MW00379696

# CLIMATE SCIENCE
# SPECIAL REPORT

Fourth National Climate Assessment | Volume I

ISBN 9781979474511

# CLIMATE SCIENCE SPECIAL REPORT (CSSR)

## TABLE OF CONTENTS

About This Report .................................................................................................................1

Guide to the Report ...............................................................................................................3

Executive Summary ...............................................................................................................12

**Chapters**

1. Our Globally Changing Climate .................................................................................35

2. Physical Drivers of Climate Change .........................................................................73

3. Detection and Attribution of Climate Change .......................................................114

4. Climate Models, Scenarios, and Projections ..........................................................133

5. Large-Scale Circulation and Climate Variability ..................................................161

6. Temperature Changes in the United States ............................................................185

7. Precipitation Change in the United States ..............................................................207

8. Droughts, Floods, and Wildfires ..............................................................................231

9. Extreme Storms ...........................................................................................................257

10. Changes in Land Cover and Terrestrial Biogeochemistry ..................................277

11. Arctic Changes and their Effects on Alaska and the Rest of the United States ................................303

12. Sea Level Rise .............................................................................................................333

13. Ocean Acidification and Other Ocean Changes ...................................................364

14. Perspectives on Climate Change Mitigation .........................................................393

15. Potential Surprises: Compound Extremes and Tipping Elements .....................411

**Appendices**

A. Observational Datasets Used in Climate Studies .................................................430

B. Model Weighting Strategy ........................................................................................436

C. Detection and Attribution Methodologies Overview ..........................................443

D. Acronyms and Units ..................................................................................................452

E. Glossary.........................................................................................................................460

C

U

# CSSR Writing Team

## Coordinating Lead Authors

**Donald J. Wuebbles,** National Science Foundation and U.S. Global Change Research Program – University of Illinois

**David W. Fahey,** NOAA Earth System Research Laboratory

**Kathy A. Hibbard,** NASA Headquarters

## Lead Authors

**Jeff R. Arnold,** U.S. Army Corps of Engineers

**Benjamin DeAngelo,** NOAA Climate Program Office

**Sarah Doherty,** University of Washington

**David R. Easterling,** NOAA National Centers for Environmental Information

**James Edmonds,** Pacific Northwest National Laboratory

**Timothy Hall,** NASA Goddard Institute for Space Studies

**Katharine Hayhoe,** Texas Tech University

**Forrest M. Hoffman,** Oak Ridge National Laboratory

**Radley Horton,** Columbia University

**Deborah Huntzinger,** Northern Arizona University

**Libby Jewett,** NOAA Ocean Acidification Program

**Thomas Knutson,** NOAA Geophysical Fluid Dynamics Lab

**Robert E. Kopp,** Rutgers University

**James P. Kossin,** NOAA National Centers for Environmental Information

**Kenneth E. Kunkel,** North Carolina State University

**Allegra N. LeGrande,** NASA Goddard Institute for Space Studies

**L. Ruby Leung,** Pacific Northwest National Laboratory

**Wieslaw Maslowski,** Naval Postgraduate School

**Carl Mears,** Remote Sensing Systems

**Judith Perlwitz,** NOAA Earth System Research Laboratory

**Anastasia Romanou,** Columbia University

**Benjamin M. Sanderson,** National Center for Atmospheric Research

**William V. Sweet,** NOAA National Ocean Service

**Patrick C. Taylor,** NASA Langley Research Center

**Robert J. Trapp,** University of Illinois at Urbana-Champaign

**Russell S. Vose,** NOAA National Centers for Environmental Information

**Duane E. Waliser,** NASA Jet Propulsion Laboratory

**Michael F. Wehner,** Lawrence Berkeley National Laboratory

**Tristram O. West,** DOE Office of Science

## Review Editors

**Linda O. Mearns,** National Center for Atmospheric Research

**Ross J. Salawitch,** University of Maryland

**Christopher P. Weaver,** USEPA

## Contributing Authors

**Richard Alley,** Pennsylvania State University

**C. Taylor Armstrong,** NOAA Ocean Acidification Program

**John Bruno,** University of North Carolina

**Shallin Busch,** NOAA Ocean Acidification Program

**Sarah Champion,** North Carolina State University

**Imke Durre,** NOAA National Centers for Environmental Information

**Dwight Gledhill,** NOAA Ocean Acidification Program

**Justin Goldstein,** U.S. Global Change Research Program – ICF

**Boyin Huang,** NOAA National Centers for Environmental Information

**Hari Krishnan,** Lawrence Berkeley National Laboratory

**Lisa Levin,** University of California – San Diego

**Frank Muller-Karger,** University of South Florida

**Alan Rhoades,** University of California – Davis

**Laura Stevens,** North Carolina State University

**Liqiang Sun,** North Carolina State University

**Eugene Takle,** Iowa State University

**Paul Ullrich,** University of California – Davis

**Eugene Wahl,** NOAA National Centers for Environmental Information

**John Walsh,** University of Alaska – Fairbanks

*Volume Editors*

**David J. Dokken,** U.S. Global Change Research Program – ICF

**David W. Fahey,** National Oceanic and Atmospheric Administration

**Kathy A. Hibbard,** National Aeronautics and Space Administration

**Thomas K. Maycock,** Cooperative Institute for Climate and Satellites – North Carolina

**Brooke C. Stewart,** Cooperative Institute for Climate and Satellites – North Carolina

**Donald J. Wuebbles,** National Science Foundation and U.S. Global Change Research Program – University of Illinois

*Science Steering Committee*

**Benjamin DeAngelo,** National Oceanic and Atmospheric Administration

**David W. Fahey,** National Oceanic and Atmospheric Administration

**Kathy A. Hibbard,** National Aeronautics and Space Administration

**Wayne Higgins,** Department of Commerce

**Jack Kaye,** National Aeronautics and Space Administration

**Dorothy Koch,** Department of Energy

**Russell S. Vose,** National Oceanic and Atmospheric Administration

**Donald J. Wuebbles,** National Science Foundation and U.S. Global Change Research Program – University of Illinois

*Subcommittee on Global Change Research*

**Ann Bartuska,** Chair, Department of Agriculture

**Virginia Burkett,** Co-Chair, Department of the Interior

**Gerald Geernaert,** Vice-Chair, Department of Energy

**Michael Kuperberg,** Executive Director, U.S. Global Change Research Program

**John Balbus,** Department of Health and Human Services

**Bill Breed,** U.S. Agency for International Development

**Pierre Comizzoli,** Smithsonian Institution

**Wayne Higgins,** Department of Commerce

**Scott Harper,** Department of Defense (Acting)

**William Hohenstein,** Department of Agriculture

**Jack Kaye,** National Aeronautics and Space Administration

**Dorothy Koch,** Department of Energy

**Andrew Miller,** U.S. Environmental Protection Agency

**David Reidmiller,** U.S. Global Change Research Program

**Trigg Talley,** Department of State

**Michael Van Woert,** National Science Foundation

*Liaison to the Executive Office of the President*

**Kimberly Miller,** Office of Management and Budget

*Report Production Team*

**Bradley Akamine,** U.S. Global Change Research Program – ICF

**Jim Biard,** Cooperative Institute for Climate and Satellites – North Carolina

**Andrew Buddenberg,** Cooperative Institute for Climate and Satellites – North Carolina

**Sarah Champion,** Cooperative Institute for Climate and Satellites – North Carolina

**David J. Dokken,** U.S. Global Change Research Program – ICF

**Amrutha Elamparuthy,** U.S. Global Change Research Program – Straughan Environmental, Inc.

**Jennifer Fulford,** TeleSolv Consulting

**Jessicca Griffin,** Cooperative Institute for Climate and Satellites – North Carolina

**Kate Johnson,** ERT Inc.

**Angel Li,** Cooperative Institute for Climate and Satellites – North Carolina

**Liz Love-Brotak,** NOAA National Centers for Environmental Information

**Thomas K. Maycock,** Cooperative Institute for Climate and Satellites – North Carolina

**Deborah Misch,** TeleSolv Consulting

**Katie Reeves,** U.S. Global Change Research Program – ICF

**Deborah Riddle,** NOAA National Centers for Environmental Information

**Reid Sherman,** U.S. Global Change Research Program – Straughan Environmental, Inc.

**Mara Sprain,** LAC Group

**Laura Stevens,** Cooperative Institute for Climate and Satellites – North Carolina

**Brooke C. Stewart,** Cooperative Institute for Climate and Satellites – North Carolina

**Liqiang Sun,** Cooperative Institute for Climate and Satellites – North Carolina

**Kathryn Tipton,** U.S. Global Change Research Program – ICF

**Sara Veasey,** NOAA National Centers for Environmental Information

*Administrative Lead Agency*

Department of Commerce / National Oceanic and Atmospheric Administration

# About This Report

As a key part of the Fourth National Climate Assessment (NCA4), the U.S. Global Change Research Program (USGCRP) oversaw the production of this stand-alone report of the state of science relating to climate change and its physical impacts.

The Climate Science Special Report (CSSR) is designed to be an authoritative assessment of the science of climate change, with a focus on the United States, to serve as the foundation for efforts to assess climate-related risks and inform decision-making about responses. In accordance with this purpose, it does not include an assessment of literature on climate change mitigation, adaptation, economic valuation, or societal responses, nor does it include policy recommendations.

As Volume I of NCA4, CSSR serves several purposes, including providing 1) an updated detailed analysis of the findings of how climate change is affecting weather and climate across the United States; 2) an executive summary and other CSSR materials that provide the basis for the discussion of climate science found in the second volume of the NCA4; and 3) foundational information and projections for climate change, including extremes, to improve "end-to-end" consistency in sectoral, regional, and resilience analyses within the second volume. CSSR integrates and evaluates the findings on climate science and discusses the uncertainties associated with these findings. It analyzes current trends in climate change, both human-induced and natural, and projects major trends to the end of this century. As an assessment and analysis of the science, this report provides important input to the development of other parts of NCA4, and their primary focus on the human welfare, societal, economic, and environmental elements of climate change.

Much of this report is written at a level more appropriate for a scientific audience, though the Executive Summary is intended to be accessible to a broader audience.

## Report Development, Review, and Approval Process

The National Oceanic and Atmospheric Administration (NOAA) serves as the administrative lead agency for the preparation of NCA4. The CSSR Federal Science Steering Committee (SSC)[1] has representatives from three agencies (NOAA, the National Aeronautics and Space Administration [NASA], and the Department of Energy [DOE]); USGCRP;[2] and three Coordinating Lead Authors, all of whom were Federal employees during the development of this report. Following a public notice for author nominations in March 2016, the SSC selected the writing team, consisting of scientists representing Federal agencies, national laboratories, universities, and the private sector. Contributing Authors were requested to provide special input to the Lead Authors to help with specific issues of the assessment.

The first Lead Author Meeting was held in Washington, DC, in April 2016, to refine the outline contained in the SSC-endorsed prospectus and to make writing assignments. Over the course of 18 months before final

---

[1] The CSSR SSC was charged with overseeing the development and production of the report. SSC membership was open to all USGCRP agencies.

[2] The USGCRP is made up of 13 Federal departments and agencies that carry out research and support the Nation's response to global change. The USGCRP is overseen by the Subcommittee on Global Change Research (SGCR) of the National Science and Technology Council's Committee on Environment, Natural Resources, and Sustainability (CENRS), which in turn is overseen by the White House Office of Science and Technology Policy (OSTP). The agencies within USGCRP are the Department of Agriculture, the Department of Commerce (NOAA), the Department of Defense, the Department of Energy, the Department of Health and Human Services, the Department of the Interior, the Department of State, the Department of Transportation, the Environmental Protection Agency, the National Aeronautics and Space Administration, the National Science Foundation, the Smithsonian Institution, and the U.S. Agency for International Development.

1

publication, seven CSSR drafts were generated, with each successive iteration—from zero- to sixth-order drafts—undergoing additional expert review, as follows: (i) by the writing team itself (13–20 June 2016); (ii) by the SSC convened to oversee report development (29 July–18 August 2016); (iii) by the technical agency representatives (and designees) comprising the Subcommittee on Global Change Research (SGCR, 3–14 October 2016); (iv) by the SSC and technical liaisons again (5–13 December 2016); (v) by the general public during the Public Comment Period (15 December 2016–3 February 2017) and an expert panel convened by the National Academies of Sciences, Engineering, and Medicine (NAS, 21 December 2016–13 March 2017);[3] and (vi) by the SGCR again (3–24 May 2017) to confirm the Review Editor conclusions that all public and NAS comments were adequately addressed. In October 2016, an 11-member core writing team was tasked with capturing the most important CSSR key findings and generating an Executive Summary. Two additional Lead Authors Meetings were held after major review milestones to facilitate chapter team deliberations and consistency: 2–4 November 2016 (Boulder, CO) and 21–22 April 2017 (Asheville, NC). Literature cutoff dates were enforced, with all cited material published by June 2017. The fifth-order draft including the Executive Summary was compiled in June 2017, and submitted to the Office of Science and Technology Policy (OSTP). OSTP is responsible for the Federal clearance process prior to final report production and public release. This published report represents the final (sixth-order) draft.

---

[3] Author responses to comments submitted as part of the Public Comment Period and a USGCRP response to the review conducted by NAS can be found on <science2017.globalchange.gov/downloads>.

## The Sustained National Climate Assessment

The Climate Science Special Report has been developed as part of the USGCRP's sustained National Climate Assessment (NCA) process. This process facilitates continuous and transparent participation of scientists and stakeholders across regions and sectors, enabling new information and insights to be assessed as they emerge. The Climate Science Special Report is aimed at a comprehensive assessment of the science underlying the changes occurring in Earth's climate system, with a special focus on the United States.

## Sources Used in this Report

The findings in this report are based on a large body of scientific, peer-reviewed research, as well as a number of other publicly available sources, including well-established and carefully evaluated observational and modeling datasets. The team of authors carefully reviewed these sources to ensure a reliable assessment of the state of scientific understanding. Each source of information was determined to meet the four parts of the quality assurance guidance provided to authors (following the approach from NCA3): 1) utility, 2) transparency and traceability, 3) objectivity, and 4) integrity and security. Report authors assessed and synthesized information from peer-reviewed journal articles, technical reports produced by Federal agencies, scientific assessments (such as the rigorously-reviewed international assessments from the Intergovernmental Panel on Climate Change,[1] reports of the National Academy of Sciences and its associated National Research Council, and various regional climate impact assessments, conference proceedings, and government statistics (such as population census and energy usage).

# Guide to the Report

The following subsections describe the format of the Climate Science Special Report and the overall structure and features of the chapters.

## Executive Summary

The Executive Summary describes the major findings from the Climate Science Special Report. It summarizes the overall findings and includes some key figures and additional bullet points covering overarching and especially noteworthy conclusions. The Executive Summary and the majority of the Key Findings are written to be accessible to a wide range of audiences.

## Chapters

### Key Findings and Traceable Accounts

Each topical chapter includes Key Findings, which are based on the authors' expert judgment of the synthesis of the assessed literature. Each Key Finding includes a confidence statement and, as appropriate, framing of key scientific uncertainties, so as to better support assessment of climate-related risks. (See "Documenting Uncertainty" below).

Each Key Finding is also accompanied by a Traceable Account that documents the supporting evidence, process, and rationale the authors used in reaching these conclusions and provides additional information on sources of uncertainty through confidence and likelihood statements. The Traceable Accounts can be found at the end of each chapter.

### Regional Analyses

Throughout the report, the regional analyses of climate changes for the United States are structured on 10 different regions as shown in Figure 1. There are differences from the regions used in the Third National Climate Assessment[2]: 1) the Great Plains are split into the Northern Great Plains and Southern Great Plains; and 2) The U.S. islands in the Caribbean are analyzed as a separate region apart from the Southeast.

### Chapter Text

Each chapter assesses the state of the science for a particular aspect of the changing climate. The first chapter gives a summary of the global changes occurring in the Earth's climate system. This is followed in Chapter 2 by a summary of the scientific basis for climate change. Chapter 3 gives an overview of the processes used in the detection and attribution of climate change and associated studies using those techniques. Chapter 4 then discusses the scenarios for greenhouse gases and particles and the modeling tools used to study future projections. Chapters 5 through 9 primarily focus on physical changes in climate occurring in the United States, including those projected to occur in the future. Chapter 10 provides a focus on land use change and associated feedbacks on climate. Chapter 11 addresses changes in Alaska in the Arctic, and how the latter affects the United States. Chapters 12 and 13 discuss key issues connected with sea level rise and ocean changes, including ocean acidification, and their potential effects on the United States. Finally, Chapters 14 and 15 discuss some important perspectives on how mitigation activities could affect future changes in climate and provide perspectives on what surprises could be in store for the changing climate beyond the analyses already covered in the rest of the assessment.

Throughout the report, results are presented in United States customary units (e.g., degrees Fahrenheit) as well as in the International System of Units (e.g., degrees Celsius).

Caribbean

**Figure 1.** Map of the ten regions of the United States used throughout the Climate Science Special Report. Regions are similar to that used in the Third National Climate Assessment except that 1) the Great Plains are split into the Northern Great Plains and Southern Great Plains, and 2) the Caribbean islands have been split from the Southeast region. (Figure source: adapted from Melillo et al. 2014[2]).

**Reference Time Periods for Graphics**

There are many different types of graphics in the Climate Science Special Report. Some of the graphs in this report illustrate historical changes and future trends in climate compared to some reference period, with the choice of this period determined by the purpose of the graph and the availability of data. The scientific community does not have a standard set of reference time periods for assessing the science, and these tend to be chosen differently for different reports and assessments. Some graphics are pulled from other studies using different time periods.

Where graphs were generated for this report (those not based largely on prior publications), they are mostly based on one of two reference periods. The 1901–1960 reference period is particularly used for graphs that illustrate past changes in climate conditions, whether in observations or in model simulations. This 60-year time period was also used for analyses in the Third National Climate Assessment (NCA3[2]). The beginning date was chosen because earlier historical observations are generally considered to be less reliable. While a 30-year base period is often used for climate analyses, the choice of 1960 as the ending date of this period was based on past changes in human influences on the climate system. Human-induced forcing exhibited a slow rise during the early part of the last century but then accelerated after 1960. Thus, these graphs highlight observed changes in climate during the period of rapid increase in human-caused

4

forcing and also reveal how well climate models simulate these observed changes.

Thus, a number of the graphs in the report are able to highlight the recent, more rapid changes relative to the early part of the century (the reference period) and also reveal how well the climate models simulate observed changes. In this report, this time period is used as the base period in most maps of observed trends and all time-varying, area-weighted averages that show both observed and projected quantities. For the observed trends, 1986–2015 is generally chosen as the most recent 30-year period (2016 data was not fully available until late in our development of the assessment).

The other commonly used reference period in this report is 1976–2005. The choice of a 30-year period is chosen to account for natural variations and to have a reasonable sampling in order to estimate likelihoods of trends in extremes. This period is consistent with the World Meteorological Organization's recommendation for climate statistics. This period is used for graphs that illustrate projected changes simulated by climate models. The purpose of these graphs is to show projected changes compared to a period that allows stakeholders and decision makers to base fundamental planning and decisions on average and extreme climate conditions in a non-stationary climate; thus, a recent available 30-year period was chosen.[3] The year 2005 was chosen as an end date because the historical period simulated by the models used in this assessment ends in that year.

For future projections, 30-year periods are again used for consistency. Projections are centered around 2030, 2050, and 2085 with an interval of plus and minus 15 years (for example, results for 2030 cover the period 2015–2045); Most model runs used here only project out to 2100 for future scenarios, but where

possible, results beyond 2100 are shown. Note that these time periods are different than those used in some of the graphics in NCA3. There are also exceptions for graphics that are based on existing publications.

For global results that may be dependent on findings from other assessments (such as those produced by the Intergovernmental Panel on Climate Change, or IPCC), and for other graphics that depend on specific published work, the use of other time periods was also allowed, but an attempt was made to keep them as similar to the selected periods as possible. For example, in the discussion of radiative forcing, the report uses the standard analyses from IPCC for the industrial era (1750 to 2011) (following IPCC 2013a[1]). And, of course, the paleoclimatic discussion of past climates goes back much further in time.

**Model Results: Past Trends and Projected Futures**
The NCA3 included global modeling results from both the CMIP3 (Coupled Model Intercomparison Project, 3rd phase) models used in the 2007 international assessment[4] and the CMIP5 (Coupled Model Intercomparison Project, Phase 5) models used in the more recent international assessment.[1] Here, the primary resource for this assessment is the more recent global model results and associated downscaled products from CMIP5. The CMIP5 models and the associated downscaled products are discussed in Chapter 4: Projections.

**Treatment of Uncertainties: Likelihoods, Confidence, and Risk Framing**
Throughout this report's assessment of the scientific understanding of climate change, the authors have assessed to the fullest extent possible the state-of-the-art understanding of the science resulting from the information in the scientific literature to arrive at a series of findings referred to as Key Findings. The approach used to represent the extent of un-

5

derstanding represented in the Key Findings is done through two metrics:

- **Confidence** in the validity of a finding based on the type, amount, quality, strength, and consistency of evidence (such as mechanistic understanding, theory, data, models, and expert judgment); the skill, range, and consistency of model projections; and the degree of agreement within the body of literature.

- **Likelihood,** or probability of an effect or impact occurring, is based on measures of uncertainty expressed probabilistically (based on the degree of understanding or knowledge, e.g., resulting from evaluating statistical analyses of observations or model results or on expert judgment).

The terminology used in the report associated with these metrics is shown in Figure 2. This language is based on that used in NCA3,[2] the IPCC's Fifth Assessment Report,[1] and most recently the USGCRP Climate and Health assessment.[5] Wherever used, the confidence and likelihood statements are italicized.

Assessments of confidence in the Key Findings are based on the expert judgment of the author team. Authors provide supporting evidence for each of the chapter's Key Findings in the Traceable Accounts. Confidence is expressed qualitatively and ranges from low confidence (inconclusive evidence or disagreement among experts) to very high confidence (strong evidence and high consensus) (see Figure 2). Confidence should not be interpreted probabilistically, as it is distinct from statistical likelihood. See chapter 1 in IPCC[1] for further discussion of this terminology.

In this report, likelihood is the chance of occurrence of an effect or impact based on measures of uncertainty expressed probabilis-

tically (based on statistical analysis of observations or model results or on expert judgment). The authors used expert judgment based on the synthesis of the literature assessed to arrive at an estimation of the likelihood that a particular observed effect was related to human contributions to climate change or that a particular impact will occur within the range of possible outcomes. Model uncertainty is an important contributor to uncertainty in climate projections, and includes, but is not restricted to, the uncertainties introduced by errors in the model's representation of the physical and bio-geochemical processes affecting the climate system as well as in the model's response to external forcing.[1]

Where it is considered justified to report the likelihood of particular impacts within the range of possible outcomes, this report takes a plain-language approach to expressing the expert judgment of the chapter team, based on the best available evidence. For example, an outcome termed "likely" has at least a 66% chance of occurring (a likelihood greater than about 2 of 3 chances); an outcome termed "very likely," at least a 90% chance (more than 9 out of 10 chances). See Figure 2 for a complete list of the likelihood terminology used in this report.

Traceable Accounts for each Key Finding 1) document the process and rationale the authors used in reaching the conclusions in their Key Finding, 2) provide additional information to readers about the quality of the information used, 3) allow traceability to resources and data, and 4) describe the level of likelihood and confidence in the Key Finding. Thus, the Traceable Accounts represent a synthesis of the chapter author team's judgment of the validity of findings, as determined through evaluation of evidence and agreement in the scientific literature. The Traceable Accounts also identify areas where data are

| Confidence Level | Likelihood | |
|---|---|---|
| **Very High** | **Virtually Certain** | 99%–100% |
| Strong evidence (established theory, multiple sources, consistent results, well documented and accepted methods, etc.), high consensus | **Extremely Likely** | 95%–100% |
| **High** | **Very Likely** | 90%–100% |
| Moderate evidence (several sources, some consistency, methods vary and/or documentation limited, etc.), medium consensus | **Likely** | 66%–100% |
| **Medium** | **About as Likely as Not** | 33%–66% |
| Suggestive evidence (a few sources, limited consistency, models incomplete, methods emerging, etc.), competing schools of thought | **Unlikely** | 0%–33% |
| **Low** | **Very Unlikely** | 0%–10% |
| Inconclusive evidence (limited sources, extrapolations, inconsistent findings, poor documentation and/or methods not tested, etc.), disagreement or lack of opinions among experts | **Extremely Unlikely** | 0%–5% |
| | **Exceptionally Unlikely** | 0%–1% |

**Figure 2.** Confidence levels and likelihood statements used in the report. (Figure source: adapted from USGCRP 2016[5] and IPCC 2013[1]; likelihoods use the broader range from the IPCC assessment). As an example, regarding "likely," a 66%–100% probability can be interpreted as a likelihood of greater than 2 out of 3 chances for the statement to be certain or true. Not all likelihoods are used in the report.

limited or emerging. Each Traceable Account includes 1) a description of the evidence base, 2) major uncertainties, and 3) an assessment of confidence based on evidence.

All Key Findings include a description of confidence. Where it is considered scientifically justified to report the likelihood of particular impacts within the range of possible outcomes, Key Findings also include a likelihood designation.

Confidence and likelihood levels are based on the expert judgment of the author team. They determined the appropriate level of confidence or likelihood by assessing the available literature, determining the quality and quantity of available evidence, and evaluating the level of agreement across different studies. Often, the underlying studies provided their own estimates of uncertainty and confidence intervals. When available, these confidence intervals were assessed by the authors in

7

making their own expert judgments. For specific descriptions of the process by which the author team came to agreement on the Key Findings and the assessment of confidence and likelihood, see the Traceable Accounts in each chapter.

In addition to the use of systematic language to convey confidence and likelihood information, this report attempts to highlight aspects of the science that are most relevant for supporting other parts of the Fourth National Climate Assessment and its analyses of key societal risks posed by climate change. This includes attention to trends and changes in the tails of the probability distribution of future climate change and its proximate impacts (for example, on sea level or temperature and precipitation extremes) and on defining plausible bounds for the magnitude of future changes, since many key risks are disproportionately determined by plausible low-probability, high-consequence outcomes. Therefore, in addition to presenting the expert judgment on the "most likely" range of projected future climate outcomes, where appropriate, this report also provides information on the outcomes lying outside this range, which nevertheless cannot be ruled out and may therefore be relevant for assessing overall risk. In some cases, this involves an evaluation of the full range of information contained in the ensemble of climate models used for this report, and in other cases this involves the consideration of additional lines of scientific evidence beyond the models.

Complementing this use of risk-focused language and presentation around specific scientific findings in the report, Chapter 15: Potential Surprises provides an overview of potential low probability/high consequence "surprises" resulting from climate change. This includes its analyses of thresholds, also called tipping points, in the climate system and the compounding effects of multiple, interacting climate change impacts whose consequences may be much greater than the sum of the individual impacts. Chapter 15 also highlights critical knowledge gaps that determine the degree to which such high-risk tails and bounding scenarios can be precisely defined, including missing processes and feedbacks.

8

# REFERENCES

1. IPCC, 2013a: *Climate Change 2013: The Physical Science Basis. Contribution of Working Group I to the Fifth Assessment Report of the Intergovernmental Panel on Climate Change.* Cambridge University Press, Cambridge, UK and New York, NY, 1535 pp. http://www.climatechange2013.org/report/

2. Melillo, J.M., T.C. Richmond, and G.W. Yohe, eds., 2014a: *Climate Change Impacts in the United States: The Third National Climate Assessment.* U.S. Global Change Research Program: Washington, D.C., 841 pp. http://dx.doi.org/10.7930/J0Z31WJ2

3. **Arguez, A. and R.S. Vose, 2011:** The definition of the standard WMO climate normal: The key to deriving alternative climate normals. *Bulletin of the American Meteorological Society,* **92,** 699-704. http://dx.doi.org/10.1175/2010BAMS2955.1

4. IPCC, 2007: *Climate Change 2007: The Physical Science Basis. Contribution of Working Group I to the Fourth Assessment Report of the Intergovernmental Panel on Climate Change.* Solomon, S., D. Qin, M. Manning, Z. Chen, M. Marquis, K.B. Averyt, M. Tignor, and H.L. Miller, Eds. Cambridge University Press, Cambridge. U.K, New York, NY, USA, 996 pp. http://www.ipcc.ch/publications_and_data/publications_ipcc_fourth_assessment_report_wg1_report_the_physical_science_basis.htm

5. USGCRP, 2016: *The Impacts of Climate Change on Human Health in the United States: A Scientific Assessment.* Crimmins, A., J. Balbus, J.L. Gamble, C.B. Beard, J.E. Bell, D. Dodgen, R.J. Eisen, N. Fann, M.D. Hawkins, S.C. Herring, L. Jantarasami, D.M. Mills, S. Saha, **M.C. Sarofim, J. Trtanj, and L. Ziska, Eds.** U.S. Global Change Research Program, Washington, DC, 312 pp. http://dx.doi.org/10.7930/J0R49NQX

# Highlights of the U.S. Global Change Research Program
# Climate Science Special Report

The climate of the United States is strongly connected to the changing global climate. The statements below highlight past, current, and projected climate changes for the United States and the globe.

Global annually averaged surface air temperature has increased by about 1.8°F (1.0°C) over the last 115 years (1901–2016). **This period is now the warmest in the history of modern civilization.** The last few years have also seen record-breaking, climate-related weather extremes, and the last three years have been the warmest years on record for the globe. These trends are expected to continue over climate timescales.

This assessment concludes, based on extensive evidence, that it is extremely likely that **human activities, especially emissions of greenhouse gases, are the dominant cause of the observed warming since the mid-20th century.** For the warming over the last century, there is no convincing alternative explanation supported by the extent of the observational evidence.

In addition to warming, many other aspects of global climate are changing, primarily in response to human activities. **Thousands of studies conducted by researchers around the world have documented changes in surface, atmospheric, and oceanic temperatures; melting glaciers; diminishing snow cover; shrinking sea ice; rising sea levels; ocean acidification; and increasing atmospheric water vapor.**

For example, **global average sea level has risen by about 7–8 inches** since 1900, with almost half (about 3 inches) of that rise occurring since 1993. Human-caused climate change has made a substantial contribution to this rise since 1900, contributing to a rate of rise that is greater than during any preceding century in at least 2,800 years. Global sea level rise has already affected the United States; **the incidence of daily tidal flooding is accelerating in more than 25 Atlantic and Gulf Coast cities.**

**Global average sea levels are expected to continue to rise—by at least several inches in the next 15 years and by 1–4 feet by 2100. A rise of as much as 8 feet by 2100 cannot be ruled out.** Sea level rise will be higher than the global average on the East and Gulf Coasts of the United States.

Changes in the characteristics of extreme events are particularly important for human safety, infrastructure, agriculture, water quality and quantity, and natural ecosystems. **Heavy rainfall is increasing in intensity and frequency across the United States and globally and is expected to continue to increase.** The largest observed changes in the United States have occurred in the Northeast.

**Heatwaves have become more frequent in the United States since the 1960s, while extreme cold temperatures and cold waves are less frequent.** Recent record-setting hot years are projected to become common in the near future for the United States, as annual average temperatures continue to rise. Annual average temperature over the contiguous United States has increased by 1.8°F (1.0°C) for the period 1901–2016; **over the next few decades (2021–2050), annual average temperatures are expected to rise by about 2.5°F for the United States, relative to the recent past (average from 1976–2005), under all plausible future climate scenarios.**

**The incidence of large forest fires in the western United States and Alaska has increased since the early 1980s and is projected to further increase** in those regions as the climate changes, with profound changes to regional ecosystems.

**Annual trends toward earlier spring melt and reduced snowpack are already affecting water resources in the western United States** and these trends are expected to continue. Under higher scenarios, and assuming no change to current water resources management, **chronic, long-duration hydrological drought is increasingly possible before the end of this century.**

**The magnitude of climate change beyond the next few decades will depend primarily on the amount of greenhouse gases (especially carbon dioxide) emitted globally.** Without major reductions in emissions, the increase in annual average global temperature relative to preindustrial times could reach 9°F (5°C) or more by the end of this century. **With significant reductions in emissions, the increase in annual average global temperature could be limited to 3.6°F (2°C) or less.**

**The global atmospheric carbon dioxide ($CO$ ) concentration has now passed 400 parts per million (ppm), a level that last occurred about 3 million years ago, when both global average temperature and sea level were significantly higher than today.** Continued growth in $CO_2$ emissions over this century and beyond would lead to an atmospheric concentration not experienced in tens to hundreds of millions of years. There is broad consensus that the further and the faster the Earth system is pushed towards warming, the greater the risk of unanticipated changes and impacts, some of which are potentially large and irreversible.

The observed increase in carbon emissions over the past 15–20 years has been consistent with higher emissions pathways. **In 2014 and 2015, emission growth rates slowed as economic growth became less carbon-intensive.** Even if this slowing trend continues, however, it is not yet at a rate that would limit global average temperature change to well below 3.6°F (2°C) above preindustrial levels.

**Recommended Citation for Chapter**
**Wuebbles**, D.J., D.W. Fahey, K.A. Hibbard, B. DeAngelo, S. Doherty, K. Hayhoe, R. Horton, J.P. Kossin, P.C. Taylor, A.M. Waple, and C.P. Weaver, 2017: Executive summary. In: *Climate Science Special Report: Fourth National Climate Assessment, Volume I* [Wuebbles, D.J., D.W. Fahey, K.A. Hibbard, D.J. Dokken, B.C. Stewart, and T.K. Maycock (eds.)]. U.S. Global Change Research Program, Washington, DC, USA, pp. 12-34, doi: 10.7930/J0DJ5CTG.

# Executive Summary

## Introduction

New observations and new research have increased our understanding of past, current, and future climate change since the Third U.S. National Climate Assessment (NCA3) was published in May 2014. This Climate Science Special Report (CSSR) is designed to capture that new information and build on the existing body of science in order to summarize the current state of knowledge and provide the scientific foundation for the Fourth National Climate Assessment (NCA4).

Since NCA3, stronger evidence has emerged for continuing, rapid, human-caused warming of the global atmosphere and ocean. This report concludes that "it is *extremely likely* that human influence has been the dominant cause of the observed warming since the mid-20th century. For the warming over the last century, there is no convincing alternative explanation supported by the extent of the observational evidence."

The last few years have also seen record-breaking, climate-related weather extremes, the three warmest years on record for the globe, and continued decline in arctic sea ice. These trends are expected to continue in the future over climate (multidecadal) timescales. Significant advances have also been made in our understanding of extreme weather events and how they relate to increasing global temperatures and associated climate changes. Since 1980, the cost of extreme events for the United States has exceeded $1.1 trillion; therefore, better understanding of the frequency and severity of these events in the context of a changing climate is warranted.

Periodically taking stock of the current state of knowledge about climate change and putting new weather extremes, changes in sea ice, increases in ocean temperatures, and ocean acidification into context ensures that rigorous, scientifically-based information is available to inform dialogue and decisions at every level. This climate science report serves as the climate science foundation of the NCA4 and is generally intended for those who have a technical background in climate science. In this Executive Summary, gray boxes present highlights of the main report. These are followed by related points and selected figures providing more scientific details. The summary material on each topic presents the most salient points of chapter findings and therefore represents only a subset of the report's content. For more details, the reader is referred to the individual chapters. This report discusses climate trends and findings at several scales: global, nationwide for the United States, and for ten specific U.S. regions (shown in Figure 1 in the Guide to the Report). A statement of scientific confidence also follows each point in the Executive Summary. The confidence scale is described in the Guide to the Report. At the end of the Executive Summary and in Chapter 1: Our Globally Changing Climate, there is also a summary box highlighting the most notable advances and topics since NCA3 and since the 2013 Intergovernmental Panel on Climate Change (IPCC) Fifth Assessment Report.

## Global and U.S. Temperatures Continue to Rise

Long-term temperature observations are among the most consistent and widespread evidence of a warming planet. Temperature (and, above all, its local averages and extremes) affects agricultural productivity, energy use, human health, water resources, infrastructure, natural ecosystems, and many other essential aspects of society and the natural environment. Recent data add to the weight of evidence for rapid global-scale warming, the dominance of human causes, and the expected continuation of increasing temperatures, including more record-setting extremes. (Ch. 1)

### Changes in Observed and Projected Global Temperature

> **The global, long-term, and unambiguous warming trend has continued during recent years. Since the last National Climate Assessment was published, 2014 became the warmest year on record globally; 2015 surpassed 2014 by a wide margin; and 2016 surpassed 2015. Sixteen of the warmest years on record for the globe occurred in the last 17 years (1998 was the exception). (Ch. 1; Fig. ES.1)**

- Global annual average temperature (as calculated from instrumental records over both land and oceans) has increased by more than 1.2°F (0.65°C) for the period 1986–2016 relative to 1901–1960; the linear regression change over the entire period from 1901–2016 is 1.8°F (1.0°C) (*very high confidence*; Fig. ES.1). Longer-term climate records over past centuries and millennia indicate that average temperatures in recent decades over much of the world have been much higher, and have risen faster during this time period than at any time in the past 1,700 years or more, the time period for which the global distribution of surface temperatures can be reconstructed (*high confidence*). (Ch. 1)

## Global Temperatures Continue to Rise

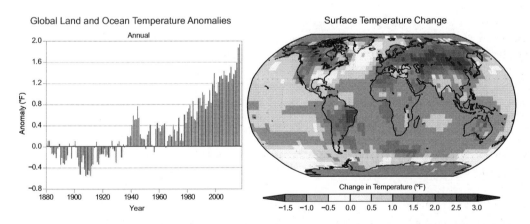

**Figure ES.1:** (left) Global annual average temperature has increased by more than 1.2°F (0.7°C) for the period 1986–2016 relative to 1901–1960. Red bars show temperatures that were above the 1901–1960 average, and blue bars indicate temperatures below the average. (right) Surface temperature change (in °F) for the period 1986–2016 relative to 1901–1960. Gray indicates missing data. *From Figures 1.2. and 1.3 in Chapter 1.*

- Many lines of evidence demonstrate that it is *extremely likely* that human influence has been the dominant cause of the observed warming since the mid-20th century. Over the last century, there are no convincing alternative explanations supported by the extent of the observational evidence. Solar output changes and internal natural variability can only contribute marginally to the observed changes in climate over the last century, and there is no convincing evidence for natural cycles in the observational record that could explain the observed changes in climate. (*Very high confidence*) (Ch. 1)

- The *likely* range of the human contribution to the global mean temperature increase over the period 1951–2010 is 1.1° to 1.4°F (0.6° to 0.8°C), and the central estimate of the observed warming of 1.2°F (0.65°C) lies within this range (*high confidence*). This translates to a *likely* human contribution of 92%–123% of the observed 1951–2010 change. The *likely* contributions of natural forcing and internal variability to global temperature change over that period are minor (*high confidence*). (Ch. 3; Fig. ES.2)

- Natural variability, including El Niño events and other recurring patterns of ocean–atmosphere interactions, impact temperature and precipitation, especially regionally, over timescales of months to years. The global influence of natural variability, however, is limited to a small fraction of observed climate trends over decades. (*Very high confidence*) (Ch. 1)

### Human Activities Are the Primary Driver of Recent Global Temperature Rise

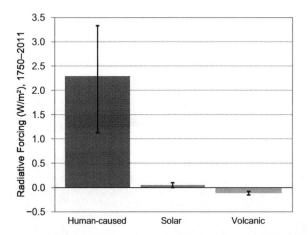

Figure ES.2: Global annual average radiative forcing change from 1750 to 2011 due to human activities, changes in total solar irradiance, and volcanic emissions. Black bars indicate the uncertainty in each. Radiative forcing is a measure of the influence a factor (such as greenhouse gas emissions) has in changing the global balance of incoming and outgoing energy. Radiative forcings greater than zero (positive forcings) produce climate warming; forcings less than zero (negative forcings) produce climate cooling. Over this time period, solar forcing has oscillated on approximately an 11-year cycle between −0.11 and +0.19 W/m². Radiative forcing due to volcanic emissions is always negative (cooling) and can be very large immediately following significant eruptions but is short-lived. Over the industrial era, the largest volcanic forcing followed the eruption of Mt. Tambora in 1815 (−11.6 W/m²). This forcing declined to −4.5 W/m² in 1816, and to near-zero by 1820. Forcing due to human activities, in contrast, has becoming increasingly positive (warming) since about 1870, and has grown at an accelerated rate since about 1970. There are also natural variations in temperature and other climate variables which operate on annual to decadal timescales. This natural variability contributes very little to climate trends over decades and longer. *Simplified from Figure 2.6 in Chapter 2. See Chapter 2 for more details.*

- Global climate is projected to continue to change over this century and beyond. The magnitude of climate change beyond the next few decades will depend primarily on the amount of greenhouse (heat-trapping) gases emitted globally and on the remaining uncertainty in the sensitivity of Earth's climate to those emissions (*very high confidence*). With significant reductions in the emissions of greenhouse gases, the global annually averaged temperature rise could be limited to 3.6°F (2°C) or less. Without major reductions in these emissions, the increase in annual average global temperatures relative to preindustrial times could reach 9°F (5°C) or more by the end of this century. (Ch. 1; Fig. ES.3)

- If greenhouse gas concentrations were stabilized at their current level, existing concentrations would commit the world to at least an additional 1.1°F (0.6°C) of warming over this century relative to the last few decades (*high confidence* in continued warming, *medium confidence* in amount of warming. (Ch. 4)

## Scenarios Used in this Assessment

Projections of future climate conditions use a range of plausible future scenarios. Consistent with previous practice, this assessment relies on scenarios generated for the Intergovernmental Panel on Climate Change (IPCC). The IPCC completed its last assessment in 2013–2014, and its projections were based on updated scenarios, namely four "representative concentration pathways" (RCPs). The RCP scenarios are numbered according to changes in radiative forcing in 2100 relative to preindustrial conditions: +2.6, +4.5, +6.0 and +8.5 watts per square meter (W/m²). Radiative forcing is a measure of the influence a factor (such as greenhouse gas emissions) has in changing the global balance of incoming and outgoing energy. Absorption by greenhouse gases (GHGs) of infrared energy radiated from the surface leads to warming of the surface and atmosphere. Though multiple emissions pathways could lead to the same 2100 radiative forcing value, an associated pathway of $CO_2$ and other human-caused emissions of greenhouse gases, aerosols, and air pollutants has been selected for each RCP. RCP8.5 implies a future with continued high emissions growth, whereas the other RCPs represent different pathways of mitigating emissions. Figure ES.3 shows these emissions pathways and the corresponding projected changes in global temperature.

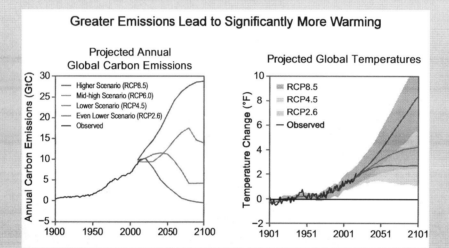

**Figure ES.3:** The two panels above show annual historical and a range of plausible future carbon emissions in units of gigatons of carbon (GtC) per year (left) and the historical observed and future temperature change that would result for a range of future scenarios relative to the 1901–1960 average, based on the central estimate (lines) and a range (shaded areas, two standard deviations) as simulated by the full suite of CMIP5 global climate models (right). By 2081–2100, the projected range in global mean temperature change is 1.1°–4.3°F under the even lower scenario (RCP2.6; 0.6°–2.4°C, green), 2.4°–5.9°F under the lower scenario (RCP4.5; 1.3°–3.3°C, blue), 3.0°–6.8°F under the mid-high scenario (RCP6.0; 1.6°–3.8°C, not shown) and 5.0°–10.2°F under the higher scenario (RCP8.5; 2.8°–5.7°C, orange). See the main report for more details on these scenarios and implications. *Based on Figure 4.1 in Chapter 4.*

**Changes in Observed and Projected U.S. Temperature**

Annual average temperature over the contiguous United States has increased by 1.8°F (1.0°C) for the period 1901–2016 and is projected to continue to rise. (*Very high confidence*). (Ch. 6; Fig. ES.4)

- Annual average temperature over the contiguous United States has increased by 1.2°F (0.7°C) for the period 1986–2016 relative to 1901–1960 and by 1.8°F (1.0°C) based on a linear regression for the period 1901–2016 (*very high confidence*). Surface and satellite data are consistent in their depiction of rapid warming since 1979 (*high confidence*). Paleo-temperature evidence shows that recent decades are the warmest of the past 1,500 years (*medium confidence*). (Ch. 6)

- Annual average temperature over the contiguous United States is projected to rise (*very high confidence*). Increases of about 2.5°F (1.4°C) are projected for the period 2021–2050 relative to the average from 1976–2005 in all RCP scenarios, implying recent record-setting years may be "common" in the next few decades (*high confidence*). Much larger rises are projected by late century (2071–2100): 2.8°–7.3°F (1.6°–4.1°C) in a lower scenario (RCP4.5) and 5.8°–11.9°F (3.2°–6.6°C) in a higher scenario (RCP8.5) (*high confidence*). (Ch. 6; Fig. ES.4)

- In the United States, the urban heat island effect results in daytime temperatures 0.9°–7.2°F (0.5°–4.0°C) higher and nighttime temperatures 1.8°– 4.5°F (1.0°–2.5°C) higher in urban areas than in rural areas, with larger temperature differences in humid regions (primarily in the eastern United States) and in cities with larger and denser populations. The urban heat island effect will strengthen in the future as the structure and spatial extent as well as population density of urban areas change and grow (*high confidence*). (Ch. 10)

## Significantly More Warming Occurs Under
## Higher Greenhouse Gas Concentration Scenarios

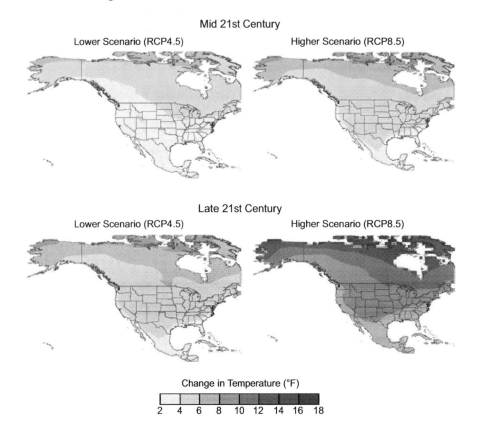

Figure ES.4: These maps show the projected changes in annual average temperatures for mid- and late-21st century for two future pathways. Changes are the differences between the average projected temperatures for mid-century (2036–2065; top), and late-century (2070–2099; bottom), and those observed for the near-present (1976–2005). *See Figure 6.7 in Chapter 6 for more details.*

### Many Temperature and Precipitation Extremes Are Becoming More Common

Temperature and precipitation extremes can affect water quality and availability, agricultural productivity, human health, vital infrastructure, iconic ecosystems and species, and the likelihood of disasters. Some extremes have already become more frequent, intense, or of longer duration, and many extremes are expected to continue to increase or worsen, presenting substantial challenges for built, agricultural, and natural systems. Some storm types such as hurricanes, tornadoes, and winter storms are also exhibiting changes that have been linked to climate change, although the current state of the science does not yet permit detailed understanding.

**Observed Changes in Extremes**

> There have been marked changes in temperature extremes across the contiguous United States. The number of high temperature records set in the past two decades far exceeds the number of low temperature records. (*Very high confidence*) (Ch. 6, Fig. ES.5)

- The frequency of cold waves has decreased since the early 1900s, and the frequency of heat waves has increased since the mid-1960s (the Dust Bowl era of the 1930s remains the peak period for extreme heat in the United States). (*Very high confidence*). (Ch. 6)

- The frequency and intensity of extreme heat and heavy precipitation events are increasing in most continental regions of the world (*very high confidence*). These trends are consistent with expected physical responses to a warming climate. Climate model studies are also consistent with these trends, although models tend to underestimate the observed trends, especially for the increase in extreme precipitation events (*very high confidence* for temperature, *high confidence* for extreme precipitation). (Ch. 1)

## Record Warm Daily Temperatures Are Occurring More Often

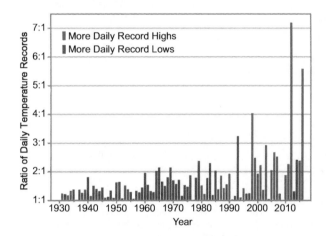

Figure ES.5: Observed changes in the occurrence of record-setting daily temperatures in the contiguous United States. Red bars indicate a year with more daily record highs than daily record lows, while blue bars indicate a year with more record lows than highs. The height of the bar indicates the ratio of record highs to lows (red) or of record lows to highs (blue). For example, a ratio of 2:1 for a blue bar means that there were twice as many record daily lows as daily record highs that year. (Figure source: NOAA/NCEI). *From Figure 6.5 in Chapter 6.*

Heavy precipitation events in most parts of the United States have increased in both intensity and frequency since 1901 (*high confidence*). There are important regional differences in trends, with the largest increases occurring in the northeastern United States (*high confidence*). (Ch. 7; Fig. ES.6)

## Extreme Precipitation Has Increased Across Much of the United States

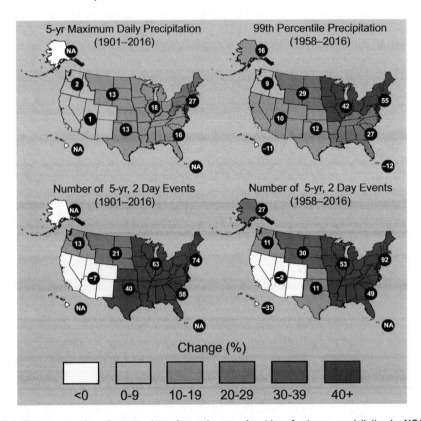

Figure ES.6: These maps show the percentage change in several metrics of extreme precipitation by NCA4 region, including (upper left) the maximum daily precipitation in consecutive 5-year periods; (upper right) the amount of precipitation falling in daily events that exceed the 99th percentile of all non-zero precipitation days (top 1% of all daily precipitation events); (lower left) the number of 2-day events with a precipitation total exceeding the largest 2-day amount that is expected to occur, on average, only once every 5 years, as calculated over 1901–2016; and (lower right) the number of 2-day events with a precipitation total exceeding the largest 2-day amount that is expected to occur, on average, only once every 5 years, as calculated over 1958–2016. The number in each black circle is the percent change over the entire period, either 1901–2016 or 1958–2016. Note that Alaska and Hawai'i are not included in the 1901–2016 maps owing to a lack of observations in the earlier part of the 20th century. *(Figure source: CICS-NC / NOAA NCEI). Based on figure 7.4 in Chapter 7.*

- Recent droughts and associated heat waves have reached record intensity in some regions of the United States; however, by geographical scale and duration, the Dust Bowl era of the 1930s remains the benchmark drought and extreme heat event in the historical record. (*Very high confidence*) (Ch. 8)

- Northern Hemisphere spring snow cover extent, North America maximum snow depth, snow water equivalent in the western United States, and extreme snowfall years in the southern and western United States have all declined, while extreme snowfall years in parts of the northern United States have increased. (*Medium confidence*). (Ch. 7)

- There has been a trend toward earlier snowmelt and a decrease in snowstorm frequency on the southern margins of climatologically snowy areas (*medium confidence*). Winter storm tracks have shifted northward since 1950 over the Northern Hemisphere (*medium confidence*). Potential linkages between the frequency and intensity of severe winter storms in the United States and accelerated warming in the Arctic have been postulated, but they are complex, and, to some extent, contested, and confidence in the connection is currently *low*. (Ch. 9)

- Tornado activity in the United States has become more variable, particularly over the 2000s, with a decrease in the number of days per year with tornadoes and an increase in the number of tornadoes on these days (*medium confidence*). Confidence in past trends for hail and severe thunderstorm winds, however, is *low* (Ch. 9)

**Projected Changes in Extremes**

- The frequency and intensity of extreme high temperature events are *virtually certain* to increase in the future as global temperature increases (*high confidence*). Extreme precipitation events will *very likely* continue to increase in frequency and intensity throughout most of the world (*high confidence*). Observed and projected trends for some other types of extreme events, such as floods, droughts, and severe storms, have more variable regional characteristics. (Ch. 1)

> **Extreme temperatures in the contiguous United States are projected to increase even more than average temperatures (*very high confidence*). (Ch. 6)**

- Both extremely cold days and extremely warm days are expected to become warmer. Cold waves are predicted to become less intense while heat waves will become more intense. The number of days below freezing is projected to decline while the number above 90°F will rise. (*Very high confidence*) (Ch. 6)

- The frequency and intensity of heavy precipitation events in the United States are projected to continue to increase over the 21st century (*high confidence*). There are, however, important regional and seasonal differences in projected changes in total precipitation: the northern United States, including Alaska, is projected to receive more precipitation in the winter and spring, and parts of the southwestern United States are projected to receive less precipitation in the winter and spring (*medium confidence*). (Ch. 7)

- The frequency and severity of landfalling "atmospheric rivers" on the U.S. West Coast (narrow streams of moisture that account for 30%–40% of the typical snowpack and annual precipitation in the region and are associated with severe flooding events) will increase as a result of increasing evaporation and resulting higher atmospheric water vapor that occurs with increasing temperature. (*Medium confidence*) (Ch. 9)

- Projections indicate large declines in snowpack in the western United States and shifts to more precipitation falling as rain than snow in the cold season in many parts of the central and eastern United States (*high confidence*). (Ch. 7)

- Substantial reductions in western U.S. winter and spring snowpack are projected as the climate warms. Earlier spring melt and reduced snow water equivalent have been formally attributed to human-induced warming (*high confidence*) and will *very likely* be exacerbated as the climate continues to warm (*very high confidence*). Under higher scenarios, and assuming no change to current water resources management, chronic, long-duration hydrological drought is increasingly possible by the end of this century (*very high confidence*). (Ch. 8)

> **Future decreases in surface soil moisture from human activities over most of the United States are *likely* as the climate warms under the higher scenarios. (*Medium confidence*) (Ch. 8)**

- The human effect on recent major U.S. droughts is complicated. Little evidence is found for a human influence on observed precipitation deficits, but much evidence is found for a human influence on surface soil moisture deficits due to increased evapotranspiration caused by higher temperatures. (*High confidence*) (Ch. 8)

- The incidence of large forest fires in the western United States and Alaska has increased since the early 1980s (*high confidence*) and is projected to further increase in those regions as the climate warms, with profound changes to certain ecosystems (*medium confidence*). (Ch. 8)

- Both physics and numerical modeling simulations generally indicate an increase in tropical cyclone intensity in a warmer world, and the models generally show an increase in the number of very intense tropical cyclones. For Atlantic and eastern North Pacific hurricanes and western North Pacific typhoons, increases are projected in precipitation rates (*high confidence*) and intensity (*medium confidence*). The frequency of the most intense of these storms is projected to increase in the Atlantic and western North Pacific (*low confidence*) and in the eastern North Pacific (*medium confidence*). (Ch. 9)

## Box ES.1: The Connected Climate System: Distant Changes Affect the United States

Weather conditions and the ways they vary across regions and over the course of the year are influenced, in the United States as elsewhere, by a range of factors, including local conditions (such as topography and urban heat islands), global trends (such as human-caused warming), and global and regional circulation patterns, including cyclical and chaotic patterns of natural variability within the climate system. For example, during an El Niño year, winters across the southwestern United States are typically wetter than average, and global temperatures are higher than average. During a La Niña year, conditions across the southwestern United States are typically dry, and there tends to be a lowering of global temperatures (Fig. ES.7).

El Niño is not the only repeating pattern of natural variability in the climate system. Other important patterns include the North Atlantic Oscillation (NAO)/Northern Annular Mode (NAM), which particularly affects conditions on the U.S. East Coast, and the North Pacific Oscillation (NPO) and Pacific North American Pattern (PNA), which especially affect conditions in Alaska and the U.S. West Coast. These patterns are closely linked to other atmospheric circulation phenomena like the position of the jet streams. Changes in the occurrence of these patterns or their properties have contributed to recent U.S. temperature and precipitation trends (*medium confidence*) although *confidence is low* regarding the size of the role of human activities in these changes. (Ch. 5)

Understanding the full scope of human impacts on climate requires a global focus because of the interconnected nature of the climate system. For example, the climate of the Arctic and the climate of the continental United States are connected through atmospheric circulation patterns. While the Arctic may seem remote to most Americans, the climatic effects of perturbations to arctic sea ice, land ice, surface temperature, snow cover, and permafrost affect the amount of warming, sea level change, carbon cycle impacts, and potentially even weather patterns in the lower 48 states. The Arctic is warming at a rate approximately twice as fast as the global average and, if it continues to warm at the same rate, Septembers will be nearly ice-free in the Arctic Ocean sometime between now and the 2040s (see Fig. ES.10). The important influence of arctic climate change on Alaska is apparent; the influence of arctic changes on U.S. weather over the coming decades remains an open question with the potential for significant impact. (Ch. 11)

Changes in the Tropics can also impact the rest of the globe, including the United States. There is growing evidence that the Tropics have expanded poleward by about 70 to 200 miles in each hemisphere over the period 1979–2009, with an accompanying shift of the subtropical dry zones, midlatitude jets, and storm tracks (*medium to high confidence*). Human activities have played a role in the change (*medium confidence*), although confidence is presently low regarding the magnitude of the human contribution relative to natural variability (Ch. 5).

(continued on next page)

**Box ES.1 (*continued*)**

## Large-Scale Patterns of Natural Variability Affect U.S. Climate

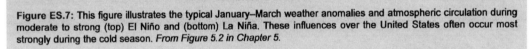

**Figure ES.7:** This figure illustrates the typical January–March weather anomalies and atmospheric circulation during moderate to strong (top) El Niño and (bottom) La Niña. These influences over the United States often occur most strongly during the cold season. *From Figure 5.2 in Chapter 5.*

## Oceans Are Rising, Warming, and Becoming More Acidic

Oceans occupy two-thirds of the planet's surface and host unique ecosystems and species, including those important for global commercial and subsistence fishing. Understanding climate impacts on the ocean and the ocean's feedbacks to the climate system is critical for a comprehensive understanding of current and future changes in climate.

### Global Ocean Heat

> The world's oceans have absorbed about 93% of the excess heat caused by greenhouse gas warming since the mid-20th century, making them warmer and altering global and regional climate feedbacks. (*Very high confidence*) (Ch. 13)

- Ocean heat content has increased at all depths since the 1960s and surface waters have warmed by about $1.3° \pm 0.1°F$ ($0.7° \pm 0.08°C$) per century globally since 1900 to 2016. Under higher scenarios, a global increase in average sea surface temperature of $4.9° \pm 1.3°F$ ($2.7° \pm 0.7°C$) is projected by 2100. (*Very high confidence*). (Ch. 13)

### Global and Regional Sea Level Rise

> Global mean sea level (GMSL) has risen by about 7–8 inches (about 16–21 cm) since 1900, with about 3 of those inches (about 7 cm) occurring since 1993 (*very high confidence*). (Ch. 12)

- Human-caused climate change has made a substantial contribution to GMSL rise since 1900 (*high confidence*), contributing to a rate of rise that is greater than during any preceding century in at least 2,800 years (*medium confidence*). (Ch. 12; Fig. ES.8)

- Relative to the year 2000, GMSL is *very likely* to rise by 0.3–0.6 feet (9–18 cm) by 2030, 0.5–1.2 feet (15–38 cm) by 2050, and 1.0–4.3 feet (30–130 cm) by 2100 (*very high confidence* in lower bounds; *medium confidence* in upper bounds for 2030 and 2050; *low* confidence in upper bounds for 2100). Future emissions pathways have little effect on projected GMSL rise in the first half of the century, but significantly affect projections for the second half of the century (*high confidence*). (Ch. 12)

## Recent Sea Level Rise Fastest for Over 2,000 Years

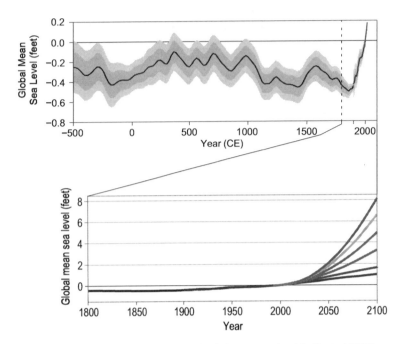

**Figure ES.8:** The top panel shows observed and reconstructed mean sea level for the last 2,500 years. The bottom panel shows projected mean sea level for six future scenarios. The six scenarios—spanning a range designed to inform a variety of decision makers—extend from a low scenario, consistent with continuation of the rate of sea level rise over the last quarter century, to an extreme scenario, assuming rapid mass loss from the Antarctic ice sheet. Note that the range on the vertical axis in the bottom graph is approximately ten times greater than in the top graph. *Based on Figure 12.2 and 12.4 in Chapter 12. See the main report for more details.*

- Emerging science regarding Antarctic ice sheet stability suggests that, for higher scenarios, a GMSL rise exceeding 8 feet (2.4 m) by 2100 is physically possible, although the probability of such an extreme outcome cannot currently be assessed. Regardless of emission pathway, it is *extremely likely* that GMSL rise will continue beyond 2100 (*high confidence*). (Ch. 12)

- Relative sea level rise in this century will vary along U.S. coastlines due, in part, to changes in Earth's gravitational field and rotation from melting of land ice, changes in ocean circulation, and vertical land motion (*very high confidence*). For almost all future GMSL rise scenarios, relative sea level rise is *likely* to be greater than the global average in the U.S. Northeast and the western Gulf of Mexico. In intermediate and low GMSL rise scenarios, relative sea level rise is *likely* to be less than the global average in much of the Pacific Northwest and Alaska. For high GMSL rise scenarios, relative sea level rise is *likely* to be higher than the global average along all U.S. coastlines outside Alaska. Almost all U.S. coastlines experience more than global mean sea level rise in response to Antarctic ice loss, and thus would be particularly affected under extreme GMSL rise scenarios involving substantial Antarctic mass loss (*high confidence*). (Ch. 12)

### Coastal Flooding

- As sea levels have risen, the number of tidal floods each year that cause minor impacts (also called "nuisance floods") have increased 5- to 10-fold since the 1960s in several U.S. coastal cities (*very high confidence*). Rates of increase are accelerating in over 25 Atlantic and Gulf Coast cities (*very high confidence*). Tidal flooding will continue increasing in depth, frequency, and extent this century (*very high confidence*). (Ch. 12)

## "Nuisance Flooding" Increases Across the United States

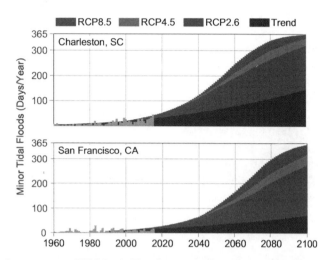

Figure ES. 9: Annual occurrences of tidal floods (days per year), also called sunny-day or nuisance flooding, have increased for some U.S. coastal cities. The figure shows historical exceedances (orange bars) for two of the locations—Charleston, SC and San Francisco, CA—and future projections through 2100. The projections are based upon the continuation of the historical trend (blue) and under median RCP2.6, 4.5 and 8.5 conditions. *From Figure 12.5, Chapter 12.*

- Assuming storm characteristics do not change, sea level rise will increase the frequency and extent of extreme flooding associated with coastal storms, such as hurricanes and nor'easters (*very high confidence*). A projected increase in the intensity of hurricanes in the North Atlantic (*medium confidence*) could increase the probability of extreme flooding along most of the U.S. Atlantic and Gulf Coast states beyond what would be projected based solely on relative sea level rise. However, there is *low confidence* in the projected increase in frequency of intense Atlantic hurricanes, and the associated flood risk amplification, and flood effects could be offset or amplified by such factors, such as changes in overall storm frequency or tracks. (Ch.12; Fig. ES. 9)

### Global Ocean Circulation

- The potential slowing of the Atlantic meridional overturning circulation (AMOC; of which the Gulf Stream is one component)—as a result of increasing ocean heat content and freshwater-driven buoyancy changes—could have dramatic climate feedbacks as the ocean absorbs less heat and $CO_2$ from the atmosphere. This slowing would also affect the climates of

North America and Europe. Any slowing documented to date cannot be directly tied to human-caused forcing, primarily due to lack of adequate observational data and to challenges in modeling ocean circulation changes. Under a higher scenario (RCP8.5), models show that the AMOC weakens over the 21st century (*low confidence*). (Ch. 13)

### Global and Regional Ocean Acidification

The world's oceans are currently absorbing more than a quarter of the $CO_2$ emitted to the atmosphere annually from human activities, making them more acidic (*very high confidence*), with potential detrimental impacts to marine ecosystems. (Ch. 13)

- Higher-latitude systems typically have a lower buffering capacity against changing acidity, exhibiting seasonally corrosive conditions sooner than low-latitude systems. The rate of acidification is unparalleled in at least the past 66 million years (*medium confidence*). Under the higher scenario (RCP8.5), the global average surface ocean acidity is projected to increase by 100% to 150% (*high confidence*). (Ch. 13)

- Acidification is regionally greater than the global average along U.S. coastal systems as a result of upwelling (e.g., in the Pacific Northwest) (*high confidence*), changes in freshwater inputs (e.g., in the Gulf of Maine) (*medium confidence*), and nutrient input (e.g., in urbanized estuaries) (*high confidence*). (Ch. 13)

### Ocean Oxygen

- Increasing sea surface temperatures, rising sea levels, and changing patterns of precipitation, winds, nutrients, and ocean circulation are contributing to overall declining oxygen concentrations at intermediate depths in various ocean locations and in many coastal areas. Over the last half century, major oxygen losses have occurred in inland seas, estuaries, and in the coastal and open ocean (*high confidence*). Ocean oxygen levels are projected to decrease by as much as 3.5% under the higher scenario (RCP8.5) by 2100 relative to preindustrial values (*high confidence*). (Ch. 13)

## Climate Change in Alaska and across the Arctic Continues to Outpace Global Climate Change

Residents of Alaska are on the front lines of climate change. Crumbling buildings, roads, and bridges and eroding shorelines are commonplace. Accelerated melting of multiyear sea ice cover, mass loss from the Greenland Ice Sheet, reduced snow cover, and permafrost thawing are stark examples of the rapid changes occurring in the Arctic. Furthermore, because elements of the climate system are interconnected (see Box ES.1), changes in the Arctic influence climate conditions outside the Arctic.

### Arctic Temperature Increases

> **Annual average near-surface air temperatures across Alaska and the Arctic have increased over the last 50 years at a rate more than twice as fast as the global average temperature. (*Very high confidence*) (Ch. 11)**

- Rising Alaskan permafrost temperatures are causing permafrost to thaw and become more discontinuous; this process releases additional carbon dioxide and methane resulting in additional warming (*high confidence*). The overall magnitude of the permafrost-carbon feedback is uncertain (Ch.2); however, it is clear that these emissions have the potential to compromise the ability to limit global temperature increases. (Ch. 11)

- Atmospheric circulation patterns connect the climates of the Arctic and the contiguous United States. Evidenced by recent record warm temperatures in the Arctic and emerging science, the midlatitude circulation has influenced observed arctic temperatures and sea ice (*high confidence*). However, confidence is low regarding whether or by what mechanisms observed arctic warming may have influenced the midlatitude circulation and weather patterns over the continental United States. The influence of arctic changes on U.S. weather over the coming decades remains an open question with the potential for significant impact. (Ch. 11)

### Arctic Land Ice Loss

- Arctic land ice loss observed in the last three decades continues, in some cases accelerating (*very high confidence*). It is *virtually certain* that Alaska glaciers have lost mass over the last 50 years, with each year since 1984 showing an annual average ice mass less than the previous year. Over the satellite record, average ice mass loss from Greenland was −269 Gt per year between April 2002 and April 2016, accelerating in recent years (*high confidence*). (Ch. 11)

### Arctic Sea Ice Loss

> **Since the early 1980s, annual average arctic sea ice has decreased in extent between 3.5% and 4.1% per decade, has become thinner by between 4.3 and 7.5 feet, and is melting at least 15 more days each year. September sea ice extent has decreased between 10.7% and 15.9% per decade. (*Very high confidence*) (Ch. 11)**

- Arctic sea ice loss is expected to continue through the 21st century, *very likely* resulting in nearly sea ice-free late summers by the 2040s (*very high confidence*). (Ch. 11)

- It is *very likely* that human activities have contributed to observed arctic surface temperature warming, sea ice loss, glacier mass loss, and northern hemisphere snow extent decline (*high confidence*). (Ch. 11)

## Multiyear Sea Ice Has Declined Dramatically

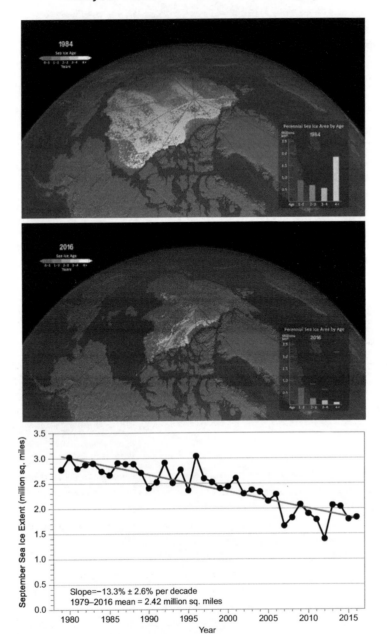

**Figure ES.10:** September sea ice extent and age shown for (top) 1984 and (middle) 2016, illustrating significant reductions in sea ice extent and age (thickness). The bar graph in the lower right of each panel illustrates the sea ice area (unit: million km²) covered within each age category (> 1 year), and the green bars represent the maximum extent for each age range during the record. The year 1984 is representative of September sea ice characteristics during the 1980s. The years 1984 and 2016 are selected as endpoints in the time series; a movie of the complete time series is available at http://svs.gsfc.nasa.gov/cgi-bin/details.cgi?aid=4489. (bottom) The satellite-era arctic sea ice areal extent trend from 1979 to 2016 for September (unit: million mi²). *From Figure 11.1 in Chapter 11.*

## Limiting Globally Averaged Warming to 2°C (3.6°F) Will Require Major Reductions in Emissions

Human activities are now the dominant cause of the observed trends in climate. For that reason, future climate projections are based on scenarios of how human activities will continue to affect the climate over the remainder of this century and beyond (see Sidebar: Scenarios Used in this Assessment). There remains significant uncertainty about future emissions due to changing economic, political, and demographic factors. For that reason, this report quantifies possible climate changes for a broad set of plausible future scenarios through the end of the century. (Ch. 2, 4, 10, 14)

> The observed increase in global carbon emissions over the past 15–20 years has been consistent with higher scenarios (e.g., RCP8.5) (*very high confidence*). In 2014 and 2015, emission growth rates slowed as economic growth became less carbon-intensive (*medium confidence*). Even if this slowing trend continues, however, it is not yet at a rate that would limit the increase in the global average temperature to well below 3.6°F (2°C) above preindustrial levels (*high confidence*). (Ch. 4)

- Global mean atmospheric carbon dioxide ($CO_2$) concentration has now passed 400 ppm, a level that last occurred about 3 million years ago, when global average temperature and sea level were significantly higher than today (*high confidence*). Continued growth in $CO_2$ emissions over this century and beyond would lead to an atmospheric concentration not experienced in tens of millions of years (*medium confidence*). The present-day emissions rate of nearly 10 GtC per year suggests that there is no climate analog for this century any time in at least the last 50 million years (*medium confidence*). (Ch. 4)

- Warming and associated climate effects from $CO_2$ emissions persist for decades to millennia. In the near-term, changes in climate are determined by past and present greenhouse gas emissions modified by natural variability. Reducing net emissions of $CO_2$ is necessary to limit near-term climate change and long-term warming. Other greenhouse gases (e.g., methane) and black carbon aerosols exert stronger warming effects than $CO_2$ on a per ton basis, but they do not persist as long in the atmosphere (Ch. 2); therefore, mitigation of non-$CO_2$ species contributes substantially to near-term cooling benefits but cannot be relied upon for ultimate stabilization goals. (*Very high confidence*) (Ch. 14)

> Choices made today will determine the magnitude of climate change risks beyond the next few decades. (Ch. 4, 14)

- Stabilizing global mean temperature to less than 3.6°F (2°C) above preindustrial levels requires substantial reductions in net global $CO_2$ emissions prior to 2040 relative to present-day values and likely requires net emissions to become zero or possibly negative later in the century. After accounting for the temperature effects of non-$CO_2$ species, cumulative global $CO_2$ emissions must stay below about 800 GtC in order to provide a two-thirds likelihood of preventing 3.6°F (2°C) of warming. Given estimated cumulative emissions since 1870, no more than approximately 230 GtC may be emitted in the future in order to remain under this temperature limit. Assuming global emissions are equal to or greater than those consistent with the RCP4.5 scenario, this cumulative carbon threshold would be exceeded in approximately two decades. (Ch. 14)

- Achieving global greenhouse gas emissions reductions before 2030 consistent with targets and actions announced by governments in the lead up to the 2015 Paris climate conference would hold open the possibility of meeting the long-term temperature goal of limiting global warming to 3.6°F (2°C) above preindustrial levels, whereas there would be virtually no chance if net global emissions followed a pathway well above those implied by country announcements. Actions in the announcements are, by themselves, insufficient to meet a 3.6°F (2°C) goal; the likelihood of achieving that depends strongly on the magnitude of global emissions reductions after 2030. (*High confidence*) (Ch. 14)

- Climate intervention or geoengineering strategies such as solar radiation management are measures that attempt to limit or reduce global temperature increases. Further assessments of the technical feasibilities, costs, risks, co-benefits, and governance challenges of climate intervention or geoengineering strategies, which are as yet unproven at scale, are a necessary step before judgments about the benefits and risks of these approaches can be made with high confidence. (*High confidence*) (Ch. 14)

- In recent decades, land-use and land-cover changes have turned the terrestrial biosphere (soil and plants) into a net "sink" for carbon (drawing down carbon from the atmosphere), and this sink has steadily increased since 1980 (*high confidence*). Because of the uncertainty in the trajectory of land cover, the possibility of the land becoming a net carbon source cannot be excluded (*very high confidence*). (Ch. 10)

### There is a Significant Possibility for Unanticipated Changes

Humanity's effect on the Earth system, through the large-scale combustion of fossil fuels and widespread deforestation and the resulting release of carbon dioxide ($CO_2$) into the atmosphere, as well as through emissions of other greenhouse gases and radiatively active substances from human activities, is unprecedented. There is significant potential for humanity's effect on the planet to result in unanticipated surprises and a broad consensus that the further and faster the Earth system is pushed towards warming, the greater the risk of such surprises.

There are at least two types of potential surprises: *compound events*, where multiple extreme climate events occur simultaneously or sequentially (creating greater overall impact), and *critical threshold* or *tipping point events*, where some threshold is crossed in the climate system (that leads to large impacts). The probability of such surprises—some of which may be abrupt and/or irreversible—as well as other more predictable but difficult-to-manage impacts, increases as the influence of human activities on the climate system increases. (Ch. 15)

Unanticipated and difficult or impossible-to-manage changes in the climate system are possible throughout the next century as critical thresholds are crossed and/or multiple climate-related extreme events occur simultaneously. (Ch. 15)

- Positive feedbacks (self-reinforcing cycles) within the climate system have the potential to accelerate human-induced climate change and even shift the Earth's climate system, in part or in whole, into new states that are very different from those experienced in the recent past (for example, ones with greatly diminished ice sheets or different large-scale patterns of atmosphere or ocean circulation). Some feedbacks and potential state shifts can be modeled and quantified; others can be modeled or identified but not quantified; and some are probably still unknown. (*Very high confidence* in the potential for state shifts and in the incompleteness of knowledge about feedbacks and potential state shifts). (Ch. 15)

- The physical and socioeconomic impacts of compound extreme events (such as simultaneous heat and drought, wildfires associated with hot and dry conditions, or flooding associated with high precipitation on top of snow or waterlogged ground) can be greater than the sum of the parts (*very high confidence*). Few analyses consider the spatial or temporal correlation between extreme events. (Ch. 15)

- While climate models incorporate important climate processes that can be well quantified, they do not include all of the processes that can contribute to feedbacks (Ch. 2), compound extreme events, and abrupt and/or irreversible changes. For this reason, future changes outside the range projected by climate models cannot be ruled out (*very high confidence*). Moreover, the systematic tendency of climate models to underestimate temperature change during warm paleoclimates suggests that climate models are more likely to underestimate than to overestimate the amount of long-term future change (*medium confidence*). (Ch. 15)

## Box ES.2: A Summary of Advances Since NCA3

Advances in scientific understanding and scientific approach, as well as developments in global policy, have occurred since NCA3. A detailed summary of these advances can be found at the end of Chapter 1: Our Globally Changing Climate. Highlights of what aspects are either especially strengthened or are emerging in the current findings include

- *Detection and attribution*: Significant advances have been made in the attribution of the human influence for individual climate and weather extreme events since NCA3. (Chapters 3, 6, 7, 8).

- *Atmospheric circulation and extreme events*: The extent to which atmospheric circulation in the midlatitudes is changing or is projected to change, possibly in ways not captured by current climate models, is a new important area of research. (Chapters 5, 6, 7).

- *Increased understanding of specific types of extreme events*: How climate change may affect specific types of extreme events in the United States is another key area where scientific understanding has advanced. (Chapter 9).

- *High-resolution global climate model simulations*: As computing resources have grown, multidecadal simulations of global climate models are now being conducted at horizontal resolutions on the order of 15 miles (25 km) that provide more realistic characterization of intense weather systems, including hurricanes. (Chapter 9).

- *Oceans and coastal waters*: Ocean acidification, warming, and oxygen loss are all increasing, and scientific understanding of the severity of their impacts is growing. Both oxygen loss and acidification may be magnified in some U.S. coastal waters relative to the global average, raising the risk of serious ecological and economic consequences. (Chapters 2, 13).

- *Local sea level change projections*: For the first time in the NCA process, sea level rise projections incorporate geographic variation based on factors such as local land subsidence, ocean currents, and changes in Earth's gravitational field. (Chapter 12).

- *Accelerated ice-sheet loss*: New observations from many different sources confirm that ice-sheet loss is accelerating. Combining observations with simultaneous advances in the physical understanding of ice sheets leads to the conclusion that up to 8.5 feet of global sea level rise is possible by 2100 under a higher scenario (RCP8.5), up from 6.6 feet in NCA3. (Chapter 12).

- *Low sea-ice areal extent*: The annual arctic sea ice extent minimum for 2016 relative to the long-term record was the second lowest on record. The arctic sea ice minimums in 2014 and 2015 were also amongst the lowest on record. Since 1981, the sea ice minimum has decreased by 13.3% per decade, more than 46% over the 35 years. The annual arctic sea ice maximum in March 2017 was the lowest maximum areal extent on record. (Chapter 11).

- *Potential surprises*: Both large-scale state shifts in the climate system (sometimes called "tipping points") and compound extremes have the potential to generate unanticipated climate surprises. The further the Earth system departs from historical climate forcings, and the more the climate changes, the greater the potential for these surprises. (Chapter 15).

- *Mitigation*: This report discusses some important aspects of climate science that are relevant to long-term temperature goals and different mitigation scenarios, including those implied by government announcements for the Paris Agreement. (Chapters 4, 14).

# 1

# Our Globally Changing Climate

**KEY FINDINGS**

1. The global climate continues to change rapidly compared to the pace of the natural variations in climate that have occurred throughout Earth's history. Trends in globally averaged temperature, sea level rise, upper-ocean heat content, land-based ice melt, arctic sea ice, depth of seasonal permafrost thaw, and other climate variables provide consistent evidence of a warming planet. These observed trends are robust and have been confirmed by multiple independent research groups around the world. (*Very high confidence*)

2. The frequency and intensity of extreme heat and heavy precipitation events are increasing in most continental regions of the world (*very high confidence*). These trends are consistent with expected physical responses to a warming climate. Climate model studies are also consistent with these trends, although models tend to underestimate the observed trends, especially for the increase in extreme precipitation events (*very high confidence* for temperature, *high confidence* for extreme precipitation). The frequency and intensity of extreme high temperature events are *virtually certain* to increase in the future as global temperature increases (*high confidence*). Extreme precipitation events will *very likely* continue to increase in frequency and intensity throughout most of the world (*high confidence*). Observed and projected trends for some other types of extreme events, such as floods, droughts, and severe storms, have more variable regional characteristics.

3. Many lines of evidence demonstrate that it is *extremely likely* that human influence has been the dominant cause of the observed warming since the mid-20th century. Formal detection and attribution studies for the period 1951 to 2010 find that the observed global mean surface temperature warming lies in the middle of the range of likely human contributions to warming over that same period. We find no convincing evidence that natural variability can account for the amount of global warming observed over the industrial era. For the period extending over the last century, there are no convincing alternative explanations supported by the extent of the observational evidence. Solar output changes and internal variability can only contribute marginally to the observed changes in climate over the last century, and we find no convincing evidence for natural cycles in the observational record that could explain the observed changes in climate. (*Very high confidence*)

4. Global climate is projected to continue to change over this century and beyond. The magnitude of climate change beyond the next few decades will depend primarily on the amount of greenhouse (heat-trapping) gases emitted globally and on the remaining uncertainty in the sensitivity of Earth's climate to those emissions (*very high confidence*). With significant reductions in the emissions of greenhouse gases, the global annually averaged temperature rise could be limited to 3.6°F (2°C) or less. Without major reductions in these emissions, the increase in annual average global temperatures relative to preindustrial times could reach 9°F (5°C) or more by the end of this century (*high confidence*).

(continued on next page)

**KEY FINDINGS (continued)**

5. Natural variability, including El Niño events and other recurring patterns of ocean–atmosphere interactions, impact temperature and precipitation, especially regionally, over months to years. The global influence of natural variability, however, is limited to a small fraction of observed climate trends over decades. (*Very high confidence*)

6. Longer-term climate records over past centuries and millennia indicate that average temperatures in recent decades over much of the world have been much higher, and have risen faster during this time period, than at any time in the past 1,700 years or more, the time period for which the global distribution of surface temperatures can be reconstructed. (*High confidence*)

**Recommended Citation for Chapter**

**Wuebbles**, D.J., D.R. Easterling, K. Hayhoe, T. Knutson, R.E. Kopp, J.P. Kossin, K.E. Kunkel, A.N. LeGrande, C. Mears, W.V. Sweet, P.C. Taylor, R.S. Vose, and M.F. Wehner, 2017: Our globally changing climate. In: *Climate Science Special Report: Fourth National Climate Assessment, Volume I* [Wuebbles, D.J., D.W. Fahey, K.A. Hibbard, D.J. Dokken, B.C. Stewart, and T.K. Maycock (eds.)]. U.S. Global Change Research Program, Washington, DC, USA, pp. 35-72, doi: 10.7930/J08S4N35.

## 1.1 Introduction

Since the Third U.S. National Climate Assessment (NCA3) was published in May 2014, new observations along multiple lines of evidence have strengthened the conclusion that Earth's climate is changing at a pace and in a pattern not explainable by natural influences. While this report focuses especially on observed and projected future changes for the United States, it is important to understand those changes in the global context (this chapter).

The world has warmed over the last 150 years, especially over the last six decades, and that warming has triggered many other changes to Earth's climate. Evidence for a changing climate abounds, from the top of the atmosphere to the depths of the oceans. Thousands of studies conducted by tens of thousands of scientists around the world have documented changes in surface, atmospheric, and oceanic temperatures; melting glaciers; disappearing snow cover; shrinking sea ice; rising sea level; and an increase in atmospheric water vapor.

Rainfall patterns and storms are changing, and the occurrence of droughts is shifting.

Many lines of evidence demonstrate that human activities, especially emissions of greenhouse gases, are primarily responsible for the observed climate changes in the industrial era, especially over the last six decades (see attribution analysis in Ch. 3: Detection and Attribution). Formal detection and attribution studies for the period 1951 to 2010 find that the observed global mean surface temperature warming lies in the middle of the range of likely human contributions to warming over that same period. The Intergovernmental Panel on Climate Change concluded that it is extremely likely that human influence has been the dominant cause of the observed warming since the mid-20th century.[1] Over the last century, there are no alternative explanations supported by the evidence that are either credible or that can contribute more than marginally to the observed patterns. There is no convincing evidence that natural variability can account for the amount of and the pattern of global warming observed over the industrial era.[2, 3, 4, 5] Solar flux variations over the last six decades have been too small to explain the observed changes in climate.[6, 7, 8] There are no apparent natural cycles in the observational record that can explain the recent changes in climate (e.g., PAGES 2k Consortium 2013;[9] Marcott et al. 2013;[10] Otto-Bliesner et al. 2016[11]). In addition, natural cycles within Earth's climate system can only redistribute heat; they cannot be responsible for the observed increase in the overall heat content of the climate system.[12] Any explanations for the observed changes in climate must be grounded in understood physical mechanisms, appropriate in scale, and consistent in timing and direction with the long-term observed trends. Known human activities quite reasonably explain what has happened without the need for other factors. Internal variability and forcing factors other than human activities cannot explain what is happening, and there are no suggested factors, even speculative ones, that can explain the timing or magnitude and that would somehow cancel out the role of human factors.[3, 13] The science underlying this evidence, along with the observed and projected changes in climate, is discussed in later chapters, starting with the basis for a human influence on climate in Chapter 2: Physical Drivers of Climate Change.

Throughout this report, we also analyze projections of future changes in climate. As discussed in Chapter 4, beyond the next few decades, the magnitude of climate change depends primarily on cumulative emissions of greenhouse gases and aerosols and the sensitivity of the climate system to those emissions. Predicting how climate will change in future decades is a different scientific issue from predicting weather a few weeks from now. Local weather is short term, with limited predictability, and is determined by the complicated movement and interaction of high pressure and low pressure systems in the atmosphere; thus, it is difficult to forecast day-to-day

changes beyond about two weeks into the future. Climate, on the other hand, is the statistics of weather—meaning not just average values but also the prevalence and intensity of extremes—as observed over a period of decades. Climate emerges from the interaction, over time, of rapidly changing local weather and more slowly changing regional and global influences, such as the distribution of heat in the oceans, the amount of energy reaching Earth from the sun, and the composition of the atmosphere. See Chapter 4: Projections and later chapters for more on climate projections.

Throughout this report, we include many findings that further strengthen or add to the understanding of climate change relative to those found in NCA3 and other assessments of the science. Several of these are highlighted in an "Advances Since NCA3" box at the end of this chapter.

## 1.2 Indicators of a Globally Changing Climate

Highly diverse types of direct measurements made on land, sea, and in the atmosphere over many decades have allowed scientists to conclude with high confidence that global mean temperature is increasing. Observational datasets for many other climate variables support the conclusion with high confidence that the global climate is changing (also see EPA 2016[14]).[15, 16] Figure 1.1 depicts several of the observational indicators that demonstrate trends consistent with a warming planet over the last century. Temperatures in the lower atmosphere and ocean have increased, as have near-surface humidity and sea level. Not only has ocean heat content increased dramatically (Figure 1.1), but more than 90% of the energy gained in the combined ocean–atmosphere system over recent decades has gone into the ocean.[17, 18] Five different observational datasets show the heat content of the oceans is increasing.

## Indicators of Warming from Multiple Datasets

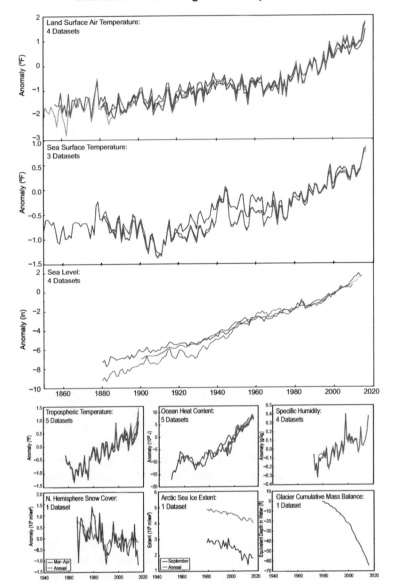

**Figure 1.1:** This image shows observations globally from nine different variables that are key indicators of a warming climate. The indicators (listed below) all show long-term trends that are consistent with global warming. In parentheses are the number of datasets shown in each graph, the length of time covered by the combined datasets and their anomaly reference period (where applicable), and the direction of the trend: land surface air temperature (4 datasets, 1850–2016 relative to 1976–2005, increase); sea surface temperature (3 datasets, 1850–2016 relative to 1976–2005, increase); sea level (4 datasets, 1880–2014 relative to 1996–2005, increase); tropospheric temperature (5 datasets, 1958–2016 relative to 1981–2005, increase); ocean heat content, upper 700m (5 datasets, 1950–2016 relative to 1996–2005, increase); specific humidity (4 datasets, 1973–2016 relative to 1980–2003, increase); Northern Hemisphere snow cover, March–April and annual (1 dataset, 1967–2016 relative to 1976–2005, decrease); arctic sea ice extent, September and annual (1 dataset, 1979–2016, decrease); glacier cumulative mass balance (1 dataset, 1980–2016, decrease). More information on the datasets can be found in the accompanying metadata. (Figure source: NOAA NCEI and CICS-NC, updated from Melillo et al. 2014;[144] Blunden and Arndt 2016[15]).

Basic physics tells us that a warmer atmosphere can hold more water vapor; this is exactly what is measured from satellite data. At the same time, a warmer world means higher evaporation rates and major changes to the hydrological cycle (e.g., Kundzewicz 2008;[19] IPCC 2013[1]), including increases in the prevalence of torrential downpours. In addition, arctic sea ice, mountain glaciers, and Northern Hemisphere spring snow cover have all decreased. The relatively small increase in Antarctic sea ice in the 15-year period from 2000 through early 2016 appears to be best explained as being due to localized natural variability (see e.g., Meehl et al. 2016;[16] Ramsayer 2014[20]); while possibly also related to natural variability, the 2017 Antarctic sea ice minimum reached in early March was the lowest measured since reliable records began in 1979. The vast majority of the glaciers in the world are losing mass at significant rates. The two largest ice sheets on our planet—on the land masses of Greenland and Antarctica—are shrinking.

Many other indicators of the changing climate have been determined from other observations—for example, changes in the growing season and the allergy season (see e.g., EPA 2016;[14] USGCRP 2017[21]). In general, the indicators demonstrate continuing changes in climate since the publication of NCA3. As with temperature, independent researchers have analyzed each of these indicators and come to the same conclusion: all of these changes paint a consistent and compelling picture of a warming planet.

### 1.3 Trends in Global Temperatures

Global annual average temperature (as calculated from instrumental records over both land and oceans; used interchangeably with global average temperature in the discussion below) has increased by more than 1.2°F (0.7°C) for the period 1986–2016 relative to

1901–1960 (Figure 1.2); see Vose et al.[22] for discussion on how global annual average temperature is derived by scientists. The linear regression change over the entire period from 1901–2016 is 1.8°F (1.0°C). Global average temperature is not expected to increase smoothly over time in response to the human warming influences, because the warming trend is superimposed on natural variability associated with, for example, the El Niño/La Niña ocean-heat oscillations and the cooling effects of particles emitted by volcanic eruptions. Even so, 16 of the 17 warmest years in the instrumental record (since the late 1800s) occurred in the period from 2001 to 2016 (1998 was the exception). Global average temperature for 2016 has now surpassed 2015 by a small amount as the warmest year on record. The year 2015 far surpassed 2014 by 0.29°F (0.16°C), four times greater than the difference between 2014 and the next warmest year, 2010.[23] Three of the four warmest years on record have occurred since the analyses through 2012 were reported in NCA3.

A strong El Niño contributed to 2015's record warmth.[15] Though an even more powerful El Niño occurred in 1998, the global temperature in that year was significantly lower (by 0.49°F [0.27°C]) than that in 2015. This suggests that human-induced warming now has a stronger influence on the occurrence of record temperatures than El Niño events. In addition, the El Niño/La Niña cycle may itself be affected by the human influence on Earth's climate system.[3, 24] It is the complex interaction of natural sources of variability with the continuously growing human warming influence that is now shaping Earth's weather and, as a result, its climate.

Globally, the persistence of the warming over the past 60 years far exceeds what can be accounted for by natural variability alone.[1] That does not mean, of course, that natural sources

## Global Land and Ocean Temperature Anomalies

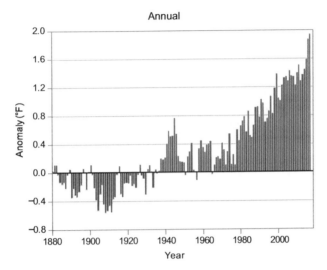

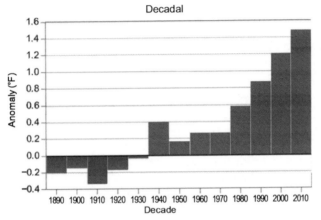

**Figure 1.2:** Top: Global annual average temperatures (as measured over both land and oceans) for 1880–2016 relative to the reference period of 1901–1960; red bars indicate temperatures above the average over 1901–1960, and blue bars indicate temperatures below the average. Global annual average temperature has increased by more than 1.2°F (0.7°C) for the period 1986–2016 relative to 1901–1960. While there is a clear long-term global warming trend, some years do not show a temperature increase relative to the previous year, and some years show greater changes than others. These year-to-year fluctuations in temperature are mainly due to natural sources of variability, such as the effects of El Niños, La Niñas, and volcanic eruptions. Based on the NCEI (NOAAGlobalTemp) dataset (updated from Vose et al.[22]) Bottom: Global average temperature averaged over decadal periods (1886–1895, 1896–1905, ..., 1996–2005, except for the 11 years in the last period, 2006–2016). Horizontal label indicates midpoint year of decadal period. Every decade since 1966–1975 has been warmer than the previous decade. (Figure source: [top] adapted from NCEI 2016,[23] [bottom] NOAA NCEI and CICS-NC).

of variability have become insignificant. They can be expected to continue to contribute a degree of "bumpiness" in the year-to-year global average temperature trajectory, as well as exert influences on the average rate of warming that can last a decade or more (see Box 1.1).[25, 26, 27]

Warming during the first half of the 1900s occurred mostly in the Northern Hemisphere.[28] Recent decades have seen greater warming in response to accelerating increases in green-house gas concentrations, particularly at high northern latitudes, and over land as compared to the ocean (see Figure 1.3). In general, winter is warming faster than summer (especially in northern latitudes). Also, nights are warming faster than days.[29, 30] There is also some evidence of faster warming at higher elevations.[31]

Most ocean areas around Earth are warming (see Ch. 13: Ocean Changes). Even in the absence of significant ice melt, the ocean is expected to warm more slowly given its larger

## Surface Temperature Change

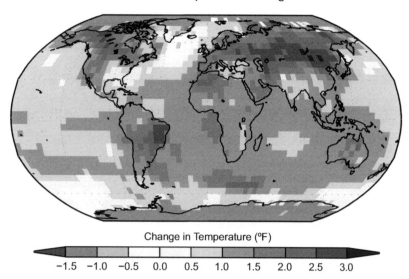

Change in Temperature (°F)

-1.5  -1.0  -0.5  0.0  0.5  1.0  1.5  2.0  2.5  3.0

Figure 1.3: Surface temperature change (in °F) for the period 1986–2015 relative to 1901–1960 from the NOAA National Centers for Environmental Information's (NCEI) surface temperature product. For visual clarity, statistical significance is not depicted on this map. Changes are generally significant (at the 90% level) over most land and ocean areas. Changes are not significant in parts of the North Atlantic Ocean, the South Pacific Ocean, and the southeastern United States. There is insufficient data in the Arctic Ocean and Antarctica for computing long-term changes (those sections are shown in gray because no trend can be derived). The relatively coarse resolution (5.0° × 5.0°) of these maps does not capture the finer details associated with mountains, coastlines, and other small-scale effects (see Ch. 6: Temperature Changes for a focus on the United States). (Figure source: updated from Vose et al. 2012[22]).

heat capacity, leading to land–ocean differences in warming (as seen in Figure 1.3). As a result, the climate for land areas often responds more rapidly than the ocean areas, even though the forcing driving a change in climate occurs equally over land and the oceans.[1] A few regions, such as the North Atlantic Ocean, have experienced cooling over the last century, though these areas have warmed over recent decades. Regional climate variability is important to determining potential effects of climate change on the ocean circulation (e.g., Hurrell and Deser 2009;[32] Hoegh-Guldberg et al. 2014[33]) as are the effects of the increasing freshwater in the North Atlantic from melting of sea and land ice.[34]

Figure 1.4 shows the projected changes in globally averaged temperature for a range of future pathways that vary from assuming strong continued dependence on fossil fuels in energy and transportation systems over the 21st century (the high scenario is Representative Concentration Pathway 8.5, or RCP8.5) to assuming major emissions reduction (the even lower scenario, RCP2.6). Chapter 4: Projections describes the future scenarios and the models of Earth's climate system being used to quantify the impact of human choices and natural variability on future climate. These analyses also suggest that global surface temperature increases for the end of the 21st century are *very likely* to exceed 1.5°C (2.7°F) relative to the 1850–1900 average for all projections, with the exception of the lowest part of the uncertainty range for RCP2.6.[1, 35, 36, 37]

## Projected Global Temperatures

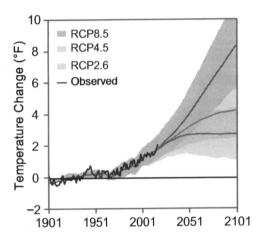

Figure 1.4: Multimodel simulated time series from 1900 to 2100 for the change in global annual mean surface temperature relative to 1901–1960 for a range of the Representative Concentration Pathways (RCPs; see Ch. 4: Projections for more information). These scenarios account for the uncertainty in future emissions from human activities (as analyzed with the 20+ models from around the world used in the most recent international assessment[1]). The mean (solid lines) and associated uncertainties (shading, showing ±2 standard deviations [5%–95%] across the distribution of individual models based on the average over 2081–2100) are given for all of the RCP scenarios as colored vertical bars. The numbers of models used to calculate the multimodel means are indicated. (Figure source: adapted from Walsh et al. 2014[201]).

## Box 1.1: Was there a "Hiatus" in Global Warming?

Natural variability in the climate system leads to year-to-year and decade-to-decade changes in global mean temperature. For short enough periods of time, this variability can lead to temporary slowdowns or even reversals in the globally-averaged temperature increase. Focusing on overly short periods can lead to incorrect conclusions about longer-term changes. Over the past decade, such a slowdown led to numerous assertions about a "hiatus" (a period of zero or negative temperature trend) in global warming over the previous 1.5 decades, which is not found when longer periods are analyzed (see Figure 1.5).[38] Thus the surface and tropospheric temperature records do not support the assertion that long-term (time periods of 25 years or longer) global warming has ceased or substantially slowed,[39, 40] a conclusion further reinforced by recently updated and improved datasets.[26, 41, 42, 43]

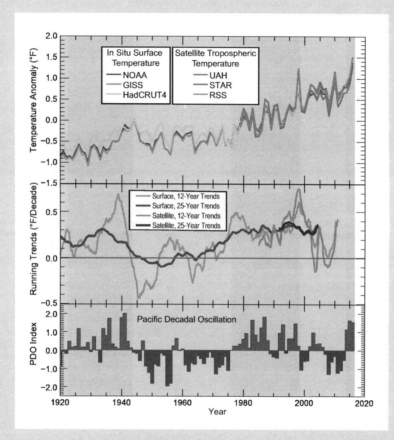

Figure 1.5: Panel A shows the annual mean temperature anomalies relative to a 1901–1960 baseline for global mean surface temperature and global mean tropospheric temperature. Short-term variability is superposed on a long-term warming signal, particularly since the 1960s. Panel B shows the linear trend of short (12-year) and longer (25-year) overlapping periods plotted at the time of the center of the trend period. For the longer period, trends are positive and nearly constant since about 1975. Panel C shows the annual mean Pacific Decadal Oscillation (PDO) index. Short-term temperature trends show a marked tendency to be lower during periods of generally negative PDO index, shown by the blue shading. (Figure source: adapted and updated from Trenberth 2015[3] and Santer et al. 2017;[38] Panel B, © American Meteorological Society. Used with permission.)

(continued on next page)

## Box 1.1 *(continued)*

For the 15 years following the 1997–1998 El Niño–Southern Oscillation (ENSO) event, the observed rate of temperature increase was smaller than the underlying long-term increasing trend on 30-year climate time scales,[44] even as other measures of global warming such as ocean heat content (see Ch. 13: Ocean Changes) and arctic sea ice extent (see Ch. 12: Sea Level Rise) continued to change.[45] Variation in the rate of warming on this time scale is not unexpected and can be the result of long-term internal variability in the climate system, or short-term changes in climate forcings such as aerosols or solar irradiance. Temporary periods similar or larger in magnitude to the current slowdown have occurred earlier in the historical record.

Even though such slowdowns are not unexpected, the slowdown of the early 2000s has been used as informal evidence to cast doubt on the accuracy of climate projections from CMIP5 models, since the measured rate of warming in all surface and tropospheric temperature datasets from 2000 to 2014 was less than expected given the results of the CMIP3 and CMIP5 historical climate simulations.[38] Thus, it is important to explore a physical explanation of the recent slowdown and to identify the relative contributions of different factors.

Numerous studies have investigated the role of natural modes of variability and how they affected the flow of energy in the climate system of the post-2000 period.[16, 46, 47, 48, 49] For the 2000–2013 time period, they find

- In the Pacific Ocean, a number of interrelated features, including cooler than expected tropical ocean surface temperatures, stronger than normal trade winds, and a shift to the cool phase of the Pacific Decadal Oscillation (PDO) led to cooler than expected surface temperatures in the Eastern Tropical Pacific, a region that has been shown to have an influence on global-scale climate.[49]

- For most of the world's oceans, heat was transferred from the surface into the deeper ocean,[46, 47, 50, 51] causing a reduction in surface warming worldwide.

- Other studies attributed part of the cause of the measurement/model discrepancy to natural fluctuations in radiative forcings, such as volcanic aerosols, stratospheric water vapor, or solar output.[52, 53, 54, 55, 56]

When comparing model predictions with measurements, it is important to note that the CMIP5 runs used an assumed representation of these factors for time periods after 2000, possibly leading to errors, especially in the year-to-year simulation of internal variability in the oceans. It is *very likely* that the early 2000s slowdown was caused by a combination of short-term variations in forcing and internal variability in the climate system, though the relative contribution of each is still an area of active research .

Although 2014 already set a new high in globally averaged temperature record up to that time, in 2015–2016, the situation changed dramatically. A switch of the PDO to the positive phase, combined with a strong El Niño event during the fall and winter of 2015–2016, led to months of record-breaking globally averaged temperatures in both the surface and satellite temperature records (see Figure 1.5),[3] bringing observed temperature trends into better agreement with model expectations (see Figure 1.6).

(continued on next page)

## Box 1.1 *(continued)*

On longer time scales, observed temperature changes and model simulations are more consistent. The observed temperature changes on longer time scales have also been attributed to anthropogenic causes with high confidence (see Ch. 3: Detection and Attribution for further discussion).[6] The pronounced globally averaged surface temperature record of 2015 and 2016 appear to make recent observed temperature changes more consistent with model simulations—including with CMIP5 projections that were (notably) developed in advance of occurrence of the 2015–2016 observed anomalies (Figure 1.6). A second important point illustrated by Figure 1.6 is the broad overall agreement between observations and models on the century time scale, which is robust to the shorter-term variations in trends in the past decade or so. Continued global warming and the frequent setting of new high global mean temperature records or near-records is consistent with expectations based on model projections of continued anthropogenic forcing toward warmer global mean conditions.

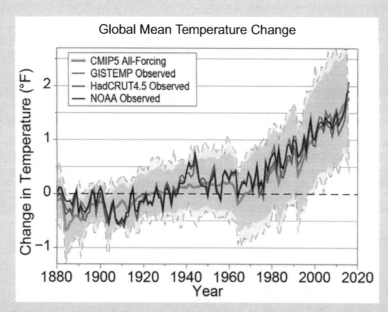

Figure 1.6: Comparison of global mean temperature anomalies (°F) from observations (through 2016) and the CMIP5 multimodel ensemble (through 2016), using the reference period 1901–1960. The CMIP5 multimodel ensemble (orange range) is constructed from blended surface temperature (ocean regions) and surface air temperature (land regions) data from the models, masked where observations are not available in the GISTEMP dataset.[27] The importance of using blended model data is shown in Richardson et al.[42] The thick solid orange curve is the model ensemble mean, formed from the ensemble across 36 models of the individual model ensemble means. The shaded region shows the +/- two standard deviation range of the individual ensemble member annual means from the 36 CMIP5 models. The dashed lines show the range from maximum to minimum values for each year among these ensemble members. The sources for the three observational indices are: HadCRUT4.5 (red): http://www.metoffice.gov.uk/hadobs/hadcrut4/data/current/download.html; NOAA (black): https://www.ncdc.noaa.gov/monitoring-references/faq/anomalies.php; and GISTEMP (blue): https://data.giss.nasa.gov/pub/gistemp/gistemp1200_ERSSTv4.nc. (NOAA and HadCRUT4 downloaded on Feb. 15, 2017; GISTEMP downloaded on Feb. 10, 2017). (Figure source: adapted from Knutson et al. 2016[27]).

## 1.4 Trends in Global Precipitation

Annual averaged precipitation across global land areas exhibits a slight rise (that is not statistically significant because of a lack of data coverage early in the record) over the past century (see Figure 1.7) along with ongoing increases in atmospheric moisture levels. Interannual and interdecadal variability is clearly found in all precipitation evaluations, owing to factors such as the North Atlantic Oscillation (NAO) and ENSO—note that precipitation reconstructions are updated operationally by NOAA NCEI on a monthly basis.[57, 58]

The hydrological cycle and the amount of global mean precipitation is primarily controlled by the atmosphere's energy budget and its interactions with clouds.[59] The amount of global mean precipitation also changes as a result of a mix of fast and slow atmospheric responses to the changing climate.[60] In the long term, increases in tropospheric radiative effects from increasing amounts of atmospheric $CO_2$ (i.e., increasing $CO_2$ leads to greater energy absorbed by the atmosphere and

re-emitted to the surface, with the additional transport to the atmosphere coming by convection) must be balanced by increased latent heating, resulting in precipitation increases of approximately 0.55% to 0.72% per °F (1% to 3% per °C).[1, 61] Global atmospheric water vapor should increase by about 6%–7% per °C of warming based on the Clausius–Clapeyron relationship (see Ch. 2: Physical Drivers of Climate Change); satellite observations of changes in precipitable water over oceans have been detected at about this rate and attributed to human-caused changes in the atmosphere.[62] Similar observed changes in land-based measurements have also been attributed to the changes in climate from greenhouse gases.[63]

Earlier studies suggested a climate change pattern of wet areas getting wetter and dry areas getting drier (e.g., Greve et al. 2014[64]). While Hadley Cell expansion should lead to more drying in the subtropics, the poleward shift of storm tracks should lead to enhanced wet regions. While this high/low rainfall behavior appears to be valid over ocean areas,

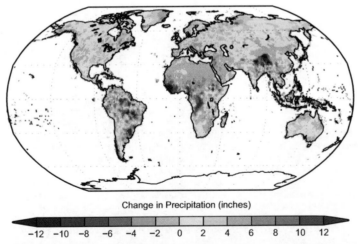

Change in Precipitation (inches)

−12 −10 −8 −6 −4 −2 0 2 4 6 8 10 12

**Figure 1.7:** Surface annually averaged precipitation change (in inches) for the period 1986–2015 relative to 1901–1960. The data is from long-term stations, so precipitation changes over the ocean and Antarctica cannot be evaluated. The trends are not considered to be statistically significant because of a lack of data coverage early in the record. The relatively coarse resolution (0.5° × 0.5°) of these maps does not capture the finer details associated with mountains, coastlines, and other small-scale effects. (Figure source: NOAA NCEI and CICS-NC).

changes over land are more complicated. The wet versus dry pattern in observed precipitation has only been attributed for the zonal mean[65, 66] and not regionally due to the large amount of spatial variation in precipitation changes as well as significant natural variability. The detected signal in zonal mean precipitation is largest in the Northern Hemisphere, with decreases in the subtropics and increases at high latitudes. As a result, the observed increase (about 5% since the 1950s[67, 68]) in annual averaged arctic precipitation have been detected and attributed to human activities.[69]

## 1.5 Trends in Global Extreme Weather Events

A change in the frequency, duration, and/or magnitude of extreme weather events is one of the most important consequences of a warming climate. In statistical terms, a small shift in the mean of a weather variable, with or without this shift occurring in concert with a change in the shape of its probability distribution, can cause a large change in the probability of a value relative to an extreme threshold (see Figure 1.8 in IPCC 2013[1]).[70] Examples include extreme high temperature events and heavy precipitation events. Some of the other extreme events, such as intense tropical cyclones, midlatitude cyclones, lightning, and hail and tornadoes associated with thunderstorms can occur as more isolated events and generally have more limited temporal and spatial observational datasets, making it more difficult to study their long-term trends. Detecting trends in the frequency and intensity of extreme weather events is challenging.[71] The most intense events are rare by definition, and observations may be incomplete and suffer from reporting biases. Further discussion on trends and projections of extreme events for the United States can be found in Chapters 6–9 and 11.

An emerging area in the science of detection and attribution has been the attribution of

extreme weather and climate events. Extreme event attribution generally addresses the question of whether climate change has altered the odds of occurrence of an extreme event like one just experienced. Attribution of extreme weather events under a changing climate is now an important and highly visible aspect of climate science. As discussed in a recent National Academy of Sciences (NAS) report,[72] the science of event attribution is rapidly advancing, including the understanding of the mechanisms that produce extreme events and the development of methods that are used for event attribution. Several other reports and papers have reviewed the topic of extreme event attribution.[73, 74, 75] This report briefly reviews extreme event attribution methodologies in practice (Ch. 3: Detection and Attribution) and provides a number of examples within the chapters on various climate phenomena (especially relating to the United States in Chapters 6–9).

### Extreme Heat and Cold
The frequency of multiday heat waves and extreme high temperatures at both daytime and nighttime hours is increasing over many of the global land areas.[1] There are increasing areas of land throughout our planet experiencing an excess number of daily highs above given thresholds (for example, the 90th percentile), with an approximate doubling of the world's land area since 1998 with 30 extreme heat days per year.[76] At the same time, frequencies of cold waves and extremely low temperatures are decreasing over the United States and much of the earth. In the United States, the number of record daily high temperatures has been about double the number of record daily low temperatures in the 2000s,[77] and much of the United States has experienced decreases of 5%–20% per decade in cold wave frequency.[1, 75]

The enhanced radiative forcing caused by greenhouse gases has a direct influence on heat extremes by shifting distributions of daily temperature.[78] Recent work indicates changes in atmospheric circulation may also play a significant role (see Ch. 5: Circulation and Variability). For example, a recent study found that increasing anticyclonic circulations partially explain observed trends in heat events over North America and Eurasia, among other effects.[79] Observed changes in circulation may also be the result of human influences on climate, though this is still an area of active research.

*Extreme Precipitation*

A robust consequence of a warming climate is an increase in atmospheric water vapor, which exacerbates precipitation events under similar meteorological conditions, meaning that when rainfall occurs, the amount of rain falling in that event tends to be greater. As a result, what in the past have been considered to be extreme precipitation events are becoming more frequent.[1, 80, 81, 82] On a global scale, the observational annual-maximum daily precipitation has increased by 8.5% over the last 110 years; global climate models also derive an increase in extreme precipitation globally but tend to underestimate the rate of the observed increase.[80, 82, 83] Extreme precipitation events are increasing in frequency globally over both wet and dry regions.[82] Although more spatially heterogeneous than heat extremes, numerous studies have found increases in precipitation extremes on many regions using a variety of methods and threshold definitions,[84] and those increases can be attributed to human-caused changes to the atmosphere.[85, 86] Finally, extreme precipitation associated with tropical cyclones (TCs) is expected to increase in the future,[87] but current trends are not clear.[84]

The impact of extreme precipitation trends on flooding globally is complex because additional factors like soil moisture and changes in land cover are important.[88] Globally, due to limited data, there is low confidence for any significant current trends in river-flooding associated with climate change,[89] but the magnitude and intensity of river flooding is projected to increase in the future.[90] More on flooding trends in the United States is in Chapter 8: Droughts, Floods, and Wildfires.

*Tornadoes and Thunderstorms*

Increasing air temperature and moisture increase the risk of extreme convection, and there is evidence for a global increase in severe thunderstorm conditions.[91] Strong convection, along with wind shear, represents favorable conditions for tornadoes. Thus, there is reason to expect increased tornado frequency and intensity in a warming climate.[92] Inferring current changes in tornado activity is hampered by changes in reporting standards, and trends remain highly uncertain (see Ch. 9: Extreme Storms).[84]

*Winter Storms*

Winter storm tracks have shifted slightly northward (by about 0.4 degrees latitude) in recent decades over the Northern Hemisphere.[93] More generally, extratropical cyclone activity is projected to change in complex ways under future climate scenarios, with increases in some regions and seasons and decreases in others. There are large model-to-model differences among CMIP5 climate models, with some models underestimating the current cyclone track density.[94, 95]

Enhanced arctic warming (arctic amplification), due in part to sea ice loss, reduces lower tropospheric meridional temperature gradients, diminishing baroclinicity (a measure of how misaligned the gradient of pressure is from the gradient of air density)—an important energy source for extratropical cyclones. At the same time, upper-level meridional temperature gradients will increase due to a warming tropical upper troposphere and a cooling high-latitude lower stratosphere.

While these two effects counteract each other with respect to a projected change in midlatitude storm tracks, the simulations indicate that the magnitude of arctic amplification may modulate some aspects (e.g., jet stream position, wave extent, and blocking frequency) of the circulation in the North Atlantic region in some seasons.[96]

### Tropical Cyclones

Detection and attribution of trends in past tropical cyclone (TC) activity is hampered by uncertainties in the data collected prior to the satellite era and by uncertainty in the relative contributions of natural variability and anthropogenic influences. Theoretical arguments and numerical modeling simulations support an expectation that radiative forcing by greenhouse gases and anthropogenic aerosols can affect TC activity in a variety of ways, but robust formal detection and attribution for past observed changes has not yet been realized. Since the IPCC AR5,[1] there is new evidence that the locations where tropical cyclones reach their peak intensity have migrated poleward in both the Northern and Southern Hemispheres, in concert with the independently measured expansion of the tropics.[97] In the western North Pacific, this migration has substantially changed the tropical cyclone hazard exposure patterns in the region and appears to have occurred outside of the historically measured modes of regional natural variability.[98]

Whether global trends in high-intensity tropical cyclones are already observable is a topic of active debate. Some research suggests positive trends,[99, 100] but significant uncertainties remain (see Ch. 9: Extreme Storms).[100] Other studies have suggested that aerosol pollution has masked the increase in TC intensity expected otherwise from enhanced greenhouse warming.[101, 102]

Tropical cyclone intensities are expected to increase with warming, both on average and at the high end of the scale, as the range of achievable intensities expands, so that the most intense storms will exceed the intensity of any in the historical record.[102] Some studies have projected an overall increase in tropical cyclone activity.[103] However, studies with high-resolution models are giving a different result. For example, a high-resolution dynamical downscaling study of global TC activity under the lower scenario (RCP4.5) projects an increased occurrence of the highest-intensity tropical cyclones (Saffir–Simpson Categories 4 and 5), along with a reduced overall tropical cyclone frequency, though there are considerable basin-to-basin differences.[87] Chapter 9: Extreme Storms covers more on extreme storms affecting the United States.

## 1.6 Global Changes in Land Processes

Changes in regional land cover have had important effects on climate, while climate change also has important effects on land cover (also see Ch. 10: Land Cover).[1] In some cases, there are changes in land cover that are both consequences of and influences on global climate change (e.g., declines in land ice and snow cover, thawing permafrost, and insect damage to forests).

Northern Hemisphere snow cover extent has decreased, especially in spring, primarily due to earlier spring snowmelt (by about 0.2 million square miles [0.5 million square km][104, 105]), and this decrease since the 1970s is at least partially driven by anthropogenic influences.[106] Snow cover reductions, especially in the Arctic region in summer, have led to reduced seasonal albedo.[107]

While global-scale trends in drought are uncertain due to insufficient observations, regional trends indicate increased frequency and intensity of drought and aridification on land cover in the Mediterranean[108, 109] and West

Africa[110, 111] and decreased frequency and intensity of droughts in central North America[112] and northwestern Australia.[110, 111, 113]

Anthropogenic land-use changes, such as deforestation and growing cropland extent, have increased the global land surface albedo, resulting in a small cooling effect. Effects of other land-use changes, including modifications of surface roughness, latent heat flux, river runoff, and irrigation, are difficult to quantify, but may offset the direct land-use albedo changes.[114, 115]

Globally, land-use change since 1750 has been typified by deforestation, driven by the growth in intensive farming and urban development. Global land-use change is estimated to have released 190 ± 65 GtC (gigatonnes of carbon) through 2015.[116, 117] Over the same period, cumulative fossil fuel and industrial emissions are estimated to have been 410 ± 20 GtC, yielding total anthropogenic emissions of 600 ± 70 GtC, of which cumulative land-use change emissions were about 32%.[116, 117] Tropical deforestation is the dominant driver of land-use change emissions, estimated at 0.1–1.7 GtC per year, primarily from biomass burning. Global deforestation emissions of about 3 GtC per year are compensated by around 2 GtC per year of forest regrowth in some regions, mainly from abandoned agricultural land.[118, 119]

Natural terrestrial ecosystems are gaining carbon through uptake of $CO_2$ by enhanced photosynthesis due to higher $CO_2$ levels, increased nitrogen deposition, and longer growing seasons in mid- and high latitudes. Anthropogenic atmospheric $CO_2$ absorbed by land ecosystems is stored as organic matter in live biomass (leaves, stems, and roots), dead biomass (litter and woody debris), and soil carbon.

Many studies have documented a lengthening growing season, primarily due to the changing climate,[120, 121, 122, 123] and elevated $CO_2$ is expected to further lengthen the growing season in places where the length is water limited.[124] In addition, a recent study has shown an overall increase in greening of Earth in vegetated regions,[125] while another has demonstrated evidence that the greening of Northern Hemisphere extratropical vegetation is attributable to anthropogenic forcings, particularly rising atmospheric greenhouse gas levels.[126] However, observations[127, 128, 129] and models[130, 131, 132] indicate that nutrient limitations and land availability will constrain future land carbon sinks.

Modifications to the water, carbon, and biogeochemical cycles on land result in both positive and negative feedbacks to temperature increases.[114, 133, 134] Snow and ice albedo feedbacks are positive, leading to increased temperatures with loss of snow and ice extent. While land ecosystems are expected to have a net positive feedback due to reduced natural sinks of $CO_2$ in a warmer world, anthropogenically increased nitrogen deposition may reduce the magnitude of the net feedback.[131, 135, 136] Increased temperature and reduced precipitation increase wildfire risk and susceptibility of terrestrial ecosystems to pests and disease, with resulting feedbacks on carbon storage. Increased temperature and precipitation, particularly at high latitudes, drives up soil decomposition, which leads to increased $CO_2$ and $CH_4$ (methane) emissions.[137, 138, 139, 140, 141, 142, 143] While some of these feedbacks are well known, others are not so well quantified and yet others remain unknown; the potential for surprise is discussed further in Chapter 15: Potential Surprises.

## 1.7 Global Changes in Sea Ice, Glaciers, and Land Ice

Since NCA3,[144] there have been significant advances in the understanding of changes in the cryosphere. Observations continue to show declines in arctic sea ice extent and thickness, Northern Hemisphere snow cover, and the volume of mountain glaciers and continental ice sheets.[1, 145, 146, 147, 148, 149] Evidence suggests in many cases that the net loss of mass from the global cryosphere is accelerating indicating significant climate feedbacks and societal consequences.[150, 151, 152, 153, 154, 155]

Arctic sea ice areal extent, thickness, and volume have declined since 1979.[1, 146, 147, 148, 156] The annual arctic sea ice extent minimum for 2016 relative to the long-term record was the second lowest (2012 was the lowest) (http://nsidc.org/arcticseaicenews/). The arctic sea ice minimum extents in 2014 and 2015 were also among the lowest on record. Annually averaged arctic sea ice extent has decreased by 3.5%–4.1% per decade since 1979 with much larger reductions in summer and fall.[1, 146, 148, 157] For example, September sea ice extent decreased by 13.3% per decade between 1979 and 2016. At the same time, September multi-year sea ice has melted faster than perennial sea ice (13.5% ± 2.5% and 11.5% ± 2.1% per decade, respectively, relative to the 1979–2012 average) corresponding to 4–7.5 feet (1.3–2.3 meter) declines in winter sea ice thickness.[1, 156] October 2016 serves as a recent example of the observed lengthening of the arctic sea ice melt season marking the slowest recorded arctic sea ice growth rate for that month.[146, 158, 159] The annual arctic sea ice maximum in March 2017 was the lowest maximum areal extent on record (http://nsidc.org/arcticseaicenews/).

While current generation climate models project a nearly ice-free Arctic Ocean in late summer by mid-century, they still simulate weaker reductions in volume and extent than observed, suggesting that projected changes are too conservative.[1, 147, 160, 161] See Chapter 11: Arctic Changes for further discussion of the implications of changes in the Arctic.

In contrast to the Arctic, sea ice extent around Antarctica has increased since 1979 by 1.2% to 1.8% per decade.[1] Strong regional differences in the sea ice growth rates are found around Antarctica but most regions (about 75%) show increases over the last 30 years.[162] The gain in antarctic sea ice is much smaller than the decrease in arctic sea ice. Changes in wind patterns, ice–ocean feedbacks, and freshwater flux have contributed to antarctic sea ice growth.[162, 163, 164, 165]

Since the NCA3,[144] the Gravity Recovery and Climate Experiment (GRACE) constellation (e.g., Velicogna and Wahr 2013[166]) has provided a record of gravimetric land ice measurements, advancing knowledge of recent mass loss from the global cryosphere. These measurements indicate that mass loss from the Antarctic Ice Sheet, Greenland Ice Sheet, and mountain glaciers around the world continues accelerating in some cases.[151, 152, 154, 155, 167, 168] The annually averaged ice mass from 37 global reference glaciers has decreased every year since 1984, a decline expected to continue even if climate were to stabilize.[1, 153, 169, 170]

Ice sheet dynamics in West Antarctica are characterized by land ice that transitions to coastal and marine ice sheet systems. Recent observed rapid mass loss from West Antarctica's floating ice shelves is attributed to increased glacial discharge rates due to diminishing ice shelves from the surrounding ocean becoming warmer.[171, 172] Recent evidence suggests that the Amundsen Sea sector is expected to disintegrate entirely[151, 168, 172] raising sea level by at least 1.2 meters (about 4 feet) and potentially an additional foot or more on top of current sea level rise projections during this century (see Section 1.2.7 and Ch. 12: Sea Level Rise for further details).[173] The potential for unanticipated rapid ice sheet melt and/or disintegration is discussed further in Chapter 15: Potential Surprises.

Over the last decade, the Greenland Ice Sheet mass loss has accelerated, losing 244 ± 6 Gt per year on average between January 2003 and May 2013.[1, 155, 174, 175] The portion of the Greenland Ice Sheet experiencing annual melt has increased since 1980 including significant events.[1, 176, 177, 178] A recent example, an unprecedented 98.6% of the Greenland Ice Sheet surface experienced melt on a single day in July 2012.[179, 180] Encompassing this event, GRACE data indicate that Greenland lost 562 Gt of mass between April 2012 and April 2013—more than double the average annual mass loss.

In addition, permafrost temperatures and active layer thicknesses have increased across much of the Arctic (also see Ch. 11: Arctic Changes).[1, 181, 182] Rising permafrost temperatures causing permafrost to thaw and become more discontinuous raises concerns about potential emissions of carbon dioxide and methane.[1] The potentially large contribution of carbon and methane emissions from permafrost and the continental shelf in the Arctic to overall warming is discussed further in Chapter 15: Potential Surprises.

## 1.8 Global Changes in Sea Level

Statistical analyses of tide gauge data indicate that global mean sea level has risen about 8–9 inches (20–23 cm) since 1880, with a rise rate of approximately 0.5–0.6 inches/decade from 1901 to1990 (about 12–15 mm/decade; also see Ch. 12: Sea Level Rise).[183, 184] However, since the early 1990s, both tide gauges and satellite altimeters have recorded a faster rate of sea level rise of about 1.2 inches/decade (approximately 3 cm/decade),[183, 184, 185] resulting in

about 3 inches (about 8 cm) of the global rise since the early 1990s. Nearly two-thirds of the sea level rise measured since 2005 has resulted from increases in ocean mass, primarily from land-based ice melt; the remaining one-third of the rise is in response to changes in density from increasing ocean temperatures.[186]

Global sea level rise and its regional variability forced by climatic and ocean circulation patterns are contributing to significant increases in annual tidal-flood frequencies, which are measured by NOAA tide gauges and associated with minor infrastructure impacts to date; along some portions of the U.S. coast, frequency of the impacts from such events appears to be accelerating (also see Ch. 12: Sea-Level Rise).[187, 188]

Future projections show that by 2100, global mean sea level is *very likely* to rise by 1.6–4.3 feet (0.5–1.3 m) under the higher scenario (RCP8.5), 1.1–3.1 feet (0.35–0.95 m) under a lower scenario (RCP4.5), and 0.8–2.6 feet (0.24–0.79 m) under and even lower scenario (RCP2.6) (see Ch. 4: Projections for a description of the scenarios).[189] Sea level will not rise uniformly around the coasts of the United States and its oversea territories. Local sea level rise is *likely* to be greater than the global average along the U.S. Atlantic and Gulf Coasts and less than the global average in most of the Pacific Northwest. Emerging science suggests these projections may be underestimates, particularly for higher scenarios; a global mean sea level rise exceeding 8 feet (2.4 m) by 2100 cannot be excluded (see Ch. 12: Sea Level Rise), and even higher amounts are possible as a result of marine ice sheet instability (see Ch. 15: Potential Surprises). We have updated the global sea level rise scenarios for 2100 of Parris et al.[190] accordingly,[191] and also extended to year 2200 in Chapter 12: Sea Level Rise. The scenarios are regionalized to better match the decision context needed for local risk framing purposes.

## 1.9 Recent Global Changes Relative to Paleoclimates

Paleoclimate records demonstrate long-term natural variability in the climate and overlap the records of the last two millennia, referred to here as the "Common Era." Before the emissions of greenhouse gases from fossil fuels and other human-related activities became a major factor over the last few centuries, the strongest drivers of climate during the last few thousand years had been volcanoes and land-use change (which has both albedo and greenhouse gas emissions effects).[192] Based on a number of proxies for temperature (for example, from tree rings, fossil pollen, corals, ocean and lake sediments, and ice cores), temperature records are available for the last 2,000 years on hemispherical and continental scales (Figures 1.8 and 1.9).[9, 193] High-resolution temperature records for North America extend back less than half of this period, with temperatures in the early parts of the Common Era inferred from analyses of pollen and other archives. For this era, there is a general cooling trend, with a relatively rapid increase in temperature over the last 150–200 years (Figure 1.9, ). For context, global annual averaged temperatures for 1986–2015 are likely much higher, and appear to have risen at a more rapid rate during the last 3 decades, than any similar period possibly over the past 2,000 years or longer (IPCC[1] makes a similar statement, but for the last 1,400 years because of data quality issues before that time).

Global temperatures of the magnitude observed recently (and projected for the rest of this century) are related to very different forcings than past climates, but studies of past climates suggest that such global temperatures were *likely* last observed during the Eemian period—the last interglacial—125,000 years ago; at that time, global temperatures were, at their peak, about 1.8°F–3.6°F (1°C–2°C) warmer than preindustrial temperatures.[194] Coincident with these higher temperatures, sea levels during that period were about 16–30 feet (6–9 meters) higher than modern levels[195, 196] (for further discussion on sea levels in the past, see Ch. 12: Sea Level Rise).

Modeling studies suggest that the Eemian period warming can be explained in part by the hemispheric changes in solar insolation from orbital forcing as a result of cyclic changes in the shape of Earth's orbit around the sun (e.g., Kaspar et al. 2005[197]), even though greenhouse gas concentrations were similar to preindustrial levels. Equilibrium climate with modern greenhouse gas concentrations (about 400 ppm $CO_2$) most recently occurred 3 million years ago during the Pliocene. During the warmest parts of this period, global temperatures were 5.4°F–7.2°F (3°C–4°C) higher than today, and sea levels were about 82 feet (25 meters) higher.[198]

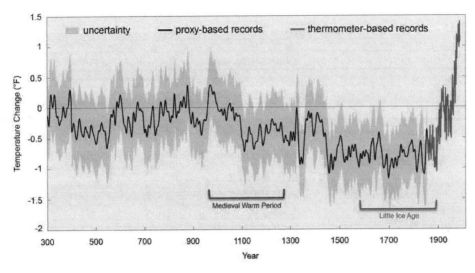

**Figure 1.8:** Changes in the temperature of the Northern Hemisphere from surface observations (in red) and from proxies (in black; uncertainty range represented by shading) relative to 1961–1990 average temperature. If this graph were plotted relative to 1901–1960 instead of 1961–1990, the temperature changes would be 0.47°F (0.26°C) higher. These analyses suggest that current temperatures are higher than seen in the Northern Hemisphere, and likely globally, in at least the last 1,700 years, and that the last decade (2006–2015) was the warmest decade on record. (Figure source: adapted from Mann et al. 2008[193]).

## Proxy Temperature Reconstructions

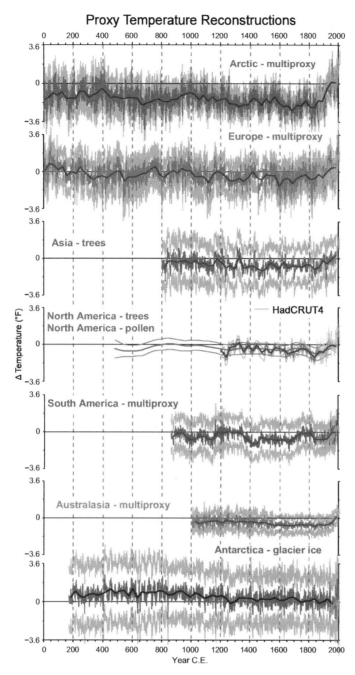

Figure 1.9: Proxy temperatures reconstructions for the seven regions of the PAGES 2k Network. Temperature anomalies are relative to the 1961–1990 reference period. If this graph were plotted relative to 1901–1960 instead of 1961–1990, the temperature changes would 0.47°F (0.26°C) higher. Gray lines around expected-value estimates indicate uncertainty ranges as defined by each regional group (see PAGE 2k Consortium[9] and related Supplementary Information). Note that the changes in temperature over the last century tend to occur at a much faster rate than found in the previous time periods. The teal values are from the HadCRUT4 surface observation record for land and ocean for the 1800s to 2000.[202] (Figure source: adapted from PAGES 2k Consortium 2013[9]).

## Box 1.2: Advances Since NCA3

This assessment reflects both advances in scientific understanding and approach since NCA3, as well as global policy developments. Highlights of what aspects are either especially strengthened or are emerging in the findings include

- *Spatial downscaling*: Projections of climate changes are downscaled to a finer resolution than the original global climate models using the Localized Constructed Analogs (LOCA) empirical statistical downscaling model. The downscaling generates temperature and precipitation on a 1/16th degree latitude/longitude grid for the contiguous United States. LOCA, one of the best statistical downscaling approaches, produces downscaled estimates using a multi-scale spatial matching scheme to pick appropriate analog days from observations (Chapters 4, 6, 7).

- *Risk-based framing*: Highlighting aspects of climate science most relevant to assessment of key societal risks are included more here than in prior national climate assessments. This approach allows for emphasis of possible outcomes that, while relatively unlikely to occur or characterized by high uncertainty, would be particularly consequential, and thus associated with large risks (Chapters 6, 7, 8, 9, 12, 15).

- *Detection and attribution*: Significant advances have been made in the attribution of the human influence for individual climate and weather extreme events since NCA3. This assessment contains extensive discussion of new and emerging findings in this area (Chapters 3, 6, 7, 8).

- *Atmospheric circulation and extreme events:* The extent to which atmospheric circulation in the midlatitudes is changing or is projected to change, possibly in ways not captured by current climate models, is a new important area of research. While still in its formative stages, this research is critically important because of the implications of such changes for climate extremes including extended cold air outbreaks, long-duration heat waves, and changes in storms and drought patterns (Chapters 5, 6, 7).

- *Increased understanding of specific types of extreme events:* How climate change may affect specific types of extreme events in the United States is another key area where scientific understanding has advanced. For example, this report highlights how intense flooding associated with atmospheric rivers could increase dramatically as the atmosphere and oceans warm or how tornadoes could be concentrated into a smaller number of high-impact days over the average severe weather season (Chapter 9).

- *Model weighting*: For the first time, maps and plots of climate projections will not show a straight average of all available climate models. Rather, each model is given a weight based on their 1) historical performance relative to observations and 2) independence relative to other models. Although this is a more accurate way of representing model output, it does not significantly alter the key findings: the weighting produces very similar trends and spatial patterns to the equal-weighting-of-models approach used in prior assessments (Chapters 4, 6, 7, Appendix B).

- *High-resolution global climate model simulations*: As computing resources have grown, multidecadal simulations of global climate models are now being conducted at horizontal resolutions on the order of 15 miles (25 km) that provide more realistic characterization of intense weather systems, including hurricanes. Even the limited number of high-resolution models currently available have increased confidence in projections of extreme weather (Chapter 9).

## Box 1.2 *(continued)*

- *The so-called "global warming hiatus"*: Since NCA3, many studies have investigated causes for the reported slowdown in the rate of increase in near-surface global mean temperature from roughly 2000 through 2013. The slowdown, which ended with the record warmth in 2014–2016, is understood to have been caused by a combination of internal variability, mostly in the heat exchange between the ocean and the atmosphere, and short-term variations in external forcing factors, both human and natural. On longer time scales, relevant to human-induced climate change, there is no hiatus, and the planet continues to warm at a steady pace as predicted by basic atmospheric physics and the well-documented increase in heat-trapping gases (Chapter 1).

- *Oceans and coastal waters*: Ocean acidification, warming, and oxygen loss are all increasing, and scientific understanding of the severity of their impacts is growing. Both oxygen loss and acidification may be magnified in some U.S. coastal waters relative to the global average, raising the risk of serious ecological and economic consequences. There is some evidence, still highly uncertain, that the Atlantic Meridional Circulation (AMOC), sometimes referred to as the ocean's conveyor belt, may be slowing down (Chapters 2, 13).

- *Local sea level change projections*: For the first time in the NCA process, sea level rise projections incorporate geographic variation based on factors such as local land subsidence, ocean currents, and changes in Earth's gravitational field (Chapter 12).

- *Accelerated ice-sheet loss*: New observations from many different sources confirm that ice-sheet loss is accelerating. Combining observations with simultaneous advances in the physical understanding of ice sheets, scientists are now concluding that up to 8.5 feet of global sea level rise is possible by 2100 under a higher scenario, up from 6.6 feet in NCA3 (Chapter 12).

- *Low sea-ice areal extent*: The annual arctic sea ice extent minimum for 2016 relative to the long-term record was the second lowest on record. The arctic sea ice minimums in 2014 and 2015 were also amongst the lowest on record. Since 1981, the sea ice minimum has decreased by 13.3% per decade, more than 46% over the 35 years. The annual arctic sea ice maximum in March 2017 was the lowest maximum areal extent on record. (Chapter 11).

- *Potential surprises*: Both large-scale state shifts in the climate system (sometimes called "tipping points") and compound extremes have the potential to generate unanticipated surprises. The further Earth system departs from historical climate forcings, and the more the climate changes, the greater the potential for these surprises. For the first time in the NCA process we include an extended discussion of these potential surprises (Chapter 15).

- *Mitigation*: This report discusses some important aspects of climate science that are relevant to long-term temperature goals and different mitigation scenarios, including those implied by government announcements for the Paris Agreement. (Chapters 4, 14).

# TRACEABLE ACCOUNTS

### Key Finding 1

The global climate continues to change rapidly compared to the pace of the natural variations in climate that have occurred throughout Earth's history. Trends in globally averaged temperature, sea level rise, upper-ocean heat content, land-based ice melt, arctic sea ice, depth of seasonal permafrost thaw, and other climate variables provide consistent evidence of a warming planet. These observed trends are robust and have been confirmed by multiple independent research groups around the world.

### Description of evidence base

The Key Finding and supporting text summarize extensive evidence documented in the climate science literature. Similar to statements made in previous national (NCA3)[144] and international[1] assessments.

Evidence for changes in global climate arises from multiple analyses of data from in-situ, satellite, and other records undertaken by many groups over several decades. These observational datasets are used throughout this chapter and are discussed further in Appendix 1 (e.g., updates of prior uses of these datasets by Vose et al. 2012;[22] Karl et al. 2015[26]). Changes in the mean state have been accompanied by changes in the frequency and nature of extreme events (e.g., Kunkel and Frankson 2015;[81] Donat et al. 2016[82]). A substantial body of analysis comparing the observed changes to a broad range of climate simulations consistently points to the necessity of invoking human-caused changes to adequately explain the observed climate system behavior. The influence of human impacts on the climate system has also been observed in a number of individual climate variables (attribution studies are discussed in Ch. 3: Detection and Attribution and in other chapters).

### Major uncertainties

Key remaining uncertainties relate to the precise magnitude and nature of changes at global, and particularly regional, scales, and especially for extreme events and our ability to observe these changes at sufficient resolution and to simulate and attribute such changes using climate models. Innovative new approaches to instigation and maintenance of reference quality observation networks such as the U.S. Climate Reference Network (http://www.ncei.noaa.gov/crn/), enhanced climate observational and data analysis capabilities, and continued improvements in climate modeling all have the potential to reduce uncertainties.

### Assessment of confidence based on evidence and agreement, including short description of nature of evidence and level of agreement

There is *very high confidence* that global climate is changing and this change is apparent across a wide range of observations, given the evidence base and remaining uncertainties. All observational evidence is consistent with a warming climate since the late 1800s. There is *very high confidence* that the global climate change of the past 50 years is primarily due to human activities, given the evidence base and remaining uncertainties.[1] Recent changes have been consistently attributed in large part to human factors across a very broad range of climate system characteristics.

### Summary sentence or paragraph that integrates the above information

The key message and supporting text summarizes extensive evidence documented in the climate science peer-reviewed literature. The trends described in NCA3 have continued and our understanding of the observations related to climate and the ability to evaluate the many facets of the climate system have increased substantially.

### Key Finding 2

The frequency and intensity of extreme heat and heavy precipitation events are increasing in most continental regions of the world (*very high confidence*). These trends are consistent with expected physical responses to a warming climate. Climate model studies are also consistent with these trends, although models tend to underestimate the observed trends, especially for the increase in extreme precipitation events (*very high confidence* for temperature, *high confidence* for extreme precipitation). The frequency and intensity of extreme high temperature events are *virtually certain* to increase in the future as global temperature increases (*high confidence*). Extreme precipitation events will *very likely* continue to increase in frequency and intensity throughout most of the world (*high confidence*). Observed and projected trends for some other types of extreme events, such as floods, droughts, and severe storms, have more variable regional characteristics.

### Description of evidence base

The Key Finding and supporting text summarizes extensive evidence documented in the climate science literature and are similar to statements made in previous national (NCA3)[144] and international[1] assessments. The analyses of past trends and future projections in extreme events and the fact that models tend to underestimate the observed trends are also well substantiated through more recent peer-reviewed literature as well.[75, 76, 81, 82, 83, 88, 90, 199]

### Major uncertainties

Key remaining uncertainties relate to the precise magnitude and nature of changes at global, and particularly regional, scales, and especially for extreme events and our ability to simulate and attribute such changes using climate models. Innovative new approaches to climate data analysis, continued improvements in climate modeling, and instigation and maintenance of reference quality observation networks such as the U.S. Climate Reference Network (http://www.ncei.noaa. gov/crn/) all have the potential to reduce uncertainties.

### Assessment of confidence based on evidence and agreement, including short description of nature of evidence and level of agreement

There is *very high confidence* for the statements about past extreme changes in temperature and precipitation and *high confidence* for future projections, based on the observational evidence and physical understanding, that there are major trends in extreme events and significant projected changes for the future.

### Summary sentence or paragraph that integrates the above information

The Key Finding and supporting text summarizes extensive evidence documented in the climate science

peer-reviewed literature. The trends for extreme events that were described in the NCA3 and IPCC assessments have continued, and our understanding of the data and ability to evaluate the many facets of the climate system have increased substantially.

### Key Finding 3

Many lines of evidence demonstrate that it is *extremely likely* that human influence has been the dominant cause of the observed warming since the mid-20th century. Formal detection and attribution studies for the period 1951 to 2010 find that the observed global mean surface temperature warming lies in the middle of the range of likely human contributions to warming over that same period. We find no convincing evidence that natural variability can account for the amount of global warming observed over the industrial era. For the period extending over the last century, there are no convincing alternative explanations supported by the extent of the observational evidence. Solar output changes and internal variability can only contribute marginally to the observed changes in climate over the last century, and we find no convincing evidence for natural cycles in the observational record that could explain the observed changes in climate. (*Very high confidence*)

### Description of evidence base

The Key Finding and supporting text summarizes extensive evidence documented in the climate science literature and are similar to statements made in previous national (NCA3)[144] and international[1] assessments. The human effects on climate have been well documented through many papers in the peer-reviewed scientific literature (e.g., see Ch. 2: Physical Drivers of Climate Change and Ch. 3: Detection and Attribution for more discussion of supporting evidence).

### Major uncertainties

Key remaining uncertainties relate to the precise magnitude and nature of changes at global, and particularly regional, scales, and especially for extreme events and our ability to simulate and attribute such changes using climate models. The exact effects from land use changes relative to the effects from greenhouse gas emissions need to be better understood.

**Assessment of confidence based on evidence and agreement, including short description of nature of evidence and level of agreement**

There is *very high confidence* for a major human influence on climate.

**Summary sentence or paragraph that integrates the above information**

The key message and supporting text summarizes extensive evidence documented in the climate science peer-reviewed literature. The analyses described in the NCA3 and IPCC assessments support our findings, and new observations and modeling studies have further substantiated these conclusions.

**Key Finding 4**

Global climate is projected to continue to change over this century and beyond. The magnitude of climate change beyond the next few decades will depend primarily on the amount of greenhouse (heat-trapping) gases emitted globally and on the remaining uncertainty in the sensitivity of Earth's climate to those emissions (*very high confidence*). With significant reductions in the emissions of greenhouse gases, the global annually averaged temperature rise could be limited to 3.6°F (2°C) or less. Without major reductions in these emissions, the increase in annual average global temperatures relative to preindustrial times could reach 9°F (5°C) or more by the end of this century (*high confidence*).

**Description of evidence base**

The Key Finding and supporting text summarizes extensive evidence documented in the climate science literature and are similar to statements made in previous national (NCA3)[144] and international[1] assessments. The projections for future climate have been well documented through many papers in the peer-reviewed scientific literature (e.g., see Ch. 4: Projections for descriptions of the scenarios and the models used).

**Major uncertainties**

Key remaining uncertainties relate to the precise magnitude and nature of changes at global, and particularly regional, scales, and especially for extreme events and our ability to simulate and attribute such changes using climate models. Of particular importance are remaining uncertainties in the understanding of feedbacks in the climate system, especially in ice–albedo and cloud cover feedbacks. Continued improvements in climate modeling to represent the physical processes affecting Earth's climate system are aimed at reducing uncertainties. Monitoring and observation programs also can help improve the understanding needed to reduce uncertainties.

**Assessment of confidence based on evidence and agreement, including short description of nature of evidence and level of agreement**

There is *very high confidence* for continued changes in climate and *high confidence* for the levels shown in the Key Finding.

**Summary sentence or paragraph that integrates the above information**

The Key Finding and supporting text summarizes extensive evidence documented in the climate science peer-reviewed literature. The projections that were described in the NCA3 and IPCC assessments support our findings, and new modeling studies have further substantiated these conclusions.

**Key Finding 5**

Natural variability, including El Niño events and other recurring patterns of ocean–atmosphere interactions, impact temperature and precipitation, especially regionally, over months to years. The global influence of natural variability, however, is limited to a small fraction of observed climate trends over decades.

**Description of evidence base**

The Key Finding and supporting text summarizes extensive evidence documented in the climate science literature and are similar to statements made in previous national (NCA3)[144] and international[1] (IPCC 2013) assessments. The role of natural variability in climate trends has been extensively discussed in the peer-reviewed literature (e.g., Karl et al. 2015;[26] Rahmstorf et al. 2015;[34] Lewandowsky et al. 2016;[39] Mears and Wentz 2016;[41] Trenberth et al. 2014;[200] Santer et al. 2017[38, 40, 68]).

**Major uncertainties**

Uncertainties still exist in the precise magnitude and nature of the full effects of individual ocean cycles and other aspects of natural variability on the climate system. Increased emphasis on monitoring should reduce this uncertainty significantly over the next few decades.

**Assessment of confidence based on evidence and agreement, including short description of nature of evidence and level of agreement**

There is *very high confidence*, affected to some degree by limitations in the observational record, that the role of natural variability on future climate change is limited.

**Summary sentence or paragraph that integrates the above information**

The Key Finding and supporting text summarizes extensive evidence documented in the climate science peer-reviewed literature. There has been an extensive increase in the understanding of the role of natural variability on the climate system over the last few decades, including a number of new findings since NCA3.

**Key Finding 6**

Longer-term climate records over past centuries and millennia indicate that average temperatures in recent decades over much of the world have been much higher, and have risen faster during this time period, than at any time in the past 1,700 years or more, the time period for which the global distribution of surface temperatures can be reconstructed.

**Description of evidence base**

The Key Finding and supporting text summarizes extensive evidence documented in the climate science literature and are similar to statements made in previous national (NCA3)[144] and international[1] assessments. There are many recent studies of the paleoclimate leading to this conclusion including those cited in the report (e.g., Mann et al. 2008;[193] PAGE 2k Consortium 2013[9]).

**Major uncertainties**

Despite the extensive increase in knowledge in the last few decades, there are still many uncertainties in understanding the hemispheric and global changes in climate over Earth's history, including that of the last few millennia. Additional research efforts in this direction can help reduce those uncertainties.

**Assessment of confidence based on evidence and agreement, including short description of nature of evidence and level of agreement**

There is *high confidence* for current temperatures to be higher than they have been in at least 1,700 years and perhaps much longer.

**Summary sentence or paragraph that integrates the above information**

The Key Finding and supporting text summarizes extensive evidence documented in the climate science peer-reviewed literature. There has been an extensive increase in the understanding of past climates on our planet, including a number of new findings since NCA3.

# REFERENCES

1.  IPCC, 2013: *Climate Change 2013: The Physical Science Basis. Contribution of Working Group I to the Fifth Assessment Report of the Intergovernmental Panel on Climate Change.* Cambridge University Press, Cambridge, UK and New York, NY, 1535 pp. http://www.climatechange2013.org/report/

2.  Trenberth, K.E. and J.T. Fasullo, 2013: An apparent hiatus in global warming? *Earth's Future*, **1**, 19-32. http://dx.doi.org/10.1002/2013EF000165

3.  Trenberth, K.E., 2015: Has there been a hiatus? *Science*, **349**, 691-692. http://dx.doi.org/10.1126/science.aac9225

4.  Marotzke, J. and P.M. Forster, 2015: Forcing, feedback and internal variability in global temperature trends. *Nature*, **517**, 565-570. http://dx.doi.org/10.1038/nature14117

5.  Lehmann, J., D. Coumou, and K. Frieler, 2015: Increased record-breaking precipitation events under global warming. *Climatic Change*, **132**, 501-515. http://dx.doi.org/10.1007/s10584-015-1434-y

6.  **Bindoff, N.L., P.A. Stott, K.M. AchutaRao, M.R. Allen, N. Gillett, D. Gutzler, K. Hansingo, G. Hegerl, Y. Hu, S. Jain, I.I. Mokhov, J. Overland, J. Perlwitz, R. Sebbari, and X. Zhang, 2013: Detection and attribution of climate change: From global to regional.** *Climate Change 2013: The Physical Science Basis. Contribution of Working Group I to the Fifth Assessment Report of the Intergovernmental Panel on Climate Change.* Stocker, T.F., D. Qin, G.-K. Plattner, M. Tignor, S.K. Allen, J. Boschung, A. Nauels, Y. Xia, V. Bex, and P.M. Midgley, Eds. Cambridge University Press, Cambridge, United Kingdom and New York, NY, USA, 867–952. http://www.climatechange2013.org/report/full-report/

7.  Schurer, A.P., S.F.B. Tett, and G.C. Hegerl, 2014: Small influence of solar variability on climate over the past millennium. *Nature Geoscience*, **7**, 104-108. http://dx.doi.org/10.1038/ngeo2040

8.  Kopp, G., 2014: An assessment of the solar irradiance record for climate studies. *Journal of Space Weather and Space Climate*, **4**, A14. http://dx.doi.org/10.1051/swsc/2014012

9.  PAGES 2K Consortium, 2013: Continental-scale temperature variability during the past two millennia. *Nature Geoscience*, **6**, 339-346. http://dx.doi.org/10.1038/ngeo1797

10. Marcott, S.A., J.D. Shakun, P.U. Clark, and A.C. Mix, 2013: A reconstruction of regional and global temperature for the past 11,300 years. *Science*, **339**, 1198-1201. http://dx.doi.org/10.1126/science.1228026

11. Otto-Bliesner, B.L., E.C. Brady, J. Fasullo, A. Jahn, L. Landrum, S. Stevenson, N. Rosenbloom, A. Mai, and G. Strand, 2016: Climate Variability and Change since 850 CE: An Ensemble Approach with the Community Earth System Model. *Bulletin of the American Meteorological Society*, **97**, 735-754. http://dx.doi.org/10.1175/bams-d-14-00233.1

12. Church, J.A., N.J. White, L.F. Konikow, C.M. Domingues, J.G. Cogley, E. Rignot, J.M. Gregory, M.R. van den Broeke, A.J. Monaghan, and I. Velicogna, 2011: Revisiting the Earth's sea-level and energy budgets from 1961 to 2008. *Geophysical Research Letters*, **38**, L18601. http://dx.doi.org/10.1029/2011GL048794

13. Anderson, B.T., J.R. Knight, M.A. Ringer, J.-H. Yoon, and A. Cherchi, 2012: Testing for the possible influence of unknown climate forcings upon global temperature increases from 1950 to 2000. *Journal of Climate*, **25**, 7163-7172. http://dx.doi.org/10.1175/jcli-d-11-00645.1

14. EPA, 2016: Climate Change Indicators in the United States, 2016. 4th edition. EPA 430-R-16-004. U.S. Environmental Protection Agency, Washington, D.C., 96 pp. https://www.epa.gov/sites/production/files/2016-08/documents/climate_indicators_2016.pdf

15. Blunden, J. and D.S. Arndt, 2016: State of the climate in 2015. *Bulletin of the American Meteorological Society*, **97**, Si-S275. http://dx.doi.org/10.1175/2016BAMSStateoftheClimate.1

16. Meehl, G.A., A. Hu, B.D. Santer, and S.-P. Xie, 2016: **Contribution of the Interdecadal Pacific Oscillation** to twentieth-century global surface temperature trends. *Nature Climate Change*, **6**, 1005-1008. http://dx.doi.org/10.1038/nclimate3107

17. **Johnson, G.C., J.M. Lyman, J. Antonov, N. Bindoff, T. Boyer, C.M. Domingues, S.A. Good, M. Ishii, and J.K. Willis, 2015: Ocean heat content** [in "State of the Climate in 2014"]. *Bulletin of the American Meteorological Society*, **96 (7)**, S64–S66, S68. http://dx.doi.org/10.1175/2014BAMSStateoftheClimate.1

18. Rhein, M., S.R. Rintoul, S. Aoki, E. Campos, D. Chambers, R.A. Feely, S. Gulev, G.C. Johnson, S.A. Josey, A. Kostianoy, C. Mauritzen, D. Roemmich, L.D. Talley, and F. Wang, 2013: Observations: Ocean. *Climate Change 2013: The Physical Science Basis. Contribution of Working Group I to the Fifth Assessment Report of the Intergovernmental Panel on Climate Change.* Stocker, T.F., D. Qin, G.-K. Plattner, M. Tignor, S.K. Allen, J. Boschung, A. Nauels, Y. Xia, V. Bex, and P.M. Midgley, Eds. Cambridge University Press, Cambridge, United Kingdom and New York, NY, USA, 255–316. http://www.climatechange2013.org/report/full-report/

19. Kundzewicz, Z.W., 2008: Climate change impacts on the hydrological cycle. *Ecohydrology & Hydrobiology,* **8**, 195-203. http://dx.doi.org/10.2478/v10104-009-0015-y

20. Ramsayer, K., 2014: Antarctic sea ice reaches new record maximum. https://www.nasa.gov/content/goddard/antarctic-sea-ice-reaches-new-record-maximum

21. USGCRP, 2017: [National Climate Assessment] Indicators. U.S. Global Change Research Program. http://www.globalchange.gov/browse/indicators

22. Vose, R.S., D. Arndt, V.F. Banzon, D.R. Easterling, B. Gleason, B. Huang, E. Kearns, J.H. Lawrimore, M.J. Menne, T.C. Peterson, R.W. Reynolds, T.M. Smith, C.N. Williams, and D.L. Wuertz, 2012: NOAA's merged land-ocean surface temperature analysis. *Bulletin of the American Meteorological Society,* **93**, 1677-1685. http://dx.doi.org/10.1175/BAMS-D-11-00241.1

23. NCEI, 2016: Climate at a Glance: Global Temperature Anomalies. http://www.ncdc.noaa.gov/cag/time-series/global/globe/land_ocean/ytd/12/1880-2015

24. Steinman, B.A., M.B. Abbott, M.E. Mann, N.D. Stansell, and B.P. Finney, 2012: 1,500 year quantitative reconstruction of winter precipitation in the Pacific Northwest. *Proceedings of the National Academy of Sciences,* **109**, 11619-11623. http://dx.doi.org/10.1073/pnas.1201083109

25. Deser, C., R. Knutti, S. Solomon, and A.S. Phillips, 2012: Communication of the role of natural variability in future North American climate. *Nature Climate Change,* **2**, 775-779. http://dx.doi.org/10.1038/nclimate1562

26. Karl, T.R., A. Arguez, B. Huang, J.H. Lawrimore, J.R. McMahon, M.J. Menne, T.C. Peterson, R.S. Vose, and H.-M. Zhang, 2015: Possible artifacts of data biases in the recent global surface warming hiatus. *Science,* **348**, 1469-1472. http://dx.doi.org/10.1126/science.aaa5632

27. Knutson, T.R., R. Zhang, and L.W. Horowitz, 2016: Prospects for a prolonged slowdown in global warming in the early 21st century. *Nature Communcations,* **7**, 13676. http://dx.doi.org/10.1038/ncomms13676

28. Delworth, T.L. and T.R. Knutson, 2000: Simulation of early 20th century global warming. *Science,* **287**, 2246-2250. http://dx.doi.org/10.1126/science.287.5461.2246

29. Alexander, L.V., X. Zhang, T.C. Peterson, J. Caesar, B. Gleason, A.M.G. Klein Tank, M. Haylock, D. Collins, B. Trewin, F. Rahimzadeh, A. Tagipour, K. Rupa Kumar, J. Revadekar, G. Griffiths, L. Vincent, D.B. Stephenson, J. Burn, E. Aguilar, M. Brunet, M. Taylor, M. New, P. Zhai, M. Rusticucci, and J.L. Vazquez-Aguirre, 2006: Global observed changes in daily climate extremes of temperature and precipitation. *Journal of Geophysical Research,* **111**, D05109. http://dx.doi.org/10.1029/2005JD006290

30. Davy, R., I. Esau, A. Chernokulsky, S. Outten, and S. Zilitinkevich, 2016: Diurnal asymmetry to the observed global warming. *International Journal of Climatology,* **37**, 79-93. http://dx.doi.org/10.1002/joc.4688

31. Mountain Research Initiative, 2015: Elevation-dependent warming in mountain regions of the world. *Nature Climate Change,* **5**, 424-430. http://dx.doi.org/10.1038/nclimate2563

32. Hurrell, J.W. and C. Deser, 2009: North Atlantic climate variability: The role of the North Atlantic oscillation. *Journal of Marine Systems,* **78**, 28-41. http://dx.doi.org/10.1016/j.jmarsys.2008.11.026

33. Hoegh-Guldberg, O., R. Cai, E.S. Poloczanska, P.G. Brewer, S. Sundby, K. Hilmi, V.J. Fabry, and S. Jung, 2014: The Ocean. *Climate Change 2014: Impacts, Adaptation, and Vulnerability. Part B: Regional Aspects. Contribution of Working Group II to the Fifth Assessment Report of the Intergovernmental Panel of Climate Change.* Barros, V.R., C.B. Field, D.J. Dokken, M.D. Mastrandrea, K.J. Mach, T.E. Bilir, M. Chatterjee, K.L. Ebi, Y.O. Estrada, R.C. Genova, B. Girma, E.S. Kissel, A.N. Levy, S. MacCracken, P.R. Mastrandrea, and L.L. White, Eds. Cambridge University Press, Cambridge, United Kingdom and New York, NY, USA, 1655-1731. http://www.ipcc.ch/pdf/assessment-report/ar5/wg2/WGIIAR5-Chap30_FINAL.pdf

34. Rahmstorf, S., J.E. Box, G. Feulner, M.E. Mann, A. Robinson, S. Rutherford, and E.J. Schaffernicht, 2015: Exceptional twentieth-century slowdown in Atlantic Ocean overturning circulation. *Nature Climate Change,* **5**, 475-480. http://dx.doi.org/10.1038/nclimate2554

35. Knutti, R., J. Rogelj, J. Sedlacek, and E.M. Fischer, 2016: A scientific critique of the two-degree climate change target. *Nature Geoscience,* **9**, 13-18. http://dx.doi.org/10.1038/ngeo2595

36. Peters, G.P., R.M. Andrew, T. Boden, J.G. Canadell, P. Ciais, C. Le Quere, G. Marland, M.R. Raupach, and C. Wilson, 2013: The challenge to keep global warming below 2°C. *Nature Climate Change,* **3**, 4-6. http://dx.doi.org/10.1038/nclimate1783

37. Schellnhuber, H.J., S. Rahmstorf, and R. Winkelmann, 2016: Why the right climate target was agreed in Paris. *Nature Climate Change,* **6**, 649-653. http://dx.doi.org/10.1038/nclimate3013

38. Santer, B.D., S. Solomon, G. Pallotta, C. Mears, S. Po-Chedley, Q. Fu, F. Wentz, C.-Z. Zou, J. Painter, I. Cvijanovic, and C. Bonfils, 2017: Comparing tropospheric warming in climate models and satellite data. *Journal of Climate,* **30**, 373-392. http://dx.doi.org/10.1175/JCLI-D-16-0333.1

39. Lewandowsky, S., J.S. Risbey, and N. Oreskes, 2016: The "pause" in global warming: Turning a routine fluctuation into a problem for science. *Bulletin of the American Meteorological Society,* **97**, 723-733. http://dx.doi.org/10.1175/BAMS-D-14-00106.1

40. Santer, B.D., S. Soloman, F.J. Wentz, Q. Fu, S. Po-Chedley, C. Mears, J.F. Painter, and C. Bonfils, 2017: Tropospheric warming over the past two decades. *Scientific Reports,* **7**, 2336. http://dx.doi.org/10.1038/s41598-017-02520-7

41. Mears, C.A. and F.J. Wentz, 2016: Sensitivity of satellite-derived tropospheric temperature trends to the diurnal cycle adjustment. *Journal of Climate,* **29**, 3629-3646. http://dx.doi.org/10.1175/JCLI-D-15-0744.1

42. Richardson, M., K. Cowtan, E. Hawkins, and M.B. Stolpe, 2016: Reconciled climate response estimates from climate models and the energy budget of Earth. *Nature Climate Change,* **6**, 931-935. http://dx.doi.org/10.1038/nclimate3066

43. Hausfather, Z., K. Cowtan, D.C. Clarke, P. Jacobs, M. Richardson, and R. Rohde, 2017: Assessing recent warming using instrumentally homogeneous sea surface temperature records. *Science Advances,* **3**, e1601207. http://dx.doi.org/10.1126/sciadv.1601207

44. Fyfe, J.C., G.A. Meehl, M.H. England, M.E. Mann, B.D. Santer, G.M. Flato, E. Hawkins, N.P. Gillett, S.-P. Xie, Y. Kosaka, and N.C. Swart, 2016: Making sense of the early-2000s warming slowdown. *Nature Climate Change,* **6**, 224-228. http://dx.doi.org/10.1038/nclimate2938

45. Benestad, R.E., 2017: A mental picture of the greenhouse effect. *Theoretical and Applied Climatology,* **128**, 679-688. http://dx.doi.org/10.1007/s00704-016-1732-y

46. Balmaseda, M.A., K.E. Trenberth, and E. Källén, 2013: Distinctive climate signals in reanalysis of global ocean heat content. *Geophysical Research Letters,* **40**, 1754-1759. http://dx.doi.org/10.1002/grl.50382

47. England, M.H., S. McGregor, P. Spence, G.A. Meehl, A. Timmermann, W. Cai, A.S. Gupta, M.J. McPhaden, A. Purich, and A. Santoso, 2014: Recent intensification of wind-driven circulation in the Pacific and the ongoing warming hiatus. *Nature Climate Change,* **4**, 222-227. http://dx.doi.org/10.1038/nclimate2106

48. Meehl, G.A., J.M. Arblaster, J.T. Fasullo, A. Hu, and K.E. Trenberth, 2011: Model-based evidence of deep-ocean heat uptake during surface-temperature hiatus periods. *Nature Climate Change,* **1**, 360-364. http://dx.doi.org/10.1038/nclimate1229

49. Kosaka, Y. and S.-P. Xie, 2013: Recent global-warming hiatus tied to equatorial Pacific surface cooling. *Nature,* **501**, 403-407. http://dx.doi.org/10.1038/nature12534

50. Chen, X. and K.-K. Tung, 2014: Varying planetary heat sink led to global-warming slowdown and acceleration. *Science,* **345**, 897-903. http://dx.doi.org/10.1126/science.1254937

51. Nieves, V., J.K. Willis, and W.C. Patzert, 2015: Recent hiatus caused by decadal shift in Indo-Pacific heating. *Science,* **349**, 532-535. http://dx.doi.org/10.1126/science.aaa4521

52. Solomon, S., K.H. Rosenlof, R.W. Portmann, J.S. Daniel, S.M. Davis, T.J. Sanford, and G.-K. Plattner, 2010: Contributions of stratospheric water vapor to decadal changes in the rate of global warming. *Science,* **327**, 1219-1223. http://dx.doi.org/10.1126/science.1182488

53. Schmidt, G.A., D.T. Shindell, and K. Tsigaridis, 2014: Reconciling warming trends. *Nature Geoscience,* **7**, 158-160. http://dx.doi.org/10.1038/ngeo2105

54. Huber, M. and R. Knutti, 2014: Natural variability, radiative forcing and climate response in the recent hiatus reconciled. *Nature Geoscience,* **7**, 651-656. http://dx.doi.org/10.1038/ngeo2228

55. Ridley, D.A., S. Solomon, J.E. Barnes, V.D. Burlakov, T. Deshler, S.I. Dolgii, A.B. Herber, T. Nagai, R.R. Neely, A.V. Nevzorov, C. Ritter, T. Sakai, B.D. Santer, M. Sato, A. Schmidt, O. Uchino, and J.P. Vernier, 2014: Total volcanic stratospheric aerosol optical depths and implications for global climate change. *Geophysical Research Letters,* **41**, 7763-7769. http://dx.doi.org/10.1002/2014GL061541

56. Santer, B.D., C. Bonfils, J.F. Painter, M.D. Zelinka, C. Mears, S. Solomon, G.A. Schmidt, J.C. Fyfe, J.N.S. Cole, L. Nazarenko, K.E. Taylor, and F.J. Wentz, 2014: Volcanic contribution to decadal changes in tropospheric temperature. *Nature Geoscience,* **7**, 185-189. http://dx.doi.org/10.1038/ngeo2098

57. Becker, A., P. Finger, A. Meyer-Christoffer, B. Rudolf, K. Schamm, U. Schneider, and M. Ziese, 2013: A description of the global land-surface precipitation data products of the Global Precipitation Climatology Centre with sample applications including centennial (trend) analysis from 1901–present. *Earth System Science Data,* **5**, 71-99. http://dx.doi.org/10.5194/essd-5-71-2013

58. Adler, R.F., G.J. Huffman, A. Chang, R. Ferraro, P.-P. Xie, J. Janowiak, B. Rudolf, U. Schneider, S. Curtis, D. Bolvin, A. Gruber, J. Susskind, P. Arkin, and E. Nelkin, 2003: The version-2 Global Precipitation Climatology Project (GPCP) monthly precipitation analysis (1979–present). *Journal of Hydrometeorology,* **4**, 1147-1167. http://dx.doi.org/10.1175/1525-7541(2003)004<1147:TVGPCP>2.0.CO;2

59. Allen, M.R. and W.J. Ingram, 2002: Constraints on future changes in climate and the hydrologic cycle. *Nature*, **419**, 224-232. http://dx.doi.org/10.1038/nature01092

60. Collins, M., R. Knutti, J. Arblaster, J.-L. Dufresne, T. Fichefet, P. Friedlingstein, X. Gao, W.J. Gutowski, T. Johns, G. Krinner, M. Shongwe, C. Tebaldi, A.J. Weaver, and M. Wehner, 2013: Long-term climate change: Projections, commitments and irreversibility. *Climate Change 2013: The Physical Science Basis. Contribution of Working Group I to the Fifth Assessment Report of the Intergovernmental Panel on Climate Change.* Stocker, T.F., D. Qin, G.-K. Plattner, M. Tignor, S.K. Allen, J. Boschung, A. Nauels, Y. Xia, V. Bex, and P.M. Midgley, Eds. Cambridge University Press, Cambridge, United Kingdom and New York, NY, USA, 1029–1136. http://www.climatechange2013.org/report/full-report/

61. Held, I.M. and B.J. Soden, 2006: Robust responses of the hydrological cycle to global warming. *Journal of Climate*, **19**, 5686-5699. http://dx.doi.org/10.1175/jcli3990.1

62. Santer, B.D., C. Mears, F.J. Wentz, K.E. Taylor, P.J. Gleckler, T.M.L. Wigley, T.P. Barnett, J.S. Boyle, W. Brüggemann, N.P. Gillett, S.A. Klein, G.A. Meehl, T. Nozawa, D.W. Pierce, P.A. Stott, W.M. Washington, and M.F. Wehner, 2007: Identification of human-induced changes in atmospheric moisture content. *Proceedings of the National Academy of Sciences*, **104**, 15248-15253. http://dx.doi.org/10.1073/pnas.0702872104

63. Willett, K.M., D.J. Philip, W.T. Peter, and P.G. Nathan, 2010: A comparison of large scale changes in surface humidity over land in observations and CMIP3 general circulation models. *Environmental Research Letters*, **5**, 025210. http://dx.doi.org/10.1088/1748-9326/5/2/025210

64. Greve, P., B. Orlowsky, B. Mueller, J. Sheffield, M. Reichstein, and S.I. Seneviratne, 2014: Global assessment of trends in wetting and drying over land. *Nature Geoscience*, **7**, 716-721. http://dx.doi.org/10.1038/ngeo2247

65. Zhang, X., F.W. Zwiers, G.C. Hegerl, F.H. Lambert, N.P. Gillett, S. Solomon, P.A. Stott, and T. Nozawa, 2007: Detection of human influence on twentieth-century precipitation trends. *Nature*, **448**, 461-465. http://dx.doi.org/10.1038/nature06025

66. Marvel, K. and C. Bonfils, 2013: Identifying external influences on global precipitation. *Proceedings of the National Academy of Sciences*, **110**, 19301-19306. http://dx.doi.org/10.1073/pnas.1314382110

67. Walsh, J.E., J.E. Overland, P.Y. Groisman, and B. Rudolf, 2011: Ongoing climate change in the Arctic. *Ambio*, **40**, 6-16. http://dx.doi.org/10.1007/s13280-011-0211-z

68. Vihma, T., J. Screen, M. Tjernström, B. Newton, X. Zhang, V. Popova, C. Deser, M. Holland, and T. Prowse, 2016: The atmospheric role in the Arctic water cycle: A review on processes, past and future changes, and their impacts. *Journal of Geophysical Research Biogeosciences*, **121**, 586-620. http://dx.doi.org/10.1002/2015JG003132

69. Min, S.-K., X. Zhang, and F. Zwiers, 2008: Human-induced Arctic moistening. *Science*, **320**, 518-520. http://dx.doi.org/10.1126/science.1153468

70. Katz, R.W. and B.G. Brown, 1992: Extreme events in a changing climate: Variability is more important than averages. *Climatic Change*, **21**, 289-302. http://dx.doi.org/10.1007/bf00139728

71. Sardeshmukh, P.D., G.P. Compo, and C. Penland, 2015: Need for caution in interpreting extreme weather statistics. *Journal of Climate*, **28**, 9166-9187. http://dx.doi.org/10.1175/JCLI-D-15-0020.1

72. NAS, 2016: *Attribution of Extreme Weather Events in the Context of Climate Change.* The National Academies Press, Washington, DC, 186 pp. http://dx.doi.org/10.17226/21852

73. Hulme, M., 2014: Attributing weather extremes to 'climate change'. *Progress in Physical Geography*, **38**, 499-511. http://dx.doi.org/10.1177/0309133314538644

74. Stott, P., 2016: How climate change affects extreme weather events. *Science*, **352**, 1517-1518. http://dx.doi.org/10.1126/science.aaf7271

75. Easterling, D.R., K.E. Kunkel, M.F. Wehner, and L. Sun, 2016: Detection and attribution of climate extremes in the observed record. *Weather and Climate Extremes*, **11**, 17-27. http://dx.doi.org/10.1016/j.wace.2016.01.001

76. Seneviratne, S.I., M.G. Donat, B. Mueller, and L.V. Alexander, 2014: No pause in the increase of hot temperature extremes. *Nature Climate Change*, **4**, 161-163. http://dx.doi.org/10.1038/nclimate2145

77. Meehl, G.A., C. Tebaldi, G. Walton, D. Easterling, and L. McDaniel, 2009: Relative increase of record high maximum temperatures compared to record low minimum temperatures in the US. *Geophysical Research Letters*, **36**, L23701. http://dx.doi.org/10.1029/2009GL040736

78. Min, S.-K., X. Zhang, F. Zwiers, H. Shiogama, Y.-S. Tung, and M. Wehner, 2013: Multimodel detection and attribution of extreme temperature changes. *Journal of Climate*, **26**, 7430-7451. http://dx.doi.org/10.1175/JCLI-D-12-00551.1

79. Horton, D.E., N.C. Johnson, D. Singh, D.L. Swain, B. Rajaratnam, and N.S. Diffenbaugh, 2015: Contribution of changes in atmospheric circulation patterns to extreme temperature trends. *Nature*, **522**, 465-469. http://dx.doi.org/10.1038/nature14550

80. Asadieh, B. and N.Y. Krakauer, 2015: Global trends in extreme precipitation: climate models versus observations. *Hydrology and Earth System Sciences*, **19**, 877-891. http://dx.doi.org/10.5194/hess-19-877-2015

81. Kunkel, K.E. and R.M. Frankson, 2015: Global land surface extremes of precipitation: Data limitations and trends. *Journal of Extreme Events*, **02**, 1550004. http://dx.doi.org/10.1142/S2345737615500049

82. Donat, M.G., A.L. Lowry, L.V. Alexander, P.A. Ogorman, and N. Maher, 2016: More extreme precipitation in the world's dry and wet regions. *Nature Climate Change*, **6**, 508-513. http://dx.doi.org/10.1038/nclimate2941

83. Fischer, E.M. and R. Knutti, 2016: Observed heavy precipitation increase confirms theory and early models. *Nature Climate Change*, **6**, 986-991. http://dx.doi.org/10.1038/nclimate3110

84. Kunkel, K.E., T.R. Karl, H. Brooks, J. Kossin, J. Lawrimore, D. Arndt, L. Bosart, D. Changnon, S.L. Cutter, N. Doesken, K. Emanuel, P.Y. Groisman, R.W. Katz, T. Knutson, J. O'Brien, C.J. Paciorek, T.C. Peterson, K. Redmond, D. Robinson, J. Trapp, R. Vose, S. Weaver, M. Wehner, K. Wolter, and D. Wuebbles, 2013: Monitoring and understanding trends in extreme storms: State of knowledge. *Bulletin of the American Meteorological Society*, **94**, 499–514. http://dx.doi.org/10.1175/BAMS-D-11-00262.1

85. Min, S.K., X. Zhang, F.W. Zwiers, and G.C. Hegerl, 2011: Human contribution to more-intense precipitation extremes. *Nature*, **470**, 378-381. http://dx.doi.org/10.1038/nature09763

86. Zhang, X., H. Wan, F.W. Zwiers, G.C. Hegerl, and S.-K. Min, 2013: Attributing intensification of precipitation extremes to human influence. *Geophysical Research Letters*, **40**, 5252-5257. http://dx.doi.org/10.1002/grl.51010

87. Knutson, T.R., J.J. Sirutis, M. Zhao, R.E. Tuleya, M. Bender, G.A. Vecchi, G. Villarini, and D. Chavas, 2015: Global projections of intense tropical cyclone activity for the late twenty-first century from dynamical downscaling of CMIP5/RCP4.5 scenarios. *Journal of Climate*, **28**, 7203-7224. http://dx.doi.org/10.1175/JCLI-D-15-0129.1

88. Berghuijs, W.R., R.A. Woods, C.J. Hutton, and M. Sivapalan, 2016: Dominant flood generating mechanisms across the United States. *Geophysical Research Letters*, **43**, 4382-4390. http://dx.doi.org/10.1002/2016GL068070

89. Kundzewicz, Z.W., S. Kanae, S.I. Seneviratne, J. Handmer, N. Nicholls, P. Peduzzi, R. Mechler, L.M. Bouwer, N. Arnell, K. Mach, R. Muir-Wood, G.R. Brakenridge, W. Kron, G. Benito, Y. Honda, K. Takahashi, and B. Sherstyukov, 2014: Flood risk and climate change: Global and regional perspectives. *Hydrological Sciences Journal*, **59**, 1-28. http://dx.doi.org/10.1080/02626667.2013.857411

90. Arnell, N.W. and S.N. Gosling, 2016: The impacts of climate change on river flood risk at the global scale. *Climatic Change*, **134**, 387-401. http://dx.doi.org/10.1007/s10584-014-1084-5

91. Sander, J., J.F. Eichner, E. Faust, and M. Steuer, 2013: Rising variability in thunderstorm-related U.S. losses as a reflection of changes in large-scale thunderstorm forcing. *Weather, Climate, and Society*, **5**, 317-331. http://dx.doi.org/10.1175/WCAS-D-12-00023.1

92. Diffenbaugh, N.S., M. Scherer, and R.J. Trapp, 2013: Robust increases in severe thunderstorm environments in response to greenhouse forcing. *Proceedings of the National Academy of Sciences*, **110**, 16361-16366. http://dx.doi.org/10.1073/pnas.1307758110

93. Bender, F.A.-M., V. Ramanathan, and G. Tselioudis, 2012: Changes in extratropical storm track cloudiness 1983–2008: Observational support for a poleward shift. *Climate Dynamics*, **38**, 2037-2053. http://dx.doi.org/10.1007/s00382-011-1065-6

94. Chang, E.K.M., 2013: CMIP5 projection of significant reduction in extratropical cyclone activity over North America. *Journal of Climate*, **26**, 9903-9922. http://dx.doi.org/10.1175/JCLI-D-13-00209.1

95. Colle, B.A., Z. Zhang, K.A. Lombardo, E. Chang, P. Liu, and M. Zhang, 2013: Historical evaluation and future prediction of eastern North American and western Atlantic extratropical cyclones in the CMIP5 models during the cool season. *Journal of Climate*, **26**, 6882-6903. http://dx.doi.org/10.1175/JCLI-D-12-00498.1

96. Barnes, E.A. and L.M. Polvani, 2015: CMIP5 projections of Arctic amplification, of the North American/North Atlantic circulation, and of their relationship. *Journal of Climate*, **28**, 5254-5271. http://dx.doi.org/10.1175/JCLI-D-14-00589.1

97. Kossin, J.P., K.A. Emanuel, and G.A. Vecchi, 2014: The poleward migration of the location of tropical cyclone maximum intensity. *Nature*, **509**, 349-352. http://dx.doi.org/10.1038/nature13278

98. Kossin, J.P., K.A. Emanuel, and S.J. Camargo, 2016: Past and projected changes in western North Pacific tropical cyclone exposure. *Journal of Climate*, **29**, 5725-5739. http://dx.doi.org/10.1175/JCLI-D-16-0076.1

99. Elsner, J.B., J.P. Kossin, and T.H. Jagger, 2008: The increasing intensity of the strongest tropical cyclones. *Nature*, **455**, 92-95. http://dx.doi.org/10.1038/nature07234

100. Kossin, J.P., T.L. Olander, and K.R. Knapp, 2013: Trend analysis with a new global record of tropical cyclone intensity. *Journal of Climate*, **26**, 9960-9976. http://dx.doi.org/10.1175/JCLI-D-13-00262.1

101. Wang, C., L. Zhang, S.-K. Lee, L. Wu, and C.R. Mechoso, 2014: A global perspective on CMIP5 climate model biases. *Nature Climate Change*, **4**, 201-205. http://dx.doi.org/10.1038/nclimate2118

102. Sobel, A.H., S.J. Camargo, T.M. Hall, C.-Y. Lee, M.K. Tippett, and A.A. Wing, 2016: Human influence on tropical cyclone intensity. *Science*, **353**, 242-246. http://dx.doi.org/10.1126/science.aaf6574

103. Emanuel, K.A., 2013: Downscaling CMIP5 climate models shows increased tropical cyclone activity over the 21st century. *Proceedings of the National Academy of Sciences*, **110**, 12219-12224. http://dx.doi.org/10.1073/pnas.1301293110

104. NSIDC, 2017: SOTC (State of the Cryosphere): Northern Hemisphere Snow. National Snow and Ice Data Center. https://nsidc.org/cryosphere/sotc/snow_extent.html

105. Kunkel, K.E., D.A. Robinson, S. Champion, X. Yin, T. Estilow, and R.M. Frankson, 2016: Trends and extremes in Northern Hemisphere snow characteristics. *Current Climate Change Reports*, **2**, 65-73. http://dx.doi.org/10.1007/s40641-016-0036-8

106. Rupp, D.E., P.W. Mote, N.L. Bindoff, P.A. Stott, and D.A. Robinson, 2013: Detection and attribution of observed changes in Northern Hemisphere spring snow cover. *Journal of Climate*, **26**, 6904-6914. http://dx.doi.org/10.1175/JCLI-D-12-00563.1

107. Callaghan, T.V., M. Johansson, R.D. Brown, P.Y. Groisman, N. Labba, V. Radionov, R.S. Bradley, S. Blangy, O.N. Bulygina, T.R. Christensen, J.E. Colman, R.L.H. Essery, B.C. Forbes, M.C. Forchhammer, V.N. Golubev, R.E. Honrath, G.P. Juday, A.V. Meshcherskaya, G.K. Phoenix, J. Pomeroy, A. Rautio, D.A. Robinson, N.M. Schmidt, M.C. Serreze, V.P. Shevchenko, A.I. Shiklomanov, A.B. Shmakin, P. Sköld, M. Sturm, M.-k. Woo, and E.F. Wood, 2011: Multiple effects of changes in Arctic snow cover. *Ambio*, **40**, 32-45. http://dx.doi.org/10.1007/s13280-011-0213-x

108. Sousa, P.M., R.M. Trigo, P. Aizpurua, R. Nieto, L. Gimeno, and R. Garcia-Herrera, 2011: Trends and extremes of drought indices throughout the 20th century in the Mediterranean. *Natural Hazards and Earth System Sciences*, **11**, 33-51. http://dx.doi.org/10.5194/nhess-11-33-2011

109. Hoerling, M., M. Chen, R. Dole, J. Eischeid, A. Kumar, J.W. Nielsen-Gammon, P. Pegion, J. Perlwitz, X.-W. Quan, and T. Zhang, 2013: Anatomy of an extreme event. *Journal of Climate*, **26**, 2811–2832. http://dx.doi.org/10.1175/JCLI-D-12-00270.1

110. Sheffield, J., E.F. Wood, and M.L. Roderick, 2012: Little change in global drought over the past 60 years. *Nature*, **491**, 435-438. http://dx.doi.org/10.1038/nature11575

111. Dai, A., 2013: Increasing drought under global warming in observations and models. *Nature Climate Change*, **3**, 52-58. http://dx.doi.org/10.1038/nclimate1633

112. Peterson, T.C., R.R. Heim, R. Hirsch, D.P. Kaiser, H. Brooks, N.S. Diffenbaugh, R.M. Dole, J.P. Giovannetone, K. Guirguis, T.R. Karl, R.W. Katz, K. Kunkel, D. Lettenmaier, G.J. McCabe, C.J. Paciorek, K.R. Ryberg, S. Schubert, V.B.S. Silva, B.C. Stewart, A.V. Vecchia, G. Villarini, R.S. Vose, J. Walsh, M. Wehner, D. Wolock, K. Wolter, C.A. Woodhouse, and D. Wuebbles, 2013: Monitoring and understanding changes in heat waves, cold waves, floods and droughts in the United States: State of knowledge. *Bulletin of the American Meteorological Society*, **94**, 821-834. http://dx.doi.org/10.1175/BAMS-D-12-00066.1

113. Jones, D.A., W. Wang, and R. Fawcett, 2009: High-quality spatial climate data-sets for Australia. *Australian Meteorological and Oceanographic Journal*, **58**, 233-248. http://dx.doi.org/10.22499/2.5804.003

114. Bonan, G.B., 2008: Forests and climate change: Forcings, feedbacks, and the climate benefits of forests. *Science*, **320**, 1444-1449. http://dx.doi.org/10.1126/science.1155121

115. de Noblet-Ducoudré, N., J.-P. Boisier, A. Pitman, G.B. Bonan, V. Brovkin, F. Cruz, C. Delire, V. Gayler, B.J.J.M.v.d. Hurk, P.J. Lawrence, M.K.v.d. Molen, C. Müller, C.H. Reick, B.J. Strengers, and A. Voldoire, 2012: Determining robust impacts of land-use-induced land cover changes on surface climate over North America and Eurasia: Results from the first set of LUCID experiments. *Journal of Climate*, **25**, 3261-3281. http://dx.doi.org/10.1175/JCLI-D-11-00338.1

116. Le Quéré, C., R. Moriarty, R.M. Andrew, J.G. Canadell, S. Sitch, J.I. Korsbakken, P. Friedlingstein, G.P. Peters, R.J. Andres, T.A. Boden, R.A. Houghton, J.I. House, R.F. Keeling, P. Tans, A. Arneth, D.C.E. Bakker, L. Barbero, L. Bopp, J. Chang, F. Chevallier, L.P. Chini, P. Ciais, M. Fader, R.A. Feely, T. Gkritzalis, I. Harris, J. Hauck, T. Ilyina, A.K. Jain, E. Kato, V. Kitidis, K. Klein Goldewijk, C. Koven, P. Landschützer, S.K. Lauvset, N. Lefèvre, A. Lenton, I.D. Lima, N. Metzl, F. Millero, D.R. Munro, A. Murata, J.E.M.S. Nabel, S. Nakaoka, Y. Nojiri, K. O'Brien, A. Olsen, T. Ono, F.F. Pérez, B. Pfeil, D. Pierrot, B. Poulter, G. Rehder, C. Rödenbeck, S. Saito, U. Schuster, J. Schwinger, R. Séférian, T. Steinhoff, B.D. Stocker, A.J. Sutton, T. Takahashi, B. Tilbrook, I.T. van der Laan-Luijkx, G.R. van der Werf, S. van Heuven, N. Vandemark, N. Viovy, A. Wiltshire, S. Zaehle, and N. Zeng, 2015: Global carbon budget 2015. *Earth System Science Data*, **7**, 349-396. http://dx.doi.org/10.5194/essd-7-349-2015

117. Le Quéré, C., R.M. Andrew, J.G. Canadell, S. Sitch, J.I. Korsbakken, G.P. Peters, A.C. Manning, T.A. Boden, P.P. Tans, R.A. Houghton, R.F. Keeling, S. Alin, O.D. Andrews, P. Anthoni, L. Barbero, L. Bopp, F. Chevallier, L.P. Chini, P. Ciais, K. Currie, C. Delire, S.C. Doney, P. Friedlingstein, T. Gkritzalis, I. Harris, J. Hauck, V. Haverd, M. Hoppema, K. Klein Goldewijk, A.K. Jain, E. Kato, A. Körtzinger, P. Landschützer, N. Lefèvre, A. Lenton, S. Lienert, D. Lombardozzi, J.R. Melton, N. Metzl, F. Millero, P.M.S. Monteiro, D.R. Munro, J.E.M.S. Nabel, S.I. Nakaoka, K. O'Brien, A. Olsen, A.M. Omar, T. Ono, D. Pierrot, B. Poulter, C. Rödenbeck, J. Salisbury, U. Schuster, J. Schwinger, R. Séférian, I. Skjelvan, B.D. Stocker, A.J. Sutton, T. Takahashi, H. Tian, B. Tilbrook, I.T. van der Laan-Luijkx, G.R. van der Werf, N. Viovy, A.P. Walker, A.J. Wiltshire, and S. Zaehle, 2016: Global carbon budget 2016. *Earth System Science Data*, **8**, 605-649. http://dx.doi.org/10.5194/essd-8-605-2016

118. Houghton, R.A., J.I. House, J. Pongratz, G.R. van der Werf, R.S. DeFries, M.C. Hansen, C. Le Quéré, and N. Ramankutty, 2012: Carbon emissions from land use and land-cover change. *Biogeosciences*, **9**, 5125-5142. http://dx.doi.org/10.5194/bg-9-5125-2012

119. Pan, Y., R.A. Birdsey, J. Fang, R. Houghton, P.E. Kauppi, W.A. Kurz, O.L. Phillips, A. Shvidenko, S.L. Lewis, J.G. Canadell, P. Ciais, R.B. Jackson, S.W. Pacala, A.D. McGuire, S. Piao, A. Rautiainen, S. Sitch, and D. Hayes, 2011: A large and persistent carbon sink in the world's forests. *Science*, **333**, 988-93. http://dx.doi.org/10.1126/science.1201609

120. Myneni, R.B., C.D. Keeling, C.J. Tucker, G. Asrar, and R.R. Nemani, 1997: Increased plant growth in the northern high latitudes from 1981 to 1991. *Nature*, **386**, 698-702. http://dx.doi.org/10.1038/386698a0

121. Menzel, A., T.H. Sparks, N. Estrella, E. Koch, A. Aasa, R. Ahas, K. Alm-Kübler, P. Bissolli, O.G. Braslavská, A. Briede, F.M. Chmielewski, Z. Crepinsek, Y. Curnel, Å. Dahl, C. Defila, A. Donnelly, Y. Filella, K. Jatczak, F. Måge, A. Mestre, Ø. Nordli, J. Peñuelas, P. Pirinen, V. Remišvá, H. Scheifinger, M. Striz, A. Susnik, A.J.H. Van Vliet, F.-E. Wielgolaski, S. Zach, and A.N.A. Zust, 2006: European phenological response to climate change matches the warming pattern. *Global Change Biology*, **12**, 1969-1976. http://dx.doi.org/10.1111/j.1365-2486.2006.01193.x

122. Schwartz, M.D., R. Ahas, and A. Aasa, 2006: Onset of spring starting earlier across the Northern Hemisphere. *Global Change Biology*, **12**, 343-351. http://dx.doi.org/10.1111/j.1365-2486.2005.01097.x

123. Kim, Y., J.S. Kimball, K. Zhang, and K.C. McDonald, 2012: Satellite detection of increasing Northern Hemisphere non-frozen seasons from 1979 to 2008: Implications for regional vegetation growth. *Remote Sensing of Environment*, **121**, 472-487. http://dx.doi.org/10.1016/j.rse.2012.02.014

124. Reyes-Fox, M., H. Steltzer, M.J. Trlica, G.S. McMaster, A.A. Andales, D.R. LeCain, and J.A. Morgan, 2014: Elevated CO2 further lengthens growing season under warming conditions. *Nature*, **510**, 259-262. http://dx.doi.org/10.1038/nature13207

125. Zhu, Z., S. Piao, R.B. Myneni, M. Huang, Z. Zeng, J.G. Canadell, P. Ciais, S. Sitch, P. Friedlingstein, A. Arneth, C. Cao, L. Cheng, E. Kato, C. Koven, Y. Li, X. Lian, Y. Liu, R. Liu, J. Mao, Y. Pan, S. Peng, J. Penuelas, B. Poulter, T.A.M. Pugh, B.D. Stocker, N. Viovy, X. Wang, Y. Wang, Z. Xiao, H. Yang, S. Zaehle, and N. Zeng, 2016: Greening of the Earth and its drivers. *Nature Climate Change*, **6**, 791-795. http://dx.doi.org/10.1038/nclimate3004

126. Mao, J., A. Ribes, B. Yan, X. Shi, P.E. Thornton, R. Seferian, P. Ciais, R.B. Myneni, H. Douville, S. Piao, Z. Zhu, R.E. Dickinson, Y. Dai, D.M. Ricciuto, M. Jin, F.M. Hoffman, B. Wang, M. Huang, and X. Lian, 2016: Human-induced greening of the northern extratropical land surface. *Nature Climate Change*, **6**, 959-963. http://dx.doi.org/10.1038/nclimate3056

127. Finzi, A.C., D.J.P. Moore, E.H. DeLucia, J. Lichter, K.S. Hofmockel, R.B. Jackson, H.-S. Kim, R. Matamala, H.R. McCarthy, R. Oren, J.S. Pippen, and W.H. Schlesinger, 2006: Progressive nitrogen limitation of ecosystem processes under elevated CO2 in a warm-temperate forest. *Ecology*, **87**, 15-25. http://dx.doi.org/10.1890/04-1748

128. Palmroth, S., R. Oren, H.R. McCarthy, K.H. Johnsen, A.C. Finzi, J.R. Butnor, M.G. Ryan, and W.H. Schlesinger, 2006: Aboveground sink strength in forests controls the allocation of carbon below ground and its [CO2]-induced enhancement. *Proceedings of the National Academy of Sciences*, **103**, 19362-19367. http://dx.doi.org/10.1073/pnas.0609492103

129. Norby, R.J., J.M. Warren, C.M. Iversen, B.E. Medlyn, and R.E. McMurtrie, 2010: CO2 enhancement of forest productivity constrained by limited nitrogen availability. *Proceedings of the National Academy of Sciences*, **107**, 19368-19373. http://dx.doi.org/10.1073/pnas.1006463107

130. Sokolov, A.P., D.W. Kicklighter, J.M. Melillo, B.S. Felzer, C.A. Schlosser, and T.W. Cronin, 2008: Consequences of considering carbon–nitrogen interactions on the feedbacks between climate and the terrestrial carbon cycle. *Journal of Climate*, **21**, 3776-3796. http://dx.doi.org/10.1175/2008JCLI2038.1

131. Thornton, P.E., S.C. Doney, K. Lindsay, J.K. Moore, N. Mahowald, J.T. Randerson, I. Fung, J.F. Lamarque, J.J. Feddema, and Y.H. Lee, 2009: Carbon-nitrogen interactions regulate climate-carbon cycle feedbacks: Results from an atmosphere-ocean general circulation model. *Biogeosciences*, **6**, 2099-2120. http://dx.doi.org/10.5194/bg-6-2099-2009

132. Zaehle, S. and A.D. Friend, 2010: Carbon and nitrogen cycle dynamics in the O-CN land surface model: 1. Model description, site-scale evaluation, and sensitivity to parameter estimates. *Global Biogeochemical Cycles*, **24**, GB1005. http://dx.doi.org/10.1029/2009GB003521

133. Betts, R.A., O. Boucher, M. Collins, P.M. Cox, P.D. Falloon, N. Gedney, D.L. Hemming, C. Huntingford, C.D. Jones, D.M.H. Sexton, and M.J. Webb, 2007: Projected increase in continental runoff due to plant responses to increasing carbon dioxide. *Nature*, **448**, 1037-1041. http://dx.doi.org/10.1038/nature06045

134. Bernier, P.Y., R.L. Desjardins, Y. Karimi-Zindashty, D. Worth, A. Beaudoin, Y. Luo, and S. Wang, 2011: Boreal lichen woodlands: A possible negative feedback to climate change in eastern North America. *Agricultural and Forest Meteorology*, **151**, 521-528. http://dx.doi.org/10.1016/j.agrformet.2010.12.013

135. Churkina, G., V. Brovkin, W. von Bloh, K. Trusilova, M. Jung, and F. Dentener, 2009: Synergy of rising nitrogen depositions and atmospheric CO2 on land carbon uptake moderately offsets global warming. *Global Biogeochemical Cycles*, **23**, GB4027. http://dx.doi.org/10.1029/2008GB003291

136. Zaehle, S., P. Friedlingstein, and A.D. Friend, 2010: Terrestrial nitrogen feedbacks may accelerate future climate change. *Geophysical Research Letters*, **37**, L01401. http://dx.doi.org/10.1029/2009GL041345

137. Page, S.E., F. Siegert, J.O. Rieley, H.-D.V. Boehm, A. Jaya, and S. Limin, 2002: The amount of carbon released from peat and forest fires in Indonesia during 1997. *Nature*, **420**, 61-65. http://dx.doi.org/10.1038/nature01131

138. Ciais, P., M. Reichstein, N. Viovy, A. Granier, J. Ogee, V. Allard, M. Aubinet, N. Buchmann, C. Bernhofer, A. Carrara, F. Chevallier, N. De Noblet, A.D. Friend, P. Friedlingstein, T. Grunwald, B. Heinesch, P. Keronen, A. Knohl, G. Krinner, D. Loustau, G. Manca, G. Matteucci, F. Miglietta, J.M. Ourcival, D. Papale, K. Pilegaard, S. Rambal, G. Seufert, J.F. Soussana, M.J. Sanz, E.D. Schulze, T. Vesala, and R. Valentini, 2005: Europe-wide reduction in primary productivity caused by the heat and drought in 2003. *Nature*, **437**, 529-533. http://dx.doi.org/10.1038/nature03972

139. Chambers, J.Q., J.I. Fisher, H. Zeng, E.L. Chapman, D.B. Baker, and G.C. Hurtt, 2007: Hurricane Katrina's carbon footprint on U.S. Gulf Coast forests. *Science*, **318**, 1107-1107. http://dx.doi.org/10.1126/science.1148913

140. Kurz, W.A., G. Stinson, G.J. Rampley, C.C. Dymond, and E.T. Neilson, 2008: Risk of natural disturbances makes future contribution of Canada's forests to the global carbon cycle highly uncertain. *Proceedings of the National Academy of Sciences*, **105**, 1551-1555. http://dx.doi.org/10.1073/pnas.0708133105

141. Clark, D.B., D.A. Clark, and S.F. Oberbauer, 2010: Annual wood production in a tropical rain forest in NE Costa Rica linked to climatic variation but not to increasing CO2. *Global Change Biology*, **16**, 747-759. http://dx.doi.org/10.1111/j.1365-2486.2009.02004.x

142. van der Werf, G.R., J.T. Randerson, L. Giglio, G.J. Collatz, M. Mu, P.S. Kasibhatla, D.C. Morton, R.S. DeFries, Y. Jin, and T.T. van Leeuwen, 2010: Global fire emissions and the contribution of deforestation, savanna, forest, agricultural, and peat fires (1997–2009). *Atmospheric Chemistry and Physics*, **10**, 11707-11735. http://dx.doi.org/10.5194/acp-10-11707-2010

143. Lewis, S.L., P.M. Brando, O.L. Phillips, G.M.F. van der Heijden, and D. Nepstad, 2011: The 2010 Amazon drought. *Science*, **331**, 554-554. http://dx.doi.org/10.1126/science.1200807

144. Melillo, J.M., T.C. Richmond, and G.W. Yohe, eds., 2014: *Climate Change Impacts in the United States: The Third National Climate Assessment*. U.S. Global Change Research Program: Washington, D.C., 841 pp. http://dx.doi.org/10.7930/J0Z31WJ2

145. Derksen, C. and R. Brown, 2012: Spring snow cover extent reductions in the 2008–2012 period exceeding climate model projections. *Geophysical Research Letters*, **39**, L19504. http://dx.doi.org/10.1029/2012gl053387

146. Stroeve, J., A. Barrett, M. Serreze, and A. Schweiger, 2014: Using records from submarine, aircraft and satellites to evaluate climate model simulations of Arctic sea ice thickness. *The Cryosphere*, **8**, 1839-1854. http://dx.doi.org/10.5194/tc-8-1839-2014

147. Stroeve, J.C., T. Markus, L. Boisvert, J. Miller, and A. Barrett, 2014: Changes in Arctic melt season and implications for sea ice loss. *Geophysical Research Letters*, **41**, 1216-1225. http://dx.doi.org/10.1002/2013GL058951

148. Comiso, J.C. and D.K. Hall, 2014: Climate trends in the Arctic as observed from space. *Wiley Interdisciplinary Reviews: Climate Change*, **5**, 389-409. http://dx.doi.org/10.1002/wcc.277

149. Derksen, D., R. Brown, L. Mudryk, and K. Loujus, 2015: [The Arctic] Terrestrial snow cover [in "State of the Climate in 2014"]. *Bulletin of the American Meteorological Society*, **96 (12)**, S133-S135. http://dx.doi.org/10.1175/2015BAMSStateoftheClimate.1

150. Rignot, E., I. Velicogna, M.R. van den Broeke, A. Monaghan, and J.T.M. Lenaerts, 2011: Acceleration of the contribution of the Greenland and Antarctic ice sheets to sea level rise. *Geophysical Research Letters*, **38**, L05503. http://dx.doi.org/10.1029/2011GL046583

151. Rignot, E., J. Mouginot, M. Morlighem, H. Seroussi, and B. Scheuchl, 2014: Widespread, rapid grounding line retreat of Pine Island, Thwaites, Smith, and Kohler Glaciers, West Antarctica, from 1992 to 2011. *Geophysical Research Letters*, **41**, 3502-3509. http://dx.doi.org/10.1002/2014GL060140

152. Williams, S.D.P., P. Moore, M.A. King, and P.L. Whitehouse, 2014: Revisiting GRACE Antarctic ice mass trends and accelerations considering autocorrelation. *Earth and Planetary Science Letters*, **385**, 12-21. http://dx.doi.org/10.1016/j.epsl.2013.10.016

153. Zemp, M., H. Frey, I. Gärtner-Roer, S.U. Nussbaumer, M. Hoelzle, F. Paul, W. Haeberli, F. Denzinger, A.P. Ahlstrøm, B. Anderson, S. Bajracharya, C. Baroni, L.N. Braun, B.E. Cáceres, G. Casassa, G. Cobos, L.R. Dávila, H. Delgado Granados, M.N. Demuth, L. Espizua, A. Fischer, K. Fujita, B. Gadek, A. Ghazanfar, J.O. Hagen, P. Holmlund, N. Karimi, Z. Li, M. Pelto, P. Pitte, V.V. Popovnin, C.A. Portocarrero, R. Prinz, C.V. Sangewar, I. Severskiy, O. Sigurðsson, A. Soruco, R. Usubaliev, and C. Vincent, 2015: Historically unprecedented global glacier decline in the early 21st century. *Journal of Glaciology*, **61**, 745-762. http://dx.doi.org/10.3189/2015JoG15J017

154. Seo, K.-W., C.R. Wilson, T. Scambos, B.-M. Kim, D.E. Waliser, B. Tian, B.-H. Kim, and J. Eom, 2015: Surface mass balance contributions to acceleration of Antarctic ice mass loss during 2003–2013. *Journal of Geophysical Research Solid Earth*, **120**, 3617-3627. http://dx.doi.org/10.1002/2014JB011755

155. Harig, C. and F.J. Simons, 2016: Ice mass loss in Greenland, the Gulf of Alaska, and the Canadian Archipelago: Seasonal cycles and decadal trends. *Geophysical Research Letters*, **43**, 3150-3159. http://dx.doi.org/10.1002/2016GL067759

156. Perovich, D., S. Gerlnad, S. Hendricks, W. Meier, M. Nicolaus, and M. Tschudi, 2015: [The Arctic] Sea ice cover [in "State of the Climate in 2014"]. *Bulletin of the American Meteorological Society*, **96 (12)**, S145-S146. http://dx.doi.org/10.1175/2015BAMSStateoftheClimate.1

157. Stroeve, J.C., M.C. Serreze, M.M. Holland, J.E. Kay, J. Malanik, and A.P. Barrett, 2012: The Arctic's rapidly shrinking sea ice cover: A research synthesis. *Climatic Change*, **110**, 1005-1027. http://dx.doi.org/10.1007/s10584-011-0101-1

158. Parkinson, C.L., 2014: Spatially mapped reductions in the length of the Arctic sea ice season. *Geophysical Research Letters*, **41**, 4316-4322. http://dx.doi.org/10.1002/2014GL060434

159. NSIDC, 2016: Sluggish Ice Growth in the Arctic. Arctic Sea Ice News and Analysis, National Snow and Ice Data Center. http://nsidc.org/arcticseaicenews/2016/11/sluggish-ice-growth-in-the-arctic/

160. Stroeve, J.C., V. Kattsov, A. Barrett, M. Serreze, T. Pavlova, M. Holland, and W.N. Meier, 2012: Trends in Arctic sea ice extent from CMIP5, CMIP3 and observations. *Geophysical Research Letters*, **39**, L16502. http://dx.doi.org/10.1029/2012GL052676

161. Zhang, R. and T.R. Knutson, 2013: The role of global climate change in the extreme low summer Arctic sea ice extent in 2012 [in "Explaining Extreme Events of 2012 from a Climate Perspective"]. *Bulletin of the American Meteorological Society*, **94 (9)**, S23-S26. http://dx.doi.org/10.1175/BAMS-D-13-00085.1

162. Zunz, V., H. Goosse, and F. Massonnet, 2013: How does internal variability influence the ability of CMIP5 models to reproduce the recent trend in Southern Ocean sea ice extent? *The Cryosphere*, **7**, 451-468. http://dx.doi.org/10.5194/tc-7-451-2013

163. Eisenman, I., W.N. Meier, and J.R. Norris, 2014: A spurious jump in the satellite record: Has Antarctic sea ice expansion been overestimated? *The Cryosphere*, **8**, 1289-1296. http://dx.doi.org/10.5194/tc-8-1289-2014

164. Pauling, A.G., C.M. Bitz, I.J. Smith, and P.J. Langhorne, 2016: The response of the Southern Ocean and Antarctic sea ice to freshwater from ice shelves in an Earth system model. *Journal of Climate*, **29**, 1655-1672. http://dx.doi.org/10.1175/JCLI-D-15-0501.1

165. Meehl, G.A., J.M. Arblaster, C.M. Bitz, C.T.Y. Chung, and H. Teng, 2016: Antarctic sea-ice expansion between 2000 and 2014 driven by tropical Pacific decadal climate variability. *Nature Geoscience*, **9**, 590-595. http://dx.doi.org/10.1038/ngeo2751

166. Velicogna, I. and J. Wahr, 2013: Time-variable gravity observations of ice sheet mass balance: Precision and limitations of the GRACE satellite data. *Geophysical Research Letters*, **40**, 3055-3063. http://dx.doi.org/10.1002/grl.50527

167. Harig, C. and F.J. Simons, 2015: Accelerated West Antarctic ice mass loss continues to outpace East Antarctic gains. *Earth and Planetary Science Letters*, **415**, 134-141. http://dx.doi.org/10.1016/j.epsl.2015.01.029

168. Joughin, I., B.E. Smith, and B. Medley, 2014: Marine ice sheet collapse potentially under way for the Thwaites Glacier Basin, West Antarctica. *Science*, **344**, 735-738. http://dx.doi.org/10.1126/science.1249055

169. Pelto, M.S., 2015: [Global Climate] Alpine glaciers [in "State of the Climate in 2014"]. *Bulletin of the American Meteorological Society*, **96 (12)**, S19-S20. http://dx.doi.org/10.1175/2015BAMSStateoftheClimate.1

170. Mengel, M., A. Levermann, K. Frieler, A. Robinson, B. Marzeion, and R. Winkelmann, 2016: Future sea level rise constrained by observations and long-term commitment. *Proceedings of the National Academy of Sciences*, **113**, 2597-2602. http://dx.doi.org/10.1073/pnas.1500515113

171. Jenkins, A., P. Dutrieux, S.S. Jacobs, S.D. McPhail, J.R. Perrett, A.T. Webb, and D. White, 2010: Observations beneath Pine Island Glacier in West Antarctica and implications for its retreat. *Nature Geoscience*, **3**, 468-472. http://dx.doi.org/10.1038/ngeo890

172. Feldmann, J. and A. Levermann, 2015: Collapse of the West Antarctic Ice Sheet after local destabilization of the Amundsen Basin. *Proceedings of the National Academy of Sciences*, **112**, 14191-14196. http://dx.doi.org/10.1073/pnas.1512482112

173. DeConto, R.M. and D. Pollard, 2016: Contribution of Antarctica to past and future sea-level rise. *Nature*, **531**, 591-597. http://dx.doi.org/10.1038/nature17145

174. Harig, C. and F.J. Simons, 2012: Mapping Greenland's mass loss in space and time. *Proceedings of the National Academy of Sciences*, **109**, 19934-19937. http://dx.doi.org/10.1073/pnas.1206785109

175. Jacob, T., J. Wahr, W.T. Pfeffer, and S. Swenson, 2012: Recent contributions of glaciers and ice caps to sea level rise. *Nature*, **482**, 514-518. http://dx.doi.org/10.1038/nature10847

176. Tedesco, M., X. Fettweis, M.R.v.d. Broeke, R.S.W.v.d. Wal, C.J.P.P. Smeets, W.J.v.d. Berg, M.C. Serreze, and J.E. Box, 2011: The role of albedo and accumulation in the 2010 melting record in Greenland. *Environmental Research Letters*, **6**, 014005. http://dx.doi.org/10.1088/1748-9326/6/1/014005

177. Fettweis, X., M. Tedesco, M. van den Broeke, and J. Ettema, 2011: Melting trends over the Greenland ice sheet (1958–2009) from spaceborne microwave data and regional climate models. *The Cryosphere*, **5**, 359-375. http://dx.doi.org/10.5194/tc-5-359-2011

178. Tedesco, M., E. Box, J. Cappelen, R.S. Fausto, X. Fettweis, K. Hansen, T. Mote, C.J.P.P. Smeets, D.V. As, R.S.W.V.d. Wal, and J. Wahr, 2015: [The Arctic] Greenland ice sheet [in "State of the Climate in 2014"]. *Bulletin of the American Meteorological Society*, **96 (12)**, S137-S139. http://dx.doi.org/10.1175/2015BAMS-StateoftheClimate.1

179. Nghiem, S.V., D.K. Hall, T.L. Mote, M. Tedesco, M.R. Albert, K. Keegan, C.A. Shuman, N.E. DiGirolamo, and G. Neumann, 2012: The extreme melt across the Greenland ice sheet in 2012. *Geophysical Research Letters*, **39**, L20502. http://dx.doi.org/10.1029/2012GL053611

180. Tedesco, M., X. Fettweis, T. Mote, J. Wahr, P. Alexander, J.E. Box, and B. Wouters, 2013: Evidence and analysis of 2012 Greenland records from spaceborne observations, a regional climate model and reanalysis data. *The Cryosphere*, **7**, 615-630. http://dx.doi.org/10.5194/tc-7-615-2013

181. Romanovsky, V.E., S.L. Smith, H.H. Christiansen, N.I. Shiklomanov, D.A. Streletskiy, D.S. Drozdov, G.V. Malkova, N.G. Oberman, A.L. Kholodov, and S.S. Marchenko, 2015: [The Arctic] Terrestrial permafrost [in "State of the Climate in 2014"]. *Bulletin of the American Meteorological Society*, **96 (12)**, S139-S141. http://dx.doi.org/10.1175/2015BAMSStateoftheClimate.1

182. Shiklomanov, N.E., D.A. Streletskiy, and F.E. Nelson, 2012: Northern Hemisphere component of the global Circumpolar Active Layer Monitory (CALM) program. In *Proceedings of the 10th International Conference on Permafrost*, Salekhard, Russia. Kane, D.L. and K.M. Hinkel, Eds., 377-382. http://research.iarc.uaf.edu/NICOP/proceedings/10th/TICOP_vol1.pdf

183. Church, J.A. and N.J. White, 2011: Sea-level rise from the late 19th to the early 21st century. *Surveys in Geophysics*, **32**, 585-602. http://dx.doi.org/10.1007/s10712-011-9119-1

184. Hay, C.C., E. Morrow, R.E. Kopp, and J.X. Mitrovica, 2015: Probabilistic reanalysis of twentieth-century sea-level rise. *Nature*, **517**, 481-484. http://dx.doi.org/10.1038/nature14093

185. Nerem, R.S., D.P. Chambers, C. Choe, and G.T. Mitchum, 2010: Estimating mean sea level change from the TOPEX and Jason altimeter missions. *Marine Geodesy*, **33**, 435-446. http://dx.doi.org/10.1080/01490419.2010.491031

186. Merrifield, M.A., P. Thompson, E. Leuliette, G.T. Mitchum, D.P. Chambers, S. Jevrejeva, R.S. Nerem, M. Menéndez, W. Sweet, B. Hamlington, and J.J. Marra, 2015: [Global Oceans] Sea level variability and change [in "State of the Climate in 2014"]. *Bulletin of the American Meteorological Society*, **96 (12)**, S82-S85. http://dx.doi.org/10.1175/2015BAMSStateoftheClimate.1

187. Ezer, T. and L.P. Atkinson, 2014: Accelerated flooding along the U.S. East Coast: On the impact of sea-level rise, tides, storms, the Gulf Stream, and the North Atlantic Oscillations. *Earth's Future*, **2**, 362-382. http://dx.doi.org/10.1002/2014EF000252

188. Sweet, W.V. and J. Park, 2014: From the extreme to the mean: Acceleration and tipping points of coastal inundation from sea level rise. *Earth's Future*, **2**, 579-600. http://dx.doi.org/10.1002/2014EF000272

189. Kopp, R.E., R.M. Horton, C.M. Little, J.X. Mitrovica, M. Oppenheimer, D.J. Rasmussen, B.H. Strauss, and C. Tebaldi, 2014: Probabilistic 21st and 22nd century sea-level projections at a global network of tide-gauge sites. *Earth's Future*, **2**, 383-406. http://dx.doi.org/10.1002/2014EF000239

190. Parris, A., P. Bromirski, V. Burkett, D. Cayan, M. Culver, J. Hall, R. Horton, K. Knuuti, R. Moss, J. Obeysekera, A. Sallenger, and J. Weiss, 2012: Global Sea Level Rise Scenarios for the United States National Climate Assessment. National Oceanic and Atmospheric Administration, Silver Spring, MD. 37 pp. http://scenarios.globalchange.gov/sites/default/files/NOAA_SLR_r3_0.pdf

191. Sweet, W.V., R.E. Kopp, C.P. Weaver, J. Obeysekera, R.M. Horton, E.R. Thieler, and C. Zervas, 2017: Global and Regional Sea Level Rise Scenarios for the United States. National Oceanic and Atmospheric Administration, National Ocean Service, Silver Spring, MD. 75 pp. https://tidesandcurrents.noaa.gov/publications/techrpt83_Global_and_Regional_SLR_Scenarios_for_the_US_final.pdf

192. Schmidt, G.A., J.H. Jungclaus, C.M. Ammann, E. Bard, P. Braconnot, T.J. Crowley, G. Delaygue, F. Joos, N.A. Krivova, R. Muscheler, B.L. Otto-Bliesner, J. Pongratz, D.T. Shindell, S.K. Solanki, F. Steinhilber, and L.E.A. Vieira, 2011: Climate forcing reconstructions for use in PMIP simulations of the last millennium (v1.0). *Geoscientific Model Development*, **4**, 33-45. http://dx.doi.org/10.5194/gmd-4-33-2011

193. Mann, M.E., Z. Zhang, M.K. Hughes, R.S. Bradley, S.K. Miller, S. Rutherford, and F. Ni, 2008: Proxy-based reconstructions of hemispheric and global surface temperature variations over the past two millennia. *Proceedings of the National Academy of Sciences*, **105**, 13252-13257. http://dx.doi.org/10.1073/pnas.0805721105

194. Turney, C.S.M. and R.T. Jones, 2010: Does the Agulhas Current amplify global temperatures during super-interglacials? *Journal of Quaternary Science*, **25**, 839-843. http://dx.doi.org/10.1002/jqs.1423

195. Dutton, A. and K. Lambeck, 2012: Ice volume and sea level during the Last Interglacial. *Science*, **337**, 216-219. http://dx.doi.org/10.1126/science.1205749

196. Kopp, R.E., F.J. Simons, J.X. Mitrovica, A.C. Maloof, and M. Oppenheimer, 2009: Probabilistic assessment of sea level during the last interglacial stage. *Nature*, **462**, 863-867. http://dx.doi.org/10.1038/nature08686

197. Kaspar, F., N. Kühl, U. Cubasch, and T. Litt, 2005: A model-data comparison of European temperatures in the Eemian interglacial. *Geophysical Research Letters*, **32**, L11703. http://dx.doi.org/10.1029/2005GL022456

198. Haywood, A.M., D.J. Hill, A.M. Dolan, B.L. Otto-Bliesner, F. Bragg, W.L. Chan, M.A. Chandler, C. Contoux, H.J. Dowsett, A. Jost, Y. Kamae, G. Lohmann, D.J. Lunt, A. Abe-Ouchi, S.J. Pickering, G. Ramstein, N.A. Rosenbloom, U. Salzmann, L. Sohl, C. Stepanek, H. Ueda, Q. Yan, and Z. Zhang, 2013: Large-scale features of Pliocene climate: Results from the Pliocene Model Intercomparison Project. *Climate of the Past*, **9**, 191-209. http://dx.doi.org/10.5194/cp-9-191-2013

199. Wuebbles, D., G. Meehl, K. Hayhoe, T.R. Karl, K. Kunkel, B. Santer, M. Wehner, B. Colle, E.M. Fischer, R. Fu, A. Goodman, E. Janssen, V. Kharin, H. Lee, W. Li, L.N. Long, S.C. Olsen, Z. Pan, A. Seth, J. Sheffield, and L. Sun, 2014: CMIP5 climate model analyses: Climate extremes in the United States. *Bulletin of the American Meteorological Society*, **95**, 571-583. http://dx.doi.org/10.1175/BAMS-D-12-00172.1

200. Trenberth, K.E., A. Dai, G. van der Schrier, P.D. Jones, J. Barichivich, K.R. Briffa, and J. Sheffield, 2014: Global warming and changes in drought. *Nature Climate Change*, **4**, 17-22. http://dx.doi.org/10.1038/nclimate2067

201. Walsh, J., D. Wuebbles, K. Hayhoe, J. Kossin, K. Kunkel, G. Stephens, P. Thorne, R. Vose, M. Wehner, J. Willis, D. Anderson, S. Doney, R. Feely, P. Hennon, V. Kharin, T. Knutson, F. Landerer, T. Lenton, J. Kennedy, and R. Somerville, 2014: Ch. 2: Our changing climate. *Climate Change Impacts in the United States: The Third National Climate Assessment*. Melillo, J.M., T.C. Richmond, and G.W. Yohe, Eds. U.S. Global Change Research Program, Washington, D.C., 19-67. http://dx.doi.org/10.7930/J0KW5CXT

202. Jones, P.D., D.H. Lister, T.J. Osborn, C. Harpham, M. Salmon, and C.P. Morice, 2012: Hemispheric and large-scale land surface air temperature variations: An extensive revision and an update to 2010. *Journal of Geophysical Research*, **117**, D05127. http://dx.doi.org/10.1029/2011JD017139

# 2
# Physical Drivers of Climate Change

## KEY FINDINGS

1. Human activities continue to significantly affect Earth's climate by altering factors that change its radiative balance. These factors, known as radiative forcings, include changes in greenhouse gases, small airborne particles (aerosols), and the reflectivity of the Earth's surface. In the industrial era, human activities have been, and are increasingly, the dominant cause of climate warming. The increase in radiative forcing due to these activities has far exceeded the relatively small net increase due to natural factors, which include changes in energy from the sun and the cooling effect of volcanic eruptions. (*Very high confidence*)

2. Aerosols caused by human activity play a profound and complex role in the climate system through radiative effects in the atmosphere and on snow and ice surfaces and through effects on cloud formation and properties. The combined forcing of aerosol–radiation and aerosol–cloud interactions is negative (cooling) over the industrial era (*high confidence*), offsetting a substantial part of greenhouse gas forcing, which is currently the predominant human contribution. The magnitude of this offset, globally averaged, has declined in recent decades, despite increasing trends in aerosol emissions or abundances in some regions (*medium* to *high confidence*).

3. The interconnected Earth–atmosphere–ocean system includes a number of positive and negative feedback processes that can either strengthen (positive feedback) or weaken (negative feedback) the system's responses to human and natural influences. These feedbacks operate on a range of time scales from very short (essentially instantaneous) to very long (centuries). Global warming by net radiative forcing over the industrial era includes a substantial amplification from these feedbacks (approximately a factor of three) (*high confidence*). While there are large uncertainties associated with some of these feedbacks, the net feedback effect over the industrial era has been positive (amplifying warming) and will continue to be positive in coming decades (*very high confidence*).

**Recommended Citation for Chapter**

**Fahey**, D.W., S.J. Doherty, K.A. Hibbard, A. Romanou, and P.C. Taylor, 2017: Physical drivers of climate change. In: *Climate Science Special Report: Fourth National Climate Assessment, Volume I* [Wuebbles, D.J., D.W. Fahey, K.A. Hibbard, D.J. Dokken, B.C. Stewart, and T.K. Maycock (eds.)]. U.S. Global Change Research Program, Washington, DC, USA, pp. 73-113, doi: 10.7930/J0513WCR.

## 2.0 Introduction

Earth's climate is undergoing substantial change due to anthropogenic activities (Ch. 1: Our Globally Changing Climate). Understanding the causes of past and present climate change and confidence in future projected changes depend directly on our ability to understand and model the physical drivers of climate change.[1] Our understanding is challenged by the complexity and interconnectedness of the components of the climate system (that is, the atmosphere, land, ocean, and cryosphere). This chapter lays out the foundation of climate change by describing its physical drivers, which are primarily associat-

ed with atmospheric composition (gases and aerosols) and cloud effects. We describe the principle radiative forcings and the variety of feedback responses which serve to amplify these forcings.

## 2.1 Earth's Energy Balance and the Greenhouse Effect

The temperature of the Earth system is determined by the amounts of incoming (short-wavelength) and outgoing (both short- and long-wavelength) radiation. In the modern era, radiative fluxes are well-constrained by satellite measurements (Figure 2.1). About a third (29.4%) of incoming, short-wavelength

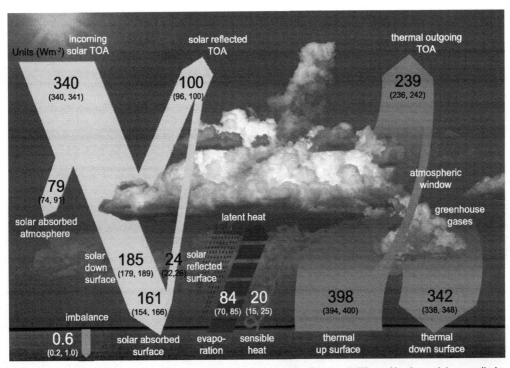

Figure 2.1: Global mean energy budget of Earth under present-day climate conditions. Numbers state magnitudes of the individual energy fluxes in watts per square meter (W/m²) averaged over Earth's surface, adjusted within their uncertainty ranges to balance the energy budgets of the atmosphere and the surface. Numbers in parentheses attached to the energy fluxes cover the range of values in line with observational constraints. Fluxes shown include those resulting from feedbacks. Note the net imbalance of 0.6 W/m² in the global mean energy budget. The observational constraints are largely provided by satellite-based observations, which have directly measured solar and infrared fluxes at the top of the atmosphere over nearly the whole globe since 1984.[217, 218] More advanced satellite-based measurements focusing on the role of clouds in Earth's radiative fluxes have been available since 1998.[219, 220] Top of Atmosphere (TOA) reflected solar values given here are based on observations 2001–2010; TOA outgoing longwave is based on 2005–2010 observations. (Figure source: Hartmann et al. 2013,[221] Figure 2-11; © IPCC, used with permission).

energy from the sun is reflected back to space, and the remainder is absorbed by Earth's system. The fraction of sunlight scattered back to space is determined by the reflectivity (albedo) of clouds, land surfaces (including snow and ice), oceans, and particles in the atmosphere. The amount and albedo of clouds, snow cover, and ice cover are particularly strong determinants of the amount of sunlight reflected back to space because their albedos are much higher than that of land and oceans.

In addition to reflected sunlight, Earth loses energy through infrared (long-wavelength) radiation from the surface and atmosphere. Absorption by greenhouse gases (GHGs) of infrared energy radiated from the surface leads to warming of the surface and atmosphere. Figure 2.1 illustrates the importance of greenhouse gases in the energy balance of Earth's system. The naturally occurring GHGs in Earth's atmosphere—principally water vapor and carbon dioxide—keep the near-surface air temperature about 60°F (33°C) warmer than it would be in their absence, assuming albedo is held constant.[2] Geothermal heat from Earth's interior, direct heating from energy production, and frictional heating through tidal flows also contribute to the amount of energy available for heating Earth's surface and atmosphere, but their total contribution is an extremely small fraction (< 0.1%) of that due to net solar (shortwave) and infrared (longwave) radiation (e.g., see Davies and Davies 2010;[3] Flanner 2009;[4] Munk and Wunsch 1998,[5] where these forcings are quantified).

Thus, Earth's equilibrium temperature in the modern era is controlled by a short list of factors: incoming sunlight, absorbed and reflected sunlight, emitted infrared radiation, and infrared radiation absorbed and re-emitted in the atmosphere, primarily by GHGs. Changes in these factors affect Earth's radiative balance and therefore its climate, including

but not limited to the average, near-surface air temperature. Anthropogenic activities have changed Earth's radiative balance and its albedo by adding GHGs, particles (aerosols), and aircraft contrails to the atmosphere, and through land-use changes. Changes in the radiative balance (or forcings) produce changes in temperature, precipitation, and other climate variables through a complex set of physical processes, many of which are coupled (Figure 2.2). These changes, in turn, trigger feedback processes which can further amplify and/or dampen the changes in radiative balance (Sections 2.5 and 2.6).

In the following sections, the principal components of the framework shown in Figure 2.2 are described. Climate models are structured to represent these processes; climate models and their components and associated uncertainties, are discussed in more detail in Chapter 4: Projections.

The processes and feedbacks connecting changes in Earth's radiative balance to a climate response (Figure 2.2) operate on a large range of time scales. Reaching an equilibrium temperature distribution in response to anthropogenic activities takes decades or longer because some components of Earth's system—in particular the oceans and cryosphere—are slow to respond due to their large thermal masses and the long time scale of circulation between the ocean surface and the deep ocean. Of the substantial energy gained in the combined ocean–atmosphere system over the previous four decades, over 90% of it has gone into ocean warming (see Box 3.1 Figure 1 of Rhein et al. 2013).[6] Even at equilibrium, internal variability in Earth's climate system causes limited annual- to decadal-scale variations in regional temperatures and other climate parameters that do not contribute to long-term trends. For example, it is *likely* that natural variability has contributed between

## Simplified Conceptual Framework of the Climate System

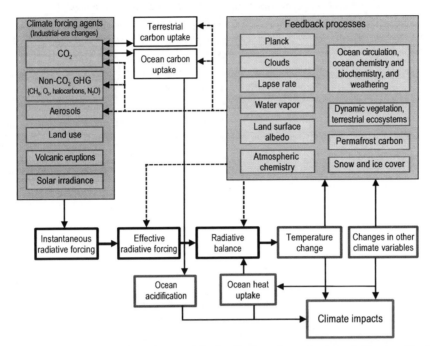

**Figure 2.2:** Simplified conceptual modeling framework for the climate system as implemented in many climate models (Ch. 4: Projections). Modeling components include forcing agents, feedback processes, carbon uptake processes, and radiative forcing and balance. The lines indicate physical interconnections (solid lines) and feedback pathways (dashed lines). Principal changes (blue boxes) lead to climate impacts (red box) and feedbacks. (Figure source: adapted from Knutti and Rugenstein 2015[82]).

−0.18°F (−0.1°C) and 0.18°F (0.1°C) to changes in surface temperatures from 1951 to 2010; by comparison, anthropogenic GHGs have *likely* contributed between 0.9°F (0.5°C) and 2.3°F (1.3°C) to observed surface warming over this same period.[7] Due to these longer time scale responses and natural variability, changes in Earth's radiative balance are not realized immediately as changes in climate, and even in equilibrium there will always be variability around mean conditions.

### 2.2 Radiative Forcing (RF) and Effective Radiative Forcing (ERF)

Radiative forcing (RF) is widely used to quantify a radiative imbalance in Earth's atmosphere resulting from either natural changes or anthropogenic activities over the industrial era. It is expressed as a change in net radiative flux (W/m²) either at the tropopause or top of the atmosphere,[8] with the latter nominally defined at 20 km altitude to optimize observation/model comparisons.[9] The instantaneous RF is defined as the immediate change in net radiative flux following a change in a climate driver. RF can also be calculated after allowing different types of system response: for example, after allowing stratospheric temperatures to adjust, after allowing both stratospheric and surface temperature to adjust, or after allowing temperatures to adjust everywhere (the equilibrium RF) (Figure 8.1 of Myhre et al. 2013[8]).

In this report, we follow the Intergovernmental Panel on Climate Change (IPCC) recommendation that the RF caused by a forcing agent be evaluated as the net radiative flux change at the tropopause after stratospheric temperatures have adjusted to a new radiative equilibrium while assuming all other variables (for example, temperatures and cloud cover) are held fixed (Box 8.1 of Myhre et al. 2013[8]). A change that results in a net increase in the downward flux (shortwave plus longwave) constitutes a positive RF, normally resulting in a warming of the surface and/or atmosphere and potential changes in other climate parameters. Conversely, a change that yields an increase in the net upward flux constitutes a negative RF, leading to a cooling of the surface and/or atmosphere and potential changes in other climate parameters.

RF serves as a metric to compare present, past, or future perturbations to the climate system (e.g., Boer and Yu 2003;[10] Gillett et al. 2004;[11] Matthews et al. 2004;[12] Meehl et al. 2004;[13] Jones et al. 2007;[14] Mahajan et al. 2013;[15] Shiogama et al. 2013[16]). For clarity and consistency, RF calculations require that a time period be defined over which the forcing occurs. Here, this period is the industrial era, defined as beginning in 1750 and extending to 2011, unless otherwise noted. The 2011 end date is that adopted by the CMIP5 calculations, which are the basis of RF evaluations by the IPCC.[8]

A refinement of the RF concept introduced in the latest IPCC assessment[17] is the use of effective radiative forcing (ERF). ERF for a climate driver is defined as its RF plus rapid adjustment(s) to that RF.[8] These rapid adjustments occur on time scales much shorter than, for example, the response of ocean temperatures. For an important subset of climate drivers, ERF is more reliably correlated with the climate response to the forcing than is RF; as such, it is an increasingly used metric when discussing forcing. For atmospheric components, ERF includes rapid adjustments due

to direct warming of the troposphere, which produces horizontal temperature variations, variations in the vertical lapse rate, and changes in clouds and vegetation, and it includes the microphysical effects of aerosols on cloud lifetime. Rapid changes in land surface properties (temperature, snow and ice cover, and vegetation) are also included. Not included in ERF are climate responses driven by changes in sea surface temperatures or sea ice cover. For forcing by aerosols in snow (Section 2.3.2), ERF includes the effects of direct warming of the snowpack by particulate absorption (for example, snow-grain size changes). Changes in all of these parameters in response to RF are quantified in terms of their impact on radiative fluxes (for example, albedo) and included in the ERF. The largest differences between RF and ERF occur for forcing by light-absorbing aerosols because of their influence on clouds and snow (Section 2.3.2). For most non-aerosol climate drivers, the differences between RF and ERF are small.

## 2.3 Drivers of Climate Change over the Industrial Era

Climate drivers of significance over the industrial era include both those associated with anthropogenic activity and, to a lesser extent, those of natural origin. The only significant natural climate drivers in the industrial era are changes in solar irradiance, volcanic eruptions, and the El Niño–Southern Oscillation. Natural emissions and sinks of GHGs and tropospheric aerosols have varied over the industrial era but have not contributed significantly to RF. The effects of cosmic rays on cloud formation have been studied, but global radiative effects are not considered significant.[18] There are other known drivers of natural origin that operate on longer time scales (for example, changes in Earth's orbit [Milankovitch cycles] and changes in atmospheric $CO_2$ via chemical weathering of rock). Anthropogenic drivers can be divided into a

## Radiative Forcing of Climate Between 1750 and 2011

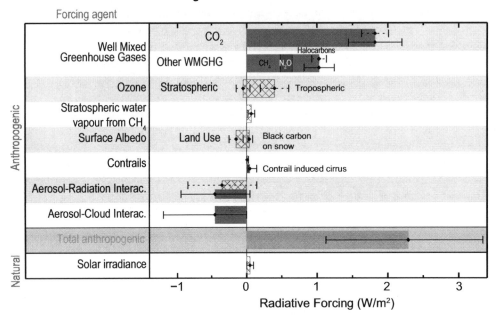

**Figure 2.3:** Bar chart for radiative forcing (RF; hatched) and effective radiative forcing (ERF; solid) for the period 1750–2011, where the total ERF is derived from the Intergovernmental Panel on Climate Change's Fifth Assessment Report. Uncertainties (5% to 95% confidence range) are given for RF (dotted lines) and ERF (solid lines). Volcanic forcing is not shown because this forcing is intermittent, exerting forcing over only a few years for eruptions during the industrial era; the net forcing over the industrial era is negligible. (Figure source: Myhre et al. 2013,[8] Figure 8-15; © IPCC, used with permission).

number of categories, including well-mixed greenhouse gases (WMGHGs), short-lived climate forcers (SLCFs, which include methane, some hydrofluorocarbons [HFCs], ozone, and aerosols), contrails, and changes in albedo (for example, land-use changes). Some WMGHGs are also considered SLCFs (for example, methane). Figures 2.3–2.7 summarize features of the principal climate drivers in the industrial era. Each is described briefly in the following.

### 2.3.1 Natural Drivers

*Solar Irradiance*

Changes in solar irradiance directly impact the climate system because the irradiance is Earth's primary energy source.[19] In the industrial era, the largest variations in total solar irradiance follow an 11-year cycle.[20, 21] Direct solar observations have been available since

1978,[22] though proxy indicators of solar cycles are available back to the early 1600s.[23] Although these variations amount to only 0.1% of the total solar output of about 1360 W/m²,[24] relative variations in irradiance at specific wavelengths can be much larger (tens of percent). Spectral variations in solar irradiance are highest at near-ultraviolet (UV) and shorter wavelengths,[25] which are also the most important wavelengths for driving changes in ozone.[26, 27] By affecting ozone concentrations, variations in total and spectral solar irradiance induce discernible changes in atmospheric heating and changes in circulation.[21, 28, 29] The relationships between changes in irradiance and changes in atmospheric composition, heating, and dynamics are such that changes in total solar irradiance are not directly correlated with the resulting radiative flux changes.[26, 30, 31]

The IPCC estimate of the RF due to changes in total solar irradiance over the industrial era is 0.05 W/m² (range: 0.0 to 0.10 W/m²).[8] This forcing does not account for radiative flux changes resulting from changes in ozone driven by changes in the spectral irradiance. Understanding of the links between changes in spectral irradiance, ozone concentrations, heating rates, and circulation changes has recently improved using, in particular, satellite data starting in 2002 that provide solar spectral irradiance measurements through the UV[26] along with a series of chemistry–climate modeling studies.[26, 27, 32, 33, 34] At the regional scale, circulation changes driven by solar spectral irradiance variations may be significant for some locations and seasons but are poorly quantified.[28] Despite remaining uncertainties, there is *very high confidence* that solar radiance-induced changes in RF are small relative to RF from anthropogenic GHGs over the industrial era (Figure 2.3).[8]

*Volcanoes*

Most volcanic eruptions are minor events with the effects of emissions confined to the troposphere and only lasting for weeks to months. In contrast, explosive volcanic eruptions inject substantial amounts of sulfur dioxide ($SO_2$) and ash into the stratosphere, which lead to significant short-term climate effects (Myhre et al. 2013,[8] and references therein). $SO_2$ oxidizes to form sulfuric acid ($H_2SO_4$) which condenses, forming new particles or adding mass to preexisting particles, thereby substantially enhancing the attenuation of sunlight transmitted through the stratosphere (that is, increasing aerosol optical depth). These aerosols increase Earth's albedo by scattering sunlight back to space, creating a negative RF that cools the planet.[35, 36] The RF persists for the lifetime of aerosol in the stratosphere, which is a few years, far exceeding that in the troposphere (about a week). The oceans respond to a negative volcanic RF through cooling and chang-

es in ocean circulation patterns that last for decades after major eruptions (for example, Mt. Tambora in 1815).[37, 38, 39, 40] In addition to the direct RF, volcanic aerosol heats the stratosphere, altering circulation patterns, and depletes ozone by enhancing surface reactions, which further changes heating and circulation. The resulting impacts on advective heat transport can be larger than the temperature impacts of the direct forcing.[36] Aerosol from both explosive and non-explosive eruptions also affects the troposphere through changes in diffuse radiation and through aerosol–cloud interactions. It has been proposed that major eruptions might "fertilize" the ocean with sufficient iron to affect phyotoplankton production and, therefore, enhance the ocean carbon sink.[41] Volcanoes also emit $CO_2$ and water vapor, although in small quantities relative to other emissions. At present, conservative estimates of annual $CO_2$ emissions from volcanoes are less than 1% of $CO_2$ emissions from all anthropogenic activities.[42] The magnitude of volcanic effects on climate depends on the number and strength of eruptions, the latitude of injection and, for ocean temperature and circulation impacts, the timing of the eruption relative to ocean temperature and circulation patterns.[39, 40]

Volcanic eruptions represent the largest natural forcing within the industrial era. In the last millennium, eruptions caused several multiyear, transient episodes of negative RF of up to several W/m² (Figure 2.6). The RF of the last major volcanic eruption, Mt. Pinatubo in 1991, decayed to negligible values later in the 1990s, with the temperature signal lasting about twice as long due to the effects of changes in ocean heat uptake.[37] A net volcanic RF has been omitted from the drivers of climate change in the industrial era in Figure 2.3 because the value from multiple, episodic eruptions is negligible compared with the other climate drivers. While future explosive volcanic eruptions have the potential to again alter Earth's climate for periods of several years, predictions of occurrence, intensity, and location remain elusive. If a sufficient number of non-explosive eruptions occur over an extended time period in the future, average changes in tropospheric composition or circulation could yield a significant RF.[36]

### 2.3.2 Anthropogenic Drivers

*Principal Well-mixed Greenhouse Gases (WMGHGs)*

The principal WMGHGs are carbon dioxide ($CO_2$), methane ($CH_4$), and nitrous oxide

($N_2O$). With atmospheric lifetimes of a decade to a century or more, these gases have modest-to-small regional variabilities and are circulated and mixed around the globe to yield small interhemispheric gradients. The atmospheric abundances and associated radiative forcings of WMGHGs have increased substantially over the industrial era (Figures 2.4–2.6). Contributions from natural sources of these constituents are accounted for in the industrial-era RF calculations shown in Figure 2.6.

$CO_2$ has substantial global sources and sinks (Figure 2.7). $CO_2$ emission sources have grown

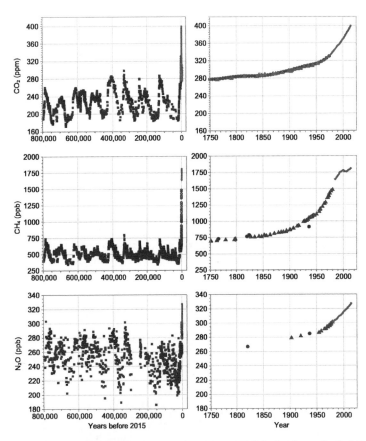

**Figure 2.4:** Atmospheric concentrations of $CO_2$ (top), $CH_4$ (middle), and $N_2O$ (bottom) over the last 800,000 years (left panels) and for 1750–2015 (right panels). Measurements are shown from ice cores (symbols with different colors for different studies) and for direct atmospheric measurements (red lines). (Adapted from IPCC 2007,[88] Figure SPM.1, © IPCC, used with permission; data are from https://www.epa.gov/climate-indicators/climate-change-indicators-atmospheric-concentrations-greenhouse-gases).

in the industrial era primarily from fossil fuel combustion (that is, coal, gas, and oil), cement manufacturing, and land-use change from activities such as deforestation.[43] Carbonation of finished cement products is a sink of atmospheric $CO_2$, offsetting a substantial fraction (0.43) of the industrial-era emissions from

cement production.[44] A number of processes act to remove $CO_2$ from the atmosphere, including uptake in the oceans, residual land uptake, and rock weathering. These combined processes yield an effective atmospheric lifetime for emitted $CO_2$ of many decades to millennia, far greater than any other major

### Radiative Forcing of Well-mixed Greenhouse Gases

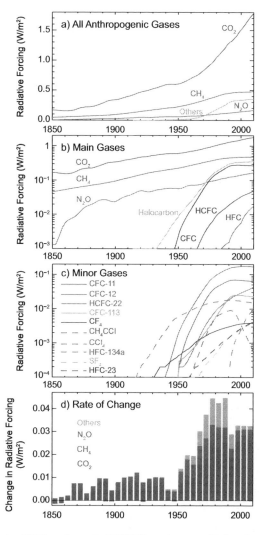

Figure 2.5: (a) Radiative forcing (RF) from the major WMGHGs and groups of halocarbons (Others) from 1850 to 2011; (b) the data in (a) with a logarithmic scale; (c) RFs from the minor WMGHGs from 1850 to 2011 (logarithmic scale); (d) the annual rate of change ([W/m²]/year) in forcing from the major WMGHGs and halocarbons from 1850 to 2011. (Figure source: Myhre et al. 2013,[8] Figure 8-06; © IPCC, used with permission).

## Time Evolution of Forcings

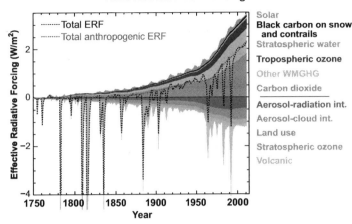

**Figure 2.6:** Time evolution in effective radiative forcings (ERFs) across the industrial era for anthropogenic and natural forcing mechanisms. The terms contributing to cumulative totals of positive and negative ERF are shown with colored regions. The terms are labeled in order on the right-hand side with positive ERFs above the zero line and negative ERFs below the zero line. The forcings from black-carbon-on-snow and contrail terms are grouped together into a single term in the plot. Also shown are the cumulative sum of all forcings (Total; black dashed line) and of anthropogenic-only forcings (Total Anthropogenic; red dashed line). Uncertainties in 2011 ERF values are shown in the original figure (Myhre et al. 2013,[8] Figure 8-18). See the Intergovernmental Panel on Climate Change Fifth Assessment Report (IPCC AR5) Supplementary Material Table 8.SM.8[8] for further information on the forcing time evolutions. Forcing numbers are provided in Annex II of IPCC AR5. The total anthropogenic forcing was 0.57 (0.29 to 0.85) W/m² in 1950, 1.25 (0.64 to 1.86) W/m² in 1980, and 2.29 (1.13 to 3.33) W/m² in 2011. (Figure source: Myhre et al. 2013,[8] Figure 8-18; © IPCC, used with permission).

GHG. Seasonal variations in $CO_2$ atmospheric concentrations occur in response to seasonal changes in photosynthesis in the biosphere, and to a lesser degree to seasonal variations in anthropogenic emissions. In addition to fossil fuel reserves, there are large natural reservoirs of carbon in the oceans, in vegetation and soils, and in permafrost.

In the industrial era, the $CO_2$ atmospheric growth rate has been exponential (Figure 2.4), with the increase in atmospheric $CO_2$ approximately twice that absorbed by the oceans. Over at least the last 50 years, $CO_2$ has shown the largest annual RF increases among all GHGs (Figures 2.4 and 2.5). The global average $CO_2$ concentration has increased by 40% over the industrial era, increasing from 278 parts per million (ppm) in 1750 to 390 ppm in 2011;[43] it now exceeds 400 ppm (as of 2016) (http://www.esrl.noaa.gov/gmd/ccgg/trends/). $CO_2$ has been chosen as the refer-

ence in defining the global warming potential (GWP) of other GHGs and climate agents. The GWP of a GHG is the integrated RF over a specified time period (for example, 100 years) from the emission of a given mass of the GHG divided by the integrated RF from the same mass emission of $CO_2$.

The global mean methane concentration and RF have also grown substantially in the industrial era (Figures 2.4 and 2.5). Methane is a stronger GHG than $CO_2$ for the same emission mass and has a shorter atmospheric lifetime of about 12 years. Methane also has indirect climate effects through induced changes in $CO_2$, stratospheric water vapor, and ozone.[45] The 100-year GWP of methane is 28–36, depending on whether oxidation into $CO_2$ is included and whether climate-carbon feedbacks are accounted for; its 20-year GWP is even higher (84–86) (Myhre et al. 2013[8] Table 8.7). With a current global mean value near 1840 parts per

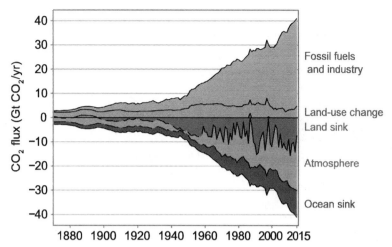

**Figure 2.7:** $CO_2$ sources and sinks (Gt$CO_2$/yr) over the period 1870–2015. The partitioning of atmospheric emissions among the atmosphere, land, and ocean is shown as equivalent negative emissions in the lower panel; of these, the land and ocean terms are sinks of atmospheric $CO_2$. $CO_2$ emissions from net land-use changes are mainly from deforestation. The atmospheric $CO_2$ growth rate is derived from atmospheric observations and ice core data. The ocean $CO_2$ sink is derived from a combination of models and observations. The land sink is the residual of the other terms in a balanced $CO_2$ budget and represents the sink of anthropogenic $CO_2$ in natural land ecosystems. These terms only represent changes since 1750 and do not include natural $CO_2$ fluxes (for example, from weathering and outgassing from lakes and rivers). **(Figure source: Le Quére et al. 2016,**[135] **Figure 3).**

billion by volume (ppb), the methane concentration has increased by a factor of about 2.5 over the industrial era. The annual growth rate for methane has been more variable than that for $CO_2$ and $N_2O$ over the past several decades, and has occasionally been negative for short periods.

Methane emissions, which have a variety of natural and anthropogenic sources, totaled $556 \pm 56$ Tg $CH_4$ in 2011 based on top-down analyses, with about 60% from anthropogenic sources.[43] The methane budget is complicated by the variety of natural and anthropogenic sources and sinks that influence its atmospheric concentration. These include the global abundance of the hydroxyl radical (OH), which controls the methane atmospheric lifetime; changes in large-scale anthropogenic activities such as mining, natural gas extraction, animal husbandry, and agricultural practices; and natural wetland emissions (Table 6.8, Ciais et al. 2013[43]). The remaining uncertainty in the cause(s) of the approximately 20-year

negative trend in the methane annual growth rate starting in the mid-1980s and the rapid increases in the annual rate in the last decade (Figure 2.4) reflect the complexity of the methane budget.[43, 46, 47]

Growth rates in the global mean nitrous oxide ($N_2O$) concentration and RF over the industrial era are smaller than for $CO_2$ and methane (Figures 2.4 and 2.5). $N_2O$ is emitted in the nitrogen cycle in natural ecosystems and has a variety of anthropogenic sources, including the use of synthetic fertilizers in agriculture, motor vehicle exhaust, and some manufacturing processes. The current global value near 330 ppb reflects steady growth over the industrial era with average increases in recent decades of 0.75 ppb per year (Figure 2.4).[43] Fertilization in global food production is responsible for about 80% of the growth rate. Anthropogenic sources account for approximately 40% of the annual $N_2O$ emissions of 17.9 (8.1 to 30.7) TgN.[43] $N_2O$ has an atmospheric lifetime of about 120 years and a GWP in the range 265–298 (Myhre et al. 2013[8] Table 8.7). The primary sink of $N_2O$ is photochemical destruction in the stratosphere, which produces nitrogen oxides ($NO_x$) that catalytically destroy ozone (e.g., Skiba and Rees 2014[48]). Small indirect climate effects, such as the response of stratospheric ozone, are generally not included in the $N_2O$ RF.

$N_2O$ is a component of the larger global budget of total reactive nitrogen (N) comprising $N_2O$, ammonia ($NH_3$), and nitrogen oxides ($NO_x$) and other compounds. Significant uncertainties are associated with balancing this budget over oceans and land while accounting for deposition and emission processes.[43, 49] Furthermore, changes in climate parameters such as temperature, moisture, and $CO_2$ concentrations are expected to affect the $N_2O$ budget in the future, and perhaps atmospheric concentrations.

*Other Well-mixed Greenhouse Gases*
Other WMGHGs include several categories of synthetic (i.e., manufactured) gases, including chlorofluorocarbons (CFCs), halons, hydrochlorofluorocarbons (HCFCs), hydrofluorocarbons (HFCs), perfluorocarbons (PFCs), and sulfur hexafluoride ($SF_6$), collectively known as halocarbons. Natural sources of these gases in the industrial era are small compared to anthropogenic sources. Important examples are the expanded use of CFCs as refrigerants and in other applications beginning in the mid-20th century. The atmospheric abundances of principal CFCs began declining in the 1990s after their regulation under the Montreal Protocol as substances that deplete stratospheric ozone (Figure 2.5). All of these gases are GHGs covering a wide range of GWPs, atmospheric concentrations, and trends. PFCs, $SF_6$, and HFCs are in the basket of gases covered under the United Nations Framework Convention on Climate Change. The United States joined other countries in proposing that

HFCs be controlled as a WMGHGs under the Montreal Protocol because of their large projected future abundances.[50] In October 2016, the Montreal Protocol adopted an amendment to phase down global HFC production and consumption, avoiding emissions equivalent to approximately 105 Gt $CO_2$ by 2100 based on earlier projections.[50] The atmospheric growth rates of some halocarbon concentrations are significant at present (for example, $SF_6$ and HFC-134a), although their RF contributions remain small (Figure 2.5).

*Water Vapor*
Water vapor in the atmosphere acts as a powerful natural GHG, significantly increasing Earth's equilibrium temperature. In the stratosphere, water vapor abundances are controlled by transport from the troposphere and from oxidation of methane. Increases in methane from anthropogenic activities therefore increase stratospheric water vapor, producing a positive RF (e.g., Solomon et al. 2010;[51] Hegglin et al. 2014[52]). Other less-important anthropogenic sources of stratospheric water vapor are hydrogen oxidation,[53] aircraft exhaust,[54, 55] and explosive volcanic eruptions.[56]

In the troposphere, the amount of water vapor is controlled by temperature.[57] Atmospheric circulation, especially convection, limits the buildup of water vapor in the atmosphere such that the water vapor from direct emissions, for example by combustion of fossil fuels or by large power plant cooling towers, does not accumulate in the atmosphere but actually offsets water vapor that would otherwise evaporate from the surface. Direct changes in atmospheric water vapor are negligible in comparison to the indirect changes caused by temperature changes resulting from radiative forcing. As such, changes in tropospheric water vapor are considered a feedback in the climate system (see Section 2.6.1 and Figure 2.2). As increasing GHG concentrations warm the atmosphere, tropospheric water vapor concentrations increase, thereby amplifying the warming effect.[57]

*Ozone*

Ozone is a naturally occurring GHG in the troposphere and stratosphere and is produced and destroyed in response to a variety of anthropogenic and natural emissions. Ozone abundances have high spatial and temporal variability due to the nature and variety of the production, loss, and transport processes controlling ozone abundances, which adds complexity to the ozone RF calculations. In the global troposphere, emissions of methane, $NO_x$, carbon monoxide (CO), and non-methane volatile organic compounds (VOCs) form ozone photochemically both near and far downwind of these precursor source emissions, leading to regional and global positive RF contributions (e.g., Dentener et al. 2005[58]). Stratospheric ozone is destroyed photochemically in reactions involving the halogen species chlorine and bromine. Halogens are released in the stratosphere from the decomposition of some halocarbons emitted at the surface as a result of natural processes and human activities.[59] Stratospheric ozone depletion, which is most notable in the polar regions, yields a net negative RF.[8]

*Aerosols*

Atmospheric aerosols are perhaps the most complex and most uncertain component of forcing due to anthropogenic activities.[8] Aerosols have diverse natural and anthropogenic sources, and emissions from these sources interact in non-linear ways.[60] Aerosol types are categorized by composition; namely, sulfate, black carbon, organic, nitrate, dust, and sea salt. Individual particles generally include a mix of these components due to chemical and physical transformations of aerosols and aerosol precursor gases following emission. Aerosol tropospheric lifetimes are days to

weeks due to the general hygroscopic nature of primary and secondary particles and the ubiquity of cloud and precipitation systems in the troposphere. Particles that act as cloud condensation nuclei (CCN) or are scavenged by cloud droplets are removed from the troposphere in precipitation. The heterogeneity of aerosol sources and locations combined with short aerosol lifetimes leads to the high spatial and temporal variabilities observed in the global aerosol distribution and their associated forcings.

Aerosols from anthropogenic activities influence RF in three primary ways: through aerosol–radiation interactions, through aerosol–cloud interactions, and through albedo changes from absorbing-aerosol deposition on snow and ice.[60] RF from aerosol–radiation interactions, also known as the aerosol "direct effect," involves absorption and scattering of longwave and shortwave radiation. RF from aerosol-cloud interactions, also known as the cloud albedo "indirect effect," results from changes in cloud droplet number and size due to changes in aerosol (cloud condensation nuclei) number and composition. The RF for the global net aerosol–radiation and aerosol–cloud interaction is negative.[8] However, the RF is not negative for all aerosol types. Light-absorbing aerosols, such as black carbon, absorb sunlight, producing a positive RF. This absorption warms the atmosphere; on net, this response is assessed to increase cloud cover and therefore increase planetary albedo (the "semi-direct" effect). This "rapid response" lowers the ERF of atmospheric black carbon by approximately 15% relative to its RF from direct absorption alone.[61] ERF for aerosol–cloud interactions includes this rapid adjustment for absorbing aerosol (that is, the cloud response to atmospheric heating) and it includes cloud lifetime effects (for example, glaciation and thermodynamic effects).[60] Light-absorbing aerosols also affect climate when present in surface snow by lowering surface albedo, yielding a positive RF (e.g., Flanner et al. 2009[62]). For black carbon deposited on snow, the ERF is a factor of three higher than the RF because of positive feedbacks that reduce snow albedo and accelerate snow melt (e.g., Flanner et al. 2009;[62] Bond et al. 2013[61]). There is *very high confidence* that the RF from snow and ice albedo is positive.[61]

*Land Surface*
Land-cover changes (LCC) due to anthropogenic activities in the industrial era have changed the land surface brightness (albedo), principally through deforestation and afforestation. There is strong evidence that these changes have increased Earth's global surface albedo, creating a negative (cooling) RF of $-0.15 \pm 0.10$ W/m$^2$.[8] In specific regions, however, LCC has lowered surface albedo producing a positive RF (for example, through afforestation and pasture abandonment). In addition to the direct radiative forcing through albedo changes, LCC also have indirect forcing effects on climate, such as altering carbon cycles and altering dust emissions through effects on the hydrologic cycle. These effects are generally not included in the direct LCC RF calculations and are instead included in the net GHG and aerosol RFs over the industrial era. These indirect forcings may be of opposite sign to that of the direct LCC albedo forcing and may constitute a significant fraction of industrial-era RF driven by human activities.[63] Some of these effects, such as alteration of the carbon cycle, constitute climate feedbacks (Figure 2.2) and are discussed more extensively in Chapter 10: Land Cover. The increased use of satellite observations to quantify LCC has resulted in smaller negative LCC RF values (e.g., Ju and Masek 2016[64]). In areas with significant irrigation, surface temperatures and precipitation are affected by a change in energy partitioning from sensible to latent heating. Direct RF due to irrigation is generally small and can be positive or nega-

tive, depending on the balance of longwave (surface cooling or increases in water vapor) and shortwave (increased cloudiness) effects.[65]

*Contrails*
Line-shaped (linear) contrails are a special type of cirrus cloud that forms in the wake of jet-engine aircraft operating in the mid- to upper troposphere under conditions of high ambient humidity. Persistent contrails, which can last for many hours, form when ambient humidity conditions are supersaturated with respect to ice. As persistent contrails spread and drift with the local winds after formation, they lose their linear features, creating additional cirrus cloudiness that is indistinguishable from background cloudiness. Contrails and contrail cirrus are additional forms of cirrus cloudiness that interact with solar and thermal radiation to provide a global net positive RF and thus are visible evidence of an anthropogenic contribution to climate change.[66]

## 2.4 Industrial-era Changes in Radiative Forcing Agents

The IPCC best-estimate values of present day RFs and ERFs from principal anthropogenic and natural climate drivers are shown in Figure 2.3 and in Table 2.1. The past changes in the industrial era leading up to present day RF are shown for anthropogenic gases in Figure 2.5 and for all climate drivers in Figure 2.6.

The combined figures have several striking features. First, there is a large range in the magnitudes of RF terms, with contrails, stratospheric ozone, black carbon on snow, and stratospheric water vapor being small fractions of the largest term ($CO_2$). The sum of ERFs from $CO_2$ and non-$CO_2$ GHGs, tropospheric ozone, stratospheric water, contrails, and black carbon on snow shows a gradual increase from 1750 to the mid-1960s and accelerated annual growth in the subsequent 50 years (Figure 2.6). The sum of aerosol effects, strato-

**Table 2.1.** Global mean RF and ERF values in 2011 for the industrial era. [a]

| Radiative Forcing Term | Radiative forcing (W/m²) | Effective radiative forcing (W/m²) [b] |
|---|---|---|
| Well-mixed greenhouse gases (CO₂, CH₄, N₂O, and halocarbons) | +2.83 (2.54 to 3.12) | +2.83 (2.26 to 3.40) |
| Tropospheric ozone | +0.40 (0.20 to 0.60) | |
| Stratospheric ozone | −0.05 (−0.15 to +0.05) | |
| Stratospheric water vapor from CH₄ | +0.07 (+0.02 to +0.12) | |
| Aerosol–radiation interactions | −0.35 (−0.85 to +0.15) | −0.45 (−0.95 to +0.05) |
| Aerosol–cloud interactions | Not quantified | −0.45 (−1.2 to 0.0) |
| Surface albedo (land use) | −0.15 (−0.25 to −0.05) | |
| Surface albedo (black carbon aerosol on snow and ice) | +0.04 (+0.02 to +0.09) | |
| Contrails | +0.01 (+0.005 to +0.03) | |
| Combined contrails and contrail-induced cirrus | Not quantified | +0.05 (0.02 to 0.15) |
| Total anthropogenic | Not quantified | +2.3 (1.1 to 3.3) |
| Solar irradiance | +0.05 (0.0 to +0.10) | |

[a] From IPCC[8]

[b] RF is a good estimate of ERF for most forcing agents except black carbon on snow and ice and aerosol–cloud interactions.

spheric ozone depletion, and land use show a monotonically increasing cooling trend for the first two centuries of the depicted time series. During the past several decades, however, this combined cooling trend has leveled off due to reductions in the emissions of aerosols and aerosol precursors, largely as a result of legislation designed to improve air quality.[67, 68] In contrast, the volcanic RF reveals its episodic, short-lived characteristics along with large values that at times dominate the total RF. Changes in total solar irradiance over the industrial era are dominated by the 11-year solar cycle and other short-term variations. The solar irradiance RF between 1745 and 2005 is 0.05 (range of 0.0–0.1) W/m²,[8] a very small fraction of total anthropogenic forcing in 2011. The large relative uncertainty derives from inconsistencies among solar models, which all rely on proxies of solar irradiance to fit the industrial era. In total, ERF has increased substantially in the industrial era, driven almost completely by anthropogenic activities, with

annual growth in ERF notably higher after the mid-1960s.

The principal anthropogenic activities that have increased ERF are those that increase net GHG emissions. The atmospheric concentrations of CO₂, CH₄, and N₂O are higher now than they have been in at least the past 800,000 years.[69] All have increased monotonically over the industrial era (Figure 2.4), and are now 40%, 250%, and 20%, respectively, above their preindustrial concentrations as reflected in the RF time series in Figure 2.5. Tropospheric ozone has increased in response to growth in precursor emissions in the industrial era. Emissions of synthetic GHGs have grown rapidly beginning in the mid-20th century, with many bringing halogens to the stratosphere and causing ozone depletion in subsequent decades. Aerosol RF effects are a sum over aerosol–radiation and aerosol–cloud interactions; this RF has increased in the industrial era due to increased emissions of aerosol and aerosol precursors (Figure 2.6). These global aerosol RF trends average across disparate trends at the regional scale. The recent leveling off of global aerosol concentrations is the result of declines in many regions that were driven by enhanced air quality regulations, particularly starting in the 1980s (e.g., Philipona et al. 2009;[70] Liebensperger et al. 2012;[71] Wild 2016[72]). These declines are partially offset by increasing trends in other regions, such as much of Asia and possibly the Arabian Peninsula.[73, 74, 75] In highly polluted regions, negative aerosol RF may fully offset positive GHG RF, in contrast to global annual averages in which positive GHG forcing fully offsets negative aerosol forcing (Figures 2.3 and 2.6).

### 2.5 The Complex Relationship between Concentrations, Forcing, and Climate Response

Climate changes occur in response to ERFs, which generally include certain rapid responses to the underlying RF terms (Figure 2.2). Responses within Earth's system to forcing can act to either amplify (positive feedback) or reduce (negative feedback) the original forcing. These feedbacks operate on a range of time scales, from days to centuries. Thus, in general, the full climate impact of a given forcing is not immediately realized. Of interest are the climate response at a given point in time under continuously evolving forcings and the total climate response realized for a given forcing. A metric for the former, which approximates near-term climate change from a GHG forcing, is the transient climate response (TCR), defined as the change in global mean surface temperature when the atmospheric $CO_2$ concentration has doubled in a scenario of concentration increasing at 1% per year. The latter is given by the equilibrium climate sensitivity (ECS), defined as the change at equilibrium in annual and global mean surface temperature following a doubling of the atmospheric $CO_2$ concentration.[76] TCR is more

representative of near-term climate change from a GHG forcing. To estimate ECS, climate model runs have to simulate thousands of years in order to allow sufficient time for ocean temperatures to reach equilibrium.

In the IPCC's Fifth Assessment Report, ECS is assessed to be a factor of 1.5 or more greater than the TCR (ECS is 2.7°F to 8.1°F [1.5°C to 4.5°C] and TCR is 1.8°F to 4.5°F [1.0°C to 2.5°C][76]), exemplifying that longer time-scale feedbacks are both significant and positive. Confidence in the model-based TCR and ECS values is increased by their agreement, within respective uncertainties, with other methods of calculating these metrics (Box 12.2 of Collins et al. 2013)[77]. The alternative methods include using reconstructed temperatures from paleoclimate archives, the forcing/response relationship from past volcanic eruptions, and observed surface and ocean temperature changes over the industrial era.[77]

While TCR and ECS are defined specifically for the case of doubled $CO_2$, the climate sensitivity factor, λ, more generally relates the equilibrium surface temperature response (ΔT) to a constant forcing (ERF) as given by ΔT = λERF.[76, 78] The λ factor is highly dependent on feedbacks within Earth's system; all feedbacks are quantified themselves as radiative forcings, since each one acts by affecting Earth's albedo or its greenhouse effect. Models in which feedback processes are more positive (that is, more strongly amplify warming) tend to have a higher climate sensitivity (see Figure 9.43 of Flato et al.[76]). In the absence of feedbacks, λ would be equal to 0.54°F/(W/m²) (0.30°C/[W/m²]). The magnitude of λ for ERF over the industrial era varies across models, but in all cases λ is greater than 0.54°F/(W/m²), indicating the sum of all climate feedbacks tends to be positive. Overall, the global warming response to ERF includes a substantial amplification from feedbacks, with a model mean $\lambda$ of 0.86°F/(W/m²) (0.48°C/[W/m²]) with a 90% uncertainty range of ±0.23°F/(W/m²) (±0.13°C/[W/m²]) (as derived from climate sensitivity parameter in Table 9.5 of Flato et al.[76] combined with methodology of Bony et al.[79]). Thus, there is *high confidence* that the response of Earth's system to the industrial-era net positive forcing is to amplify that forcing (Figure 9.42 of Flato et al.[76]).

The models used to quantify $\lambda$ account for the near-term feedbacks described below (Section 2.6.1), though with mixed levels of detail regarding feedbacks to atmospheric composition. Feedbacks to the land and ocean carbon sink, land albedo and ocean heat uptake, most of which operate on longer time scales (Section 2.6.2), are currently included on only a limited basis, or in some cases not at all, in climate models. Climate feedbacks are the largest source of uncertainty in quantifying climate sensitivity;[76] namely, the responses of clouds, the carbon cycle, ocean circulation and, to a lesser extent, land and sea ice to surface temperature and precipitation changes.

The complexity of mapping forcings to climate responses on a global scale is enhanced by geographic and seasonal variations in these forcings and responses, driven in part by similar variations in anthropogenic emissions and concentrations. Studies show that the spatial pattern and timing of climate responses are not always well correlated with the spatial pattern and timing of a radiative forcing, since adjustments within the climate system can determine much of the response (e.g., Shindell and Faluvegi 2009;[80] Crook and Forster 2011;[81] Knutti and Rugenstein 2015[82]). The RF patterns of short-lived climate drivers with inhomogeneous source distributions, such as aerosols, tropospheric ozone, contrails, and land cover change, are leading examples of highly inhomogeneous forcings. Spatial and temporal variability in aerosol and

aerosol precursor emissions is enhanced by in-atmosphere aerosol formation and chemical transformations, and by aerosol removal in precipitation and surface deposition. Even for relatively uniformly distributed species (for example, WMGHGs), RF patterns are less homogenous than their concentrations. The RF of a uniform $CO_2$ distribution, for example, depends on latitude and cloud cover.[83] With the added complexity and variability of regional forcings, the global mean RFs are known with more confidence than the regional RF patterns. Forcing feedbacks in response to spatially variable forcings also have variable geographic and temporal patterns.

Quantifying the relationship between spatial RF patterns and regional and global climate responses in the industrial era is difficult because it requires distinguishing forcing responses from the inherent internal variability of the climate system, which acts on a range of time scales. The ability to test the accuracy of modeled responses to forcing patterns is limited by the sparsity of long-term observational records of regional climate variables. As a result, there is generally *very low confidence* in our understanding of the qualitative and quantitative forcing–response relationships at the regional scale. However, there is *medium to high confidence* in other features, such as aerosol effects altering the location of the Inter Tropical Convergence Zone (ITCZ) and the positive feedback to reductions of snow and ice and albedo changes at high latitudes.[8, 60]

## 2.6 Radiative-forcing Feedbacks

### 2.6.1 Near-term Feedbacks
*Planck Feedback*

When the temperatures of Earth's surface and atmosphere increase in response to RF, more infrared radiation is emitted into the lower atmosphere; this serves to restore radiative balance at the tropopause. This radiative feedback, defined as the Planck feedback, only partially offsets the positive RF while triggering other feedbacks that affect radiative balance. The Planck feedback magnitude is −3.20 ± 0.04 W/m² per 1.8°F (1°C) of warming and is the strongest and primary stabilizing feedback in the climate system.[84]

*Water Vapor and Lapse Rate Feedbacks*
Warmer air holds more moisture (water vapor) than cooler air—about 7% more per degree Celsius—as dictated by the Clausius–Clapeyron relationship.[85] Thus, as global temperatures increase, the total amount of water vapor in the atmosphere increases, adding further to greenhouse warming—a positive feedback—with a mean value derived from a suite of atmosphere/ocean global climate models (AOGCM) of 1.6 ± 0.3 W/m² per 1.8°F (1°C) of warming (Table 9.5 of Flato et al. 2013).[76] The water vapor feedback is responsible for more than doubling the direct climate warming from $CO_2$ emissions alone.[57, 79, 84, 86] Observations confirm that global tropospheric water vapor has increased commensurate with measured warming (FAQ 3.2 and its Figure 1a in IPCC 2013).[17] Interannual variations and trends in stratospheric water vapor, while influenced by tropospheric abundances, are controlled largely by tropopause temperatures and dynamical processes.[87] Increases in tropospheric water vapor have a larger warming effect in the upper troposphere (where it is cooler) than in the lower troposphere, thereby decreasing the rate at which temperatures decrease with altitude (the lapse rate). Warmer temperatures aloft increase outgoing infrared radiation—a negative feedback—with a mean value derived from the same AOGCM suite of −0.6 ± 0.4 W/m² per 1.8°F (1°C) warming. These feedback values remain largely unchanged between recent IPCC assessments.[17, 88] Recent advances in both observations and models have increased confidence that the net effect of the water vapor and lapse rate feedbacks is a significant positive RF.[76]

*Cloud Feedbacks*
An increase in cloudiness has two direct impacts on radiative fluxes: first, it increases scattering of sunlight, which increases Earth's albedo and cools the surface (the shortwave cloud radiative effect); second, it increases trapping of infrared radiation, which warms the surface (the longwave cloud radiative effect). A decrease in cloudiness has the opposite effects. Clouds have a relatively larger shortwave effect when they form over dark surfaces (for example, oceans) than over higher albedo surfaces, such as sea ice and deserts. For clouds globally, the shortwave cloud radiative effect is about −50 W/m², and the longwave effect is about +30 W/m², yielding a net cooling influence.[89, 90] The relative magnitudes of both effects vary with cloud type as well as with location. For low-altitude, thick clouds (for example, stratus and stratocumulus) the shortwave radiative effect dominates, so they cause a net cooling. For high-altitude, thin clouds (for example, cirrus) the longwave effect dominates, so they cause a net warming (e.g., Hartmann et al. 1992;[91] Chen et al. 2000[92]). Therefore, an increase in low clouds is a negative feedback to RF, while an increase in high clouds is a positive feedback. The potential magnitude of cloud feedbacks is large compared with global RF (see Section 2.4). Cloud feedbacks also influence natural variability within the climate system and may amplify atmospheric circulation patterns and the El Niño–Southern Oscillation.[93]

The net radiative effect of cloud feedbacks is positive over the industrial era, with an assessed value of +0.27 ± 0.42 W/m² per 1.8°F (1°C) warming.[84] The net cloud feedback can be broken into components, where the longwave cloud feedback is positive (+0.24 ± 0.26 W/m² per 1.8°F [1°C] warming) and the shortwave feedback is near-zero (+0.14 ± 0.40 W/m² per 1.8°F [1°C] warming[84]), though the two do not add linearly. The value of the shortwave cloud feedback shows a significant sensitivity to computation methodology.[84, 94, 95] Uncertainty in cloud feedback remains the largest source of inter-model differences in calculated climate sensitivity.[60, 84]

### Snow, Ice, and Surface Albedo

Snow and ice are highly reflective to solar radiation relative to land surfaces and the ocean. Loss of snow cover, glaciers, ice sheets, or sea ice resulting from climate warming lowers Earth's surface albedo. The losses create the snow–albedo feedback because subsequent increases in absorbed solar radiation lead to further warming as well as changes in turbulent heat fluxes at the surface.[96] For seasonal snow, glaciers, and sea ice, a positive albedo feedback occurs where light-absorbing aerosols are deposited to the surface, darkening the snow and ice and accelerating the loss of snow and ice mass (e.g., Hansen and Nazarenko 2004;[97] Jacobson 2004;[98] Flanner et al. 2009;[62] Skeie et al. 2011;[99] Bond et al. 2013;[61] Yang et al. 2015[100]).

For ice sheets (for example, on Antarctica and Greenland—see Ch. 11: Arctic Changes), the positive radiative feedback is further amplified by dynamical feedbacks on ice-sheet mass loss. Specifically, since continental ice shelves limit the discharge rates of ice sheets into the ocean; any melting of the ice shelves accelerates the discharge rate, creating a positive feedback on the ice-stream flow rate and total mass loss (e.g., Holland et al. 2008;[101] Schoof 2010;[102] Rignot et al. 2010;[103] Joughin et al. 2012[104]). Warming oceans also lead to accelerated melting of basal ice (ice at the base of a glacier or ice sheet) and subsequent ice-sheet loss (e.g., Straneo et al. 2013;[105] Thoma et al. 2015;[106] Alley et al. 2016;[107] Silvano et al. 2016[108]). Feedbacks related to ice sheet dynamics occur on longer time scales than other feedbacks—many centuries or longer. Significant ice-sheet melt can also lead to changes in freshwater input to the oceans, which in turn can affect ocean temperatures and circulation, ocean–atmosphere heat exchange and moisture fluxes, and atmospheric circulation.[69]

The complete contribution of ice-sheet feedbacks on time scales of millennia are not generally included in CMIP5 climate simulations. These slow feedbacks are also not thought to change in proportion to global mean surface temperature change, implying that the apparent climate sensitivity changes with time, making it difficult to fully understand climate sensitivity considering only the industrial age. This slow response increases the likelihood for tipping points, as discussed further in Chapter 15: Potential Surprises.

The surface-albedo feedback is an important influence on interannual variations in sea ice as well as on long-term climate change. While there is a significant range in estimates of the snow-albedo feedback, it is assessed as positive,[84, 109, 110] with a best estimate of $0.27 \pm 0.06$ W/m$^2$ per 1.8°F (1°C) of warming globally. Within the cryosphere, the surface-albedo feedback is most effective in polar regions;[94, 111] there is also evidence that polar surface-albedo feedbacks might influence the tropical climate as well.[112]

Changes in sea ice can also influence arctic cloudiness. Recent work indicates that arctic clouds have responded to sea ice loss in fall but not summer.[113, 114, 115, 116, 117] This has important implications for future climate change, as an increase in summer clouds could offset a portion of the amplifying surface-albedo feedback, slowing down the rate of arctic warming.

### Atmospheric Composition

Climate change alters the atmospheric abundance and distribution of some radiatively active species by changing natural emissions, atmospheric photochemical reaction rates, atmospheric lifetimes, transport patterns, or deposition rates. These changes in turn alter the associated ERFs, forming a feedback.[118, 119, 120] Atmospheric composition feedbacks occur through a variety of processes. Important examples include climate-driven changes in temperature and precipitation that affect 1) natural sources of $NO_x$ from soils and lightning and VOC sources from vegetation, all of which affect ozone abundances;[120, 121, 122] 2) regional aridity, which influences surface dust sources as well as susceptibility to wildfires; and 3) surface winds, which control the emission of dust from the land surface and the emissions of sea salt and dimethyl sulfide—a natural precursor to sulfate aerosol—from the ocean surface.

Climate-driven ecosystem changes that alter the carbon cycle potentially impact atmospheric $CO_2$ and $CH_4$ abundances (Section 2.6.2). Atmospheric aerosols affect clouds and precipitation rates, which in turn alter aerosol removal rates, lifetimes, and atmospheric abundances. Longwave radiative feedbacks and climate-driven circulation changes also alter stratospheric ozone abundance.[123] Investigation of these and other composition–climate interactions is an active area of research (e.g., John et al. 2012;[124] Pacifico et al. 2012;[125] Morgenstern et al. 2013;[126] Holmes et al. 2013;[127] Naik et al. 2013;[128] Voulgarakis et al. 2013;[129] Isaksen et al. 2014;[130] Dietmuller et al. 2014;[131] Banerjee et al. 2014[132]). While understanding of key processes is improving, atmospheric composition feedbacks are absent or limited in many global climate modeling studies used to project future climate, though this is rapidly changing.[133] For some composition–climate feedbacks involving shorter-lived constituents, the net effects may be near zero at the global scale while significant at local to regional scales (e.g., Raes et al. 2010;[120] Han et al. 2013[134]).

### 2.6.2 Long-term Feedbacks

*Terrestrial Ecosystems and Climate Change Feedbacks*

The cycling of carbon through the climate system is an important long-term climate feedback that affects atmospheric $CO_2$ concentrations. The global mean atmospheric $CO_2$ concentration is determined by emissions from burning fossil fuels, wildfires, and permafrost thaw balanced against $CO_2$ uptake by the oceans and terrestrial biosphere (Figures 2.2 and 2.7).[43, 135] During the past decade, just less than a third of anthropogenic $CO_2$ has been taken up by the terrestrial environment, and another quarter by the oceans (Le Quéré et al.[135] Table 8) through photosynthesis and through direct absorption by ocean surface waters. The capacity of the land to continue uptake of $CO_2$ is uncertain and depends on land-use management and on responses of the biosphere to climate change (see Ch. 10: Land Cover). Altered uptake rates affect atmospheric $CO_2$ abundance, forcing, and rates of climate change. Such changes are expected to evolve on the decadal and longer time scale, though abrupt changes are possible.

Significant uncertainty exists in quantification of carbon-cycle feedbacks, with large differences in the assumed characteristics of the land carbon-cycle processes in current models. Ocean carbon-cycle changes in future climate scenarios are also highly uncertain. Both of these contribute significant uncertainty to longer-term (century-scale) climate projections. Basic principles of carbon cycle dynamics in terrestrial ecosystems suggest that increased atmospheric $CO_2$ concentrations can directly enhance plant growth rates and, therefore, increase carbon uptake (the "$CO_2$ fertilization" effect), nominally sequestering much of the added carbon from fossil-fuel combustion (e.g., Wenzel et al. 2016[136]). However, this effect is variable; sometimes plants acclimate so that higher $CO_2$ concentrations no longer enhance growth (e.g., Franks et al. 2013[137]). In addition, $CO_2$ fertilization is often offset by other factors limiting plant growth, such as water and or nutrient availability and temperature and incoming solar radiation that can be modified by changes in vegetation structure. Large-scale plant mortality through fire, soil moisture drought, and/or temperature changes also impact successional processes that contribute to reestablishment and revegetation (or not) of disturbed ecosystems, altering the amount and distribution of plants available to uptake $CO_2$. With sufficient disturbance, it has been argued that forests could, on net, turn into a source rather than a sink of $CO_2$.[138]

Climate-induced changes in the horizontal (for example, landscape to biome) and vertical (soils to canopy) structure of terrestrial ecosystems also alter the physical surface roughness and albedo, as well as biogeochemical (carbon and nitrogen) cycles and biophysical evapotranspiration and water demand. Combined, these responses constitute climate feedbacks by altering surface albedo and atmospheric GHG abundances. Drivers of these changes in terrestrial ecosystems include changes in the biophysical growing season, altered seasonality, wildfire patterns, and multiple additional interacting factors (Ch. 10: Land Cover).

Accurate determination of future $CO_2$ stabilization scenarios depends on accounting for the significant role that the land biosphere plays in the global carbon cycle and feedbacks between climate change and the terrestrial carbon cycle.[139] Earth System Models (ESMs) are increasing the representation of terrestrial carbon cycle processes, including plant photosynthesis, plant and soil respiration and decomposition, and $CO_2$ fertilization, with the latter based on the assumption that an increased atmospheric $CO_2$ concentration provides more substrate for photosynthesis

and productivity. Recent advances in ESMs are beginning to account for other important factors such as nutrient limitations.[140, 141, 142] ESMs that do include carbon-cycle feedbacks appear, on average, to overestimate terrestrial $CO_2$ uptake under the present-day climate[143, 144] and underestimate nutrient limitations to $CO_2$ fertilization.[142] The sign of the land carbon-cycle feedback through 2100 remains unclear in the newest generation of ESMs.[142, 145, 146] Eleven CMIP5 ESMs forced with the same $CO_2$ emissions scenario—one consistent with RCP8.5 concentrations—produce a range of 795 to 1145 ppm for atmospheric $CO_2$ concentration in 2100. The majority of the ESMs (7 out of 11) simulated a $CO_2$ concentration larger (by 44 ppm on average) than their equivalent non-interactive carbon cycle counterpart.[146] This difference in $CO_2$ equates to about 0.4°F (0.2°C) more warming by 2100. The inclusion of carbon-cycle feedbacks does not alter the lower-end bound on climate sensitivity, but, in most climate models, inclusion pushes the upper bound higher.[146]

*Ocean Chemistry, Ecosystem, and Circulation Changes*
The ocean plays a significant role in climate change by playing a critical role in controlling the amount of GHGs (including $CO_2$, water vapor, and $N_2O$) and heat in the atmosphere (Figure 2.7). To date most of the net energy increase in the climate system from anthropogenic RF is in the form of ocean heat (see Box 3.1 Figure 1 of Rhein et al. 2013).[6] This additional heat is stored predominantly (about 60%) in the upper 700 meters of the ocean (see Ch. 12: Sea Level Rise and Ch. 13: Ocean Changes).[147] Ocean warming and climate-driven changes in ocean stratification and circulation alter oceanic biological productivity and therefore $CO_2$ uptake; combined, these feedbacks affect the rate of warming from radiative forcing.

Marine ecosystems take up $CO_2$ from the atmosphere in the same way that plants do on land. About half of the global net primary production (NPP) is by marine plants (approximately $50 \pm 28$ GtC/year[148, 149, 150]). Phytoplankton NPP supports the biological pump, which transports 2–12 GtC/year of organic carbon to the deep sea,[151, 152] where it is sequestered away from the atmospheric pool of carbon for 200–1,500 years. Since the ocean is an important carbon sink, climate-driven changes in NPP represent an important feedback because they potentially change atmospheric $CO_2$ abundance and forcing.

There are multiple links between RF-driven changes in climate, physical changes to the ocean, and feedbacks to ocean carbon and heat uptake. Changes in ocean temperature, circulation, and stratification driven by climate change alter phytoplankton NPP. Absorption of $CO_2$ by the ocean also increases its acidity, which can also affect NPP and therefore the carbon sink (see Ch. 13: Ocean Changes for a more detailed discussion of ocean acidification).

In addition to being an important carbon sink, the ocean dominates the hydrological cycle, since most surface evaporation and rainfall occur over the ocean.[153, 154] The ocean component of the water vapor feedback derives from the rate of evaporation, which depends on surface wind stress and ocean temperature. Climate warming from radiative forcing also is associated with intensification of the water cycle (Ch. 7: Precipitation Change). Over decadal time scales the surface ocean salinity has increased in areas of high salinity, such as the subtropical gyres, and decreased in areas of low salinity, such as the Warm Pool region (see Ch. 13: Ocean Changes).[155, 156] This increase in stratification in select regions and mixing in other regions are feedback processes because

they lead to altered patterns of ocean circulation, which impacts uptake of anthropogenic heat and $CO_2$.

Increased stratification inhibits surface mixing, high-latitude convection, and deep-water formation, thereby potentially weakening ocean circulations, in particular the Atlantic Meridional Overturning Circulation (AMOC) (see also Ch. 13: Ocean Changes).[157, 158] Reduced deep-water formation and slower overturning are associated with decreased heat and carbon sequestration at greater depths. Observational evidence is mixed regarding whether the AMOC has slowed over the past decades to century (see Sect. 13.2.1 of Ch. 13: Ocean Changes). Future projections show that the strength of AMOC may significantly decrease as the ocean warms and freshens and as upwelling in the Southern Ocean weakens due to the storm track moving poleward (see also Ch. 13: Ocean Changes).[159] Such a slowdown of the ocean currents will impact the rate at which the ocean absorbs $CO_2$ and heat from the atmosphere.

Increased ocean temperatures also accelerate ice sheet melt, particularly for the Antarctic Ice Sheet where basal sea ice melting is important relative to surface melting due to colder surface temperatures.[160] For the Greenland Ice Sheet, submarine melting at tidewater margins is also contributing to volume loss.[161] In turn, changes in ice sheet melt rates change cold- and freshwater inputs, also altering ocean stratification. This affects ocean circulation and the ability of the ocean to absorb more GHGs and heat.[162] Enhanced sea ice export to lower latitudes gives rise to local salinity anomalies (such as the Great Salinity Anomaly[163]) and therefore to changes in ocean circulation and air–sea exchanges of momentum, heat, and freshwater, which in turn affect the atmospheric distribution of heat and GHGs.

Remote sensing of sea surface temperature and chlorophyll as well as model simulations and sediment records suggest that global phytoplankton NPP may have increased recently as a consequence of decadal-scale natural climate variability, such as the El Niño–Southern Oscillation, which promotes vertical mixing and upwelling of nutrients.[150, 164, 165] Analyses of longer trends, however, suggest that phytoplankton NPP has decreased by about 1% per year over the last 100 years.[166, 167, 168] The latter results, although controversial,[169] are the only studies of the global rate of change over this period. In contrast, model simulations show decreases of only 6.6% in NPP and 8% in the biological pump over the last five decades.[170] Total NPP is complex to model, as there are still areas of uncertainty on how multiple physical factors affect phytoplankton growth, grazing, and community composition, and as certain phytoplankton species are more efficient at carbon export.[171, 172] As a result, model uncertainty is still significant in NPP projections.[173] While there are variations across climate model projections, there is good agreement that in the future there will be increasing stratification, decreasing NPP, and a decreasing sink of $CO_2$ to the ocean via biological activity.[172] Overall, compared to the 1990s, in 2090 total NPP is expected to decrease by 2%–16% and export production (that is, particulate flux to the deep ocean) could decline by 7%–18% under the higher scenario (RCP8.5).[172] Consistent with this result, carbon cycle feedbacks in the ocean were positive (that is, higher $CO_2$ concentrations leading to a lower rate of $CO_2$ sequestration to the ocean, thereby accelerating the growth of atmospheric $CO_2$ concentrations) across the suite of CMIP5 models.

*Permafrost and Hydrates*
Permafrost and methane hydrates contain large stores of methane and (for permafrost) carbon in the form of organic materials, mostly at northern high latitudes. With warming, this organic material can thaw, making previously frozen organic matter available for microbial decomposition, releasing $CO_2$ and methane to the atmosphere, providing additional radiative forcing and accelerating warming. This process defines the permafrost–carbon feedback. Combined data and modeling studies suggest that this feedback is *very likely* positive.[174, 175, 176] This feedback was not included in recent IPCC projections but is an active area of research. Meeting stabilization or mitigation targets in the future will require limits on total GHG abundances in the atmosphere. Accounting for additional permafrost-carbon release reduces the amount of anthropogenic emissions that can occur and still meet these limits.[177]

The permafrost–carbon feedback in the higher scenario (RCP8.5; Section 1.2.2 and Figure 1.4) contributes $120 \pm 85$ Gt of additional carbon by 2100; this represents 6% of the total anthropogenic forcing for 2100 and corresponds to a global temperature increase of $+0.52° \pm 0.38°$F ($+0.29° \pm 0.21°$C).[174] Considering the broader range of forcing scenarios (Figure 1.4), it is *likely* that the permafrost–carbon feedback increases carbon emissions between 2% and 11% by 2100. A key feature of the permafrost feedback is that, once initiated, it will continue for an extended period because emissions from decomposition occur slowly over decades and longer. In the coming few decades, enhanced plant growth at high latitudes and its associated $CO_2$ sink[145] are expected to partially offset the increased emissions from permafrost thaw;[174, 176] thereafter, decomposition will dominate uptake. Recent evidence indicates that permafrost thaw is occurring faster than expected; poorly understood deep-soil carbon decomposition and ice wedge processes *likely* contribute.[178, 179] Chapter 11: Arctic Changes includes a more detailed discussion of permafrost and methane hydrates in the Arctic. Future changes in permafrost emissions and the potential for even greater emissions from methane hydrates in the continental shelf are discussed further in Chapter 15: Potential Surprises.

# TRACEABLE ACCOUNTS

### Key Finding 1

Human activities continue to significantly affect Earth's climate by altering factors that change its radiative balance. These factors, known as radiative forcings, include changes in greenhouse gases, small airborne particles (aerosols), and the reflectivity of Earth's surface. In the industrial era, human activities have been, and are increasingly, the dominant cause of climate warming. The increase in radiative forcing due to these activities has far exceeded the relatively small net increase due to natural factors, which include changes in energy from the sun and the cooling effect of volcanic eruptions. (*Very high confidence*)

### Description of evidence base

The Key Finding and supporting text summarizes extensive evidence documented in the climate science literature, including in previous national (NCA3)[180] and international[17] assessments. The assertion that Earth's climate is controlled by its radiative balance is a well-established physical property of the planet. Quantification of the changes in Earth's radiative balance come from a combination of observations and calculations. Satellite data are used directly to observe changes in Earth's outgoing visible and infrared radiation. Since 2002, observations of incoming sunlight include both total solar irradiance and solar spectral irradiance.[26] Extensive in situ and remote sensing data are used to quantify atmospheric concentrations of radiative forcing agents (greenhouse gases [e.g., Ciais et al. 2013;[43] Le Quéré et al. 2016[135]] and aerosols [e.g., Bond et al. 2013;[61] Boucher et al. 2013;[60] Myhre et al. 2013;[8] Jiao et al. 2014;[181] Tsigaridis et al. 2014;[182] Koffi et al. 2016[183]]) and changes in land cover,[64, 184, 185] as well as the relevant properties of these agents (for example, aerosol microphysical and optical properties). Climate models are constrained by these observed concentrations and properties. Concentrations of long-lived greenhouse gases in particular are well-quantified with observations because of their relatively high spatial homogeneity. Climate model calculations of radiative forcing by greenhouse gases and aerosols are supported by observations of radiative fluxes from the surface, from airborne research platforms, and from satellites. Both direct observations and modeling studies show large, explosive eruptions affect climate parameters for years to decades.[36, 186] Over the industrial era, radiative forcing by volcanoes has been episodic and currently does not contribute significantly to forcing trends. Observations indicate a positive but small increase in solar input over the industrial era.[8, 22, 23] Relatively higher variations in solar input at shorter (UV) wavelengths[25] may be leading to indirect changes in Earth's radiative balance through their impact on ozone concentrations that are larger than the radiative impact of changes in total solar irradiance,[21, 26, 27, 28, 29] but these changes are also small in comparison to anthropogenic greenhouse gas and aerosol forcing.[8] The finding of an increasingly strong positive forcing over the industrial era is supported by observed increases in atmospheric temperatures (see Ch. 1: Our Globally Changing Climate) and by observed increases in ocean temperatures (Ch. 1: Our Globally Changing Climate and Ch. 13: Ocean Changes). The attribution of climate change to human activities is supported by climate models, which are able to reproduce observed temperature trends when RF from human activities is included and considerably deviate from observed trends when only natural forcings are included (Ch. 3: Detection and Attribution, Figure 3.1).

### Major uncertainties

The largest source of uncertainty in radiative forcing (both natural and anthropogenic) over the industrial era is quantifying forcing by aerosols. This finding is consistent across previous assessments (e.g., IPCC 2007;[88] IPCC 2013[17]). The major uncertainties associated with aerosol forcing is discussed below in the Traceable Accounts for Key Finding 2.

Recent work has highlighted the potentially larger role of variations in UV solar irradiance, versus total solar irradiance, in solar forcing. However, this increase in solar forcing uncertainty is not sufficiently large to reduce confidence that anthropogenic activities dominate industrial-era forcing.

**Assessment of confidence based on evidence and agreement, including short description of nature of evidence and level of agreement**

There is *very high confidence* that anthropogenic radiative forcing exceeds natural forcing over the industrial era based on quantitative assessments of known radiative forcing components. Assessments of the natural forcings of solar irradiance changes and volcanic activity show with *very high confidence* that both forcings are small over the industrial era relative to total anthropogenic forcing. Total anthropogenic forcing is assessed to have become larger and more positive during the industrial era, while natural forcings show no similar trend.

**Summary sentence or paragraph that integrates the above information**

This key finding is consistent with that in the IPCC Fourth Assessment Report (AR4)[88] and Fifth Assessment Report (AR5);[17] namely, anthropogenic radiative forcing is positive (climate warming) and substantially larger than natural forcing from variations in solar input and volcanic emissions. Confidence in this finding has increased from AR4 to AR5, as anthropogenic GHG forcings have continued to increase, whereas solar forcing remains small and volcanic forcing near-zero over decadal time scales.

**Key Finding 2**

Aerosols caused by human activity play a profound and complex role in the climate system through radiative effects in the atmosphere and on snow and ice surfaces and through effects on cloud formation and properties. The combined forcing of aerosol–radiation and aerosol–cloud interactions is negative (cooling) over the industrial era (*high confidence*), offsetting a substantial part of greenhouse gas forcing, which is currently the predominant human contribution. The magnitude of this offset, globally averaged, has declined in recent decades, despite increasing trends in aerosol emissions or abundances in some regions. (*Medium to high confidence*)

**Description of evidence base**

The Key Finding and supporting text summarize extensive evidence documented in the climate science literature, including in previous national (NCA3)[180] and international[17] assessments. Aerosols affect Earth's albedo by directly interacting with solar radiation (scattering and absorbing sunlight) and by affecting cloud properties (albedo and lifetime).

Fundamental physical principles show how atmospheric aerosols scatter and absorb sunlight (aerosol–radiation interaction), and thereby directly reduce incoming solar radiation reaching the surface. Extensive in situ and remote sensing data are used to measure emission of aerosols and aerosol precursors from specific source types, the concentrations of aerosols in the atmosphere, aerosol microphysical and optical properties, and, via remote sensing, their direct impacts on radiative fluxes. Atmospheric models used to calculate aerosol forcings are constrained by these observations (see Key Finding 1).

In addition to their direct impact on radiative fluxes, aerosols also act as cloud condensation nuclei. Aerosol–cloud interactions are more complex, with a strong theoretical basis supported by observational evidence. Multiple observational and modeling studies have concluded that increasing the number of aerosols in the atmosphere increases cloud albedo and lifetime, adding to the negative forcing (aerosol–cloud microphysical interactions) (e.g., Twohy 2005;[187] Lohmann and Feichter 2005;[188] Quaas et al. 2009;[189] Rosenfeld et al. 2014[190]). Particles that absorb sunlight increase atmospheric heating; if they are sufficiently absorbing, the net effect of scattering plus absorption is a positive radiative forcing. Only a few source types (for example, from diesel engines) produce aerosols that are sufficiently absorbing that they have a positive radiative forcing.[61] Modeling studies, combined with observational inputs, have investigated the thermodynamic response to aerosol absorption in the atmosphere. Averaging over aerosol locations relative to the clouds and other factors, the resulting changes in cloud properties represent a negative forcing, offsetting approximately 15% of the positive radiative forcing from heating by absorbing aerosols (specifically, black carbon).[61]

Modeling and observational evidence both show that annually averaged global aerosol ERF increased until the 1980s and since then has flattened or slightly declined,[191, 192, 193, 194] driven by the introduction of stronger air quality regulations (Smith and Bond 2014; Fiore et al. 2015). In one recent study,[195] global mean aerosol RF has become less negative since IPCC AR5,[8] due to a combination of declining sulfur dioxide emissions (which produce negative RF) and increasing black carbon emissions (which produce positive RF). Within these global trends there are significant regional variations (e.g., Mao et al. 2014[196]), driven by both changes in aerosol abundance and changes in the relative contributions of primarily light-scattering and light-absorbing aerosols.[68, 195] In Europe and North America, aerosol ERF has significantly declined (become less negative) since the 1980s.[70, 71, 197, 198, 199, 200] In contrast, observations show significant increases in aerosol abundances over India,[201, 202] and these increases are expected to continue into the near future.[203] Several modeling and observational studies point to aerosol ERF for China peaking around 1990,[204, 205, 206] though in some regions of China aerosol abundances and ERF have continued to increase.[206] The suite of scenarios used for future climate projection (i.e., the scenarios shown in Ch. 1: Our Globally Changing Climate, Figure 1.4) includes emissions for aerosols and aerosol precursors. Across this range of scenarios, globally averaged ERF of aerosols is expected to decline (become less negative) in the coming decades,[67, 192] reducing the current aerosol offset to the increasing RF from GHGs.

### Major uncertainties

Aerosol–cloud interactions are the largest source of uncertainty in both aerosol and total anthropogenic radiative forcing. These include the microphysical effects of aerosols on clouds and changes in clouds that result from the rapid response to absorption of sunlight by aerosols. This finding, consistent across previous assessments (e.g., Forster et al. 2007;[207] Myhre et al. 2013[8]), is due to poor understanding of how both natural and anthropogenic aerosol emissions have changed and how changing aerosol concentrations and composition affect cloud properties (albedo and lifetime).[60, 208] From a theoretical standpoint, aerosol–cloud interactions are complex, and using observations to isolate the effects of aerosols on clouds is complicated by the fact that other factors (for example, the thermodynamic state of the atmosphere) also strongly influence cloud properties. Further, changes in aerosol properties and the atmospheric thermodynamic state are often correlated and interact in non-linear ways.[209]

### Assessment of confidence based on evidence and agreement, including short description of nature of evidence and level of agreement

There is *very high confidence* that aerosol radiative forcing is negative on a global, annually averaged basis, *medium confidence* in the magnitude of the aerosol RF, *high* confidence that aerosol ERF is also, on average, negative, and *low to medium confidence* in the magnitude of aerosol ERF. Lower confidence in the magnitude of aerosol ERF is due to large uncertainties in the effects of aerosols on clouds. Combined, we assess a *high level of confidence* that aerosol ERF is negative and sufficiently large to be substantially offsetting positive GHG forcing. Improvements in the quantification of emissions, in observations (from both surface-based networks and satellites), and in modeling capability give *medium* to *high confidence* in the finding that aerosol forcing trends are decreasing in recent decades.

### Summary sentence or paragraph that integrates the above information

This key finding is consistent with the findings of IPCC AR5[8] that aerosols constitute a negative radiative forcing. While significant uncertainty remains in the quantification of aerosol ERF, we assess with *high confidence* that aerosols offset about half of the positive forcing by anthropogenic $CO_2$ and about a third of the forcing by all well-mixed anthropogenic GHGs. The fraction of GHG forcing that is offset by aerosols has been decreasing over recent decades, as aerosol forcing has leveled off while GHG forcing continues to increase.

### Key Finding 3

The interconnected Earth–atmosphere–ocean climate system includes a number of positive and negative feedback processes that can either strengthen (positive feedback) or weaken (negative feedback) the system's responses to human and natural influences. These feedbacks operate on a range of time scales from very short (essentially instantaneous) to very long (centuries). Global warming by net radiative forcing over the industrial era includes a substantial amplification from these feedbacks (approximately a factor of three) (*high confidence*). While there are large uncertainties associated with some of these feedbacks, the net feedback effect over the industrial era has been positive (amplifying warming) and will continue to be positive in coming decades. (*Very high confidence*)

### Description of evidence base

The variety of climate system feedbacks all depend on fundamental physical principles and are known with a range of uncertainties. The Planck feedback is based on well-known radiative transfer models. The largest positive feedback is the water vapor feedback, which derives from the dependence of vapor pressure on temperature. There is *very high confidence* that this feedback is positive, approximately doubling the direct forcing due to $CO_2$ emissions alone. The lapse rate feedback derives from thermodynamic principles. There is *very high confidence* that this feedback is negative and partially offsets the water vapor feedback. The water vapor and lapse-rate feedbacks are linked by the fact that both are driven by increases in atmospheric water vapor with increasing temperature. Estimates of the magnitude of these two feedbacks have changed little across recent assessments.[60, 210] The snow- and ice-albedo feedback is positive in sign, with the magnitude of the feedback dependent in part on the time scale of interest.[109, 110] The assessed strength of this feedback has also not changed significantly since IPCC 2007.[88] Cloud feedbacks modeled using microphysical principles are either positive or negative, depending on the sign of the change in clouds with warming (increase or decrease) and the type of cloud that changes (low or high clouds). Recent international assessments[60, 210] and a separate feedback assessment[84] all give best

estimates of the cloud feedback as net positive. Feedback via changes in atmospheric composition is not well-quantified but is expected to be small relative to water-vapor-plus-lapse-rate, snow, and cloud feedbacks at the global scale.[120] Carbon cycle feedbacks through changes in the land biosphere are currently of uncertain sign and have asymmetric uncertainties: they might be small and negative but could also be large and positive.[138] Recent best estimates of the ocean carbon-cycle feedback are that it is positive with significant uncertainty that includes the possibility of a negative feedback for present-day $CO_2$ levels.[170, 211] The permafrost–carbon feedback is *very likely* positive, and as discussed in Chapter 15: Potential Surprises, could be a larger positive feedback in the longer term. Thus, in the balance of multiple negative and positive feedback processes, the preponderance of evidence is that positive feedback processes dominate the overall radiative forcing feedback from anthropogenic activities.

### Major uncertainties

Uncertainties in cloud feedbacks are the largest source of uncertainty in the net climate feedback (and therefore climate sensitivity) on the decadal to century time scale.[60, 84] This results from the fact that cloud feedbacks can be either positive or negative, depending not only on the direction of change (more or less cloud) but also on the type of cloud affected and, to a lesser degree, the location of the cloud.[84] On decadal and longer time scales, the biological and physical responses of the ocean and land to climate change, and the subsequent changes in land and oceanic sinks of $CO_2$, contribute significant uncertainty to the net climate feedback (Ch. 13: Ocean Changes). Changes in the Brewer–Dobson atmospheric circulation driven by climate change and subsequent effects on stratosphere–troposphere coupling also contribute to climate feedback uncertainty.[77, 212, 213, 214, 215, 216]

### Assessment of confidence based on evidence and agreement, including short description of nature of evidence and level of agreement

There is *high confidence* that the net effect of all feedback processes in the climate system is positive, thereby amplifying warming. This confidence is based on consistency across multiple assessments, including IPCC AR5 (IPCC 2013[17] and references therein), of the magnitude of, in particular, the largest feedbacks in the climate system, two of which (water vapor feedback and snow/ice albedo feedback) are definitively positive in sign. While significant increases in low cloud cover with climate warming would be a large negative feedback to warming, modeling and observational studies do not support the idea of increases, on average, in low clouds with climate warming.

**Summary sentence or paragraph that integrates the above information**

The net effect of all identified feedbacks to forcing is positive based on the best current assessments and therefore amplifies climate warming. Feedback uncertainties, which are large for some processes, are included in these assessments. The various feedback processes operate on different time scales with carbon cycle and snow– and ice–albedo feedbacks operating on longer timelines than water vapor, lapse rate, cloud, and atmospheric composition feedbacks.

# REFERENCES

1. Clark, P.U., J.D. Shakun, S.A. Marcott, A.C. Mix, M. Eby, S. Kulp, A. Levermann, G.A. Milne, P.L. Pfister, B.D. Santer, D.P. Schrag, S. Solomon, T.F. Stocker, B.H. Strauss, A.J. Weaver, R. Winkelmann, D. Archer, E. Bard, A. Goldner, K. Lambeck, R.T. Pierrehumbert, and G.-K. Plattner, 2016: Consequences of twenty-first-century policy for multi-millennial climate and sea-level change. *Nature Climate Change*, **6**, 360-369. http://dx.doi.org/10.1038/nclimate2923

2. Lacis, A.A., G.A. Schmidt, D. Rind, and R.A. Ruedy, 2010: Atmospheric $CO_2$: Principal control knob governing Earth's temperature. *Science*, **330**, 356-359. http://dx.doi.org/10.1126/science.1190653

3. Davies, J.H. and D.R. Davies, 2010: Earth's surface heat flux. *Solid Earth*, **1**, 5-24. http://dx.doi.org/10.5194/se-1-5-2010

4. Flanner, M.G., 2009: Integrating anthropogenic heat flux with global climate models. *Geophysical Research Letters*, **36**, L02801. http://dx.doi.org/10.1029/2008gl036465

5. Munk, W. and C. Wunsch, 1998: Abyssal recipes II: Energetics of tidal and wind mixing. *Deep Sea Research Part I: Oceanographic Research Papers*, **45**, 1977-2010. http://dx.doi.org/10.1016/S0967-0637(98)00070-3

6. Rhein, M., S.R. Rintoul, S. Aoki, E. Campos, D. Chambers, R.A. Feely, S. Gulev, G.C. Johnson, S.A. Josey, A. Kostianoy, C. Mauritzen, D. Roemmich, L.D. Talley, and F. Wang, 2013: Observations: Ocean. *Climate Change 2013: The Physical Science Basis. Contribution of Working Group I to the Fifth Assessment Report of the Intergovernmental Panel on Climate Change*. Stocker, T.F., D. Qin, G.-K. Plattner, M. Tignor, S.K. Allen, J. Boschung, A. Nauels, Y. Xia, V. Bex, and P.M. Midgley, Eds. Cambridge University Press, Cambridge, United Kingdom and New York, NY, USA, 255–316. http://www.climatechange2013.org/report/full-report/

7. **Bindoff, N.L., P.A. Stott, K.M. AchutaRao, M.R. Allen, N. Gillett, D. Gutzler, K. Hansingo, G. Hegerl, Y. Hu, S. Jain, I.I. Mokhov, J. Overland, J. Perlwitz, R. Sebbari, and X. Zhang, 2013: Detection and attribution of climate change: From global to regional.** *Climate Change 2013: The Physical Science Basis. Contribution of Working Group I to the Fifth Assessment Report of the Intergovernmental Panel on Climate Change*. Stocker, T.F., D. Qin, G.-K. Plattner, M. Tignor, S.K. Allen, J. Boschung, A. Nauels, Y. Xia, V. Bex, and P.M. Midgley, Eds. Cambridge University Press, Cambridge, United Kingdom and New York, NY, USA, 867–952. http://www.climatechange2013.org/report/full-report/

8. Myhre, G., D. Shindell, F.-M. Bréon, W. Collins, J. Fuglestvedt, J. Huang, D. Koch, J.-F. Lamarque, D. Lee, B. Mendoza, T. Nakajima, A. Robock, G. Stephens, T. Takemura, and H. Zhang, 2013: Anthropogenic and natural radiative forcing. *Climate Change 2013: The Physical Science Basis. Contribution of Working Group I to the Fifth Assessment Report of the Intergovernmental Panel on Climate Change*. Stocker, T.F., D. Qin, G.-K. Plattner, M. Tignor, S.K. Allen, J. Boschung, A. Nauels, Y. Xia, V. Bex, and P.M. Midgley, Eds. Cambridge University Press, Cambridge, United Kingdom and New York, NY, USA, 659–740. http://www.climatechange2013.org/report/full-report/

9. Loeb, N.G., S. Kato, and B.A. Wielicki, 2002: Defining top-of-the-atmosphere flux reference level for earth radiation budget studies. *Journal of Climate*, **15**, 3301-3309. http://dx.doi.org/10.1175/1520-0442(2002)015<3301:dtotaf>2.0.co;2

10. Boer, G. and B. Yu, 2003: Climate sensitivity and response. *Climate Dynamics*, **20**, 415-429. http://dx.doi.org/10.1007/s00382-002-0283-3

11. Gillett, N.P., M.F. Wehner, S.F.B. Tett, and A.J. Weaver, 2004: Testing the linearity of the response to combined greenhouse gas and sulfate aerosol forcing. *Geophysical Research Letters*, **31**, L14201. http://dx.doi.org/10.1029/2004GL020111

12. Matthews, H.D., A.J. Weaver, K.J. Meissner, N.P. Gillett, and M. Eby, 2004: Natural and anthropogenic climate change: Incorporating historical land cover change, vegetation dynamics and the global carbon cycle. *Climate Dynamics*, **22**, 461-479. http://dx.doi.org/10.1007/s00382-004-0392-2

13. Meehl, G.A., W.M. Washington, C.M. Ammann, J.M. Arblaster, T.M.L. Wigley, and C. Tebaldi, 2004: Combinations of natural and anthropogenic forcings in twentieth-century climate. *Journal of Climate*, **17**, 3721-3727. http://dx.doi.org/10.1175/1520-0442(2004)017<3721:CONAAF>2.0.CO;2

14. Jones, A., J.M. Haywood, and O. Boucher, 2007: Aerosol forcing, climate response and climate sensitivity in the Hadley Centre climate model. *Journal of Geophysical Research*, **112**, D20211. http://dx.doi.org/10.1029/2007JD008688

15. Mahajan, S., K.J. Evans, J.J. Hack, and J.E. Truesdale, 2013: Linearity of climate response to increases in black carbon aerosols. *Journal of Climate*, **26**, 8223-8237. http://dx.doi.org/10.1175/JCLI-D-12-00715.1

16. Shiogama, H., D.A. Stone, T. Nagashima, T. Nozawa, and S. Emori, 2013: On the linear additivity of climate forcing-response relationships at global and continental scales. *International Journal of Climatology*, **33**, 2542-2550. http://dx.doi.org/10.1002/joc.3607

17. IPCC, 2013: *Climate Change 2013: The Physical Science Basis. Contribution of Working Group I to the Fifth Assessment Report of the Intergovernmental Panel on Climate Change.* Cambridge University Press, Cambridge, UK and New York, NY, 1535 pp. http://www.climatechange2013.org/report/

18. Krissansen-Totton, J. and R. Davies, 2013: Investigation of cosmic ray–cloud connections using MISR. *Geophysical Research Letters*, **40**, 5240-5245. http://dx.doi.org/10.1002/grl.50996

19. Lean, J., 1997: The sun's variable radiation and its relevance for earth. *Annual Review of Astronomy and Astrophysics*, **35**, 33-67. http://dx.doi.org/10.1146/annurev.astro.35.1.33

20. Fröhlich, C. and J. Lean, 2004: Solar radiative output and its variability: Evidence and mechanisms. *The Astronomy and Astrophysics Review*, **12**, 273-320. http://dx.doi.org/10.1007/s00159-004-0024-1

21. Gray, L.J., J. Beer, M. Geller, J.D. Haigh, M. Lockwood, K. Matthes, U. Cubasch, D. Fleitmann, G. Harrison, L. Hood, J. Luterbacher, G.A. Meehl, D. Shindell, B. van Geel, and W. White, 2010: Solar influences on climate. *Reviews of Geophysics*, **48**, RG4001. http://dx.doi.org/10.1029/2009RG000282

22. Kopp, G., 2014: An assessment of the solar irradiance record for climate studies. *Journal of Space Weather and Space Climate*, **4**, A14. http://dx.doi.org/10.1051/swsc/2014012

23. Kopp, G., N. Krivova, C.J. Wu, and J. Lean, 2016: The impact of the revised sunspot record on solar irradiance reconstructions. *Solar Physics*, **291**, 2951-1965. http://dx.doi.org/10.1007/s11207-016-0853-x

24. Kopp, G. and J.L. Lean, 2011: A new, lower value of total solar irradiance: Evidence and climate significance. *Geophysical Research Letters*, **38**, L01706. http://dx.doi.org/10.1029/2010GL045777

25. Floyd, L.E., J.W. Cook, L.C. Herring, and P.C. Crane, 2003: SUSIM'S 11-year observational record of the solar UV irradiance. *Advances in Space Research*, **31**, 2111-2120. http://dx.doi.org/10.1016/S0273-1177(03)00148-0

26. Ermolli, I., K. Matthes, T. Dudok de Wit, N.A. Krivova, K. Tourpali, M. Weber, Y.C. Unruh, L. Gray, U. Langematz, P. Pilewskie, E. Rozanov, W. Schmutz, A. Shapiro, S.K. Solanki, and T.N. Woods, 2013: Recent variability of the solar spectral irradiance and its impact on climate modelling. *Atmospheric Chemistry and Physics*, **13**, 3945-3977. http://dx.doi.org/10.5194/acp-13-3945-2013

27. Bolduc, C., M.S. Bourqui, and P. Charbonneau, 2015: A comparison of stratospheric photochemical response to different reconstructions of solar ultraviolet radiative variability. *Journal of Atmospheric and Solar-Terrestrial Physics*, **132**, 22-32. http://dx.doi.org/10.1016/j.jastp.2015.06.008

28. Lockwood, M., 2012: Solar influence on global and regional climates. *Surveys in Geophysics*, **33**, 503-534. http://dx.doi.org/10.1007/s10712-012-9181-3

29. Seppälä, A., K. Matthes, C.E. Randall, and I.A. Mironova, 2014: What is the solar influence on climate? Overview of activities during CAWSES-II. *Progress in Earth and Planetary Science*, **1**, 1-12. http://dx.doi.org/10.1186/s40645-014-0024-3

30. Xu, J. and A.M. Powell, 2013: What happened to surface temperature with sunspot activity in the past 130 years? *Theoretical and Applied Climatology*, **111**, 609-622. http://dx.doi.org/10.1007/s00704-012-0694-y

31. Gao, F.-L., L.-R. Tao, G.-M. Cui, J.-L. Xu, and T.-C. Hua, 2015: The influence of solar spectral variations on global radiative balance. *Advances in Space Research*, **55**, 682-687. http://dx.doi.org/10.1016/j.asr.2014.10.028

32. Swartz, W.H., R.S. Stolarski, L.D. Oman, E.L. Fleming, and C.H. Jackman, 2012: Middle atmosphere response to different descriptions of the 11-yr solar cycle in spectral irradiance in a chemistry-climate model. *Atmospheric Chemistry and Physics*, **12**, 5937-5948. http://dx.doi.org/10.5194/acp-12-5937-2012

33. Chiodo, G., D.R. Marsh, R. Garcia-Herrera, N. Calvo, and J.A. García, 2014: On the detection of solar signal in the tropical stratosphere. *Atmospheric Chemistry and Physics*, **14**, 5251-5269. http://dx.doi.org/10.5194/acp-14-5251-2014

34. Dhomse, S.S., M.P. Chipperfield, W. Feng, W.T. Ball, Y.C. Unruh, J.D. Haigh, N.A. Krivova, S.K. Solanki, and A.K. Smith, 2013: Stratospheric O$_3$ changes during 2001–2010: The small role of solar flux variations in a chemical transport model. *Atmospheric Chemistry and Physics*, **13**, 10113-10123. http://dx.doi.org/10.5194/acp-13-10113-2013

35. Andronova, N.G., E.V. Rozanov, F. Yang, M.E. Schlesinger, and G.L. Stenchikov, 1999: Radiative forcing by volcanic aerosols from 1850 to 1994. *Journal of Geophysical Research*, **104**, 16807-16826. http://dx.doi.org/10.1029/1999JD900165

36. Robock, A., 2000: Volcanic eruptions and climate. *Reviews of Geophysics*, **38**, 191-219. http://dx.doi.org/10.1029/1998RG000054

37. Stenchikov, G., T.L. Delworth, V. Ramaswamy, R.J. Stouffer, A. Wittenberg, and F. Zeng, 2009: Volcanic signals in oceans. *Journal of Geophysical Research*, **114**, D16104. http://dx.doi.org/10.1029/2008JD011673

38. Otterå, O.H., M. Bentsen, H. Drange, and L. Suo, 2010: External forcing as a metronome for Atlantic multidecadal variability. *Nature Geoscience*, **3**, 688-694. http://dx.doi.org/10.1038/ngeo955

39. Zanchettin, D., C. Timmreck, H.-F. Graf, A. Rubino, S. Lorenz, K. Lohmann, K. Krüger, and J.H. Jungclaus, 2012: Bi-decadal variability excited in the coupled ocean–atmosphere system by strong tropical volcanic eruptions. *Climate Dynamics*, **39**, 419-444. http://dx.doi.org/10.1007/s00382-011-1167-1

40. Zhang, D., R. Blender, and K. Fraedrich, 2013: Volcanoes and ENSO in millennium simulations: Global impacts and regional reconstructions in East Asia. *Theoretical and Applied Climatology*, **111**, 437-454. http://dx.doi.org/10.1007/s00704-012-0670-6

41. Langmann, B., 2014: On the role of climate forcing by volcanic sulphate and volcanic ash. *Advances in Meteorology*, **2014**, 17. http://dx.doi.org/10.1155/2014/340123

42. Gerlach, T., 2011: Volcanic versus anthropogenic carbon dioxide. *Eos, Transactions, American Geophysical Union*, **92**, 201-202. http://dx.doi.org/10.1029/2011EO240001

43. Ciais, P., C. Sabine, G. Bala, L. Bopp, V. Brovkin, J. Canadell, A. Chhabra, R. DeFries, J. Galloway, M. Heimann, C. Jones, C. Le Quéré, R.B. Myneni, S. Piao, and P. Thornton, 2013: Carbon and other biogeochemical cycles. *Climate Change 2013: The Physical Science Basis. Contribution of Working Group I to the Fifth Assessment Report of the Intergovernmental Panel on Climate Change*. Stocker, T.F., D. Qin, G.-K. Plattner, M. Tignor, S.K. Allen, J. Boschung, A. Nauels, Y. Xia, V. Bex, and P.M. Midgley, Eds. Cambridge University Press, Cambridge, United Kingdom and New York, NY, USA, 465–570. http://www.climatechange2013.org/report/full-report/

44. Xi, F., S.J. Davis, P. Ciais, D. Crawford-Brown, D. Guan, C. Pade, T. Shi, M. Syddall, J. Lv, L. Ji, L. Bing, J. Wang, W. Wei, K.-H. Yang, B. Lagerblad, I. Galan, C. Andrade, Y. Zhang, and Z. Liu, 2016: Substantial global carbon uptake by cement carbonation. *Nature Geoscience*, **9**, 880-883. http://dx.doi.org/10.1038/ngeo2840

45. Lelieveld, J. and P.J. Crutzen, 1992: Indirect chemical effects of methane on climate warming. *Nature*, **355**, 339-342. http://dx.doi.org/10.1038/355339a0

46. Saunois, M., R.B. Jackson, P. Bousquet, B. Poulter, and J.G. Canadell, 2016: The growing role of methane in anthropogenic climate change. *Environmental Research Letters*, **11**, 120207. http://dx.doi.org/10.1088/1748-9326/11/12/120207

47. Nisbet, E.G., E.J. Dlugokencky, M.R. Manning, D. Lowry, R.E. Fisher, J.L. France, S.E. Michel, J.B. Miller, J.W.C. White, B. Vaughn, P. Bousquet, J.A. Pyle, N.J. Warwick, M. Cain, R. Brownlow, G. Zazzeri, M. Lanoisellé, A.C. Manning, E. Gloor, D.E.J. Worthy, E.G. Brunke, C. Labuschagne, E.W. Wolff, and A.L. Ganesan, 2016: Rising atmospheric methane: 2007–2014 growth and isotopic shift. *Global Biogeochemical Cycles*, **30**, 1356-1370. http://dx.doi.org/10.1002/2016GB005406

48. Skiba, U.M. and R.M. Rees, 2014: Nitrous oxide, climate change and agriculture. *CAB Reviews*, **9**, 7. http://dx.doi.org/10.1079/PAVSNNR20149010

49. Fowler, D., M. Coyle, U. Skiba, M.A. Sutton, J.N. Cape, S. Reis, L.J. Sheppard, A. Jenkins, B. Grizzetti, J.N. Galloway, P. Vitousek, A. Leach, A.F. Bouwman, K. Butterbach-Bahl, F. Dentener, D. Stevenson, M. Amann, and M. Voss, 2013: The global nitrogen cycle in the twenty-first century. *Philosophical Transactions of the Royal Society B: Biological Sciences*, **368**, 20130164. http://dx.doi.org/10.1098/rstb.2013.0164

50. Velders, G.J.M., D.W. Fahey, J.S. Daniel, S.O. Andersen, and M. McFarland, 2015: Future atmospheric abundances and climate forcings from scenarios of global and regional hydrofluorocarbon (HFC) emissions. *Atmospheric Environment*, **123, Part A**, 200-209. http://dx.doi.org/10.1016/j.atmosenv.2015.10.071

51. Solomon, S., K.H. Rosenlof, R.W. Portmann, J.S. Daniel, S.M. Davis, T.J. Sanford, and G.-K. Plattner, 2010: Contributions of stratospheric water vapor to decadal changes in the rate of global warming. *Science*, **327**, 1219-1223. http://dx.doi.org/10.1126/science.1182488

52. Hegglin, M.I., D.A. Plummer, T.G. Shepherd, J.F. Scinocca, J. Anderson, L. Froidevaux, B. Funke, D. Hurst, A. Rozanov, J. Urban, T. von Clarmann, K.A. Walker, H.J. Wang, S. Tegtmeier, and K. Weigel, 2014: Vertical structure of stratospheric water vapour trends derived from merged satellite data. *Nature Geoscience*, **7**, 768-776. http://dx.doi.org/10.1038/ngeo2236

53. le Texier, H., S. Solomon, and R.R. Garcia, 1988: The role of molecular hydrogen and methane oxidation in the water vapour budget of the stratosphere. *Quarterly Journal of the Royal Meteorological Society*, **114**, 281-295. http://dx.doi.org/10.1002/qj.49711448002

54. Rosenlof, K.H., S.J. Oltmans, D. Kley, J.M. Russell, E.W. Chiou, W.P. Chu, D.G. Johnson, K.K. Kelly, H.A. Michelsen, G.E. Nedoluha, E.E. Remsberg, G.C. Toon, and M.P. McCormick, 2001: Stratospheric water vapor increases over the past half-century. *Geophysical Research Letters*, **28**, 1195-1198. http://dx.doi.org/10.1029/2000GL012502

55. Morris, G.A., J.E. Rosenfield, M.R. Schoeberl, and C.H. Jackman, 2003: Potential impact of subsonic and supersonic aircraft exhaust on water vapor in the lower stratosphere assessed via a trajectory model. *Journal of Geophysical Research*, **108**, 4103. http://dx.doi.org/10.1029/2002JD002614

56. Löffler, M., S. Brinkop, and P. Jöckel, 2016: Impact of major volcanic eruptions on stratospheric water vapour. *Atmospheric Chemistry and Physics*, **16**, 6547-6562. http://dx.doi.org/10.5194/acp-16-6547-2016

57. Held, I.M. and B.J. Soden, 2000: Water vapor feedback and global warming. *Annual Review of Energy and the Environment*, **25**, 441-475. http://dx.doi.org/10.1146/annurev.energy.25.1.441

58. Dentener, F., D. Stevenson, J. Cofala, R. Mechler, M. Amann, P. Bergamaschi, F. Raes, and R. Derwent, 2005: The impact of air pollutant and methane emission controls on tropospheric ozone and radiative forcing: CTM calculations for the period 1990-2030. *Atmospheric Chemistry and Physics*, **5**, 1731-1755. http://dx.doi.org/10.5194/acp-5-1731-2005

59. WMO, 2014: Scientific Asssessment of Ozone Depletion: 2014. World Meteorological Organization Geneva, Switzerland. 416 pp. http://www.esrl.noaa.gov/csd/assessments/ozone/2014/

60. Boucher, O., D. Randall, P. Artaxo, C. Bretherton, G. Feingold, P. Forster, V.-M. Kerminen, Y. Kondo, H. Liao, U. Lohmann, P. Rasch, S.K. Satheesh, S. Sherwood, B. Stevens, and X.Y. Zhang, 2013: Clouds and aerosols. *Climate Change 2013: The Physical Science Basis. Contribution of Working Group I to the Fifth Assessment Report of the Intergovernmental Panel on Climate Change*. Stocker, T.F., D. Qin, G.-K. Plattner, M. Tignor, S.K. Allen, J. Boschung, A. Nauels, Y. Xia, V. Bex, and P.M. Midgley, Eds. Cambridge University Press, Cambridge, United Kingdom and New York, NY, USA, 571–658. http://www.climatechange2013.org/report/full-report/

61. Bond, T.C., S.J. Doherty, D.W. Fahey, P.M. Forster, T. Berntsen, B.J. DeAngelo, M.G. Flanner, S. Ghan, B. Kärcher, D. Koch, S. Kinne, Y. Kondo, P.K. Quinn, M.C. Sarofim, M.G. Schultz, M. Schulz, C. Venkataraman, H. Zhang, S. Zhang, N. Bellouin, S.K. Guttikunda, P.K. Hopke, M.Z. Jacobson, J.W. Kaiser, Z. Klimont, U. Lohmann, J.P. Schwarz, D. Shindell, T. Storelvmo, S.G. Warren, and C.S. Zender, 2013: Bounding the role of black carbon in the climate system: A scientific assessment. *Journal of Geophysical Research Atmospheres*, **118**, 5380-5552. http://dx.doi.org/10.1002/jgrd.50171

62. Flanner, M.G., C.S. Zender, P.G. Hess, N.M. Mahowald, T.H. Painter, V. Ramanathan, and P.J. Rasch, 2009: Springtime warming and reduced snow cover from carbonaceous particles. *Atmospheric Chemistry and Physics*, **9**, 2481-2497. http://dx.doi.org/10.5194/acp-9-2481-2009

63. Ward, D.S., N.M. Mahowald, and S. Kloster, 2014: Potential climate forcing of land use and land cover change. *Atmospheric Chemistry and Physics*, **14**, 12701-12724. http://dx.doi.org/10.5194/acp-14-12701-2014

64. Ju, J. and J.G. Masek, 2016: The vegetation greenness trend in Canada and US Alaska from 1984–2012 Landsat data. *Remote Sensing of Environment*, **176**, 1-16. http://dx.doi.org/10.1016/j.rse.2016.01.001

65. Cook, B.I., S.P. Shukla, M.J. Puma, and L.S. Nazarenko, 2015: Irrigation as an historical climate forcing. *Climate Dynamics*, **44**, 1715-1730. http://dx.doi.org/10.1007/s00382-014-2204-7

66. Burkhardt, U. and B. Kärcher, 2011: Global radiative forcing from contrail cirrus. *Nature Climate Change*, **1**, 54-58. http://dx.doi.org/10.1038/nclimate1068

67. Smith, S.J. and T.C. Bond, 2014: Two hundred fifty years of aerosols and climate: The end of the age of aerosols. *Atmospheric Chemistry and Physics*, **14**, 537-549. http://dx.doi.org/10.5194/acp-14-537-2014

68. Fiore, A.M., V. Naik, and E.M. Leibensperger, 2015: Air quality and climate connections. *Journal of the Air & Waste Management Association*, **65**, 645-686. http://dx.doi.org/10.1080/10962247.2015.1040526

69. Masson-Delmotte, V., M. Schulz, A. Abe-Ouchi, J. Beer, A. Ganopolski, J.F. González Rouco, E. Jansen, K. Lambeck, J. Luterbacher, T. Naish, T. Osborn, B. Otto-Bliesner, T. Quinn, R. Ramesh, M. Rojas, X. Shao, and A. Timmermann, 2013: Information from paleoclimate archives. *Climate Change 2013: The Physical Science Basis. Contribution of Working Group I to the Fifth Assessment Report of the Intergovernmental Panel on Climate Change*. Stocker, T.F., D. Qin, G.-K. Plattner, M. Tignor, S.K. Allen, J. Boschung, A. Nauels, Y. Xia, V. Bex, and P.M. Midgley, Eds. Cambridge University Press, Cambridge, United Kingdom and New York, NY, USA, 383–464. http://www.climatechange2013.org/report/full-report/

70. Philipona, R., K. Behrens, and C. Ruckstuhl, 2009: How declining aerosols and rising greenhouse gases forced rapid warming in Europe since the 1980s. *Geophysical Research Letters*, **36**, L02806. http://dx.doi.org/10.1029/2008GL036350

71. Leibensperger, E.M., L.J. Mickley, D.J. Jacob, W.T. Chen, J.H. Seinfeld, A. Nenes, P.J. Adams, D.G. Streets, N. Kumar, and D. Rind, 2012: Climatic effects of 1950-2050 changes in US anthropogenic aerosols – Part 1: Aerosol trends and radiative forcing. *Atmospheric Chemistry and Physics* **12**, 3333-3348. http://dx.doi.org/10.5194/acp-12-3333-2012

72. Wild, M., 2016: Decadal changes in radiative fluxes at land and ocean surfaces and their relevance for global warming. *Wiley Interdisciplinary Reviews: Climate Change*, **7**, 91-107. http://dx.doi.org/10.1002/wcc.372

73. Hsu, N.C., R. Gautam, A.M. Sayer, C. Bettenhausen, C. Li, M.J. Jeong, S.C. Tsay, and B.N. Holben, 2012: Global and regional trends of aerosol optical depth over land and ocean using SeaWiFS measurements from 1997 to 2010. *Atmospheric Chemistry and Physics*, **12**, 8037-8053. http://dx.doi.org/10.5194/acp-12-8037-2012

74. Chin, M., T. Diehl, Q. Tan, J.M. Prospero, R.A. Kahn, L.A. Remer, H. Yu, A.M. Sayer, H. Bian, I.V. Geogdzhayev, B.N. Holben, S.G. Howell, B.J. Huebert, N.C. Hsu, D. Kim, T.L. Kucsera, R.C. Levy, M.I. Mishchenko, X. Pan, P.K. Quinn, G.L. Schuster, D.G. Streets, S.A. Strode, O. Torres, and X.P. Zhao, 2014: Multi-decadal aerosol variations from 1980 to 2009: A perspective from observations and a global model. *Atmospheric Chemistry and Physics*, **14**, 3657-3690. http://dx.doi.org/10.5194/acp-14-3657-2014

75. Lynch, P., J.S. Reid, D.L. Westphal, J. Zhang, T.F. Hogan, E.J. Hyer, C.A. Curtis, D.A. Hegg, Y. Shi, J.R. Campbell, J.I. Rubin, W.R. Sessions, F.J. Turk, and A.L. Walker, 2016: An 11-year global gridded aerosol optical thickness reanalysis (v1.0) for atmospheric and climate sciences. *Geoscientific Model Development*, **9**, 1489-1522. http://dx.doi.org/10.5194/gmd-9-1489-2016

76. Flato, G., J. Marotzke, B. Abiodun, P. Braconnot, S.C. Chou, W. Collins, P. Cox, F. Driouech, S. Emori, V. Eyring, C. Forest, P. Gleckler, E. Guilyardi, C. Jakob, V. Kattsov, C. Reason, and M. Rummukainen, 2013: Evaluation of climate models. *Climate Change 2013: The Physical Science Basis. Contribution of Working Group I to the Fifth Assessment Report of the Intergovernmental Panel on Climate Change*. Stocker, T.F., D. Qin, G.-K. Plattner, M. Tignor, S.K. Allen, J. Boschung, A. Nauels, Y. Xia, V. Bex, and P.M. Midgley, Eds. Cambridge University Press, Cambridge, United Kingdom and New York, NY, USA, 741–866. http://www.climatechange2013.org/report/full-report/

77. Collins, M., R. Knutti, J. Arblaster, J.-L. Dufresne, T. Fichefet, P. Friedlingstein, X. Gao, W.J. Gutowski, T. Johns, G. Krinner, M. Shongwe, C. Tebaldi, A.J. Weaver, and M. Wehner, 2013: Long-term climate change: Projections, commitments and irreversibility. *Climate Change 2013: The Physical Science Basis. Contribution of Working Group I to the Fifth Assessment Report of the Intergovernmental Panel on Climate Change*. Stocker, T.F., D. Qin, G.-K. Plattner, M. Tignor, S.K. Allen, J. Boschung, A. Nauels, Y. Xia, V. Bex, and P.M. Midgley, Eds. Cambridge University Press, Cambridge, United Kingdom and New York, NY, USA, 1029–1136. http://www.climatechange2013.org/report/full-report/

78. Knutti, R. and G.C. Hegerl, 2008: The equilibrium sensitivity of the Earth's temperature to radiation changes. *Nature Geoscience*, **1**, 735-743. http://dx.doi.org/10.1038/ngeo337

79. Bony, S., R. Colman, V.M. Kattsov, R.P. Allan, C.S. Bretherton, J.-L. Dufresne, A. Hall, S. Hallegatte, M.M. Holland, W. Ingram, D.A. Randall, B.J. Soden, G. Tselioudis, and M.J. Webb, 2006: How well do we understand and evaluate climate change feedback processes? *Journal of Climate*, **19**, 3445-3482. http://dx.doi.org/10.1175/JCLI3819.1

80. Shindell, D. and G. Faluvegi, 2009: Climate response to regional radiative forcing during the twentieth century. *Nature Geoscience*, **2**, 294-300. http://dx.doi.org/10.1038/ngeo473

81. Crook, J.A. and P.M. Forster, 2011: A balance between radiative forcing and climate feedback in the modeled 20th century temperature response. *Journal of Geophysical Research*, **116**, D17108. http://dx.doi.org/10.1029/2011JD015924

82. Knutti, R. and M.A.A. Rugenstein, 2015: Feedbacks, climate sensitivity and the limits of linear models. *Philosophical Transactions of the Royal Society A: Mathematical, Physical and Engineering Sciences*, **373**, 20150146. http://dx.doi.org/10.1098/rsta.2015.0146

83. Ramanathan, V., M.S. Lian, and R.D. Cess, 1979: Increased atmospheric $CO_2$: Zonal and seasonal estimates of the effect on the radiation energy balance and surface temperature. *Journal of Geophysical Research*, **84**, 4949-4958. http://dx.doi.org/10.1029/JC084iC08p04949

84. Vial, J., J.-L. Dufresne, and S. Bony, 2013: On the interpretation of inter-model spread in CMIP5 climate sensitivity estimates. *Climate Dynamics*, **41**, 3339-3362. http://dx.doi.org/10.1007/s00382-013-1725-9

85. Allen, M.R. and W.J. Ingram, 2002: Constraints on future changes in climate and the hydrologic cycle. *Nature*, **419**, 224-232. http://dx.doi.org/10.1038/nature01092

86. Soden, B.J. and I.M. Held, 2006: An assessment of climate feedbacks in coupled ocean–atmosphere models. *Journal of Climate*, **19**, 3354-3360. http://dx.doi.org/10.1175/JCLI3799.1

87. Dessler, A.E., M.R. Schoeberl, T. Wang, S.M. Davis, K.H. Rosenlof, and J.P. Vernier, 2014: Variations of stratospheric water vapor over the past three decades. *Journal of Geophysical Research Atmospheres*, **119**, 12,588-12,598. http://dx.doi.org/10.1002/2014JD021712

88. IPCC, 2007: *Climate Change 2007: The Physical Science Basis. Contribution of Working Group I to the Fourth Assessment Report of the Intergovernmental Panel on Climate Change*. Solomon, S., D. Qin, M. Manning, Z. Chen, M. Marquis, K.B. Averyt, M. Tignor, and H.L. Miller, Eds. Cambridge University Press, Cambridge. U.K, New York, NY, USA, 996 pp. http://www.ipcc.ch/publications_and_data/publications_ipcc_fourth_assessment_report_wg1_report_the_physical_science_basis.htm

89. Loeb, N.G., B.A. Wielicki, D.R. Doelling, G.L. Smith, D.F. Keyes, S. Kato, N. Manalo-Smith, and T. Wong, 2009: Toward optimal closure of the Earth's top-of-atmosphere radiation budget. *Journal of Climate*, **22**, 748-766. http://dx.doi.org/10.1175/2008JCLI2637.1

90. Sohn, B.J., T. Nakajima, M. Satoh, and H.S. Jang, 2010: Impact of different definitions of clear-sky flux on the determination of longwave cloud radiative forcing: NICAM simulation results. *Atmospheric Chemistry and Physics,* **10,** 11641-11646. http://dx.doi.org/10.5194/acp-10-11641-2010

91. Hartmann, D.L., M.E. Ockert-Bell, and M.L. Michelsen, 1992: The effect of cloud type on Earth's energy balance: Global analysis. *Journal of Climate,* **5,** 1281-1304. http://dx.doi.org/10.1175/1520-0442(1992)005<1281:teocto>2.0.co;2

92. Chen, T., W.B. Rossow, and Y. Zhang, 2000: Radiative effects of cloud-type variations. *Journal of Climate,* **13,** 264-286. http://dx.doi.org/10.1175/1520-0442(2000)013<0264:reoctv>2.0.co;2

93. Rädel, G., T. Mauritsen, B. Stevens, D. Dommenget, D. Matei, K. Bellomo, and A. Clement, 2016: Amplification of El Niño by cloud longwave coupling to atmospheric circulation. *Nature Geoscience,* **9,** 106-110. http://dx.doi.org/10.1038/ngeo2630

94. Taylor, P.C., R.G. Ellingson, and M. Cai, 2011: Geographical distribution of climate feedbacks in the NCAR CCSM3.0. *Journal of Climate,* **24,** 2737-2753. http://dx.doi.org/10.1175/2010JCLI3788.1

95. Klocke, D., J. Quaas, and B. Stevens, 2013: Assessment of different metrics for physical climate feedbacks. *Climate Dynamics,* **41,** 1173-1185. http://dx.doi.org/10.1007/s00382-013-1757-1

96. Sejas, S.A., M. Cai, A. Hu, G.A. Meehl, W. Washington, and P.C. Taylor, 2014: Individual feedback contributions to the seasonality of surface warming. *Journal of Climate,* **27,** 5653-5669. http://dx.doi.org/10.1175/JCLI-D-13-00658.1

97. Hansen, J. and L. Nazarenko, 2004: Soot climate forcing via snow and ice albedos. *Proceedings of the National Academy of Sciences of the United States of America,* **101,** 423-428. http://dx.doi.org/10.1073/pnas.2237157100

98. Jacobson, M.Z., 2004: Climate response of fossil fuel and biofuel soot, accounting for soot's feedback to snow and sea ice albedo and emissivity. *Journal of Geophysical Research,* **109,** D21201. http://dx.doi.org/10.1029/2004JD004945

99. Skeie, R.B., T. Berntsen, G. Myhre, C.A. Pedersen, J. Ström, S. Gerland, and J.A. Ogren, 2011: Black carbon in the atmosphere and snow, from pre-industrial times until present. *Atmospheric Chemistry and Physics,* **11,** 6809-6836. http://dx.doi.org/10.5194/acp-11-6809-2011

100. Yang, S., B. Xu, J. Cao, C.S. Zender, and M. Wang, 2015: Climate effect of black carbon aerosol in a Tibetan Plateau glacier. *Atmospheric Environment,* **111,** 71-78. http://dx.doi.org/10.1016/j.atmosenv.2015.03.016

101. Holland, D.M., R.H. Thomas, B. de Young, M.H. Ribergaard, and B. Lyberth, 2008: Acceleration of Jakobshavn Isbrae triggered by warm subsurface ocean waters. *Nature Geoscience,* **1,** 659-664. http://dx.doi.org/10.1038/ngeo316

102. Schoof, C., 2010: Ice-sheet acceleration driven by melt supply variability. *Nature,* **468,** 803-806. http://dx.doi.org/10.1038/nature09618

103. Rignot, E., M. Koppes, and I. Velicogna, 2010: Rapid submarine melting of the calving faces of West Greenland glaciers. *Nature Geoscience,* **3,** 187-191. http://dx.doi.org/10.1038/ngeo765

104. Joughin, I., R.B. Alley, and D.M. Holland, 2012: Ice-sheet response to oceanic forcing. *Science,* **338,** 1172-1176. http://dx.doi.org/10.1126/science.1226481

105. Straneo, F. and P. Heimbach, 2013: North Atlantic warming and the retreat of Greenland's outlet glaciers. *Nature,* **504,** 36-43. http://dx.doi.org/10.1038/nature12854

106. Thoma, M., J. Determann, K. Grosfeld, S. Goeller, and H.H. Hellmer, 2015: Future sea-level rise due to projected ocean warming beneath the Filchner Ronne Ice Shelf: A coupled model study. *Earth and Planetary Science Letters,* **431,** 217-224. http://dx.doi.org/10.1016/j.epsl.2015.09.013

107. Alley, K.E., T.A. Scambos, M.R. Siegfried, and H.A. Fricker, 2016: Impacts of warm water on Antarctic ice shelf stability through basal channel formation. *Nature Geoscience,* **9,** 290-293. http://dx.doi.org/10.1038/ngeo2675

108. Silvano, A., S.R. Rintoul, and L. Herraiz-Borreguero, 2016: Ocean-ice shelf interaction in East Antarctica. *Oceanography,* **29,** 130-143. http://dx.doi.org/10.5670/oceanog.2016.105

109. Hall, A. and X. Qu, 2006: Using the current seasonal cycle to constrain snow albedo feedback in future climate change. *Geophysical Research Letters,* **33,** L03502. http://dx.doi.org/10.1029/2005GL025127

110. Fernandes, R., H. Zhao, X. Wang, J. Key, X. Qu, and A. Hall, 2009: Controls on Northern Hemisphere snow albedo feedback quantified using satellite Earth observations. *Geophysical Research Letters,* **36,** L21702. http://dx.doi.org/10.1029/2009GL040057

111. Winton, M., 2006: Surface albedo feedback estimates for the AR4 climate models. *Journal of Climate,* **19,** 359-365. http://dx.doi.org/10.1175/JCLI3624.1

112. Hall, A., 2004: The role of surface albedo feedback in climate. *Journal of Climate,* **17,** 1550-1568. http://dx.doi.org/10.1175/1520-0442(2004)017<1550:TROSAF>2.0.CO;2

113. Kay, J.E. and A. Gettelman, 2009: Cloud influence on and response to seasonal Arctic sea ice loss. *Journal of Geophysical Research*, **114**, D18204. http://dx.doi.org/10.1029/2009JD011773

114. Kay, J.E., K. Raeder, A. Gettelman, and J. Anderson, 2011: The boundary layer response to recent Arctic sea ice loss and implications for high-latitude climate feedbacks. *Journal of Climate*, **24**, 428-447. http://dx.doi.org/10.1175/2010JCLI3651.1

115. Kay, J.E. and T. L'Ecuyer, 2013: Observational constraints on Arctic Ocean clouds and radiative fluxes during the early 21st century. *Journal of Geophysical Research Atmospheres*, **118**, 7219-7236. http://dx.doi.org/10.1002/jgrd.50489

116. Pistone, K., I. Eisenman, and V. Ramanathan, 2014: Observational determination of albedo decrease caused by vanishing Arctic sea ice. *Proceedings of the National Academy of Sciences*, **111**, 3322-3326. http://dx.doi.org/10.1073/pnas.1318201111

117. Taylor, P.C., S. Kato, K.-M. Xu, and M. Cai, 2015: Covariance between Arctic sea ice and clouds within atmospheric state regimes at the satellite footprint level. *Journal of Geophysical Research Atmospheres*, **120**, 12656-12678. http://dx.doi.org/10.1002/2015JD023520

118. Liao, H., Y. Zhang, W.-T. Chen, F. Raes, and J.H. Seinfeld, 2009: Effect of chemistry-aerosol-climate coupling on predictions of future climate and future levels of tropospheric ozone and aerosols. *Journal of Geophysical Research*, **114**, D10306. http://dx.doi.org/10.1029/2008JD010984

119. Unger, N., S. Menon, D.M. Koch, and D.T. Shindell, 2009: Impacts of aerosol-cloud interactions on past and future changes in tropospheric composition. *Atmospheric Chemistry and Physics*, **9**, 4115-4129. http://dx.doi.org/10.5194/acp-9-4115-2009

120. Raes, F., H. Liao, W.-T. Chen, and J.H. Seinfeld, 2010: Atmospheric chemistry-climate feedbacks. *Journal of Geophysical Research*, **115**, D12121. http://dx.doi.org/10.1029/2009JD013300

121. Tai, A.P.K., L.J. Mickley, C.L. Heald, and S. Wu, 2013: Effect of CO2 inhibition on biogenic isoprene emission: Implications for air quality under 2000 to 2050 changes in climate, vegetation, and land use. *Geophysical Research Letters*, **40**, 3479-3483. http://dx.doi.org/10.1002/grl.50650

122. Yue, X., L.J. Mickley, J.A. Logan, R.C. Hudman, M.V. Martin, and R.M. Yantosca, 2015: Impact of 2050 climate change on North American wildfire: consequences for ozone air quality. *Atmospheric Chemistry and Physics*, **15**, 10033-10055. http://dx.doi.org/10.5194/acp-15-10033-2015

123. Nowack, P.J., N. Luke Abraham, A.C. Maycock, P. Braesicke, J.M. Gregory, M.M. Joshi, A. Osprey, and J.A. Pyle, 2015: A large ozone-circulation feedback and its implications for global warming assessments. *Nature Climate Change*, **5**, 41-45. http://dx.doi.org/10.1038/nclimate2451

124. John, J.G., A.M. Fiore, V. Naik, L.W. Horowitz, and J.P. Dunne, 2012: Climate versus emission drivers of methane lifetime against loss by tropospheric OH from 1860–2100. *Atmospheric Chemistry and Physics*, **12**, 12021-12036. http://dx.doi.org/10.5194/acp-12-12021-2012

125. Pacifico, F., G.A. Folberth, C.D. Jones, S.P. Harrison, and W.J. Collins, 2012: Sensitivity of biogenic isoprene emissions to past, present, and future environmental conditions and implications for atmospheric chemistry. *Journal of Geophysical Research*, **117**, D22302. http://dx.doi.org/10.1029/2012JD018276

126. Morgenstern, O., G. Zeng, N. Luke Abraham, P.J. Telford, P. Braesicke, J.A. Pyle, S.C. Hardiman, F.M. O'Connor, and C.E. Johnson, 2013: Impacts of climate change, ozone recovery, and increasing methane on surface ozone and the tropospheric oxidizing capacity. *Journal of Geophysical Research Atmospheres*, **118**, 1028-1041. http://dx.doi.org/10.1029/2012JD018382

127. Holmes, C.D., M.J. Prather, O.A. Søvde, and G. Myhre, 2013: Future methane, hydroxyl, and their uncertainties: Key climate and emission parameters for future predictions. *Atmospheric Chemistry and Physics*, **13**, 285-302. http://dx.doi.org/10.5194/acp-13-285-2013

128. Naik, V., A. Voulgarakis, A.M. Fiore, L.W. Horowitz, J.F. Lamarque, M. Lin, M.J. Prather, P.J. Young, D. Bergmann, P.J. Cameron-Smith, I. Cionni, W.J. Collins, S.B. Dalsøren, R. Doherty, V. Eyring, G. Faluvegi, G.A. Folberth, B. Josse, Y.H. Lee, I.A. MacKenzie, T. Nagashima, T.P.C. van Noije, D.A. Plummer, M. Righi, S.T. Rumbold, R. Skeie, D.T. Shindell, D.S. Stevenson, S. Strode, K. Sudo, S. Szopa, and G. Zeng, 2013: Preindustrial to present-day changes in tropospheric hydroxyl radical and methane lifetime from the Atmospheric Chemistry and Climate Model Intercomparison Project (ACCMIP). *Atmospheric Chemistry and Physics*, **13**, 5277-5298. http://dx.doi.org/10.5194/acp-13-5277-2013

129. Voulgarakis, A., V. Naik, J.F. Lamarque, D.T. Shindell, P.J. Young, M.J. Prather, O. Wild, R.D. Field, D. Bergmann, P. Cameron-Smith, I. Cionni, W.J. Collins, S.B. Dalsøren, R.M. Doherty, V. Eyring, G. Faluvegi, G.A. Folberth, L.W. Horowitz, B. Josse, I.A. MacKenzie, T. Nagashima, D.A. Plummer, M. Righi, S.T. Rumbold, D.S. Stevenson, S.A. Strode, K. Sudo, S. Szopa, and G. Zeng, 2013: Analysis of present day and future OH and methane lifetime in the ACCMIP simulations. *Atmospheric Chemistry and Physics*, **13**, 2563-2587. http://dx.doi.org/10.5194/acp-13-2563-2013

130. Isaksen, I., T. Berntsen, S. Dalsøren, K. Eleftheratos, Y. Orsolini, B. Rognerud, F. Stordal, O. Søvde, C. Zerefos, and C. Holmes, 2014: Atmospheric ozone and methane in a changing climate. *Atmosphere*, **5**, 518. http://dx.doi.org/10.3390/atmos5030518

131. Dietmüller, S., M. Ponater, and R. Sausen, 2014: Interactive ozone induces a negative feedback in CO2-driven climate change simulations. *Journal of Geophysical Research Atmospheres*, **119**, 1796-1805. http://dx.doi.org/10.1002/2013JD020575

132. Banerjee, A., A.T. Archibald, A.C. Maycock, P. Telford, N.L. Abraham, X. Yang, P. Braesicke, and J.A. Pyle, 2014: Lightning NO$_x$, a key chemistry–climate interaction: Impacts of future climate change and consequences for tropospheric oxidising capacity. *Atmospheric Chemistry and Physics*, **14**, 9871-9881. http://dx.doi.org/10.5194/acp-14-9871-2014

133. ACC-MIP, 2017: Atmospheric Chemistry and Climate MIP. WCRP Working Group on Coupled Modeling. https://www.wcrp-climate.org/modelling-wgcm-mip-catalogue/modelling-wgcm-mips-2/226-modelling-wgcm-acc-mip

134. Han, Z., J. Li, W. Guo, Z. Xiong, and W. Zhang, 2013: A study of dust radiative feedback on dust cycle and meteorology over East Asia by a coupled regional climate-chemistry-aerosol model. *Atmospheric Environment*, **68**, 54-63. http://dx.doi.org/10.1016/j.atmosenv.2012.11.032

135. Le Quéré, C., R.M. Andrew, J.G. Canadell, S. Sitch, J.I. Korsbakken, G.P. Peters, A.C. Manning, T.A. Boden, P.P. Tans, R.A. Houghton, R.F. Keeling, S. Alin, O.D. Andrews, P. Anthoni, L. Barbero, L. Bopp, F. Chevallier, L.P. Chini, P. Ciais, K. Currie, C. Delire, S.C. Doney, P. Friedlingstein, T. Gkritzalis, I. Harris, J. Hauck, V. Haverd, M. Hoppema, K. Klein Goldewijk, A.K. Jain, E. Kato, A. Körtzinger, P. Landschützer, N. Lefèvre, A. Lenton, S. Lienert, D. Lombardozzi, J.R. Melton, N. Metzl, F. Millero, P.M.S. Monteiro, D.R. Munro, J.E.M.S. Nabel, S.I. Nakaoka, K. O'Brien, A. Olsen, A.M. Omar, T. Ono, D. Pierrot, B. Poulter, C. Rödenbeck, J. Salisbury, U. Schuster, J. Schwinger, R. Séférian, I. Skjelvan, B.D. Stocker, A.J. Sutton, T. Takahashi, H. Tian, B. Tilbrook, I.T. van der Laan-Luijkx, G.R. van der Werf, N. Viovy, A.P. Walker, A.J. Wiltshire, and S. Zaehle, 2016: Global carbon budget 2016. *Earth System Science Data*, **8**, 605-649. http://dx.doi.org/10.5194/essd-8-605-2016

136. Wenzel, S., P.M. Cox, V. Eyring, and P. Friedlingstein, 2016: Projected land photosynthesis constrained by changes in the seasonal cycle of atmospheric CO2. *Nature*, **538**, 499-501. http://dx.doi.org/10.1038/nature19772

137. Franks, P.J., M.A. Adams, J.S. Amthor, M.M. Barbour, J.A. Berry, D.S. Ellsworth, G.D. Farquhar, O. Ghannoum, J. Lloyd, N. McDowell, R.J. Norby, D.T. Tissue, and S. von Caemmerer, 2013: Sensitivity of plants to changing atmospheric CO2 concentration: From the geological past to the next century. *New Phytologist*, **197**, 1077-1094. http://dx.doi.org/10.1111/nph.12104

138. Seppälä, R., 2009: A global assessment on adaptation of forests to climate change. *Scandinavian Journal of Forest Research*, **24**, 469-472. http://dx.doi.org/10.1080/02827580903378626

139. Hibbard, K.A., G.A. Meehl, P.M. Cox, and P. Friedlingstein, 2007: A strategy for climate change stabilization experiments. *Eos, Transactions, American Geophysical Union*, **88**, 217-221. http://dx.doi.org/10.1029/2007EO200002

140. Thornton, P.E., J.-F. Lamarque, N.A. Rosenbloom, and N.M. Mahowald, 2007: Influence of carbon-nitrogen cycle coupling on land model response to CO2 fertilization and climate variability. *Global Biogeochemical Cycles*, **21**, GB4018. http://dx.doi.org/10.1029/2006GB002868

141. Brzostek, E.R., J.B. Fisher, and R.P. Phillips, 2014: Modeling the carbon cost of plant nitrogen acquisition: Mycorrhizal trade-offs and multipath resistance uptake improve predictions of retranslocation. *Journal of Geophysical Research Biogeosciences*, **119**, 1684-1697. http://dx.doi.org/10.1002/2014JG002660

142. Wieder, W.R., C.C. Cleveland, W.K. Smith, and K. Todd-Brown, 2015: Future productivity and carbon storage limited by terrestrial nutrient availability. *Nature Geoscience*, **8**, 441-444. http://dx.doi.org/10.1038/ngeo2413

143. Anav, A., P. Friedlingstein, M. Kidston, L. Bopp, P. Ciais, P. Cox, C. Jones, M. Jung, R. Myneni, and Z. Zhu, 2013: Evaluating the land and ocean components of the global carbon cycle in the CMIP5 earth system models. *Journal of Climate*, **26**, 6801-6843. http://dx.doi.org/10.1175/jcli-d-12-00417.1

144. Smith, W.K., S.C. Reed, C.C. Cleveland, A.P. Ballantyne, W.R.L. Anderegg, W.R. Wieder, Y.Y. Liu, and S.W. Running, 2016: Large divergence of satellite and Earth system model estimates of global terrestrial CO2 fertilization. *Nature Climate Change*, **6**, 306-310. http://dx.doi.org/10.1038/nclimate2879

145. Friedlingstein, P., P. Cox, R. Betts, L. Bopp, W.v. Bloh, V. Brovkin, P. Cadule, S. Doney, M. Eby, I. Fung, G. Bala, J. John, C. Jones, F. Joos, T. Kato, M. Kawamiya, W. Knorr, K. Lindsay, H.D. Matthews, T. Raddatz, P. Rayner, C. Reick, E. Roeckner, K.-G. Schnitzler, R. Schnur, K. Strassmann, A.J. Weaver, C. Yoshikawa, and N. Zeng, 2006: Climate–carbon cycle feedback analysis: Results from the C⁴MIP model intercomparison. *Journal of Climate*, **19**, 3337-3353. http://dx.doi.org/10.1175/JCLI3800.1

146. Friedlingstein, P., M. Meinshausen, V.K. Arora, C.D. Jones, A. Anav, S.K. Liddicoat, and R. Knutti, 2014: Uncertainties in CMIP5 climate projections due to carbon cycle feedbacks. *Journal of Climate*, **27**, 511-526. http://dx.doi.org/10.1175/JCLI-D-12-00579.1

147. Johnson, G.C., J.M. Lyman, T. Boyer, C.M. Domingues, M. Ishii, R. Killick, D. Monselesan, and S.E. Wijffels, 2016: [Global Oceans] Ocean heat content [in "State of the Climate in 2015"]. *Bulletin of the American Meteorological Society*, **97**, S66-S70. http://dx.doi.org/10.1175/2016BAMSStateoftheClimate.1

148. Falkowski, P.G., M.E. Katz, A.H. Knoll, A. Quigg, J.A. Raven, O. Schofield, and F.J.R. Taylor, 2004: The evolution of modern eukaryotic phytoplankton. *Science*, **305**, 354-360. http://dx.doi.org/10.1126/science.1095964

149. Carr, M.-E., M.A.M. Friedrichs, M. Schmeltz, M. Noguchi Aita, D. Antoine, K.R. Arrigo, I. Asanuma, O. Aumont, R. Barber, M. Behrenfeld, R. Bidigare, E.T. Buitenhuis, J. Campbell, A. Ciotti, H. Dierssen, M. Dowell, J. Dunne, W. Esaias, B. Gentili, W. Gregg, S. Groom, N. Hoepffner, J. Ishizaka, T. Kameda, C. Le Quéré, S. Lohrenz, J. Marra, F. Mélin, K. Moore, A. Morel, T.E. Reddy, J. Ryan, M. Scardi, T. Smyth, K. Turpie, G. Tilstone, K. Waters, and Y. Yamanaka, 2006: A comparison of global estimates of marine primary production from ocean color. *Deep Sea Research Part II: Topical Studies in Oceanography*, **53**, 741-770. http://dx.doi.org/10.1016/j.dsr2.2006.01.028

150. Chavez, F.P., M. Messié, and J.T. Pennington, 2011: Marine primary production in relation to climate variability and change. *Annual Review of Marine Science*, **3**, 227-260. http://dx.doi.org/10.1146/annurev.marine.010908.163917

151. Doney, S.C., 2010: The growing human footprint on coastal and open-ocean biogeochemistry. *Science*, **328**, 1512-6. http://dx.doi.org/10.1126/science.1185198

152. Passow, U. and C.A. Carlson, 2012: The biological pump in a high CO2 world. *Marine Ecology Progress Series*, **470**, 249-271. http://dx.doi.org/10.3354/meps09985

153. Trenberth, K.E., P.D. Jones, P. Ambenje, R. Bojariu, D. Easterling, A.K. Tank, D. Parker, F. Rahimzadeh, J.A. Renwick, M. Rusticucci, B. Soden, and P. Zhai, 2007: Observations: Surface and atmospheric climate change. *Climate Change 2007: The Physical Science Basis. Contribution of Working Group I to the Fourth Assessment Report of the Intergovernmental Panel on Climate Change*. Solomon, S., D. Qin, M. Manning, Z. Chen, M. Marquis, K.B. Averyt, M. Tignor, and H.L. Miller, Eds. Cambridge University Press, Cambridge, United Kingdom and New York, NY, USA. http://www.ipcc.ch/publications_and_data/ar4/wg1/en/ch3.html

154. Schanze, J.J., R.W. Schmitt, and L.L. Yu, 2010: The global oceanic freshwater cycle: A state-of-the-art quantification. *Journal of Marine Research*, **68**, 569-595. http://dx.doi.org/10.1357/002224010794657164

155. Durack, P.J. and S.E. Wijffels, 2010: Fifty-year trends in global ocean salinities and their relationship to broad-scale warming. *Journal of Climate*, **23**, 4342-4362. http://dx.doi.org/10.1175/2010jcli3377.1

156. Good, P., J.M. Gregory, J.A. Lowe, and T. Andrews, 2013: Abrupt CO2 experiments as tools for predicting and understanding CMIP5 representative concentration pathway projections. *Climate Dynamics*, **40**, 1041-1053. http://dx.doi.org/10.1007/s00382-012-1410-4

157. Andrews, T., J.M. Gregory, M.J. Webb, and K.E. Taylor, 2012: Forcing, feedbacks and climate sensitivity in CMIP5 coupled atmosphere-ocean climate models. *Geophysical Research Letters*, **39**, L09712. http://dx.doi.org/10.1029/2012GL051607

158. Kostov, Y., K.C. Armour, and J. Marshall, 2014: Impact of the Atlantic meridional overturning circulation on ocean heat storage and transient climate change. *Geophysical Research Letters*, **41**, 2108-2116. http://dx.doi.org/10.1002/2013GL058998

159. Rahmstorf, S., J.E. Box, G. Feulner, M.E. Mann, A. Robinson, S. Rutherford, and E.J. Schaffernicht, 2015: Exceptional twentieth-century slowdown in Atlantic Ocean overturning circulation. *Nature Climate Change*, **5**, 475-480. http://dx.doi.org/10.1038/nclimate2554

160. Rignot, E. and R.H. Thomas, 2002: Mass balance of polar ice sheets. *Science*, **297**, 1502-1506. http://dx.doi.org/10.1126/science.1073888

161. van den Broeke, M., J. Bamber, J. Ettema, E. Rignot, E. Schrama, W.J. van de Berg, E. van Meijgaard, I. Velicogna, and B. Wouters, 2009: Partitioning recent Greenland mass loss. *Science*, **326**, 984-986. http://dx.doi.org/10.1126/science.1178176

162. Enderlin, E.M. and G.S. Hamilton, 2014: Estimates of iceberg submarine melting from high-resolution digital elevation models: Application to Sermilik Fjord, East Greenland. *Journal of Glaciology*, **60**, 1084-1092. http://dx.doi.org/10.3189/2014JoG14J085

163. Gelderloos, R., F. Straneo, and C.A. Katsman, 2012: Mechanisms behind the temporary shutdown of deep convection in the Labrador Sea: Lessons from the great salinity anomaly years 1968–71. *Journal of Climate*, **25**, 6743-6755. http://dx.doi.org/10.1175/jcli-d-11-00549.1

164. Bidigare, R.R., F. Chai, M.R. Landry, R. Lukas, C.C.S. Hannides, S.J. Christensen, D.M. Karl, L. Shi, and Y. Chao, 2009: Subtropical ocean ecosystem structure changes forced by North Pacific climate variations. *Journal of Plankton Research*, **31**, 1131-1139. http://dx.doi.org/10.1093/plankt/fbp064

165. Zhai, P.-W., Y. Hu, C.A. Hostetler, B. Cairns, R.A. Ferrare, K.D. Knobelspiesse, D.B. Josset, C.R. Trepte, P.L. Lucker, and J. Chowdhary, 2013: Uncertainty and interpretation of aerosol remote sensing due to vertical inhomogeneity. *Journal of Quantitative Spectroscopy and Radiative Transfer,* **114,** 91-100. http://dx.doi.org/10.1016/j.jqsrt.2012.08.006

166. Behrenfeld, M.J., R.T. O'Malley, D.A. Siegel, C.R. McClain, J.L. Sarmiento, G.C. Feldman, A.J. Milligan, P.G. Falkowski, R.M. Letelier, and E.S. Boss, 2006: Climate-driven trends in contemporary ocean productivity. *Nature,* **444,** 752-755. http://dx.doi.org/10.1038/nature05317

167. Boyce, D.G., M.R. Lewis, and B. Worm, 2010: Global phytoplankton decline over the past century. *Nature,* **466,** 591-596. http://dx.doi.org/10.1038/nature09268

168. Capotondi, A., M.A. Alexander, N.A. Bond, E.N. Curchitser, and J.D. Scott, 2012: Enhanced upper ocean stratification with climate change in the CMIP3 models. *Journal of Geophysical Research,* **117,** C04031. http://dx.doi.org/10.1029/2011JC007409

169. Rykaczewski, R.R. and J.P. Dunne, 2011: A measured look at ocean chlorophyll trends. *Nature,* **472,** E5-E6. http://dx.doi.org/10.1038/nature09952

170. Laufkötter, C., M. Vogt, N. Gruber, M. Aita-Noguchi, O. Aumont, L. Bopp, E. Buitenhuis, S.C. Doney, J. Dunne, T. Hashioka, J. Hauck, T. Hirata, J. John, C. Le Quéré, I.D. Lima, H. Nakano, R. Seferian, I. Totterdell, M. Vichi, and C. Völker, 2015: Drivers and uncertainties of future global marine primary production in marine ecosystem models. *Biogeosciences,* **12,** 6955-6984. http://dx.doi.org/10.5194/bg-12-6955-2015

171. Jin, X., N. Gruber, J.P. Dunne, J.L. Sarmiento, and R.A. Armstrong, 2006: Diagnosing the contribution of phytoplankton functional groups to the production and export of particulate organic carbon, CaCO3, and opal from global nutrient and alkalinity distributions. *Global Biogeochemical Cycles,* **20,** GB2015. http://dx.doi.org/10.1029/2005GB002532

172. Fu, W., J.T. Randerson, and J.K. Moore, 2016: Climate change impacts on net primary production (NPP) and export production (EP) regulated by increasing stratification and phytoplankton community structure in the CMIP5 models. *Biogeosciences,* **13,** 5151-5170. http://dx.doi.org/10.5194/bg-13-5151-2016

173. Frölicher, T.L., K.B. Rodgers, C.A. Stock, and W.W.L. Cheung, 2016: Sources of uncertainties in 21st century projections of potential ocean ecosystem stressors. *Global Biogeochemical Cycles,* **30,** 1224-1243. http://dx.doi.org/10.1002/2015GB005338

174. Schaefer, K., H. Lantuit, E.R. Vladimir, E.A.G. Schuur, and R. Witt, 2014: The impact of the permafrost carbon feedback on global climate. *Environmental Research Letters,* **9,** 085003. http://dx.doi.org/10.1088/1748-9326/9/8/085003

175. Koven, C.D., E.A.G. Schuur, C. Schädel, T.J. Bohn, E.J. Burke, G. Chen, X. Chen, P. Ciais, G. Grosse, J.W. Harden, D.J. Hayes, G. Hugelius, E.E. Jafarov, G. Krinner, P. Kuhry, D.M. Lawrence, A.H. MacDougall, S.S. Marchenko, A.D. McGuire, S.M. Natali, D.J. Nicolsky, D. Olefeldt, S. Peng, V.E. Romanovsky, K.M. Schaefer, J. Strauss, C.C. Treat, and M. Turetsky, 2015: A simplified, data-constrained approach to estimate the permafrost carbon–climate feedback. *Philosophical Transactions of the Royal Society A: Mathematical, Physical and Engineering Sciences,* **373,** 20140423. http://dx.doi.org/10.1098/rsta.2014.0423

176. Schuur, E.A.G., A.D. McGuire, C. Schadel, G. Grosse, J.W. Harden, D.J. Hayes, G. Hugelius, C.D. Koven, P. Kuhry, D.M. Lawrence, S.M. Natali, D. Olefeldt, V.E. Romanovsky, K. Schaefer, M.R. Turetsky, C.C. Treat, and J.E. Vonk, 2015: Climate change and the permafrost carbon feedback. *Nature,* **520,** 171-179. http://dx.doi.org/10.1038/nature14338

177. González-Eguino, M. and M.B. Neumann, 2016: Significant implications of permafrost thawing for climate change control. *Climatic Change,* **136,** 381-388. http://dx.doi.org/10.1007/s10584-016-1666-5

178. Koven, C.D., D.M. Lawrence, and W.J. Riley, 2015: Permafrost carbon–climate feedback is sensitive to deep soil carbon decomposability but not deep soil nitrogen dynamics. *Proceedings of the National Academy of Sciences,* **112,** 3752-3757. http://dx.doi.org/10.1073/pnas.1415123112

179. Liljedahl, A.K., J. Boike, R.P. Daanen, A.N. Fedorov, G.V. Frost, G. Grosse, L.D. Hinzman, Y. Iijma, J.C. Jorgenson, N. Matveyeva, M. Necsoiu, M.K. Raynolds, V.E. Romanovsky, J. Schulla, K.D. Tape, D.A. Walker, C.J. Wilson, H. Yabuki, and D. Zona, 2016: Pan-Arctic ice-wedge degradation in warming permafrost and its influence on tundra hydrology. *Nature Geoscience,* **9,** 312-318. http://dx.doi.org/10.1038/ngeo2674

180. Melillo, J.M., T.C. Richmond, and G.W. Yohe, eds., 2014: *Climate Change Impacts in the United States: The Third National Climate Assessment.* U.S. Global Change Research Program: Washington, D.C., 841 pp. http://dx.doi.org/10.7930/J0Z31WJ2

181. Jiao, C., M.G. Flanner, Y. Balkanski, S.E. Bauer, N. Bellouin, T.K. Berntsen, H. Bian, K.S. Carslaw, M. Chin, N. De Luca, T. Diehl, S.J. Ghan, T. Iversen, A. Kirkevåg, D. Koch, X. Liu, G.W. Mann, J.E. Penner, G. Pitari, M. Schulz, Ø. Seland, R.B. Skeie, S.D. Steenrod, P. Stier, T. Takemura, K. Tsigaridis, T. van Noije, Y. Yun, and K. Zhang, 2014: An AeroCom assessment of black carbon in Arctic snow and sea ice. *Atmospheric Chemistry and Physics,* **14,** 2399-2417. http://dx.doi.org/10.5194/acp-14-2399-2014

182. Tsigaridis, K., N. Daskalakis, M. Kanakidou, P.J. Adams, P. Artaxo, R. Bahadur, Y. Balkanski, S.E. Bauer, N. Bellouin, A. Benedetti, T. Bergman, T.K. Berntsen, J.P. Beukes, H. Bian, K.S. Carslaw, M. Chin, G. Curci, T. Diehl, R.C. Easter, S.J. Ghan, S.L. Gong, A. Hodzic, C.R. Hoyle, T. Iversen, S. Jathar, J.L. Jimenez, J.W. Kaiser, A. Kirkevåg, D. Koch, H. Kokkola, Y.H. Lee, G. Lin, X. Liu, G. Luo, X. Ma, G.W. Mann, N. Mihalopoulos, J.J. Morcrette, J.F. Müller, G. Myhre, S. Myriokefalitakis, N.L. Ng, D. O'Donnell, J.E. Penner, L. Pozzoli, K.J. Pringle, L.M. Russell, M. Schulz, J. Sciare, Ø. Seland, D.T. Shindell, S. Sillman, R.B. Skeie, D. Spracklen, T. Stavrakou, S.D. Steenrod, T. Takemura, P. Tiitta, S. Tilmes, H. Tost, T. van Noije, P.G. van Zyl, K. von Salzen, F. Yu, Z. Wang, Z. Wang, R.A. Zaveri, H. Zhang, K. Zhang, Q. Zhang, and X. Zhang, 2014: The AeroCom evaluation and intercomparison of organic aerosol in global models. *Atmospheric Chemistry and Physics*, **14**, 10845-10895. http://dx.doi.org/10.5194/acp-14-10845-2014

183. Koffi, B., M. Schulz, F.-M. Bréon, F. Dentener, B.M. Steensen, J. Griesfeller, D. Winker, Y. Balkanski, S.E. Bauer, N. Bellouin, T. Berntsen, H. Bian, M. Chin, T. Diehl, R. Easter, S. Ghan, D.A. Hauglustaine, T. Iversen, A. Kirkevåg, X. Liu, U. Lohmann, G. Myhre, P. Rasch, Ø. Seland, R.B. Skeie, S.D. Steenrod, P. Stier, J. Tackett, T. Takemura, K. Tsigaridis, M.R. Vuolo, J. Yoon, and K. Zhang, 2016: Evaluation of the aerosol vertical distribution in global aerosol models through comparison against CALIOP measurements: AeroCom phase II results. *Journal of Geophysical Research Atmospheres*, **121**, 7254-7283. http://dx.doi.org/10.1002/2015JD024639

184. Zhu, Z., S. Piao, R.B. Myneni, M. Huang, Z. Zeng, J.G. Canadell, P. Ciais, S. Sitch, P. Friedlingstein, A. Arneth, C. Cao, L. Cheng, E. Kato, C. Koven, Y. Li, X. Lian, Y. Liu, R. Liu, J. Mao, Y. Pan, S. Peng, J. Penuelas, B. Poulter, T.A.M. Pugh, B.D. Stocker, N. Viovy, X. Wang, Y. Wang, Z. Xiao, H. Yang, S. Zaehle, and N. Zeng, 2016: Greening of the Earth and its drivers. *Nature Climate Change*, **6**, 791-795. http://dx.doi.org/10.1038/nclimate3004

185. Mao, J., A. Ribes, B. Yan, X. Shi, P.E. Thornton, R. Seferian, P. Ciais, R.B. Myneni, H. Douville, S. Piao, Z. Zhu, R.E. Dickinson, Y. Dai, D.M. Ricciuto, M. Jin, F.M. Hoffman, B. Wang, M. Huang, and X. Lian, 2016: Human-induced greening of the northern extratropical land surface. *Nature Climate Change*, **6**, 959-963. http://dx.doi.org/10.1038/nclimate3056

186. Raible, C.C., S. Brönnimann, R. Auchmann, P. Brohan, T.L. Frölicher, H.-F. Graf, P. Jones, J. Luterbacher, S. Muthers, R. Neukom, A. Robock, S. Self, A. Sudrajat, C. Timmreck, and M. Wegmann, 2016: Tambora 1815 as a test case for high impact volcanic eruptions: Earth system effects. *Wiley Interdisciplinary Reviews: Climate Change*, **7**, 569-589. http://dx.doi.org/10.1002/wcc.407

187. Twohy, C.H., M.D. Petters, J.R. Snider, B. Stevens, W. Tahnk, M. Wetzel, L. Russell, and F. Burnet, 2005: Evaluation of the aerosol indirect effect in marine stratocumulus clouds: Droplet number, size, liquid water path, and radiative impact. *Journal of Geophysical Research*, **110**, D08203. http://dx.doi.org/10.1029/2004JD005116

188. Lohmann, U. and J. Feichter, 2005: Global indirect aerosol effects: A review. *Atmospheric Chemistry and Physics*, **5**, 715-737. http://dx.doi.org/10.5194/acp-5-715-2005

189. Quaas, J., Y. Ming, S. Menon, T. Takemura, M. Wang, J.E. Penner, A. Gettelman, U. Lohmann, N. Bellouin, O. Boucher, A.M. Sayer, G.E. Thomas, A. McComiskey, G. Feingold, C. Hoose, J.E. Kristjánsson, X. Liu, Y. Balkanski, L.J. Donner, P.A. Ginoux, P. Stier, B. Grandey, J. Feichter, I. Sednev, S.E. Bauer, D. Koch, R.G. Grainger, Kirkev, aring, A. g, T. Iversen, Ø. Seland, R. Easter, S.J. Ghan, P.J. Rasch, H. Morrison, J.F. Lamarque, M.J. Iacono, S. Kinne, and M. Schulz, 2009: Aerosol indirect effects – general circulation model intercomparison and evaluation with satellite data. *Atmospheric Chemistry and Physics*, **9**, 8697-8717. http://dx.doi.org/10.5194/acp-9-8697-2009

190. Rosenfeld, D., M.O. Andreae, A. Asmi, M. Chin, G. de Leeuw, D.P. Donovan, R. Kahn, S. Kinne, N. Kivekäs, M. Kulmala, W. Lau, K.S. Schmidt, T. Suni, T. Wagner, M. Wild, and J. Quaas, 2014: Global observations of aerosol–cloud–precipitation–climate interactions. *Reviews of Geophysics*, **52**, 750-808. http://dx.doi.org/10.1002/2013RG000441

191. Wild, M., 2009: Global dimming and brightening: A review. *Journal of Geophysical Research*, **114**, D00D16. http://dx.doi.org/10.1029/2008JD011470

192. Szopa, S., Y. Balkanski, M. Schulz, S. Bekki, D. Cugnet, A. Fortems-Cheiney, S. Turquety, A. Cozic, C. Déandreis, D. Hauglustaine, A. Idelkadi, J. Lathière, F. Lefevre, M. Marchand, R. Vuolo, N. Yan, and J.-L. Dufresne, 2013: Aerosol and ozone changes as forcing for climate evolution between 1850 and 2100. *Climate Dynamics*, **40**, 2223-2250. http://dx.doi.org/10.1007/s00382-012-1408-y

193. Stjern, C.W. and J.E. Kristjánsson, 2015: Contrasting influences of recent aerosol changes on clouds and precipitation in Europe and East Asia. *Journal of Climate*, **28**, 8770-8790. http://dx.doi.org/10.1175/jcli-d-14-00837.1

194. Wang, Y., J.H. Jiang, and H. Su, 2015: Atmospheric responses to the redistribution of anthropogenic aerosols. *Journal of Geophysical Research Atmospheres*, **120**, 9625-9641. http://dx.doi.org/10.1002/2015JD023665

195. Myhre, G., W. Aas, R. Cherian, W. Collins, G. Faluvegi, M. Flanner, P. Forster, Ø. Hodnebrog, Z. Klimont, M.T. Lund, J. Mülmenstädt, C. Lund Myhre, D. Olivié, M. Prather, J. Quaas, B.H. Samset, J.L. Schnell, M. Schulz, D. Shindell, R.B. Skeie, T. Takemura, and S. Tsyro, 2017: Multi-model simulations of aerosol and ozone radiative forcing due to anthropogenic emission changes during the period 1990–2015. *Atmospheric Chemistry and Physics*, **17**, 2709-2720. http://dx.doi.org/10.5194/acp-17-2709-2017

196. Mao, K.B., Y. Ma, L. Xia, W.Y. Chen, X.Y. Shen, T.J. He, and T.R. Xu, 2014: Global aerosol change in the last decade: An analysis based on MODIS data. *Atmospheric Environment*, **94**, 680-686. http://dx.doi.org/10.1016/j.atmosenv.2014.04.053

197. Marmer, E., B. Langmann, H. Fagerli, and V. Vestreng, 2007: Direct shortwave radiative forcing of sulfate aerosol over Europe from 1900 to 2000. *Journal of Geophysical Research*, **112**, D23S17. http://dx.doi.org/10.1029/2006JD008037

198. Murphy, D.M., J.C. Chow, E.M. Leibensperger, W.C. Malm, M. Pitchford, B.A. Schichtel, J.G. Watson, and W.H. White, 2011: Decreases in elemental carbon and fine particle mass in the United States. *Atmospheric Chemistry and Physics*, **11**, 4679-4686. http://dx.doi.org/10.5194/acp-11-4679-2011

199. Kühn, T., A.I. Partanen, A. Laakso, Z. Lu, T. Bergman, S. Mikkonen, H. Kokkola, H. Korhonen, P. Räisänen, D.G. Streets, S. Romakkaniemi, and A. Laaksonen, 2014: Climate impacts of changing aerosol emissions since 1996. *Geophysical Research Letters*, **41**, 4711-4718. http://dx.doi.org/10.1002/2014GL060349

200. Turnock, S.T., D.V. Spracklen, K.S. Carslaw, G.W. Mann, M.T. Woodhouse, P.M. Forster, J. Haywood, C.E. Johnson, M. Dalvi, N. Bellouin, and A. Sanchez-Lorenzo, 2015: Modelled and observed changes in aerosols and surface solar radiation over Europe between 1960 and 2009. *Atmospheric Chemistry and Physics*, **15**, 9477-9500. http://dx.doi.org/10.5194/acp-15-9477-2015

201. Babu, S.S., M.R. Manoj, K.K. Moorthy, M.M. Gogoi, V.S. Nair, S.K. Kompalli, S.K. Satheesh, K. Niranjan, K. Ramagopal, P.K. Bhuyan, and D. Singh, 2013: Trends in aerosol optical depth over Indian region: Potential causes and impact indicators. *Journal of Geophysical Research Atmospheres*, **118**, 11,794-11,806. http://dx.doi.org/10.1002/2013JD020507

202. Krishna Moorthy, K., S. Suresh Babu, M.R. Manoj, and S.K. Satheesh, 2013: Buildup of aerosols over the Indian Region. *Geophysical Research Letters*, **40**, 1011-1014. http://dx.doi.org/10.1002/grl.50165

203. Pietikäinen, J.P., K. Kupiainen, Z. Klimont, R. Makkonen, H. Korhonen, R. Karinkanta, A.P. Hyvärinen, N. Karvosenoja, A. Laaksonen, H. Lihavainen, and V.M. Kerminen, 2015: Impacts of emission reductions on aerosol radiative effects. *Atmospheric Chemistry and Physics*, **15**, 5501-5519. http://dx.doi.org/10.5194/acp-15-5501-2015

204. Streets, D.G., C. Yu, Y. Wu, M. Chin, Z. Zhao, T. Hayasaka, and G. Shi, 2008: Aerosol trends over China, 1980–2000. *Atmospheric Research*, **88**, 174-182. http://dx.doi.org/10.1016/j.atmosres.2007.10.016

205. Li, J., Z. Han, and Z. Xie, 2013: Model analysis of long-term trends of aerosol concentrations and direct radiative forcings over East Asia. *Tellus B*, **65**, 20410. http://dx.doi.org/10.3402/tellusb.v65i0.20410

206. Wang, Y., Y. Yang, S. Han, Q. Wang, and J. Zhang, 2013: Sunshine dimming and brightening in Chinese cities (1955-2011) was driven by air pollution rather than clouds. *Climate Research*, **56**, 11-20. http://dx.doi.org/10.3354/cr01139

207. Forster, P., V. Ramaswamy, P. Artaxo, T. Berntsen, R. Betts, D.W. Fahey, J. Haywood, J. Lean, D.C. Lowe, G. Myhre, J. Nganga, R. Prinn, G. Raga, M. Schulz, and R. Van Dorland, 2007: Ch. 2: Changes in atmospheric constituents and in radiative forcing. *Climate Change 2007: The Physical Science Basis. Contribution of Working Group I to the Fourth Assessment Report (AR4) of the Intergovernmental Panel on Climate Change*. Solomon, S., D. Qin, M. Manning, Z. Chen, M. Marquis, K.B. Averyt, M. Tignor, and H.L. Miller, Eds. Cambridge University Press, Cambridge, UK. http://www.ipcc.ch/publications_and_data/ar4/wg1/en/ch2.html

208. Carslaw, K.S., L.A. Lee, C.L. Reddington, K.J. Pringle, A. Rap, P.M. Forster, G.W. Mann, D.V. Spracklen, M.T. Woodhouse, L.A. Regayre, and J.R. Pierce, 2013: Large contribution of natural aerosols to uncertainty in indirect forcing. *Nature*, **503**, 67-71. http://dx.doi.org/10.1038/nature12674

209. Stevens, B. and G. Feingold, 2009: Untangling aerosol effects on clouds and precipitation in a buffered system. *Nature*, **461**, 607-613. http://dx.doi.org/10.1038/nature08281

210. Randall, D.A., R.A. Wood, S. Bony, R. Colman, T. Fichefet, J. Fyfe, V. Kattsov, A. Pitman, J. Shukla, J. Srinivasan, R.J. Stouffer, A. Sumi, and K.E. Taylor, 2007: Ch. 8: Climate models and their evaluation. *Climate Change 2007: The Physical Science Basis. Contribution of Working Group I to the Fourth Assessment Report of the Intergovernmental Panel on Climate Change*. Solomon, S., D. Qin, M. Manning, Z. Chen, M. Marquis, K.B. Averyt, M. Tignor, and H.L. Miller, Eds. Cambridge University Press, Cambridge, United Kingdom and New York, NY, USA, 589-662. www.ipcc.ch/pdf/assessment-report/ar4/wg1/ar4-wg1-chapter8.pdf

211. Steinacher, M., F. Joos, T.L. Frölicher, L. Bopp, P. Cadule, V. Cocco, S.C. Doney, M. Gehlen, K. Lindsay, and J.K. Moore, 2010: Projected 21st century decrease in marine productivity: A multi-model analysis. *Biogeosciences*, **7**, 979-1005. http://dx.doi.org/10.5194/bg-7-979-2010

212. Hauglustaine, D.A., J. Lathière, S. Szopa, and G.A. Folberth, 2005: Future tropospheric ozone simulated with a climate-chemistry-biosphere model. *Geophysical Research Letters*, **32**, L24807. http://dx.doi.org/10.1029/2005GL024031

213. Jiang, X., S.J. Eichelberger, D.L. Hartmann, R. Shia, and Y.L. Yung, 2007: Influence of doubled CO2 on ozone via changes in the Brewer–Dobson circulation. *Journal of the Atmospheric Sciences*, **64**, 2751-2755. http://dx.doi.org/10.1175/jas3969.1

214. Li, F., J. Austin, and J. Wilson, 2008: The strength of the Brewer–Dobson circulation in a changing climate: Coupled chemistry–climate model simulations. *Journal of Climate*, **21**, 40-57. http://dx.doi.org/10.1175/2007jcli1663.1

215. Shepherd, T.G. and C. McLandress, 2011: A robust mechanism for strengthening of the Brewer–Dobson circulation in response to climate change: Critical-layer control of subtropical wave breaking. *Journal of the Atmospheric Sciences*, **68**, 784-797. http://dx.doi.org/10.1175/2010jas3608.1

216. McLandress, C., T.G. Shepherd, M.C. Reader, D.A. Plummer, and K.P. Shine, 2014: The climate impact of past changes in halocarbons and $CO_2$ in the tropical UTLS region. *Journal of Climate*, **27**, 8646-8660. http://dx.doi.org/10.1175/jcli-d-14-00232.1

217. Barkstrom, B.R., 1984: The Earth Radiation Budget Experiment (ERBE). *Bulletin of the American Meteorological Society*, **65**, 1170-1185. http://dx.doi.org/10.1175/1520-0477(1984)065<1170:terbe>2.0.co;2

218. Smith, G.L., B.R. Barkstrom, E.F. Harrison, R.B. Lee, and B.A. Wielicki, 1994: Radiation budget measurements for the eighties and nineties. *Advances in Space Research*, **14**, 81-84. http://dx.doi.org/10.1016/0273-1177(94)90351-4

219. Wielicki, B.A., E.F. Harrison, R.D. Cess, M.D. King, and D.A. Randall, 1995: Mission to planet Earth: Role of clouds and radiation in climate. *Bulletin of the American Meteorological Society*, **76**, 2125-2153. http://dx.doi.org/10.1175/1520-0477(1995)076<2125:mtpero>2.0.co;2

220. Wielicki, B.A., B.R. Barkstrom, E.F. Harrison, R.B. Lee, III, G.L. Smith, and J.E. Cooper, 1996: Clouds and the Earth's Radiant Energy System (CERES): An Earth observing system experiment. *Bulletin of the American Meteorological Society*, **77**, 853-868. http://dx.doi.org/10.1175/1520-0477(1996)077<0853:catere>2.0.co;2

221. Hartmann, D.L., A.M.G. Klein Tank, M. Rusticucci, L.V. Alexander, S. Brönnimann, Y. Charabi, F.J. Dentener, E.J. Dlugokencky, D.R. Easterling, A. Kaplan, B.J. Soden, P.W. Thorne, M. Wild, and P.M. Zhai, 2013: Observations: Atmosphere and surface. *Climate Change 2013: The Physical Science Basis. Contribution of Working Group I to the Fifth Assessment Report of the Intergovernmental Panel on Climate Change*. Stocker, T.F., D. Qin, G.-K. Plattner, M. Tignor, S.K. Allen, J. Boschung, A. Nauels, Y. Xia, V. Bex, and P.M. Midgley, Eds. Cambridge University Press, Cambridge, United Kingdom and New York, NY, USA, 159–254. http://www.climatechange2013.org/report/full-report/

# 3
# Detection and Attribution of Climate Change

**KEY FINDINGS**

1. The *likely* range of the human contribution to the global mean temperature increase over the period 1951–2010 is 1.1° to 1.4°F (0.6° to 0.8°C), and the central estimate of the observed warming of 1.2°F (0.65°C) lies within this range *(high confidence)*. This translates to a *likely* human contribution of 93%–123% of the observed 1951–2010 change. It is *extremely likely* that more than half of the global mean temperature increase since 1951 was caused by human influence on climate *(high confidence)*. The *likely* contributions of natural forcing and internal variability to global temperature change over that period are minor *(high confidence)*.

2. The science of event attribution is rapidly advancing through improved understanding of the mechanisms that produce extreme events and the marked progress in development of methods that are used for event attribution *(high confidence)*.

**Recommended Citation for Chapter**

**Knutson**, T., J.P. Kossin, C. Mears, J. Perlwitz, and M.F. Wehner, 2017: Detection and attribution of climate change. In: *Climate Science Special Report: Fourth National Climate Assessment, Volume I* [Wuebbles, D.J., D.W. Fahey, K.A. Hibbard, D.J. Dokken, B.C. Stewart, and T.K. Maycock (eds.)]. U.S. Global Change Research Program, Washington, DC, USA, pp. 114-132, doi: 10.7930/J01834ND.

## 3.1 Introduction

Detection and attribution of climate change involves assessing the causes of observed changes in the climate system through systematic comparison of climate models and observations using various statistical methods. Detection and attribution studies are important for a number of reasons. For example, such studies can help determine whether a human influence on climate variables (for example, temperature) can be distinguished from natural variability. Detection and attribution studies can help evaluate whether model simulations are consistent with observed trends or other changes in the climate system. Results from detection and attribution studies

can inform decision making on climate policy and adaptation.

There are several general types of detection and attribution studies, including: attribution of trends or long-term changes in climate variables; attribution of changes in extremes; attribution of weather or climate events; attribution of climate-related impacts; and the estimation of climate sensitivity using observational constraints. Paleoclimate proxies can also be useful for detection and attribution studies, particularly to provide a longer-term perspective on climate variability as a baseline on which to compare recent climate changes of the past century or so (for example, see Figure

12.2 from Ch. 12: Sea Level Rise). Detection and attribution studies can be done at various scales, from global to regional.

Since the Intergovernmental Panel on Climate Change (IPCC) Fifth Assessment Report (AR5) chapter on detection and attribution[1] and the Third National Climate Assessment (NCA3[2]), the science of detection and attribution has advanced, with a major scientific question being the issue of attribution of extreme events.[3, 4, 5, 6] Therefore, the methods used in this developing area of the science are briefly reviewed in Appendix C: Detection and Attribution Methods, along with a brief overview of the various general detection and attribution methodologies, including some recent developments in these areas. Detection and attribution of changes in extremes in general presents a number of challenges,[7] including limitations of observations, models, statistical methods, process understanding for extremes, and uncertainties about the natural variability of extremes. Although the present report does not focus on climate impacts on ecosystems or human systems, a relatively new and developing area of detection and attribution science (reviewed in Stone et al. 2013[8]), concerns detecting and attributing the impacts of climate change on natural or human systems. Many new developments in detection and attribution science have been fostered by the International Detection and Attribution Group (IDAG; http://www.image.ucar.edu/idag/ and http://www.clivar.org/clivar-panels/ etccdi/idag/international-detection-attribution-group-idag) which is an international group of scientists who have collaborated since 1995 on "assessing and reducing uncertainties in the estimates of climate change."

In the remainder of this chapter, we review highlights of detection and attribution science, particularly key attribution findings for the rise in global mean temperature. However, as this is a U.S.-focused assessment, the report as a whole will focus more on the detection and attribution findings for particular regional phenomena (for example, regional temperature, precipitation) or at least global-scale phenomena that are directly affecting the United States (for example, sea level rise). Most of these findings are contained in the individual phenomena chapters, rather than in this general overview chapter on detection and attribution. We provide summary links to the chapters where particular detection and attribution findings are presented in more detail.

## 3.2 Detection and Attribution of Global Temperature Changes

The concept of detection and attribution is illustrated in Figure 3.1, which shows a very simple example of detection and attribution of global mean temperature. While more powerful pattern-based detection and attribution methods (discussed later), and even greater use of time averaging, can result in much stronger statements about detection and attribution, the example in Figure 3.1 serves to illustrate the general concept. In the figure, observed global mean temperature anomalies (relative to a 1901–1960 baseline) are compared with anomalies from historical simulations of CMIP5 models. The spread of different individual model simulations (the blue and orange shading) arises both from differences between the models in their responses to the different specified climate forcing agents (natural and anthropogenic) and from internal (unforced) climate variability. Observed annual temperatures after about 1980 are shown to be inconsistent with models that include only natural forcings (blue shading) and are consistent with the model simulations that include both anthropogenic and natural forcing (orange shading). This implies that the observed global warming is attributable in large part to anthropogenic forcing. A key aspect of a detection and attribution finding will be the

## Global Mean Temperature Change

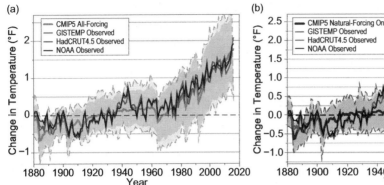

**Figure 3.1:** Comparison of observed global mean temperature anomalies from three observational datasets to CMIP5 climate model historical experiments using: (a) anthropogenic and natural forcings combined, or (b) natural forcings only. In (a) the thick orange curve is the CMIP5 grand ensemble mean across 36 models while the orange shading and outer dashed lines depict the ±2 standard deviation and absolute ranges of annual anomalies across all individual sim-ulations of the 36 models. Model data are a masked blend of surface air temperature over land regions and sea surface temperature over ice-free ocean regions to be more consistent with observations than using surface air temperature alone. All time series (°F) are referenced to a 1901–1960 baseline value. The simulations in (a) have been extended from 2006 through 2016 using projections under the higher scenario (RCP8.5). (b) As in (a) but the blue curves and shading are based on 18 CMIP5 models using natural forcings only. See legends to identify observational datasets. Observations after about 1980 are shown to be inconsistent with the natural forcing-only models (indicating detectable warming) and also consistent with the models that include both anthropogenic and natural forcing, implying that the warming is attributable in part to anthropogenic forcing according to the models. (Figure source: adapted from Melillo et al.[2] and Knutson et al.[19]).

assessment of the adequacy of the models and observations used for these conclusions, as discussed and assessed in Flato et al.,[9] Bindoff et al.,[1] and IPCC.[10]

The detection and attribution of global tem-perature change to human causes has been one of the most important and visible find-ings over the course of the past global climate change scientific assessments by the IPCC. The first IPCC report[11] concluded that a human in-fluence on climate had not yet been detected, but judged that "the unequivocal detection of the enhanced greenhouse effect from obser-vations is not likely for a decade or more." The second IPCC report[12] concluded that "the balance of evidence suggests a discernible human influence on climate." The third IPCC report[13] strengthened this conclusion to: "most of the observed warming over the last 50 years is likely to have been due to the increase of greenhouse gas concentrations." The fourth

IPCC report[14] further strengthened the con-clusion to: "Most of the observed increase in global average temperatures since the mid-20th century is very likely due to the observed increase in anthropogenic greenhouse gas con-centrations." The fifth IPCC report[10] further strengthened this to: "It is extremely likely that more than half of the observed increase in global average surface temperature from 1951 to 2010 was caused by the anthropogenic increase in greenhouse gas concentrations and other anthropogenic forcings together." These increasingly confident statements have result-ed from scientific advances, including better observational datasets, improved models and detection/attribution methods, and improved estimates of climate forcings. Importantly, the continued long-term warming of the glob-al climate system since the time of the first IPCC report and the broad-scale agreement of the spatial pattern of observed temperature changes with climate model projections of greenhouse gas-induced changes as published in the late 1980s (e.g., Stouffer and Manabe 2017[15]) give more confidence in the attribution of observed warming since 1951 as being due primarily to human activity.

The IPCC AR5 presented an updated assessment of detection and attribution research at the global to regional scale[1] which is briefly summarized here. Key attribution assessment results from IPCC AR5 for global mean temperature are summarized in Figure 3.2, which shows assessed *likely* ranges and midpoint estimates for several factors contributing to increases in global mean temperature. According to Bindoff et al.,[1] the *likely* range of the anthropogenic contribution to global mean temperature increases over 1951–2010 was 0.6° to 0.8°C (1.1° to 1.4°F), compared with the

observed warming 5th to 95th percentile range of 0.59° to 0.71°C (1.1° to 1.3°F). The estimated *likely* contribution ranges for natural forcing and internal variability were both much smaller (−0.1° to 0.1°C, or −0.2° to 0.2°F) than the observed warming. The confidence intervals that encompass the *extremely likely* range for the anthropogenic contribution are wider than the *likely* range. Using these wider confidence limits, the lower limit of attributable warming contribution range still lies above 50% of the observed warming rate, and thus Bindoff et al.[1] concluded that it is *extremely likely* that more than half of the global mean temperature increase since 1951 was caused by human influence on climate. This assessment concurs with the Bindoff et al.[1] assessment of attributable warming and cooling influences.

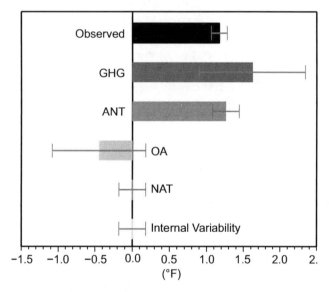

HG - well-mixed greenhouse gases  OA - other anthropogenic forcings

NT - all anthropogenic forcings combined  NAT - natural forcings

**Figure 3.2:** Observed global mean temperature trend (black bar) and attributable warming or cooling influences of anthropogenic and natural forcings over 1951–2010. Observations are from HadCRUT4, along with observational uncertainty (5% to 95%) error bars.[62] Likely ranges (bar-whisker plots) and midpoint values (colored bars) for attributable forcings are from IPCC AR5.[1]. GHG refers to well-mixed greenhouse gases, OA to other anthropogenic forcings, NAT to natural forcings, and ANT to all anthropogenic forcings combined. Likely ranges are broader for contributions from well-mixed greenhouse gases and for other anthropogenic forcings, assessed separately, than for the contributions from all anthropogenic forcings combined, as it is more difficult to quantitatively constrain the separate contributions of the various anthropogenic forcing agents. (Figure source: redrawn from Bindoff et al.;[1] © IPCC. Used with permission.)

Apart from formal detection attribution studies such as those underlying the results above, which use global climate model output and pattern-based regression methods, anthropogenic influences on global mean temperature can also be estimated using simpler empirical models, such as multiple linear regression/energy balance models (e.g., Canty et al. 2013[16]; Zhou and Tung 2013[17]). For example, Figure 3.3 illustrates how the global mean surface temperature changes since the late 1800s can be decomposed into components linearly related to several forcing variables (anthropogenic forcing, solar variability, volcanic forcing, plus an internal variability component, here related to El Niño–Southern Oscillation). Using this approach, Canty et al.[16] also infer a substantial contribution of anthropogenic forcing to the rise in global mean temperature since the late 1800s. Stern and Kaufmann[18] use another method—Granger causality tests—and again infer that "human activity is partially responsible for the observed rise in global temperature and that this rise in temperature also has an effect on the global carbon cycle." They also conclude that anthropogenic sulfate aerosol effects may only be about half as large as inferred in a number of previous studies.

Multi-century to multi-millennial-scale climate model integrations with unchanging external forcing provide a means of estimating potential contributions of internal climate variability to observed trends. Bindoff et al.[1] conclude, based on multimodel assessments, that the likely range contribution of internal variability to observed trends over 1951–2010 is about ±0.2°F, compared to the observed warming of about 1.2°F over that period. A recent 5,200-year integration of the CMIP5 model having apparently the largest global mean temperature variability among CMIP5 models shows rare instances of multidecadal global warming approaching the observed 1951–2010 warming trend.[19] However, even that most

extreme model cannot simulate century-scale warming trends from internal variability that approach the observed global mean warming over the past century. According to a multimodel analysis of observed versus CMIP5 modeled global temperature trends (Knutson et al. 2013[20], Fig. 7a), the modeled natural fluctuations (forced plus internal) would need to be larger by about a factor of three for even an unusual natural variability episode (95th percentile) to approach the observed trend since 1900. Thus, using present models there is no known source of internal climate variability that can reproduce the observed warming over the past century without including strong positive forcing from anthropogenic greenhouse gas emissions (Figure 3.1). The modeled century-scale trend due to natural forcings (solar and volcanic) is also minor (Figure 3.1), so that, using present models, there is no known source of natural variability that can reproduce the observed global warming over the past century. One study[21] comparing paleoclimate data with models concluded that current climate models may substantially underestimate regional sea surface temperature variability on multidecadal to multi-centennial timescales, especially at low latitudes. The causes of this apparent discrepancy--whether due to data issues, external forcings/response, or simulated internal variability issues--and its implications for simulations of global temperature variability in climate models remain unresolved. Since Laepple and Huybers[21] is a single paleoclimate-based study and focuses on regional, not global mean, temperature variability, we have consequently not modified our conclusions regarding global temperature attribution from those contained in Bindoff et al.,[1] although further research on this issue is warranted. In summary, we are not aware of any convincing evidence that natural variability alone could have accounted for the amount and timing of global warming that was observed over the industrial era.

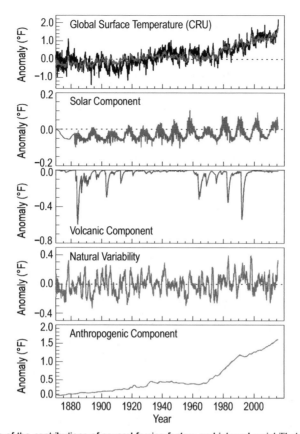

**Figure 3.3:** Estimates of the contributions of several forcing factors and internal variability to global mean temperature change since 1870, based on an empirical approach using multiple linear regression and energy balance models. The top panel shows global temperature anomalies (°F) from the observations[62] in black with the multiple linear regression result in red (1901–1960 base period). The lower four panels show the estimated contribution to global mean temperature anomalies from four factors: solar variability; volcanic eruptions; internal variability related to El Niño/Southern Oscillation; and anthropogenic forcing. The anthropogenic contribution includes a warming component from greenhouse gases concentrations and a cooling component from anthropogenic aerosols. (Figure source: adapted from Canty et al.[16]).

While most detection and attribution studies focus on changes in temperature and other variables in the historical record since about 1860 or later, some studies relevant to detection and attribution focus on changes over much longer periods. For example, geological and tide-based reconstructions of global mean sea level (Ch. 12: Sea Level Rise, Figure 12.2b) suggest that the rate of sea level rise in the last century was faster than during any century over the past ~2,800 years. As an example, for Northern Hemisphere annual mean temperatures, Schurer et al.[22] use detection and attribution fingerprinting methods along with paleoclimate reconstructions and millennial-scale climate model simulations from eight models to explore causes for temperature variations from 850 AD to the present, including the Medieval Climate Anomaly (MCA, around 900 to 1200 AD) and the Little Ice Age (LIA, around 1450 to 1800 AD). They conclude that solar variability and volcanic eruptions were the main causal factors for changes in Northern Hemisphere temperatures from 1400 to 1900, but that greenhouse gas changes of uncertain origin apparently contributed to the cool conditions during 1600–1800. Their study provides further support for previous IPCC report conclusions (e.g., IPCC 2007[14]) that internal variability alone was extremely unlikely to have been the cause of the recent observed 50- and 100-year warming trends. Andres and Peltier[23] also inferred from millennial-scale climate model simulations that volcanoes, solar variability, greenhouse gases, and orbital variations all contributed significantly to the transition from the MCA to the LIA.

An active and important area of climate research that involves detection and attribution science is the estimation of global climate sensitivity, based on past observational constraints. An important measure of climate sensitivity, with particular relevance for climate projections over the coming decades, is the transient climate response (TCR), defined as the rise in global mean surface temperature at the time of $CO_2$ doubling for a 1% per year transient increase of atmospheric $CO_2$. (Equilibrium climate sensitivity is discussed in Ch. 2: Physical Drivers of Climate Change). The TCR of the climate system has an estimated range of 0.9° to 2.0°C (1.6° to 3.6°F) and 0.9° to 2.5°C (1.6° to 4.5°F), according to two recent assessments (Otto et al.[24] and Lewis and Curry[25], respectively). Marvel et al.[26] suggest, based on experiments with a single climate model, that after accounting for the different efficacies of various historical climate forcing agents, the TCR could be adjusted upward from the Otto et al.[24] and Lewis and Curry[25] estimates. Richardson et al.[27] report a best estimate for TCR of 1.66°C (2.99 °F), with a 5% to 95% confidence range of 1.0° to 3.3°C (1.8° to 5.9°F). Furthermore, Richardson et al. conclude that the earlier studies noted above may underestimate TCR because the surface temperature dataset they used undersamples rapidly warming regions due to limited coverage and because surface water warms less than surface air. Gregory et al.[28] note, within CMIP5 models, that the TCR to the second doubling of $CO_2$ (that is, from doubling to quadrupling)

is 40% higher than that for the first doubling. They explore the various physical reasons for this finding and conclude this may also lead to an underestimate of TCR in the empirical observation-based studies. In summary, estimation of TCR from observations continues to be an active area of research with considerable remaining uncertainties, as discussed above. Even the low-end estimates for TCR cited above from some recent studies (about 0.9°C or 1.6°F) imply that the climate will continue to warm substantially if atmospheric $CO_2$ concentrations continue to increase over the coming century as projected under a number of future scenarios.

## 3.3 Detection and Attribution with a United States Regional Focus

Detection and attribution at regional scales is generally more challenging than at the global scale for a number of reasons. At the regional scale, the magnitude of natural variability swings are typically larger than for global means. If the climate change signal is similar in magnitude at the regional and global scales, this makes it more difficult to detect anthropogenic climate changes at the regional scale. Furthermore, there is less spatial pattern information at the regional scale that can be used to distinguish contributions from various forcings. Other forcings that have typically received less attention than greenhouse gases, such as land-use change, could be more important at regional scales than globally.[29] Also, simulated internal variability at regional scales may be less reliable than at global scales (Bindoff et al.[1]). While detection and attribution of changes in extremes (including at the regional scale) presents a number of key challenges,[7] previous studies (e.g., Zwiers et al. 2011[30]) have demonstrated how detection and attribution methods, combined with generalized extreme value distributions, can be used to detect a human influence on extreme temperatures at the regional scale, including over North America.

In IPCC AR5,[1] which had a broader global focus than this report, attributable human contributions were reported for warming over all continents except Antarctica. Changes in daily temperature extremes throughout the world; ocean surface and subsurface temperature and salinity sea level pressure patterns; arctic sea ice loss; northern hemispheric snow cover decrease; global mean sea level rise; and ocean acidification were all associated with human activity in AR5.[1] IPCC AR5 also reported medium confidence in anthropogenic contributions to increased atmospheric specific humidity, zonal mean precipitation over Northern Hemisphere mid to high latitudes, and intensification of heavy precipitation over land regions. IPCC AR5 had weaker attribution conclusions than IPCC AR4 on some phenomena, including tropical cyclone and drought changes.

Although the present assessment follows most of the IPCC AR5 conclusions on detection and attribution of relevance to the United States, we make some additional attribution assessment statements in the relevant chapters of this report. Among the notable detection and attribution-relevant findings in this report are the following (refer to the listed chapters for further details):

- Ch. 5: Circulation and Variability: The tropics have expanded poleward by about 70 to 200 miles in each hemisphere over the period 1979–2009, with an accompanying shift of the subtropical dry zones, midlatitude jets, and storm tracks (*medium to high confidence*). Human activities have played a role in this change (*medium confidence*), although confidence is presently *low* regarding the magnitude of the human contribution relative to natural variability.

- Ch. 6: Temperature Change: Detectable anthropogenic warming since 1901 has

occurred over the western and northern regions of the contiguous United States according to observations and CMIP5 models (*medium confidence*), although over the southeastern United States there has been no detectable warming trend since 1901. The combined influence of natural and anthropogenic forcings on temperature *extremes* have been detected over large subregions of North America (*medium confidence*).

- Ch. 7: Precipitation Change: For the continental United States, there is *high confidence* in the detection of extreme precipitation increases, while there is *low confidence* in attributing the extreme precipitation changes purely to anthropogenic forcing. There is stronger evidence for a human contribution (*medium confidence*) when taking into account process-based understanding (for example, increased water vapor in a warmer atmosphere).

- Ch. 8: Drought, Floods, and Wildfire: While by some measures drought has decreased over much of the continental United States in association with long-term increases in precipitation, neither the precipitation increases nor inferred drought decreases have been confidently attributed to anthropogenic forcing. Detectable changes—a mix of increases and decreases—in some classes of flood frequency have occurred in parts of the United States, although attribution studies have not established a robust connection between increased riverine flooding and human-induced climate change. There is *medium confidence* for a human-caused climate change contribution to increased forest fire activity in Alaska in recent decades and *low* to *medium confidence* for a detectable human climate change contribution in the western United States.

- Ch. 9: Extreme Storms: There is broad agreement in the literature that human factors (greenhouse gases and aerosols) have had a measurable impact on the observed oceanic and atmospheric variability in the North Atlantic, and there is *medium confidence* that this has contributed to the observed increase in Atlantic hurricane activity since the 1970s. There is no consensus on the relative magnitude of human and natural influences on past changes in hurricane activity.

- Ch. 10: Land Cover: Modifications to land use and land cover due to human activities produce changes in surface albedo, latent and sensible heat, and atmospheric aerosol and greenhouse gas concentrations, accounting for an estimated $40\% \pm 16\%$ of the human-caused global radiative forcing from 1850 to 2010 *(high confidence)*.

- Ch. 11: Arctic Changes: It is *very likely* that human activities have contributed to observed arctic surface temperature warming, sea ice loss, glacier mass loss, and Northern Hemisphere snow extent decline *(high confidence)*.

- Ch. 12: Sea Level Rise: Human-caused climate change has made a substantial contribution to global mean sea level rise since 1900 *(high confidence)*, contributing to a rate of rise that is greater than during any preceding century in at least 2,800 years *(medium confidence)*.

- Ch. 13: Ocean Changes: The world's oceans have absorbed about 93% of the excess heat caused by greenhouse warming since the mid-20th Century. The world's oceans are currently absorbing more than a quarter of the carbon dioxide emitted to the atmosphere annually from human activities, making them more acidic *(very high confidence)*.

### 3.4 Extreme Event Attribution

Since the IPCC AR5 and NCA3,[2] the attribution of extreme weather and climate events has been an emerging area in the science of detection and attribution. Attribution of extreme weather events under a changing climate is now an important and highly visible aspect of climate science. As discussed in the recent National Academy of Sciences report,[5] the science of event attribution is rapidly advancing, including the understanding of the mechanisms that produce extreme events and the rapid progress in development of methods used for event attribution.

When an extreme weather event occurs, the question is often asked: was this event caused by climate change? A generally more appropriate framing for the question is whether climate change has altered the odds of occurrence of an extreme event like the one just experienced. Extreme event attribution studies to date have generally been concerned with answering the latter question. In recent developments, Hannart et al.[31] discuss the application of causal theory to event attribution, including discussion of conditions under which stronger causal statements can be made, in principle, based on theory of causality and distinctions between necessary and sufficient causality.

Several recent studies, including NAS,[5] have reviewed aspects of extreme event attribution.[3, 4, 6] Hulme[4] and NAS[5] discuss the motivations for scientists to be pursuing extreme event attribution, including the need to inform risk management and adaptation planning. Hulme[4] categorizes event attribution studies/statements into general types, including those based on: physical reasoning, statistical analysis of time series, fraction of attributable risk (FAR) estimation (discussed in the Appendix), or those that rely on the philosophical argument that there are no longer any purely natural weather events. The NAS[5] report outlines two general approaches to event attribution: 1) using observations to estimate a change in probability of magnitude of events, or 2) using model simulations to compare an event in the current climate versus that in a hypothetical "counterfactual" climate not influenced by human activities. As discussed by Trenberth et al.,[32] Shepherd,[33] and Horton et al.,[34] an ingredients-based or conditional attribution approach can also be used, when one examines the impact of certain environmental changes (for example, greater atmospheric moisture) on the character of an extreme event using model experiments, all else being equal. Further discussion of methodologies is given in Appendix C.

Examples of extreme event attribution studies are numerous. Many are cited by Hulme,[4] NAS,[5] Easterling et al.,[3] and there are many further examples in an annual collection of studies of extreme events of the previous year, published in the *Bulletin of the American Meteorological Society*.[35, 36, 37, 38, 39]

While an extensive review of extreme event attribution is beyond the scope of this report, particularly given the recent publication of several assessments or review papers on the topic, some general findings from the more comprehensive NAS[5] report are summarized here:

- Confidence in attribution findings of anthropogenic influence is greatest for extreme events that are related to an aspect of temperature, followed by hydrological drought and heavy precipitation, with little or no confidence for severe convective storms or extratropical storms.

- Event attribution is more reliable when based on sound physical principles, consistent evidence from observations, and

numerical models that can replicate the event.

- Statements about attribution are sensitive to the way the questions are posed (that is, framing).

- Assumptions used in studies must be clearly stated and uncertainties estimated in order for a clear, unambiguous interpretation of an event attribution to be possible.

The NAS report noted that uncertainties about the roles of low-frequency natural variability and confounding factors (for example, the effects of dams on flooding) could be sources of difficulties in event attribution studies. In addition, the report noted that attribution conclusions would be more robust in cases where observed changes in the event being examined are consistent with expectations from model-based attribution studies. The report endorsed the need for more research to improve understanding of a number of important aspects of event attribution studies, including physical processes, models and their capabilities, natural variability, reliable long-term observational records, statistical methods, confounding factors, and future projections of the phenomena of interest.

As discussed in Appendix C: Detection and Attribution Methodologies, confidence is typically lower for an attribution-without-detection statement than for an attribution statement accompanied by an established, detectable anthropogenic influence (for example, a detectable and attributable long-term trend or increase in variability) for the phenomenon itself. An example of the former would be stating that a change in the probability or magnitude of a heat wave in the southeastern United States was attributable to rising greenhouse gases, because there has not been a detectable century-scale trend in either temperature or temperature variability in this region (e.g., Ch. 6: Temperature Change; Knutson et al. 2013[20]).

To our knowledge, no extreme weather event observed to date has been found to have zero probability of occurrence in a preindustrial climate, according to climate model simulations. Therefore, the causes of attributed extreme events are a combination of natural variations in the climate system compounded (or alleviated) by the anthropogenic change to the climate system. Event attribution statements quantify the relative contribution of these human and natural causal factors. In the future, as the climate change signal gets stronger compared to natural variability, humans may experience weather events which are essentially impossible to simulate in a preindustrial climate. This is already becoming the case at large time and spatial scales, where for example the record global mean surface temperature anomaly observed in 2016 (relative to a 1901–1960 baseline) is essentially impossible for global climate models to reproduce under preindustrial climate forcing conditions (for example, see Figure 3.1).

The European heat wave of 2003[40] and Australia's extreme temperatures and heat indices of 2013 (e.g., Arblaster et al. 2014[41]; King et al. 2014[42]; Knutson et al. 2014[43]; Lewis and Karoly 2014[44]; Perkins et al. 2014[45]) are examples of extreme weather or climate events where relatively strong evidence for a human contribution to the event has been found. Similarly, in the United States, the science of event attribution for weather and climate extreme events has been actively pursued since the NCA3. For example, for the case of the recent California drought, investigators have attempted to determine, using various methods discussed in this chapter, whether human-caused climate change contributed to the event (see discussion in Ch. 8: Droughts, Floods, and Wildfires).

As an example, illustrating different methods of attribution for an event in the United States, Hoerling et al.[46] concluded that the 2011 Texas heat wave/meteorological drought was primarily caused by antecedent and concurrent negative rainfall anomalies due mainly to natural variability and the La Niña conditions at the time of the event, but with a relatively small (not detected) warming contribution from anthropogenic forcing. The anthropogenic contribution nonetheless doubled the chances of reaching a new temperature record in 2011 compared to the 1981–2010 reference period, according to their study. Rupp et al.,[47] meanwhile, concluded that extreme heat events in Texas were about 20 times more likely for 2008 La Niña conditions than similar conditions during the 1960s. This pair of studies illustrates how the framing of the attribution question can matter. For example, the studies used different baseline reference periods to determine the magnitude of anomalies, which can also affect quantitative conclusions, since using an earlier baseline period typically results in larger magnitude anomalies (in a generally warming climate). The Hoerling et al. analysis focused on both what caused most of the magnitude of the anomalies as well as changes in probability of the event, whereas Rupp et al. focused on the changes in the probability of the event. Otto et al.[48] showed for the case of the Russian heat wave of 2010 how a different focus of attribution (fraction of anomaly explained vs. change in probability of occurrence over a threshold) can give seemingly conflicting results, yet have no real fundamental contradiction. In the illustrative case for the 2011 Texas heat/drought, we conclude that there is *medium* confidence that anthropogenic forcing contributed to the heat wave, both in terms of a small contribution to the anomaly magnitude and a significant increase in the probability of occurrence of the event.

In this report, we do not assess or compile all individual weather or climate extreme events for which an attributable anthropogenic climate change has been claimed in a published study, as there are now many such studies that provide this information. Some event attribution-related studies that focus on the United States are discussed in more detail in Chapters 6–9, which primarily examine phenomena such as precipitation extremes, droughts, floods, severe storms, and temperature extremes. For example, as discussed in Chapter 6: Temperature Change (Table 6.3), a number of extreme temperature events (warm anomalies) in the United States have been partly attributed to anthropogenic influence on climate.

## Traceable Accounts

### Key Finding 1

The *likely* range of the human contribution to the global mean temperature increase over the period 1951–2010 is 1.1° to 1.4°F (0.6° to 0.8°C), and the central estimate of the observed warming of 1.2°F (0.65°C) lies within this range (*high confidence*). This translates to a *likely* human contribution of 93%–123% of the observed 1951-2010 change. It is *extremely likely* that more than half of the global mean temperature increase since 1951 was caused by human influence on climate (*high confidence*). The *likely* contributions of natural forcing and internal variability to global temperature change over that period are minor (*high confidence*).

### Description of evidence base

This Key Finding summarizes key detection and attribution evidence documented in the climate science literature and in the IPCC AR5,[1] and references therein. The Key Finding is essentially the same as the summary assessment of IPCC AR5.

According to Bindoff et al.,[1] the *likely* range of the anthropogenic contribution to global mean temperature increases over 1951–2010 was 1.1° to 1.4°F (0.6° to 0.8°C, compared with the observed warming 5th to 95th percentile range of 1.1° to 1.3°F (0.59° to 0.71°C). The estimated likely contribution ranges for natural forcing and internal variability were both much smaller (–0.2° to 0.2°F, or –0.1° to 0.1°C) than the observed warming. The confidence intervals that encompass the *extremely likely* range for the anthropogenic contribution are wider than the *likely* range, but nonetheless allow for the conclusion that it is *extremely likely* that more than half of the global mean temperature increase since 1951 was caused by human influence on climate (*high confidence*).

The attribution of temperature increases since 1951 is based largely on the detection and attribution analyses of Gillett et al.,[49] Jones et al.,[50] and consideration of Ribes and Terray,[51] Huber and Knutti,[52] Wigley and Santer,[53] and IPCC AR4.[54] The IPCC finding receives further support from alternative approaches, such as multiple linear regression/energy balance modeling[16] and a new methodological approach to detection and attribution that uses additive decomposition and hypothesis testing,[55] which infer similar attributable warming results. Individual study results used to derive the IPCC finding are summarized in Figure 10.4 of Bindoff et al.,[1] which also assesses model dependence by comparing results obtained from several individual CMIP5 models. The estimated potential influence of internal variability is based on Knutson et al.[20] and Huber and Knutti,[52] with consideration of the above references. Moreover, simulated global temperature multidecadal variability is assessed to be adequate,[1] with *high confidence* that models reproduce global and Northern Hemisphere temperature variability across a range of timescales.[9] Further support for these assessments comes from assessments of paleoclimate data[56] and increased confidence in physical understanding and models of the climate system.[10, 15] A more detailed traceable account is contained in Bindoff et al.[1] Post-IPCC AR5 supporting evidence includes additional analyses showing the unusual nature of observed global warming since the late 1800s compared to simulated internal climate variability,[19] and the recent occurrence of new record high global mean temperatures are consistent with model projections of continued warming on multidecadal scales (for example, Figure 3.1).

### Major uncertainties

As discussed in the main text, estimation of the transient climate response (TCR), defined as the global mean surface temperature change at the time of $CO_2$ doubling in a 1% per year $CO_2$ transient increase experiment, continues to be an active area of research with considerable remaining uncertainties. Some detection attribution methods use model-based methods together with observations to attempt to infer scaling magnitudes of the forced responses based on regression methods (that is, they do not use the models' climate sensitivities directly). However, if climate models are significantly more sensitive to $CO_2$ increases than the real world, as suggested by the studies of Otto et al.[24] and Lewis and Curry[25] (though see differing conclusions from other studies in the main text), this could lead to an overestimate of attributable warming estimates, at least as obtained using some detection and attribution methods. In any case, it is important to better constrain the TCR to have higher confidence in general in attributable warming estimates obtained using various methods.

The global temperature change since 1951 attributable to anthropogenic forcings other than greenhouse gases has a wide estimated *likely* range (−1.1° to +0.2°F in Fig. 3.1). This wide range is largely due to the considerable uncertainty of estimated total radiative forcing due to aerosols (i.e., the direct effect combined with the effects of aerosols on clouds[57]). Although more of the relevant physical processes are being included in models, confidence in these model representations remains low.[58] In detection/attribution studies there are substantial technical challenges in quantifying the separate attributable contributions to temperature change from greenhouse gases and aerosols.[1] Finally, there is a range of estimates of the potential contributions of internal climate variability, and some sources of uncertainty around modeled estimates (e.g., Laepple and Huybers 2014[21]). However, current CMIP5 multimodel estimates (*likely* range of ±0.2°F, or 0.1°C, over 60 years) would have to increase by a factor of about three for even half of the observed 60-year trend to lie within a revised *likely* range of potential internal variability (e.g., Knutson et al. 2013;[20] Huber and Knutti 2012[52]). Recently, Knutson et al.[19] examined a 5,000-year integration of the CMIP5 model having the strongest internal multidecadal variability among 25 CMIP5 models they examined. While the internal variability within this strongly varying model can on rare occasions produce 60-year warmings approaching that observed from 1951–2010, even this most extreme model did not produce any examples of centennial-scale internal variability warming that could match the observed global warming since the late 1800s, even in a 5,000-year integration.

### Assessment of confidence based on evidence and agreement, including short description of nature of evidence and level of agreement

There is *very high confidence* that global temperature has been increasing and that anthropogenic forcings have played a major role in the increase observed over

the past 60 years, with strong evidence from several studies using well-established detection and attribution techniques. There is *high confidence* that the role of internal variability is minor, as the CMIP5 climate models as a group simulate only a minor role for internal variability over the past 60 years, and the models have been assessed by IPCC AR5 as adequate for the purpose of estimating the potential role of internal variability.

### If appropriate, estimate likelihood of impact or consequence, including short description of basis of estimate

The amount of historical warming attributable to anthropogenic forcing has a very high likelihood of consequence, as it is related to the amount of future warming to be expected under various emission scenarios, and the impacts of global warming are generally larger for higher warming rates and higher warming amounts.

### Summary sentence or paragraph that integrates the above information

Detection and attribution studies, climate models, observations, paleoclimate data, and physical understanding lead to *high confidence* (*extremely likely*) that more than half of the observed global mean warming since 1951 was caused by humans, and *high confidence* that internal climate variability played only a minor role (and possibly even a negative contribution) in the observed warming since 1951. The key message and supporting text summarizes extensive evidence documented in the peer-reviewed detection and attribution literature, including in the IPCC AR5.

### Key Finding 2

The science of event attribution is rapidly advancing through improved understanding of the mechanisms that produce extreme events and the marked progress in development of methods that are used for event attribution (*high confidence*).

### Description of evidence base

This Key Finding paraphrases a conclusion of the National Academy of Sciences report[5] on attribution of extreme weather events in the context of climate change. That report discusses advancements in event attribution in more detail than possible here due to space limitations. Weather and climate science in general continue to seek improved physical understanding of extreme weather events. One aspect of improved understanding is the ability to more realistically simulate extreme weather events in models, as the models embody current physical understanding in a simulation framework that can be tested on sample cases. NAS[5] provides references to studies that evaluate weather and climate models used to simulated extreme events in a climate context. Such models can include coupled climate models (e.g., Taylor et al. 2012;[59] Flato et al. 2013[9]), atmospheric models with specified sea surface temperatures, regional models for dynamical downscaling, weather forecasting models, or statistical downscaling models. Appendix C includes a brief description of the evolving set of methods used for event attribution, discussed in more detail in references such as NAS,[5] Hulme,[4] Trenberth et al.,[32] Shepherd,[33] Horton et al.,[34] Hannart,[60] and Hannart et al.[31, 61] Most of this methodology as applied to extreme weather and climate event attribution, has evolved since the European heat wave study of Stott et al.[40]

### Major uncertainties

While the science of event attribution is rapidly advancing, studies of individual events will typically contain caveats. In some cases, attribution statements are made without a clear detection of an anthropogenic influence on observed occurrences of events similar to the one in question, so that there is reliance on models to assess probabilities of occurrence. In such cases there will typically be uncertainties in the model-based estimations of the anthropogenic influence, in the estimation of the influence of natural variability on the event's occurrence, and even in the observational records related to the event (e.g., long-term records of hurricane occurrence). Despite these uncertainties in individual attribution studies, the science of event attribution is advancing through increased physical understanding and development of new methods of attribution and evaluation of models.

### Assessment of confidence based on evidence and agreement, including short description of nature of evidence and level of agreement

There is *very high confidence* that weather and climate science are advancing in their understanding of the physical mechanisms that produce extreme events. For example, hurricane track forecasts have improved in part due to improved models. There is *high confidence* that new methods being developed will help lead to further advances in the science of event attribution.

### If appropriate, estimate likelihood of impact or consequence, including short description of basis of estimate

Improving science of event attribution has a high likelihood of impact, as it is one means by which scientists can better understand the relationship between occurrence of extreme events and long-term climate change. A further impact will be the improved ability to communicate this information to the public and to policymakers for various uses, including improved adaptation planning.[4, 5]

### Summary sentence or paragraph that integrates the above information

Owing to the improved physical understanding of extreme weather and climate events as the science in these fields progress, and owing to the high promise of newly developed methods for exploring the roles of different influences on occurrence of extreme events, there is *high confidence* that the science of event attribution is rapidly advancing.

# References

1. Bindoff, N.L., P.A. Stott, K.M. AchutaRao, M.R. Allen, N. Gillett, D. Gutzler, K. Hansingo, G. Hegerl, Y. Hu, S. Jain, I.I. Mokhov, J. Overland, J. Perlwitz, R. Sebbari, and X. Zhang, 2013: Detection and attribution of climate change: From global to regional. *Climate Change 2013: The Physical Science Basis. Contribution of Working Group I to the Fifth Assessment Report of the Intergovernmental Panel on Climate Change.* Stocker, T.F., D. Qin, G.-K. Plattner, M. Tignor, S.K. Allen, J. Boschung, A. Nauels, Y. Xia, V. Bex, and P.M. Midgley, Eds. Cambridge University Press, Cambridge, United Kingdom and New York, NY, USA, 867–952. http://www.climatechange2013.org/report/full-report/

2. Melillo, J.M., T.C. Richmond, and G.W. Yohe, eds., 2014: *Climate Change Impacts in the United States: The Third National Climate Assessment.* U.S. Global Change Research Program: Washington, D.C., 841 pp. http://dx.doi.org/10.7930/J0Z31WJ2

3. Easterling, D.R., K.E. Kunkel, M.F. Wehner, and L. Sun, 2016: Detection and attribution of climate extremes in the observed record. *Weather and Climate Extremes,* **11,** 17-27. http://dx.doi.org/10.1016/j.wace.2016.01.001

4. Hulme, M., 2014: Attributing weather extremes to 'climate change'. *Progress in Physical Geography,* **38,** 499-511. http://dx.doi.org/10.1177/0309133314538644

5. NAS, 2016: *Attribution of Extreme Weather Events in the Context of Climate Change.* The National Academies Press, Washington, DC, 186 pp. http://dx.doi.org/10.17226/21852

6. Stott, P., 2016: How climate change affects extreme weather events. *Science,* **352,** 1517-1518. http://dx.doi.org/10.1126/science.aaf7271

7. Zwiers, F.W., L.V. Alexander, G.C. Hegerl, T.R. Knutson, J.P. Kossin, P. Naveau, N. Nicholls, C. Schär, S.I. Seneviratne, and X. Zhang, 2013: Climate extremes: Challenges in estimating and understanding recent changes in the frequency and intensity of extreme climate and weather events. *Climate Science for Serving Society: Research, Modeling and Prediction Priorities.* Asrar, G.R. and J.W. Hurrell, Eds. Springer Netherlands, Dordrecht, 339-389. http://dx.doi.org/10.1007/978-94-007-6692-1_13

8. Stone, D., M. Auffhammer, M. Carey, G. Hansen, C. Huggel, W. Cramer, D. Lobell, U. Molau, A. Solow, L. Tibig, and G. Yohe, 2013: The challenge to detect and attribute effects of climate change on human and natural systems. *Climatic Change,* **121,** 381-395. http://dx.doi.org/10.1007/s10584-013-0873-6

9. Flato, G., J. Marotzke, B. Abiodun, P. Braconnot, S.C. Chou, W. Collins, P. Cox, F. Driouech, S. Emori, V. Eyring, C. Forest, P. Gleckler, E. Guilyardi, C. Jakob, V. Kattsov, C. Reason, and M. Rummukainen, 2013: Evaluation of climate models. *Climate Change 2013: The Physical Science Basis. Contribution of Working Group I to the Fifth Assessment Report of the Intergovernmental Panel on Climate Change.* Stocker, T.F., D. Qin, G.-K. Plattner, M. Tignor, S.K. Allen, J. Boschung, A. Nauels, Y. Xia, V. Bex, and P.M. Midgley, Eds. Cambridge University Press, Cambridge, United Kingdom and New York, NY, USA, 741–866. http://www.climatechange2013.org/report/full-report/

10. IPCC, 2013: *Climate Change 2013: The Physical Science Basis. Contribution of Working Group I to the Fifth Assessment Report of the Intergovernmental Panel on Climate Change.* Cambridge University Press, Cambridge, UK and New York, NY, 1535 pp. http://www.climatechange2013.org/report/

11. IPCC, 1990: *Climate Change: The IPCC Scientific Assessment.* Houghton, J.T., G.J. Jenkins, and J.J. Ephraums, Eds. Cambridge University Press, Cambridge, United Kingdom and New York, NY, USA, 212 pp.

12. IPCC, 1996: *Climate Change 1995: The Science of Climate Change. Contribution of Working Group I to the Second Assessment Report of the Intergovernmental Panel on Climate Change.* Houghton, J.T., L.G. Meira Filho, B.A. Callander, N. Harris, A. Kattenberg, and K. Maskell, Eds. Cambridge University Press, Cambridge, United Kingdom and New York, NY, USA, 584 pp.

13. IPCC, 2001: *Climate Change 2001: The Scientific Basis. Contribution of Working Group I to the Third Assessment Report of the Intergovernmental Panel on Climate Change.* Houghton, J.T., Y. Ding, D.J. Griggs, M. Noguer, P.J. van der Linden, X. Dai, K. Maskell, and C.A. Johnson, Eds. Cambridge University Press, Cambridge, United Kingdom and New York, NY, USA, 881 pp.

14. IPCC, 2007: *Climate Change 2007: The Physical Science Basis. Contribution of Working Group I to the Fourth Assessment Report of the Intergovernmental Panel on Climate Change.* Solomon, S., D. Qin, M. Manning, Z. Chen, M. Marquis, K.B. Averyt, M. Tignor, and H.L. Miller, Eds. Cambridge University Press, Cambridge. U.K, New York, NY, USA, 996 pp. http://www.ipcc.ch/publications_and_data/publications_ipcc_fourth_assessment_report_wg1_report_the_physical_science_basis.htm

15. Stouffer, R.J. and S. Manabe, 2017: Assessing temperature pattern projections made in 1989. *Nature Climate Change,* **7,** 163-165. http://dx.doi.org/10.1038/nclimate3224

16. Canty, T., N.R. Mascioli, M.D. Smarte, and R.J. Salawitch, 2013: An empirical model of global climate – Part 1: A critical evaluation of volcanic cooling. *Atmospheric Chemistry and Physics,* **13,** 3997-4031. http://dx.doi.org/10.5194/acp-13-3997-2013

17. Zhou, J. and K.-K. Tung, 2013: Deducing multidecadal anthropogenic global warming trends using multiple regression analysis. *Journal of the Atmospheric Sciences,* **70,** 3-8. http://dx.doi.org/10.1175/jas-d-12-0208.1

18. Stern, D.I. and R.K. Kaufmann, 2014: Anthropogenic and natural causes of climate change. *Climatic Change,* **122,** 257-269. http://dx.doi.org/10.1007/s10584-013-1007-x

19. Knutson, T.R., R. Zhang, and L.W. Horowitz, 2016: Prospects for a prolonged slowdown in global warming in the early 21st century. *Nature Communcations,* **7,** 13676. http://dx.doi.org/10.1038/ncomms13676

20. Knutson, T.R., F. Zeng, and A.T. Wittenberg, 2013: Multimodel assessment of regional surface temperature trends: CMIP3 and CMIP5 twentieth-century simulations. *Journal of Climate,* **26,** 8709-8743. http://dx.doi.org/10.1175/JCLI-D-12-00567.1

21. Laepple, T. and P. Huybers, 2014: Ocean surface temperature variability: Large model–data differences at decadal and longer periods. *Proceedings of the National Academy of Sciences,* **111,** 16682-16687. http://dx.doi.org/10.1073/pnas.1412077111

22. Schurer, A.P., G.C. Hegerl, M.E. Mann, S.F.B. Tett, and S.J. Phipps, 2013: Separating forced from chaotic climate variability over the past millennium. *Journal of Climate,* **26,** 6954-6973. http://dx.doi.org/10.1175/jcli-d-12-00826.1

23. Andres, H.J. and W.R. Peltier, 2016: Regional influences of natural external forcings on the transition from the Medieval Climate Anomaly to the Little Ice Age. *Journal of Climate,* **29,** 5779-5800. http://dx.doi.org/10.1175/jcli-d-15-0599.1

24. Otto, A., F.E.L. Otto, O. Boucher, J. Church, G. Hegerl, P.M. Forster, N.P. Gillett, J. Gregory, G.C. Johnson, R. Knutti, N. Lewis, U. Lohmann, J. Marotzke, G. Myhre, D. Shindell, B. Stevens, and M.R. Allen, 2013: Energy budget constraints on climate response. *Nature Geoscience,* **6,** 415-416. http://dx.doi.org/10.1038/ngeo1836

25. Lewis, N. and J.A. Curry, 2015: The implications for climate sensitivity of AR5 forcing and heat uptake estimates. *Climate Dynamics,* **45,** 1009-1023. http://dx.doi.org/10.1007/s00382-014-2342-y

26. Marvel, K., G.A. Schmidt, R.L. Miller, and L.S. Nazarenko, 2016: Implications for climate sensitivity from the response to individual forcings. *Nature Climate Change,* **6,** 386-389. http://dx.doi.org/10.1038/nclimate2888

27. Richardson, M., K. Cowtan, E. Hawkins, and M.B. Stolpe, 2016: Reconciled climate response estimates from climate models and the energy budget of Earth. *Nature Climate Change,* **6,** 931-935. http://dx.doi.org/10.1038/nclimate3066

28. Gregory, J.M., T. Andrews, and P. Good, 2015: The inconstancy of the transient climate response parameter under increasing $CO_2$. *Philosophical Transactions of the Royal Society A: Mathematical, Physical and Engineering Sciences,* **373,** 20140417. http://dx.doi.org/10.1098/rsta.2014.0417

29. Pielke Sr., R.A., R. Mahmood, and C. McAlpine, 2016: Land's complex role in climate change. *Physics Today,* **69,** 40-46. http://dx.doi.org/10.1063/PT.3.3364

30. Zwiers, F.W., X.B. Zhang, and Y. Feng, 2011: Anthropogenic influence on long return period daily temperature extremes at regional scales. *Journal of Climate,* **24,** 881-892. http://dx.doi.org/10.1175/2010jcli3908.1

31. Hannart, A., J. Pearl, F.E.L. Otto, P. Naveau, and M. Ghil, 2016: Causal counterfactual theory for the attribution of weather and climate-related events. *Bulletin of the American Meteorological Society,* **97,** 99-110. http://dx.doi.org/10.1175/bams-d-14-00034.1

32. Trenberth, K.E., J.T. Fasullo, and T.G. Shepherd, 2015: Attribution of climate extreme events. *Nature Climate Change,* **5,** 725-730. http://dx.doi.org/10.1038/nclimate2657

33. Shepherd, T.G., 2016: A common framework for approaches to extreme event attribution. *Current Climate Change Reports,* **2,** 28-38. http://dx.doi.org/10.1007/s40641-016-0033-y

34. Horton, R.M., J.S. Mankin, C. Lesk, E. Coffel, and C. Raymond, 2016: A review of recent advances in research on extreme heat events. *Current Climate Change Reports,* **2,** 242-259. http://dx.doi.org/10.1007/s40641-016-0042-x

35. Herring, S.C., A. Hoell, M.P. Hoerling, J.P. Kossin, C.J. Schreck III, and P.A. Stott, 2016: Explaining Extreme Events of 2015 from a Climate Perspective. *Bulletin of the American Meteorological Society,* **97,** S1-S145. http://dx.doi.org/10.1175/BAMS-ExplainingExtremeEvents2015.1

36. Herring, S.C., M.P. Hoerling, J.P. Kossin, T.C. Peterson, and P.A. Stott, 2015: Explaining Extreme Events of 2014 from a Climate Perspective. *Bulletin of the American Meteorological Society,* **96,** S1-S172. http://dx.doi.org/10.1175/BAMS-ExplainingExtremeEvents2014.1

37. Herring, S.C., M.P. Hoerling, T.C. Peterson, and P.A. Stott, 2014: Explaining Extreme Events of 2013 from a Climate Perspective. *Bulletin of the American Meteorological Society,* **95,** S1-S104. http://dx.doi.org/10.1175/1520-0477-95.9.s1.1

38. Peterson, T.C., M.P. Hoerling, P.A. Stott, and S.C. Herring, 2013: Explaining Extreme Events of 2012 from a Climate Perspective. *Bulletin of the American Meteorological Society,* **94,** S1-S74. http://dx.doi.org/10.1175/bams-d-13-00085.1

39. Peterson, T.C., P.A. Stott, and S. Herring, 2012: Explaining extreme events of 2011 from a climate perspective. *Bulletin of the American Meteorological Society*, **93**, 1041-1067. http://dx.doi.org/10.1175/BAMS-D-12-00021.1

40. Stott, P.A., D.A. Stone, and M.R. Allen, 2004: Human contribution to the European heatwave of 2003. *Nature*, **432**, 610-614. http://dx.doi.org/10.1038/nature03089

41. Arblaster, J.M., E.-P. Lim, H.H. Hendon, B.C. Trewin, M.C. Wheeler, G. Liu, and K. Braganza, 2014: Understanding Australia's hottest September on record [in "Explaining Extreme Events of 2013 from a Climate Perspective"]. *Bulletin of the American Meteorological Society*, **95 (9)**, S37-S41. http://dx.doi.org/10.1175/1520-0477-95.9.S1.1

42. King, A.D., D.J. Karoly, M.G. Donat, and L.V. Alexander, 2014: Climate change turns Australia's 2013 Big Dry into a year of record-breaking heat [in "Explaining Extreme Events of 2013 from a Climate Perspective"]. *Bulletin of the American Meteorological Society*, **95 (9)**, S41-S45. http://dx.doi.org/10.1175/1520-0477-95.9.S1.1

43. Knutson, T.R., F. Zeng, and A.T. Wittenberg, 2014: Multimodel assessment of extreme annual-mean warm anomalies during 2013 over regions of Australia and the western tropical Pacific [in "Explaining Extreme Events of 2013 from a Climate Perspective"]. *Bulletin of the American Meteorological Society*, **95 (9)**, S26-S30. http://dx.doi.org/10.1175/1520-0477-95.9.S1.1

44. Lewis, S. and D.J. Karoly, 2014: The role of anthropogenic forcing in the record 2013 Australia-wide annual and spring temperatures [in "Explaining Extreme Events of 2013 from a Climate Perspective"]. *Bulletin of the American Meteorological Society*, **95 (9)**, S31-S33. http://dx.doi.org/10.1175/1520-0477-95.9.S1.1

45. Perkins, S.E., S.C. Lewis, A.D. King, and L.V. Alexander, 2014: Increased simulated risk of the hot Australian summer of 2012/13 due to anthropogenic activity as measured by heat wave frequency and intensity [in "Explaining Extreme Events of 2013 from a Climate Perspective"]. *Bulletin of the American Meteorological Society*, **95 (9)**, S34-S37. http://dx.doi.org/10.1175/1520-0477-95.9.S1.1

46. Hoerling, M., M. Chen, R. Dole, J. Eischeid, A. Kumar, J.W. Nielsen-Gammon, P. Pegion, J. Perlwitz, X.-W. Quan, and T. Zhang, 2013: Anatomy of an extreme event. *Journal of Climate*, **26**, 2811–2832. http://dx.doi.org/10.1175/JCLI-D-12-00270.1

47. Rupp, D.E., P.W. Mote, N. Massey, C.J. Rye, R. Jones, and M.R. Allen, 2012: Did human influence on climate make the 2011 Texas drought more probable? [in "Explaining Extreme Events of 2011 from a Climate Perspective"]. *Bulletin of the American Meteorological Society*, **93**, 1052-1054. http://dx.doi.org/10.1175/BAMS-D-12-00021.1

48. Otto, F.E.L., N. Massey, G.J. van Oldenborgh, R.G. Jones, and M.R. Allen, 2012: Reconciling two approaches to attribution of the 2010 Russian heat wave. *Geophysical Research Letters*, **39**, L04702. http://dx.doi.org/10.1029/2011GL050422

49. Gillett, N.P., J.C. Fyfe, and D.E. Parker, 2013: Attribution of observed sea level pressure trends to greenhouse gas, aerosol, and ozone changes. *Geophysical Research Letters*, **40**, 2302-2306. http://dx.doi.org/10.1002/grl.50500

50. Jones, G.S., P.A. Stott, and N. Christidis, 2013: Attribution of observed historical near surface temperature variations to anthropogenic and natural causes using CMIP5 simulations. *Journal of Geophysical Research*, **118**, 4001-4024. http://dx.doi.org/10.1002/jgrd.50239

51. Ribes, A. and L. Terray, 2013: Application of regularised optimal fingerprinting to attribution. Part II: Application to global near-surface temperature. *Climate Dynamics*, **41**, 2837-2853. http://dx.doi.org/10.1007/s00382-013-1736-6

52. Huber, M. and R. Knutti, 2012: Anthropogenic and natural warming inferred from changes in Earth's energy balance. *Nature Geoscience*, **5**, 31-36. http://dx.doi.org/10.1038/ngeo1327

53. Wigley, T.M.L. and B.D. Santer, 2013: A probabilistic quantification of the anthropogenic component of twentieth century global warming. *Climate Dynamics*, **40**, 1087-1102. http://dx.doi.org/10.1007/s00382-012-1585-8

54. Hegerl, G.C., F.W. Zwiers, P. Braconnot, N.P. Gillett, Y. Luo, J.A.M. Orsini, N. Nicholls, J.E. Penner, and P.A. Stott, 2007: Understanding and attributing climate change. *Climate Change 2007: The Physical Science Basis. Contribution of Working Group I to the Fourth Assessment Report of the Intergovernmental Panel on Climate Change*. Solomon, S., D. Qin, M. Manning, Z. Chen, M. Marquis, K.B. Averyt, M. Tignor, and H.L. Miller, Eds. Cambridge University Press, Cambridge, United Kingdom and New York, NY, USA, 663-745. http://www.ipcc.ch/publications_and_data/ar4/wg1/en/ch9.html

55. Ribes, A., F.W. Zwiers, J.-M. Azaïs, and P. Naveau, 2017: A new statistical approach to climate change detection and attribution. *Climate Dynamics*, **48**, 367-386. http://dx.doi.org/10.1007/s00382-016-3079-6

56. Masson-Delmotte, V., M. Schulz, A. Abe-Ouchi, J. Beer, A. Ganopolski, J.F. González Rouco, E. Jansen, K. Lambeck, J. Luterbacher, T. Naish, T. Osborn, B. Otto-Bliesner, T. Quinn, R. Ramesh, M. Rojas, X. Shao, and A. Timmermann, 2013: Information from paleoclimate archives. *Climate Change 2013: The Physical Science Basis. Contribution of Working Group I to the Fifth Assessment Report of the Intergovernmental Panel on Climate Change*. Stocker, T.F., D. Qin, G.-K. Plattner, M. Tignor, S.K. Allen, J. Boschung, A. Nauels, Y. Xia, V. Bex, and P.M. Midgley, Eds. Cambridge University Press, Cambridge, United Kingdom and New York, NY, USA, 383–464. http://www.climatechange2013.org/report/full-report/

57. Myhre, G., D. Shindell, F.-M. Bréon, W. Collins, J. Fuglestvedt, J. Huang, D. Koch, J.-F. Lamarque, D. Lee, B. Mendoza, T. Nakajima, A. Robock, G. Stephens, T. Takemura, and H. Zhang, 2013: Anthropogenic and natural radiative forcing. *Climate Change 2013: The Physical Science Basis. Contribution of Working Group I to the Fifth Assessment Report of the Intergovernmental Panel on Climate Change*. Stocker, T.F., D. Qin, G.-K. Plattner, M. Tignor, S.K. Allen, J. Boschung, A. Nauels, Y. Xia, V. Bex, and P.M. Midgley, Eds. Cambridge University Press, Cambridge, United Kingdom and New York, NY, USA, 659–740. http://www.climatechange2013.org/report/full-report/

58. Boucher, O., D. Randall, P. Artaxo, C. Bretherton, G. Feingold, P. Forster, V.-M. Kerminen, Y. Kondo, H. Liao, U. Lohmann, P. Rasch, S.K. Satheesh, S. Sherwood, B. Stevens, and X.Y. Zhang, 2013: Clouds and aerosols. *Climate Change 2013: The Physical Science Basis. Contribution of Working Group I to the Fifth Assessment Report of the Intergovernmental Panel on Climate Change*. Stocker, T.F., D. Qin, G.-K. Plattner, M. Tignor, S.K. Allen, J. Boschung, A. Nauels, Y. Xia, V. Bex, and P.M. Midgley, Eds. Cambridge University Press, Cambridge, United Kingdom and New York, NY, USA, 571–658. http://www.climatechange2013.org/report/full-report/

59. Taylor, K.E., R.J. Stouffer, and G.A. Meehl, 2012: An overview of CMIP5 and the experiment design. *Bulletin of the American Meteorological Society*, **93**, 485-498. http://dx.doi.org/10.1175/BAMS-D-11-00094.1

60. Hannart, A., 2016: Integrated optimal fingerprinting: Method description and illustration. *Journal of Climate*, **29**, 1977-1998. http://dx.doi.org/10.1175/jcli-d-14-00124.1

61. Hannart, A., A. Carrassi, M. Bocquet, M. Ghil, P. Naveau, M. Pulido, J. Ruiz, and P. Tandeo, 2016: DADA: Data assimilation for the detection and attribution of weather and climate-related events. *Climatic Change*, **136**, 155-174. http://dx.doi.org/10.1007/s10584-016-1595-3

62. Morice, C.P., J.J. Kennedy, N.A. Rayner, and P.D. Jones, 2012: Quantifying uncertainties in global and regional temperature change using an ensemble of observational estimates: The HadCRUT4 dataset. *Journal of Geophysical Research*, **117**, D08101. http://dx.doi.org/10.1029/2011JD017187

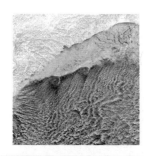

# 4

# Climate Models, Scenarios, and Projections

## KEY FINDINGS

1. If greenhouse gas concentrations were stabilized at their current level, existing concentrations would commit the world to at least an additional 1.1°F (0.6°C) of warming over this century relative to the last few decades (*high confidence* in continued warming, *medium confidence* in amount of warming).

2. Over the next two decades, global temperature increase is projected to be between 0.5°F and 1.3°F (0.3°–0.7°C) (*medium confidence*). This range is primarily due to uncertainties in natural sources of variability that affect short-term trends. In some regions, this means that the trend may not be distinguishable from natural variability (*high confidence*).

3. Beyond the next few decades, the magnitude of climate change depends primarily on cumulative emissions of greenhouse gases and aerosols and the sensitivity of the climate system to those emissions (*high confidence*). Projected changes range from 4.7°–8.6°F (2.6°–4.8°C) under the higher scenario (RCP8.5) to 0.5°–1.3°F (0.3°–1.7°C) under the much lower scenario (RCP2.6), for 2081–2100 relative to 1986–2005 (*medium confidence*).

4. Global mean atmospheric carbon dioxide ($CO_2$) concentration has now passed 400 ppm, a level that last occurred about 3 million years ago, when global average temperature and sea level were significantly higher than today (*high confidence*). Continued growth in $CO_2$ emissions over this century and beyond would lead to an atmospheric concentration not experienced in tens of millions of years (*medium confidence*). The present-day emissions rate of nearly 10 GtC per year suggests that there is no climate analog for this century any time in at least the last 50 million years (*medium confidence*).

5. The observed increase in global carbon emissions over the past 15–20 years has been consistent with higher scenarios (*very high confidence*). In 2014 and 2015, emission growth rates slowed as economic growth has become less carbon-intensive (*medium confidence*). Even if this trend continues, however, it is not yet at a rate that would limit the increase in the global average temperature to well below 3.6°F (2°C) above preindustrial levels (*high confidence*).

6. Combining output from global climate models and dynamical and statistical downscaling models using advanced averaging, weighting, and pattern scaling approaches can result in more relevant and robust future projections. For some regions, sectors, and impacts, these techniques are increasing the ability of the scientific community to provide guidance on the use of climate projections for quantifying regional-scale changes and impacts (*medium to high confidence*).

**Recommended Citation for Chapter**

**Hayhoe**, K., J. Edmonds, R.E. Kopp, A.N. LeGrande, B.M. Sanderson, M.F. Wehner, and D.J. Wuebbles, 2017: Climate models, scenarios, and projections. In: *Climate Science Special Report: Fourth National Climate Assessment, Volume I* [Wuebbles, D.J., D.W. Fahey, K.A. Hibbard, D.J. Dokken, B.C. Stewart, and T.K. Maycock (eds.)]. U.S. Global Change Research Program, Washington, DC, USA, pp. 133-160, doi: 10.7930/J0WH2N54.

## 4.1 The Human Role in Future Climate

The Earth's climate, past and future, is not static; it changes in response to both natural and anthropogenic drivers (see Ch. 2: Physical Drivers of Climate Change). Human emissions of carbon dioxide ($CO_2$), methane ($CH_4$), and other greenhouse gases now overwhelm the influence of natural drivers on the external forcing of Earth's climate (see Ch. 3: Detection and Attribution). Climate change (see Ch. 1: Our Globally Changing Climate) and ocean acidification (see Ch. 13: Ocean Changes) are already occurring due to the buildup of atmospheric $CO_2$ from human emissions in the industrial era.[1, 2]

Even if existing concentrations could be immediately stabilized, temperature would continue to increase by an estimated 1.1°F (0.6°C) over this century, relative to 1980–1999.[3] This is because of the long timescale over which some climate feedbacks act (Ch. 2: Physical Drivers of Climate Change). Over the next few decades, concentrations are projected to increase and the resulting global temperature increase is projected to range from 0.5°F to 1.3°F (0.3°C to 0.7°C). This range depends on natural variability, on emissions of short-lived species such as $CH_4$ and black carbon that contribute to warming, and on emissions of sulfur dioxide ($SO_2$) and other aerosols that have a net cooling effect (Ch. 2: Physical Drivers of Climate Change). The role of emission reductions of non-$CO_2$ gases and aerosols in achieving various global temperature targets is discussed in Chapter 14: Mitigation.

Over the past 15–20 years, the growth rate in atmospheric carbon emissions from human activities has increased from 1.5 to 2 parts per million (ppm) per year due to increasing carbon emissions from human activities that track the rate projected under higher scenarios, in large part due to growing contributions from developing economies.[4, 5, 6] One possible

analog for the rapid pace of change occurring today is the relatively abrupt warming of 9°–14°F (5°–8°C) that occurred during the Paleocene-Eocene Thermal Maximum (PETM), approximately 55–56 million years ago.[7, 8, 9, 10] However, emissions today are nearly 10 GtC per year. During the PETM, the rate of maximum sustained carbon release was less than 1.1 GtC per year, with significant differences in both background conditions and forcing relative to today. This suggests that there is no precise past analog any time in the last 66 million years for the conditions occurring today.[10, 11]

Since 2014, growth rates of global carbon emissions have declined, a trend cautiously attributed to declining coal use in China, despite large uncertainties in emissions reporting.[12, 13] Economic growth is becoming less carbon-intensive, as both developed and emerging economies begin to phase out coal and transition to natural gas and renewable, non-carbon energy.[14, 15]

Beyond the next few decades, the magnitude of future climate change will be primarily a function of future carbon emissions and the response of the climate system to those emissions. This chapter describes the scenarios that provide the basis for the range of future projections presented in this report: from those consistent with continued increases in greenhouse gas emissions, to others that can only be achieved by various levels of emission reductions (see Ch. 14: Mitigation). This chapter also describes the models used to quantify projected changes at the global to regional scale and how it is possible to estimate the range in potential climate change—as determined by climate sensitivity, which is the response of global temperature to a natural or anthropogenic forcing (see Ch. 2: Physical Drivers of Climate Change)—that would result from a given scenario.[3]

## 4.2 Future Scenarios

Climate projections are typically presented for a range of plausible pathways, scenarios, or targets that capture the relationships between human choices, emissions, concentrations, and temperature change. Some scenarios are consistent with continued dependence on fossil fuels, while others can only be achieved by deliberate actions to reduce emissions. The resulting range reflects the uncertainty inherent in quantifying human activities (including technological change) and their influence on climate.

The first Intergovernmental Panel on Climate Change Assessment Report (IPCC FAR) in 1990 discussed three types of scenarios: equilibrium scenarios, in which $CO_2$ concentration was fixed; transient scenarios, in which $CO_2$ concentration increased by a fixed percentage each year over the duration of the scenario; and four brand-new Scientific Assessment (SA90) emission scenarios based on World Bank population projections.[16] Today, that original portfolio has expanded to encompass a wide variety of time-dependent or transient scenarios that project how population, energy sources, technology, emissions, atmospheric concentrations, radiative forcing, and/or global temperature change over time.

Other scenarios are simply expressed in terms of an end-goal or target, such as capping cumulative carbon emissions at a specific level or stabilizing global temperature at or below a certain threshold such as 3.6°F (2°C), a goal that is often cited in a variety of scientific and policy discussions, most recently the Paris Agreement.[17] To stabilize climate at any particular temperature level, however, it is not enough to halt the growth in annual carbon emissions. Global net carbon emissions will eventually need to reach zero[3] and negative emissions may be needed for a greater-than-50% chance of limiting warming below 3.6°F (2°C) (see also Ch. 14: Mitigation for a discussion of negative emissions).[18]

Finally, some scenarios, like the "commitment" scenario in Key Finding 1 and the fixed-$CO_2$ equilibrium scenarios described above, continue to explore hypothetical questions such as, "what would the world look like, long-term, if humans were able to stabilize atmospheric $CO_2$ concentration at a given level?" This section describes the different types of scenarios used today and their relevance to assessing impacts and informing policy targets.

### 4.2.1 Emissions Scenarios, Representative Concentration Pathways, and Shared Socioeconomic Pathways

The standard sets of time-dependent scenarios used by the climate modeling community as input to global climate model simulations provide the basis for the majority of the future projections presented in IPCC assessment reports and U.S. National Climate Assessments (NCAs). Developed by the integrated assessment modeling community, these sets of standard scenarios have become more comprehensive with each new generation, as the original SA90 scenarios[19] were replaced by the IS92 emission scenarios of the 1990s,[20] which were in turn succeeded by the Special Report on Emissions Scenarios in 2000 (SRES)[21] and by the Representative Concentration Pathways in 2010 (RCPs).[22]

SA90, IS92, and SRES are all emission-based scenarios. They begin with a set of storylines that were based on population projections initially. By SRES, they had become much more complex, laying out a consistent picture of demographics, international trade, flow of information and technology, and other social, technological, and economic characteristics of future worlds. These assumptions were then fed through socioeconomic and Integrated Assessment Models (IAMs) to derive emissions. For SRES, the use of various IAMs resulted in multiple emissions scenarios corresponding to each storyline; however, one scenario for each storyline was selected as the representative "marker" scenario to be used as input to global models to calculate the resulting atmospheric concentrations, radiative forcing, and climate change for the higher A1fi (fossil-intensive), mid-high A2, mid-low B2, and lower B1 storylines. IS92-based projections were used in the IPCC Second and Third Assessment Reports (SAR and TAR)[23, 24] and the first NCA.[25] Projections based on SRES scenarios were used in the second and third NCAs[26, 27] as well as the IPCC TAR and Fourth Assessment Reports (AR4).[24, 28]

The most recent set of time-dependent scenarios, RCPs, builds on these two decades of scenario development. However, RCPs differ from previous sets of standard scenarios in at least four important ways. First, RCPs are not emissions scenarios; they are radiative forcing scenarios. Each scenario is tied to one value: the change in radiative forcing at the tropopause by 2100 relative to preindustrial levels. The four RCPs are numbered according to the change in radiative forcing by 2100: +2.6, +4.5, +6.0 and +8.5 watts per square meter (W/m$^2$).[29, 30, 31, 32]

The second difference is that, starting from these radiative forcing values, IAMs are used to work backwards to derive a range of emissions trajectories and corresponding policies and technological strategies for each RCP that would achieve the same ultimate impact on radiative forcing. From the multiple emissions pathways that could lead to the same 2100 radiative forcing value, an associated pathway of annual carbon dioxide and other anthropogenic emissions of greenhouse gases, aerosols, air pollutants, and other short-lived species has been selected for each RCP to use as input to future climate model simulations

(e.g., Meinshausen et al. 2011;[33] Cubasch et al. 2013[34]). In addition, RCPs provide climate modelers with gridded trajectories of land use and land cover.

A third difference between the RCPs and previous scenarios is that while none of the SRES scenarios included a scenario with explicit policies and measures to limit climate forcing, all of the three lower RCP scenarios (2.6, 4.5, and 6.0) are climate-policy scenarios. At the higher end of the range, the RCP8.5 scenario corresponds to a future where carbon dioxide and methane emissions continue to rise as a result of fossil fuel use, albeit with significant declines in emission growth rates over the second half of the century (Figure 4.1), significant reduction in aerosols, and modest improvements in energy intensity and technology.[32] Atmospheric carbon dioxide levels for RCP8.5 are similar to those of the SRES A1FI scenario: they rise from current-day levels of 400 up to 936 ppm by the end of this century. CO$_2$-equivalent levels (including emissions of other non-CO$_2$ greenhouse gases, aerosols, and other substances that affect climate) reach more than 1200 ppm by 2100, and global temperature is projected to increase by 5.4°–9.9°F (3°–5.5°C) by 2100 relative to the 1986–2005 average. RCP8.5 reflects the upper range of the open literature on emissions, but is not intended to serve as an upper limit on possible emissions nor as a business-as-usual or reference scenario for the other three scenarios.

Under the lower scenarios (RCP4.5 and RCP2.6),[29, 30] atmospheric CO$_2$ levels remain below 550 and 450 ppm by 2100, respectively. Emissions of other substances are also lower; by 2100, CO$_2$-equivalent concentrations that include all emissions from human activities reach 580 ppm under RCP4.5 and 425 ppm under RCP2.6. RCP4.5 is similar to SRES B1, but the RCP2.6 scenario is much lower than any SRES scenario because it includes the option of using policies to achieve net negative carbon dioxide emissions before the end of the century, while SRES scenarios do not. RCP-based projections were used in the most recent IPCC Fifth Assessment Report (AR5)[3] and the third NCA[27] and are used in this fourth NCA as well.

Within the RCP family, individual scenarios have not been assigned a formal likelihood. Higher-numbered scenarios correspond to higher emissions and a larger and more rapid global temperature change (Figure 4.1); the range of values covered by the scenarios was chosen to reflect the then-current range in the open literature. Since the choice of scenario constrains the magnitudes of future changes, most assessments (including this one; see Ch. 6: Temperature Change) quantify future change and corresponding impacts under a range of future scenarios that reflect the uncertainty in the consequences of human choices over the coming century.

Fourth, a broad range of socioeconomic scenarios were developed independently from the RCPs and a subset of these were constrained, using emissions limitations policies consistent with their underlying storylines, to create five Shared Socioeconomic Pathways (SSPs) with climate forcing that matches the RCP values. This pairing of SSPs and RCPs is designed to

meet the needs of the impacts, adaptation, and vulnerability (IAV) communities, enabling them to couple alternative socioeconomic scenarios with the climate scenarios developed using RCPs to explore the socioeconomic challenges to climate mitigation and adaptation.[35] The five SSPs consist of SSP1 ("Sustainability"; low challenges to mitigation and adaptation), SSP2 ("Middle of the Road"; middle challenges to mitigation and adaptation), SSP3 ("Regional Rivalry"; high challenges to mitigation and adaptation), SSP4 ("Inequality"; low challenges to mitigation, high challenges to adaptation), and SSP5 ("Fossil-fueled Development"; high challenges to mitigation, low challenges to adaptation). Each scenario has an underlying SSP narrative, as well as consistent assumptions regarding demographics, urbanization, economic growth, and technology development. Only SSP5 produces a reference scenario that is consistent with RCP8.5; climate forcing in the other SSPs' reference scenarios that don't include climate policy remains below 8.5 W/m². In addition, the nature of SSP3 makes it impossible for that scenario to produce a climate forcing as low as 2.6 W/m². While new research is under way to explore scenarios that limit climate forcing to 2.0 W/m², neither the RCPs nor the SSPs have produced scenarios in that range.

## Emissions, Concentrations, and Temperature Projections

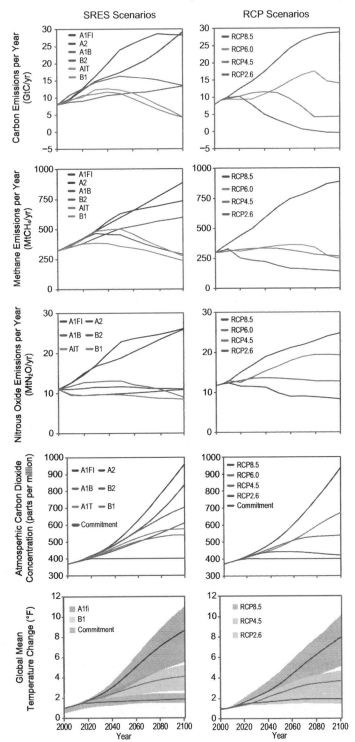

**Figure 4.1:** The climate projections used in this report are based on the 2010 Representative Concentration Pathways (RCP, right). They are largely consistent with scenarios used in previous assessments, the 2000 Special Report on Emission Scenarios (SRES, left). This figure compares SRES and RCP annual carbon emissions (GtC per year, first row), annual methane emissions (MtCH$_4$ per year, second row), annual nitrous oxide emissions (MtN$_2$O per year, third row), carbon dioxide concentration in the atmosphere (ppm, fourth row), and global mean temperature change relative to 1900–1960 as simulated by CMIP3 models for the SRES scenarios and CMIP5 models for the RCP scenarios (°F, fifth row). Note that global mean temperature from SRES A1FI simulations are only available from four global climate models. (Data from IPCC-DDC, IIASA, CMIP3, and CMIP5).

### 4.2.2 Alternative Scenarios

The emissions and radiative forcing scenarios described above include a component of time: how much will climate change, and by when? Ultimately, however, the magnitude of human-induced climate change depends less on the year-to-year emissions than it does on the net amount of carbon, or cumulative carbon, emitted into the atmosphere. The lower the atmospheric concentrations of $CO_2$, the greater the chance that eventual global temperature change will not reach the high end temperature projections, or possibly remain below 3.6°F (2°C) relative to preindustrial levels.

Cumulative carbon targets offer an alternative approach to expressing a goal designed to limit global temperature to a certain level. As discussed in Chapter 14: Mitigation, it is possible to quantify the expected amount of carbon that can be emitted globally in order to meet a specific global warming target such as 3.6°F (2°C) or even 2.7°F (1.5°C)—although if current carbon emission rates of just under 10 GtC per year were to continue, the lower target would be reached in a matter of years. The higher target would be reached in a matter of decades (see Ch. 14: Mitigation).

Under a lower scenario (RCP4.5), global temperature change is more likely than not to exceed 3.6°F (2°C),[3, 36] whereas under the even lower scenario (RCP2.6), it is likely to remain below 3.6°F (2°C).[3, 37] While new research is under way to explore scenarios consistent with limiting climate forcing to 2.0 W/m[2], a level consistent with limiting global mean surface temperature change to 2.7°F (1.5°C), neither the RCPs nor the SSPs have produced scenarios that allow for such a small amount of temperature change (see also Ch. 14: Mitigation). [37]

Future projections are most commonly summarized for a given future scenario (for example, RCP8.5 or 4.5) over a range of future climatological time periods (for example, temperature change in 2040–2079 or 2070–2099 relative to 1980–2009). While this approach has the advantage of developing projections for a specific time horizon, uncertainty in future projections is relatively high, incorporating both the uncertainty due to multiple scenarios as well as uncertainty regarding the response of the climate system to human emissions. These uncertainties increase the further out in time the projections go. Using these same transient, scenario-based simulations, however, it is possible to analyze the projected changes for a given global mean temperature (GMT) threshold by extracting a time slice (typically 20 years) centered around the point in time at which that change is reached (Figure 4.2).

Derived GMT scenarios offer a way for the public and policymakers to understand the impacts for any given temperature threshold, as many physical changes and impacts have been shown to scale with global mean surface temperature, including shifts in average precipitation, extreme heat, runoff, drought risk, wildfire, temperature-related crop yield changes, and even risk of coral bleaching (e.g., NRC 2011;[38] Collins et al. 2013;[3] Frieler et al. 2013;[39] Swain and Hayhoe 2015[40]). They also allow scientists to highlight the effect of global mean temperature on projected regional change by de-emphasizing the uncertainty due to both climate sensitivity and future scenarios.[40, 41] This approach is less useful for those impacts that vary based on rate of change, such as species migrations, or where equilibrium changes are very different from transient effects, such as sea level rise.

Pattern scaling techniques[42] are based on a similar assumption to GMT scenarios, namely that large-scale patterns of regional change will scale with global temperature change. These techniques can be used to quantify regional projections for scenarios that are not

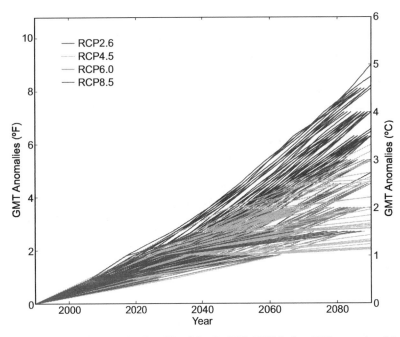

**Figure 4.2:** Global mean temperature anomalies (°F) relative to 1976–2005 for four RCP scenarios, 2.6 (green), 4.5 (yellow), 6.0 (orange), and 8.5 (red). Each line represents an individual simulation from the CMIP5 archive. Every RCP-based simulation with annual or monthly temperature outputs available was used here. The values shown here were calculated in 0.5°C increments; since not every simulation reaches the next 0.5°C increment before end of century, many lines terminate before 2100. (Figure source: adapted from Swain and Hayhoe 2015[40]).

readily available in preexisting databases of global climate model simulations, including changes in both mean and extremes (e.g., Fix et al. 2016[43]). A comprehensive assessment both confirms and constrains the validity of applying pattern scaling to quantify climate response to a range of projected future changes.[44] For temperature-based climate targets, these pattern scaling frames or GMT scenarios offer the basis for more consistent comparisons across studies examining regional change or potential risks and impacts.

### 4.2.3 Analogs from the Paleoclimate Record

Most CMIP5 simulations project transient changes in climate through 2100; a few simulations extend to 2200, 2300, or beyond. However, as discussed in Chapter 2: Physical Drivers of Climate Change, the long-term impact of human activities on the carbon cycle and Earth's climate over the next few decades and for the remainder of this century can only be assessed by considering changes that occur over multiple centuries and even millennia.[38]

In the past, there have been several examples of "hothouse" climates where carbon dioxide concentrations and/or global mean temperatures were similar to preindustrial, current, or plausible future levels. These periods are sometimes referenced as analogs, albeit imperfect and incomplete, of future climate (e.g., Crowley 1990[10]), though comparing climate model simulations to geologic reconstructions of temperature and carbon dioxide during these periods suggests that today's global climate models tend to underestimate the magnitude of change in response to higher $CO_2$ (see Ch. 15: Potential Surprises).

The last interglacial period, approximately 125,000 years ago, is known as the Eemian. During that time, $CO_2$ concentration was similar to preindustrial concentrations, around 280 ppm.[45] Global mean temperature was approximately 1.8°–3.6°F (1°–2°C) higher than preindustrial temperatures,[46, 47] although the poles were significantly warmer [48, 49] and sea level was 6 to 9 meters (20 to 30 feet) higher than today.[50] During the Pliocene, approximately 3 million years ago, long-term $CO_2$ concentration was similar to today's, around 400 ppm[51]—although this level was sustained over long periods of time, whereas today the global $CO_2$ concentration is increasing rapidly. At that time, global mean temperature was approximately 3.6°–6.3°F (2°–3.5°C) above preindustrial, and sea level was somewhere between 66 ± 33 feet (20 ± 10 meters) higher than today.[52, 53, 54]

Under the higher scenario (RCP8.5), $CO_2$ concentrations are projected to reach 936 ppm by 2100. During the Eocene, 35 to 55 million years ago, $CO_2$ levels were between 680 and 1260 ppm, or somewhere between two and a half to four and a half times higher than preindustrial levels.[55] If Eocene conditions are used as an analog, this suggests that if the $CO_2$ concentrations projected to occur under the RCP8.5 scenario by 2100 were sustained over long periods of time, global temperatures would be approximately 9°–14°F (5°–8°C) above preindustrial temperatures.[56] During the Eocene, there were no permanent land-based ice sheets; Antarctic glaciation did not begin until approximately 34 million years ago.[57] Calibrating sea level rise models against past climate suggests that, under the RCP8.5 scenario, Antarctica could contribute 3 feet (1 meter) of sea level rise by 2100 and 50 feet (15 meters) by 2500.[58] If atmospheric $CO_2$ were sustained at levels approximately two to three times above preindustrial for tens of thousands of years, it is estimated that Greenland

and Antarctic ice sheets could melt entirely,[59] resulting in approximately 215 feet (65 meters) of sea level rise.[60]

## 4.3 Modeling Tools

Using transient scenarios such as SRES and RCP as input, global climate models (GCMs) produce trajectories of future climate change, including global and regional changes in temperature, precipitation, and other physical characteristics of the climate system (see also Ch. 6: Temperature Change and Ch. 7: Precipitation Change).[3, 61] The resolution of global models has increased significantly since IPCC FAR.[19] However, even the latest experimental high-resolution simulations, at 15–30 miles (25–50 km) per gridbox, are unable to simulate all of the important fine-scale processes occurring at regional to local scales. Instead, downscaling methods are often used to correct systematic biases, or offsets relative to observations, in global projections and translate them into the higher-resolution information typically required for impact assessments.

Dynamical downscaling with regional climate models (RCMs) directly simulates the response of regional climate processes to global change, while empirical statistical downscaling models (ESDMs) tend to be more flexible and computationally efficient. Comparing the ability of dynamical and statistical methods to reproduce observed climate shows that the relative performance of the two approaches depends on the assessment criteria.[62] Although dynamical and statistical methods can be combined into a hybrid framework, many assessments still tend to rely on one or the other type of downscaling, where the choice is based on the needs of the assessment. The projections shown in this report, for example, are either based on the original GCM simulations or on simulations that have been statistically downscaled using the LOcalized Constructed Analogs method (LOCA).[63] This section describes the global climate models used today, briefly summarizes their development over the past few decades, and explains the general characteristics and relative strengths and weaknesses of the dynamical and statistical downscaling.

### 4.3.1 Global Climate Models

Global climate models are mathematical frameworks that were originally built on fundamental equations of physics. They account for the conservation of energy, mass, and momentum and how these are exchanged among different components of the climate system. Using these fundamental relationships, GCMs are able to simulate many important aspects of Earth's climate: large-scale patterns of temperature and precipitation, general characteristics of storm tracks and extratropical cyclones, and observed changes in global mean temperature and ocean heat content as a result of human emissions.[64]

The complexity of climate models has grown over time, as they incorporate additional components of Earth's climate system (Figure 4.3). For example, GCMs were previously referred to as "general circulation models" when they included only the physics needed to simulate the general circulation of the atmosphere. Today, global climate models simulate many more aspects of the climate system: atmospheric chemistry and aerosols, land surface interactions including soil and vegetation, land and sea ice, and increasingly even an interactive carbon cycle and/or biogeochemistry. Models that include this last component are also referred to as Earth system models (ESMs).

In addition to expanding the number of processes in the models and improving the treatment of existing processes, the total number of GCMs and the average horizontal spatial resolution of the models have increased over time, as computers become more powerful, and with each successive version of the World Cli-

mate Research Programme's (WCRP's) Coupled Model Intercomparison Project (CMIP). CMIP5 provides output from over 50 GCMs with spatial resolutions ranging from about 30 to 200 miles (50 to 300 km) per horizontal size and variable vertical resolution on the order of hundreds of meters in the troposphere or lower atmosphere.

It is often assumed that higher-resolution, more complex, and more up-to-date models will perform better and/or produce more robust projections than previous-generation models. However, a large body of research comparing CMIP3 and CMIP5 simulations concludes that, although the spatial resolution of CMIP5 has improved relative to CMIP3, the overall improvement in performance is relatively minor. For certain variables, regions, and seasons, there is some improvement; for others, there is little difference or even sometimes degradation in performance, as greater complexity does not necessarily imply improved performance.[65, 66, 67, 68] CMIP5 simulations do show modest improvement in model ability to simulate ENSO,[69] some aspects of cloud characteristics,[70] and the rate of arctic sea ice loss,[71] as well as greater consensus regarding projected drying in the southwestern United States and Mexico.[68]

Projected changes in hurricane rainfall rates and the reduction in tropical storm frequency are similar, but CMIP5-based projections of increases in the frequency of the strongest hurricanes are generally smaller than CMIP3-based projections.[72] On the other hand, many studies find little to no significant difference in large-scale patterns of changes in both mean and extreme temperature and precipitation from CMIP3 to CMIP5.[65, 68, 73, 74] Also, CMIP3 simulations are driven by SRES scenarios, while CMIP5 simulations are driven by RCP scenarios. Although some scenarios have comparable $CO_2$ concentration pathways (Figure 4.1), differences in non-$CO_2$ species and aerosols

## A Climate Modeling Timeline
### (When Various Components Became Commonly Used)

| 1890s | 1960s | Hydrological | 1970s | 1990s | 2000s | 2010s |
|---|---|---|---|---|---|---|
| Radiative Transfer | Non-Linear Fluid Dynamics | Cycle | Sea Ice and Land Surface | Atmospheric Chemistry | Aerosols and Vegetation | Biogeochemical Cycles and Carbon |

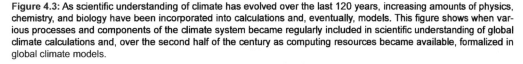

Energy Balance Models    Atmosphere-Ocean General Circulation Models    Earth System Models

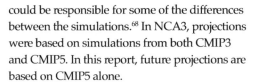

**Figure 4.3:** As scientific understanding of climate has evolved over the last 120 years, increasing amounts of physics, chemistry, and biology have been incorporated into calculations and, eventually, models. This figure shows when various processes and components of the climate system became regularly included in scientific understanding of global climate calculations and, over the second half of the century as computing resources became available, formalized in global climate models.

could be responsible for some of the differences between the simulations.[68] In NCA3, projections were based on simulations from both CMIP3 and CMIP5. In this report, future projections are based on CMIP5 alone.

GCMs are constantly being expanded to include more physics, chemistry, and, increasingly, even the biology and biogeochemistry at work in the climate system (Figure 4.3). Interactions within and between the various components of the climate system result in positive and negative feedbacks that can act to enhance or dampen the effect of human emissions on the climate system. The extent to which models explicitly resolve or incorporate these processes determines their climate sensitivity, or response to external forcing (see Ch. 2: Physical Drivers of Climate Change, Section 2.5 on climate sensitivity, and Ch. 15: Potential Surprises on the importance of processes not included in present-day GCMs).

Confidence in the usefulness of the future projections generated by global climate models is based on multiple factors. These include the fundamental nature of the physical processes they represent, such as radiative transfer or geophysical fluid dynamics, which can be

tested directly against measurements or theoretical calculations to demonstrate that model approximations are valid (e.g., IPCC 1990[19]). They also include the vast body of literature dedicated to evaluating and assessing model abilities to simulate observed features of the earth system, including large-scale modes of natural variability, and to reproduce their net response to external forcing that captures the interaction of many processes which produce observable climate system feedbacks (e.g., Flato et al. 2013[64]). There is no better framework for integrating our knowledge of the physical processes in a complex coupled system like Earth's climate.

Given their complexities, GCMs typically build on previous generations and therefore many models are not fully independent from each other. Many share both ideas and model components or code, complicating the interpretation of multimodel ensembles that often are assumed to be independent.[75, 76] Consideration of the independence of different models is one of the key pieces of information going into the weighting approach used in this report (see Appendix B: Weighting Strategy).

### 4.3.2 Regional Climate Models

Dynamical downscaling models are often referred to as regional climate models, since they include many of the same physical processes that make up a global climate model, but simulate these processes at higher spatial resolution over smaller regions, such as the western or eastern United States (Figure 4.4).[77] Most RCM simulations use GCM fields from pre-computed global simulations as boundary conditions. This approach allows RCMs to draw from a broad set of GCM simulations, such as CMIP5, but does not allow for possible two-way feedbacks and interactions between the regional and global scales. Dynamical downscaling can also be conducted interactively through nesting a higher-resolution regional grid or model into a global model during a simulation. Both approaches directly simulate the dynamics of the regional climate system, but only the second allows for two-way interactions between regional and global change.

RCMs are computationally intensive, providing a broad range of output variables that resolve regional climate features important for assessing climate impacts. The size of individual grid cells can be as fine as 0.6 to 1.2 miles (1 to 2 km) per gridbox in some studies, but more commonly range from about 6 to 30 miles (10 to 50 km). At smaller spatial scales, and for specific variables and areas with complex terrain, such as coastlines or mountains, regional climate models have been shown to add value.[78] As model resolution increases, RCMs are also able to explicitly resolve some processes that are parameterized in global models. For example, some models with spatial scales below 2.5 miles (4 km) are able to dispense with the parameterization of convective precipitation, a significant source of error and uncertainty in coarser models.[79] RCMs can also incorporate changes in land use, land cover, or hydrology into local climate at spatial scales relevant to planning and decision-making at the regional level.

Despite the differences in resolution, RCMs are still subject to many of the same types of uncertainty as GCMs. Even the highest-resolution RCM cannot explicitly model physical processes that occur at even smaller scales than the model is able to resolve; instead, parameterizations are required. Similarly, RCMs might not include a process or an interaction that is not yet well understood, even if it is able to be resolved at the spatial scale of the model. One additional source of uncertainty unique to RCMs arises from the fact that at their boundaries RCMs require output from GCMs to provide large-scale circulation such as winds, temperature, and moisture; the degree to which the driving GCM correctly captures large-scale circulation and climate will affect the performance of the RCM.[80] RCMs can be evaluated by directly comparing their output to observations; although this process can be challenging and time-consuming, it is often necessary to quantify the appropriate level of confidence that can be placed in their output.[77]

Studies have also highlighted the importance of large ensemble simulations when quantifying regional change.[81] However, due to their computational demand, extensive ensembles of RCM-based projections are rare. The largest ensembles of RCM simulations for North America are hosted by the North American Regional Climate Change Assessment Program (NARCCAP) and the North American CORDEX project (NA-CORDEX). These simulations are useful for examining patterns of change over North America and providing a broad suite of surface and upper-air variables to characterize future impacts. Since these ensembles are based on four simulations from four CMIP3 GCMs for a mid-high SRES scenario (NARCCAP) and six CMIP5 GCMs for two RCP scenarios (NA-CORDEX), they do not encompass the full range of uncertainty in future projections due to human activities, natural variability, and climate sensitivity.

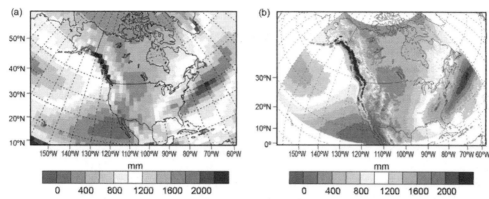

Figure 4.4: CMIP5 global climate models typically operate at coarser horizontal spatial scales on the order of 30 to 200 miles (50 to 300 km), while regional climate models have much finer resolutions, on the order of 6 to 30 miles (10 to 50 km). This figure compares annual average precipitation (in millimeters) for the historical period 1979–2008 using (a) a resolution of 250 km or 150 miles with (b) a resolution of 15 miles or 25 km to illustrate the importance of spatial scale in resolving key topographical features, particularly along the coasts and in mountainous areas. In this case, both simulations are by the GFDL HIRAM, an experimental high-resolution model. (Figure source: adapted from Dixon et al. 2016[86]).

### 4.3.3 Empirical Statistical Downscaling Models

Empirical statistical downscaling models (ESDMs) combine GCM output with historical observations to translate large-scale predictors or patterns into high-resolution projections at the scale of observations. The observations used in an ESDM can range from individual weather stations to gridded datasets. As output, ESDMs can generate a range of products, from large grids to analyses optimized for a specific location, variable, or decision-context.

Statistical techniques are varied, from the simple difference or delta approaches used in the first NCA (subtracting historical simulated values from future values, and adding the resulting delta to historical observations)[25] to the parametric quantile mapping approach used in NCA2 and 3.[26, 27, 82] Even more complex clustering and advanced mathematical modeling techniques can rival dynamical downscaling in their demand for computational resources (e.g., Vrac et al. 2007[83]).

Statistical models are generally flexible and less computationally demanding than RCMs. A number of databases using a variety of

methods, including the LOcalized Constructed Analogs method (LOCA), provide statistically downscaled projections for a continuous period from 1960 to 2100 using a large ensemble of global models and a range of higher and lower future scenarios to capture uncertainty due to human activities. ESDMs are also effective at removing biases in historical simulated values, leading to a good match between the average (multidecadal) statistics of observed and statistically downscaled climate at the spatial scale and over the historical period of the observational data used to train the statistical model. Unless methods can simultaneously downscale multiple variables, however, statistical downscaling carries the risk of altering some of the physical interdependences between variables. ESDMs are also limited in that they require observational data as input; the longer and more complete the record, the greater the confidence that the ESDM is being trained on a representative sample of climatic conditions for that location. Application of ESDMs to remote locations with sparse temporal and/or spatial records is challenging, though in many cases reanalysis[84] or even monthly satellite data[85] can be used in lieu of in situ observations. Lack of data availability can also limit the use of ESDMs in applications that require more variables than temperature and precipitation. Finally, statistical models are based on the key assumption that the relationship between large-scale weather systems and local climate or the spatial pattern of surface climate will remain stationary over the time horizon of the projections. This assumption may not hold if climate change alters local feedback processes that affect these relationships.

ESDMs can be evaluated in three different ways, each of which provides useful insight into model performance.[77] First, the model's goodness-of-fit can be quantified by comparing downscaled simulations for the historical period with the identical observations used to train the model. Second, the generalizability of the model can be determined by comparing downscaled historical simulations with observations from a different time period than was used to train the model; this is often accomplished via cross-validation. Third and most importantly, the stationarity of the model can be evaluated through a "perfect model" experiment using coarse-resolution GCM simulations to generate future projections, then comparing these with high-resolution GCM simulations for the same future time period. Initial analyses using the perfect model approach have demonstrated that the assumption of stationarity can vary significantly by ESDM method, by quantile, and by the time scale (daily or monthly) of the GCM input.[86]

ESDMs are best suited for analyses that require a broad range of future projections of standard, near-surface variables such as temperature and precipitation, at the scale of observations that may already be used for planning purposes. If the study needs to evaluate the full range of projected changes pro-vided by multiple models and scenarios, then statistical downscaling may be more appropriate than dynamical downscaling. However, even within statistical downscaling, selecting an appropriate method for any given study depends on the questions being asked (see Kotamarthi et al. 2016[77] for further discussion on selection of appropriate downscaling methods). This report uses projections generated by LOCA,[63] which spatially matches model-simulated days, past and future, to analogs from observations.

### 4.3.4 Averaging, Weighting, and Selection of Global Models

The results of individual climate model simulations using the same inputs can differ from each other over shorter time scales ranging from several years to several decades.[87, 88] These differences are the result of normal, natural variability, as well as the various ways models characterize various small-scale processes. Although decadal predictability is an active research area,[89] the timing of specific natural variations is largely unpredictable beyond several seasons. For this reason, multimodel simulations are generally averaged to remove the effects of randomly occurring natural variations from long-term trends and make it easier to discern the impact of external drivers, both human and natural, on Earth's climate. Multimodel averaging is typically the last stage in any analysis, used to prepare figures showing projected changes in quantities such as annual or seasonal temperature or precipitation (see Ch. 6: Temperature Change and Ch. 7: Precipitation Change). While the effect of averaging on the systematic errors depends on the extent to which models have similar errors or offsetting errors, there is growing recognition of the value of large ensembles of climate model simulations in addressing uncertainty in both natural variability and scientific modeling (e.g., Deser et al. 2012[87]).

Previous assessments have used a simple average to calculate the multimodel ensemble. This approach implicitly assumes each climate model is independent from the others and of equal ability. Neither of these assumptions, however, are completely valid. Some models share many components with other models in the CMIP5 archive, whereas others have been developed largely in isolation.[75, 76] Also, some models are more successful than others at replicating observed climate and trends over the past century, at simulating the large-scale dynamical features responsible for creating or affecting the average climate conditions over a certain region, such as the Arctic or the Caribbean (e.g., Wang et al. 2007;[90] Wang et al. 2014;[91] Ryu and Hayhoe 2014[92]), or at simulating past climates with very different states than present day.[93] Evaluation of the success of a specific model often depends on the variable or metric being considered in the analysis, with some models performing better than others for certain regions or variables. However, all future simulations agree that both global and regional temperatures will increase over this century in response to increasing emissions of greenhouse gases from human activities.

Can more sophisticated weighting or model selection schemes improve the quality of future projections? In the past, model weights were often based on historical performance; yet performance varies by region and variable, and may not equate to improved future projections.[65] For example, ranking GCMs based on their average biases in temperature gives a very different result than when the same models are ranked based on their ability to simulate observed temperature trends.[94, 95] If GCMs are weighted in a way that does not accurately capture the true uncertainty in regional change, the result can be less robust than an equally-weighted mean.[96] Although the intent of weighting models is to increase the robustness of the projections, by giving lesser weight to outliers a weighting scheme may increase the risk of underestimating the range of uncertainty, a tendency that has already been noted in multi-model ensembles (see Ch. 15: Potential Surprises).

Despite these challenges, for the first time in an official U.S. Global Change Research Program report, this assessment uses model weighting to refine future climate change projections (see also Appendix B: Weighting Strategy).[97] The weighting approach is unique: it takes into account the interdependence of individual climate models as well as their relative abilities in simulating North American climate. Understanding of model history, together with the fingerprints of particular model biases, has been used to identify model pairs that are not independent. In this report, model independence and selected global and North American model quality metrics are considered in order to determine the weighting parameters.[97] Evaluation of this approach shows improved performance of the weighted ensemble over the Arctic, a region where model-based trends often differ from observations, but little change in global-scale temperature response and in other regions where modeled and observed trends are similar, although there are small regional differences in the statistical significance of projected changes. The choice of metric used to evaluate models has very little effect on the independence weighting, and some moderate influence on the skill weighting if only a small number of variables are used to assess model quality. Because a large number of variables are combined to produce a comprehensive "skill metric," the metric is not highly sensitive to any single variable. All multimodel figures in this report use the approach described in Appendix B: Weighting Strategy.

## 4.4 Uncertainty in Future Projections

The timing and magnitude of projected future climate change is uncertain due to the ambiguity introduced by human choices (as discussed in Section 4.2), natural variability, and scientific uncertainty,[87, 98, 99] which includes uncertainty in both scientific modeling and climate sensitivity (see Ch. 2: Physical Drivers of Climate Change). Confidence in projections of specific aspects of future climate change increases if formal detection and attribution analyses (Ch. 3: Detection and Attribution) indicate that an observed change has been influenced by human activities, and the projection is consistent with attribution. However, in many cases, especially at the regional scales considered in this assessment, a human-forced response may not yet have emerged from the noise of natural climate variability but may be expected to in the future (e.g., Hawkins and Sutton 2009[98], 2011[99]). In such cases, confidence in such "projections without attribution" may still be significant under higher scenarios, if the relevant physical mechanisms of change are well understood.

Scientific uncertainty encompasses multiple factors. The first is parametric uncertainty—the ability of GCMs to simulate processes that occur on spatial or temporal scales smaller than they can resolve. The second is structural uncertainty—whether GCMs include and accurately represent all the important physical processes occurring on scales they can resolve. Structural uncertainty can arise because a process is not yet recognized—such as "tipping points" or mechanisms of abrupt change—or because it is known but is not yet understood well enough to be modeled accurately—such as dynamical mechanisms that are important to melting ice sheets (see Ch. 15: Potential Surprises). The third is climate sensitivity—a measure of the response of the planet to increasing levels of $CO_2$, which is formally defined in Chapter 2: Physical Drivers of Cli-

mate Change as the equilibrium temperature change resulting from a doubling of $CO_2$ levels in the atmosphere relative to preindustrial levels. Various lines of evidence constrain the likely value of climate sensitivity to between 2.7°F and 8.1°F (1.5°C and 4.5°C;[100] see Ch. 2: Physical Drivers of Climate Change for further discussion).

Which of these sources of uncertainty—human, natural, and scientific—is most important depends on the time frame and the variable considered. As future scenarios diverge (Figure 4.1), so too do projected changes in global and regional temperatures.[98] Uncertainty in the magnitude and sign of projected changes in precipitation and other aspects of climate is even greater. The processes that lead to precipitation happen at scales smaller than what can be resolved by even high-resolution models, requiring significant parameterization. Precipitation also depends on many large-scale aspects of climate, including atmospheric circulation, storm tracks, and moisture convergence. Due to the greater level of complexity associated with modeling precipitation, scientific uncertainty tends to dominate in precipitation projections throughout the entire century, affecting both the magnitude and sometimes (depending on location) the sign of the projected change in precipitation.[99]

Over the next few decades, the greater part of the range or uncertainty in projected global and regional change will be the result of a combination of natural variability (mostly related to uncertainty in specifying the initial conditions of the state of the ocean)[88] and scientific limitations in our ability to model and understand the Earth's climate system (Figure 4.5, Ch. 5: Circulation & Variability). Differences in future scenarios, shown in orange in Figure 4.5, represent the difference between scenarios, or human activity. Over the short term, this uncertainty is relatively small. As time progresses, however, differences in various possible future pathways become larger and the delayed ocean response to these differences begins to be realized. By about 2030, the human source of uncertainty becomes increasingly important in determining the magnitude and patterns of future global warming. Even though natural variability will continue to occur, most of the difference between present and future climates will be determined by choices that society makes today and over the next few decades. The further out in time we look, the greater the influence of these human choices are on the magnitude of future warming.

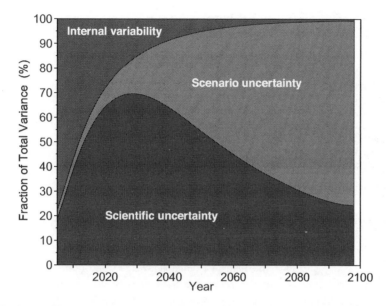

Figure 4.5: The fraction of total variance in decadal mean surface air temperature predictions explained by the three components of total uncertainty is shown for the lower 48 states (similar results are seen for Hawai'i and Alaska, not shown). Orange regions represent human or scenario uncertainty, blue regions represent scientific uncertainty, and green regions represent the internal variability component. As the size of the region is reduced, the relative importance of internal variability increases. In interpreting this figure, it is important to remember that it shows the fractional sources of uncertainty. Total uncertainty increases as time progresses. (Figure source: adapted from Hawkins and Sutton 2009[98]).

# TRACEABLE ACCOUNTS

### Key Finding 1

If greenhouse gas concentrations were stabilized at their current level, existing concentrations would commit the world to at least an additional 1.1°F (0.6°C) of warming over this century relative to the last few decades (*high confidence* in continued warming, *medium confidence* in amount of warming).

### Description of evidence base

The basic physics underlying the impact of human emissions on global climate, and the role of climate sensitivity in moderating the impact of those emissions on global temperature, has been documented since the 1800s in a series of peer-reviewed journal articles that is summarized in a collection titled, "The Warming Papers: The Scientific Foundation for the Climate Change Forecast".[101]

The estimate of committed warming at constant atmospheric concentrations is based on IPCC AR5 WG1, Chapter 12, section 12.5.2,[3] page 1103 which is in turn derived from AR4 WG1, Chapter 10, section 10.7.1,[28] page 822.

### Major uncertainties

The uncertainty in projected change under a commitment scenario is low and primarily the result of uncertainty in climate sensitivity. This key finding describes a hypothetical scenario that assumes all human-caused emissions cease and the Earth system responds only to what is already in the atmosphere.

### Assessment of confidence based on evidence and agreement, including short description of nature of evidence and level of agreement

The statement has *high confidence* in the sign of future change and *medium confidence* in the amount of warming, based on the estimate of committed warming at constant atmospheric concentrations from Collins et al.[3] based on Meehl et al.[28] for a hypothetical scenario where concentrations in the atmosphere were fixed at a known level.

### Summary sentence or paragraph that integrates the above information

The key finding is based on the basic physical principles of radiative transfer that have been well established for decades to centuries; the amount of estimated warming for this hypothetical scenario is derived from Collins et al.[3] which is in turn based on Meehl et al.[28] using CMIP3 models.

### Key Finding 2

Over the next two decades, global temperature increase is projected to be between 0.5°F and 1.3°F (0.3°–0.7°C) (*medium confidence*). This range is primarily due to uncertainties in natural sources of variability that affect short-term trends. In some regions, this means that the trend may not be distinguishable from natural variability (*high confidence*).

### Description of evidence base

The estimate of projected near-term warming under continued emissions of carbon dioxide and other greenhouse gases and aerosols was obtained directly from IPCC AR5 WG1.[61]

The statement regarding the sources of uncertainty in near-term projections and regional uncertainty is based on Hawkins and Sutton[98, 99] and Deser et al.[87, 88]

### Major uncertainties

As stated in the key finding, natural variability is the primary uncertainty in quantifying the amount of global temperature change over the next two decades.

### Assessment of confidence based on evidence and agreement, including short description of nature of evidence and level of agreement

The first statement regarding projected warming over the next two decades has *medium confidence* in the amount of warming due to the uncertainties described in the key finding. The second statement has *high confidence*, as the literature strongly supports the statement that natural variability is the primary source of uncertainty over time scales of years to decades.[87, 88, 89]

**Summary sentence or paragraph that integrates the above information**

The estimated warming presented in this Key Finding is based on calculations reported by Kirtman et al.[61] The key finding that natural variability is the most important uncertainty over the near-term is based on multiple peer reviewed publications.

**Key Finding 3**

Beyond the next few decades, the magnitude of climate change depends primarily on cumulative emissions of greenhouse gases and aerosols and the sensitivity of the climate system to those emissions (*high confidence*). Projected changes range from 4.7°–8.6°F (2.6°–4.8°C) under the higher scenario (RCP8.5) to 0.5°–1.3°F (0.3°–1.7°C) under the much lower scenario (RCP2.6), for 2081–2100 relative to 1986–2005 (*medium confidence*).

**Description of evidence base**

The estimate of projected long-term warming under continued emissions of carbon dioxide and other greenhouse gases and aerosols under the RCP scenarios was obtained directly from IPCC AR5 WG1.[3]

All credible climate models assessed in Chapter 9 of the IPCC WG1 AR5[64] from the simplest to the most complex respond with elevated global mean temperature, the simplest indicator of climate change, when atmospheric concentrations of greenhouse gases increase. It follows then that an emissions pathway that tracks or exceeds the higher scenario (RCP8.5) would lead to larger amounts of climate change.

The statement regarding the sources of uncertainty in long-term projections is based on Hawkins and Sutton.[98, 99]

**Major uncertainties**

As stated in the key finding, the magnitude of climate change over the long term is uncertain due to human emissions of greenhouse gases and climate sensitivity.

**Assessment of confidence based on evidence and agreement, including short description of nature of evidence and level of agreement**

The first statement regarding additional warming and its dependence on human emissions and climate sensitivity has *high confidence*, as understanding of the radiative properties of greenhouse gases and the existence of both positive and negative feedbacks in the climate system is basic physics, dating to the 19th century. The second has *medium confidence* in the specific magnitude of warming, due to the uncertainties described in the key finding.

**Summary sentence or paragraph that integrates the above information**

The estimated warming presented in this key finding is based on calculations reported by Collins et al.[3] The key finding that human emissions and climate sensitivity are the most important sources of uncertainty over the long-term is based on both basic physics regarding the radiative properties of greenhouse gases, as well as a large body of peer reviewed publications.

**Key Finding 4**

Global mean atmospheric carbon dioxide ($CO_2$) concentration has now passed 400 ppm, a level that last occurred about 3 million years ago, when global average temperature and sea level were significantly higher than today (*high confidence*). Continued growth in $CO_2$ emissions over this century and beyond would lead to an atmospheric concentration not experienced in tens of millions of years (*medium confidence*). The present-day emissions rate of nearly 10 GtC per year suggests that there is no climate analog for this century any time in at least the last 50 million years (*medium confidence*).

**Description of evidence base**

The key finding is based on a large body of research including Crowley,[10] Schneider et al.,[45] Lunt et al.,[46] Otto-Bleisner et al.,[47] NEEM,[48] Jouzel et al.,[49] Dutton et al.,[53] Seki et al.,[51] Haywood et al.,[52] Miller et al.,[54] Royer,[56] Bowen et al.,[7] Kirtland Turner et al.,[8] Penman et al.,[9] Zeebe et al.,[11] and summarized in NRC[38] and Masson-Delmotte et al.[102]

**Major uncertainties**

The largest uncertainty is the measurement of past sea

level, given the contributions of not only changes in land ice mass, but also in solid earth, mantle, isostatic adjustments, etc. that occur on timescales of millions of years. This uncertainty increases the further back in time we go; however, the signal (and forcing) size is also much greater. There are also associated uncertainties in precise quantification of past global mean temperature and carbon dioxide levels. There is uncertainty in the age models used to determine rates of change and coincidence of response at shorter, sub-millennial timescales.

**Assessment of confidence based on evidence and agreement, including short description of nature of evidence and level of agreement**

*High confidence* in the likelihood statement that past global mean temperature and sea level rise were higher with similar or higher $CO_2$ concentrations is based on Masson-Delmotte et al.[102] in IPCC AR5. *Medium confidence* that no precise analog exists in 66 million years is based on Zeebe et al.[11] as well as the larger body of literature summarized in Masson-Delmotte et al.[102]

**Summary sentence or paragraph that integrates the above information**

The key finding is based on a vast body of literature that summarizes the results of observations, paleoclimate analyses, and paleoclimate modeling over the past 50 years and more.

**Key Finding 5**

The observed increase in global carbon emissions over the past 15–20 years has been consistent with higher scenarios (*very high confidence*). In 2014 and 2015, emission growth rates slowed as economic growth has become less carbon-intensive (*medium confidence*). Even if this trend continues, however, it is not yet at a rate that would limit the increase in the global average temperature to well below 3.6°F (2°C) above preindustrial levels (*high confidence*).

**Description of Evidence Base**

Observed emissions for 2014 and 2015 and estimated emissions for 2016 suggest a decrease in the growth rate and possibly even emissions of carbon; this shift is attributed primarily to decreased coal use in China although with significant uncertainty as noted in the

references in the text. This statement is based on Tans and Keeling 2017;[4] Raupach et al. 2007;[5] Le Quéré et al. 2009;[6] Jackson et al. 2016;[12] Korsbakken et al. 2016[13] and personal communication with Le Quéré (2017).

The statement that the growth rate of carbon dioxide increased over the past 15–20 years is based on the data available here: https://www.esrl.noaa.gov/gmd/ccgg/trends/gr.html

The evidence that actual emission rates track or exceed the higher scenario (RCP8.5) is as follows. The actual emission of $CO_2$ from fossil fuel consumption and concrete manufacture over the period 2005–2014 is 90.11 Pg.[104] The emissions consistent with RCP8.5 over the same period assuming linear trends between years 2000, 2005, 2010, and 2020 in the specification is 99.24 Pg.

Actual emissions:

http://www.globalcarbonproject.org/ and Le Quéré et al.[103]

Emissions consistent with RCP8.5

http://tntcat.iiasa.ac.at:8787/RcpDb/dsd?Action=htmlpage&page=compare

The numbers for fossil fuel and industrial emissions (RCP) compared to fossil fuel and cement emissions (observed) in units of GtC are

|  | RCP8.5 | Actual | Difference |
|---|---|---|---|
| 2005 | 7.97 | 8.23 | 0.26 |
| 2006 | 8.16 | 8.53 | 0.36 |
| 2007 | 8.35 | 8.78 | 0.42 |
| 2008 | 8.54 | 8.96 | 0.42 |
| 2009 | 8.74 | 8.87 | 0.14 |
| 2010 | 8.93 | 9.21 | 0.28 |
| 2011 | 9.19 | 9.54 | 0.36 |
| 2012 | 9.45 | 9.69 | 0.24 |
| 2013 | 9.71 | 9.82 | 0.11 |
| 2014 | 9.97 | 9.89 | -0.08 |
| 2015 | 10.23 | 9.90 | -0.34 |
| total | 99.24 | 101.41 | 2.18 |

**Major Uncertainties**

None

**Assessment of confidence based on evidence and agreement, including short description of nature of evidence and level of agreement**

*Very high confidence* in increasing emissions over the last 20 years and *high confidence* in the fact that recent emission trends will not be sufficient to avoid 3.6°F (2°C). *Medium confidence* in recent findings that the growth rate is slowing. Climate change scales with the amount of anthropogenic greenhouse gas in the atmosphere. If emissions exceed those consistent with RCP8.5, the likely range of changes in temperatures and climate variables will be larger than projected.

**Summary sentence or paragraph that integrates the above information**

The key finding is based on basic physics relating emissions to concentrations, radiative forcing, and resulting change in global mean temperature, as well as on IEA data on national emissions as reported in the peer-reviewed literature.

**Key Finding 6**

Combining output from global climate models and dynamical and statistical downscaling models using advanced averaging, weighting, and pattern scaling approaches can result in more relevant and robust future projections. For some regions, sectors, and impacts, these techniques are increasing the ability of the scientific community to provide guidance on the use of climate projections for quantifying regional-scale changes and impacts (*medium to high confidence*).

**Description of evidence base**

The contribution of weighting and pattern scaling to improving the robustness of multimodel ensemble projections is described and quantified by a large body of literature as summarized in the text, including Sanderson et al.[76] and Knutti et al.[97] The state of the art of dynamical and statistical downscaling and the scientific community's ability to provide guidance regarding the application of climate projections to regional impact

assessments is summarized in Kotamarthi et al.[77] and supported by Feser et al.[78] and Prein et al.[79]

**Major uncertainties**

Regional climate models are subject to the same structural and parametric uncertainties as global models, as well as the uncertainty due to incorporating boundary conditions. The primary source of error in application of empirical statistical downscaling methods is inappropriate application, followed by stationarity.

**Assessment of confidence based on evidence and agreement, including short description of nature of evidence and level of agreement**

Advanced weighting techniques have significantly improved over previous Bayesian approaches; confidence in their ability to improve the robustness of multimodel ensembles, while currently rated as *medium*, is likely to grow in coming years. Downscaling has evolved significantly over the last decade and is now broadly viewed as a robust source for high-resolution climate projections that can be used as input to regional impact assessments.

**Summary sentence or paragraph that integrates the above information**

Scientific understanding of climate projections, downscaling, multimodel ensembles, and weighting has evolved significantly over the last decades to the extent that appropriate methods are now broadly viewed as robust sources for climate projections that can be used as input to regional impact assessments.

# REFERENCES

1. Hartmann, D.L., A.M.G. Klein Tank, M. Rusticucci, L.V. Alexander, S. Brönnimann, Y. Charabi, F.J. Dentener, E.J. Dlugokencky, D.R. Easterling, A. Kaplan, B.J. Soden, P.W. Thorne, M. Wild, and P.M. Zhai, 2013: Observations: Atmosphere and surface. *Climate Change 2013: The Physical Science Basis. Contribution of Working Group I to the Fifth Assessment Report of the Intergovernmental Panel on Climate Change.* Stocker, T.F., D. Qin, G.-K. Plattner, M. Tignor, S.K. Allen, J. Boschung, A. Nauels, Y. Xia, V. Bex, and P.M. Midgley, Eds. Cambridge University Press, Cambridge, United Kingdom and New York, NY, USA, 159–254. http://www.climatechange2013.org/report/full-report/

2. Rhein, M., S.R. Rintoul, S. Aoki, E. Campos, D. Chambers, R.A. Feely, S. Gulev, G.C. Johnson, S.A. Josey, A. Kostianoy, C. Mauritzen, D. Roemmich, L.D. Talley, and F. Wang, 2013: Observations: Ocean. *Climate Change 2013: The Physical Science Basis. Contribution of Working Group I to the Fifth Assessment Report of the Intergovernmental Panel on Climate Change.* Stocker, T.F., D. Qin, G.-K. Plattner, M. Tignor, S.K. Allen, J. Boschung, A. Nauels, Y. Xia, V. Bex, and P.M. Midgley, Eds. Cambridge University Press, Cambridge, United Kingdom and New York, NY, USA, 255–316. http://www.climatechange2013.org/report/full-report/

3. Collins, M., R. Knutti, J. Arblaster, J.-L. Dufresne, T. Fichefet, P. Friedlingstein, X. Gao, W.J. Gutowski, T. Johns, G. Krinner, M. Shongwe, C. Tebaldi, A.J. Weaver, and M. Wehner, 2013: Long-term climate change: Projections, commitments and irreversibility. *Climate Change 2013: The Physical Science Basis. Contribution of Working Group I to the Fifth Assessment Report of the Intergovernmental Panel on Climate Change.* Stocker, T.F., D. Qin, G.-K. Plattner, M. Tignor, S.K. Allen, J. Boschung, A. Nauels, Y. Xia, V. Bex, and P.M. Midgley, Eds. Cambridge University Press, Cambridge, United Kingdom and New York, NY, USA, 1029–1136. http://www.climatechange2013.org/report/full-report/

4. Tans, P. and R. Keeling, 2017: Trends in Atmospheric Carbon Dioxide. Annual Mean Growth Rate of CO2 at Mauna Loa. NOAA Earth System Research Laboratory. https://www.esrl.noaa.gov/gmd/ccgg/trends/gr.html

5. Raupach, M.R., G. Marland, P. Ciais, C. Le Quéré, J.G. Canadell, G. Klepper, and C.B. Field, 2007: Global and regional drivers of accelerating CO2 emissions. *Proceedings of the National Academy of Sciences*, **104**, 10288-10293. http://dx.doi.org/10.1073/pnas.0700609104

6. Le Quéré, C., M.R. Raupach, J.G. Canadell, G. Marland, L. Bopp, P. Ciais, T.J. Conway, S.C. Doney, R.A. Feely, P. Foster, P. Friedlingstein, K. Gurney, R.A. Houghton, J.I. House, C. Huntingford, P.E. Levy, M.R. Lomas, J. Majkut, N. Metzl, J.P. Ometto, G.P. Peters, I.C. Prentice, J.T. Randerson, S.W. Running, J.L. Sarmiento, U. Schuster, S. Sitch, T. Takahashi, N. Viovy, G.R. van der Werf, and F.I. Woodward, 2009: Trends in the sources and sinks of carbon dioxide. *Nature Geoscience*, **2**, 831-836. http://dx.doi.org/10.1038/ngeo689

7. Bowen, G.J., B.J. Maibauer, M.J. Kraus, U. Rohl, T. Westerhold, A. Steimke, P.D. Gingerich, S.L. Wing, and W.C. Clyde, 2015: Two massive, rapid releases of carbon during the onset of the Palaeocene-Eocene thermal maximum. *Nature Geoscience*, **8**, 44-47. http://dx.doi.org/10.1038/ngeo2316

8. Kirtland Turner, S., P.F. Sexton, C.D. Charles, and R.D. Norris, 2014: Persistence of carbon release events through the peak of early Eocene global warmth. *Nature Geoscience*, **7**, 748-751. http://dx.doi.org/10.1038/ngeo2240

9. Penman, D.E., B. Hönisch, R.E. Zeebe, E. Thomas, and J.C. Zachos, 2014: Rapid and sustained surface ocean acidification during the Paleocene-Eocene Thermal Maximum. *Paleoceanography*, **29**, 357-369. http://dx.doi.org/10.1002/2014PA002621

10. Crowley, T.J., 1990: Are there any satisfactory geologic analogs for a future greenhouse warming? *Journal of Climate*, **3**, 1282-1292. http://dx.doi.org/10.1175/1520-0442(1990)003<1282:atasga>2.0.co;2

11. Zeebe, R.E., A. Ridgwell, and J.C. Zachos, 2016: Anthropogenic carbon release rate unprecedented during the past 66 million years. *Nature Geoscience*, **9**, 325-329. http://dx.doi.org/10.1038/ngeo2681

12. Jackson, R.B., J.G. Canadell, C. Le Quere, R.M. Andrew, J.I. Korsbakken, G.P. Peters, and N. Nakicenovic, 2016: Reaching peak emissions. *Nature Climate Change*, **6**, 7-10. http://dx.doi.org/10.1038/nclimate2892

13. Korsbakken, J.I., G.P. Peters, and R.M. Andrew, 2016: Uncertainties around reductions in China's coal use and CO2 emissions. *Nature Climate Change*, **6**, 687-690. http://dx.doi.org/10.1038/nclimate2963

14. IEA, 2016: Decoupling of global emissions and economic growth confirmed. International Energy Agency, March 16. https://www.iea.org/newsroomandevents/pressreleases/2016/march/decoupling-of-global-emissions-and-economic-growth-confirmed.html

15. Green, F. and N. Stern, 2016: China's changing economy: Implications for its carbon dioxide emissions. *Climate Policy*, **17**, 423-442. http://dx.doi.org/10.1080/14693062.2016.1156515

16. Bretherton, F., K. Bryan, J. Woods, J. Hansen, M. Hoffert, X. Jiang, S. Manabe, G. Meehl, S. Raper, D. Rind, M. Schlesinger, R. Stouffer, T. Volk, and T. Wigley, 1990: Time-dependent greenhouse-gas-induced climate change. *Climate Change: The IPCC Scientific Assessment Report prepared for Intergovernmental Panel on Climate Change by Working Group I* Houghton, J.T., G.J. Jenkins, and J.J. Ephraums, Eds. Cambridge University Press, Cambridge, United Kingdom and New York, NY, USA, 173-193. https://www.ipcc.ch/publications_and_data/publications_ipcc_first_assessment_1990_wg1.shtml

17. UNFCCC, 2015: Paris Agreement. United Nations Framework Convention on Climate Change, [Bonn, Germany]. 25 pp. http://unfccc.int/files/essential_background/convention/application/pdf/english_paris_agreement.pdf

18. Smith, P., S.J. Davis, F. Creutzig, S. Fuss, J. Minx, B. Gabrielle, E. Kato, R.B. Jackson, A. Cowie, E. Kriegler, D.P. van Vuuren, J. Rogelj, P. Ciais, J. Milne, J.G. Canadell, D. McCollum, G. Peters, R. Andrew, V. Krey, G. Shrestha, P. Friedlingstein, T. Gasser, A. Grubler, W.K. Heidug, M. Jonas, C.D. Jones, F. Kraxner, E. Littleton, J. Lowe, J.R. Moreira, N. Nakicenovic, M. Obersteiner, A. Patwardhan, M. Rogner, E. Rubin, A. Sharifi, A. Torvanger, Y. Yamagata, J. Edmonds, and C. Yongsung, 2016: Biophysical and economic limits to negative CO2 emissions. *Nature Climate Change*, **6**, 42-50. http://dx.doi.org/10.1038/nclimate2870

19. IPCC, 1990: *Climate Change: The IPCC Scientific Assessment*. Houghton, J.T., G.J. Jenkins, and J.J. Ephraums, Eds. Cambridge University Press, Cambridge, United Kingdom and New York, NY, USA, 212 pp. https://www.ipcc.ch/publications_and_data/publications_ipcc_first_assessment_1990_wg1.shtml

20. Leggett, J., W.J. Pepper, R.J. Swart, J. Edmonds, L.G.M. Filho, I. Mintzer, M.X. Wang, and J. Watson, 1992: Emissions scenarios for the IPCC: An update. *Climate Change 1992: The Supplementary Report to the IPCC Scientific Assessment*. Houghton, J.T., B.A. Callander, and S.K. Varney, Eds. Cambridge University Press, Cambridge, United Kingdom, New York, NY, USA, and Victoria, Australia, 73-95. https://www.ipcc.ch/ipccreports/1992%20IPCC%20Supplement/IPCC_Suppl_Report_1992_wg_I/ipcc_wg_I_1992_suppl_report_section_a3.pdf

21. Nakicenovic, N., J. Alcamo, G. Davis, B.d. Vries, J. Fenhann, S. Gaffin, K. Gregory, A. Grübler, T.Y. Jung, T. Kram, E.L.L. Rovere, L. Michaelis, S. Mori, T. Morita, W. Pepper, H. Pitcher, L. Price, K. Riahi, A. Roehrl, H.-H. Rogner, A. Sankovski, M. Schlesinger, P. Shukla, S. Smith, R. Swart, S.v. Rooijen, N. Victor, and Z. Dadi, 2000: IPCC Special Report on Emissions Scenarios. Nakicenovic, N. and R. Swart (Eds.). Cambridge University Press. http://www.ipccreports/sres/emission/index.php?idp=0

22. Moss, R.H., J.A. Edmonds, K.A. Hibbard, M.R. Manning, S.K. Rose, D.P. van Vuuren, T.R. Carter, S. Emori, M. Kainuma, T. Kram, G.A. Meehl, J.F.B. Mitchell, N. Nakicenovic, K. Riahi, S.J. Smith, R.J. Stouffer, A.M. Thomson, J.P. Weyant, and T.J. Wilbanks, 2010: The next generation of scenarios for climate change research and assessment. *Nature*, **463**, 747-756. http://dx.doi.org/10.1038/nature08823

23. Kattenberg, A., F. Giorgi, H. Grassl, G. Meehl, J. Mitchell, R. Stouffer, T. Tokioka, A. Weaver, and T. Wigley, 1996: Climate models - projections of future climate. *Climate Change 1995: The Science of Climate Change. Contribution of Working Group I to the Second Assessment Report of the Intergovernmental Panel on Climate Change*. Houghton, J.T., L.G. Meira Filho, B.A. Callander, N. Harris, A. Kattenberg, and K. Maskell, Eds. Cambridge University Press, Cambridge, United Kingdom and New York, NY, USA, 285-358. https://www.ipcc.ch/ipccreports/sar/wg_I/ipcc_sar_wg_I_full_report.pdf

24. Cubasch, U., G. Meehl, G. Boer, R. Stouffer, M. Dix, A. Noda, C. Senior, S. Raper, and K. Yap, 2001: Projections of future climate change. *Climate Change 2001: The Scientific Basis. Contribution of Working Group I to the Third Assessment Report of the Intergovernmental Panel on Climate Change*. Houghton, J.T., Y. Ding, D.J. Griggs, M. Noquer, P.J. van der Linden, X. Dai, K. Maskell, and C.A. Johnson, Eds. Cambridge University Press, Cambridge, United Kingdom and New York, NY, USA, 525-582. https://www.ipccreports/tar/wg1/pdf/TAR-09.PDF

25. NAST, 2001: Climate Change Impacts on the United States: The Potential Consequences of Climate Variability and Change, Report for the US Global Change Research Program. U.S. Global Climate Research Program, National Assessment Synthesis Team, Cambridge, UK. 620 pp. http://www.globalchange.gov/browse/reports/climate-change-impacts-united-states-potential-consequences-climate-variability-and-3

26. Karl, T.R., J.T. Melillo, and T.C. Peterson, eds., 2009: *Global Climate Change Impacts in the United States*. Cambridge University Press: New York, NY, 189 pp. http://downloads.globalchange.gov/usimpacts/pdfs/climate-impacts-report.pdf

27. Melillo, J.M., T.C. Richmond, and G.W. Yohe, eds., 2014: *Climate Change Impacts in the United States: The Third National Climate Assessment*. U.S. Global Change Research Program: Washington, D.C., 841 pp. http://dx.doi.org/10.7930/J0Z31WJ2

28. Meehl, G.A., T.F. Stocker, W.D. Collins, P. Friedling-stein, A.T. Gaye, J.M. Gregory, A. Kitoh, R. Knutti, J.M. Murphy, A. Noda, S.C.B. Raper, I.G. Watterson, A.J. Weaver, and Z.-C. Zhao, 2007: Ch. 10: Global climate projections. *Climate Change 2007: The Physical Science basis: Contribution of Working Group I to the Fourth Assessment Report of the Intergovernmental Panel on Climate Change.* Solomon, S., D. Qin, M. Manning, Z. Chen, M. Marquis, K.B. Averyt, M. Tignor, and H.L. Miller, Eds. Cambridge University Press, Cambridge, UK and New York, NY, 747-845. http://www.ipcc.ch/pdf/assessment-report/ar4/wg1/ar4-wg1-chapter10.pdf

29. van Vuuren, D.P., S. Deetman, M.G.J. den Elzen, A. Hof, M. Isaac, K. Klein Goldewijk, T. Kram, A. Mendoza Beltran, E. Stehfest, and J. van Vliet, 2011: RCP2.6: Exploring the possibility to keep global mean temperature increase below 2°C. *Climatic Change*, **109**, 95-116. http://dx.doi.org/10.1007/s10584-011-0152-3

30. Thomson, A.M., K.V. Calvin, S.J. Smith, G.P. Kyle, A. Volke, P. Patel, S. Delgado-Arias, B. Bond-Lamberty, M.A. Wise, and L.E. Clarke, 2011: RCP4.5: A pathway for stabilization of radiative forcing by 2100. *Climatic Change*, **109**, 77-94. http://dx.doi.org/10.1007/s10584-011-0151-4

31. Masui, T., K. Matsumoto, Y. Hijioka, T. Kinoshita, T. Nozawa, S. Ishiwatari, E. Kato, P.R. Shukla, Y. Yamagata, and M. Kainuma, 2011: An emission pathway for stabilization at 6 Wm⁻² radiative forcing. *Climatic Change*, **109**, 59. http://dx.doi.org/10.1007/s10584-011-0150-5

32. Riahi, K., S. Rao, V. Krey, C. Cho, V. Chirkov, G. Fischer, G. Kindermann, N. Nakicenovic, and P. Rafaj, 2011: RCP 8.5—A scenario of comparatively high greenhouse gas emissions. *Climatic Change*, **109**, 33-57. http://dx.doi.org/10.1007/s10584-011-0149-y

33. Meinshausen, M., S.J. Smith, K. Calvin, J.S. Daniel, M.L.T. Kainuma, J.-F. Lamarque, K. Matsumoto, S.A. Montzka, S.C.B. Raper, K. Riahi, A. Thomson, G.J.M. Velders, and D.P.P. van Vuuren, 2011: The RCP greenhouse gas concentrations and their extensions from 1765 to 2300. *Climatic Change*, **109**, 213-241. http://dx.doi.org/10.1007/s10584-011-0156-z

34. Cubasch, U., D. Wuebbles, D. Chen, M.C. Facchini, D. Frame, N. Mahowald, and J.-G. Winther, 2013: Introduction. *Climate Change 2013: The Physical Science Basis. Contribution of Working Group I to the Fifth Assessment Report of the Intergovernmental Panel on Climate Change.* Stocker, T.F., D. Qin, G.-K. Plattner, M. Tignor, S.K. Allen, J. Boschung, A. Nauels, Y. Xia, V. Bex, and P.M. Midgley, Eds. Cambridge University Press, Cambridge, United Kingdom and New York, NY, USA, 119–158. http://www.climatechange2013.org/report/full-report/

35. O'Neill, B.C., E. Kriegler, K. Riahi, K.L. Ebi, S. Hallegatte, T.R. Carter, R. Mathur, and D.P. van Vuuren, 2014: A new scenario framework for climate change research: The concept of shared socioeconomic pathways. *Climatic Change*, **122**, 387-400. http://dx.doi.org/10.1007/s10584-013-0905-2

36. IIASA, 2016: RCP Database. Version 2.0.5. International Institute for Applied Systems Analysis. https://tntcat.iiasa.ac.at/RcpDb/dsd?Action=htmlpage&page=compare

37. Sanderson, B.M., B.C. O'Neill, and C. Tebaldi, 2016: What would it take to achieve the Paris temperature targets? *Geophysical Research Letters*, **43**, 7133-7142. http://dx.doi.org/10.1002/2016GL069563

38. NRC, 2011: *Climate Stabilization Targets: Emissions, Concentrations, and Impacts over Decades to Millennia.* National Research Council. The National Academies Press, Washington, D.C., 298 pp. http://dx.doi.org/10.17226/12877

39. Frieler, K., M. Meinshausen, A. Golly, M. Mengel, K. Lebek, S.D. Donner, and O. Hoegh-Guldberg, 2013: Limiting global warming to 2°C is unlikely to save most coral reefs. *Nature Climate Change*, **3**, 165-170. http://dx.doi.org/10.1038/nclimate1674

40. Swain, S. and K. Hayhoe, 2015: CMIP5 projected changes in spring and summer drought and wet conditions over North America. *Climate Dynamics*, **44**, 2737-2750. http://dx.doi.org/10.1007/s00382-014-2255-9

41. Herger, N., B.M. Sanderson, and R. Knutti, 2015: Improved pattern scaling approaches for the use in climate impact studies. *Geophysical Research Letters*, **42**, 3486-3494. http://dx.doi.org/10.1002/2015GL063569

42. Mitchell, T.D., 2003: Pattern scaling: An examination of the accuracy of the technique for describing future climates. *Climatic Change*, **60**, 217-242. http://dx.doi.org/10.1023/a:1026035305597

43. Fix, M.J., D. Cooley, S.R. Sain, and C. Tebaldi, 2016: A comparison of U.S. precipitation extremes under RCP8.5 and RCP4.5 with an application of pattern scaling. *Climatic Change*, **First online**, 1-13. http://dx.doi.org/10.1007/s10584-016-1656-7

44. Tebaldi, C. and J.M. Arblaster, 2014: Pattern scaling: Its strengths and limitations, and an update on the latest model simulations. *Climatic Change*, **122**, 459-471. http://dx.doi.org/10.1007/s10584-013-1032-9

45. Schneider, R., J. Schmitt, P. Köhler, F. Joos, and H. Fischer, 2013: A reconstruction of atmospheric carbon dioxide and its stable carbon isotopic composition from the penultimate glacial maximum to the last glacial inception. *Climate of the Past*, **9**, 2507-2523. http://dx.doi.org/10.5194/cp-9-2507-2013

46. Lunt, D.J., T. Dunkley Jones, M. Heinemann, M. Huber, A. LeGrande, A. Winguth, C. Loptson, J. Marotzke, C.D. Roberts, J. Tindall, P. Valdes, and C. Winguth, 2012: A model–data comparison for a multi-model ensemble of early Eocene atmosphere–ocean simulations: EoMIP. *Climate of the Past*, **8**, 1717-1736. http://dx.doi.org/10.5194/cp-8-1717-2012

47. Otto-Bliesner, B.L., N. Rosenbloom, E.J. Stone, N.P. McKay, D.J. Lunt, E.C. Brady, and J.T. Overpeck, 2013: How warm was the last interglacial? New model–data comparisons. *Philosophical Transactions of the Royal Society A: Mathematical, Physical and Engineering Sciences*, **371**, 20130097. http://dx.doi.org/10.1098/rsta.2013.0097

48. NEEM, 2013: Eemian interglacial reconstructed from a Greenland folded ice core. *Nature*, **493**, 489-494. http://dx.doi.org/10.1038/nature11789

49. Jouzel, J., V. Masson-Delmotte, O. Cattani, G. Dreyfus, S. Falourd, G. Hoffmann, B. Minster, J. Nouet, J.M. Barnola, J. Chappellaz, H. Fischer, J.C. Gallet, S. Johnsen, M. Leuenberger, L. Loulergue, D. Luethi, H. Oerter, F. Parrenin, G. Raisbeck, D. Raynaud, A. Schilt, J. Schwander, E. Selmo, R. Souchez, R. Spahni, B. Stauffer, J.P. Steffensen, B. Stenni, T.F. Stocker, J.L. Tison, M. Werner, and E.W. Wolff, 2007: Orbital and millennial Antarctic climate variability over the past 800,000 years. *Science*, **317**, 793-796. http://dx.doi.org/10.1126/science.1141038

50. Kopp, R.E., F.J. Simons, J.X. Mitrovica, A.C. Maloof, and M. Oppenheimer, 2009: Probabilistic assessment of sea level during the last interglacial stage. *Nature*, **462**, 863-867. http://dx.doi.org/10.1038/nature08686

51. Seki, O., G.L. Foster, D.N. Schmidt, A. Mackensen, K. Kawamura, and R.D. Pancost, 2010: Alkenone and boron-based Pliocene pCO2 records. *Earth and Planetary Science Letters*, **292**, 201-211. http://dx.doi.org/10.1016/j.epsl.2010.01.037

52. Haywood, A.M., D.J. Hill, A.M. Dolan, B.L. Otto-Bliesner, F. Bragg, W.L. Chan, M.A. Chandler, C. Contoux, H.J. Dowsett, A. Jost, Y. Kamae, G. Lohmann, D.J. Lunt, A. Abe-Ouchi, S.J. Pickering, G. Ramstein, N.A. Rosenbloom, U. Salzmann, L. Sohl, C. Stepanek, H. Ueda, Q. Yan, and Z. Zhang, 2013: Large-scale features of Pliocene climate: Results from the Pliocene Model Intercomparison Project. *Climate of the Past*, **9**, 191-209. http://dx.doi.org/10.5194/cp-9-191-2013

53. Dutton, A., A.E. Carlson, A.J. Long, G.A. Milne, P.U. Clark, R. DeConto, B.P. Horton, S. Rahmstorf, and M.E. Raymo, 2015: Sea-level rise due to polar ice-sheet mass loss during past warm periods. *Science*, **349**, aaa4019. http://dx.doi.org/10.1126/science.aaa4019

54. Miller, K.G., J.D. Wright, J.V. Browning, A. Kulpecz, M. Kominz, T.R. Naish, B.S. Cramer, Y. Rosenthal, W.R. Peltier, and S. Sosdian, 2012: High tide of the warm Pliocene: Implications of global sea level for Antarctic deglaciation. *Geology*, **40**, 407-410. http://dx.doi.org/10.1130/g32869.1

55. Jagniecki, E.A., T.K. Lowenstein, D.M. Jenkins, and R.V. Demicco, 2015: Eocene atmospheric CO2 from the nahcolite proxy. *Geology*, **43**, 1075-1078. http://dx.doi.org/10.1130/g36886.1

56. Royer, D.L., 2014: 6.11 - Atmospheric $CO_2$ and $O_2$ during the Phanerozoic: Tools, patterns, and impacts. *Treatise on Geochemistry (Second Edition)*. Holland, H.D. and K.K. Turekian, Eds. Elsevier, Amsterdam, Netherlands, 251-267. http://dx.doi.org/10.1016/B978-0-08-095975-7.01311-5

57. Pagani, M., M. Huber, Z. Liu, S.M. Bohaty, J. Henderiks, W. Sijp, S. Krishnan, and R.M. DeConto, 2011: The role of carbon dioxide during the onset of Antarctic glaciation. *Science*, **334**, 1261-1264. http://dx.doi.org/10.1126/science.1203909

58. DeConto, R.M. and D. Pollard, 2016: Contribution of Antarctica to past and future sea-level rise. *Nature*, **531**, 591-597. http://dx.doi.org/10.1038/nature17145

59. Gasson, E., D.J. Lunt, R. DeConto, A. Goldner, M. Heinemann, M. Huber, A.N. LeGrande, D. Pollard, N. Sagoo, M. Siddall, A. Winguth, and P.J. Valdes, 2014: Uncertainties in the modelled $CO_2$ threshold for Antarctic glaciation. *Climate of the Past*, **10**, 451-466. http://dx.doi.org/10.5194/cp-10-451-2014

60. Vaughan, D.G., J.C. Comiso, I. Allison, J. Carrasco, G. Kaser, R. Kwok, P. Mote, T. Murray, F. Paul, J. Ren, E. Rignot, O. Solomina, K. Steffen, and T. Zhang, 2013: Observations: Cryosphere. *Climate Change 2013: The Physical Science Basis. Contribution of Working Group I to the Fifth Assessment Report of the Intergovernmental Panel on Climate Change.* Stocker, T.F., D. Qin, G.-K. Plattner, M. Tignor, S.K. Allen, J. Boschung, A. Nauels, Y. Xia, V. Bex, and P.M. Midgley, Eds. Cambridge University Press, Cambridge, United Kingdom and New York, NY, USA, 317–382. http://www.climatechange2013.org/report/full-report/

61. Kirtman, B., S.B. Power, J.A. Adedoyin, G.J. Boer, R. Bojariu, I. Camilloni, F.J. Doblas-Reyes, A.M. Fiore, M. Kimoto, G.A. Meehl, M. Prather, A. Sarr, C. Schär, R. Sutton, G.J. van Oldenborgh, G. Vecchi, and H.J. Wang, 2013: Near-term climate change: Projections and predictability. *Climate Change 2013: The Physical Science Basis. Contribution of Working Group I to the Fifth Assessment Report of the Intergovernmental Panel on Climate Change.* Stocker, T.F., D. Qin, G.-K. Plattner, M. Tignor, S.K. Allen, J. Boschung, A. Nauels, Y. Xia, V. Bex, and P.M. Midgley, Eds. Cambridge University Press, Cambridge, UK and New York, NY, USA, 953–1028. http://www.climatechange2013.org/report/full-report/

62. Vaittinada Ayar, P., M. Vrac, S. Bastin, J. Carreau, M. Déqué, and C. Gallardo, 2016: Intercomparison of statistical and dynamical downscaling models under the EURO- and MED-CORDEX initiative framework: Present climate evaluations. *Climate Dynamics*, **46**, 1301-1329. http://dx.doi.org/10.1007/s00382-015-2647-5

63. Pierce, D.W., D.R. Cayan, and B.L. Thrasher, 2014: Statistical downscaling using Localized Constructed Analogs (LOCA). *Journal of Hydrometeorology*, **15**, 2558-2585. http://dx.doi.org/10.1175/jhm-d-14-0082.1

64. Flato, G., J. Marotzke, B. Abiodun, P. Braconnot, S.C. Chou, W. Collins, P. Cox, F. Driouech, S. Emori, V. Eyring, C. Forest, P. Gleckler, E. Guilyardi, C. Jakob, V. Kattsov, C. Reason, and M. Rummukainen, 2013: Evaluation of climate models. *Climate Change 2013: The Physical Science Basis. Contribution of Working Group I to the Fifth Assessment Report of the Intergovernmental Panel on Climate Change*. Stocker, T.F., D. Qin, G.-K. Plattner, M. Tignor, S.K. Allen, J. Boschung, A. Nauels, Y. Xia, V. Bex, and P.M. Midgley, Eds. Cambridge University Press, Cambridge, United Kingdom and New York, NY, USA, 741–866. http://www.climatechange2013.org/report/full-report/

65. Knutti, R. and J. Sedláček, 2013: Robustness and uncertainties in the new CMIP5 climate model projections. *Nature Climate Change*, **3**, 369-373. http://dx.doi.org/10.1038/nclimate1716

66. Kumar, D., E. Kodra, and A.R. Ganguly, 2014: Regional and seasonal intercomparison of CMIP3 and CMIP5 climate model ensembles for temperature and precipitation. *Climate Dynamics*, **43**, 2491-2518. http://dx.doi.org/10.1007/s00382-014-2070-3

67. Sheffield, J., A.P. Barrett, B. Colle, D.N. Fernando, R. Fu, K.L. Geil, Q. Hu, J. Kinter, S. Kumar, B. Langenbrunner, K. Lombardo, L.N. Long, E. Maloney, A. Mariotti, J.E. Meyerson, K.C. Mo, J.D. Neelin, S. Nigam, Z. Pan, T. Ren, A. Ruiz-Barradas, Y.L. Serra, A. Seth, J.M. Thibeault, J.C. Stroeve, Z. Yang, and L. Yin, 2013: North American climate in CMIP5 experiments. Part I: Evaluation of historical simulations of continental and regional climatology. *Journal of Climate*, **26**, 9209-9245. http://dx.doi.org/10.1175/jcli-d-12-00592.1

68. Sheffield, J., A. Barrett, D. Barrie, S.J. Camargo, E.K.M. Chang, B. Colle, D.N. Fernando, R. Fu, K.L. Geil, Q. Hu, X. Jiang, N. Johnson, K.B. Karnauskas, S.T. Kim, J. Kinter, S. Kumar, B. Langenbrunner, K. Lombardo, L.N. Long, E. Maloney, A. Mariotti, J.E. Meyerson, K.C. Mo, J.D. Neelin, S. Nigam, Z. Pan, T. Ren, A. Ruiz-Barradas, R. Seager, Y.L. Serra, A. Seth, D.-Z. Sun, J.M. Thibeault, J.C. Stroeve, C. Wang, S.-P. Xie, Z. Yang, L. Yin, J.-Y. Yu, T. Zhang, and M. Zhao, 2014: Regional Climate Processes and Projections for North America: CMIP3/CMIP5 Differences, Attribution and Outstanding Issues. NOAA Technical Report OAR CPO-2. NOAA Climate Program Office, Silver Spring, MD. 47 pp. http://dx.doi.org/10.7289/V5D-B7ZRC

69. Bellenger, H., E. Guilyardi, J. Leloup, M. Lengaigne, and J. Vialard, 2014: ENSO representation in climate models: From CMIP3 to CMIP5. *Climate Dynamics*, **42**, 1999-2018. http://dx.doi.org/10.1007/s00382-013-1783-z

70. Lauer, A. and K. Hamilton, 2013: Simulating clouds with global climate models: A comparison of CMIP5 results with CMIP3 and satellite data. *Journal of Climate*, **26**, 3823-3845. http://dx.doi.org/10.1175/jcli-d-12-00451.1

71. Wang, M. and J.E. Overland, 2012: A sea ice free summer Arctic within 30 years: An update from CMIP5 models. *Geophysical Research Letters*, **39**, L18501. http://dx.doi.org/10.1029/2012GL052868

72. Knutson, T.R., J.J. Sirutis, G.A. Vecchi, S. Garner, M. Zhao, H.-S. Kim, M. Bender, R.E. Tuleya, I.M. Held, and G. Villarini, 2013: Dynamical downscaling projections of twenty-first-century Atlantic hurricane activity: CMIP3 and CMIP5 model-based scenarios. *Journal of Climate*, **27**, 6591-6617. http://dx.doi.org/10.1175/jcli-d-12-00539.1

73. Kharin, V.V., F.W. Zwiers, X. Zhang, and M. Wehner, 2013: Changes in temperature and precipitation extremes in the CMIP5 ensemble. *Climatic Change*, **119**, 345-357. http://dx.doi.org/10.1007/s10584-013-0705-8

74. Sun, L., K.E. Kunkel, L.E. Stevens, A. Buddenberg, J.G. Dobson, and D.R. Easterling, 2015: Regional Surface Climate Conditions in CMIP3 and CMIP5 for the United States: Differences, Similarities, and Implications for the U.S. National Climate Assessment. NOAA Technical Report NESDIS 144. National Oceanic and Atmospheric Administration, National Environmental Satellite, Data, and Information Service, 111 pp. http://dx.doi.org/10.7289/V5RB72KG

75. Knutti, R., D. Masson, and A. Gettelman, 2013: Climate model genealogy: Generation CMIP5 and how we got there. *Geophysical Research Letters*, **40**, 1194-1199. http://dx.doi.org/10.1002/grl.50256

76. Sanderson, B.M., R. Knutti, and P. Caldwell, 2015: A representative democracy to reduce interdependency in a multimodel ensemble. *Journal of Climate*, **28**, 5171-5194. http://dx.doi.org/10.1175/JCLI-D-14-00362.1

77. Kotamarthi, R., L. Mearns, K. Hayhoe, C. Castro, and D. Wuebbles, 2016: Use of Climate Information for Decision-Making and Impact Research. U.S. Department of Defense, Strategic Environment Research and Development Program Report, 55 pp. http://dx.doi.org/10.13140/RG.2.1.1986.0085

78. Feser, F., B. Rockel, H.v. Storch, J. Winterfeldt, and M. Zahn, 2011: Regional climate models add value to global model data: A review and selected examples. *Bulletin of the American Meteorological Society*, **92**, 1181-1192. http://dx.doi.org/10.1175/2011BAMS3061.1

79. Prein, A.F., W. Langhans, G. Fosser, A. Ferrone, N. Ban, K. Goergen, M. Keller, M. Tölle, O. Gutjahr, F. Feser, E. Brisson, S. Kollet, J. Schmidli, N.P.M. van Lipzig, and R. Leung, 2015: A review on regional convection-permitting climate modeling: Demonstrations, prospects, and challenges. *Reviews of Geophysics*, **53**, 323-361. http://dx.doi.org/10.1002/2014RG000475

80. Wang, Y., L.R. Leung, J.L. McGregor, D.-K. Lee, W.-C. Wang, Y. Ding, and F. Kimura, 2004: Regional climate modeling: Progress, challenges, and prospects. *Journal of the Meteorological Society of Japan. Series II*, **82**, 1599-1628. http://dx.doi.org/10.2151/jmsj.82.1599

81. Xie, S.-P., C. Deser, G.A. Vecchi, M. Collins, T.L. Delworth, A. Hall, E. Hawkins, N.C. Johnson, C. Cassou, A. Giannini, and M. Watanabe, 2015: Towards predictive understanding of regional climate change. *Nature Climate Change*, **5**, 921-930. http://dx.doi.org/10.1038/nclimate2689

82. Stoner, A.M.K., K. Hayhoe, X. Yang, and D.J. Wuebbles, 2012: An asynchronous regional regression model for statistical downscaling of daily climate variables. *International Journal of Climatology*, **33**, 2473-2494. http://dx.doi.org/10.1002/joc.3603

83. Vrac, M., M. Stein, and K. Hayhoe, 2007: Statistical downscaling of precipitation through nonhomogeneous stochastic weather typing. *Climate Research*, **34**, 169-184. http://dx.doi.org/10.3354/cr00696

84. Brands, S., J.M. Gutiérrez, S. Herrera, and A.S. Cofiño, 2012: On the use of reanalysis data for downscaling. *Journal of Climate*, **25**, 2517-2526. http://dx.doi.org/10.1175/jcli-d-11-00251.1

85. Thrasher, B., J. Xiong, W. Wang, F. Melton, A. Michaelis, and R. Nemani, 2013: Downscaled climate projections suitable for resource management. *Eos, Transactions, American Geophysical Union*, **94**, 321-323. http://dx.doi.org/10.1002/2013EO370002

86. Dixon, K.W., J.R. Lanzante, M.J. Nath, K. Hayhoe, A. Stoner, A. Radhakrishnan, V. Balaji, and C.F. Gaitán, 2016: Evaluating the stationarity assumption in statistically downscaled climate projections: Is past performance an indicator of future results? *Climatic Change*, **135**, 395-408. http://dx.doi.org/10.1007/s10584-016-1598-0

87. Deser, C., A. Phillips, V. Bourdette, and H. Teng, 2012: Uncertainty in climate change projections: The role of internal variability. *Climate Dynamics*, **38**, 527-546. http://dx.doi.org/10.1007/s00382-010-0977-x

88. Deser, C., R. Knutti, S. Solomon, and A.S. Phillips, 2012: Communication of the role of natural variability in future North American climate. *Nature Climate Change*, **2**, 775-779. http://dx.doi.org/10.1038/nclimate1562

89. Deser, C., A.S. Phillips, M.A. Alexander, and B.V. Smoliak, 2014: Projecting North American climate over the next 50 years: Uncertainty due to internal variability. *Journal of Climate*, **27**, 2271-2296. http://dx.doi.org/10.1175/JCLI-D-13-00451.1

90. Wang, M., J.E. Overland, V. Kattsov, J.E. Walsh, X. Zhang, and T. Pavlova, 2007: Intrinsic versus forced variation in coupled climate model simulations over the Arctic during the twentieth century. *Journal of Climate*, **20**, 1093-1107. http://dx.doi.org/10.1175/JCLI4043.1

91. Wang, C., L. Zhang, S.-K. Lee, L. Wu, and C.R. Mechoso, 2014: A global perspective on CMIP5 climate model biases. *Nature Climate Change*, **4**, 201-205. http://dx.doi.org/10.1038/nclimate2118

92. Ryu, J.-H. and K. Hayhoe, 2014: Understanding the sources of Caribbean precipitation biases in CMIP3 and CMIP5 simulations. *Climate Dynamics*, **42**, 3233-3252. http://dx.doi.org/10.1007/s00382-013-1801-1

93. Braconnot, P., S.P. Harrison, M. Kageyama, P.J. Bartlein, V. Masson-Delmotte, A. Abe-Ouchi, B. Otto-Bliesner, and Y. Zhao, 2012: Evaluation of climate models using palaeoclimatic data. *Nature Climate Change*, **2**, 417-424. http://dx.doi.org/10.1038/nclimate1456

94. Jun, M., R. Knutti, and D.W. Nychka, 2008: Local eigenvalue analysis of CMIP3 climate model errors. *Tellus A*, **60**, 992-1000. http://dx.doi.org/10.1111/j.1600-0870.2008.00356.x

95. Giorgi, F. and E. Coppola, 2010: Does the model regional bias affect the projected regional climate change? An analysis of global model projections. *Climatic Change*, **100**, 787-795. http://dx.doi.org/10.1007/s10584-010-9864-z

96. Weigel, A.P., R. Knutti, M.A. Liniger, and C. Appenzeller, 2010: Risks of model weighting in multimodel climate projections. *Journal of Climate*, **23**, 4175-4191. http://dx.doi.org/10.1175/2010jcli3594.1

97. Knutti, R., J. Sedláček, B.M. Sanderson, R. Lorenz, E.M. Fischer, and V. Eyring, 2017: A climate model projection weighting scheme accounting for performance and interdependence. *Geophysical Research Letters*, **44**, 1909-1918. http://dx.doi.org/10.1002/2016GL072012

98. Hawkins, E. and R. Sutton, 2009: The potential to narrow uncertainty in regional climate predictions. *Bulletin of the American Meteorological Society*, **90**, 1095-1107. http://dx.doi.org/10.1175/2009BAMS2607.1

99. Hawkins, E. and R. Sutton, 2011: The potential to narrow uncertainty in projections of regional precipitation change. *Climate Dynamics*, **37**, 407-418. http://dx.doi.org/10.1007/s00382-010-0810-6

100. IPCC, 2013: Summary for policymakers. *Climate Change 2013: The Physical Science Basis. Contribution of Working Group I to the Fifth Assessment Report of the Intergovernmental Panel on Climate Change.* Stocker, T.F., D. Qin, G.-K. Plattner, M. Tignor, S.K. Allen, J. Boschung, A. Nauels, Y. Xia, V. Bex, and P.M. Midgley, Eds. Cambridge University Press, Cambridge, United Kingdom and New York, NY, USA, 1–30. http://www.climatechange2013.org/report/

101. Archer, D. and R. Pierrehumbert, eds., 2011: *The Warming Papers: The Scientific Foundation for the Climate Change Forecast.* Wiley-Blackwell: Oxford, UK, 432 pp. http://www.wiley.com/WileyCDA/WileyTitle/productCd-1405196165.html

102. Masson-Delmotte, V., M. Schulz, A. Abe-Ouchi, J. Beer, A. Ganopolski, J.F. González Rouco, E. Jansen, K. Lambeck, J. Luterbacher, T. Naish, T. Osborn, B. Otto-Bliesner, T. Quinn, R. Ramesh, M. Rojas, X. Shao, and A. Timmermann, 2013: Information from paleoclimate archives. *Climate Change 2013: The Physical Science Basis. Contribution of Working Group I to the Fifth Assessment Report of the Intergovernmental Panel on Climate Change.* Stocker, T.F., D. Qin, G.-K. Plattner, M. Tignor, S.K. Allen, J. Boschung, A. Nauels, Y. Xia, V. Bex, and P.M. Midgley, Eds. Cambridge University Press, Cambridge, United Kingdom and New York, NY, USA, 383–464. http://www.climatechange2013.org/report/full-report/

103. Le Quéré, C., R. Moriarty, R.M. Andrew, J.G. Canadell, S. Sitch, J.I. Korsbakken, P. Friedlingstein, G.P. Peters, R.J. Andres, T.A. Boden, R.A. Houghton, J.I. House, R.F. Keeling, P. Tans, A. Arneth, D.C.E. Bakker, L. Barbero, L. Bopp, J. Chang, F. Chevallier, L.P. Chini, P. Ciais, M. Fader, R.A. Feely, T. Gkritzalis, I. Harris, J. Hauck, T. Ilyina, A.K. Jain, E. Kato, V. Kitidis, K. Klein Goldewijk, C. Koven, P. Landschützer, S.K. Lauvset, N. Lefèvre, A. Lenton, I.D. Lima, N. Metzl, F. Millero, D.R. Munro, A. Murata, J.E.M.S. Nabel, S. Nakaoka, Y. Nojiri, K. O'Brien, A. Olsen, T. Ono, F.F. Pérez, B. Pfeil, D. Pierrot, B. Poulter, G. Rehder, C. Rödenbeck, S. Saito, U. Schuster, J. Schwinger, R. Séférian, T. Steinhoff, B.D. Stocker, A.J. Sutton, T. Takahashi, B. Tilbrook, I.T. van der Laan-Luijkx, G.R. van der Werf, S. van Heuven, D. Vandemark, N. Viovy, A. Wiltshire, S. Zaehle, and N. Zeng, 2015: Global carbon budget 2015. *Earth System Science Data*, **7**, 349-396. http://dx.doi.org/10.5194/essd-7-349-2015

# 5

# Large-Scale Circulation and Climate Variability

## KEY FINDINGS

1. The tropics have expanded poleward by about 70 to 200 miles in each hemisphere over the period 1979–2009, with an accompanying shift of the subtropical dry zones, midlatitude jets, and storm tracks (*medium to high confidence*). Human activities have played a role in this change (*medium confidence*), although confidence is presently *low* regarding the magnitude of the human contribution relative to natural variability.

2. Recurring patterns of variability in large-scale atmospheric circulation (such as the North Atlantic Oscillation and Northern Annular Mode) and the atmosphere–ocean system (such as El Niño–Southern Oscillation) cause year-to-year variations in U.S. temperatures and precipitation (*high confidence*). Changes in the occurrence of these patterns or their properties have contributed to recent U.S. temperature and precipitation trends (*medium confidence*), although confidence is *low* regarding the size of the role of human activities in these changes.

**Recommended Citation for Chapter**

**Perlwitz**, J., T. Knutson, J.P. Kossin, and A.N. LeGrande, 2017: Large-scale circulation and climate variability. In: *Climate Science Special Report: Fourth National Climate Assessment, Volume I* [Wuebbles, D.J., D.W. Fahey, K.A. Hibbard, D.J. Dokken, B.C. Stewart, and T.K. Maycock (eds.)]. U.S. Global Change Research Program, Washington, DC, USA, pp. 161-184, doi: 10.7930/J0RV0KVQ.

## 5.1 Introduction

The causes of regional climate trends cannot be understood without considering the impact of variations in large-scale atmospheric circulation and an assessment of the role of internally generated climate variability. There are contributions to regional climate trends from changes in large-scale latitudinal circulation, which is generally organized into three cells in each hemisphere—Hadley cell, Ferrell cell and Polar cell—and which determines the location of subtropical dry zones and midlatitude jet streams (Figure 5.1). These circulation cells are expected to shift poleward during warmer periods,[1, 2, 3, 4] which could result in poleward shifts in precipitation patterns, affecting natural ecosystems, agriculture, and water resources.[5, 6]

In addition, regional climate can be strongly affected by non-local responses to recurring patterns (or modes) of variability of the atmospheric circulation or the coupled atmosphere–ocean system. These modes of variability represent preferred spatial patterns and their temporal variation. They account for gross features in variance and for teleconnections which describe climate links between geographically separated regions. Modes of variability are often described as a product of a spatial climate pattern and an associated climate index time series that are identified based on statistical methods like Principal Component Analysis (PC analysis), which is also called Empirical Orthogonal Function Analysis (EOF analysis), and cluster analysis.

## Atmospheric Circulation

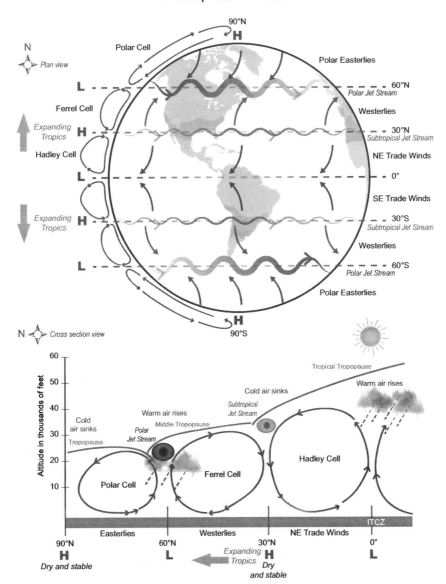

**Figure 5.1:** (top) Plan and (bottom) cross-section schematic view representations of the general circulation of the atmosphere. Three main circulations exist between the equator and poles due to solar heating and Earth's rotation: 1) **Hadley cell** – Low-latitude air moves toward the equator. Due to solar heating, air near the equator rises vertically and moves poleward in the upper atmosphere. 2) **Ferrel cell** – A midlatitude mean atmospheric circulation cell. In this cell, the air flows poleward and eastward near the surface and equatorward and westward at higher levels. 3) **Polar cell** – Air rises, diverges, and travels toward the poles. Once over the poles, the air sinks, forming the polar highs. At the surface, air diverges outward from the polar highs. Surface winds in the polar cell are easterly (polar easterlies). A high pressure band is located at about 30° N/S latitude, leading to dry/hot weather due to descending air motion (subtropical dry zones are indicated in orange in the schematic views). Expanding tropics (indicted by orange arrows) are associated with a poleward shift of the subtropical dry zones. A low pressure band is found at 50°–60° N/S, with rainy and stormy weather in relation to the polar jet stream bands of strong westerly wind in the upper levels of the atmosphere. (Figure source: adapted from NWS 2016[177]).

On intraseasonal to interannual time scales, the climate of the United States is strongly affected by modes of atmospheric circulation variability like the North Atlantic Oscillation (NAO)/Northern Annular Mode (NAM), North Pacific Oscillation (NPO), and Pacific/North American Pattern (PNA).[7, 8, 9] These modes are closely linked to other atmospheric circulation phenomena like blocking and quasi-stationary wave patterns and jet streams that can lead to weather and climate extremes.[10] On an interannual time scale, coupled atmosphere–ocean phenomena like El Niño–Southern Oscillation (ENSO) have a prominent effect.[11] On longer time scales, U.S. climate anomalies are linked to slow variations of sea surface temperature related to the Pacific Decadal Oscillation (PDO) and the Atlantic Multidecadal Oscillation (AMO).[12, 13, 14]

These modes of variability can affect the local-to-regional climate response to external forcing in various ways. The climate response may be altered by the forced response of these existing, recurring modes of variability.[15] Further, the structure and strength of regional temperature and precipitation impacts of these recurring modes of variability may be modified due to a change in the background climate.[16] Modes of internal variability of the climate system also contribute to observed decadal and multidecadal temperature and precipitation trends on local to regional scales, masking possible systematic changes due to an anthropogenic influence.[17] However, there are still large uncertainties in our understanding of the impact of human-induced climate change on atmospheric circulation.[4, 18] Furthermore, the confidence in any specific projected change in ENSO variability in the 21st century remains *low*.[19]

## 5.2 Modes of Variability: Past and Projected Changes

### 5.2.1 Width of the Tropics and Global Circulation

Evidence continues to mount for an expansion of the tropics over the past several decades, with a poleward expansion of the Hadley cell and an associated poleward shift of the subtropical dry zones and storm tracks in each hemisphere.[5, 20, 21, 22, 23, 24, 25, 26, 27, 28, 29] The rate of expansion is uncertain and depends on the metrics and data sources that are used. Recent estimates of the widening of the global tropics for the period 1979–2009 range between 1° and 3° latitude (between about 70 and 200 miles) in each hemisphere, an average trend of between approximately 0.5° and 1.0° per decade.[26] While the roles of increasing greenhouse gases in both hemispheres,[4, 30] stratospheric ozone depletion in the Southern Hemisphere,[31] and anthropogenic aerosols in the Northern Hemisphere[32, 33] have been implicated as contributors to the observed expansion, there is uncertainty in the relative contributions of natural and anthropogenic factors, and natural variability may currently be dominating.[23, 34, 35]

Most of the previous work on tropical expansion to date has focused on zonally averaged changes. There are only a few recent studies that diagnose regional characteristics of tropical expansion. The findings depend on analysis methods and datasets. For example, a northward expansion of the tropics in most regions of the Northern Hemisphere, including the Eastern Pacific with impact on drying in the American Southwest, is found based on diagnosing outgoing longwave radiation.[36] However, other studies do not find a significant poleward expansion of the tropics over the Eastern Pacific and North America.[37, 38] Thus, while some studies associate the observed drying of the U.S. Southwest with the poleward expansion of the tropics,[5, 39] regional impacts of the observed zonally averaged changes in the width of the tropics are not understood.

Due to human-induced greenhouse gas increases, the Hadley cell is *likely* to widen in the future, with an accompanying poleward shift in the subtropical dry zones, midlatitude jets, and storm tracks.[2, 4, 5, 40, 41, 42, 43] Large uncertainties remain in projected changes in non-zonal to regional circulation components and related changes in precipitation patterns.[18, 40, 44, 45] Uncertainties in projected changes in midlatitude jets are also related to the projected rate of arctic amplification and variations in the stratospheric polar vortex. Both factors could shift the midlatitude jet equatorward, especially in the North Atlantic region.[46, 47, 48, 49]

### 5.2.2 El Niño–Southern Oscillation

El Niño–Southern Oscillation (ENSO) is a main source of climate variability, with a two- to seven-year timescale, originating from coupled ocean–atmosphere interactions in the tropical Pacific. Major ENSO events affect weather patterns over many parts of the globe through atmospheric teleconnections. ENSO strongly affects precipitation and temperature in the United States with impacts being most pronounced during the cold season (Figure 5.2).[11, 50, 51, 52, 53] A cooling trend of the tropical Pacific Ocean that resembles La Niña conditions contributed to drying in southwestern North America from 1979 to 2006[54] and is found to explain most of the decrease in heavy daily precipitation events in the southern United States from 1979 to 2013.[55]

El Niño teleconnections are modulated by the location of maximum anomalous tropical Pacific sea surface temperatures (SST). Eastern Pacific (EP) El Niño events affect winter temperatures primarily over the Great Lakes, Northeast, and Southwest, while Central Pacific (CP) events influence temperatures primarily over the northwestern and southeastern United States.[56] The CP El Niño also enhances the drying effect, but weakens the wetting effect, typically produced by traditional EP El Niño events on U.S. winter

precipitation.[57] It is not clear whether observed decadal-scale modulations of ENSO properties, including an increase in ENSO amplitude[58] and an increase in frequency of CP El Niño events,[59, 60] are due to internal variability or anthropogenic forcing. Uncertainties in both the diagnosed distinct U.S. climate effects of EP and CP events and causes for the decadal scale changes result from the limited sample size of observed ENSO events in each category[61, 62] and the relatively short record of the comprehensive observations (since late 1970s) that would allow the investigation of ENSO-related coupled atmosphere–ocean feedbacks.[19] Furthermore, unforced global climate model simulations show that decadal to centennial modulations of ENSO can be generated without any change in external forcing.[63] A model study based on large, single-model ensembles of atmospheric and coupled atmosphere–ocean models finds that external radiative forcing resulted in an atmospheric teleconnection pattern that is independent of ENSO-like variations during the 1979–2014 period and is characterized by a hemisphere-scale increasing trend in heights.[53]

The representation of ENSO in climate models has improved from CMIP3 to CMIP5 models, especially in relation to ENSO amplitude.[64, 65] However, CMIP5 models still cannot capture the seasonal timing of ENSO events.[66] Furthermore, they still exhibit errors in simulating key atmospheric feedbacks, and the improvement in ENSO amplitudes might therefore result from error compensations.[64] Limited observational records and the nonstationarity of tropical Pacific teleconnections to North America on multidecadal time scales pose challenges for evaluating teleconnections between ENSO and U.S. climate in coupled atmosphere–ocean models.[61, 67] For a given SST forcing, however, the atmospheric component of CMIP5 models simulate the sign of the precipitation change over the southern section of North America.[68]

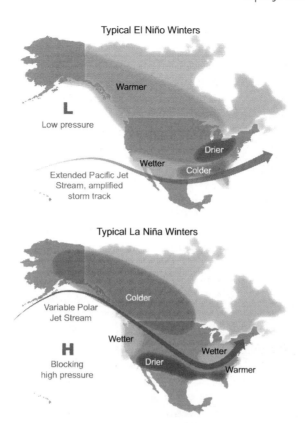

Typical El Niño Winters

Typical La Niña Winters

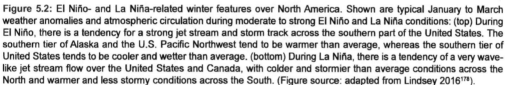

**Figure 5.2:** El Niño- and La Niña-related winter features over North America. Shown are typical January to March weather anomalies and atmospheric circulation during moderate to strong El Niño and La Niña conditions: (top) During El Niño, there is a tendency for a strong jet stream and storm track across the southern part of the United States. The southern tier of Alaska and the U.S. Pacific Northwest tend to be warmer than average, whereas the southern tier of United States tends to be cooler and wetter than average. (bottom) During La Niña, there is a tendency of a very wave-like jet stream flow over the United States and Canada, with colder and stormier than average conditions across the North and warmer and less stormy conditions across the South. (Figure source: adapted from Lindsey 2016[178]).

Climate projections suggest that ENSO will remain a primary mode of natural climate variability in the 21st century.[19] Climate models do not agree, however, on projected changes in the intensity or spatial pattern of ENSO.[19] This uncertainty is related to a model dependence of simulated changes in the zonal gradient of tropical Pacific sea surface temperature in a warming climate.[19] Model studies suggest an eastward shift of ENSO-induced teleconnection patterns due to greenhouse gas-induced climate change.[69, 70, 71, 72] However, the impact of such a shift on ENSO-induced climate anomalies in the United States is not well understood.[72, 73]

In summary, there is *high confidence* that, in the 21st century, ENSO will remain a main source of climate variability over the United States on seasonal to interannual timescales. There is *low confidence* for a specific projected change in ENSO variability.

### 5.2.3 Extra-tropical Modes of Variability and Phenomena

*North Atlantic Oscillation and Northern Annular Mode*

The North Atlantic Oscillation (NAO), the leading recurring mode of variability in the extratropical North Atlantic region, describes an opposing pattern of sea level pressure between the Atlantic subtropical high and the Iceland/Arctic low. Variations in the NAO are accompanied by changes in the location and intensity of the Atlantic midlatitude storm track and blocking activity that affect climate over the North Atlantic and surrounding continents. A negative NAO phase is related to anomalously cold conditions and an enhanced number of cold outbreaks in the eastern United States, while a strong positive phase of the NAO tends to be associated with above-normal temperatures in this region.[7, 74] The positive phase of the NAO is associated with increased precipitation frequency and positive daily rainfall anomalies, including extreme daily precipitation anomalies in the northeastern United States.[75, 76]

The Northern Annular Mode/Arctic Oscillation (NAM/AO) is closely related to the NAO. It describes a similar out-of-phase pressure variation between mid- and high latitudes but on a hemispheric rather than regional scale.[77, 78] The time series of the NAO and NAM/AO are highly correlated, with persistent NAO and NAM/AO events being indistinguishable.[79, 80]

The wintertime NAO/NAM index exhibits pronounced variability on multidecadal time scales, with an increase from the 1960s to the 1990s, a shift to a more negative phase since the 1990s due to a series of winters like 2009–2010 and 2010–2011 (which had exceptionally low index values), and a return to more positive values after 2011.[30] Decadal scale temperature trends in the eastern United States, including occurrences of cold outbreaks during recent years, are linked to these changes in the NAO/NAM.[81, 82, 83, 84]

The NAO's influence on the ocean occurs through changes in heat content, gyre circulations, mixed layer depth, salinity, high-latitude deep water formation, and sea ice cov-

er.[7, 85] Climate model simulations show that multidecadal variations in the NAO induce multidecadal variations in the strength of the Atlantic Meridional Overturning Circulation (AMOC) and poleward ocean heat transport in the Atlantic, extending to the Arctic, with potential impacts on recent arctic sea ice loss and Northern Hemisphere warming.[85] However, other model simulations suggest that the NAO and recent changes in Northern Hemisphere climate were affected by recent variations in the AMOC,[86] for which enhanced freshwater discharge from the Greenland Ice Sheet (GrIS) may have been a contributing cause.[87]

Climate models are widely analyzed for their ability to simulate the spatial patterns of the NAO/NAM and their relationship to temperature and precipitation anomalies over the United States.[9, 65, 88] Climate models reproduce the broad spatial and temporal features of the NAO, although there are large differences among the individual models in the location of the NAO centers of action and their average magnitude. These differences affect the agreement between observed and simulated climate anomalies related to the NAO.[9, 65] Climate models tend to have a NAM pattern that is more annular than observed,[65, 88] resulting in a strong bias in the Pacific center of the NAM. As a result, temperature anomalies over the northwestern United States associated with the NAM in most models are of opposite sign compared to observation.[88] Biases in the model representation of NAO/NAM features are linked to limited abilities of general circulation models to reproduce dynamical processes, including atmospheric blocking,[89] troposphere–stratosphere coupling,[90] and climatological stationary waves.[90, 91]

The CMIP5 models on average simulate a progressive shift of the NAO/NAM towards the positive phase due to human-induced climate change.[92] However, the spread between model simulations is larger than the projected multimodel increase,[19] and there are uncertainties related to future scenarios.[9] Furthermore, it is found that shifts between preferred periods of positive and negative NAO phase will continue to occur similar to those observed in the past.[19, 93] There is no consensus on the location of changes of NAO centers among the global climate models under future warming scenarios.[9] Uncertainties in future projections of the NAO/NAM in some seasons are linked to model spread in projected future arctic warming[46, 47] (Ch. 11: Arctic Changes) and to how models resolve stratospheric processes.[19, 94]

In summary, while it is *likely* that the NAO/NAM index will become slightly more positive (on average) due to increases in GHGs, there is *low confidence* in temperature and precipitation changes over the United States related to such variations in the NAO/NAM.

*North Pacific Oscillation/West Pacific Oscillation*
The North Pacific Oscillation (NPO) is a recurring mode of variability in the extratropical North Pacific region and is characterized by a north-south seesaw in sea level pressure. Effects of NPO on U.S. hydroclimate and marginal ice zone extent in the arctic seas have been reported.[8]

The NPO is linked to tropical sea surface temperature variability. Specifically, NPO contributes to the excitation of ENSO events via the "Seasonal Footprinting Mechanism".[95, 96] In turn, warm events in the central tropical Pacific Ocean are suggested to force an NPO-like circulation pattern.[97] There is *low confidence* in future projections of the NPO due to the small number of modeling studies as well as the finding that many climate models do not properly simulate the observed linkages between the NPO and tropical sea surface temperature variability.[19, 98]

*Pacific/North American Pattern*
The Pacific/North American (PNA) pattern is the leading recurring mode of internal atmospheric variability over the North Pacific and the North American continent, especially during the cold season. It describes a quadripole pattern of mid-tropospheric height anomalies, with anomalies of similar sign located over the subtropical northeastern Pacific and northwestern North America and of the opposite sign centered over the Gulf of Alaska and the southeastern United States. The PNA pattern is associated with strong fluctuations in the strength and location of the East Asian jet stream. The positive phase of the PNA pattern is associated with above average temperatures over the western and northwestern United States, and below average temperatures across the south-central and southeastern United States, including an enhanced occurrence of extreme cold temperatures.[9, 99, 100] Significant negative correlation between the PNA and winter precipitation over the Ohio River Valley has been documented.[9, 99, 101] The PNA is related to ENSO events[102] and also serves as a bridge linking ENSO and NAO variability.[103]

Climate models are able to reasonably represent the atmospheric circulation and climate anomalies associated with the PNA pattern. However, individual models exhibit differences compared to the observed relationship, due to displacements of the simulated PNA centers of action and offsets in their magnitudes.[9] Climate models do not show consistent location changes of the PNA centers due to increases in GHGs.[9, 72] Therefore, there is *low confidence* for projected changes in the PNA and the association with temperature and precipitation variations over the United States.

*Blocking and Quasi-Stationary Waves*
Anomalous atmospheric flow patterns in the extratropics that remain in place for an ex-

tended period of time (for example, blocking and quasi-stationary Rossby waves)—and thus affect a region with similar weather conditions like rain or clear sky for several days to weeks—can lead to flooding, drought, heat waves, and cold waves.[10, 104, 105] Specifically, blocking describes large-scale, persistent high pressure systems that interrupt the typical westerly flow, while planetary waves (Rossby waves) describe large-scale meandering of the atmospheric jet stream.

A persistent pattern of high pressure in the circulation off the West Coast of the United States has been associated with the recent multiyear California drought[106, 107, 108] (Ch. 8: Droughts, Floods, and Wildfire). Blocking in the Alaskan region, which is enhanced during La Niña winters (Figure 5.2),[109] is associated with higher temperatures in western Alaska but shift to lower mean and extreme surface temperatures from the Yukon southward to the southern Plains.[110] The anomalously cold winters of 2009–2010 and 2010–2011 in the United States are linked to the blocked (or negative) phase of the NAO.[111] Stationary Rossby wave patterns may have contributed to the North American temperature extremes during summers like 2011.[112] It has been suggested that arctic amplification has already led to weakened westerly winds and hence more slowly moving and amplified wave patterns and enhanced occurrence of blocking[113, 114] (Ch. 11: Arctic Changes). While some studies suggest an observed increase in the metrics of these persistent circulation patterns,[113, 115] other studies suggest that observed changes are small compared to atmospheric internal variability.[116, 117, 118]

A decrease of blocking frequency with climate change is found in CMIP3, CMIP5, and higher-resolution models.[19, 119, 120] Climate models robustly project a change in Northern Hemisphere winter quasi-stationary wave fields

that are linked to a wetting of the North American West Coast,[45, 121, 122] due to a strengthening of the zonal mean westerlies in the subtropical upper troposphere. However, CMIP5 models still underestimate observed blocking activity in the North Atlantic sector while they tend to overestimate activity in the North Pacific, although with a large intermodel spread.[19] Most climate models also exhibit biases in the representation of relevant stationary waves.[44]

In summary, there is *low confidence* in projected changes in atmospheric blocking and wintertime quasi-stationary waves. Therefore, our confidence is *low* on the association between observed and projected changes in weather and climate extremes over the United States and variations in these persistent atmospheric circulation patterns.

### 5.2.4 Modes of Variability on Decadal to Multidecadal Time Scales

*Pacific Decadal Oscillation (PDO) / Interdecadal Pacific Oscillation (IPO)*

The Pacific Decadal Oscillation (PDO) was first introduced by Mantua et al. 1997[123] as the leading empirical orthogonal function of North Pacific (20°–70°N) monthly averaged sea surface temperature anomalies.[14] Interdecadal Pacific Oscillation (IPO) refers to the same phenomenon and is based on Pacific-wide sea surface temperatures. PDO/IPO lacks a characteristic timescale and represents a combination of physical processes that span the tropics and extratropics, including both remote tropical forcing and local North Pacific atmosphere–ocean interactions.[14] Consequently, PDO-related variations in temperature and precipitation in the United States are very similar to (and indeed may be caused by) variations associated with ENSO and the strength of the Aleutian low (North Pacific Index, NPI), as shown in Figure 5.3. A PDO-related temperature variation in Alaska is also apparent.[124, 125]

## Cold Season Relationship
## between Climate Indices and Precipitation/Temperature Anomalies

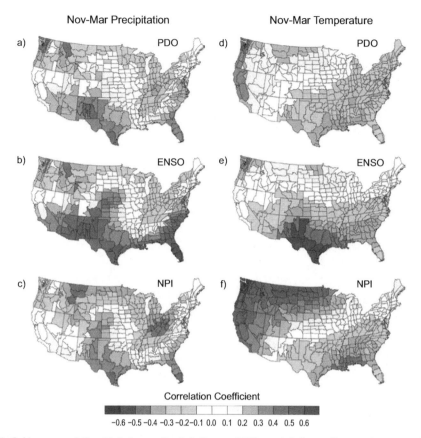

Nov–Mar Precipitation          Nov–Mar Temperature

Correlation Coefficient

–0.6 –0.5 –0.4 –0.3 –0.2 –0.1  0.0  0.1  0.2  0.3  0.4  0.5  0.6

**Figure 5.3:** Cold season relationship between climate indices and U.S. precipitation and temperature anomalies determined from U.S. climate division data,[179] for the years 1901–2014. November–March mean U.S. precipitation anomalies correlated with (a) the Pacific Decadal Oscillation (PDO) index, (b) the El Niño–Southern Oscillation (ENSO) index, and (c) the North Pacific Index (NPI). November–March U.S. temperature anomalies correlated with (d) the PDO index, (e) the ENSO index, and (f) the NPI. United States temperature and precipitation related to the Pacific Decadal Oscillation are very similar to (and indeed may be caused by) variations associated with ENSO and the Aleutian low strength (North Pacific Index). (Figure source: Newman et al. 2016[14]; ©American Meteorological Society, used with permission).

The PDO does not show a long-term trend either in SST reconstructions or in the ensemble mean of historical CMIP3 and CMIP5 simulations.[14] Emerging science suggests that externally forced natural and anthropogenic factors have contributed to the observed PDO-like variability. For example, a model study finds that the observed PDO phase is affected by large volcanic events and the variability in incoming solar radiation.[126] Aerosols from anthropogenic sources could change the temporal variability of the North Pacific SST through modifications of the atmospheric circulation.[127, 128] Furthermore, some studies show that periods with near-zero warming trends of global mean temperature and periods of accelerated temperatures could result from the interplay between internally generated PDO/ IPO-like temperature variations in the tropical Pacific Ocean and greenhouse gas-induced ocean warming.[129, 130]

Future changes in the spatial and temporal characteristics of PDO/IPO are uncertain. Based on CMIP3 models, one study finds that most of these models do not exhibit significant changes,[98] while another study points out that the PDO/IPO becomes weaker and more frequent by the end of the 21st century in some models.[131] Furthermore, future changes in ENSO variability, which strongly contributes to the PDO/IPO,[132] are also uncertain (Section 5.2.2). Therefore, there is *low confidence* in projected future changes in the PDO/IPO.

*Atlantic Multidecadal Variability (AMV) / Atlantic Multidecadal Oscillation (AMO)*
The North Atlantic Ocean region exhibits coherent multidecadal variability that exerts measurable impacts on regional climate for variables such as U.S. precipitation[12, 133, 134, 135] and Atlantic hurricane activity.[13, 136, 137, 138, 139, 140] This observed Atlantic multidecadal variability, or AMV, is generally understood to be driven by a combination of internal and external factors.[12, 141, 142, 143, 144, 145, 146, 147, 148] The AMV manifests in SST variability and patterns as well as synoptic-scale variability of atmospheric conditions. The internal part of the observed AMV is often referred to as the Atlantic Multidecadal Oscillation (AMO) and is putatively driven by changes in the strength of the Atlantic Meridional Overturning Circulation (AMOC).[142, 143, 149, 150] It is important to understand the distinction between the AMO, which is often assumed to be natural (because of its putative relationship with natural AMOC variability), and AMV, which simply represents the observed multidecadal variability as a whole.

The relationship between observed AMV and the AMOC has recently been called into question and arguments have been made that AMV can occur in the absence of the AMOC via stochastic forcing of the ocean by coherent atmospheric circulation variability, but this is

presently a topic of debate.[151, 152, 153, 154] Despite the ongoing debates, it is generally acknowledged that observed AMV, as a whole, represents a complex conflation of natural internal variability of the AMOC, natural red-noise stochastic forcing of the ocean by the atmosphere,[146] natural external variability from volcanic events[155, 156] and mineral aerosols,[157] and anthropogenic forcing from greenhouse gases and pollution aerosols.[158, 159, 160, 161]

As also discussed in Chapter 9: Extreme Storms (in the context of Atlantic hurricanes), determining the relative contributions of each mechanism to the observed multidecadal variability in the Atlantic is presently an active area of research and debate, and no consensus has yet been reached.[146, 161, 162, 163, 164, 165, 166] Still, despite the level of disagreement about the relative magnitude of human influences (particularly whether natural or anthropogenic factors are dominating), there is broad agreement in the literature of the past decade or so that human factors have had a measurable impact on the observed AMV. Furthermore, the AMO, as measured by indices constructed from environmental data (e.g., Enfield et al. 2001[12]), is generally based on detrended SST data and is then, by construction, segregated from the century-scale linear SST trends that are likely forced by increasing greenhouse gas concentrations. In particular, removal of a linear trend is not expected to account for all of the variability forced by changes in sulfate aerosol concentrations that have occurred over the past century. In this case, increasing sulfate aerosols are argued to cause cooling of Atlantic SST, thus offsetting the warming caused by increasing greenhouse gas concentration. After the Clean Air Act and Amendments of the 1970s, however, a steady reduction of sulfate aerosols is argued to have caused SST warming that compounds the warming from the ongoing increases in greenhouse gas concentrations.[160, 161] This combination of greenhouse gas and sulfate aerosol forcing, by itself, can lead to Atlantic multidecadal SST variability that would not be removed by removing a linear trend.[155]

In summary, it is unclear what the statistically derived AMO indices represent, and it is not readily supportable to treat AMO index variability as tacitly representing natural variability, nor is it clear that the observed AMV is truly oscillatory in nature.[167] There is a physical basis for treating the AMOC as oscillatory (via thermohaline circulation arguments),[168] but there is no expectation of true oscillatory behavior in the hypothesized external forcing agents for the remaining variability. Detrending the SST data used to construct the AMO indices may partially remove the century-scale trends forced by increasing greenhouse gas concentrations, but it is not adequate for removing multidecadal variability forced by aerosol concentration variability. There is evidence that natural AMOC variability has been occurring for hundreds of years,[149, 169, 170, 171, 172] and this has apparently played some role in the observed AMV as a whole, but a growing body of evidence shows that external factors, both natural and anthropogenic, have played a substantial additional role in the past century.

## 5.3 Quantifying the Role of Internal Variability on Past and Future U.S. Climate Trends

The role of internal variability in masking trends is substantially increased on regional and local scales relative to the global scale, and in the extratropics relative to the tropics (Ch. 4: Projections). Approaches have been developed to better quantify the externally forced and internally driven contributions to observed and future climate trends and variability and further separate these contributions into thermodynamically and dynamically driven factors.[17] Specifically, large "initial condition" climate model ensembles with 30 ensemble members and more[93, 173, 174] and long control runs[175] have been shown to be useful tools to characterize uncertainties in climate change projections at local/regional scales.

North American temperature and precipitation trends on timescales of up to a few decades are strongly affected by intrinsic atmospheric circulation variability.[17, 173] For example, it is estimated that internal circulation trends account for approximately one-third of the observed wintertime warming over North America during the past 50 years. In a few areas, such as the central Rocky Mountains and far western Alaska, internal dynamics have offset the warming trend by 10%–30%.[17] Natural climate variability superimposed upon forced climate change will result in a large range of possible trends for surface air temperature and precipitation in the United States over the next 50 years (Figure 5.4).[173]

Climate models are evaluated with respect to their proper simulation of internal decadal variability. Comparing observed and simulated variability estimates at timescales longer than 10 years suggest that models tend to overestimate the internal variability in the northern extratropics, including over the continental United States, but underestimate it over much of the tropics and subtropical ocean regions.[93, 176] Such biases affect signal-to-noise estimates of regional scale climate change response and thus assessment of internally driven contributions to regional/local trends.

a) Winter surface air temperature and sea level pressure

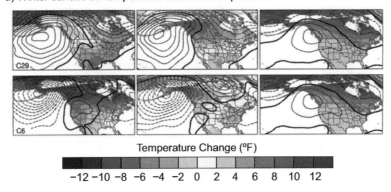

Temperature Change (°F)

−12 −10 −8 −6 −4 −2 0 2 4 6 8 10 12

b) Winter precipitation and sea level pressure

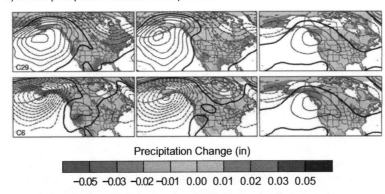

Precipitation Change (in)

−0.05 −0.03 −0.02 −0.01 0.00 0.01 0.02 0.03 0.05

**Figure 5.4:** (left) Total 2010–2060 winter trends decomposed into (center) internal and (right) forced components for two contrasting CCSM3 ensemble members (runs 29 and 6) for (a) surface air temperature [color shading; °F/(51 years)] and sea level pressure (SLP; contours) and (b) precipitation [color shading; inches per day/(51 years)] and SLP (contours). SLP contour interval is 1 hPa/(51 years), with solid (dashed) contours for positive (negative) values; the zero contour is thickened. The same climate model (CCSM3) simulates a large range of possible trends in North American climate over the 2010–2060 period because of the influence of internal climate variability superposed upon forced climate trends. (Figure source: adapted from Deser et al. 2014;[173] © American Meteorological Society, used with permission).

# TRACEABLE ACCOUNTS

### Key Finding 1

The tropics have expanded poleward by about 70 to 200 miles in each hemisphere over the period 1979–2009, with an accompanying shift of the subtropical dry zones, midlatitude jets, and storm tracks (*medium to high confidence*). Human activities have played a role in this change (*medium confidence*), although confidence is presently *low* regarding the magnitude of the human contribution relative to natural variability

### Description of evidence base

The Key Finding is supported by statements of the Intergovernmental Panel on Climate Change's Fifth Assessment Report[24] and a large number of more recent studies that examined the magnitude of the observed tropical widening and various causes.[5, 20, 22, 23, 25, 26, 27, 28, 29, 31] Additional evidence for an impact of greenhouse gas increases on the widening of the tropical belt and poleward shifts of the midlatitude jets is provided by the diagnosis of CMIP5 simulations.[4, 40] There is emerging evidence for an impact of anthropogenic aerosols on the tropical expansion in the Northern Hemisphere.[32, 33] Recent studies provide new evidence on the significance of internal variability on recent changes in the tropical width.[23, 34, 35]

### Major uncertainties

The rate of observed expansion of tropics depends on which metric is used. The linkages between different metrics are not fully explored. Uncertainties also result from the utilization of reanalysis to determine trends and from limited observational records of free atmosphere circulation, precipitation, and evaporation. The dynamical mechanisms behind changes in the width of the tropical belt (e.g., tropical–extratropical interactions and baroclinic eddies) are not fully understood. There is also a limited understanding of how various climate forcings, such as anthropogenic aerosols, affect the width of tropics. The coarse horizontal and vertical resolution of global climate models may limit the ability of these models to properly resolve latitudinal changes in the atmospheric circulation. Limited observational records affect the ability to accurately estimate

the contribution of natural decadal to multi-decadal variability on observed expansion of the tropics.

### Assessment of confidence based on evidence and agreement, including short description of nature of evidence and level of agreement

*Medium to high confidence* that the tropics and related features of the global circulation have expanded poleward is based upon the results of a large number of observational studies, using a wide variety of metrics and data sets, which reach similar conclusions. A large number of studies utilizing modeling of different complexity and theoretical considerations provide compounding evidence that human activities, including increases in greenhouse gases, ozone depletion, and anthropogenic aerosols, contributed to the observed poleward expansion of the tropics. Climate models forced with these anthropogenic drivers cannot explain the observed magnitude of tropical expansion and some studies suggest a possibly large contribution of internal variability. These multiple lines of evidence lead to the conclusion of *medium confidence* that human activities contributed to observed expansion of the tropics.

### Summary sentence or paragraph that integrates the above information

The tropics have expanded poleward in each hemisphere over the period 1979–2009 (*medium to high confidence*) as shown by a large number of studies using a variety of metrics, observations and reanalysis. Modeling studies and theoretical considerations illustrate that human activities, including increases in greenhouse gases, ozone depletion, and anthropogenic aerosols, cause a widening of the tropics. There is *medium confidence* that human activities have contributed to the observed poleward expansion, taking into account uncertainties in the magnitude of observed trends and a possible large contribution of natural climate variability.

### Key Finding 2

Recurring patterns of variability in large-scale atmospheric circulation (such as the North Atlantic Oscillation and Northern Annular Mode) and the atmosphere–ocean system (such as El Niño–Southern Oscillation) cause year-to-year variations in U.S. temperatures and precipitation (*high confidence*). Changes in the occurrence of these patterns or their properties have contributed to recent U.S. temperature and precipitation trends (*medium confidence*), although confidence is *low* regarding the size of the role of human activities in these changes.

**Description of evidence base**

The Key Finding is supported by a large number of studies that diagnose recurring patterns of variability and their changes, as well as their impact on climate over the United States. Regarding year-to-year variations, a large number of studies based on models and observations show statistically significant associations between North Atlantic Oscillation/Northern Annular Mode and United States temperature and precipitation,[7, 9, 74, 75, 76, 88] as well as El Niño–Southern Oscillation and related U.S. climate teleconnections.[11, 50, 51, 52, 53, 56, 57] Regarding recent decadal trends, several studies provide evidence for concurrent changes in the North Atlantic Oscillation/Northern Annular Mode and climate anomalies over the United States.[81, 82, 83, 84] Modeling studies provide evidence for a linkage between cooling trends of the tropical Pacific Ocean that resemble La Niña and precipitation changes in the southern United States.[54, 55] Several studies describe a decadal modification of ENSO.[58, 59, 60, 63] Modeling evidence is provided that such decadal modifications can be due to internal variability.[63] Climate models are widely analyzed for their ability to simulate recurring patterns of variability and teleconnections over the United States.[9, 64, 65, 68, 88, 98] Climate model projections are also widely analyzed to diagnose the impact of human activities on NAM/NAO, ENSO teleconnections, and other recurring modes of variability associated with climate anomalies.[9, 19, 72, 92]

**Major uncertainties**

A key uncertainty is related to limited observational records and our capability to properly simulate climate variability on decadal to multidecadal timescales, as well as to properly simulate recurring patterns of climate variability, underlying physical mechanisms, and

associated variations in temperature and precipitation over the United States.

**Assessment of confidence based on evidence and agreement, including short description of nature of evidence and level of agreement**

There is *high confidence* that preferred patterns of variability affect U.S. temperature on a year-to-year timescale, based on a large number of studies that diagnose observational data records and long simulations. There is *medium confidence* that changes in the occurrence of these patterns or their properties have contributed to recent U.S. temperature and precipitation trends. Several studies agree on a linkage between decadal changes in the NAO/NAM and climate trends over the United States, and there is some modeling evidence for a linkage between a La Niña-like cooling trend over the tropical Pacific and precipitation changes in the southwestern United States. There is no robust evidence for observed decadal changes in the properties of ENSO and related United States climate impacts. Confidence is *low* regarding the size of the role of human influences in these changes because models do not agree on the impact of human activity on preferred patterns of variability or because projected changes are small compared to internal variability.

**Summary sentence or paragraph that integrates the above information**

Recurring modes of variability strongly affect temperature and precipitation over the United States on interannual timescales (*high confidence*) as supported by a very large number of observational and modeling studies. Changes in some recurring patterns of variability have contributed to recent trends in U.S. temperature and precipitation (*medium confidence*). The causes of these changes are uncertain due to the limited observational record and because models exhibit some difficulties simulating these recurring patterns of variability and their underlying physical mechanisms.

# References

1.	Frierson, D.M.W., J. Lu, and G. Chen, 2007: Width of the Hadley cell in simple and comprehensive general circulation models. *Geophysical Research Letters*, **34**, L18804. http://dx.doi.org/10.1029/2007GL031115

2.	Mbengue, C. and T. Schneider, 2017: Storm-track shifts under climate change: Toward a mechanistic understanding using baroclinic mean available potential energy. *Journal of the Atmospheric Sciences*, **74**, 93-110. http://dx.doi.org/10.1175/jas-d-15-0267.1

3.	Sun, Y., G. Ramstein, C. Contoux, and T. Zhou, 2013: A comparative study of large-scale atmospheric circulation in the context of a future scenario (RCP4.5) and past warmth (mid-Pliocene). *Climate of the Past*, **9**, 1613-1627. http://dx.doi.org/10.5194/cp-9-1613-2013

4.	Vallis, G.K., P. Zurita-Gotor, C. Cairns, and J. Kidston, 2015: Response of the large-scale structure of the atmosphere to global warming. *Quarterly Journal of the Royal Meteorological Society*, **141**, 1479-1501. http://dx.doi.org/10.1002/qj.2456

5.	Feng, S. and Q. Fu, 2013: Expansion of global drylands under a warming climate. *Atmospheric Chemistry and Physics*, **13**, 10081-10094. http://dx.doi.org/10.5194/acp-13-10081-2013

6.	Seidel, D.J., Q. Fu, W.J. Randel, and T.J. Reichler, 2008: Widening of the tropical belt in a changing climate. *Nature Geoscience*, **1**, 21-24. http://dx.doi.org/10.1038/ngeo.2007.38

7.	Hurrell, J.W. and C. Deser, 2009: North Atlantic climate variability: The role of the North Atlantic oscillation. *Journal of Marine Systems*, **78**, 28-41. http://dx.doi.org/10.1016/j.jmarsys.2008.11.026

8.	Linkin, M.E. and S. Nigam, 2008: The North Pacific Oscillation–West Pacific teleconnection pattern: Mature-phase structure and winter impacts. *Journal of Climate*, **21**, 1979-1997. http://dx.doi.org/10.1175/2007JCLI2048.1

9.	Ning, L. and R.S. Bradley, 2016: NAO and PNA influences on winter temperature and precipitation over the eastern United States in CMIP5 GCMs. *Climate Dynamics*, **46**, 1257-1276. http://dx.doi.org/10.1007/s00382-015-2643-9

10.	Grotjahn, R., R. Black, R. Leung, M.F. Wehner, M. Barlow, M. Bosilovich, A. Gershunov, W.J. Gutowski, J.R. Gyakum, R.W. Katz, Y.-Y. Lee, Y.-K. Lim, and Prabhat, 2016: North American extreme temperature events and related large scale meteorological patterns: A review of statistical methods, dynamics, modeling, and trends. *Climate Dynamics*, **46**, 1151-1184. http://dx.doi.org/10.1007/s00382-015-2638-6

11.	Halpert, M.S. and C.F. Ropelewski, 1992: Surface temperature patterns associated with the Southern Oscillation. *Journal of Climate*, **5**, 577-593. http://dx.doi.org/10.1175/1520-0442(1992)005<0577:STPAWT>2.0.CO;2

12.	Enfield, D.B., A.M. Mestas-Nuñez, and P.J. Trimble, 2001: The Atlantic Multidecadal Oscillation and its relation to rainfall and river flows in the continental U.S. *Geophysical Research Letters*, **28**, 2077-2080. http://dx.doi.org/10.1029/2000GL012745

13.	Goldenberg, S.B., C.W. Landsea, A.M. Mestas-Nuñez, and W.M. Gray, 2001: The recent increase in Atlantic hurricane activity: Causes and implications. *Science*, **293**, 474-479. http://dx.doi.org/10.1126/science.1060040

14.	Newman, M., M.A. Alexander, T.R. Ault, K.M. Cobb, C. Deser, E.D. Lorenzo, N.J. Mantua, A.J. Miller, S. Minobe, H. Nakamura, N. Schneider, D.J. Vimont, A.S. Phillips, J.D. Scott, and C.A. Smith, 2016: The Pacific Decadal Oscillation, revisited. *Journal of Climate*, **29**, 4399-4427. http://dx.doi.org/10.1175/JCLI-D-15-0508.1

15.	Perlwitz, J., S. Pawson, R.L. Fogt, J.E. Nielsen, and W.D. Neff, 2008: Impact of stratospheric ozone hole recovery on Antarctic climate. *Geophysical Research Letters*, **35**, L08714. http://dx.doi.org/10.1029/2008GL033317

16.	Palmer, T.N., F.J. Doblas-Reyes, A. Weisheimer, and M.J. Rodwell, 2008: Toward seamless prediction: Calibration of climate change projections using seasonal forecasts. *Bulletin of the American Meteorological Society*, **89**, 459-470. http://dx.doi.org/10.1175/bams-89-4-459

17.	Deser, C., L. Terray, and A.S. Phillips, 2016: Forced and internal components of winter air temperature trends over North America during the past 50 years: Mechanisms and implications. *Journal of Climate*, **29**, 2237-2258. http://dx.doi.org/10.1175/JCLI-D-15-0304.1

18.	Shepherd, T.G., 2014: Atmospheric circulation as a source of uncertainty in climate change projections. *Nature Geoscience*, **7**, 703-708. http://dx.doi.org/10.1038/ngeo2253

19. Christensen, J.H., K. Krishna Kumar, E. Aldrian, S.-I. An, I.F.A. Cavalcanti, M. de Castro, W. Dong, P. Goswami, A. Hall, J.K. Kanyanga, A. Kitoh, J. Kossin, N.-C. Lau, J. Renwick, D.B. Stephenson, S.-P. Xie, and T. Zhou, 2013: Climate phenomena and their relevance for future regional climate change. *Climate Change 2013: The Physical Science Basis. Contribution of Working Group I to the Fifth Assessment Report of the Intergovernmental Panel on Climate Change*. Stocker, T.F., D. Qin, G.-K. Plattner, M. Tignor, S.K. Allen, J. Boschung, A. Nauels, Y. Xia, V. Bex, and P.M. Midgley, Eds. Cambridge University Press, Cambridge, United Kingdom and New York, NY, USA, 1217–1308. http://www.climatechange2013.org/report/full-report/

20. Birner, T., S.M. Davis, and D.J. Seidel, 2014: The changing width of Earth's tropical belt. *Physcis Today*, **67**, 38-44. http://dx.doi.org/10.1063/PT.3.2620

21. Brönnimann, S., A.M. Fischer, E. Rozanov, P. Poli, G.P. Compo, and P.D. Sardeshmukh, 2015: Southward shift of the northern tropical belt from 1945 to 1980. *Nature Geoscience*, **8**, 969-974. http://dx.doi.org/10.1038/ngeo2568

22. Davis, N.A. and T. Birner, 2013: Seasonal to multidecadal variability of the width of the tropical belt. *Journal of Geophysical Research Atmospheres*, **118**, 7773-7787. http://dx.doi.org/10.1002/jgrd.50610

23. **Garfinkel, C.I., D.W. Waugh, and L.M. Polvani,** 2015: Recent Hadley cell expansion: The role of internal atmospheric variability in reconciling modeled and observed trends. *Geophysical Research Letters*, **42**, 10,824-10,831. http://dx.doi.org/10.1002/2015GL066942

24. Hartmann, D.L., A.M.G. Klein Tank, M. Rusticucci, L.V. Alexander, S. Brönnimann, Y. Charabi, F.J. Dentener, E.J. Dlugokencky, D.R. Easterling, A. Kaplan, B.J. Soden, P.W. Thorne, M. Wild, and P.M. Zhai, 2013: Observations: Atmosphere and surface. *Climate Change 2013: The Physical Science Basis. Contribution of Working Group I to the Fifth Assessment Report of the Intergovernmental Panel on Climate Change*. Stocker, T.F., D. Qin, G.-K. Plattner, M. Tignor, S.K. Allen, J. Boschung, A. Nauels, Y. Xia, V. Bex, and P.M. Midgley, Eds. Cambridge University Press, Cambridge, United Kingdom and New York, NY, USA, 159–254. http://www.climatechange2013.org/report/full-report/

25. Karnauskas, K.B. and C.C. Ummenhofer, 2014: On the dynamics of the Hadley circulation and subtropical drying. *Climate Dynamics*, **42**, 2259-2269. http://dx.doi.org/10.1007/s00382-014-2129-1

26. Lucas, C., B. Timbal, and H. Nguyen, 2014: The expanding tropics: A critical assessment of the observational and modeling studies. *Wiley Interdisciplinary Reviews: Climate Change*, **5**, 89-112. http://dx.doi.org/10.1002/wcc.251

27. Norris, J.R., R.J. Allen, A.T. Evan, M.D. Zelinka, C.W. O'Dell, and S.A. Klein, 2016: Evidence for climate change in the satellite cloud record. *Nature*, **536**, 72-75. http://dx.doi.org/10.1038/nature18273

28. Quan, X.-W., M.P. Hoerling, J. Perlwitz, H.F. Diaz, and T. Xu, 2014: How fast are the tropics expanding? *Journal of Climate*, **27**, 1999-2013. http://dx.doi.org/10.1175/JCLI-D-13-00287.1

29. Reichler, T., 2016: Chapter 6 - Poleward expansion of the atmospheric circulation *Climate Change (Second Edition)*. Letcher, T.M., Ed. Elsevier, Boston, 79-104. http://dx.doi.org/10.1016/B978-0-444-63524-2.00006-3

30. **Bindoff, N.L., P.A. Stott, K.M. AchutaRao, M.R. Allen,** N. Gillett, D. Gutzler, K. Hansingo, G. Hegerl, Y. Hu, S. Jain, I.I. Mokhov, J. Overland, J. Perlwitz, R. Sebbari, and X. Zhang, 2013: Detection and attribution of climate change: From global to regional. *Climate Change 2013: The Physical Science Basis. Contribution of Working Group I to the Fifth Assessment Report of the Intergovernmental Panel on Climate Change*. Stocker, T.F., D. Qin, G.-K. Plattner, M. Tignor, S.K. Allen, J. Boschung, A. Nauels, Y. Xia, V. Bex, and P.M. Midgley, Eds. Cambridge University Press, Cambridge, United Kingdom and New York, NY, USA, 867–952. http://www.climatechange2013.org/report/full-report/

31. **Waugh, D.W., C.I. Garfinkel, and L.M. Polvani,** 2015: Drivers of the recent tropical expansion in the Southern Hemisphere: Changing SSTs or ozone depletion? *Journal of Climate*, **28**, 6581-6586. http://dx.doi.org/10.1175/JCLI-D-15-0138.1

32. Allen, R.J., S.C. Sherwood, J.R. Norris, and C.S. Zender, 2012: Recent Northern Hemisphere tropical expansion primarily driven by black carbon and tropospheric ozone. *Nature*, **485**, 350-354. http://dx.doi.org/10.1038/nature11097

33. Kovilakam, M. and S. Mahajan, 2015: Black carbon aerosol-induced Northern Hemisphere tropical expansion. *Geophysical Research Letters*, **42**, 4964-4972. http://dx.doi.org/10.1002/2015GL064559

34. Adam, O., T. Schneider, and N. Harnik, 2014: Role of changes in mean temperatures versus temperature gradients in the recent widening of the Hadley circulation. *Journal of Climate*, **27**, 7450-7461. http://dx.doi.org/10.1175/JCLI-D-14-00140.1

35. Allen, R.J., J.R. Norris, and M. Kovilakam, 2014: Influence of anthropogenic aerosols and the Pacific Decadal Oscillation on tropical belt width. *Nature Geoscience*, **7**, 270-274. http://dx.doi.org/10.1038/ngeo2091

36. Chen, S., K. Wei, W. Chen, and L. Song, 2014: Regional changes in the annual mean Hadley circulation in recent decades. *Journal of Geophysical Research Atmospheres*, **119**, 7815-7832. http://dx.doi.org/10.1002/2014JD021540

37. Lucas, C. and H. Nguyen, 2015: Regional characteristics of tropical expansion and the role of climate variability. *Journal of Geophysical Research Atmospheres*, **120**, 6809-6824. http://dx.doi.org/10.1002/2015JD023130

38. Schwendike, J., G.J. Berry, M.J. Reeder, C. Jakob, P. Govekar, and R. Wardle, 2015: Trends in the local Hadley and local Walker circulations. *Journal of Geophysical Research Atmospheres*, **120**, 7599-7618. http://dx.doi.org/10.1002/2014JD022652

39. Prein, A.F., G.J. Holland, R.M. Rasmussen, M.P. Clark, and M.R. Tye, 2016: Running dry: The U.S. Southwest's drift into a drier climate state. *Geophysical Research Letters*, **43**, 1272-1279. http://dx.doi.org/10.1002/2015GL066727

40. Barnes, E.A. and L. Polvani, 2013: Response of the midlatitude jets, and of their variability, to increased greenhouse gases in the CMIP5 models. *Journal of Climate*, **26**, 7117-7135. http://dx.doi.org/10.1175/JCLI-D-12-00536.1

41. Collins, M., R. Knutti, J. Arblaster, J.-L. Dufresne, T. Fichefet, P. Friedlingstein, X. Gao, W.J. Gutowski, T. Johns, G. Krinner, M. Shongwe, C. Tebaldi, A.J. Weaver, and M. Wehner, 2013: Long-term climate change: Projections, commitments and irreversibility. *Climate Change 2013: The Physical Science Basis. Contribution of Working Group I to the Fifth Assessment Report of the Intergovernmental Panel on Climate Change*. Stocker, T.F., D. Qin, G.-K. Plattner, M. Tignor, S.K. Allen, J. Boschung, A. Nauels, Y. Xia, V. Bex, and P.M. Midgley, Eds. Cambridge University Press, Cambridge, United Kingdom and New York, NY, USA, 1029–1136. http://www.climatechange2013.org/report/full-report/

42. Scheff, J. and D. Frierson, 2012: Twenty-first-century multimodel subtropical precipitation declines are mostly midlatitude shifts. *Journal of Climate*, **25**, 4330-4347. http://dx.doi.org/10.1175/JCLI-D-11-00393.1

43. Scheff, J. and D.M.W. Frierson, 2012: Robust future precipitation declines in CMIP5 largely reflect the poleward expansion of model subtropical dry zones. *Geophysical Research Letters*, **39**, L18704. http://dx.doi.org/10.1029/2012GL052910

44. Simpson, I.R., R. Seager, M. Ting, and T.A. Shaw, 2016: Causes of change in Northern Hemisphere winter meridional winds and regional hydroclimate. *Nature Climate Change*, **6**, 65-70. http://dx.doi.org/10.1038/nclimate2783

45. Simpson, I.R., T.A. Shaw, and R. Seager, 2014: A diagnosis of the seasonally and longitudinally varying midlatitude circulation response to global warming. *Journal of the Atmospheric Sciences*, **71**, 2489-2515. http://dx.doi.org/10.1175/JAS-D-13-0325.1

46. Barnes, E.A. and L.M. Polvani, 2015: CMIP5 projections of Arctic amplification, of the North American/North Atlantic circulation, and of their relationship. *Journal of Climate*, **28**, 5254-5271. http://dx.doi.org/10.1175/JCLI-D-14-00589.1

47. Cattiaux, J. and C. Cassou, 2013: Opposite CMIP3/CMIP5 trends in the wintertime Northern Annular Mode explained by combined local sea ice and remote tropical influences. *Geophysical Research Letters*, **40**, 3682-3687. http://dx.doi.org/10.1002/grl.50643

48. Karpechko, A.Y. and E. Manzini, 2012: Stratospheric influence on tropospheric climate change in the Northern Hemisphere. *Journal of Geophysical Research*, **117**, D05133. http://dx.doi.org/10.1029/2011JD017036

49. Scaife, A.A., T. Spangehl, D.R. Fereday, U. Cubasch, U. Langematz, H. Akiyoshi, S. Bekki, P. Braesicke, N. Butchart, M.P. Chipperfield, A. Gettelman, S.C. Hardiman, M. Michou, E. Rozanov, and T.G. Shepherd, 2012: Climate change projections and stratosphere–troposphere interaction. *Climate Dynamics*, **38**, 2089-2097. http://dx.doi.org/10.1007/s00382-011-1080-7

50. Hoerling, M.P., A. Kumar, and T. Xu, 2001: Robustness of the nonlinear climate response to ENSO's extreme phases. *Journal of Climate*, **14**, 1277-1293. http://dx.doi.org/10.1175/1520-0442(2001)014<1277:ROTNCR>2.0.CO;2

51. Kiladis, G.N. and H.F. Diaz, 1989: Global climatic anomalies associated with extremes in the Southern Oscillation. *Journal of Climate*, **2**, 1069-1090. http://dx.doi.org/10.1175/1520-0442(1989)002<1069:GCAAWE>2.0.CO;2

52. Ropelewski, C.F. and M.S. Halpert, 1987: Global and regional scale precipitation patterns associated with the El Niño/Southern Oscillation. *Monthly Weather Review*, **115**, 1606-1626. http://dx.doi.org/10.1175/1520-0493(1987)115<1606:GARSPP>2.0.CO;2

53.	Zhang, T., M.P. Hoerling, J. Perlwitz, and T. Xu, 2016: Forced atmospheric teleconnections during 1979–2014. *Journal of Climate*, **29**, 2333-2357. http://dx.doi.org/10.1175/jcli-d-15-0226.1

54.	Hoerling, M., J. Eischeid, and J. Perlwitz, 2010: Regional precipitation trends: Distinguishing natural variability from anthropogenic forcing. *Journal of Climate*, **23**, 2131-2145. http://dx.doi.org/10.1175/2009jcli3420.1

55.	Hoerling, M., J. Eischeid, J. Perlwitz, X.-W. Quan, K. Wolter, and L. Cheng, 2016: Characterizing recent trends in U.S. heavy precipitation. *Journal of Climate*, **29**, 2313-2332. http://dx.doi.org/10.1175/jcli-d-15-0441.1

56.	Yu, J.-Y., Y. Zou, S.T. Kim, and T. Lee, 2012: The changing impact of El Niño on US winter temperatures. *Geophysical Research Letters*, **39**, L15702. http://dx.doi.org/10.1029/2012GL052483

57.	Yu, J.-Y. and Y. Zou, 2013: The enhanced drying effect of Central-Pacific El Niño on US winter. *Environmental Research Letters*, **8**, 014019. http://dx.doi.org/10.1088/1748-9326/8/1/014019

58.	Li, J., S.-P. Xie, E.R. Cook, G. Huang, R. D'Arrigo, F. Liu, J. Ma, and X.-T. Zheng, 2011: Interdecadal modulation of El Niño amplitude during the past millennium. *Nature Climate Change*, **1**, 114-118. http://dx.doi.org/10.1038/nclimate1086

59.	Lee, T. and M.J. McPhaden, 2010: Increasing intensity of El Niño in the central-equatorial Pacific. *Geophysical Research Letters*, **37**, L14603. http://dx.doi.org/10.1029/2010GL044007

60.	Yeh, S.-W., J.-S. Kug, B. Dewitte, M.-H. Kwon, B.P. Kirtman, and F.-F. Jin, 2009: El Niño in a changing climate. *Nature*, **461**, 511-514. http://dx.doi.org/10.1038/nature08316

61.	Deser, C., I.R. Simpson, K.A. McKinnon, and A.S. Phillips, 2017: The Northern Hemisphere extratropical atmospheric circulation response to ENSO: How well do we know it and how do we evaluate models accordingly? *Journal of Climate*, **30**, 5059-5082. http://dx.doi.org/10.1175/jcli-d-16-0844.1

62.	Garfinkel, C.I., M.M. Hurwitz, D.W. Waugh, and A.H. Butler, 2013: Are the teleconnections of Central Pacific and Eastern Pacific El Niño distinct in boreal wintertime? *Climate Dynamics*, **41**, 1835-1852. http://dx.doi.org/10.1007/s00382-012-1570-2

63.	Capotondi, A., A.T. Wittenberg, M. Newman, E.D. Lorenzo, J.-Y. Yu, P. Braconnot, J. Cole, B. Dewitte, B. Giese, E. Guilyardi, F.-F. Jin, K. Karnauskas, B. Kirtman, T. Lee, N. Schneider, Y. Xue, and S.-W. Yeh, 2015: Understanding ENSO diversity. *Bulletin of the American Meteorological Society*, **96 (12)**, 921-938. http://dx.doi.org/10.1175/BAMS-D-13-00117.1

64.	Bellenger, H., E. Guilyardi, J. Leloup, M. Lengaigne, and J. Vialard, 2014: ENSO representation in climate models: From CMIP3 to CMIP5. *Climate Dynamics*, **42**, 1999-2018. http://dx.doi.org/10.1007/s00382-013-1783-z

65.	Flato, G., J. Marotzke, B. Abiodun, P. Braconnot, S.C. Chou, W. Collins, P. Cox, F. Driouech, S. Emori, V. Eyring, C. Forest, P. Gleckler, E. Guilyardi, C. Jakob, V. Kattsov, C. Reason, and M. Rummukainen, 2013: Evaluation of climate models. *Climate Change 2013: The Physical Science Basis. Contribution of Working Group I to the Fifth Assessment Report of the Intergovernmental Panel on Climate Change*. Stocker, T.F., D. Qin, G.-K. Plattner, M. Tignor, S.K. Allen, J. Boschung, A. Nauels, Y. Xia, V. Bex, and P.M. Midgley, Eds. Cambridge University Press, Cambridge, United Kingdom and New York, NY, USA, 741–866. http://www.climatechange2013.org/report/full-report/

66.	Sheffield, J., S.J. Camargo, R. Fu, Q. Hu, X. Jiang, N. Johnson, K.B. Karnauskas, S.T. Kim, J. Kinter, S. Kumar, B. Langenbrunner, E. Maloney, A. Mariotti, J.E. Meyerson, J.D. Neelin, S. Nigam, Z. Pan, A. Ruiz-Barradas, R. Seager, Y.L. Serra, D.-Z. Sun, C. Wang, S.-P. Xie, J.-Y. Yu, T. Zhang, and M. Zhao, 2013: North American climate in CMIP5 experiments. Part II: Evaluation of historical simulations of intraseasonal to decadal variability. *Journal of Climate*, **26**, 9247-9290. http://dx.doi.org/10.1175/jcli-d-12-00593.1

67.	Coats, S., J.E. Smerdon, B.I. Cook, and R. Seager, 2013: Stationarity of the tropical pacific teleconnection to North America in CMIP5/PMIP3 model simulations. *Geophysical Research Letters*, **40**, 4927-4932. http://dx.doi.org/10.1002/grl.50938

68.	Langenbrunner, B. and J.D. Neelin, 2013: Analyzing ENSO teleconnections in CMIP models as a measure of model fidelity in simulating precipitation. *Journal of Climate*, **26**, 4431-4446. http://dx.doi.org/10.1175/jcli-d-12-00542.1

69.	Kug, J.-S., S.-I. An, Y.-G. Ham, and I.-S. Kang, 2010: Changes in El Niño and La Niña teleconnections over North Pacific–America in the global warming simulations. *Theoretical and Applied Climatology*, **100**, 275-282. http://dx.doi.org/10.1007/s00704-009-0183-0

70. Meehl, G.A. and H. Teng, 2007: Multi-model changes in El Niño teleconnections over North America in a future warmer climate. *Climate Dynamics*, **29**, 779-790. http://dx.doi.org/10.1007/s00382-007-0268-3

71. **Stevenson, S.L., 2012: Significant changes to ENSO strength and impacts in the twenty-first century: Results from CMIP5.** *Geophysical Research Letters*, **39**, L17703. http://dx.doi.org/10.1029/2012GL052759

72. Zhou, Z.-Q., S.-P. Xie, X.-T. Zheng, Q. Liu, and H. Wang, 2014: Global warming–induced changes in El Niño teleconnections over the North **Pacific and North America.** *Journal of Climate*, **27**, 9050-9064. http://dx.doi.org/10.1175/JCLI-D-14-00254.1

73. Seager, R., N. Naik, and L. Vogel, 2012: Does **global warming cause intensified interannual hydroclimate variability?** *Journal of Climate*, **25**, 3355-3372. http://dx.doi.org/10.1175/JCLI-D-11-00363.1

74. Thompson, D.W.J. and J.M. Wallace, 2001: Regional climate impacts of the Northern Hemisphere annular mode. *Science*, **293**, 85-89. http://dx.doi.org/10.1126/science.1058958

75. Archambault, H.M., L.F. Bosart, D. Keyser, **and A.R. Aiyyer, 2008: Influence of large-scale flow regimes on cool-season precipitation** in the northeastern United States. *Monthly Weather Review*, **136**, 2945-2963. http://dx.doi.org/10.1175/2007MWR2308.1

76. Durkee, J.D., J.D. Frye, C.M. Fuhrmann, M.C. **Lacke, H.G. Jeong, and T.L. Mote, 2008: Effects** of the North Atlantic Oscillation on precipitation-type frequency and distribution in the eastern United States. *Theoretical and Applied Climatology*, **94**, 51-65. http://dx.doi.org/10.1007/s00704-007-0345-x

77. Thompson, D.W.J. and J.M. Wallace, 1998: The Arctic oscillation signature in the wintertime **geopotential height and temperature fields.** *Geophysical Research Letters*, **25**, 1297-1300. http://dx.doi.org/10.1029/98GL00950

78. Thompson, D.W.J. and J.M. Wallace, 2000: Annular modes in the extratropical circulation. Part I: Month-to-month variability. *Journal of Climate*, **13**, 1000-1016. http://dx.doi.org/10.1175/1520-0442(2000)013<1000:AMITEC>2.0.CO;2

79. Deser, C., 2000: On the teleconnectivity of the "Arctic Oscillation". *Geophysical Research Letters*, **27**, 779-782. http://dx.doi.org/10.1029/1999GL010945

80. Feldstein, S.B. and C. Franzke, 2006: Are the North Atlantic Oscillation and the Northern Annular Mode distinguishable? *Journal of the Atmospheric Sciences*, **63**, 2915-2930. http://dx.doi.org/10.1175/JAS3798.1

81. Cohen, J. and M. Barlow, 2005: The NAO, the AO, and global warming: How closely related? *Journal of Climate*, **18**, 4498-4513. http://dx.doi.org/10.1175/jcli3530.1

82. Hurrell, J.W., 1995: Decadal trends in the North Atlantic oscillation: Regional temperatures and precipitation. *Science*, **269**, 676-679. http://dx.doi.org/10.1126/science.269.5224.676

83. Overland, J., J.A. Francis, R. Hall, E. Hanna, S.-J. Kim, and T. Vihma, 2015: The melting Arctic and midlatitude weather patterns: Are they connected? *Journal of Climate*, **28**, 7917-7932. http://dx.doi.org/10.1175/JCLI-D-14-00822.1

84. Overland, J.E. and M. Wang, 2015: Increased variability in the early winter subarctic North American atmospheric circulation. *Journal of Climate*, **28**, 7297-7305. http://dx.doi.org/10.1175/jcli-d-15-0395.1

85. Delworth, T.L., F. Zeng, G.A. Vecchi, X. Yang, L. Zhang, and R. Zhang, 2016: The North Atlantic Oscillation as a driver of rapid climate change in the Northern Hemisphere. *Nature Geoscience*, **9**, 509-512. http://dx.doi.org/10.1038/ngeo2738

86. Peings, Y. and G. Magnusdottir, 2014: Forcing of the wintertime atmospheric circulation **by the multidecadal fluctuations of the North** Atlantic ocean. *Environmental Research Letters*, **9**, 034018. http://dx.doi.org/10.1088/1748-9326/9/3/034018

87. Yang, Q., T.H. Dixon, P.G. Myers, J. Bonin, D. Chambers, and M.R. van den Broeke, 2016: Re**cent increases in Arctic freshwater flux affects** Labrador Sea convection and Atlantic overturning circulation. *Nature Communications*, **7**, 10525. http://dx.doi.org/10.1038/ncomms10525

88. Gong, H., L. Wang, W. Chen, X. Chen, and D. Nath, 2017: Biases of the wintertime Arctic Oscillation in CMIP5 models. *Environmental Research Letters*, **12**, 014001. http://dx.doi.org/10.1088/1748-9326/12/1/014001

89. Davini, P. and C. Cagnazzo, 2014: On the misinterpretation of the North Atlantic Oscillation in CMIP5 models. *Climate Dynamics*, **43**, 1497-1511. http://dx.doi.org/10.1007/s00382-013-1970-y

90. Shaw, T.A., J. Perlwitz, and O. Weiner, 2014: Troposphere-stratosphere coupling: Links to North Atlantic weather and climate, including their representation in CMIP5 models. *Journal of Geophysical Research Atmospheres*, **119**, 5864-5880. http://dx.doi.org/10.1002/2013JD021191

91. Lee, Y.-Y. and R.X. Black, 2013: Boreal winter low-frequency variability in CMIP5 models. *Journal of Geophysical Research Atmospheres*, **118**, 6891-6904. http://dx.doi.org/10.1002/jgrd.50493

92. Gillett, N.P. and J.C. Fyfe, 2013: Annular mode changes in the CMIP5 simulations. *Geophysical Research Letters*, **40**, 1189-1193. http://dx.doi.org/10.1002/grl.50249

93. Deser, C., A. Phillips, V. Bourdette, and H. Teng, 2012: Uncertainty in climate change projections: The role of internal variability. *Climate Dynamics*, **38**, 527-546. http://dx.doi.org/10.1007/s00382-010-0977-x

94. Manzini, E., A.Y. Karpechko, J. Anstey, M.P. Baldwin, R.X. Black, C. Cagnazzo, N. Calvo, A. Charlton-Perez, B. Christiansen, P. Davini, E. Gerber, M. Giorgetta, L. Gray, S.C. Hardiman, Y.Y. Lee, D.R. Marsh, B.A. McDaniel, A. Purich, A.A. Scaife, D. Shindell, S.W. Son, S. Watanabe, and G. Zappa, 2014: Northern winter climate change: Assessment of uncertainty in CMIP5 projections related to stratosphere-troposphere coupling. *Journal of Geophysical Research Atmospheres*, **119**, 7979-7998. http://dx.doi.org/10.1002/2013JD021403

95. Alexander, M.A., D.J. Vimont, P. Chang, and J.D. Scott, 2010: The impact of extratropical atmospheric variability on ENSO: Testing the seasonal footprinting mechanism using coupled model experiments. *Journal of Climate*, **23**, 2885-2901. http://dx.doi.org/10.1175/2010jcli3205.1

96. Vimont, D.J., J.M. Wallace, and D.S. Battisti, 2003: The seasonal footprinting mechanism in the Pacific: Implications for ENSO. *Journal of Climate*, **16**, 2668-2675. http://dx.doi.org/10.1175/1520-0442(2003)016<2668:tsfmit>2.0.co;2

97. Di Lorenzo, E., K.M. Cobb, J.C. Furtado, N. Schneider, B.T. Anderson, A. Bracco, M.A. Alexander, and D.J. Vimont, 2010: Central Pacific El Niño and decadal climate change in the North Pacific Ocean. *Nature Geoscience*, **3**, 762-765. http://dx.doi.org/10.1038/ngeo984

98. Furtado, J.C., E.D. Lorenzo, N. Schneider, and N.A. Bond, 2011: North Pacific decadal variability and climate change in the IPCC AR4 models. *Journal of Climate*, **24**, 3049-3067. http://dx.doi.org/10.1175/2010JCLI3584.1

99. Leathers, D.J., B. Yarnal, and M.A. Palecki, 1991: The Pacific/North American teleconnection pattern and United States climate. Part I: Regional temperature and precipitation associations. *Journal of Climate*, **4**, 517-528. http://dx.doi.org/10.1175/1520-0442(1991)004<0517:TPATPA>2.0.CO;2

100. Loikith, P.C. and A.J. Broccoli, 2012: Characteristics of observed atmospheric circulation patterns associated with temperature extremes over North America. *Journal of Climate*, **25**, 7266-7281. http://dx.doi.org/10.1175/JCLI-D-11-00709.1

101. Coleman, J.S.M. and J.C. Rogers, 2003: Ohio River valley winter moisture conditions associated with the Pacific–North American teleconnection pattern. *Journal of Climate*, **16**, 969-981. http://dx.doi.org/10.1175/1520-0442(2003)016<0969:ORVWMC>2.0.CO;2

102. Nigam, S., 2003: Teleconnections. *Encyclopedia of Atmospheric Sciences*. Holton, J.R., Ed. Academic Press, 2243-2269.

103. Li, Y. and N.-C. Lau, 2012: Impact of ENSO on the atmospheric variability over the North Atlantic in late winter—Role of transient eddies. *Journal of Climate*, **25**, 320-342. http://dx.doi.org/10.1175/JCLI-D-11-00037.1

104. Petoukhov, V., S. Rahmstorf, S. Petri, and H.J. Schellnhuber, 2013: Quasiresonant amplification of planetary waves and recent Northern Hemisphere weather extremes. *Proceedings of the National Academy of Sciences*, **110**, 5336-5341. http://dx.doi.org/10.1073/pnas.1222000110

105. Whan, K., F. Zwiers, and J. Sillmann, 2016: The influence of atmospheric blocking on extreme winter minimum temperatures in North America. *Journal of Climate*, **29**, 4361-4381. http://dx.doi.org/10.1175/JCLI-D-15-0493.1

106. Seager, R., M. Hoerling, S. Schubert, H. Wang, B. Lyon, A. Kumar, J. Nakamura, and N. Henderson, 2015: Causes of the 2011–14 California drought. *Journal of Climate*, **28**, 6997-7024. http://dx.doi.org/10.1175/JCLI-D-14-00860.1

107. Swain, D., M. Tsiang, M. Haughen, D. Singh, A. Charland, B. Rajarthan, and N.S. Diffenbaugh, 2014: The extraordinary California drought of 2013/14: Character, context and the role of climate change [in "Explaining Extreme Events of 2013 from a Climate Perspective"]. *Bulletin of the American Meteorological Society*, **95 (9)**, S3-S6. http://dx.doi.org/10.1175/1520-0477-95.9.S1.1

108. Teng, H. and G. Branstator, 2017: Causes of extreme ridges that induce California droughts. *Journal of Climate*, **30**, 1477-1492. http://dx.doi.org/10.1175/jcli-d-16-0524.1

109. Renwick, J.A. and J.M. Wallace, 1996: Relationships between North Pacific wintertime blocking, El Niño, and the PNA pattern. *Monthly Weather Review*, **124**, 2071-2076. http://dx.doi.org/10.1175/1520-0493(1996)124<2071:RBNPWB>2.0.CO;2

110. Carrera, M.L., R.W. Higgins, and V.E. Kousky, 2004: Downstream weather impacts associated with atmospheric blocking over the northeast Pacific. *Journal of Climate*, **17**, 4823-4839. http://dx.doi.org/10.1175/JCLI-3237.1

111. Guirguis, K., A. Gershunov, R. Schwartz, and S. Bennett, 2011: Recent warm and cold daily winter temperature extremes in the Northern Hemisphere. *Geophysical Research Letters*, **38**, L17701. http://dx.doi.org/10.1029/2011GL048762

112. Wang, H., S. Schubert, R. Koster, Y.-G. Ham, and M. Suarez, 2014: On the role of SST forcing in the 2011 and 2012 extreme U.S. heat and drought: A study in contrasts. *Journal of Hydrometeorology*, **15**, 1255-1273. http://dx.doi.org/10.1175/JHM-D-13-069.1

113. Francis, J.A. and S.J. Vavrus, 2012: Evidence linking Arctic amplification to extreme weather in mid-latitudes. *Geophysical Research Letters*, **39**, L06801. http://dx.doi.org/10.1029/2012GL051000

114. Francis, J.A., S.J. Vavrus, and J. Cohen, 2017: Amplified Arctic warming and mid-latitude weather: Emerging connections. *Wiley Interdisciplinary Review: Climate Change*, **8**, e474. http://dx.doi.org/10.1002/wcc.474

115. Hanna, E., T.E. Cropper, R.J. Hall, and J. Cappelen, 2016: Greenland blocking index 1851–2015: A regional climate change signal. *International Journal of Climatology*, **36**, 4847-4861. http://dx.doi.org/10.1002/joc.4673

116. Barnes, E.A., 2013: Revisiting the evidence linking Arctic amplification to extreme weather in midlatitudes. *Geophysical Research Letters*, **40**, 4734-4739. http://dx.doi.org/10.1002/grl.50880

117. Barnes, E.A., E. Dunn-Sigouin, G. Masato, and T. Woollings, 2014: Exploring recent trends in Northern Hemisphere blocking. *Geophysical Research Letters*, **41**, 638-644. http://dx.doi.org/10.1002/2013GL058745

118. Screen, J.A. and I. Simmonds, 2013: Exploring links between Arctic amplification and mid-latitude weather. *Geophysical Research Letters*, **40**, 959-964. http://dx.doi.org/10.1002/grl.50174

119. Hoskins, B. and T. Woollings, 2015: Persistent extratropical regimes and climate extremes. *Current Climate Change Reports*, **1**, 115-124. http://dx.doi.org/10.1007/s40641-015-0020-8

120. Kennedy, D., T. Parker, T. Woollings, B. Harvey, and L. Shaffrey, 2016: The response of high-impact blocking weather systems to climate change. *Geophysical Research Letters*, **43**, 7250-7258. http://dx.doi.org/10.1002/2016GL069725

121. Brandefelt, J. and H. Körnich, 2008: Northern Hemisphere stationary waves in future climate projections. *Journal of Climate*, **21**, 6341-6353. http://dx.doi.org/10.1175/2008JCLI2373.1

122. Haarsma, R.J. and F. Selten, 2012: Anthropogenic changes in the Walker circulation and their impact on the extra-tropical planetary wave structure in the Northern Hemisphere. *Climate Dynamics*, **39**, 1781-1799. http://dx.doi.org/10.1007/s00382-012-1308-1

123. Mantua, N.J., S.R. Hare, Y. Zhang, J.M. Wallace, and R.C. Francis, 1997: A Pacific interdecadal climate oscillation with impacts on salmon production. *Bulletin of the American Meteorological Society*, **78**, 1069-1080. http://dx.doi.org/10.1175/1520-0477(1997)078<1069:APICOW>2.0.CO;2

124. Hartmann, B. and G. Wendler, 2005: The significance of the 1976 Pacific climate shift in the climatology of Alaska. *Journal of Climate*, **18**, 4824-4839. http://dx.doi.org/10.1175/JCLI3532.1

125. McAfee, S.A., 2014: Consistency and the lack thereof in Pacific Decadal Oscillation impacts on North American winter climate. *Journal of Climate*, **27**, 7410-7431. http://dx.doi.org/10.1175/JCLI-D-14-00143.1

126. Wang, T., O.H. Otterå, Y. Gao, and H. Wang, 2012: The response of the North Pacific Decadal Variability to strong tropical volcanic eruptions. *Climate Dynamics*, **39**, 2917-2936. http://dx.doi.org/10.1007/s00382-012-1373-5

127. Boo, K.-O., B.B.B. Booth, Y.-H. Byun, J. Lee, C. Cho, S. Shim, and K.-T. Kim, 2015: Influence of aerosols in multidecadal SST variability simulations over the North Pacific. *Journal of Geophysical Research Atmospheres*, **120**, 517-531. http://dx.doi.org/10.1002/2014JD021933

128. Yeh, S.-W., W.-M. Kim, Y.H. Kim, B.-K. Moon, R.J. Park, and C.-K. Song, 2013: Changes in the variability of the North Pacific sea surface temperature caused by direct sulfate aerosol forcing in China in a coupled general circulation model. *Journal of Geophysical Research Atmospheres*, **118**, 1261-1270. http://dx.doi.org/10.1029/2012JD017947

129. Meehl, G.A., A. Hu, J.M. Arblaster, J. Fasullo, and K.E. Trenberth, 2013: Externally forced and internally generated decadal climate variability associated with the Interdecadal Pacific Oscillation. *Journal of Climate*, **26**, 7298-7310. http://dx.doi.org/10.1175/JCLI-D-12-00548.1

130. Meehl, G.A., A. Hu, B.D. Santer, and S.-P. Xie, 2016: Contribution of the Interdecadal Pacific Oscillation to twentieth-century global surface temperature trends. *Nature Climate Change*, **6**, 1005-1008. http://dx.doi.org/10.1038/nclimate3107

131. Lapp, S.L., J.-M. St. Jacques, E.M. Barrow, and D.J. Sauchyn, 2012: GCM projections for the Pacific Decadal Oscillation under greenhouse forcing for the early 21st century. *International Journal of Climatology*, **32**, 1423-1442. http://dx.doi.org/10.1002/joc.2364

132. Newman, M., 2007: Interannual to decadal predictability of tropical and North Pacific sea surface temperatures. *Journal of Climate*, **20**, 2333-2356. http://dx.doi.org/10.1175/jcli4165.1

133. Feng, S., Q. Hu, and R.J. Oglesby, 2011: Influence of Atlantic sea surface temperatures on persistent drought in North America. *Climate Dynamics*, **37**, 569-586. http://dx.doi.org/10.1007/s00382-010-0835-x

134. Kavvada, A., A. Ruiz-Barradas, and S. Nigam, 2013: AMO's structure and climate footprint in observations and IPCC AR5 climate simulations. *Climate Dynamics*, **41**, 1345-1364. http://dx.doi.org/10.1007/s00382-013-1712-1

135. Seager, R., Y. Kushnir, M. Ting, M. Cane, N. Naik, and J. Miller, 2008: Would advance knowledge of 1930s SSTs have allowed prediction of the Dust Bowl drought? *Journal of Climate*, **21**, 3261-3281. http://dx.doi.org/10.1175/2007JCLI2134.1

136. Chylek, P. and G. Lesins, 2008: Multidecadal variability of Atlantic hurricane activity: 1851–2007. *Journal of Geophysical Research*, **113**, D22106. http://dx.doi.org/10.1029/2008JD010036

137. **Gray, W.M., J.D. Sheaffer, and C.W. Landsea, 1997:** Climate trends associated with multidecadal variability of Atlantic hurricane activity. *Hurricanes: Climate and Socioeconomic Impacts*. Diaz, H.F. and R.S. Pulwarty, Eds. Springer, Berlin, Heidelberg, 15-53. http://dx.doi.org/10.1007/978-3-642-60672-4_2

138. **Kossin, J.P., 2017:** Hurricane intensification along U. S. coast suppressed during active hurricane periods. *Nature*, **541**, 390-393. http://dx.doi.org/10.1038/nature20783

139. Landsea, C.W., R.A. Pielke Jr., A.M. Mestas-Nuñez, and J.A. Knaff, 1999: Atlantic basin hurricanes: Indices of climatic changes. *Climatic Change*, **42**, 89-129. http://dx.doi.org/10.1023/a:1005416332322

140. Zhang, R. and T.L. Delworth, 2009: A new method for attributing climate variations over the Atlantic hurricane basin's main development region. *Geophysical Research Letters*, **36**, L06701. http://dx.doi.org/10.1029/2009GL037260

141. Caron, L.-P., M. Boudreault, and C.L. Bruyère, 2015: Changes in large-scale controls of Atlantic tropical cyclone activity with the phases of the Atlantic Multidecadal Oscillation. *Climate Dynamics*, **44**, 1801-1821. http://dx.doi.org/10.1007/s00382-014-2186-5

142. Delworth, L.T. and E.M. Mann, 2000: Observed and simulated multidecadal variability in the Northern Hemisphere. *Climate Dynamics*, **16**, 661-676. http://dx.doi.org/10.1007/s003820000075

143. Delworth, T.L., F. Zeng, L. Zhang, R. Zhang, G.A. Vecchi, and X. Yang, 2017: The central role of ocean dynamics in connecting the North Atlantic Oscillation to the extratropical component of the Atlantic Multidecadal Oscillation. *Journal of Climate*, **30**, 3789-3805. http://dx.doi.org/10.1175/jcli-d-16-0358.1

144. Frankcombe, L.M., A.v.d. Heydt, and H.A. Dijkstra, 2010: North Atlantic multidecadal climate variability: An investigation of dominant time scales and processes. *Journal of Climate*, **23**, 3626-3638. http://dx.doi.org/10.1175/2010jcli3471.1

145. Knight, J.R., C.K. Folland, and A.A. Scaife, 2006: Climate impacts of the Atlantic Multidecadal Oscillation. *Geophysical Research Letters*, **33**, L17706. http://dx.doi.org/10.1029/2006GL026242

146. Mann, M.E., B.A. Steinman, and S.K. Miller, 2014: On forced temperature changes, internal variability, and the AMO. *Geophysical Research Letters*, **41**, 3211-3219. http://dx.doi.org/10.1002/2014GL059233

147. Moore, G.W.K., J. Halfar, H. Majeed, W. Adey, and A. Kronz, 2017: Amplification of the Atlantic Multidecadal Oscillation associated with the onset of the industrial-era warming. *Scientific Reports*, **7**, 40861. http://dx.doi.org/10.1038/srep40861

148. Terray, L., 2012: Evidence for multiple drivers of North Atlantic multi-decadal climate variability. *Geophysical Research Letters*, **39**, L19712. http://dx.doi.org/10.1029/2012GL053046

149. Miles, M.W., D.V. Divine, T. Furevik, E. Jansen, M. Moros, and A.E.J. Ogilvie, 2014: A signal of persistent Atlantic multidecadal variability in Arctic sea ice. *Geophysical Research Letters*, **41**, 463-469. http://dx.doi.org/10.1002/2013GL058084

150. Trenary, L. and T. DelSole, 2016: Does the Atlantic Multidecadal Oscillation get its predictability from the Atlantic Meridional Overturning Circulation? *Journal of Climate*, **29**, 5267-5280. http://dx.doi.org/10.1175/jcli-d-16-0030.1

151. Clement, A., K. Bellomo, L.N. Murphy, M.A. Cane, T. Mauritsen, G. Rädel, and B. Stevens, 2015: The Atlantic Multidecadal Oscillation without a role for ocean circulation. *Science*, **350**, 320-324. http://dx.doi.org/10.1126/science.aab3980

152. Clement, A., M.A. Cane, L.N. Murphy, K. Bellomo, T. Mauritsen, and B. Stevens, 2016: Response to Comment on "The Atlantic Multidecadal Oscillation without a role for ocean circulation". *Science*, **352**, 1527-1527. http://dx.doi.org/10.1126/science.aaf2575

153. Srivastava, A. and T. DelSole, 2017: Decadal predictability without ocean dynamics. *Proceedings of the National Academy of Sciences*, **114**, 2177-2182. http://dx.doi.org/10.1073/pnas.1614085114

154. Zhang, R., R. Sutton, G. Danabasoglu, T.L. Delworth, W.M. Kim, J. Robson, and S.G. Yeager, 2016: Comment on "The Atlantic Multidecadal Oscillation without a role for ocean circulation". *Science*, **352**, 1527-1527. http://dx.doi.org/10.1126/science.aaf1660

155. Canty, T., N.R. Mascioli, M.D. Smarte, and R.J. Salawitch, 2013: An empirical model of global climate – Part 1: A critical evaluation of volcanic cooling. *Atmospheric Chemistry and Physics*, **13**, 3997-4031. http://dx.doi.org/10.5194/acp-13-3997-2013

156. Evan, A.T., 2012: Atlantic hurricane activity following two major volcanic eruptions. *Journal of Geophysical Research*, **117**, D06101. http://dx.doi.org/10.1029/2011JD016716

157. Evan, A.T., D.J. Vimont, A.K. Heidinger, J.P. Kossin, and R. Bennartz, 2009: The role of aerosols in the evolution of tropical North Atlantic Ocean temperature anomalies. *Science*, **324**, 778-781. http://dx.doi.org/10.1126/science.1167404

158. Booth, B.B.B., N.J. Dunstone, P.R. Halloran, T. Andrews, and N. Bellouin, 2012: Aerosols implicated as a prime driver of twentieth-century North Atlantic climate variability. *Nature*, **484**, 228-232. http://dx.doi.org/10.1038/nature10946

159. Dunstone, N.J., D.M. Smith, B.B.B. Booth, L. Hermanson, and R. Eade, 2013: Anthropogenic aerosol forcing of Atlantic tropical storms. *Nature Geoscience*, **6**, 534-539. http://dx.doi.org/10.1038/ngeo1854

160. Mann, M.E. and K.A. Emanuel, 2006: Atlantic hurricane trends linked to climate change. *Eos, Transactions, American Geophysical Union*, **87**, 233-244. http://dx.doi.org/10.1029/2006EO240001

161. Sobel, A.H., S.J. Camargo, T.M. Hall, C.-Y. Lee, M.K. Tippett, and A.A. Wing, 2016: Human influence on tropical cyclone intensity. *Science*, **353**, 242-246. http://dx.doi.org/10.1126/science.aaf6574

162. Carslaw, K.S., L.A. Lee, C.L. Reddington, K.J. Pringle, A. Rap, P.M. Forster, G.W. Mann, D.V. Spracklen, M.T. Woodhouse, L.A. Regayre, and J.R. Pierce, 2013: Large contribution of natural aerosols to uncertainty in indirect forcing. *Nature*, **503**, 67-71. http://dx.doi.org/10.1038/nature12674

163. Stevens, B., 2015: Rethinking the lower bound on aerosol radiative forcing. *Journal of Climate*, **28**, 4794-4819. http://dx.doi.org/10.1175/jcli-d-14-00656.1

164. Ting, M., Y. Kushnir, R. Seager, and C. Li, 2009: Forced and internal twentieth-century SST trends in the North Atlantic. *Journal of Climate*, **22**, 1469-1481. http://dx.doi.org/10.1175/2008jcli2561.1

165. Tung, K.-K. and J. Zhou, 2013: Using data to attribute episodes of warming and cooling in instrumental records. *Proceedings of the National Academy of Sciences*, **110**, 2058-2063. http://dx.doi.org/10.1073/pnas.1212471110

166. Zhang, R., T.L. Delworth, R. Sutton, D.L.R. Hodson, K.W. Dixon, I.M. Held, Y. Kushnir, J. Marshall, Y. Ming, R. Msadek, J. Robson, A.J. Rosati, M. Ting, and G.A. Vecchi, 2013: Have aerosols caused the observed Atlantic multidecadal variability? *Journal of the Atmospheric Sciences*, **70**, 1135-1144. http://dx.doi.org/10.1175/jas-d-12-0331.1

167. Vincze, M. and I.M. Jánosi, 2011: Is the Atlantic Multidecadal Oscillation (AMO) a statistical phantom? *Nonlinear Processes in Geophysics*, **18**, 469-475. http://dx.doi.org/10.5194/npg-18-469-2011

168. Dima, M. and G. Lohmann, 2007: A hemispheric mechanism for the Atlantic Multidecadal Oscillation. *Journal of Climate*, **20**, 2706-2719. http://dx.doi.org/10.1175/jcli4174.1

169. Chylek, P., C.K. Folland, H.A. Dijkstra, G. Lesins, and M.K. Dubey, 2011: Ice-core data evidence for a prominent near 20 year time-scale of the Atlantic Multidecadal Oscillation. *Geophysical Research Letters*, **38**, L13704. http://dx.doi.org/10.1029/2011GL047501

170. Gray, S.T., L.J. Graumlich, J.L. Betancourt, and G.T. Pederson, 2004: A tree-ring based reconstruction of the Atlantic Multidecadal Oscillation since 1567 A.D. *Geophysical Research Letters*, **31**, L12205. http://dx.doi.org/10.1029/2004GL019932

171. Knudsen, M.F., B.H. Jacobsen, M.-S. Seidenkrantz, and J. Olsen, 2014: Evidence for external forcing of the Atlantic Multidecadal Oscillation since termination of the Little Ice Age. *Nature Communications*, **5**, 3323. http://dx.doi.org/10.1038/ncomms4323

172. **Mann, M.E., J.D. Woodruff, J.P. Donnelly, and Z. Zhang, 2009:** Atlantic hurricanes and climate over the past 1,500 years. *Nature*, **460**, 880-883. http://dx.doi.org/10.1038/nature08219

173. Deser, C., A.S. Phillips, M.A. Alexander, and B.V. Smoliak, 2014: Projecting North American climate over the next 50 years: Uncertainty due to internal variability. *Journal of Climate*, **27**, 2271-2296. http://dx.doi.org/10.1175/JCLI-D-13-00451.1

174. Wettstein, J.J. and C. Deser, 2014: Internal variability in projections of twenty-first-century Arctic sea ice loss: Role of the large-scale atmospheric circulation. *Journal of Climate*, **27**, 527-550. http://dx.doi.org/10.1175/JCLI-D-12-00839.1

175. Thompson, D.W.J., E.A. Barnes, C. Deser, W.E. Foust, and A.S. Phillips, 2015: Quantifying the role of internal climate variability in future climate trends. *Journal of Climate*, **28**, 6443-6456. http://dx.doi.org/10.1175/JCLI-D-14-00830.1

176. Knutson, T.R., F. Zeng, and A.T. Wittenberg, 2013: Multimodel assessment of regional surface temperature trends: CMIP3 and CMIP5 twentieth-century simulations. *Journal of Climate*, **26**, 8709-8743. http://dx.doi.org/10.1175/JCLI-D-12-00567.1

177. NWS, 2016: Global Circulations in NWS Jet Stream: An Online School for Weather. National Weather Service. http://www.srh.noaa.gov/jetstream/global/circ.html

178. **Lindsey, R., 2016:** How El Niño and La Niña affect the winter jet stream and U.S. climate. Climate.gov. https://www.climate.gov/news-features/featured-images/how-el-ni%C3%B1o-and-la-ni%C3%B1a-affect-winter-jet-stream-and-us-climate

179. Vose, R.S., S. Applequist, M. Squires, I. Durre, M.J. Menne, C.N. Williams, Jr., C. Fenimore, K. Gleason, and D. Arndt, 2014: Improved historical temperature and precipitation time series for U.S. climate divisions. *Journal of Applied Meteorology and Climatology*, **53**, 1232-1251. http://dx.doi.org/10.1175/JAMC-D-13-0248.1

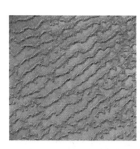

# 6

# Temperature Changes in the United States

## KEY FINDINGS

1. Annual average temperature over the contiguous United States has increased by 1.2°F (0.7°C) for the period 1986–2016 relative to 1901–1960 and by 1.8°F (1.0°C) based on a linear regression for the period 1895–2016 (*very high confidence*). Surface and satellite data are consistent in their depiction of rapid warming since 1979 (*high confidence*). Paleo-temperature evidence shows that recent decades are the warmest of the past 1,500 years (*medium confidence*).

2. There have been marked changes in temperature extremes across the contiguous United States. The frequency of cold waves has decreased since the early 1900s, and the frequency of heat waves has increased since the mid-1960s. The Dust Bowl era of the 1930s remains the peak period for extreme heat. The number of high temperature records set in the past two decades far exceeds the number of low temperature records. (*Very high confidence*)

3. Annual average temperature over the contiguous United States is projected to rise (*very high confidence*). Increases of about 2.5°F (1.4°C) are projected for the period 2021–2050 relative to 1976–2005 in all RCP scenarios, implying recent record-setting years may be "common" in the next few decades (*high confidence*). Much larger rises are projected by late century (2071–2100): 2.8°–7.3°F (1.6°–4.1°C) in a lower scenario (RCP4.5) and 5.8°–11.9°F (3.2°–6.6°C) in the higher scenario (RCP8.5) (*high confidence*).

4. Extreme temperatures in the contiguous United States are projected to increase even more than average temperatures. The temperatures of extremely cold days and extremely warm days are both expected to increase. Cold waves are projected to become less intense while heat waves will become more intense. The number of days below freezing is projected to decline while the number above 90°F will rise. (*Very high confidence*)

**Recommended Citation for Chapter**

**Vose**, R.S., D.R. Easterling, K.E. Kunkel, A.N. LeGrande, and M.F. Wehner, 2017: Temperature changes in the United States. In: *Climate Science Special Report: Fourth National Climate Assessment, Volume I* [Wuebbles, D.J., D.W. Fahey, K.A. Hibbard, D.J. Dokken, B.C. Stewart, and T.K. Maycock (eds.)]. U.S. Global Change Research Program, Washington, DC, USA, pp. 185-206, doi: 10.7930/J0N29V45.

## Introduction

Temperature is among the most important climatic elements used in decision-making. For example, builders and insurers use temperature data for planning and risk management while energy companies and regulators use temperature data to predict demand and set utility rates. Temperature is also a key indicator of climate change: recent increases are apparent over the land, ocean, and troposphere, and substantial changes are expected for this century. This chapter summarizes the major observed and projected changes in near-surface air temperature over the United States, emphasizing new data sets and model projections since the Third National Climate Assessment (NCA3). Changes are depicted using a spectrum of observations, including surface weather stations, moored ocean buoys, polar-orbiting satellites, and temperature-sensitive proxies. Projections are based on global models and downscaled products from CMIP5 (Coupled Model Intercomparison Project Phase 5) using a suite of Representative Concentration Pathways (RCPs; see Ch. 4: Projections for more on RCPs and future scenarios).

## 6.1 Historical Changes

### 6.1.1 Average Temperatures

Changes in average temperature are described using a suite of observational datasets. As in NCA3, changes in land temperature are assessed using the nClimGrid dataset.[1, 2] Along U.S. coastlines, changes in sea surface temperatures are quantified using a new reconstruction[3] that forms the ocean component of the NOAA Global Temperature dataset.[4] Changes in middle tropospheric temperature are examined using updated versions of multiple satellite datasets.[5, 6, 7]

The annual average temperature of the contiguous United States has risen since the start of the 20th century. In general, temperature increased until about 1940, decreased until

about 1970, and increased rapidly through 2016. Because the increase was not constant over time, multiple methods were evaluated in this report (as in NCA3) to quantify the trend. All methods yielded rates of warming that were significant at the 95% level. The lowest estimate of 1.2°F (0.7°C) was obtained by computing the difference between the average for 1986–2016 (i.e., present-day) and the average for 1901–1960 (i.e., the first half of the last century). The highest estimate of 1.8°F (1.0°C) was obtained by fitting a linear (least-squares) regression line through the period 1895–2016. Thus, the temperature increase cited in this assessment is 1.2°–1.8°F (0.7°–1.0°C).

This increase is about 0.1°F (0.06°C) less than presented in NCA3, and it results from the use of slightly different periods in each report. In particular, the decline in the lower bound stems from the use of different time periods to represent present-day climate (NCA3 used 1991–2012, which was slightly warmer than the 1986–2016 period used here). The decline in the upper bound stems mainly from temperature differences late in the record (e.g., the last year of data available for NCA3 was 2012, which was the warmest year on record for the contiguous United States).

Each NCA region experienced a net warming through 2016 (Table 6.1). The largest changes were in the western United States, where average temperature increased by more than 1.5°F (0.8°C) in Alaska, the Northwest, the Southwest, and also in the Northern Great Plains. As noted in NCA3, the Southeast had the least warming, driven by a combination of natural variations and human influences.[8] In most regions, average minimum temperature increased at a slightly higher rate than average maximum temperature, with the Midwest having the largest discrepancy, and the Southwest and Northwest having the smallest. This differential rate of warming resulted in a continuing

**Table 6.1.** Observed changes in annual average temperature (°F) for each National Climate Assessment region. Changes are the difference between the average for present-day (1986–2016) and the average for the first half of the last century (1901–1960 for the contiguous United States, 1925–1960 for Alaska, Hawai'i, and the Caribbean). Estimates are derived from the nClimDiv dataset[1,2].

| NCA Region | Change in Annual Average Temperature | Change in Annual Average Maximum Temperature | Change in Annual Average Minimum Temperature |
|---|---|---|---|
| Contiguous U.S. | 1.23°F | 1.06°F | 1.41°F |
| Northeast | 1.43°F | 1.16°F | 1.70°F |
| Southeast | 0.46°F | 0.16°F | 0.76°F |
| Midwest | 1.26°F | 0.77°F | 1.75°F |
| Great Plains North | 1.69°F | 1.66°F | 1.72°F |
| Great Plains South | 0.76°F | 0.56°F | 0.96°F |
| Southwest | 1.61°F | 1.61°F | 1.61°F |
| Northwest | 1.54°F | 1.52°F | 1.56°F |
| Alaska | 1.67°F | 1.43°F | 1.91°F |
| Hawaii | 1.26°F | 1.01°F | 1.49°F |
| Caribbean | 1.35°F | 1.08°F | 1.60°F |

decrease in the diurnal temperature range that is consistent with other parts of the globe.[9] Annual average sea surface temperature also increased along all regional coastlines (see Figure 1.3), though changes were generally smaller than over land owing to the higher heat capacity of water. Increases were largest in Alaska (greater than 1.0°F [0.6°C]) while increases were smallest (less than 0.5°F [0.3°C]) in coastal areas of the Southeast.

More than 95% of the land surface of the contiguous United States had an increase in annual average temperature (Figure 6.1). In contrast, only small (and somewhat dispersed) parts of the Southeast and Southern Great Plains experienced cooling. From a seasonal perspective, warming was greatest and most widespread in winter, with increases of over 1.5°F (0.8°C) in most areas. In summer, warming was less extensive (mainly along the East Coast and in the western third of the Nation), while cooling was evident in parts of the Southeast, Midwest, and Great Plains.

There has been a rapid increase in the average temperature of the contiguous United States over the past several decades. There is general consistency on this point between the surface thermometer record from NOAA[1] and the middle tropospheric satellite records from Remote Sensing Systems (RSS),[5] NOAA's Center for Satellite Applications and Research (STAR),[7] and the University of Alabama in Huntsville (UAH).[6] In particular, for the period 1979–2016, the rate of warming in the surface record was 0.512°F (0.284°C) per decade, versus trends of 0.455°F (0.253°C), 0.421°F (0.234°C), and 0.289°F (0.160 °C) per decade for RSS version 4, STAR version 3, and UAH version 6, respectively (after accounting for stratospheric influences). All trends are statistically significant at the 95% level. For the contiguous United States, the year 2016 was the second-warmest on record at the surface and in the middle troposphere (2012 was the warmest year at the surface, and 2015 was the warmest in the middle troposphere). Generally speaking, surface and satellite records

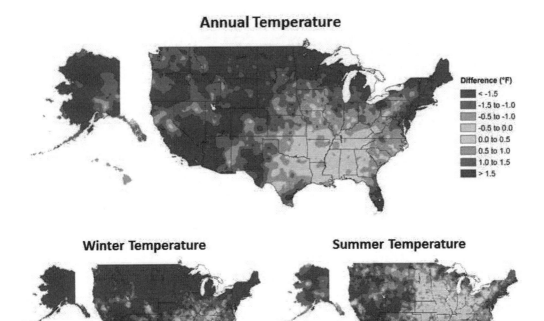

## Annual Temperature

**Difference (°F)**
- < -1.5
- -1.5 to -1.0
- -0.5 to -1.0
- -0.5 to 0.0
- 0.0 to 0.5
- 0.5 to 1.0
- 1.0 to 1.5
- > 1.5

## Winter Temperature

## Summer Temperature

**Figure 6.1.** Observed changes in annual, winter, and summer temperature (°F). Changes are the difference between the average for present-day (1986–2016) and the average for the first half of the last century (1901–1960 for the contiguous United States, 1925–1960 for Alaska and Hawai'i). Estimates are derived from the nClimDiv dataset.[1,2] (Figure source: NOAA/NCEI).

do not have identical trends because they do not represent the same physical quantity; surface measurements are made using thermometers in shelters about 1.5 meters above the ground whereas satellite measurements are mass-weighted averages of microwave emissions from deep atmospheric layers. The UAH record likely has a lower trend because it differs from the other satellite products in the treatment of target temperatures from the NOAA-9 satellite as well as in the correction for diurnal drift.[10]

Recent paleo-temperature evidence confirms the unusual character of wide-scale warming during the past few decades as determined from the instrumental record. The most important new paleoclimate study since NCA3 showed that for each of the seven continental regions, the reconstructed area-weighted average temperature for 1971–2000 was higher than for any other time in nearly 1,400 years,[11] although with significant uncertainty around the central estimate that leads to this conclusion. Recent (up to 2006) 30-year smoothed temperatures across temperate North America (including most of the continental United States) are similarly reconstructed as the warmest over the past 1,500 years[12] (Figure 6.2). Unlike the PAGES 2k seven-continent result mentioned above, this conclusion for North America is robust in relation to the estimated uncertainty range. Reconstruction data since 1500 for western temperate North America show the same conclusion at the annual time scale for 1986–2005. This time period and the running 20-year periods thereafter are warmer than all possible continuous 20-year sequences in a 1,000-member statistical reconstruction ensemble.[13]

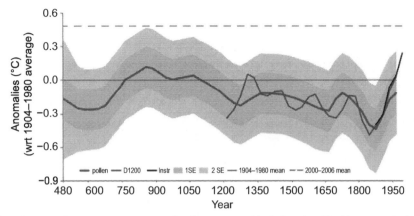

**Figure 6.2.** Pollen-based temperature reconstruction for temperate North America. The blue curve depicts the pollen-based reconstruction of 30-year averages (as anomalies from 1904 to 1980) for the temperate region (30°–55°N, 75°–130°W). The red curve shows the corresponding tree ring-based decadal average reconstruction, which was smoothed and used to calibrate the lower-frequency pollen-based estimate. Light (medium) blue zones indicate 2 standard error (1 standard error) uncertainty estimations associated with each 30-year value. The black curve shows comparably smoothed instrumental temperature values up to 1980. The dashed black line represents the average temperature anomaly of comparably smoothed instrumental data for the period 2000–2006. (Figure source: NOAA NCEI).

### 6.1.2 Temperature Extremes

Shifts in temperature extremes are examined using a suite of societally relevant climate change indices[14, 15] derived from long-term observations of daily surface temperature.[16] The coldest and warmest temperatures of the year are of particular relevance given their widespread use in engineering, agricultural, and other sectoral applications (for example, extreme annual design conditions by the American Society of Heating, Refrigeration, and Air Conditioning; plant hardiness zones by the U.S. Department of Agriculture). Cold waves and heat waves (that is, extended periods of below or above normal temperature) are likewise of great importance because of their numerous societal and environmental impacts, which span from human health to plant and animal phenology. Changes are considered for a spectrum of event frequencies and intensities, ranging from the typical annual extreme to the 1-in-10 year event (an extreme that only has a 10% chance of occurrence in any given year). The discussion focuses on the contiguous United States; Alaska, Hawai'i, and the Caribbean

do not have a sufficient number of long-term stations for a century-scale analysis.

Cold extremes have become less severe over the past century. For example, the coldest daily temperature of the year has increased at most locations in the contiguous United States (Figure 6.3). All regions experienced net increases (Table 6.2), with the largest rises in the Northern Great Plains and the Northwest (roughly 4.5°F [2.5°C]), and the smallest in the Southeast (about 1.0°F [0.6°C]). In general, there were increases throughout the record, with a slight acceleration in recent decades (Figure 6.3). The temperature of extremely cold days (1-in-10 year events) generally exhibited the same pattern of increases as the coldest daily temperature of the year. Consistent with these increases, the number of cool nights per year (those with a minimum temperature below the 10th percentile for 1961–1990) declined in all regions, with much of the West having decreases of roughly two weeks. The frequency of cold waves (6-day periods with a minimum temperature below the

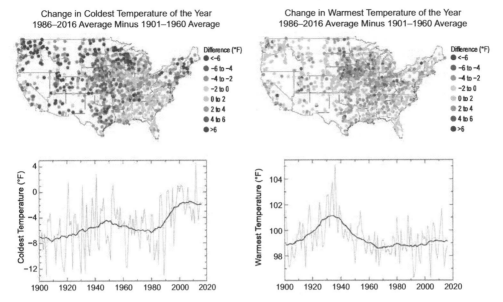

Figure 6.3. Observed changes in the coldest and warmest daily temperatures (°F) of the year in the contiguous United States. Maps (top) depict changes at stations; changes are the difference between the average for present-day (1986–2016) and the average for the first half of the last century (1901–1960). Time series (bottom) depict the area-weighted average for the contiguous United States. Estimates are derived from long-term stations with minimal missing data in the Global Historical Climatology Network–Daily dataset.[16] (Figure source: NOAA/NCEI).

**Table 6.2.** Observed changes in the coldest and warmest daily temperatures (°F) of the year for each National Climate Assessment region in the contiguous United States. Changes are the difference between the average for present-day (1986–2016) and the average for the first half of the last century (1901–1960). Estimates are derived from long-term stations with minimal missing data in the Global Historical Climatology Network–Daily dataset.[16]

| NCA Region | Change in Coldest Day of the Year | Change in Warmest Day of the Year |
|---|---|---|
| Northeast | 2.83°F | −0.92°F |
| Southeast | 1.13°F | −1.49°F |
| Midwest | 2.93°F | −2.22°F |
| Great Plains North | 4.40°F | −1.08°F |
| Great Plains South | 3.25°F | −1.07°F |
| Southwest | 3.99°F | 0.50°F |
| Northwest | 4.78°F | −0.17°F |

10th percentile for 1961–1990) has fallen over the past century (Figure 6.4). The frequency of intense cold waves (4-day, 1-in-5 year events) peaked in the 1980s and then reached record-low levels in the 2000s.[17]

Changes in warm extremes are more nuanced than changes in cold extremes. For instance, the warmest daily temperature of the year increased in some parts of the West over the past century (Figure 6.3), but there were decreases in almost all locations east of the Rocky Mountains. In fact, all eastern regions experienced a net decrease (Table 6.2), most notably the Midwest (about 2.2°F [1.2°C]) and the Southeast (roughly 1.5°F [0.8°C]). The decreases in the eastern half of Nation, particularly in the Great Plains, are mainly tied to the unprecedented summer heat of the 1930s Dust Bowl era, which was exacerbated by land-surface feedbacks driven by springtime

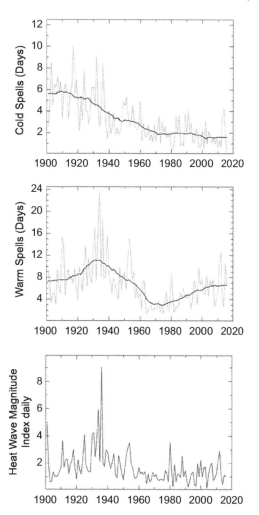

**Figure 6.4.** Observed changes in cold and heat waves in the contiguous United States. The top panel depicts changes in the frequency of cold waves; the middle panel depicts changes in the frequency of heat waves; and the bottom panel depicts changes in the intensity of heat waves. Cold and heat wave frequency indices are defined in Zhang et al.,[15] and the heat wave intensity index is defined in Russo et al.[14] Estimates are derived from long-term stations with minimal missing data in the Global Historical Climatology Network–Daily dataset.[16] (Figure source: NOAA/NCEI).

precipitation deficits and land mismanagement.[18] However, anthropogenic aerosol forcing may also have reduced summer temperatures in the Northeast and Southeast from the early 1950s to the mid-1970s,[19] and agricultural intensification may have suppressed the hottest extremes in the Midwest.[20] Since the mid-1960s, there has been only a very slight increase in the warmest daily temperature of the year (amidst large interannual variability). Heat waves (6-day periods with a maximum temperature above the 90th percentile for 1961–1990) increased in frequency until the mid-1930s, became considerably less common through the mid-1960s, and increased in frequency again thereafter (Figure 6.4). As with warm daily temperatures, heat wave magnitude reached a maximum in the 1930s. The frequency of intense heat waves (4-day, 1-in-5 year events) has generally increased since the 1960s in most regions except the Midwest and the Great

Plains.[17, 21] Since the early 1980s (Figure 6.4), there is suggestive evidence of a slight increase in the intensity of heat waves nationwide[14] as well as an increase in the concurrence of droughts and heat waves.[22]

Changes in the occurrence of record-setting daily temperatures are also apparent. Very generally, the number of record lows has been declining since the late-1970s while the number of record highs has been rising.[23] By extension, there has been an increase in the ratio of the number of record highs to record lows (Figure 6.5). Over the past two decades, the average of this ratio exceeds two (meaning that twice as many high-temperature records have been set as low-temperature records). The number of new highs has surpassed the number of new lows in 15 of the last 20 years, with 2012 and 2016 being particularly extreme (ratios of seven and five, respectively).

## 6.2 Detection and Attribution

### 6.2.1 Average Temperatures

While a confident attribution of global temperature increases to anthropogenic forcing has been made,[24] detection and attribution assessment statements for smaller regions are generally much weaker. Nevertheless, some detectable anthropogenic influences on average temperature have been reported for North America and parts of the United States (e.g., Christidis et al. 2010;[25] Bonfils et al. 2008;[26] Pierce et al. 2009[27]). Figure 6.6 shows an example for linear trends for 1901–2015, indicating a detectable anthropogenic warming since 1901 over the western and northern regions of the contiguous United States for the CMIP5 multimodel ensemble—a condition that was also met for most of the individual models.[28] The Southeast stands out as the only region with no "detectable" warming since 1901; observed trends there were inconsistent with CMIP5 All Forcing historical runs.[28] The cause

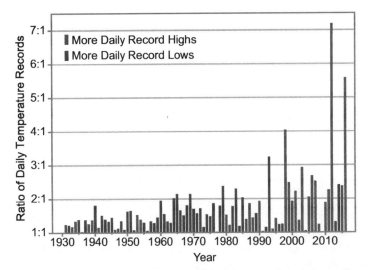

Figure 6.5. Observed changes in the occurrence of record-setting daily temperatures in the contiguous United States. Red bars indicate a year with more daily record highs than daily record lows, while blue bars indicate a year with more record lows than highs. The height of the bar indicates the ratio of record highs to lows (red) or of record lows to highs (blue). For example, a ratio of 2:1 for a blue bar means that there were twice as many record daily lows as daily record highs that year. Estimates are derived from long-term stations with minimal missing data in the Global Historical Climatology Network–Daily dataset.[16] (Figure source: NOAA/NCEI).

## Assessment of Annual Surface Temperature Trends (1901–2015)

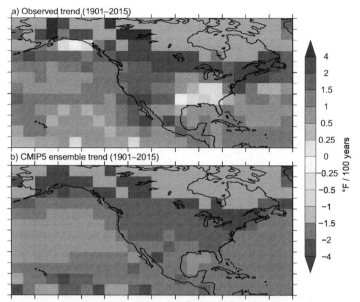

a) Observed trend (1901–2015)

b) CMIP5 ensemble trend (1901–2015)

c) Summary trend assessment

°F / 100 years

4
2
1.5
1
0.5
0.25
0
−0.25
−0.5
−1
−1.5
−2
−4

Detectable anthro. increase, greater than modeled

Detectable anthro. increase, consistent with model

Detectable increase, less than modeled

No detectable trend; white dots: consistent with model

White stippling: Obs. Consistent with All-Forcing

Insufficient data

**Figure 6.6.** Detection and attribution assessment of trends in annual average temperature (°F). Grid-box values indicate whether linear trends for 1901–2015 are detectable (that is, distinct from natural variability) and/or consistent with CMIP5 historical All-Forcing runs. If the grid-box trend is found to be both detectable and either consistent with or greater than the warming in the All-Forcing runs, then the grid box is assessed as having a detectable anthropogenic contribution to warming over the period. Gray regions represent grid boxes with data that are too sparse for detection and attribution. (Figure source: updated from Knutson et al. 2013;[28] © American Meteorological Society. Used with permission.)

of this "warming hole," or lack of a long-term warming trend, remains uncertain, though it is likely a combination of natural and human causes. Some studies conclude that changes in anthropogenic aerosols have played a crucial role (e.g., Leibensperger et al. 2012;[29, 30] Yu et al. 2014[31]), whereas other studies infer a possible large role for atmospheric circulation,[32] internal climate variability (e.g., Meehl et al. 2012;[8] Knutson et al. 2013[28]), and changes in land use (e.g., Goldstein et al. 2009;[33] Xu et al. 2015[34]). Notably, the Southeast has been warming rapidly since the early 1960s.[35, 36] In summary, there is medium confidence for detectable anthropogenic warming over the western and northern regions of the contiguous United States.

### 6.2.2 Temperature Extremes

The Intergovernmental Panel on Climate Change's (IPCC's) Fifth Assessment Report (AR5)[24] concluded that it is very likely that human influence has contributed to the observed changes in frequency and intensity of temperature extremes on the global scale since the mid-20th century. The combined influence of anthropogenic and natural forcings was also detectable (medium confidence) over large subregions of North America (e.g., Zwiers et al. 2011;[37] Min et al. 2013[38]). In general, however, results for the contiguous United States are not as compelling as for global land areas, in part because detection of changes in U.S. regional temperature extremes is affected by extreme temperature in the 1930s.[17] Table 6.3 summarizes available attribution statements for recent extreme U.S. temperature events. As an example, the recent record or near-record high March–May average temperatures occurring in 2012 over the eastern United States were attributed in part to external (natural plus anthropogenic) forcing;[39] the century-scale trend response of temperature to external forcing is typically a close approximation to the anthropogenic forcing response alone. Another study found that although the extreme March 2012 warm anomalies over the United States were mostly due to natural variability, anthropogenic warming contributed to the severity.[40] Such statements reveal that both natural and anthropogenic factors influence the severity of extreme temperature events. Nearly every modern analysis of current extreme hot and cold events reveals some degree of attributable human influence.

## 6.3 Projected Changes

### 6.3.1 Average Temperatures

Temperature projections are based on global model results and associated downscaled products from CMIP5 using a suite of Representative Concentration Pathways (RCPs). In contrast to NCA3, model weighting is employed to refine projections of temperature for each RCP (Ch. 4: Projections; Appendix B: Model Weighting). Weighting parameters are based on model independence and skill over North America for seasonal temperature and annual extremes. Unless stated otherwise, all changes presented here represent the weighted multimodel mean. The weighting scheme helps refine confidence and likelihood statements, but projections of U.S. surface air temperature remain very similar to those in NCA3. Generally speaking, extreme temperatures are projected to increase even more than average temperatures.[41]

**Table 6.3.** Extreme temperature events in the United States for which attribution statements have been made. There are three possible attribution statements: "+" shows an attributable human-induced increase in frequency or intensity, "–" shows an attributable human-induced decrease in frequency or intensity, "0" shows no attributable human contribution.

| Study | Period | Region | Type | Statement |
|---|---|---|---|---|
| Rupp et al. 2012[52]<br>Angélil et al. 2017[53] | Spring/Summer 2011 | Texas | Hot | +<br>+ |
| Hoerling et al. 2013[54] | Summer 2011 | Texas | Hot | + |
| Diffenbaugh and Scherer 2013[55]<br>Angélil et al. 2017[53] | July 2012 | Northcentral and Northeast | Hot | +<br>+ |
| Cattiaux and Yiou 2013[56]<br>Angélil et al. 2017[53] | Spring 2012 | East | Hot | 0<br>+ |
| Knutson et al. 2013b[39]<br>Angélil et al. 2017[53] | Spring 2012 | East | Hot | +<br>+ |
| Jeon et al 2016[57] | Summer 2011 | Texas/Oklahoma | Hot | + |
| Dole et al. 2014[40] | March 2012 | Upper Midwest | Hot | + |
| Seager et al. 2014[58] | 2011–2014 | California | Hot | + |
| Wolter et al. 2015[59] | Winter 2014 | Midwest | Cold | – |
| Trenary et al. 2015[60] | Winter 2014 | East | Cold | 0 |

The annual average temperature of the contiguous United States is projected to rise throughout the century. Increases for the period 2021–2050 relative to 1976–2005 are projected to be about 2.5°F (1.4°C) for a lower scenario (RCP4.5) and 2.9°F (1.6°C) for the higher scenario (RCP8.5); the similarity in warming reflects the similarity in greenhouse gas concentrations during this period (Figure 4.1). Notably, a 2.5°F (1.4°C) increase makes the near-term average comparable to the hottest year in the historical record (2012). In other words, recent record-breaking years may be "common" in the next few decades. By late-century (2071–2100), the RCPs diverge significantly, leading to different rates of warming: approximately 5.0°F (2.8°C) for RCP4.5 and 8.7°F (4.8°C) for RCP8.5. Likewise, there are different ranges of warming for each scenario: 2.8°–7.3°F (1.6°–4.1°C) for RCP4.5 and 5.8°–11.9°F (3.2°–6.6°C) for RCP8.5. (The range is defined here as the difference between the average increase in the three coolest models and the average increase in the three warmest models.) For both RCPs, slightly greater increases are projected in summer than winter (except for Alaska), and average maximums will rise slightly faster than average minimums (except in the Southeast and Southern Great Plains).

Statistically significant warming is projected for all parts of the United States throughout the century (Figure 6.7). Consistent with polar amplification, warming rates (and spatial gradients) are greater at higher latitudes. For example, warming is largest in Alaska (more than 12.0°F [6.7°C] in the northern half of the state by late-century under RCP8.5), driven in part by a decrease in snow cover and thus surface albedo. Similarly, northern regions of the contiguous United States have slightly more warming than other regions (roughly 9.0°F [5.5°C] in the Northeast, Midwest, and Northern Great Plains by late-century under RCP8.5; Table 6.4). The Southeast has slightly less warming because of latent heat release from increases in evapotranspiration (as is already evident in the observed record). Warming is smallest in Hawai'i and the Caribbean (roughly 4.0°–6.0°F [2.2°–3.3°C] by late century under RCP8.5) due to the moderating effects of surrounding oceans. From a sub-regional perspective, less warming is projected along the coasts of the contiguous United States, again due to maritime influences, although increases are still substantial. Warming at higher elevations may be underestimated because the resolution of the CMIP5 models does not capture orography in detail.

## Projected Changes in Annual Average Temperature

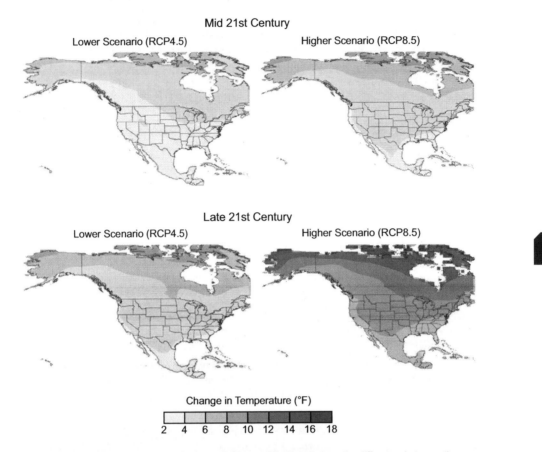

Figure 6.7. Projected changes in annual average temperatures (°F). Changes are the difference between the average for mid-century (2036–2065; top) or late-century (2070–2099, bottom) and the average for near-present (1976–2005). Each map depicts the weighted multimodel mean. Increases are statistically significant in all areas (that is, more than 50% of the models show a statistically significant change, and more than 67% agree on the sign of the change[45]). (Figure source: CICS-NC and NOAA NCEI).

Table 6.4. Projected changes in annual average temperature (°F) for each National Climate Assessment region in the contiguous United States. Changes are the difference between the average for mid-century (2036–2065) or late-century (2071–2100) and the average for near-present (1976–2005) under the higher scenario (RCP8.5) and a lower scenario (RCP4.5). Estimates are derived from 32 climate models that were statistically downscaled using the Localized Constructed Analogs technique.[51] Increases are statistically significant in all areas (that is, more than 50% of the models show a statistically significant change, and more than 67% agree on the sign of the change[45]).

| NCA Region | RCP4.5 Mid-Century (2036–2065) | RCP8.5 Mid-Century (2036–2065) | RCP4.5 Late-Century (2071–2100) | RCP8.5 Late-Century (2071–2100) |
|---|---|---|---|---|
| Northeast | 3.98°F | 5.09°F | 5.27°F | 9.11°F |
| Southeast | 3.40°F | 4.30°F | 4.43°F | 7.72°F |
| Midwest | 4.21°F | 5.29°F | 5.57°F | 9.49°F |
| Great Plains North | 4.05°F | 5.10°F | 5.44°F | 9.37°F |
| Great Plains South | 3.62°F | 4.61°F | 4.78°F | 8.44°F |
| Southwest | 3.72°F | 4.80°F | 4.93°F | 8.65°F |
| Northwest | 3.66°F | 4.67°F | 4.99°F | 8.51°F |

### 6.3.2 Temperature Extremes

Daily extreme temperatures are projected to increase substantially in the contiguous United States, particularly under the higher scenario (RCP8.5). For instance, the coldest and warmest daily temperatures of the year are expected to increase at least 5°F (2.8°C) in most areas by mid-century,[42] rising to 10°F (5.5°C) or more by late-century.[43] In general, there will be larger increases in the coldest temperatures of the year, especially in the northern half of the Nation, whereas the warmest temperatures will exhibit somewhat more uniform changes geographically (Figure 6.8). By mid-century, the upper bound for projected changes (i.e., the average of the three warmest models) is about 2°F (1.1°C) greater than the weighted multimodel mean. On a regional basis, annual extremes (Table 6.5) are consistently projected to rise faster than annual averages (Table 6.4). Future changes in "very rare" extremes are also striking; by late century, current 1-in-20 year maximums are projected to occur every year, while current 1-in-20 year minimums are not expected to occur at all.[44]

The frequency and intensity of cold waves is projected to decrease while the frequency and intensity of heat waves is projected to increase throughout the century. The frequency of cold waves (6-day periods with a minimum temperature below the 10th percentile) will decrease the most in Alaska and the least in the Northeast while the frequency of heat waves (6-day periods with a maximum temperature above the 90th percentile) will increase in all regions, particularly the Southeast, Southwest, and Alaska. By mid-century, decreases in the frequency of cold waves are similar across RCPs whereas increases in the frequency of heat waves are about 50% greater in the higher scenario (RCP8.5) than the lower scenario (RCP4.5).[45] The intensity of cold waves is projected to decrease while the intensity of heat waves is projected to increase, dramatically so under RCP8.5. By mid-century, both extreme cold waves and extreme heat waves (5-day, 1-in-10 year events) are projected to have temperature increases of at least 11.0°F (6.1°C) nationwide, with larger increases in northern regions (the Northeast, Midwest, Northern Great Plains, and Northwest; Table 6.5).

There are large projected changes in the number of days exceeding key temperature thresholds throughout the contiguous United States.

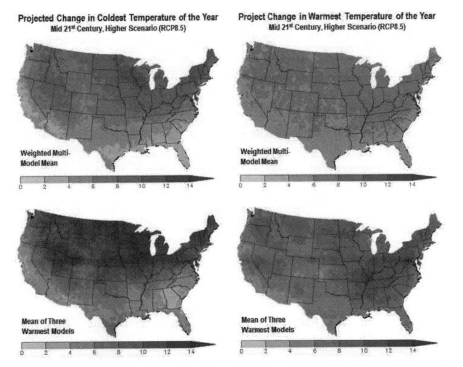

Projected Change in Coldest Temperature of the Year
Mid 21st Century, Higher Scenario (RCP8.5)

Project Change in Warmest Temperature of the Year
Mid 21st Century, Higher Scenario (RCP8.5)

**Figure 6.8.** Projected changes in the coldest and warmest daily temperatures (°F) of the year in the contiguous United States. Changes are the difference between the average for mid-century (2036–2065) and the average for near-present (1976–2005) under the higher scenario (RCP8.5). Maps in the top row depict the weighted multimodel mean whereas maps on the bottom row depict the mean of the three warmest models (that is, the models with the largest temperature increase). Maps are derived from 32 climate model projections that were statistically downscaled using the Localized Constructed Analogs technique.[51] Increases are statistically significant in all areas (that is, more than 50% of the models show a statistically significant change, and more than 67% agree on the sign of the change[45]). (Figure source: CICS-NC and NOAA NCEI).

**Table 6.5.** Projected changes in temperature extremes (°F) for each National Climate Assessment region in the contiguous United States. Changes are the difference between the average for mid-century (2036–2065) and the average for near-present (1976–2005) under the higher scenario (RCP8.5). Estimates are derived from 32 climate models that were statistically downscaled using the Localized Constructed Analogs technique.[51] Increases are statistically significant in all areas (that is, more than 50% of the models show a statistically significant change, and more than 67% agree on the sign of the change[45]).

| NCA Region | Change in Coldest Day of the Year | Change in Coldest 5-Day 1-in-10 Year Event | Change in Warmest Day of the Year | Change in Warmest 5-Day 1-in-10 Year Event |
|---|---|---|---|---|
| Northeast | 9.51°F | 15.93°F | 6.51°F | 12.88°F |
| Southeast | 4.97°F | 8.84°F | 5.79°F | 11.09°F |
| Midwest | 9.44°F | 15.52°F | 6.71°F | 13.02°F |
| Great Plains North | 8.01°F | 12.01°F | 6.48°F | 12.00°F |
| Great Plains South | 5.49°F | 9.41°F | 5.70°F | 10.73°F |
| Southwest | 6.13°F | 10.20°F | 5.85°F | 11.17°F |
| Northwest | 7.33°F | 10.95°F | 6.25°F | 12.31°F |

For instance, there are about 20–30 more days per year with a maximum over 90°F (32°C) in most areas by mid-century under RCP8.5, with increases of 40–50 days in much of the Southeast (Figure 6.9). The upper bound for projected changes is very roughly 10 days greater than the weighted multimodel mean. Consistent with widespread warming, there are 20–30 fewer days per year with a minimum temperature below freezing in the northern and eastern parts of the nation, with decreases of more than 40–50 days in much the West. The upper bound for projected changes in freezing events is very roughly 10–20 days fewer than the weighted multimodel mean in many areas.

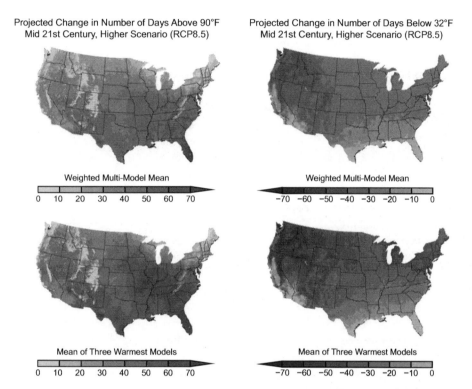

Figure 6.9. Projected changes in the number of days per year with a maximum temperature above 90°F and a minimum temperature below 32°F in the contiguous United States. Changes are the difference between the average for mid-century (2036–2065) and the average for near-present (1976–2005) under the higher scenario (RCP8.5). Maps in the top row depict the weighted multimodel mean whereas maps on the bottom row depict the mean of the three warmest models (that is, the models with the largest temperature increase). Maps are derived from 32 climate model projections that were statistically downscaled using the Localized Constructed Analogs technique.[51] Changes are statistically significant in all areas (that is, more than 50% of the models show a statistically significant change, and more than 67% agree on the sign of the change[45]). (Figure source: CICS-NC and NOAA NCEI).

# TRACEABLE ACCOUNTS

**Key Finding 1** Annual average temperature over the contiguous United States has increased by 1.2°F (0.7°C) for the period 1986–2016 relative to 1901–1960 and by 1.8°F (1.0°C) based on a linear regression for the period 1895–2016 (*very high confidence*). Surface and satellite data are consistent in their depiction of rapid warming since 1979 (*high confidence*). Paleo-temperature evidence shows that recent decades are the warmest of the past 1,500 years (*medium confidence*).

## Description of Evidence Base

The key finding and supporting text summarize extensive evidence documented in the climate science literature. Similar statements about changes exist in other reports (e.g., NCA3;[46] Global Climate Change Impacts in the United States;[47] SAP 1.1: Temperature trends in the lower atmosphere[48]).

Evidence for changes in U.S. climate arises from multiple analyses of data from in situ, satellite, and other records undertaken by many groups over several decades. The primary dataset for surface temperatures in the United States is nClimGrid,[1, 2] though trends are similar in the U.S. Historical Climatology Network, the Global Historical Climatology Network, and other datasets. Several atmospheric reanalyses (e.g., 20th Century Reanalysis, Climate Forecast System Reanalysis, ERA-Interim, Modern Era Reanalysis for Research and Applications) confirm rapid warming at the surface since 1979, with observed trends closely tracking the ensemble mean of the reanalyses. Several recently improved satellite datasets document changes in middle tropospheric temperatures.[5, 6, 7] Longer-term changes are depicted using multiple paleo analyses (e.g., Wahl and Smerdon 2012;[13] Trouet et al. 2013[12]).

## Major Uncertainties

The primary uncertainties for surface data relate to historical changes in station location, temperature instrumentation, observing practice, and spatial sampling (particularly in areas and periods with low station density, such as the intermountain West in the early 20th century). Satellite records are similarly impacted by non-climatic changes such as orbital decay, diurnal sampling, and instrument calibration to target temperatures. Several uncertainties are inherent in temperature-sensitive proxies, such as dating techniques and spatial sampling.

## Assessment of confidence based on evidence and agreement, including short description of nature of evidence and level of agreement

*Very high* (since 1895), *High* (for surface/satellite agreement since 1979), *Medium* (for paleo)

## Likelihood of Impact

*Extremely Likely*

## Summary sentence or paragraph that integrates the above information

There is *very high confidence* in observed changes in average temperature over the United States based upon the convergence of evidence from multiple data sources, analyses, and assessments.

## Key Finding 2

There have been marked changes in temperature extremes across the contiguous United States. The frequency of cold waves has decreased since the early 1900s, and the frequency of heat waves has increased since the mid-1960s. The Dust Bowl era of the 1930s remains the peak period for extreme heat. The number of high temperature records set in the past two decades far exceeds the number of low temperature records. (*Very high confidence*)

## Description of Evidence Base

The key finding and supporting text summarize extensive evidence documented in the climate science literature. Similar statements about changes have also been made in other reports (e.g., NCA3;[46] SAP 3.3: Weather and Climate Extremes in a Changing Climate;[49] IPCC Special Report on Managing the Risks of Extreme Events and Disasters to Advance Climate Change Adaptation[50]).

Evidence for changes in U.S. climate arises from multiple analyses of in situ data using widely published climate extremes indices. For the analyses presented here, the source of in situ data is the Global Historical Climatology Network–Daily dataset,[16] with changes in extremes being assessed using long-term stations with minimal missing data to avoid network-induced variability on the long-term time series. Cold wave frequency was quantified using the Cold Spell Duration Index,[15] heat wave frequency was quantified using the Warm Spell Duration Index,[15] and heat wave intensity were quantified using the Heat Wave Magnitude Index Daily.[14] Station-based index values were averaged into 4° grid boxes, which were then area-averaged into a time series for the contiguous United States. Note that a variety of other threshold and percentile-based indices were also evaluated, with consistent results (e.g., the Dust Bowl was consistently the peak period for extreme heat). Changes in record-setting temperatures were quantified as in Meehl et al. (2016).[23]

### Major Uncertainties

The primary uncertainties for in situ data relate to historical changes in station location, temperature instrumentation, observing practice, and spatial sampling (particularly the precision of estimates of change in areas and periods with low station density, such as the intermountain West in the early 20th century).

### Assessment of confidence based on evidence and agreement, including short description of nature of evidence and level of agreement

*Very high*

### Likelihood of Impact

*Extremely likely*

### Summary sentence or paragraph that integrates the above information

There is *very high confidence* in observed changes in temperature extremes over the United States based upon the convergence of evidence from multiple data sources, analyses, and assessments.

### Key Finding 3

Annual average temperature over the contiguous United States is projected to rise (*very high confidence*). Increases of about 2.5°F (1.4°C) are projected

for the period 2021–2050 relative to 1976–2005 in all RCP scenarios, implying recent record-setting years may be "common" in the next few decades (*high confidence*). Much larger rises are projected by late century (2071–2100): 2.8°–7.3°F (1.6°–4.1°C) in a lower scenario (RCP4.5) and 5.8°–11.9°F (3.2°–6.6°C) in a higher scenario (RCP8.5) (*high confidence*).

### Description of Evidence Base

The key finding and supporting text summarize extensive evidence documented in the climate science literature. Similar statements about changes have also been made in other reports (e.g., NCA3;[46] Global Climate Change Impacts in the United States[47]). The basic physics underlying the impact of human emissions on climate has also been documented in every IPCC assessment.

Projections are based on global model results and associated downscaled products from CMIP5 for RCP4.5 (lower scenario) and RCP8.5 (higher scenario). Model weighting is employed to refine projections for each RCP. Weighting parameters are based on model independence and skill over North America for seasonal temperature and annual extremes. The multimodel mean is based on 32 model projections that were statistically downscaled using the Localized Constructed Analogs technique.[51] The range is defined as the difference between the average increase in the three coolest models and the average increase in the three warmest models. All increases are significant (i.e., more than 50% of the models show a statistically significant change, and more than 67% agree on the sign of the change[45]).

### Major Uncertainties

Global climate models are subject to structural and parametric uncertainty, resulting in a range of estimates of future changes in average temperature. This is partially mitigated through the use of model weighting and pattern scaling. Furthermore, virtually every ensemble member of every model projection contains an increase in temperature by mid- and late-century. Empirical downscaling introduces additional uncertainty (e.g., with respect to stationarity).

**Assessment of confidence based on evidence and agreement, including short description of nature of evidence and level of agreement**

*Very high* for projected change in annual average temperature; *high confidence* for record-setting years becoming the norm in the near future; *high confidence* for much larger temperature increases by late century under a higher scenario (RCP8.5).

**Likelihood of Impact**

*Extremely likely*

**Summary sentence or paragraph that integrates the above information**

There is *very high confidence* in projected changes in average temperature over the United States based upon the convergence of evidence from multiple model simulations, analyses, and assessments.

**Key Finding 4**

Extreme temperatures in the contiguous United States are projected to increase even more than average temperatures. The temperatures of extremely cold days and extremely warm days are both expected to increase. Cold waves are projected to become less intense while heat waves will become more intense. The number of days below freezing is projected to decline while the number above 90°F will rise. (*Very high confidence*)

**Description of Evidence Base**

The key finding and supporting text summarize extensive evidence documented in the climate science literature (e.g., Fischer et al. 2013;[42] Sillmann et al. 2013;[43] Wuebbles et al. 2014;[44] Sun et al. 2015[45]). Similar statements about changes have also been made in other national assessments (such as NCA3) and in reports by the Climate Change Science Program (such as SAP 3.3: Weather and Climate Extremes in a Changing Climate[49]).

Projections are based on global model results and associated downscaled products from CMIP5 for RCP4.5 (lower scenario) and RCP8.5 (higher scenario). Model weighting is employed to refine projections for each RCP. Weighting parameters are based on model independence and skill over North America for seasonal temperature and annual extremes. The multimodel mean is based on 32 model projections that were statistically downscaled using the Localized Constructed Analogs technique.[51] Downscaling improves on the coarse model output, establishing a more geographically accurate baseline for changes in extremes and the number of days per year over key thresholds. The upper bound for projected changes is the average of the three warmest models. All increases are significant (i.e., more than 50% of the models show a statistically significant change, and more than 67% agree on the sign of the change[45]).

**Major Uncertainties**

Global climate models are subject to structural and parametric uncertainty, resulting in a range of estimates of future changes in temperature extremes. This is partially mitigated through the use of model weighting and pattern scaling. Furthermore, virtually every ensemble member of every model projection contains an increase in temperature by mid- and late-century. Empirical downscaling introduces additional uncertainty (e.g., with respect to stationarity).

**Assessment of confidence based on evidence and agreement, including short description of nature of evidence and level of agreement**

*Very high*

**Likelihood of Impact**

*Extremely likely*

**Summary Sentence**

There is *very high confidence* in projected changes in temperature extremes over the United States based upon the convergence of evidence from multiple model simulations, analyses, and assessments.

# REFERENCES

1. Vose, R.S., S. Applequist, M. Squires, I. Durre, M.J. Menne, C.N. Williams, Jr., C. Fenimore, K. Gleason, and D. Arndt, 2014: Improved historical temperature and precipitation time series for U.S. climate divisions. *Journal of Applied Meteorology and Climatology*, **53**, 1232-1251. http://dx.doi.org/10.1175/JAMC-D-13-0248.1

2. Vose, R.S., M. Squires, D. Arndt, I. Durre, C. Fenimore, K. Gleason, M.J. Menne, J. Partain, C.N. Williams Jr., P.A. Bieniek, and R.L. Thoman, 2017: Deriving historical temperature and precipitation time series for Alaska climate divisions via climatologically aided interpolation. *Journal of Service Climatology* **10**, 20. https://www.stateclimate.org/sites/default/files/upload/pdf/journal-articles/2017-Ross-etal.pdf

3. Huang, B., V.F. Banzon, E. Freeman, J. Lawrimore, W. Liu, T.C. Peterson, T.M. Smith, P.W. Thorne, S.D. Woodruff, and H.-M. Zhang, 2015: Extended Reconstructed Sea Surface Temperature Version 4 (ERSST. v4). Part I: Upgrades and intercomparisons. *Journal of Climate*, **28**, 911-930. http://dx.doi.org/10.1175/JCLI-D-14-00006.1

4. Vose, R.S., D. Arndt, V.F. Banzon, D.R. Easterling, B. Gleason, B. Huang, E. Kearns, J.H. Lawrimore, M.J. Menne, T.C. Peterson, R.W. Reynolds, T.M. Smith, C.N. Williams, and D.L. Wuertz, 2012: NOAA's merged land-ocean surface temperature analysis. *Bulletin of the American Meteorological Society*, **93**, 1677-1685. http://dx.doi.org/10.1175/BAMS-D-11-00241.1

5. Mears, C.A. and F.J. Wentz, 2016: Sensitivity of satellite-derived tropospheric temperature trends to the diurnal cycle adjustment. *Journal of Climate*, **29**, 3629-3646. http://dx.doi.org/10.1175/JCLI-D-15-0744.1

6. Spencer, R.W., J.R. Christy, and W.D. Braswell, 2017: UAH Version 6 global satellite temperature products: Methodology and results. *Asia-Pacific Journal of Atmospheric Sciences*, **53**, 121-130. http://dx.doi.org/10.1007/s13143-017-0010-y

7. Zou, C.-Z. and J. Li, 2014: NOAA MSU Mean Layer Temperature. National Oceanic and Atmospheric Administration, Center for Satellite Applications and Research, 35 pp. http://www.star.nesdis.noaa.gov/smcd/emb/mscat/documents/MSU_TCDR_CATBD_Zou_Li.pdf

8. Meehl, G.A., J.M. Arblaster, and G. Branstator, 2012: Mechanisms contributing to the warming hole and the consequent US east–west differential of heat extremes. *Journal of Climate*, **25**, 6394-6408. http://dx.doi.org/10.1175/JCLI-D-11-00655.1

9. Thorne, P.W., M.G. Donat, R.J.H. Dunn, C.N. Williams, L.V. Alexander, J. Caesar, I. Durre, I. Harris, Z. Hausfather, P.D. Jones, M.J. Menne, R. Rohde, R.S. Vose, R. Davy, A.M.G. Klein-Tank, J.H. Lawrimore, T.C. Peterson, and J.J. Rennie, 2016: Reassessing changes in diurnal temperature range: Intercomparison and evaluation of existing global data set estimates. *Journal of Geophysical Research Atmospheres*, **121**, 5138-5158. http://dx.doi.org/10.1002/2015JD024584

10. Po-Chedley, S., T.J. Thorsen, and Q. Fu, 2015: Removing diurnal cycle contamination in satellite-derived tropospheric temperatures: Understanding tropical tropospheric trend discrepancies. *Journal of Climate*, **28**, 2274-2290. http://dx.doi.org/10.1175/JCLI-D-13-00767.1

11. PAGES 2K Consortium, 2013: Continental-scale temperature variability during the past two millennia. *Nature Geoscience*, **6**, 339-346. http://dx.doi.org/10.1038/ngeo1797

12. Trouet, V., H.F. Diaz, E.R. Wahl, A.E. Viau, R. Graham, N. Graham, and E.R. Cook, 2013: A 1500-year reconstruction of annual mean temperature for temperate North America on decadal-to-multidecadal time scales. *Environmental Research Letters*, **8**, 024008. http://dx.doi.org/10.1088/1748-9326/8/2/024008

13. Wahl, E.R. and J.E. Smerdon, 2012: Comparative performance of paleoclimate field and index reconstructions derived from climate proxies and noise-only predictors. *Geophysical Research Letters*, **39**, L06703. http://dx.doi.org/10.1029/2012GL051086

14. Russo, S., A. Dosio, R.G. Graversen, J. Sillmann, H. Carrao, M.B. Dunbar, A. Singleton, P. Montagna, P. Barbola, and J.V. Vogt, 2014: Magnitude of extreme heat waves in present climate and their projection in a warming world. *Journal of Geophysical Research Atmospheres*, **119**, 12,500-12,512. http://dx.doi.org/10.1002/2014JD022098

15. Zhang, X., L. Alexander, G.C. Hegerl, P. Jones, A.K. Tank, T.C. Peterson, B. Trewin, and F.W. Zwiers, 2011: Indices for monitoring changes in extremes based on daily temperature and precipitation data. *Wiley Interdisciplinary Reviews: Climate Change*, **2**, 851-870. http://dx.doi.org/10.1002/wcc.147

16. Menne, M.J., I. Durre, R.S. Vose, B.E. Gleason, and T.G. Houston, 2012: An overview of the global historical climatology network-daily database. *Journal of Atmospheric and Oceanic Technology*, **29**, 897-910. http://dx.doi.org/10.1175/JTECH-D-11-00103.1

17. Peterson, T.C., R.R. Heim, R. Hirsch, D.P. Kaiser, H. Brooks, N.S. Diffenbaugh, R.M. Dole, J.P. Giovannettone, K. Guirguis, T.R. Karl, R.W. Katz, K. Kunkel, D. Lettenmaier, G.J. McCabe, C.J. Paciorek, K.R. Ryberg, S. Schubert, V.B.S. Silva, B.C. Stewart, A.V. Vecchia, G. Villarini, R.S. Vose, J. Walsh, M. Wehner, D. Wolock, K. Wolter, C.A. Woodhouse, and D. Wuebbles, 2013: Monitoring and understanding changes in heat waves, cold waves, floods and droughts in the United States: State of knowledge. *Bulletin of the American Meteorological Society*, **94**, 821-834. http://dx.doi.org/10.1175/BAMS-D-12-00066.1

18. Donat, M.G., A.D. King, J.T. Overpeck, L.V. Alexander, I. Durre, and D.J. Karoly, 2016: Extraordinary heat during the 1930s US Dust Bowl and associated large-scale conditions. *Climate Dynamics*, **46**, 413-426. http://dx.doi.org/10.1007/s00382-015-2590-5

19. Mascioli, N.R., M. Previdi, A.M. Fiore, and M. Ting, 2017: Timing and seasonality of the United States 'warming hole'. *Environmental Research Letters*, **12**, 034008. http://dx.doi.org/10.1088/1748-9326/aa5ef4

20. Mueller, N.D., E.E. Butler, K.A. McKinnon, A. Rhines, M. Tingley, N.M. Holbrook, and P. Huybers, 2016: Cooling of US Midwest summer temperature extremes from cropland intensification. *Nature Climate Change*, **6**, 317-322. http://dx.doi.org/10.1038/nclimate2825

21. Smith, T.T., B.F. Zaitchik, and J.M. Gohlke, 2013: Heat waves in the United States: Definitions, patterns and trends. *Climatic Change*, **118**, 811-825. http://dx.doi.org/10.1007/s10584-012-0659-2

22. Mazdiyasni, O. and A. AghaKouchak, 2015: Substantial increase in concurrent droughts and heatwaves in the United States. *Proceedings of the National Academy of Sciences*, **112**, 11484-11489. http://dx.doi.org/10.1073/pnas.1422945112

23. Meehl, G.A., C. Tebaldi, and D. Adams-Smith, 2016: US daily temperature records past, present, and future. *Proceedings of the National Academy of Sciences*, **113**, 13977-13982. http://dx.doi.org/10.1073/pnas.1606117113

24. Bindoff, N.L., P.A. Stott, K.M. AchutaRao, M.R. Allen, N. Gillett, D. Gutzler, K. Hansingo, G. Hegerl, Y. Hu, S. Jain, I.I. Mokhov, J. Overland, J. Perlwitz, R. Sebbari, and X. Zhang, 2013: Detection and attribution of climate change: From global to regional. *Climate Change 2013: The Physical Science Basis. Contribution of Working Group I to the Fifth Assessment Report of the Intergovernmental Panel on Climate Change*. Stocker, T.F., D. Qin, G.-K. Plattner, M. Tignor, S.K. Allen, J. Boschung, A. Nauels, Y. Xia, V. Bex, and P.M. Midgley, Eds. Cambridge University Press, Cambridge, United Kingdom and New York, NY, USA, 867–952. http://www.climatechange2013.org/report/full-report/

25. Christidis, N., P.A. Stott, F.W. Zwiers, H. Shiogama, and T. Nozawa, 2010: Probabilistic estimates of recent changes in temperature: A multi-scale attribution analysis. *Climate Dynamics*, **34**, 1139-1156. http://dx.doi.org/10.1007/s00382-009-0615-7

26. Bonfils, C., P.B. Duffy, B.D. Santer, T.M.L. Wigley, D.B. Lobell, T.J. Phillips, and C. Doutriaux, 2008: Identification of external influences on temperatures in California. *Climatic Change*, **87**, 43-55. http://dx.doi.org/10.1007/s10584-007-9374-9

27. Pierce, D.W., T.P. Barnett, B.D. Santer, and P.J. Gleckler, 2009: Selecting global climate models for regional climate change studies. *Proceedings of the National Academy of Sciences*, **106**, 8441-8446. http://dx.doi.org/10.1073/pnas.0900094106

28. Knutson, T.R., F. Zeng, and A.T. Wittenberg, 2013: Multimodel assessment of regional surface temperature trends: CMIP3 and CMIP5 twentieth-century simulations. *Journal of Climate*, **26**, 8709-8743. http://dx.doi.org/10.1175/JCLI-D-12-00567.1

29. Leibensperger, E.M., L.J. Mickley, D.J. Jacob, W.T. Chen, J.H. Seinfeld, A. Nenes, P.J. Adams, D.G. Streets, N. Kumar, and D. Rind, 2012: Climatic effects of 1950-2050 changes in US anthropogenic aerosols – Part 1: Aerosol trends and radiative forcing. *Atmospheric Chemistry and Physics* **12**, 3333-3348. http://dx.doi.org/10.5194/acp-12-3333-2012

30. Leibensperger, E.M., L.J. Mickley, D.J. Jacob, W.T. Chen, J.H. Seinfeld, A. Nenes, P.J. Adams, D.G. Streets, N. Kumar, and D. Rind, 2012: Climatic effects of 1950–2050 changes in US anthropogenic aerosols – Part 2: Climate response. *Atmospheric Chemistry and Physics*, **12**, 3349-3362. http://dx.doi.org/10.5194/acp-12-3349-2012

31. Yu, S., K. Alapaty, R. Mathur, J. Pleim, Y. Zhang, C. Nolte, B. Eder, K. Foley, and T. Nagashima, 2014: Attribution of the United States "warming hole": Aerosol indirect effect and precipitable water vapor. *Scientific Reports*, **4**, 6929. http://dx.doi.org/10.1038/srep06929

32. Abatzoglou, J.T. and K.T. Redmond, 2007: Asymmetry between trends in spring and autumn temperature and circulation regimes over western North America. *Geophysical Research Letters*, **34**, L18808. http://dx.doi.org/10.1029/2007GL030891

33. Goldstein, A.H., C.D. Koven, C.L. Heald, and I.Y. Fung, 2009: Biogenic carbon and anthropogenic pollutants combine to form a cooling haze over the southeastern United States. *Proceedings of the National Academy of Sciences*, **106**, 8835-8840. http://dx.doi.org/10.1073/pnas.0904128106

34. Xu, L., H. Guo, C.M. Boyd, M. Klein, A. Bougiatioti, K.M. Cerully, J.R. Hite, G. Isaacman-VanWertz, N.M. Kreisberg, C. Knote, K. Olson, A. Koss, A.H. Goldstein, S.V. Hering, J. de Gouw, K. Baumann, S.-H. Lee, A. Nenes, R.J. Weber, and N.L. Ng, 2015: Effects of anthropogenic emissions on aerosol formation from isoprene and monoterpenes in the southeastern United States. *Proceedings of the National Academy of Sciences*, **112**, 37-42. http://dx.doi.org/10.1073/pnas.1417609112

35. Pan, Z., X. Liu, S. Kumar, Z. Gao, and J. Kinter, 2013: Intermodel variability and mechanism attribution of central and southeastern U.S. anomalous cooling in the twentieth century as simulated by CMIP5 models. *Journal of Climate*, **26**, 6215-6237. http://dx.doi.org/10.1175/JCLI-D-12-00559.1

36. Walsh, J., D. Wuebbles, K. Hayhoe, J. Kossin, K. Kunkel, G. Stephens, P. Thorne, R. Vose, M. Wehner, J. Willis, D. Anderson, S. Doney, R. Feely, P. Hennon, V. Kharin, T. Knutson, F. Landerer, T. Lenton, J. Kennedy, and R. Somerville, 2014: Ch. 2: Our changing climate. *Climate Change Impacts in the United States: The Third National Climate Assessment*. Melillo, J.M., T.C. Richmond, and G.W. Yohe, Eds. U.S. Global Change Research Program, Washington, D.C., 19-67. http://dx.doi.org/10.7930/J0KW5CXT

37. Zwiers, F.W., X.B. Zhang, and Y. Feng, 2011: Anthropogenic influence on long return period daily temperature extremes at regional scales. *Journal of Climate*, **24**, 881-892. http://dx.doi.org/10.1175/2010jcli3908.1

38. Min, S.-K., X. Zhang, F. Zwiers, H. Shiogama, Y.-S. Tung, and M. Wehner, 2013: Multimodel detection and attribution of extreme temperature changes. *Journal of Climate*, **26**, 7430-7451. http://dx.doi.org/10.1175/JCLI-D-12-00551.1

39. Knutson, T.R., F. Zeng, and A.T. Wittenberg, 2013: The extreme March-May 2012 warm anomaly over the eastern United States: Global context and multimodel trend analysis [in "Explaining Extreme Events of 2012 from a Climate Perspective"]. *Bulletin of the American Meteorological Society*, **94 (9)**, S13-S17. http://dx.doi.org/10.1175/BAMS-D-13-00085.1

40. Dole, R., M. Hoerling, A. Kumar, J. Eischeid, J. Perlwitz, X.-W. Quan, G. Kiladis, R. Webb, D. Murray, M. Chen, K. Wolter, and T. Zhang, 2014: The making of an extreme event: Putting the pieces together. *Bulletin of the American Meteorological Society*, **95**, 427-440. http://dx.doi.org/10.1175/BAMS-D-12-00069.1

41. Collins, M., R. Knutti, J. Arblaster, J.-L. Dufresne, T. Fichefet, P. Friedlingstein, X. Gao, W.J. Gutowski, T. Johns, G. Krinner, M. Shongwe, C. Tebaldi, A.J. Weaver, and M. Wehner, 2013: Long-term climate change: Projections, commitments and irreversibility. *Climate Change 2013: The Physical Science Basis. Contribution of Working Group I to the Fifth Assessment Report of the Intergovernmental Panel on Climate Change*. Stocker, T.F., D. Qin, G.-K. Plattner, M. Tignor, S.K. Allen, J. Boschung, A. Nauels, Y. Xia, V. Bex, and P.M. Midgley, Eds. Cambridge University Press, Cambridge, United Kingdom and New York, NY, USA, 1029–1136. http://www.climatechange2013.org/report/full-report/

42. Fischer, E.M., U. Beyerle, and R. Knutti, 2013: Robust spatially aggregated projections of climate extremes. *Nature Climate Change*, **3**, 1033-1038. http://dx.doi.org/10.1038/nclimate2051

43. Sillmann, J., V.V. Kharin, F.W. Zwiers, X. Zhang, and D. Bronaugh, 2013: Climate extremes indices in the CMIP5 multimodel ensemble: Part 2. Future climate projections. *Journal of Geophysical Research Atmospheres*, **118**, 2473-2493. http://dx.doi.org/10.1002/jgrd.50188

44. Wuebbles, D., G. Meehl, K. Hayhoe, T.R. Karl, K. Kunkel, B. Santer, M. Wehner, B. Colle, E.M. Fischer, R. Fu, A. Goodman, E. Janssen, V. Kharin, H. Lee, W. Li, L.N. Long, S.C. Olsen, Z. Pan, A. Seth, J. Sheffield, and L. Sun, 2014: CMIP5 climate model analyses: Climate extremes in the United States. *Bulletin of the American Meteorological Society*, **95**, 571-583. http://dx.doi.org/10.1175/BAMS-D-12-00172.1

45. Sun, L., K.E. Kunkel, L.E. Stevens, A. Buddenberg, J.G. Dobson, and D.R. Easterling, 2015: Regional Surface Climate Conditions in CMIP3 and CMIP5 for the United States: Differences, Similarities, and Implications for the U.S. National Climate Assessment. NOAA Technical Report NESDIS 144. National Oceanic and Atmospheric Administration, National Environmental Satellite, Data, and Information Service, 111 pp. http://dx.doi.org/10.7289/V5RB72KG

46. Melillo, J.M., T.C. Richmond, and G.W. Yohe, eds., 2014: *Climate Change Impacts in the United States: The Third National Climate Assessment*. U.S. Global Change Research Program: Washington, D.C., 841 pp. http://dx.doi.org/10.7930/J0Z31WJ2

47. Karl, T.R., J.T. Melillo, and T.C. Peterson, eds., 2009: *Global Climate Change Impacts in the United States*. Cambridge University Press: New York, NY, 189 pp. http://downloads.globalchange.gov/usimpacts/pdfs/climate-impacts-report.pdf

48. CCSP, 2006: *Temperature Trends in the Lower Atmosphere: Steps for Understanding and Reconciling Differences. A Report by the U.S. Climate Change Science Program and the Subcommittee on Global Change Research.* National Oceanic and Atmospheric Administration, Washington, D.C., 164 pp. http://www.globalchange.gov/browse/reports/sap-11-temperature-trends-lower-atmosphere-steps-understanding-reconciling

49. CCSP, 2008: *Weather and Climate Extremes in a Changing Climate - Regions of Focus - North America, Hawaii, Caribbean, and U.S. Pacific Islands. A Report by the U.S. Climate Change Science Program and the Subcommittee on Global Change Research.* Karl, T.R., G.A. Meehl, C.D. Miller, S.J. Hassol, A.M. Waple, and W.L. Murray, Eds. Department of Commerce, NOAA's National Climatic Data Center, Washington, D.C., 164 pp. http://downloads.globalchange.gov/sap/sap3-3/sap3-3-final-all.pdf

50. IPCC, 2012: Managing the Risks of Extreme Events and Disasters to Advance Climate Change Adaptation. A Special Report of Working Groups I and II of the Intergovernmental Panel on Climate Change. Field, C.B., V. Barros, T.F. Stocker, D. Qin, D.J. Dokken, K.L. Ebi, M.D. Mastrandrea, K.J. Mach, G.-K. Plattner, S.K. Allen, M. Tignor, and P.M. Midgley (Eds.). Cambridge University Press, Cambridge, UK and New York, NY. 582 pp. https://www.ipcc.ch/pdf/special-reports/srex/SREX_Full_Report.pdf

51. Pierce, D.W., D.R. Cayan, and B.L. Thrasher, 2014: Statistical downscaling using Localized Constructed Analogs (LOCA). *Journal of Hydrometeorology,* **15**, 2558-2585. http://dx.doi.org/10.1175/jhm-d-14-0082.1

52. Rupp, D.E., P.W. Mote, N. Massey, C.J. Rye, R. Jones, and M.R. Allen, 2012: Did human influence on climate make the 2011 Texas drought more probable? [in "Explaining Extreme Events of 2011 from a Climate Perspective"]. *Bulletin of the American Meteorological Society,* **93**, 1052-1054. http://dx.doi.org/10.1175/BAMS-D-12-00021.1

53. Angélil, O., D. Stone, M. Wehner, C.J. Paciorek, H. Krishnan, and W. Collins, 2017: An independent assessment of anthropogenic attribution statements for recent extreme temperature and rainfall events. *Journal of Climate,* **30**, 5-16. http://dx.doi.org/10.1175/JCLI-D-16-0077.1

54. Hoerling, M., M. Chen, R. Dole, J. Eischeid, A. Kumar, J.W. Nielsen-Gammon, P. Pegion, J. Perlwitz, X.-W. Quan, and T. Zhang, 2013: Anatomy of an extreme event. *Journal of Climate,* **26**, 2811–2832. http://dx.doi.org/10.1175/JCLI-D-12-00270.1

55. Diffenbaugh, N.S. and M. Scherer, 2013: Likelihood of July 2012 U.S. temperatures in pre-industrial and current forcing regimes [in "Explaining Extreme Events of 2013 from a Climate Perspective"]. *Bulletin of the American Meteorological Society,* **94 (9),** S6-S9. http://dx.doi.org/10.1175/BAMS-D-13-00085.1

56. Cattiaux, J. and P. Yiou, 2013: U.S. heat waves of spring and summer 2012 from the flow analogue perspective [in "Explaining Extreme Events of 2012 from a Climate Perspective"]. *Bulletin of the American Meteorological Society,* **94 (9),** S10-S13. http://dx.doi.org/10.1175/BAMS-D-13-00085.1

57. Jeon, S., C.J. Paciorek, and M.F. Wehner, 2016: Quantile-based bias correction and uncertainty quantification of extreme event attribution statements. *Weather and Climate Extremes,* **12**, 24-32. http://dx.doi.org/10.1016/j.wace.2016.02.001

58. Seager, R., M. Hoerling, D.S. Siegfried, h. Wang, B. Lyon, A. Kumar, J. Nakamura, and N. Henderson, 2014: Causes and Predictability of the 2011-14 California Drought. National Oceanic and Atmospheric Administration, Drought Task Force Narrative Team, 40 pp. http://dx.doi.org/10.7289/V58K771F

59. Wolter, K., J.K. Eischeid, X.-W. Quan, T.N. Chase, M. Hoerling, R.M. Dole, G.J.V. Oldenborgh, and J.E. Walsh, 2015: How unusual was the cold winter of 2013/14 in the Upper Midwest? [in "Explaining Extreme Events of 2014 from a Climate Perspective"]. *Bulletin of the American Meteorological Society,* **96 (12),** S10-S14. http://dx.doi.org/10.1175/bams-d-15-00126.1

60. Trenary, L., T. DelSole, B. Doty, and M.K. Tippett, 2015: Was the cold eastern US Winter of 2014 due to increased variability? *Bulletin of the American Meteorological Society,* **96 (12),** S15-S19. http://dx.doi.org/10.1175/bams-d-15-00138.1

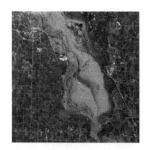

# 7
# Precipitation Change in the United States

## KEY FINDINGS

1.  Annual precipitation has decreased in much of the West, Southwest, and Southeast and increased in most of the Northern and Southern Plains, Midwest, and Northeast. A national average increase of 4% in annual precipitation since 1901 is mostly a result of large increases in the fall season. (*Medium confidence*)

2.  Heavy precipitation events in most parts of the United States have increased in both intensity and frequency since 1901 (*high confidence*). There are important regional differences in trends, with the largest increases occurring in the northeastern United States (*high confidence*). In particular, mesoscale convective systems (organized clusters of thunderstorms)—the main mechanism for warm season precipitation in the central part of the United States—have increased in occurrence and precipitation amounts since 1979 (*medium confidence*).

3.  The frequency and intensity of heavy precipitation events are projected to continue to increase over the 21st century (*high confidence*). Mesoscale convective systems in the central United States are expected to continue to increase in number and intensity in the future (*medium confidence*). There are, however, important regional and seasonal differences in projected changes in total precipitation: the northern United States, including Alaska, is projected to receive more precipitation in the winter and spring, and parts of the southwestern United States are projected to receive less precipitation in the winter and spring (*medium confidence*).

4.  Northern Hemisphere spring snow cover extent, North America maximum snow depth, snow water equivalent in the western United States, and extreme snowfall years in the southern and western United States have all declined, while extreme snowfall years in parts of the northern United States have increased (*medium confidence*). Projections indicate large declines in snowpack in the western United States and shifts to more precipitation falling as rain than snow in the cold season in many parts of the central and eastern United States (*high confidence*).

**Recommended Citation for Chapter**

**Easterling**, D.R., K.E. Kunkel, J.R. Arnold, T. Knutson, A.N. LeGrande, L.R. Leung, R.S. Vose, D.E. Waliser, and M.F. Wehner, 2017: Precipitation change in the United States. In: *Climate Science Special Report: Fourth National Climate Assessment, Volume I* [Wuebbles, D.J., D.W. Fahey, K.A. Hibbard, D.J. Dokken, B.C. Stewart, and T.K. Maycock (eds.)]. U.S. Global Change Research Program, Washington, DC, USA, pp. 207-230, doi: 10.7930/J0H993CC.

## Introduction

Changes in precipitation are one of the most important potential outcomes of a warming world because precipitation is integral to the very nature of society and ecosystems. These systems have developed and adapted to the past envelope of precipitation variations. Any large changes beyond the historical envelope may have profound societal and ecological impacts.

Historical variations in precipitation, as observed from both instrumental and proxy records, establish the context around which future projected changes can be interpreted, because it is within that context that systems have evolved. Long-term station observations from core climate networks serve as a primary source to establish observed changes in both means and extremes. Proxy records, which are used to reconstruct past climate conditions, are varied and include sources such as tree ring and ice core data. Projected changes are examined using the Coupled Model Inter-comparison Project Phase 5 (CMIP5) suite of model simulations. They establish the likelihood of distinct regional and seasonal patterns of change.

## 7.1 Historical Changes

### 7.1.1 Mean Changes

Annual precipitation averaged across the United States has increased approximately 4% over the 1901–2015 period, slightly less than the 5% increase reported in the Third National Climate Assessment (NCA3) over the 1901–2012 period.[1] There continue to be important regional and seasonal differences in precipitation changes (Figure 7.1). Seasonally, national increases are largest in the fall, while little change is observed for winter. Regional differences are apparent, as the Northeast, Midwest, and Great Plains have had increases

while parts of the Southwest and Southeast have had decreases. The slight decrease in the change in annual precipitation across the United States since NCA3 appears to be the result of the recent lingering droughts in the western and southwestern United States.[2, 3] However, the recent meteorological drought in California that began in late 2011[4, 5] now appears to be largely over, due to the substantial precipitation and snowpack the state received in the winter of 2016–2017. The year 2015 was the third wettest on record, just behind 1973 and 1983 (all of which were years marked by El Niño events). Interannual variability is substantial, as evidenced by large multiyear meteorological and agricultural droughts in the 1930s and 1950s.

Changes in precipitation differ markedly across the seasons, as do regional patterns of increases and decreases. For the contiguous United States, fall exhibits the largest (10%) and most widespread increase, exceeding 15% in much of the Northern Great Plains, Southeast, and Northeast. Winter average for the United States has the smallest increase (2%), with drying over most of the western United States as well as parts of the Southeast. In particular, a reduction in streamflow in the northwestern United States has been linked to a decrease in orographic enhancement of precipitation since 1950.[6] Spring and summer have comparable increases (about 3.5%) but substantially different patterns. In spring, the northern half of the contiguous United States has become wetter, and the southern half has become drier. In summer, there is a mixture of increases and decreases across the Nation. Alaska shows little change in annual precipitation (+1.5%); however, in all seasons, central Alaska shows declines and the panhandle shows increases. Hawai'i shows a decline of more than 15% in annual precipitation.

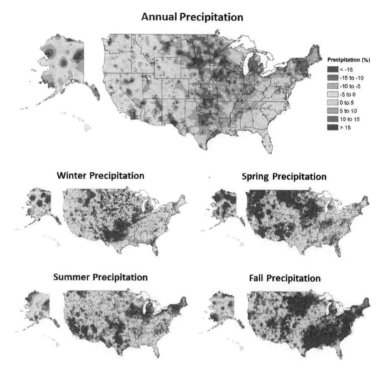

**Annual Precipitation**

Precipitation (%)
- < -15
- -15 to -10
- -10 to -5
- -5 to 0
- 0 to 5
- 5 to 10
- 10 to 15
- > 15

**Winter Precipitation**

**Spring Precipitation**

**Summer Precipitation**

**Fall Precipitation**

**Figure 7.1:** Annual and seasonal changes in precipitation over the United States. Changes are the average for present-day (1986–2015) minus the average for the first half of the last century (1901–1960 for the contiguous United States, 1925–1960 for Alaska and Hawai'i) divided by the average for the first half of the century. (Figure source: [top panel] adapted from Peterson et al. 2013,[78] © American Meteorological Society. Used with permission; [bottom four panels] NOAA NCEI, data source: nCLIMDiv].

### 7.1.2 Snow

Changes in snow cover extent (SCE) in the Northern Hemisphere exhibit a strong seasonal dependence.[7] There has been little change in winter SCE since the 1960s (when the first satellite records became available), while fall SCE has increased. However, the decline in spring SCE is larger than the increase in fall and is due in part to higher temperatures that shorten the time snow spends on the ground in the spring. This tendency is highlighted by the recent occurrences of both unusually high and unusually low monthly (October–June) SCE values, including the top 5 highest and top 5 lowest values in the 48 years of data. From 2010 onward, 7 of the 45 highest monthly SCE values occurred, all in the fall or winter (mostly in November and December), while 9 of the 10 lowest May and June values occurred.

This reflects the trend toward earlier spring snowmelt, particularly at high latitudes.[8] An analysis of seasonal maximum snow depth for 1961–2015 over North America indicates a statistically significant downward trend of 0.11 standardized anomalies per decade and a trend toward the seasonal maximum snow depth occurring earlier—approximately one week earlier on average since the 1960s.[8] There has been a statistically significant decrease over the period of 1930–2007 in the frequency of years with a large number of snowfall days (years exceeding the 90th percentile) in the southern United States and the U.S. Pacific Northwest and an increase in the northern United States.[9] In the snow belts of the Great Lakes, lake effect snowfall has increased overall since the early 20th century for Lakes Superior, Michigan-Huron, and Erie.[10] However, individual studies for

Lakes Michigan[11] and Ontario[12] indicate that this increase has not been continuous. In both cases, upward trends were observed until the 1970s/early 1980s. Since then, however, lake effect snowfall has decreased in these regions. Lake effect snows along the Great Lakes are affected greatly by ice cover extent and lake water temperatures. As ice cover diminishes in winter, the expectation is for more lake effect snow until temperatures increase enough such that much of what now falls as snow instead falls as rain.[13, 14]

End-of-season snow water equivalent (SWE)—especially important where water supply is dominated by spring snow melt (for example, in much of the American West)—has declined since 1980 in the western United States, based on analysis of in situ observations, and is associated with springtime warming.[15] Satellite measurements of SWE based on brightness temperature also show a decrease over this period.[16] The variability of western United States SWE is largely driven by the most extreme events, with the top decile of events explaining 69% of the variability.[17] The recent drought in the western United States was highlighted by the extremely dry 2014–2015 winter that followed three previous dry winters. At Donner Summit, CA, (approximate elevation of 2,100 meters) in the Sierra Nevada Mountains, end-of-season SWE on April 1, 2015, was the lowest on record, based on survey measurements back to 1910, at only 0.51 inches (1.3 cm), or less than 2% of the long-term average. This followed the previous record low in 2014. The estimated return period of this drought is at least 500 years based on paleoclimatic reconstructions.[18]

### 7.1.3 Observed changes in U.S. seasonal extreme precipitation.

Extreme precipitation events occur when the air is nearly completely saturated. Hence, extreme precipitation events are generally

observed to increase in intensity by about 6% to 7% for each degree Celsius of temperature increase, as dictated by the Clausius–Clapeyron relation. Figure 7.2 shows the observed change in the 20-year return value of the seasonal maximum 1-day precipitation totals over the period 1948–2015. A mix of increases and decreases is shown, with the Northwest showing very small changes in all seasons, the southern Great Plains showing a large increase in winter, and the Southeast showing a large increase in the fall.

A U.S. index of extreme precipitation from NCA3 was updated (Figure 7.3) through 2016. This is the number of 2-day precipitation events exceeding the threshold for a 5-year recurrence. The values were calculated by first arithmetically averaging the station data for all stations within each 1° by 1° latitude/longitude grid for each year and then averaging over the grid values across the contiguous United States for each year during the period of 1896–2015. The number of events has been well above average for the last three decades. The slight drop from 2006–2010 to 2011–2016 reflects a below-average number during the widespread severe meteorological drought year of 2012, while the other years in this pentad were well above average. The index value for 2015 was 80% above the 1901–1960 reference period average and the third highest value in the 120 years of record (after 1998 and 2008).

Maximum daily precipitation totals were calculated for consecutive 5-year blocks from 1901 (1901–1905, 1906–1910, 1911–1915, …, 2011–2016) for individual long-term stations. For each 5-year block, these values were aggregated to the regional scale by first arithmetically averaging the station 5-year maximum for all stations within each 2° by 2° latitude/longitude grid and then averaging across all grids within each region to

## Observed Change in Daily, 20-year Return Level Precipitation

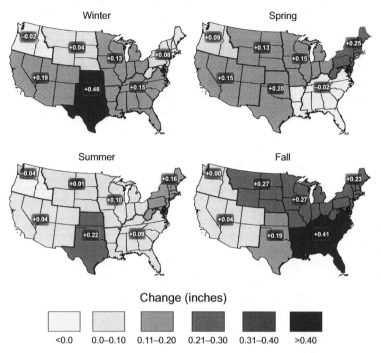

Figure 7.2: Observed changes in the 20-year return value of the seasonal daily precipitation totals for the contiguous United States over the period 1948 to 2015 using data from the Global Historical Climatology Network (GHCN) dataset. (Figure source: adapted from Kunkel et al. 2013;[61] © American Meteorological Society. Used with permission.)

## 2-Day Precipitation Events Exceeding 5-Year Recurrence Interval

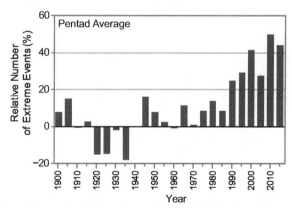

Figure 7.3: Index of the number of 2-day precipitation events exceeding the station-specific threshold for a 5-year recurrence interval in the contiguous United States, expressed as a percentage difference from the 1901–1960 mean. The annual values are averaged over 5-year periods, with the pentad label indicating the ending year of the period. Annual time series of the number of events are first calculated at individual stations. Next, the grid box time series are calculated as the average of all stations in the grid box. Finally, a national time series is calculated as the average of the grid box time series. Data source: GHCN-Daily. (Figure source: CICS-NC and NOAA NCEI).

create a regional time series. Finally, a trend was computed for the resulting regional time series. The difference between these two periods (Figure 7.4, upper left panel) indicates substantial increases over the eastern United States, particularly the northeastern United States with an increase of 27% since 1901. The increases are much smaller over the western United States, with the southwestern and northwestern United States showing little increase.

Another index of extreme precipitation from NCA3 (the total precipitation falling in the top 1% of all days with precipitation) was updated through 2016 (Figure 7.4, upper right panel). This analysis is for 1958–2016. There are increases in all regions, with the largest increases again in the northeastern United States. There are some changes in the values compared to NCA3, with small increases in some regions such as the Midwest and Southwest and small decreases in others such as the Northeast, but the overall picture of changes is the same.

## Observed Change in Heavy Precipitation

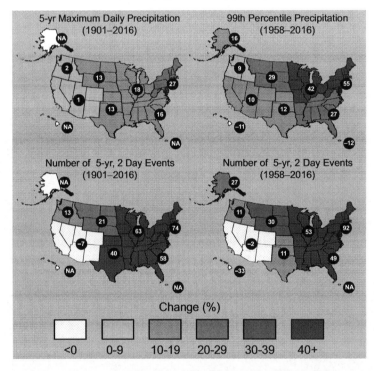

Figure 7.4: These maps show the change in several metrics of extreme precipitation by NCA4 region, including (upper left) the maximum daily precipitation in consecutive 5-year blocks, (upper right) the amount of precipitation falling in daily events that exceed the 99th percentile of all non-zero precipitation days, (lower left) the number of 2-day events with a precipitation total exceeding the largest 2-day amount that is expected to occur, on average, only once every 5 years, as calculated over 1901–2016, and (lower right) the number of 2-day events with a precipitation total exceeding the largest 2-day amount that is expected to occur, on average, only once every 5 years, as calculated over 1958–2016. The numerical value is the percent change over the entire period, either 1901–2016 or 1958–2016. The percentages are first calculated for individual stations, then averaged over 2° latitude by 2° longitude grid boxes, and finally averaged over each NCA4 region. Note that Alaska and Hawai'i are not included in the 1901–2016 maps owing to a lack of observations in the earlier part of the 20th century. (Figure source: CICS-NC and NOAA NCEI).

The national results shown in Figure 7.3 were disaggregated into regional values for two periods: 1901–2016 (Figure 7.4, lower left panel) and 1958–2016 (Figure 7.4, lower right panel) for comparison with Figure 7.4, upper right panel. As with the other metrics, there are large increases over the eastern half of the United States while the increases in the western United States are smaller and there are actually small decreases in the Southwest.

There are differences in the magnitude of changes among the four different regional metrics in Figure 7.4, but the overall picture is the same: large increases in the eastern half of the United States and smaller increases, or slight decreases, in the western United States.

### 7.1.4 Extratropical Cyclones and Mesoscale Convective Systems

As described in Chapter 9: Extreme Storms, there is uncertainty about future changes in winter extratropical cyclones (ETCs).[19] Thus, the potential effects on winter extreme precipitation events is also uncertain. Summertime ETC activity across North America has decreased since 1979, with a reduction of more than 35% in the number of strong summertime ETCs.[20] Most climate models simulate little change over this same historical period, but they project a decrease in summer ETC activity during the remainder of the 21st century.[20] This is potentially relevant to extreme precipitation in the northeastern quadrant of the United States because a large percentage of the extreme precipitation events in this region are caused by ETCs and their associated fronts.[21] This suggests that in the future there may be fewer opportunities in the summer for extreme precipitation, although increases in water vapor are likely to overcompensate for any decreases in ETCs by increasing the likelihood that an ETC will produce excessive rainfall amounts. A very idealized set of climate simulations[22] suggests that substantial projected

warming will lead to a decrease in the number of ETCs but an increase in the intensity of the strongest ETCs. One factor potentially causing this model ETC intensification is an increase in latent heat release in these storms related to a moister atmosphere. Because of the idealized nature of these simulations, the implications of these results for the real earth–atmosphere system is uncertain. However, the increased latent heat mechanism is likely to occur given the high confidence in a future moister atmosphere. For eastern North America, CMIP5 simulations of the future indicate an increase in strong ETCs.[19] Thus, it is possible that the most extreme precipitation events associated with ETCs may increase in the future.

Mesoscale convective systems (MCSs), which contribute substantially to warm season precipitation in the tropics and subtropics,[23] account for about half of rainfall in the central United States.[24] Schumacher and Johnson[25] reported that 74% of all warm season extreme rain events over the eastern two-thirds of the United States during the period 1999–2003 were associated with an MCS. Feng et al.[26] found that large regions of the central United States experienced statistically significant upward trends in April–June MCS rainfall of 0.4–0.8 mm per day (approximately 20%–40%) per decade from 1979 to 2014. They further found upward trends in MCS frequency of occurrence, lifetime, and precipitation amount, which they attribute to an enhanced west-to-east pressure gradient (enhanced Great Plains low-level jet) and enhanced specific humidity throughout the eastern Great Plains.

### 7.1.5 Detection and Attribution
*Trends*

Detectability of trends (compared to internal variability) for a number of precipitation metrics over the continental United States has been examined; however, trends identified for the U.S. regions have not been clearly attributed to anthropogenic forcing.[27, 28] One study concluded that increasing precipitation trends in some north-central U.S. regions and the extreme annual anomalies there in 2013 were at least partly attributable to the combination of anthropogenic and natural forcing.[29]

There is *medium confidence* that anthropogenic forcing has contributed to global-scale intensification of heavy precipitation over land regions with sufficient data coverage.[30] Global changes in extreme precipitation have been attributed to anthropogenically forced climate change,[31, 32] including annual maximum 1-day and 5-day accumulated precipitation over Northern Hemisphere land regions and (relevant to this report) over the North American continent.[33] Although the United States was not separately assessed, the parts of North America with sufficient data for analysis included the continental United States and parts of southern Canada, Mexico, and Central America. Since the covered region was predominantly over the United States, these detection/attribution findings are applicable to the continental United States.

Analyses of precipitation extreme changes over the United States by region (20-year return values of seasonal daily precipitation over 1948–2015, Figure 7.2) show statistically significant increases consistent with theoretical expectations and previous analyses.[34] Further, a significant increase in the area affected by precipitation extremes over North America has also been detected.[35] There is likely an anthropogenic influence on the upward trend in heavy precipitation,[36] although models underestimate the magnitude of the trend. Extreme rainfall from U.S. landfalling tropical cyclones has been higher in recent years (1994–2008) than the long-term historical average, even accounting for temporal changes in storm frequency.[10]

Based on current evidence, it is concluded that detectable but not attributable increases in mean precipitation have occurred over parts of the central United States. Formal detection-attribution studies indicate a human contribution to extreme precipitation increases over the continental United States, but confidence is *low* based on those studies alone due to the short observational period, high natural variability, and model uncertainty.

In summary, based on available studies, it is concluded that for the continental United States there is *high confidence* in the detection of extreme precipitation increases, while there is *low confidence* in attributing the extreme precipitation changes purely to anthropogenic forcing. There is stronger evidence for a human contribution (*medium confidence*) when taking into account process-based understanding (increased water vapor in a warmer atmosphere), evidence from weather and climate models, and trends in other parts of the world.

*Event Attribution*
A number of recent heavy precipitation events have been examined to determine the degree to which their occurrence and severity can be attributed to human-induced climate change. Table 7.1 summarizes available attribution statements for recent extreme U.S. precipitation events. Seasonal and annual precipitation extremes occurring in the north-central and eastern U.S. regions in 2013 were examined for evidence of an anthropogenic influence on their occurrence.[29] Increasing trends in annual precipitation were detected in the northern tier of states, March–May precipitation in the upper Midwest, and June–August precipitation in the eastern United States since 1900. These trends are attributed to external forcing (anthropogenic and natural) but could not be directly attributed to anthropogenic forcing alone. However, based on this analysis, it is

**Table 7.1.** A list of U.S. extreme precipitation events for which attribution statements have been made. In the far right column, "+" indicates that an attributable human-induced increase in frequency and/or magnitude was found, "–" indicates that an attributable human-induced decrease in frequency and/or magnitude was found, "0" indicates no attributable human contribution was identified. As in Tables 6.1 and 8.2, several of the events were originally examined in the *Bulletin of the American Meteorological Society's* (BAMS) State of the Climate Reports and reexamined by Angélil et al.[76] In these cases, both attribution statements are listed with the original authors first. Source: M. Wehner.

| Authors | Event year and duration | Region | Type | Attribution statement |
|---|---|---|---|---|
| Knutson et al. 2014[29] / Angélil et al. 2017[76] | ANN 2013 | U.S. Northern Tier | Wet | +/0 |
| Knutson et al. 2014[29] / Angélil et al. 2017[76] | MAM 2013 | U.S. Upper Midwest | Wet | +/+ |
| Knutson et al. 2014[29] / Angélil et al. 2017[76] | JJA 2013 | Eastern U.S. Region | Wet | +/– |
| Edwards et al. 2014[77] | October 4–5, 2013 | South Dakota | Blizzard | 0 |
| Hoerling et al. 2014[37] | September 10–14, 2013 | Colorado | Wet | 0 |
| Pall et al. 2017[38] | September 10–14, 2013 | Colorado | Wet | + |
| Northwest | 3.66°F | 4.67°F | 4.99°F | 8.51°F |

concluded that the probability of these kinds of extremes has increased due to anthropogenic forcing.

The human influence on individual storms has been investigated with conflicting results. For example, in examining the attribution of the 2013 Colorado floods, one study finds that despite the expected human-induced increase in available moisture, the GEOS-5 model produces fewer extreme storms in the 1983–2012 period compared to the 1871–1900 period in Colorado during the fall season; the study attributes that behavior to changes in the large-scale circulation.[37] However, another study finds that such coarse models cannot produce the observed magnitude of precipitation due to resolution constraints.[38] Based on a highly conditional set of hindcast simulations

imposing the large-scale meteorology and a substantial increase in both the probability and magnitude of the observed precipitation accumulation magnitudes in that particular meteorological situation, the study could not address the question of whether such situations have become more or less probable. Extreme precipitation event attribution is inherently limited by the rarity of the necessary meteorological conditions and the limited number of model simulations that can be performed to examine rare events. This remains an open and active area of research. However, based on these two studies, the anthropogenic contribution to the 2013 Colorado heavy rainfall-flood event is unclear.

An event attribution study of the potential influence of anthropogenic climate change on the extreme 3-day rainfall event associated with flooding in Louisiana in August 2016[39] finds that such extreme rainfall events have become more likely since 1900. Model simulations of extreme rainfall suggest that anthropogenic forcing has increased the odds of such a 3-day extreme precipitation event by 40% or more.

## 7.2 Projections

Changes in precipitation in a warmer climate are governed by many factors. Although energy constraints can be used to understand global changes in precipitation, projecting regional changes is much more difficult because of uncertainty in projecting changes in the large-scale circulation that plays an important role in the formation of clouds and precipitation.[40] For the contiguous United States (CONUS), future changes in seasonal average precipitation will include a mix of increases, decreases, or little change, depending on location and season (Figure 7.5). High-latitude regions are generally projected to become wetter while the subtropical zone is projected to become drier. As the CONUS lies between these two regions, there is significant uncertainty about the sign and magnitude of future anthropogenic changes to seasonal precipitation in much of the region, particularly in the middle latitudes of the Nation. However, because the physical mechanisms controlling extreme precipitation differ from those controlling seasonal average precipitation (Section 7.1.4), in particular atmospheric water vapor will increase with increasing temperatures, confidence is *high* that precipitation extremes will increase in frequency and intensity in the future throughout the CONUS.

Global climate models used to project precipitation changes exhibit varying degrees of fidelity in capturing the observed climatology and seasonal variations of precipitation across the United States. Global or regional climate

models with higher horizontal resolution generally achieve better skill than the CMIP5 models in capturing the spatial patterns and magnitude of winter precipitation in the western and southeastern United States (e.g., Mearns et al. 2012;[41] Wehner 2013;[42] Bacmeister et al. 2014;[43] Wehner et al. 2014[44]), leading to improved simulations of snowpack and runoff (e.g., Rauscher et al. 2008;[45] Rasmussen et al. 2011[46]). Simulation of present and future summer precipitation remains a significant challenge, as current convective parameterizations fail to properly represent the statistics of mesoscale convective systems.[47] As a result, high-resolution models that still require the parameterization of deep convection exhibit mixed results.[44, 48] Advances in computing technology are beginning to enable regional climate modeling at the higher resolutions (1–4 km), permitting the direct simulation of convective clouds systems (e.g., Ban et al. 2014[49]) and eliminating the need for this class of parameterization. However, projections from such models are not yet ready for inclusion in this report.

Important progress has been made by the climate modeling community in providing multimodel ensembles such as CMIP5[50] and NARCCAP[41] to characterize projection uncertainty arising from model differences and large ensemble simulations such as CESM-LE[51] to characterize uncertainty inherent in the climate system due to internal variability. These ensembles provide an important resource for examining the uncertainties in future precipitation projections.

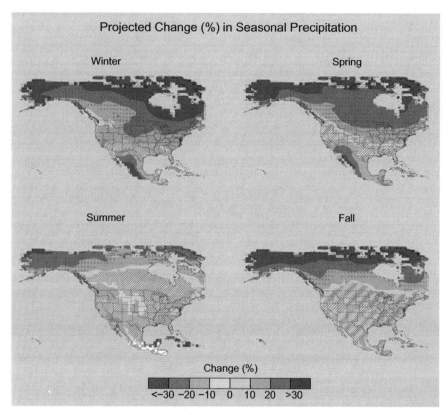

Figure 7.5: Projected change (%) in total seasonal precipitation from CMIP5 simulations for 2070–2099. The values are weighted multimodel means and expressed as the percent change relative to the 1976–2005 average. These are results for the higher scenario (RCP8.5). Stippling indicates that changes are assessed to be large compared to natural variations. Hatching indicates that changes are assessed to be small compared to natural variations. Blank regions (if any) are where projections are assessed to be inconclusive. Data source: World Climate Research Program's (WCRP's) Coupled Model Intercomparison Project. (Figure source: NOAA NCEI).

### 7.2.1 Future Changes in U.S. Seasonal Mean Precipitation

In the United States, projected changes in seasonal mean precipitation span the range from profound decreases to profound increases. In many regions and seasons, projected changes in precipitation are not large compared to natural variations. The general pattern of change is clear and consistent with theoretical expectations. Figure 7.5 shows the weighted CMIP5 multimodel average seasonal change at the end of the century compared to the present under the higher scenario (RCP8.5; see Ch. 4: Projections for discussion of RCPs).

In this figure, changes projected with high confidence to be larger than natural variations are stippled. Regions where future changes are projected with high confidence to be smaller than natural variations are hatched. In winter and spring, the northern part of the country is projected to become wetter as the global climate warms. In the early to middle parts of this century, this will likely be manifested as increases in snowfall.[52] By the latter half of the century, as temperature continues to increase, it will be too warm to snow in many current snow-producing situations, and precipitation will mostly be rainfall. In the southwestern

United States, precipitation will decrease in the spring but the changes are only a little larger than natural variations. Many other regions of the country will not experience significant changes in average precipitation. This is also the case over most of the country in the summer and fall.

This pattern of projected precipitation change arises because of changes in locally available water vapor and weather system shifts. In the northern part of the continent, increases in water vapor, together with changes in circulation that are the result of expansion of the Hadley cell, bring more moisture to these latitudes while maintaining or increasing the frequency of precipitation-producing weather systems. This change in the Hadley circulation (see Ch. 5: Circulation and Variability for discussion of circulation changes) also causes the subtropics, the region between the northern and southern edges of the tropics and the midlatitudes (about 35° of latitude), to be drier in warmer climates as well as moving the mean storm track northward and away from the subtropics, decreasing the frequency of precipitation-producing systems. The combination of these two factors results in precipitation decreases in the southwestern United States, Mexico, and the Caribbean.[53]

*Projected Changes In Snow*
The Third National Climate Assessment[54] projected reductions in annual snowpack of up to 40% in the western United States based on the SRES A2 emissions scenario in the CMIP3 suite of climate model projections. Recent research using the CMIP5 suite of climate model projections forced with a higher scenario (RCP8.5) and statistically downscaled for the western United States continues to show the expected declines in various snow metrics, including snow water equivalent, the number of extreme snowfall events, and number of snowfall days.[55] A northward shift in the rain–snow transition zone in the central and eastern United States was found using statistically downscaled CMIP5 simulations forced with RCP8.5. By the end of the 21st century, large areas that are currently snow dominated in the cold season are expected to be rainfall dominated.[56]

The Variable Infiltration Capacity (VIC) model has been used to investigate the potential effects of climate change on SWE. Declines in SWE are projected in all western U.S. mountain ranges during the 21st century with the virtual disappearance of snowpack in the southernmost mountains by the end of the 21st century under both the lower (RCP4.5) and higher (RCP8.5) scenarios.[57] The projected decreases are most robust at the lower elevations of areas where snowpack accumulation is now reliable (for example, the Cascades and northern Sierra Nevada ranges). In these areas, future decreases in SWE are largely driven by increases in temperature. At higher (colder) elevations, projections are driven more by precipitation changes and are thus more uncertain.

### 7.2.2 Extremes
*Heavy Precipitation Events*
Studies project that the observed increase in heavy precipitation events will continue in the future (e.g. Janssen et al. 2014,[58] 2016[59]). Similar to observed changes, increases are expected in all regions, even those regions where total precipitation is projected to decline, such as the southwestern United States. Under the higher scenario (RCP8.5) the number of extreme events (exceeding a 5-year return period) increases by two to three times the historical average in every region (Figure 7.6) by the end of the 21st century, with the largest increases in the Northeast. Under the lower scenario (RCP4.5), increases are 50%–100%. Research shows that there is strong evidence, both from the observed record and modeling studies, that increased water vapor resulting from high-

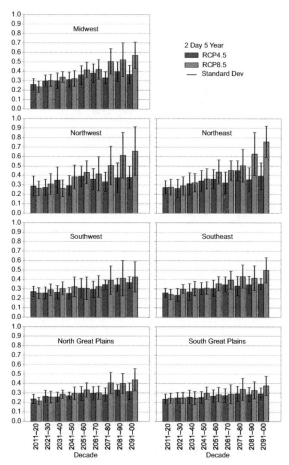

Figure 7.6: Regional extreme precipitation event frequency for a lower scenario (RCP4.5) (green; 16 CMIP5 models) and the higher scenario (RCP8.5) (blue; 14 CMIP5 models) for a 2-day duration and 5-year return. Calculated for 2006–2100 but decadal anomalies begin in 2011. Error bars are ±1 standard deviation; standard deviation is calculated from the 14 or 16 model values that represent the aggregated average over the regions, over the decades, and over the ensemble members of each model. The average frequency for the historical reference period is 0.2 by definition and the values in this graph should be interpreted with respect to a comparison with this historical average value. (Figure source: Janssen et al. 2014[58]).

er temperatures is the primary cause of the increases.[42, 60, 61] Additional effects on extreme precipitation due to changes in dynamical processes are poorly understood. However, atmospheric rivers (ARs), especially along the West Coast of the United States, are projected to increase in number and water vapor transport[62] and experience landfall at lower latitudes[63] by the end of the 21st century.

Projections of changes in the 20-year return period amount for daily precipitation (Figure 7.7) using LOcally Constructed Analogs (LOCA) downscaled data also show large percentage increases for both the middle and late 21st century. A lower scenario (RCP4.5) show increases of around 10% for mid-century and up to 14% for the late century projections. A higher scenario (RCP8.5) shows even larger increases for both mid- and late-century

## Projected Change in Daily, 20-year Extreme Precipitation

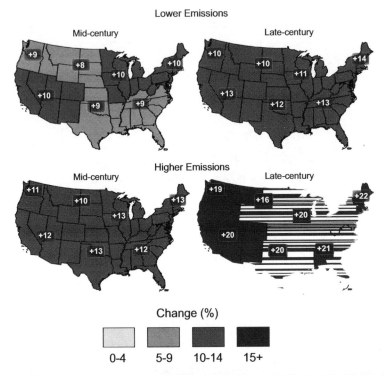

**Figure 7.7:** Projected change in the 20-year return period amount for daily precipitation for mid- (left maps) and late-21st century (right maps). Results are shown for a lower scenario (top maps; RCP4.5) and for a higher scenario (bottom maps, RCP8.5). These results are calculated from the LOCA downscaled data. (Figure source: CICS-NC and NOAA NCEI).

projections, with increases of around 20% by late 21st century. No region in either scenario shows a decline in heavy precipitation. The increases in extreme precipitation tend to increase with return level, such that increases for the 100-year return level are about 30% by the end of the century under a higher scenario (RCP8.5).

Projections of changes in the distribution of daily precipitation amounts (Figure 7.8) indicate an overall more extreme precipitation climate. Specifically, the projections indicate a slight increase in the numbers of dry days and the very lightest precipitation days and a large increase in the heaviest days. The number of days with precipitation amounts greater than the 95th percentile of all non-zero precipita-

tion days increases by more than 25%. At the same time, the number of days with precipitation amounts in the 10th–80th percentile range decreases.

Most global climate models lack sufficient resolution to project changes in mesoscale convective systems (MCSs) in a changing climate.[64] However, research by Cook et al.[65] attempted to identify clues to changes in dynamical forcing that create MCSs. To do this, they examined the ability of 18 coupled ocean–atmosphere global climate models (GCMs) to simulate potential 21st century changes in warm-season flow and the associated U.S. Midwest hydrology resulting from increases in greenhouse gases. They selected a subset of six models that best captured the

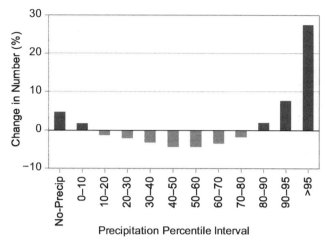

Figure 7.8: Projected change (percentage change relative to the 1976–2005 reference period average) in the number of daily zero ("No-Precip") and non-zero precipitation days (by percentile bins) for late-21st century under a higher scenario (RCP8.5). The precipitation percentile bin thresholds are based on daily non-zero precipitation amounts from the 1976–2005 reference period that have been ranked from low to high. These results are calculated from the LOCA downscaled data. (Figure source: CICS-NC and NOAA NCEI).

low-level flow and associated dynamics of the present-day climate of the central United States and then analyzed these models for changes due to enhanced greenhouse gas forcing. In each of these models, springtime precipitation increases significantly (by 20%–40%) in the upper Mississippi Valley and decreases to the south. The enhanced moisture convergence leading to modeled future climate rainfall increases in the U.S. Midwest is caused by meridional convergence at 850 hPa, connecting the rainfall changes with the Great Plains Low-Level Jet intensification.[66] This is consistent with findings from Feng et al.[26] in the observational record for the period 1979–2014 and by Pan et al.[67] by use of a regional climate model.

Changes in intense hourly precipitation events were simulated by Prein et al.[68] where they found the most intense hourly events (99.9 percentile) in the central United States increase at the expense of moderately intense (97.5 percentile) hourly events in the warm season. They also found the frequency of seasonal hourly precipitation extremes is expected to increase in all regions by up to five times in the same areas that show the highest increases in extreme precipitation rates.

*Hurricane Precipitation*

Regional model projections of precipitation from landfalling tropical cyclones over the United States, based on downscaling of CMIP5 model climate changes, suggest that the occurrence frequency of post-landfall tropical cyclones over the United States will change little compared to present day during the 21st century, as the reduced frequency of tropical cyclones over the Atlantic domain is mostly offset by a greater landfalling fraction. However, when downscaling from CMIP3 model climate changes, projections show a reduced occurrence frequency over U.S. land, indicating uncertainty about future outcomes. The average tropical cyclone rainfall rates within 500 km (about 311 miles) of the storm center increased by 8% to 17% in the simulations, which was at least as much as expected from the water vapor content increase factor alone.

Several studies have projected increases of precipitation rates within hurricanes over ocean regions,[69] particularly for the Atlantic basin.[70] The primary physical mechanism for this increase is the enhanced water vapor content in the warmer atmosphere, which enhances moisture convergence into the storm for a given circulation strength, although a more intense circulation can also contribute.[71] Since hurricanes are responsible for many of the most extreme precipitation events in the southeastern United States,[10, 21] such events are likely to be even heavier in the future. In a set of idealized forcing experiments, this effect was partly offset by differences in warming rates at the surface and at altitude.[72]

# TRACEABLE ACCOUNTS

### Key Finding 1

Annual precipitation has decreased in much of the West, Southwest, and Southeast and increased in most of the Northern and Southern Plains, Midwest, and Northeast. A national average increase of 4% in annual precipitation since 1901 is mostly a result of large increases in the fall season. (*Medium confidence*)

### Description of evidence base

The key finding and supporting text summarizes extensive evidence documented in the climate science peer-reviewed literature. Evidence of long-term changes in precipitation is based on analysis of daily precipitation observations from the U.S. Cooperative Observer Network (http://www.nws.noaa.gov/om/coop/) and shown in Figure 7.1. Published work, such as the Third National Climate Assessment,[73] and Figure 7.1 show important regional and seasonal differences in U.S. precipitation change since 1901.

### Major uncertainties

The main source of uncertainty is the sensitivity of observed precipitation trends to the spatial distribution of observing stations and to historical changes in station location, rain gauges, the local landscape, and observing practices. These issues are mitigated somewhat by new methods to produce spatial grids through time.[74]

### Assessment of confidence based on evidence and agreement, including short description of nature of evidence and level of agreement

Based on the evidence and understanding of the issues leading to uncertainties, confidence is *medium* that average annual precipitation has increased in the United States. Furthermore, confidence is also *medium* that the important regional and seasonal differences in changes documented in the text and in Figure 7.1 are robust.

### Summary sentence or paragraph that integrates the above information

Based on the patterns shown in Figure 7.1 and numerous additional studies of precipitation changes in the United States, there is *medium confidence* in the observed changes in annual and seasonal precipitation over the various regions and the United States as a whole.

### Key Finding 2

Heavy precipitation events in most parts of the United States have increased in both intensity and frequency since 1901 (*high confidence*). There are important regional differences in trends, with the largest increases occurring in the northeastern United States (*high confidence*). In particular, mesoscale convective systems (organized clusters of thunderstorms)—the main mechanism for warm season precipitation in the central part of the United States—have increased in occurrence and precipitation amounts since 1979 (*medium confidence*).

### Description of evidence base

The key finding and supporting text summarize extensive evidence documented in the climate science peer-reviewed literature. Numerous papers have been written documenting observed changes in heavy precipitation events in the United States, including those cited in the Third National Climate Assessment and in this assessment. Although station-based analyses (e.g., Westra et al. 2013[34]) do not show large numbers of statistically significant station-based trends, area averaging reduces the noise inherent in station-based data and produces robust increasing signals (see Figures 7.2 and 7.3). Evidence of long-term changes in precipitation is based on analysis of daily precipitation observations from the U.S. Cooperative Observer Network (http://www.nws.noaa.gov/om/coop/) and shown in Figures 7.2, 7.3, and 7.4.

### Major uncertainties

The main source of uncertainty is the sensitivity of observed precipitation trends to the spatial distribution of observing stations and to historical changes in station location, rain gauges, and observing practices. These issues are mitigated somewhat by methods used to produce spatial grids through gridbox averaging.

**Assessment of confidence based on evidence and agreement, including short description of nature of evidence and level of agreement**

Based on the evidence and understanding of the issues leading to uncertainties, confidence is *high* that heavy precipitation events have increased in the United States. Furthermore, confidence is also *high* that the important regional and seasonal differences in changes documented in the text and in Figures 7.2, 7.3, and 7.4 are robust.

**Summary sentence or paragraph that integrates the above information**

Based on numerous analyses of the observed record in the United States there is *high confidence* in the observed changes in heavy precipitation events, and *medium confidence* in observed changes in mesoscale convective systems.

**Key Finding 3**

The frequency and intensity of heavy precipitation events are projected to continue to increase over the 21st century (*high confidence*). Mesoscale convective systems in the central United States are expected to continue to increase in number and intensity in the future (*medium confidence*). There are, however, important regional and seasonal differences in projected changes in total precipitation: the northern United States, including Alaska, is projected to receive more precipitation in the winter and spring, and parts of the southwestern United States are projected to receive less precipitation in the winter and spring (*medium confidence*).

**Description of evidence base**

Evidence for future changes in precipitation is based on climate model projections and our understanding of the climate system's response to increasing greenhouse gases and of regional mechanisms behind the projected changes. In particular, Figure 7.7 documents projected changes in the 20-year return period amount using the LOCA data, and Figure 7.6 shows changes in 2 day totals for the 5-year return period using the CMIP5 suite of models. Each figure shows robust changes in extreme precipitation events as they are defined in

the figure. However, Figure 7.5, which shows changes in seasonal and annual precipitation, indicates where confidence in the changes is higher based on consistency between the models and that there are large areas where the projected change is uncertain.

**Major uncertainties**

A key issue is how well climate models simulate precipitation, which is one of the more challenging aspects of weather and climate simulation. In particular, comparisons of model projections for total precipitation (from both CMIP3 and CMIP5, see Sun et al. 2015[75]) by NCA3 region show a spread of responses in some regions (for example, the Southwest) such that they are opposite from the ensemble average response. The continental United States is positioned in the transition zone between expected drying in the subtropics and wetting in the mid- and higher-latitudes. There are some differences in the location of this transition between CMIP3 and CMIP5 models and thus there remains uncertainty in the exact location of the transition zone.

**Assessment of confidence based on evidence and agreement, including short description of nature of evidence and level of agreement**

Based on evidence from climate model simulations and our fundamental understanding of the relationship of water vapor to temperature, confidence is *high* that extreme precipitation will increase in all regions of the United States. However, based on the evidence and understanding of the issues leading to uncertainties, confidence is *medium* that that more total precipitation is projected for the northern U.S. and less for the Southwest.

**Summary sentence or paragraph that integrates the above information**

Based on numerous analyses of model simulations and our understanding of the climate system there is *high confidence* in the projected changes in precipitation extremes and *medium confidence* in projected changes in total precipitation over the United States.

**Key Finding 4**

Northern Hemisphere spring snow cover extent, North America maximum snow depth, snow water equivalent in the western United States, and extreme snowfall years in the southern and western United States have all declined, while extreme snowfall years in parts of the northern United States have increased (*medium confidence*). Projections indicate large declines in snowpack in the western United States and shifts to more precipitation falling as rain than snow in the cold season in many parts of the central and eastern United States (*high confidence*).

**Description of evidence base**

Evidence of historical changes in snow cover extent and a reduction in extreme snowfall years is consistent with our understanding of the climate system's response to increasing greenhouse gases. Furthermore, climate models continue to consistently show future declines in snowpack in the western United States. Recent model projections for the eastern United States also confirm a future shift from snowfall to rainfall during the cold season in colder portions of the central and eastern United States. Each of these changes is documented in the peer-reviewed literature and are cited in the main text of this chapter.

**Major uncertainties**

The main source of uncertainty is the sensitivity of observed snow changes to the spatial distribution of observing stations and to historical changes in station location, rain gauges, and observing practices, particularly for snow. Another key issue is the ability of climate models to simulate precipitation, particularly snow. Future changes in the frequency and intensity of meteorological systems causing heavy snow are less certain than temperature changes.

**Assessment of confidence based on evidence and agreement, including short description of nature of evidence and level of agreement**

Given the evidence base and uncertainties, confidence is *medium* that snow cover extent has declined in the United States and *medium* that extreme snowfall years have declined in recent years. Confidence is *high* that western United States snowpack will decline in the future, and confidence is *medium* that a shift from snow domination to rain domination will occur in the parts of the central and eastern United States cited in the text.

**Summary sentence or paragraph that integrates the above information**

Based on observational analyses of snow cover, depth, and water equivalent there is *medium confidence* in the observed changes, and based on model simulations for the future there is *high confidence* in snowpack declines in the western United States and *medium confidence* in the shift to rain from snow in the eastern United States.

# REFERENCES

1. Walsh, J., D. Wuebbles, K. Hayhoe, J. Kossin, K. Kunkel, G. Stephens, P. Thorne, R. Vose, M. Wehner, J. Willis, D. Anderson, S. Doney, R. Feely, P. Hennon, V. Kharin, T. Knutson, F. Landerer, T. Lenton, J. Kennedy, and R. Somerville, 2014: Ch. 2: Our changing climate. *Climate Change Impacts in the United States: The Third National Climate Assessment*. Melillo, J.M., T.C. Richmond, and G.W. Yohe, Eds. U.S. Global Change Research Program, Washington, D.C., 19-67. http://dx.doi.org/10.7930/J0KW5CXT

2. NOAA, 2016: Climate at a Glance: Southwest PDSI. http://www.ncdc.noaa.gov/cag/time-series/us/107/0/pdsi/12/12/1895-2016?base_prd=true&firstbaseyear=1901&lastbaseyear=2000

3. Barnston, A.G. and B. Lyon, 2016: Does the NMME capture a recent decadal shift toward increasing drought occurrence in the southwestern United States? *Journal of Climate*, **29**, 561-581. http://dx.doi.org/10.1175/JCLI-D-15-0311.1

4. Seager, R., M. Hoerling, S. Schubert, H. Wang, B. Lyon, A. Kumar, J. Nakamura, and N. Henderson, 2015: Causes of the 2011–14 California drought. *Journal of Climate*, **28**, 6997-7024. http://dx.doi.org/10.1175/JCLI-D-14-00860.1

5. NOAA, 2016: Climate at a Glance: California PDSI. http://www.ncdc.noaa.gov/cag/time-series/us/4/0/pdsi/12/9/1895-2016?base_prd=true&firstbaseyear=1901&lastbaseyear=2000

6. Luce, C.H., J.T. Abatzoglou, and Z.A. Holden, 2013: The missing mountain water: Slower westerlies decrease orographic enhancement in the Pacific Northwest USA. *Science*, **342**, 1360-1364. http://dx.doi.org/10.1126/science.1242335

7. Vaughan, D.G., J.C. Comiso, I. Allison, J. Carrasco, G. Kaser, R. Kwok, P. Mote, T. Murray, F. Paul, J. Ren, E. Rignot, O. Solomina, K. Steffen, and T. Zhang, 2013: Observations: Cryosphere. *Climate Change 2013: The Physical Science Basis. Contribution of Working Group I to the Fifth Assessment Report of the Intergovernmental Panel on Climate Change*. Stocker, T.F., D. Qin, G.-K. Plattner, M. Tignor, S.K. Allen, J. Boschung, A. Nauels, Y. Xia, V. Bex, and P.M. Midgley, Eds. Cambridge University Press, Cambridge, United Kingdom and New York, NY, USA, 317–382. http://www.climatechange2013.org/report/full-report/

8. Kunkel, K.E., D.A. Robinson, S. Champion, X. Yin, T. Estilow, and R.M. Frankson, 2016: Trends and extremes in Northern Hemisphere snow characteristics. *Current Climate Change Reports*, **2**, 65-73. http://dx.doi.org/10.1007/s40641-016-0036-8

9. Kluver, D. and D. Leathers, 2015: Regionalization of snowfall frequency and trends over the contiguous United States. *International Journal of Climatology*, **35**, 4348-4358. http://dx.doi.org/10.1002/joc.4292

10. Kunkel, K.E., D.R. Easterling, D.A.R. Kristovich, B. Gleason, L. Stoecker, and R. Smith, 2010: Recent increases in U.S. heavy precipitation associated with tropical cyclones. *Geophysical Research Letters*, **37**, L24706. http://dx.doi.org/10.1029/2010GL045164

11. Bard, L. and D.A.R. Kristovich, 2012: Trend reversal in Lake Michigan contribution to snowfall. *Journal of Applied Meteorology and Climatology*, **51**, 2038-2046. http://dx.doi.org/10.1175/jamc-d-12-064.1

12. Hartnett, J.J., J.M. Collins, M.A. Baxter, and D.P. Chambers, 2014: Spatiotemporal snowfall trends in central New York. *Journal of Applied Meteorology and Climatology*, **53**, 2685-2697. http://dx.doi.org/10.1175/jamc-d-14-0084.1

13. Wright, D.M., D.J. Posselt, and A.L. Steiner, 2013: Sensitivity of lake-effect snowfall to lake ice cover and temperature in the Great Lakes region. *Monthly Weather Review*, **141**, 670-689. http://dx.doi.org/10.1175/mwr-d-12-00038.1

14. Vavrus, S., M. Notaro, and A. Zarrin, 2013: The role of ice cover in heavy lake-effect snowstorms over the Great Lakes Basin as simulated by RegCM4. *Monthly Weather Review*, **141**, 148-165. http://dx.doi.org/10.1175/mwr-d-12-00107.1

15. Pederson, G.T., J.L. Betancourt, and G.J. McCabe, 2013: Regional patterns and proximal causes of the recent snowpack decline in the Rocky Mountains, U.S. *Geophysical Research Letters*, **40**, 1811-1816. http://dx.doi.org/10.1002/grl.50424

16. Gan, T.Y., R.G. Barry, M. Gizaw, A. Gobena, and R. Balaji, 2013: Changes in North American snowpacks for 1979–2007 detected from the snow water equivalent data of SMMR and SSM/I passive microwave and related climatic factors. *Journal of Geophysical Research Atmospheres*, **118**, 7682–7697. http://dx.doi.org/10.1002/jgrd.50507

17. Lute, A.C. and J.T. Abatzoglou, 2014: Role of extreme snowfall events in interannual variability of snowfall accumulation in the western United States. *Water Resources Research*, **50**, 2874-2888. http://dx.doi.org/10.1002/2013WR014465

18. Belmecheri, S., F. Babst, E.R. Wahl, D.W. Stahle, and V. Trouet, 2016: Multi-century evaluation of Sierra Nevada snowpack. *Nature Climate Change*, **6**, 2-3. http://dx.doi.org/10.1038/nclimate2809

19. Colle, B.A., Z. Zhang, K.A. Lombardo, E. Chang, P. Liu, and M. Zhang, 2013: Historical evaluation and future prediction of eastern North American and western Atlantic extratropical cyclones in the CMIP5 models during the cool season. *Journal of Climate*, **26**, 6882-6903. http://dx.doi.org/10.1175/JCLI-D-12-00498.1

20. Chang, E.K.M., C.-G. Ma, C. Zheng, and A.M.W. Yau, 2016: Observed and projected decrease in Northern Hemisphere extratropical cyclone activity in summer and its impacts on maximum temperature. *Geophysical Research Letters*, **43**, 2200-2208. http://dx.doi.org/10.1002/2016GL068172

21. Kunkel, K.E., D.R. Easterling, D.A. Kristovich, B. Gleason, L. Stoecker, and R. Smith, 2012: Meteorological causes of the secular variations in observed extreme precipitation events for the conterminous United States. *Journal of Hydrometeorology*, **13**, 1131-1141. http://dx.doi.org/10.1175/JHM-D-11-0108.1

22. Pfahl, S., P.A. O'Gorman, and M.S. Singh, 2015: Extratropical cyclones in idealized simulations of changed climates. *Journal of Climate*, **28**, 9373-9392. http://dx.doi.org/10.1175/JCLI-D-14-00816.1

23. Nesbitt, S.W., R. Cifelli, and S.A. Rutledge, 2006: Storm morphology and rainfall characteristics of TRMM precipitation features. *Monthly Weather Review*, **134**, 2702-2721. http://dx.doi.org/10.1175/mwr3200.1

24. Fritsch, J.M., R.J. Kane, and C.R. Chelius, 1986: The contribution of mesoscale convective weather systems to the warm-season precipitation in the United States. *Journal of Climate and Applied Meteorology*, **25**, 1333-1345. http://dx.doi.org/10.1175/1520-0450(1986)025<1333:tcomcw>2.0.co;2

25. Schumacher, R.S. and R.H. Johnson, 2006: Characteristics of U.S. extreme rain events during 1999–2003. *Weather and Forecasting*, **21**, 69-85. http://dx.doi.org/10.1175/waf900.1

26. Feng, Z., L.R. Leung, S. Hagos, R.A. Houze, C.D. Burleyson, and K. Balaguru, 2016: More frequent intense and long-lived storms dominate the springtime trend in central US rainfall. *Nature Communications*, **7**, 13429. http://dx.doi.org/10.1038/ncomms13429

27. Anderson, B.T., D.J. Gianotti, and G.D. Salvucci, 2015: Detectability of historical trends in station-based precipitation characteristics over the continental United States. *Journal of Geophysical Research Atmospheres*, **120**, 4842-4859. http://dx.doi.org/10.1002/2014JD022960

28. Easterling, D.R., K.E. Kunkel, M.F. Wehner, and L. Sun, 2016: Detection and attribution of climate extremes in the observed record. *Weather and Climate Extremes*, **11**, 17-27. http://dx.doi.org/10.1016/j.wace.2016.01.001

29. Knutson, T.R., F. Zeng, and A.T. Wittenberg, 2014: Seasonal and annual mean precipitation extremes occurring during 2013: A U.S. focused analysis [in "Explaining Extreme Events of 2013 from a Climate Perspective"]. *Bulletin of the American Meteorological Society*, **95 (9)**, S19-S23. http://dx.doi.org/10.1175/1520-0477-95.9.S1.1

30. Bindoff, N.L., P.A. Stott, K.M. AchutaRao, M.R. Allen, N. Gillett, D. Gutzler, K. Hansingo, G. Hegerl, Y. Hu, S. Jain, I.I. Mokhov, J. Overland, J. Perlwitz, R. Sebbari, and X. Zhang, 2013: Detection and attribution of climate change: From global to regional. *Climate Change 2013: The Physical Science Basis. Contribution of Working Group I to the Fifth Assessment Report of the Intergovernmental Panel on Climate Change*. Stocker, T.F., D. Qin, G.-K. Plattner, M. Tignor, S.K. Allen, J. Boschung, A. Nauels, Y. Xia, V. Bex, and P.M. Midgley, Eds. Cambridge University Press, Cambridge, United Kingdom and New York, NY, USA, 867–952. http://www.climatechange2013.org/report/full-report/

31. Min, S.K., X. Zhang, F.W. Zwiers, and G.C. Hegerl, 2011: Human contribution to more-intense precipitation extremes. *Nature*, **470**, 378-381. http://dx.doi.org/10.1038/nature09763

32. Min, S.-K., X. Zhang, F. Zwiers, H. Shiogama, Y.-S. Tung, and M. Wehner, 2013: Multimodel detection and attribution of extreme temperature changes. *Journal of Climate*, **26**, 7430-7451. http://dx.doi.org/10.1175/JCLI-D-12-00551.1

33. Zhang, X., H. Wan, F.W. Zwiers, G.C. Hegerl, and S.-K. Min, 2013: Attributing intensification of precipitation extremes to human influence. *Geophysical Research Letters*, **40**, 5252-5257. http://dx.doi.org/10.1002/grl.51010

34. Westra, S., L.V. Alexander, and F.W. Zwiers, 2013: Global increasing trends in annual maximum daily precipitation. *Journal of Climate*, **26**, 3904-3918. http://dx.doi.org/10.1175/JCLI-D-12-00502.1

35. Dittus, A.J., D.J. Karoly, S.C. Lewis, and L.V. Alexander, 2015: A multiregion assessment of observed changes in the areal extent of temperature and precipitation extremes. *Journal of Climate*, **28**, 9206-9220. http://dx.doi.org/10.1175/JCLI-D-14-00753.1

36. Dittus, A.J., D.J. Karoly, S.C. Lewis, L.V. Alexander, and M.G. Donat, 2016: A multiregion model evaluation and attribution study of historical changes in the area affected by temperature and precipitation extremes. *Journal of Climate*, **29**, 8285-8299. http://dx.doi.org/10.1175/jcli-d-16-0164.1

37. Hoerling, M., K. Wolter, J. Perlwitz, X. Quan, J. Eischeid, H. Want, S. Schubert, H. Diaz, and R. Dole, 2014: Northeast Colorado extreme rains interpreted in a climate change context [in "Explaining Extreme Events of 2013 from a Climate Perspective"]. *Bulletin of the American Meteorological Society*, **95 (9)**, S15-S18. http://dx.doi.org/10.1175/1520-0477-95.9.S1.1

38. Pall, P.C.M.P., M.F. Wehner, D.A. Stone, C.J. Paciorek, and W.D. Collins, 2017: Diagnosing anthropogenic contributions to heavy Colorado rainfall in September 2013. *Weather and Climate Extremes*, **17**, 1-6. http://dx.doi.org/10.1016/j.wace.2017.03.004

39. van der Wiel, K., S.B. Kapnick, G.J. van Oldenborgh, K. Whan, S. Philip, G.A. Vecchi, R.K. Singh, J. Arrighi, and H. Cullen, 2017: Rapid attribution of the August 2016 flood-inducing extreme precipitation in south Louisiana to climate change. *Hydrology and Earth System Sciences*, **21**, 897-921. http://dx.doi.org/10.5194/hess-21-897-2017

40. Shepherd, T.G., 2014: Atmospheric circulation as a source of uncertainty in climate change projections. *Nature Geoscience*, **7**, 703-708. http://dx.doi.org/10.1038/ngeo2253

41. Mearns, L.O., R. Arritt, S. Biner, M.S. Bukovsky, S. Stain, S. Sain, D. Caya, J. Correia, Jr., D. Flory, W. Gutowski, E.S. Takle, R. Jones, R. Leung, W. Moufouma-Okia, L. McDaniel, A.M.B. Nunes, Y. Qian, J. Roads, L. Sloan, and M. Snyder, 2012: The North American regional climate change assessment program: Overview of phase I results. *Bulletin of the American Meteorological Society*, **93**, 1337-1362. http://dx.doi.org/10.1175/BAMS-D-11-00223.1

42. Wehner, M.F., 2013: Very extreme seasonal precipitation in the NARCCAP ensemble: Model performance and projections. *Climate Dynamics*, **40**, 59-80. http://dx.doi.org/10.1007/s00382-012-1393-1

43. Bacmeister, J.T., M.F. Wehner, R.B. Neale, A. Gettelman, C. Hannay, P.H. Lauritzen, J.M. Caron, and J.E. Truesdale, 2014: Exploratory high-resolution climate simulations using the Community Atmosphere Model (CAM). *Journal of Climate*, **27**, 3073-3099. http://dx.doi.org/10.1175/JCLI-D-13-00387.1

44. Wehner, M.F., K.A. Reed, F. Li, Prabhat, J. Bacmeister, C.-T. Chen, C. Paciorek, P.J. Gleckler, K.R. Sperber, W.D. Collins, A. Gettelman, and C. Jablonowski, 2014: The effect of horizontal resolution on simulation quality in the Community Atmospheric Model, CAM5.1. *Journal of Advances in Modeling Earth Systems*, **6**, 980-997. http://dx.doi.org/10.1002/2013MS000276

45. Rauscher, S.A., J.S. Pal, N.S. Diffenbaugh, and M.M. Benedetti, 2008: Future changes in snowmelt-driven runoff timing over the western US. *Geophysical Research Letters*, **35**, L16703. http://dx.doi.org/10.1029/2008GL034424

46. Rasmussen, R., C. Liu, K. Ikeda, D. Gochis, D. Yates, F. Chen, M. Tewari, M. Barlage, J. Dudhia, W. Yu, K. Miller, K. Arsenault, V. Grubišić, G. Thompson, and E. Gutmann, 2011: High-resolution coupled climate runoff simulations of seasonal snowfall over Colorado: A process study of current and warmer climate. *Journal of Climate*, **24**, 3015-3048. http://dx.doi.org/10.1175/2010JCLI3985.1

47. Boyle, J. and S.A. Klein, 2010: Impact of horizontal resolution on climate model forecasts of tropical precipitation and diabatic heating for the TWP-ICE period. *Journal of Geophysical Research*, **115**, D23113. http://dx.doi.org/10.1029/2010JD014262

48. Sakaguchi, K., L.R. Leung, C. Zhao, Q. Yang, J. Lu, S. Hagos, S.A. Rauscher, L. Dong, T.D. Ringler, and P.H. Lauritzen, 2015: Exploring a multiresolution approach using AMIP simulations. *Journal of Climate*, **28**, 5549-5574. http://dx.doi.org/10.1175/JCLI-D-14-00729.1

49. Ban, N., J. Schmidli, and C. Schär, 2014: Evaluation of the convection-resolving regional climate modeling approach in decade-long simulations. *Journal of Geophysical Research Atmospheres*, **119**, 7889-7907. http://dx.doi.org/10.1002/2014JD021478

50. Taylor, K.E., R.J. Stouffer, and G.A. Meehl, 2012: An overview of CMIP5 and the experiment design. *Bulletin of the American Meteorological Society*, **93**, 485-498. http://dx.doi.org/10.1175/BAMS-D-11-00094.1

51. Kay, J.E., C. Deser, A. Phillips, A. Mai, C. Hannay, G. Strand, J.M. Arblaster, S.C. Bates, G. Danabasoglu, J. Edwards, M. Holland, P. Kushner, J.-F. Lamarque, D. Lawrence, K. Lindsay, A. Middleton, E. Munoz, R. Neale, K. Oleson, L. Polvani, and M. Vertenstein, 2015: The Community Earth System Model (CESM) large ensemble project: A community resource for studying climate change in the presence of internal climate variability. *Bulletin of the American Meteorological Society*, **96 (12)**, 1333-1349. http://dx.doi.org/10.1175/BAMS-D-13-00255.1

52. O'Gorman, P.A., 2014: Contrasting responses of mean and extreme snowfall to climate change. *Nature*, **512**, 416-418. http://dx.doi.org/10.1038/nature13625

53. Collins, M., R. Knutti, J. Arblaster, J.-L. Dufresne, T. Fichefet, P. Friedlingstein, X. Gao, W.J. Gutowski, T. Johns, G. Krinner, M. Shongwe, C. Tebaldi, A.J. Weaver, and M. Wehner, 2013: Long-term climate change: Projections, commitments and irreversibility. *Climate Change 2013: The Physical Science Basis. Contribution of Working Group I to the Fifth Assessment Report of the Intergovernmental Panel on Climate Change*. Stocker, T.F., D. Qin, G.-K. Plattner, M. Tignor, S.K. Allen, J. Boschung, A. Nauels, Y. Xia, V. Bex, and P.M. Midgley, Eds. Cambridge University Press, Cambridge, United Kingdom and New York, NY, USA, 1029-1136. http://www.climatechange2013.org/report/full-report/

54. Georgakakos, A., P. Fleming, M. Dettinger, C. Peters-Lidard, T.C. Richmond, K. Reckhow, K. White, and D. Yates, 2014: Ch. 3: Water resources. *Climate Change Impacts in the United States: The Third National Climate Assessment*. Melillo, J.M., T.C. Richmond, and G.W. Yohe, Eds. U.S. Global Change Research Program, Washington, D.C., 69-112. http://dx.doi.org/10.7930/J0G44N6T

55. Lute, A.C., J.T. Abatzoglou, and K.C. Hegewisch, 2015: Projected changes in snowfall extremes and interannual variability of snowfall in the western United States. *Water Resources Research*, **51**, 960-972. http://dx.doi.org/10.1002/2014WR016267

56. Ning, L. and R.S. Bradley, 2015: Snow occurrence changes over the central and eastern United States under future warming scenarios. *Scientific Reports*, **5**, 17073. http://dx.doi.org/10.1038/srep17073

57. Gergel, D.R., B. Nijssen, J.T. Abatzoglou, D.P. Lettenmaier, and M.R. Stumbaugh, 2017: Effects of climate change on snowpack and fire potential in the western USA. *Climatic Change*, **141**, 287-299. http://dx.doi.org/10.1007/s10584-017-1899-y

58. Janssen, E., D.J. Wuebbles, K.E. Kunkel, S.C. Olsen, and A. Goodman, 2014: Observational- and model-based trends and projections of extreme precipitation over the contiguous United States. *Earth's Future*, **2**, 99-113. http://dx.doi.org/10.1002/2013EF000185

59. Janssen, E., R.L. Sriver, D.J. Wuebbles, and K.E. Kunkel, 2016: Seasonal and regional variations in extreme precipitation event frequency using CMIP5. *Geophysical Research Letters*, **43**, 5385-5393. http://dx.doi.org/10.1002/2016GL069151

60. Kunkel, K.E., T.R. Karl, H. Brooks, J. Kossin, J. Lawrimore, D. Arndt, L. Bosart, D. Changnon, S.L. Cutter, N. Doesken, K. Emanuel, P.Y. Groisman, R.W. Katz, T. Knutson, J. O'Brien, C.J. Paciorek, T.C. Peterson, K. Redmond, D. Robinson, J. Trapp, R. Vose, S. Weaver, M. Wehner, K. Wolter, and D. Wuebbles, 2013: Monitoring and understanding trends in extreme storms: State of knowledge. *Bulletin of the American Meteorological Society*, **94**, 499–514. http://dx.doi.org/10.1175/BAMS-D-11-00262.1

61. Kunkel, K.E., T.R. Karl, D.R. Easterling, K. Redmond, J. Young, X. Yin, and P. Hennon, 2013: Probable maximum precipitation and climate change. *Geophysical Research Letters*, **40**, 1402-1408. http://dx.doi.org/10.1002/grl.50334

62. Dettinger, M., 2011: Climate change, atmospheric rivers, and floods in California—a multimodel analysis of storm frequency and magnitude changes. *Journal of the American Water Resources Association*, **47**, 514-523. http://dx.doi.org/10.1111/j.1752-1688.2011.00546.x

63. Shields, C.A. and J.T. Kiehl, 2016: Atmospheric river landfall-latitude changes in future climate simulations. *Geophysical Research Letters*, **43**, 8775-8782. http://dx.doi.org/10.1002/2016GL070470

64. Kooperman, G.J., M.S. Pritchard, and R.C.J. Somerville, 2013: Robustness and sensitivities of central U.S. summer convection in the super-parameterized CAM: Multi-model intercomparison with a new regional EOF index. *Geophysical Research Letters*, **40**, 3287-3291. http://dx.doi.org/10.1002/grl.50597

65. Cook, K.H., E.K. Vizy, Z.S. Launer, and C.M. Patricola, 2008: Springtime intensification of the Great Plains low-level jet and midwest precipitation in GCM simulations of the twenty-first century. *Journal of Climate*, **21**, 6321-6340. http://dx.doi.org/10.1175/2008jcli2355.1

66. Higgins, R.W., Y. Yao, E.S. Yarosh, J.E. Janowiak, and K.C. Mo, 1997: Influence of the Great Plains low-level jet on summertime precipitation and moisture transport over the central United States. *Journal of Climate*, **10**, 481-507. http://dx.doi.org/10.1175/1520-0442(1997)010<0481:iotgpl>2.0.co;2

67. Pan, Z., R.W. Arritt, E.S. Takle, W.J. Gutowski, Jr., C.J. Anderson, and M. Segal, 2004: Altered hydrologic feedback in a warming climate introduces a "warming hole". *Geophysical Research Letters*, **31**, L17109. http://dx.doi.org/10.1029/2004GL020528

68. Prein, A.F., R.M. Rasmussen, K. Ikeda, C. Liu, M.P. Clark, and G.J. Holland, 2017: The future intensification of hourly precipitation extremes. *Nature Climate Change*, **7**, 48-52. http://dx.doi.org/10.1038/nclimate3168

69. Knutson, T.R., J.L. McBride, J. Chan, K. Emanuel, G. Holland, C. Landsea, I. Held, J.P. Kossin, A.K. Srivastava, and M. Sugi, 2010: Tropical cyclones and climate change. *Nature Geoscience*, **3**, 157-163. http://dx.doi.org/10.1038/ngeo779

70. Knutson, T.R., J.J. Sirutis, G.A. Vecchi, S. Garner, M. Zhao, H.-S. Kim, M. Bender, R.E. Tuleya, I.M. Held, and G. Villarini, 2013: Dynamical downscaling projections of twenty-first-century Atlantic hurricane activity: CMIP3 and CMIP5 model-based scenarios. *Journal of Climate*, **27**, 6591-6617. http://dx.doi.org/10.1175/jcli-d-12-00539.1

71. Wang, C.-C., B.-X. Lin, C.-T. Chen, and S.-H. Lo, 2015: Quantifying the effects of long-term climate change on tropical cyclone rainfall using a cloud-resolving model: Examples of two landfall typhoons in Taiwan. *Journal of Climate*, **28**, 66-85. http://dx.doi.org/10.1175/JCLI-D-14-00044.1

72. Villarini, G., D.A. Lavers, E. Scoccimarro, M. Zhao, M.F. Wehner, G.A. Vecchi, T.R. Knutson, and K.A. Reed, 2014: Sensitivity of tropical cyclone rainfall to idealized global-scale forcings. *Journal of Climate*, **27**, 4622-4641. http://dx.doi.org/10.1175/JCLI-D-13-00780.1

73. Melillo, J.M., T.C. Richmond, and G.W. Yohe, eds., 2014: *Climate Change Impacts in the United States: The Third National Climate Assessment*. U.S. Global Change Research Program: Washington, D.C., 841 pp. http://dx.doi.org/10.7930/J0Z31WJ2

74. Vose, R.S., S. Applequist, M. Squires, I. Durre, M.J. Menne, C.N. Williams, Jr., C. Fenimore, K. Gleason, and D. Arndt, 2014: Improved historical temperature and precipitation time series for U.S. climate divisions. *Journal of Applied Meteorology and Climatology*, **53**, 1232-1251. http://dx.doi.org/10.1175/JAMC-D-13-0248.1

75. Sun, L., K.E. Kunkel, L.E. Stevens, A. Buddenberg, J.G. Dobson, and D.R. Easterling, 2015: Regional Surface Climate Conditions in CMIP3 and CMIP5 for the United States: Differences, Similarities, and Implications for the U.S. National Climate Assessment. NOAA Technical Report NESDIS 144. National Oceanic and Atmospheric Administration, National Environmental Satellite, Data, and Information Service, 111 pp. http://dx.doi.org/10.7289/V5RB72KG

76. Angélil, O., D. Stone, M. Wehner, C.J. Paciorek, H. Krishnan, and W. Collins, 2017: An independent assessment of anthropogenic attribution statements for recent extreme temperature and rainfall events. *Journal of Climate*, **30**, 5-16. http://dx.doi.org/10.1175/JCLI-D-16-0077.1

77. Edwards, L.M., M. Bunkers, J.T. Abatzoglou, D.P. Todey, and L.E. Parker, 2014: October 2013 blizzard in western South Dakota [in "Explaining Extreme Events of 2013 from a Climate Perspective"]. *Bulletin of the American Meteorological Society*, **95 (9)**, S23-S26. http://dx.doi.org/10.1175/1520-0477-95.9.S1.1

78. Peterson, T.C., R.R. Heim, R. Hirsch, D.P. Kaiser, H. Brooks, N.S. Diffenbaugh, R.M. Dole, J.P. Giovannettone, K. Guirguis, T.R. Karl, R.W. Katz, K. Kunkel, D. Lettenmaier, G.J. McCabe, C.J. Paciorek, K.R. Ryberg, S. Schubert, V.B.S. Silva, B.C. Stewart, A.V. Vecchia, G. Villarini, R.S. Vose, J. Walsh, M. Wehner, D. Wolock, K. Wolter, C.A. Woodhouse, and D. Wuebbles, 2013: Monitoring and understanding changes in heat waves, cold waves, floods and droughts in the United States: State of knowledge. *Bulletin of the American Meteorological Society*, **94**, 821-834. http://dx.doi.org/10.1175/BAMS-D-12-00066.1

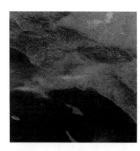

# 8

# Droughts, Floods, and Wildfires

## KEY FINDINGS

1. Recent droughts and associated heat waves have reached record intensity in some regions of the United States; however, by geographical scale and duration, the Dust Bowl era of the 1930s remains the benchmark drought and extreme heat event in the historical record (*very high confidence*). While by some measures drought has decreased over much of the continental United States in association with long-term increases in precipitation, neither the precipitation increases nor inferred drought decreases have been confidently attributed to anthropogenic forcing.

2. The human effect on recent major U.S. droughts is complicated. Little evidence is found for a human influence on observed precipitation deficits, but much evidence is found for a human influence on surface soil moisture deficits due to increased evapotranspiration caused by higher temperatures. (*High confidence*)

3. Future decreases in surface (top 10 cm) soil moisture from anthropogenic forcing over most of the United States are *likely* as the climate warms under higher scenarios. (*Medium confidence*)

4. Substantial reductions in western U.S. winter and spring snowpack are projected as the climate warms. Earlier spring melt and reduced snow water equivalent have been formally attributed to human-induced warming (*high confidence*) and will *very likely* be exacerbated as the climate continues to warm (*very high confidence*). Under higher scenarios, and assuming no change to current water resources management, chronic, long-duration hydrological drought is increasingly possible by the end of this century (*very high confidence*).

5. Detectable changes in some classes of flood frequency have occurred in parts of the United States and are a mix of increases and decreases. Extreme precipitation, one of the controlling factors in flood statistics, is observed to have generally increased and is projected to continue to do so across the United States in a warming atmosphere. However, formal attribution approaches have not established a significant connection of increased riverine flooding to human-induced climate change, and the timing of any emergence of a future detectable anthropogenic change in flooding is unclear. (*Medium confidence*)

6. The incidence of large forest fires in the western United States and Alaska has increased since the early 1980s (*high confidence*) and is projected to further increase in those regions as the climate warms, with profound changes to certain ecosystems (*medium confidence*).

**Recommended Citation for Chapter**

**Wehner**, M.F., J.R. Arnold, T. Knutson, K.E. Kunkel, and A.N. LeGrande, 2017: Droughts, floods, and wildfires. In: *Climate Science Special Report: Fourth National Climate Assessment, Volume I* [Wuebbles, D.J., D.W. Fahey, K.A. Hibbard, D.J. Dokken, B.C. Stewart, and T.K. Maycock (eds.)]. U.S. Global Change Research Program, Washington, DC, USA, pp. 231-256 doi: 10.7930/J0CJ8BNN.

## 8.1 Drought

The word "drought" brings to mind abnormally dry conditions. However, the meaning of "dry" can be ambiguous and lead to confusion in how drought is actually defined. Three different classes of droughts are defined by NOAA and describe a useful hierarchal set of water deficit characterization, each with different impacts. "Meteorological drought" describes conditions of precipitation deficit. "Agricultural drought" describes conditions of soil moisture deficit. "Hydrological drought" describes conditions of deficit in runoff.[1] Clearly these three characterizations of drought are related but are also different descriptions of water shortages with different target audiences and different time scales. In particular, agricultural drought is of concern to producers of food while hydrological drought is of concern to water system managers. Soil moisture is a function of both precipitation and evapotranspiration. Because potential evapotranspiration increases with temperature, anthropogenic climate change generally results in drier soils and often less runoff in the long term. In fact, under the higher scenario (RCP8.5; see Ch. 4: Projections for a description of the RCP scenarios) at the end of the 21st century, no region of the planet is projected to experience significantly higher levels of annual average surface soil moisture due to the sensitivity of evapotranspiration to temperature, even though much higher precipitation is projected in some regions.[2] Seasonal and annual total runoff, on the other hand, are projected to either increase or decrease, depending on location and season under the same conditions,[2] illustrating the complex relationships between the various components of the hydrological system. Meteorological drought can occur on a range of time scales, in addition to seasonal or annual time scales. "Flash droughts" can result from just a few weeks of dry weather,[3] and the paleoclimate record contains droughts of several

decades. Hence, it is vital to describe precisely the definition of drought in any public discussion to avoid confusion due to this complexity. As the climate changes, conditions currently considered "abnormally" dry may become relatively "normal" in those regions undergoing aridification, or extremely unlikely in those regions becoming wetter. Hence, the reference conditions defining drought may need to be modified from those currently in practice.

### 8.1.1 Historical Context

The United States has experienced all three types of droughts in the past, always driven, at least in some part, by natural variations in seasonal and/or annual precipitation amounts. As the climate changes, we can expect that human activities will alter the effect of these natural variations. The "Dust Bowl" drought of the 1930s is still the most significant meteorological and agricultural drought experienced in the United States in terms of its geographic and temporal extent. However, even though it happened prior to most of the current global warming, human activities exacerbated the dryness of the soil by the farming practices of the time.[4] Tree ring archives reveal that such droughts (in the agricultural sense) have occurred occasionally over the last 1,000 years.[5] Climate model simulations suggest that droughts lasting several years to decades occur naturally in the southwestern United States.[6] The Intergovernmental Panel on Climate Change Fifth Assessment Report (IPCC AR5)[7] concluded "there is low confidence in detection and attribution of changes in (meteorological) drought over global land areas since the mid-20th century, owing to observational uncertainties and difficulties in distinguishing decadal-scale variability in drought from long-term trends." As they noted, this was a weaker attribution statement than in the Fourth Assessment Report,[8] which had concluded "that an increased risk of drought was *more likely than not* due to anthropogenic forcing during the second half of the 20th century." The weaker statement in AR5 reflected additional studies with conflicting conclusions on global drought trends (e.g., Sheffield et al. 2012;[9] Dai 2013[10]). Western North America was noted as a region where determining if observed recent droughts were unusual compared to natural variability was particularly difficult. This was due to evidence from paleoclimate proxies of cases of central U.S. droughts during the past 1,000 years that were longer and more intense than historical U.S. droughts.[11] Drought is, of course, directly connected to seasonal precipitation totals. Figure 7.1 shows detectable observed recent changes in seasonal precipitation. In fact, the increases in observed summer and fall precipitation are at odds with the projections in Figure 7.5. As a consequence of this increased precipitation, drought statistics over the entire CONUS have declined.[3, 12] Furthermore, there is no detectable change in meteorological drought at the global scale.[9] However, a number of individual event attribution studies suggest that if a drought occurs, anthropogenic temperature increases can exacerbate soil moisture deficits (e.g., Seager et al. 2015;[13] Trenberth et al. 2014[14]). Future projections of the anthropogenic contribution to changes in drought risk and severity must be considered in the context of the significant role of natural variability.

### 8.1.2 Recent Major U.S. Droughts
*Meteorological and Agricultural Drought*
The United States has suffered a number of very significant droughts of all types since 2011. Each of these droughts was a result of different persistent, large-scale meteorological patterns of mostly natural origins, with varying degrees of attributable human influence. Table 8.1 summarizes available attribution statements for recent extreme U.S. droughts. Statements about meteorological drought are decidedly mixed, revealing the complexities

in interpreting the low tail of the distribution of precipitation. Statements about agricultural drought consistently maintain a human influence if only surface soil moisture measures are considered. The single agricultural drought attribution study at root depth comes to the opposite conclusion.[15] In all cases, these attribution statements are examples of attribution without detection (see Appendix C). The absence of moisture during the 2011 Texas/Oklahoma drought and heat wave was found to be an event whose likelihood was enhanced by the La Niña state of the ocean, but the human interference in the climate system still doubled the chances of reaching such high temperatures.[16] This study illustrates that the effect of human-induced climate change is combined with natural variations and can compound or inhibit the realized severity of any given extreme weather event.

The Great Plains/Midwest drought of 2012 was the most severe summer meteorological drought in the observational record for that region.[17] An unfortunate string of three different patterns of large-scale meteorology from May through August 2012 precluded the normal frequency of summer thunderstorms and was not predicted by the NOAA seasonal forecasts.[17] Little influence of the global sea surface temperature (SST) pattern on meteorological drought frequency has been found in model simulations.[17] No evidence of a human contribution to the 2012 precipitation deficit in the Great Plains and Midwest is found in numerous studies.[17, 18, 19] However, an alternative view is that the 2012 central U.S. drought can be classified as a "heat wave flash drought",[20] a type of rapidly evolving drought that has decreased in frequency over the past century.[3] Also, an increase in the chances of the unusually high temperatures seen in the United States in 2012, partly associated with resultant dry summer soil moisture anomalies, was attributed to the human interference with the climate system,[21] indicating the strong feedback between lower soil moisture and higher surface air temperatures during periods of low precipitation. One study found that most, but not all, of the 2012 surface moisture deficit in the Great Plains was attributable to the precipitation deficit.[22] That study also noted that Great Plains root depth and deeper soil mois-

**Table 8.1.** A list of U.S. droughts for which attribution statements have been made. In the last column, "+" indicates that an attributable human-induced increase in frequency and/or magnitude was found, "–" indicates that an attributable human-induced decrease in frequency and/or magnitude was found, "0" indicates no attributable human contribution was identified. As in Tables 6.2 and 7.1, several of the events were originally examined in the Bulletin of the American Meteorological Society's (BAMS) State of the Climate Reports and reexamined by Angélil et al.[18] In these cases, both attribution statements are listed with the original authors first. (Source: M. Wehner)

| Authors | Event Year and Duration | Region or State | Type | Attribution Statement |
|---|---|---|---|---|
| Rupp et al. 2012[130] / Angélil et al. 2017[18] | MAMJJA 2011 | Texas | Meteorological | +/+ |
| Hoerling et al. 2013[16] | 2012 | Texas | Meteorological | + |
| Rupp et al. 2013[19] / Angélil et al. 2017[18] | MAMJJA 2012 | CO, NE, KS, OK, IA, MO, AR & IL | Meteorological | 0/0 |
| Rupp et al. 2013[19] / Angélil et al. 2017[18] | MAM 2012 | CO, NE, KS, OK, IA, MO, AR & IL | Meteorological | 0/0 |
| Rupp et al. 2013[19] / Angélil et al. 2017[18] | JJA 2012 | CO, NE, KS, OK, IA, MO, AR & IL | Meteorological | 0/+ |
| Hoerling et al. 2014[17] | MJJA 2012 | Great Plains/Midwest | Meteorological | 0 |
| Swain et al. 2014[24] / Angélil et al. 2017[18] | ANN 2013 | California | Meteorological | +/+ |
| Wang and Schubert 2014[29] / Angélil et al. 2017[18] | JS 2013 | California | Meteorological | 0/+ |
| Knutson et al. 2014[131] / Angélil et al. 2017[18] | ANN 2013 | California | Meteorological | 0/+ |
| Knutson et al. 2014[131] / Angélil et al. 2017[18] | MAM 2013 | U.S. Southern Plains region | Meteorological | 0/+ |
| Diffenbaugh et al. 2015[28] | 2012–2014 | California | Agricultural | + |
| Seager et al. 2015[13] | 2012–2014 | California | Agricultural | + |
| Cheng et al. 2016[15] | 2011–2015 | California | Agricultural | – |
| Mote et al. 2016[31] | 2015 | Washington, Oregon, California | Hydrological (snow water equivalent) | + |

ture was higher than normal in 2012 despite the surface drying, due to wet conditions in prior years, indicating the long time scales relevant below the surface.[22]

The recent California drought, which began in 2011, is unusual in different respects. In this case, the precipitation deficit from 2011 to 2014 was a result of the "ridiculously resilient ridge" of high pressure. This very stable high pressure system steered storms towards the north, away from the highly engineered California water resource system.[13, 23, 24] A slow-moving high sea surface temperature (SST) anomaly, referred to as "The Blob"— was caused by a persistent ridge that weakened the normal cooling mechanisms for that region of the upper ocean.[25] Atmospheric modeling studies showed that the ridge that caused The Blob was favored by a pattern of persistent tropical SST anomalies that were warm in the western equatorial Pacific and simultaneously cool in the far eastern equatorial Pacific.[23, 26] It was also favored by reduced arctic sea ice and from feedbacks with The Blob's SST anomalies.[27] These studies also suggest that internal variability likely played a prominent role in the persistence of the 2013–2014 ridge off the west coast of North America. Observational records are not long enough and the anomaly was unusual enough that similarly long-lived patterns have not been often seen before. Hence, attribution statements, such as that about an increasing anthropogenic influence on the frequency of geopotential height anomalies similar to 2012–2014 (e.g., Swain et al. 2014[24]), are without associated detection (Ch. 3: Detection and Attribution). A secondary attribution question concerns the anthropogenic precipitation response in the presence of this SST anomaly. In attribution studies with a prescribed 2013 SST anomaly, a consistent increase in the human influence on the chances of very dry California conditions was found.[18]

Anthropogenic climate change did increase the risk of the high temperatures in California in the winters of 2013–2014 and 2014–2015, especially the latter,[13, 28, 29] further exacerbating the soil moisture deficit and the associated stress on irrigation systems. This raises the question, as yet unanswered, of whether droughts in the western United States are shifting from precipitation control[30] to temperature control. There is some evidence to support a relationship between mild winter and/or warm spring temperatures and drought occurrence,[31] but long-term warming trends in the tropical and North Pacific do not appear to have led to trends toward less precipitation over California.[32] An anthropogenic contribution to commonly used measures of agricultural drought, including the Palmer Drought Severity Index (PDSI), was found in California[28, 33] and is consistent with previous projections of changes in PDSI[10, 34, 35] and with an attribution study.[36] Due to its simplicity, the PDSI has been criticized as being overly sensitive to higher temperatures and thus may exaggerate the human contribution to soil dryness.[37] In fact, this study also finds that formulations of potential evaporation used in more complicated hydrologic models are similarly biased, undermining confidence in the magnitude but not the sign of projected surface soil moisture changes in a warmer climate. Seager et al.[13] analyzed climate model output directly, finding that precipitation minus evaporation in the southwestern United States is projected to experience significant decreases in surface water availability, leading to surface runoff decreases in California, Nevada, Texas, and the Colorado River headwaters even in the near term. However, the criticisms of PDSI also apply to most of the CMIP5 land surface model evapotranspiration formulations. Analysis of soil moisture in the CMIP5 models at deeper levels is complicated by the wide variety in sophistication of their component land models. A pair of studies reveals less sensitivity at depth-to-surface air temperature increases than at near-surface levels.[15, 38] Berg et al.[39] adjust for the differences in land component model vertical treatments, finding projected change in vertically integrated soil moisture down to 3 meters depth is mixed, with projected decreases in the Southwest and in the south-central United States, but increases over the northern plains. Nonetheless, the warming trend has led to declines in a number of indicators, including Sierra snow water equivalent, that are relevant to hydrological drought.[30] Attribution of the California drought and heat wave remains an interesting and controversial research topic.

In summary, there has not yet been a formal identification of a human influence on past changes in United States meteorological drought through the analysis of precipitation trends. Some, but not all, U.S. meteorological drought event attribution studies, largely in the "without detection" class, exhibit a human influence. Attribution of a human influence on past changes in U.S. agricultural drought are limited both by availability of soil moisture observations and a lack of subsurface modeling studies. While a human influence on surface soil moisture trends has been identified with *medium confidence*, its relevance to agriculture may be exaggerated.

### Runoff And Hydrological Drought

Several studies focused on the Colorado River basin in the United States that used more sophisticated runoff models driven by the CMIP3 models[40, 41, 42, 43, 44] showed that annual runoff reductions in a warmer western Unites States climate occur through a combination of evapotranspiration increases and precipitation decreases, with the overall reduction in river flow exacerbated by human water demands on the basin's supply. Reduced U.S. snowfall accumulations in much warmer

future climates are virtually certain as frozen precipitation is replaced by rain regardless of the projected changes in total precipitation amounts discussed in Chapter 7: Precipitation Change (Figure 7.6). The profound change in the hydrology of snowmelt-driven flows in the western United States is well documented. Earlier spring runoff[45] reduced the fraction of precipitation falling as snow[46] and the snow-pack water content at the end of winter,[47, 48] consistent with warmer temperatures. Formal detection and attribution (Ch. 3: Detection and Attribution) of the observed shift towards earlier snowmelt-driven flows in the western United States reveals that the shift is detectably different from natural variability and attributable to anthropogenic climate change.[49] Similarly, observed declines in the snow water equivalent in the region have been formally attributed to anthropogenic climate change[50] as have temperature, river flow, and snow-pack.[41, 51] As a harbinger, the unusually low western U.S. snowpack of 2015 may become the norm.[31]

In the northwestern United States, long-term trends in streamflow have seen declines, with the strongest trends in drought years[52] that are attributed to a decline in winter precipitation.[53] These reductions in precipitation are linked to decreased westerly wind speeds in winter over the region. Furthermore, the trends in westerlies are consistent with CMIP5-projected wind speed changes due to a decreasing meridional temperature and pressure gradients rather than low-frequency climate variability modes. Such precipitation changes have been a primary source of change in hydrological drought in the Northwest over the last 60 years[54] and are in addition to changes in snowpack properties.

We conclude with *high confidence* that these observed changes in temperature controlled aspects of western U.S. hydrology are *likely* a consequence of human changes to the climate system.

### 8.1.3 Projections of Future Droughts

The future changes in seasonal precipitation shown in Chapter 7: Precipitation Change (Figure 7.6) indicate that the southwestern United States may experience chronic future precipitation deficits, particularly in the spring. In much warmer climates, expansion of the tropics and subtropics, traceable to changes in the Hadley circulation, cause shifts in seasonal precipitation that are particularly evident in such arid and semi-arid regions and increase the risk of meteorological drought. However, uncertainty in the magnitude and timing of future southwestern drying is high. We note that the weighted and downscaled projections of Figure 7.6 exhibit significantly less drying and are assessed to be less significant in comparison to natural variations than the original unweighted CMIP5 projections.[34]

Western U.S. hydrological drought is currently controlled by the frequency and intensity of extreme precipitation events, particularly atmospheric rivers, as these events represent the source of nearly half of the annual water supply and snowpack for the western coastal states.[55, 56] Climate projections indicate greater frequency of atmospheric rivers in the future (e.g., Dettinger 2011;[55] Warner et al. 2015;[57] Gao et al. 2015;[58] see further discussion in Ch. 9: Extreme Storms). Sequences of these extreme storms have played a critical role in ending recent hydrological droughts along the U.S. West Coast.[59] However, as winter temperatures increase, the fraction of precipitation falling as snow will decrease, potentially disrupting western U.S. water management practices.

Significant U.S. seasonal precipitation deficits are not confidently projected outside of the Southwest. However, future higher tempera-

tures will *likely* lead to greater frequencies and magnitudes of agricultural droughts throughout the continental United States as the resulting increases in evapotranspiration outpace projected precipitation increases.[2] Figure 8.1 shows the weighted multimodel projection of the percent change in near-surface soil moisture at the end of the 21st century under the higher scenario (RCP8.5), indicating widespread drying over the entire continental United States. Previous National Climate Assessments[34, 60] have discussed the implication of these future drier conditions in the context of the PDSI, finding that the future normal condition would be considered drought at the present time, and that the incidence of **"extreme drought" (PDSI < −4) would be** significantly increased. However, as described below, the PDSI may overestimate future soil moisture drying.

This projection is made "without attribution" (Ch. 4: Projections), but confidence that future soils will generally be drier at the surface is *medium*, as the mechanisms leading to increased evapotranspiration in a warmer climate are elementary scientific facts. However, the land surface component models in the CMIP5 climate models vary greatly in their sophistication, causing the projected magnitude of both the average soil moisture decrease and the increased risk for agricultural drought to be less certain. The weighted projected seasonal decreases in surface soil moisture are generally towards drier conditions, even in regions and seasons where precipitation is projected to experience large increases (Figure 7.6) due to increases in the evapotranspiration associated with higher temperature. Drying is assessed to be large relative to natural variations in much of the CONUS region in the summer. Significant spring and fall drying is also projected in the mountainous western states, with potential implications for forest and wildfire risk. Also, the combination of significant summer and fall drying in the midwestern states has potential agricultural implications. The largest percent changes are projected in the southwestern United States and are consistent in magnitude with an earlier study of the Colorado River Basin using more sophisticated macroscale hydrological models.[42]

In this assessment, we limit the direct CMIP5 weighted multimodel projection of soil moisture shown in Figure 8.1 to the surface (defined as the top 10 cm of the soil), as the land surface component sub-models vary greatly in their representation of the total depth of the soil. A more relevant projection to agricultural drought would be the soil moisture at the root

depth of typical U.S. crops. Cook et al.[38] find that future drying at a depth of 30 cm will be less than at 2 cm, but still significant and comparable to a modified PDSI formulation. Few of the CMIP5 land models have detailed ecological representations of evapotranspiration processes, causing the simulation of the soil moisture budget to be less constrained than reality.[61] Over the western United States, unrealistically low elevations in the CMIP5 models due to resolution constraints present a further challenge in interpreting evapotranspiration changes. Nonetheless, Figure 8.1 shows a projected drying of surface soil moisture across nearly all of the coterminous United States in all seasons, even in regions and seasons where

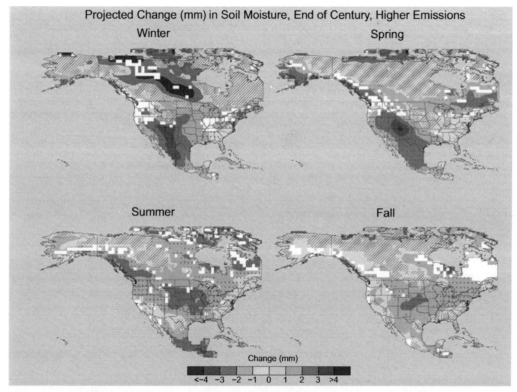

**Figure 8.1:** Projected end of the 21st century weighted CMIP5 multimodel average percent changes in near surface seasonal soil moisture (mrsos) under the higher scenario (RCP8.5). Stippling indicates that changes are assessed to be large compared to natural variations. Hashing indicates that changes are assessed to be small compared to natural variations. Blank regions (if any) are where projections are assessed to be inconclusive (Appendix B). (Figure source: NOAA NCEI and CICS-NC).

precipitation is projected to increase, consistent with increased evapotranspiration due to elevated temperatures.[38]

Widespread reductions in mean snowfall across North America are projected by the CMIP5 models.[62] Together with earlier snowmelt at altitudes high enough for snow, disruptions in western U.S. water delivery systems are expected to lead to more frequent hydrological drought conditions.[40, 41, 50, 63, 64] Due to resolution constraints, the elevation of mountains as represented in the CMIP5 models is too low to adequately represent the effects of future temperature on snowpacks. However, increased model resolution has been demonstrated to have important impacts on future projections of snowpack water content in warmer climates and is enabled by recent advances in high performance computing.[65] Figure 8.2 and Table 8.2 show a projection of changes in western U.S. mountain winter (December, January, and February) hydrology obtained from a different high-resolution atmospheric model at the middle and end of the 21st century under the higher scenario (RCP8.5). These projections indicate dramatic reductions in all aspects of snow[66] and are similar to previous statistically downscaled pro-

jections.[67, 68] Table 8.2 reveals that the reductions in snow water equivalent accelerate in the latter half of this century under the higher scenario (RCP8.5) and with substantial variations across the western United States. Changes in snow residence time, an alternative measure of snowpack relevant to the timing of runoff, is also shown to be sensitive to elevation, with widespread reductions across this region.[69] Given the larger projected increases in temperature at high altitudes compared to adjacent lower altitudes[70] and the resulting changes in both snowpack depth and melt timing in very warm future scenarios such as RCP8.5, and assuming no change to water resource management practices, several important western U.S. snowpack reservoirs effectively disappear by 2100 in this dynamical projection, resulting in chronic, long-lasting hydrological drought. This dramatic statement is also supported by two climate model studies: a multimodel statistical downscaling of the CMIP5 RCP8.5 ensemble that finds large areal reductions in snow-dominated regions of the western United States by mid-century and complete elimination of snow-dominated regions in certain watersheds,[68] and a large ensemble simulation of a global climate model.[71]

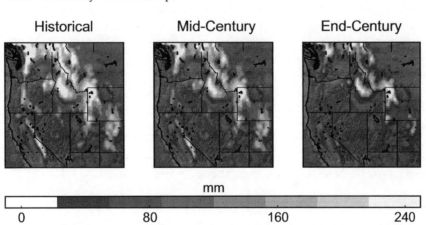

Figure 8.2: Projected changes in winter (DJF) snow water equivalent at the middle and end of this century under the higher scenario (RCP8.5) from a high-resolution version of the Community Atmospheric Model, CAM5.[66] (Figure source: H. Krishnan, LBNL).

**Table 8.2.** Projected changes in western U.S. mountain range winter (DJF) snow-related hydrology variables at the middle and end of this century. Projections are for the higher scenario (RCP8.5) from a high-resolution version of the Community Atmospheric Model, CAM5.[66]

| Mountain Range | Snow Water Equivalent (% Change) | | Snow Cover (% Change) | | Snowfall (% Change) | | Surface Temperature (change in K) | |
|---|---|---|---|---|---|---|---|---|
| | 2050 | 2100 | 2050 | 2100 | 2050 | 2100 | 2050 | 2100 |
| Cascades | −41.5 | −89.9 | −21.6 | −72.9 | −10.7 | −50.0 | 0.9 | 4.1 |
| Klamath | −50.75 | −95.8 | −38.6 | −89.0 | −23.1 | −78.7 | 0.8 | 3.5 |
| Rockies | −17.3 | −65.1 | −8.2 | −43.1 | 1.7 | −8.2 | 1.4 | 5.5 |
| Sierra Nevada | −21.8 | −89.0 | −21.9 | −77.7 | −4.7 | −66.6 | 1.1 | 4.5 |
| Wasatch and Uinta | −18.9 | −78.7 | −14.2 | −61.4 | 4.1 | −34.6 | 1.8 | 6.1 |
| **Western USA** | **−22.3** | **−70.1** | **−12.7** | **−51.5** | **−1.6** | **−21.4** | **1.3** | **5.2** |

As earlier spring melt and reduced snow water equivalent have been formally attributed to human-induced warming, substantial reductions in western U.S. winter and spring snowpack are projected (with attribution) to be *very likely* as the climate continues to warm (*very high confidence*). Under higher scenarios and assuming no change to current water-resources management, chronic, long-duration hydrological drought is increasingly possible by the end of this century (*very high confidence*).

## 8.2 Floods

Flooding damage in the United States can come from flash floods of smaller rivers and creeks, prolonged flooding along major rivers, urban flooding unassociated with proximity to a riverway, coastal flooding from storm surge which may be exacerbated by sea level rise, and the confluence of coastal storms and inland riverine flooding from the same precipitation event (Ch. 12: Sea Level Rise). Flash flooding is associated with extreme precipitation somewhere along the river which may occur upstream of the regions at risk. Flooding of major rivers in the United States with substantial winter snow accumulations usually occurs in the late winter or spring and can result from an unusually heavy seasonal snowfall followed by a "rain on snow" event or from a rapid onset of higher temperatures

that leads to rapid snow melting within the river basin. In the western coastal states, most flooding occurs in conjunction with extreme precipitation events referred to as "atmospheric rivers" (see Ch. 9: Extreme Storms),[72, 73] with mountain snowpack being vulnerable to these typically warmer-than-normal storms and their potential for rain on existing snow cover.[74] Hurricanes and tropical storms are an important driver of flooding events in the eastern United States. Changes in streamflow rates depend on many factors, both human and natural, in addition to climate change. Deforestation, urbanization, dams, floodwater management activities, and changes in agricultural practices can all play a role in past and future changes in flood statistics. Projection of future changes is thus a complex multivariate problem.[34]

The IPCC AR5[7] did not attribute changes in flooding to anthropogenic influence nor report detectable changes in flooding magnitude, duration, or frequency. Trends in extreme high values of streamflow are mixed across the United States.[34, 75, 76] Analysis of 200 U.S. stream gauges indicates areas of both increasing and decreasing flooding magnitude[77] but does not provide robust evidence that these trends are attributable to human influences. Significant increases in flood frequency have been detected in about one-third of stream gauge stations examined for the central United States, with a much stronger signal of frequency change than is found for changes in flood magnitude in these gauges.[78] This apparent disparity with ubiquitous increases in observed extreme precipitation (Figure 7.2) can be partly explained by the seasonality of the two phenomena. Extreme precipitation events in the eastern half of the CONUS are larger in the summer and fall when soil moisture and seasonal streamflow levels are low and less favorable for flooding.[79] By contrast, high streamflow events are often larger in the spring and winter when soil moisture is high and snowmelt and frozen ground can enhance runoff.[80] Furthermore, floods may be poorly explained by daily precipitation characteristics alone; the relevant mechanisms are more complex, involving processes that are seasonally and geographically variable, including the seasonal cycles of soil moisture content and snowfall/snowmelt.[81]

Recent analysis of annual maximum streamflow shows statistically significant trends in the upper Mississippi River valley (increasing) and in the Northwest (decreasing).[44] In fact, across the midwestern United States, statistically significant increases in flooding are well documented.[78, 82, 83, 84, 85, 86, 87, 88] These increases in flood risk and severity are not attributed to 20th century changes in agricultural practices[87, 89] but instead are attributed mostly to the observed increases in precipitation shown in Figures 7.1 through 7.4.[78, 84, 89, 90] Trends in maximum streamflow in the northeastern United States are less dramatic and less spatially coherent,[44, 80] although one study found mostly increasing trends[91] in that region, consistent with the increasing trends in observed extreme precipitation in the region (Ch. 6: Temperature Change).[34, 80]

The nature of the proxy archives complicates the reconstruction of past flood events in a gridded fashion as has been done with droughts. However, reconstructions of past river outflows do exist. For instance, it has been suggested that the mid-20th century water allocations from the Colorado River were made during one of the wettest periods of the past five centuries.[92] For the eastern United States, the Mississippi River has undergone century-scale variability in flood frequency—perhaps linked to the moisture availability in the central United States and the temperature structure of the Atlantic Ocean.[93]

The complex mix of processes complicates the formal attribution of observed flooding trends to anthropogenic climate change and suggests that additional scientific rigor is needed in flood attribution studies.[94] As noted above, precipitation increases have been found to strongly influence changes in flood statistics. However, in U.S. regions, no formal attribution of precipitation changes to anthropogenic forcing has been made so far, so indirect attribution of flooding changes is not possible. Hence, no formal attribution of observed flooding changes to anthropogenic forcing has been claimed.[78]

A projection study based on coupling an ensemble of regional climate model output to a hydrology model[95] finds that the magnitude of future very extreme runoff (which can lead to flooding) is decreased in most of the summer months in Washington State, Oregon, Idaho, and western Montana but substantially increases in the other seasons. Projected weighted increases in extreme runoff from the coast to the Cascade Mountains are particularly large in that study during the fall and winter which are not evident in the weighted seasonal averaged CMIP5 runoff projections.[2] For the West Coast of the United States, extremely heavy precipitation from intense atmospheric river storms is an important factor in flood frequency and severity.[55, 96] Projections indicate greater frequency of heavy atmospheric rivers in the future (e.g., Dettinger et al. 2011;[96] Warner et al. 2015;[57] Gao et al. 2015;[58] see further discussion in Ch. 9: Extreme Storms). Translating these increases in atmospheric river frequency to their impact on flood frequency requires a detailed representation of western states topography in the global projection models and/or via dynamic downscaling to regional models and is a rapidly developing science. In a report prepared for the Federal Insurance and Mitigation Administration of the Federal Emergency Management Agency, a regression-based approach of scaling river gauge data based on seven commonly used climate change indices from the CMIP3 database[97] found that at the end of the 21st century the 1% annual chance floodplain area would increase in area by about 30%, with larger changes in the Northeast and Great Lakes regions and smaller changes in the central part of the country and the Gulf Coast.[98]

Urban flooding results from heavy precipitation events that overwhelm the existing sewer infrastructure's ability to convey the resulting stormwater. Future increases in daily and sub-daily extreme precipitation rates will require significant upgrades to many communities' storm sewer systems, as will sea level rise in coastal cities and towns.[99, 100]

No studies have formally attributed (see Ch. 3: Detection and Attribution) long-term changes in observed flooding of major rivers in the United States to anthropogenic forcing. We conclude that there is *medium confidence* that detectable (though not attributable to anthropogenic forcing changes) increases in flood statistics have occurred in parts of the central United States. Key Finding 3 of Chapter 7: Precipitation Change states that the frequency and intensity of heavy precipitation events are

projected to continue to increase over the 21st century with *high confidence*. Given the connection between extreme precipitation and flooding, and the complexities of other relevant factors, we concur with the IPCC Special Report on Extremes (SREX) assessment of "medium confidence (based on physical reasoning) that projected increases in heavy rainfall would contribute to increases in local flooding in some catchments or regions".[101]

Existing studies of individual extreme flooding events are confined to changes in the locally responsible precipitation event and have not included detailed analyses of the events' hydrology. Gochis et al.[102] describe the massive floods of 2013 along the Colorado front range, estimating that the streamflow amounts ranged from 50- to 500-year return values across the region. Hoerling et al.[17] analyzed the 2013 northeastern Colorado heavy multi-day precipitation event and resulting flood, finding little evidence of an anthropogenic influence on its occurrence. However, Pall et al.[103] challenge their event attribution methodology with a more constrained study and find that the thermodynamic response of precipitation in this event due to anthropogenic forcing was substantially increased. The Pall et al.[103] approach does not rule out that the likelihood of the extremely rare large-scale meteorological pattern responsible for the flood may have changed.

## 8.3 Wildfires

A global phenomenon with natural (lightning) and human-caused ignition sources, wildfire represents a critical ecosystem process. Recent decades have seen a profound increase in forest fire activity over the western United States and Alaska.[104, 105, 106, 107] The frequency of large wildfires is influenced by a complex combination of natural and human factors. Temperature, soil moisture, relative humidity, wind speed, and vegetation (fuel density) are

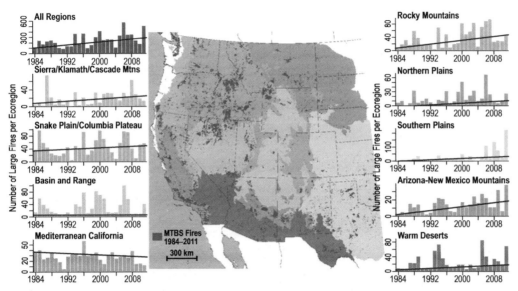

**Figure 8.3:** Trends in the annual number of large fires in the western United States for a variety of ecoregions. The black lines are fitted trend lines. Statistically significant at a 10% level for all regions except the Snake Plain/Columbian Plateau, Basin and Range, and Mediterranean California regions. (Figure source: Dennison et al.[108]).

important aspects of the relationship between fire frequency and ecosystems. Forest management and fire suppression practices can also alter this relationship from what it was in the preindustrial era. Changes in these control parameters can interact with each other in complex ways with the potential for tipping points—in both fire frequency and in ecosystem properties—that may be crossed as the climate warms.

Figure 8.3 shows that the number of large fires has increased over the period 1984–2011, with high statistical significance in 7 out of 10 western U.S. regions across a large variety of vegetation, elevation, and climatic types.[108] State-level fire data over the 20th century[109] indicates that area burned in the western United States decreased from 1916 to about 1940, was at low levels until the 1970s, then increased into the more recent period. Modeled increases in temperatures and vapor pressure deficits due to anthropogenic climate change have increased forest fire activity in the western United States by increasing the aridity of forest fu-

els during the fire season.[104] Increases in these relevant climatic drivers were found to be responsible for over half the observed increase in western U.S. forest fuel aridity from 1979 to 2015 and doubled the forest fire area over the period 1984–2015.[104] Littell et al. (2009, 2010, 2016)[109, 110, 111] found that two climatic mechanisms affect fire in the western United States: increased fuel flammability driven by warmer, drier conditions and increased fuel availability driven by antecedent moisture. Littell et al.[111] found a clear link between increased drought and increased fire risk. Yoon et al.[112] assessed the 2014 fire season, finding an increased risk of fire in California. While fire suppression practices can also lead to a significant increase in fire risk in lower-elevation and drier forest types, this is less important in higher-elevation and moister forests.[113, 114, 115] Increases in future forest fire extent, frequency, and intensity depend strongly on local ecosystem properties and will vary greatly across the United States. Westerling et al.[116] projected substantial increases in forest fire frequency in the Greater Yellowstone ecosystem by mid-century under the older SRES A2 emissions scenario, and further stated that years without large fires in the region will become extremely rare. Stavros et al.[117] projected increases in very large fires (greater than 50,000 acres) across the western United States by mid-century under both the lower and higher scenarios (RCP4.5 and RCP8.5, respectively). Likewise, Prestemon et al.[118] projected significant increases in lightning-ignited wildfire in the Southeast by mid-century but with substantial differences between ecoregions. However, other factors, related to climate change such as water scarcity or insect infestations may act to stifle future forest fire activity by reducing growth or otherwise killing trees leading to fuel reduction.[110]

Historically, wildfires have been less frequent and of smaller extent in Alaska compared to the rest of the globe.[119, 120] Shortened land snow cover seasons and higher temperatures have made the Arctic more vulnerable to wildfire.[119, 120, 121] Total area burned and the number of large fires (those with area greater than 1,000 square km or 386 square miles) in Alaska exhibits significant interannual and decadal scale variability from influences of atmospheric circulation patterns and controlled burns, but have *likely* increased since 1959.[122] The most recent decade has seen an unusually large number of severe wildfire years in Alaska, for which the risk of severe fires has *likely* increased by 33%–50% as a result of anthropogenic climate change[123] and is projected to increase by up to a factor of four by the end of the century under the mid-high scenario (RCP6.0).[121] Historically less flammable tundra and cooler boreal forest regions could shift into historically unprecedented fire risk regimes as a consequence of temperatures increasing above the minimum thresholds required for burning. Alaska's fire season is also *likely* lengthening—a trend expected to continue.[119, 124] Thresholds in temperature and precipitation shape Arctic fire regimes, and

projected increases in future lightning activity imply increased vulnerability to future climate change.[119, 121] Alaskan tundra and forest wildfires will *likely* increase under warmer and drier conditions[124, 125] and potentially result in a transition into a fire regime unprecedented in the last 10,000 years.[126] Total area burned is projected to increase between 25% and 53% by the end of the century.[127]

Boreal forests and tundra contain large stores of carbon, approximately 50% of the total global soil carbon.[128] Increased fire activity could deplete these stores, releasing them to the atmosphere to serve as an additional source of atmospheric $CO_2$ and alter the carbon cycle if ecosystems change from higher to lower carbon densities.[126, 128] Additionally, increased fires in Alaska may also enhance the degradation of Alaska's permafrost, blackening the ground, reducing surface albedo, and removing protective vegetation.

Both anthropogenic climate change and the legacy of land use/management have an influence on U.S. wildfires and are subtly and inextricably intertwined. Forest management practices have resulted in higher fuel densities in most U.S. forests, except in the Alaskan bush and the higher mountainous regions of the western United States. Nonetheless, there is *medium confidence* for a human-caused climate change contribution to increased forest fire activity in Alaska in recent decades with a *likely* further increase as the climate continues to warm, and *low to medium confidence* for a detectable human climate change contribution in the western United States based on existing studies. Recent literature does not contain a complete robust detection and attribution analysis of forest fires including estimates of natural decadal and multidecadal variability, as described in Chapter 3: Detection and Attribution, nor separate the contributions to observed trends from climate change and forest management. These assessment statements about attribution to human-induced climate change are instead multistep attribution statements (Ch. 3: Detection and Attribution) based on plausible model-based estimates of anthropogenic contributions to observed trends. The modeled contributions, in turn, are based on climate variables that are closely linked to fire risk and that, in most cases, have a detectable human influence, such as surface air temperature and snow melt timing.

# TRACEABLE ACCOUNTS

### Key Finding 1

Recent droughts and associated heat waves have reached record intensity in some regions of the United States; however, by geographical scale and duration, the Dust Bowl era of the 1930s remains the benchmark drought and extreme heat event in the historical record (*very high confidence*). While by some measures drought has decreased over much of the continental United States in association with long-term increases in precipitation, neither the precipitation increases nor inferred drought decreases have been confidently attributed to anthropogenic forcing.

### Description of evidence base

Recent droughts are well characterized and described in the literature. The Dust Bowl is not as well documented, but available observational records support the key finding. The last sentence is an "absence of evidence" statement and does not imply "evidence of absence" of future anthropogenic changes. The inferred decreases in some measures of U.S. drought or types of drought (heat wave/flash droughts) are described in Andreadis and Lettenmaier[1,2] and Mo and Lettenmaier[3].

### Major uncertainties

Record-breaking temperatures are well documented with low uncertainty.[129] The magnitude of the Dust Bowl relative to present times varies with location. Uncertainty in the key finding is affected by the quality of pre-World War II observations but is relatively low.

### Assessment of confidence based on evidence and agreement

Precipitation is well observed in the United States, leading to *very high confidence*.

### Summary sentence or paragraph that integrates the above information

The key finding is a statement that recent U.S. droughts, while sometimes long and severe, are not unprecedented in the historical record.

### Key Finding 2

The human effect on recent major U.S. droughts is complicated. Little evidence is found for a human influence on observed precipitation deficits, but much evidence is found for a human influence on surface soil moisture deficits due to increased evapotranspiration caused by higher temperatures (*high confidence*).

### Description of evidence base

Observational records of meteorological drought are not long enough to detect statistically significant trends. Additionally, paleoclimatic evidence suggests that major droughts have occurred throughout the distant past. Surface soil moisture is not well observed throughout the CONUS, but numerous event attribution studies attribute enhanced reduction of surface soil moisture during dry periods to anthropogenic warming and enhanced evapotranspiration. Sophisticated land surface models have been demonstrated to reproduce the available observations and have allowed for century scale reconstructions.

### Major uncertainties

Uncertainties stem from the length of precipitation observations and the lack of surface moisture observations.

### Assessment of confidence based on evidence and agreement

Confidence is *high* for widespread future surface soil moisture deficits, as little change is projected for future summer and fall average precipitation. In the absence of increased precipitation (and in some cases with it), evapotranspiration increases due to increased temperatures will lead to less soil moisture overall, especially near the surface.

### Summary sentence or paragraph that integrates the above information

The precipitation deficit portion of the key finding is a conservative statement reflecting the conflicting and limited event attribution literature on meteorological drought. The soil moisture portion of the key finding is limited to the surface and not the more relevant root depth and is supported by the studies cited in this chapter.

### Key Finding 3

Future decreases in surface (top 10 cm) soil moisture from anthropogenic forcing over most of the United States are *likely* as the climate warms under the higher scenarios. (*Medium confidence*)

### Description of evidence base

First principles establish that evaporation is at least linearly dependent on temperatures and accounts for much of the surface moisture decrease as temperature increases. Plant transpiration for many non-desert species controls plant temperature and responds to increased temperature by opening stomata to release more water vapor. This water comes from the soil at root depth as the plant exhausts its stored water supply (*very high confidence*). Furthermore, nearly all CMIP5 models exhibit U.S. surface soil moisture drying at the end of the century under the higher scenario (RCP8.5), and the multimodel average exhibits no significant annual soil moisture increases anywhere on the planet.[2]

### Major uncertainties

While both evaporation and transpiration changes are of the same sign as temperature increases, the relative importance of each as a function of depth is less well quantified. The amount of transpiration varies considerably among plant species, and these are treated with widely varying degrees of sophistication in the land surface components of contemporary climate models. Uncertainty in the sign of the anthropogenic change of root depth soil moisture is low in regions and seasons of projected precipitation decreases (Ch. 7: Precipitation Changes). There is moderate to high uncertainty in the magnitude of the change in soil moisture at all depths and all regions and seasons. This key finding is a "projection without attribution" statement as such a drying is not part of the observed record. Projections of summertime mean CONUS precipitation exhibit no significant change. However, recent summertime precipitation trends are positive, leading to reduced agricultural drought conditions overall.[12] While statistically significant increases in precipitation have been identified over parts of the United States, these trends have not been clearly attributed to anthropogenic forcing (Ch. 7: Precipitation Change). Furthermore, North

American summer temperature increases under the higher scenario (RCP8.5) at the end of the century are projected to be substantially more than the current observed (and modeled) temperature increase. Because of the response of evapotranspiration to temperature increases, the CMIP5 multimodel average projection is for drier surface soils even in those high latitude regions (Alaska and Canada) that are confidently projected to experience increases in precipitation. Hence, in the CONUS region, with little or no projected summertime changes in precipitation, we conclude that surface soil moisture will *likely* decrease.

### Assessment of confidence based on evidence and agreement

CMIP5 and regional models support the surface soil moisture key finding. Confidence is assessed as "medium" as this key finding—despite the high level of agreement among model projections—because of difficulties in observing long-term changes in this metric, and because, at present, there is no published evidence of detectable long-term decreases in surface soil moisture across the United States.

### Summary sentence or paragraph that integrates the above information

In the northern United States, surface soil moisture (top 10 cm) is *likely* to decrease as evaporation outpaces increases in precipitation. In the Southwest, the combination of temperature increases and precipitation decreases causes surface soil moisture decreases to be *very likely*. In this region, decreases in soil moisture at the root depth are *likely*.

### Key Finding 4

Substantial reductions in western U.S. winter and spring snowpack are projected as the climate warms. Earlier spring melt and reduced snow water equivalent have been formally attributed to human induced warming (*high confidence*) and will *very likely* be exacerbated as the climate continues to warm (*very high confidence*). Under higher scenarios, and assuming no change to current water resources management, chronic, long-duration hydrological drought is increasingly possible by the end of this century (*very high confidence*).

### Description of evidence base

First principles tell us that as temperatures rise, minimum snow levels also must rise. Certain changes in western U.S. hydrology have already been attributed to human causes in several papers following Barnett et al.[41] and are cited in the text. The CMIP3/5 models project widespread warming with future increases in atmospheric GHG concentrations, although these are underestimated in the current generation of global climate models (GCMs) at the high altitudes of the western United States due to constraints on orographic representation at current GCM spatial resolutions.

CMIP5 models were not designed or constructed for direct projection of locally relevant snowpack amounts. However, a high-resolution climate model, selected for its ability to simulate western U.S. snowpack amounts and extent, projects devastating changes in the hydrology of this region assuming constant water resource management practices.[66] This conclusion is also supported by a statistical downscaling result shown in Figure 3.1 of Walsh et al.[34] and Cayan et al.[67] and by the more recent statistical downscaling study of Klos et al.[68].

### Major uncertainties

The major uncertainty is not so much "if" but rather "how much" as changes to precipitation phases (rain or snow) are sensitive to temperature increases that in turn depend on greenhouse gas (GHG) forcing changes. Also, changes to the lower-elevation catchments will be realized prior to those at higher elevations that, even at 25 km, are not adequately resolved. Uncertainty in the final statement also stems from the usage of one model but is tempered by similar findings from statistical downscaling studies. However, this simulation is a so-called "prescribed temperature" experiment with the usual uncertainties about climate sensitivity wired in by the usage of one particular ocean temperature change. Uncertainty in the equator-to-pole differential ocean warming rate is also a factor.

### Assessment of confidence based on evidence and agreement

All CMIP5 models project large-scale warming in the western United States as GHG forcing increases. Warming is underestimated in most of the western United States due to elevation deficiencies that are a consequence of coarse model resolution.

### Summary sentence or paragraph that integrates the above information

Warmer temperatures lead to less snow and more rain if total precipitation remains unchanged. Projected winter/spring precipitation changes are a mix of increases in northern states and decreases in the Southwest. In the northern Rocky Mountains, snowpack is projected to decrease even with a projected precipitation increase due to this phase change effect. This will lead to, at the very least, profound changes to the seasonal and sub-seasonal timing of the western U.S. hydrological cycle even where annual precipitation remains nearly unchanged with a strong potential for water shortages.

### Key Finding 5

Detectable changes in some classes of flood frequency have occurred in parts of the United States and are a mix of increases and decreases. Extreme precipitation, one of the controlling factors in flood statistics, is observed to have generally increased and is projected to continue to do so across the United States in a warming atmosphere. However, formal attribution approaches have not established a significant connection of increased riverine flooding to human-induced climate change, and the timing of any emergence of a future detectible anthropogenic change in flooding is unclear. (*Medium confidence*)

### Description of evidence base

Observed changes are a mix of increases and decreases and are documented by Walsh et al.[34] and other studies cited in the text. No attribution statements have been made.

### Major uncertainties

Floods are highly variable both in space and time. The multivariate nature of floods complicates detection and attribution.

**Assessment of confidence based on evidence and agreement**

Confidence is limited to *medium* due to both the lack of an attributable change in observed flooding to date and the complicated multivariate nature of flooding. However, confidence is *high* in the projections of increased future extreme precipitation, the principal driver (among several) of many floods. It is unclear when an observed long-term increase in U.S. riverine flooding will be attributed to anthropogenic climate change. Hence, confidence is *medium* in this part of the key message at this time.

**Summary sentence or paragraph that integrates the above information**

The key finding is a relatively weak statement reflecting the lack of definitive detection and attribution of anthropogenic changes in U.S. flooding intensity, duration, and frequency.

**Key Finding 6**

The incidence of large forest fires in the western United States and Alaska has increased since the early 1980s (*high confidence*) and is projected to further increase in those regions as the climate warms with profound changes to certain ecosystems (*medium confidence*).

**Description of evidence base**

Studies by Dennison et al. (western United States)[108] and Kasischke and Turetsky (Alaska)[122] document the observed increases in fire statistics. Projections of Westerling et al. (western United States)[116] and Young et al.[121] and others (Alaska) indicate increased fire risk. These observations and projections are consistent with drying due to warmer temperatures leading to increased flammability and longer fire seasons.

**Major uncertainties**

Analyses of other regions of the United States, which also could be subject to increased fire risk, do not seem to be readily available. Likewise, projections of the western U.S. fire risk are of limited areas. In terms of attribution, there is still some uncertainty as to how well non-climatic confounding factors such as forestry management and fire suppression practices have been accounted for, particularly for the western United States. Other climate change factors, such as increased water deficits and insect infestations, could reduce fuel loads, tending towards reducing fire frequency and/or intensity.

**Assessment of confidence based on evidence and agreement**

Confidence is *high* in the observations due to solid observational evidence. Confidence in projections would be higher if there were more available studies covering a broader area of the United States and a wider range of ecosystems.

**Summary sentence or paragraph that integrates the above information**

Wildfires have increased over parts of the western United States and Alaska in recent decades and are projected to continue to increase as a result of climate change. As a result, shifts in certain ecosystem types may occur.

# REFERENCES

1. NOAA, 2008: Drought: Public Fact Sheet. National Oceanic and Atmospheric Administration, National Weather Service, Washington, D.C. http://www.nws.noaa.gov/os/brochures/climate/DroughtPublic2.pdf

2. Collins, M., R. Knutti, J. Arblaster, J.-L. Dufresne, T. Fichefet, P. Friedlingstein, X. Gao, W.J. Gutowski, T. Johns, G. Krinner, M. Shongwe, C. Tebaldi, A.J. Weaver, and M. Wehner, 2013: Long-term climate change: Projections, commitments and irreversibility. *Climate Change 2013: The Physical Science Basis. Contribution of Working Group I to the Fifth Assessment Report of the Intergovernmental Panel on Climate Change.* Stocker, T.F., D. Qin, G.-K. Plattner, M. Tignor, S.K. Allen, J. Boschung, A. Nauels, Y. Xia, V. Bex, and P.M. Midgley, Eds. Cambridge University Press, Cambridge, United Kingdom and New York, NY, USA, 1029–1136. http://www.climatechange2013.org/report/full-report/

3. Mo, K.C. and D.P. Lettenmaier, 2015: Heat wave flash droughts in decline. *Geophysical Research Letters*, **42**, 2823-2829. http://dx.doi.org/10.1002/2015GL064018

4. Bennet, H.H., F.H. Fowler, F.C. Harrington, R.C. Moore, J.C. Page, M.L. Cooke, H.A. Wallace, and R.G. Tugwell, 1936: *A Report of the Great Plains Area Drought Committee.* Hopkins Papers Box 13. Franklin D. Roosevelt Library, New Deal Network (FERI), Hyde Park, NY. http://newdeal.feri.org/hopkins/hop27.htm

5. Cook, E.R., C.A. Woodhouse, C.M. Eakin, D.M. Meko, and D.W. Stahle, 2004: Long-term aridity changes in the western United States. *Science*, **306**, 1015-1018. http://dx.doi.org/10.1126/science.1102586

6. Coats, S., B.I. Cook, J.E. Smerdon, and R. Seager, 2015: North American pancontinental droughts in model simulations of the last millennium. *Journal of Climate*, **28**, 2025-2043. http://dx.doi.org/10.1175/JCLI-D-14-00634.1

7. Bindoff, N.L., P.A. Stott, K.M. AchutaRao, M.R. Allen, N. Gillett, D. Gutzler, K. Hansingo, G. Hegerl, Y. Hu, S. Jain, I.I. Mokhov, J. Overland, J. Perlwitz, R. Sebbari, and X. Zhang, 2013: Detection and attribution of climate change: From global to regional. *Climate Change 2013: The Physical Science Basis. Contribution of Working Group I to the Fifth Assessment Report of the Intergovernmental Panel on Climate Change.* Stocker, T.F., D. Qin, G.-K. Plattner, M. Tignor, S.K. Allen, J. Boschung, A. Nauels, Y. Xia, V. Bex, and P.M. Midgley, Eds. Cambridge University Press, Cambridge, United Kingdom and New York, NY, USA, 867–952. http://www.climatechange2013.org/report/full-report/

8. Hegerl, G.C., F.W. Zwiers, P. Braconnot, N.P. Gillett, Y. Luo, J.A.M. Orsini, N. Nicholls, J.E. Penner, and P.A. Stott, 2007: Understanding and attributing climate change. *Climate Change 2007: The Physical Science Basis. Contribution of Working Group I to the Fourth Assessment Report of the Intergovernmental Panel on Climate Change.* Solomon, S., D. Qin, M. Manning, Z. Chen, M. Marquis, K.B. Averyt, M. Tignor, and H.L. Miller, Eds. Cambridge University Press, Cambridge, United Kingdom and New York, NY, USA, 663-745. http://www.ipcc.ch/publications_and_data/ar4/wg1/en/ch9.html

9. Sheffield, J., E.F. Wood, and M.L. Roderick, 2012: Little change in global drought over the past 60 years. *Nature*, **491**, 435-438. http://dx.doi.org/10.1038/nature11575

10. Dai, A., 2013: Increasing drought under global warming in observations and models. *Nature Climate Change*, **3**, 52-58. http://dx.doi.org/10.1038/nclimate1633

11. Masson-Delmotte, V., M. Schulz, A. Abe-Ouchi, J. Beer, A. Ganopolski, J.F. González Rouco, E. Jansen, K. Lambeck, J. Luterbacher, T. Naish, T. Osborn, B. Otto-Bliesner, T. Quinn, R. Ramesh, M. Rojas, X. Shao, and A. Timmermann, 2013: Information from paleoclimate archives. *Climate Change 2013: The Physical Science Basis. Contribution of Working Group I to the Fifth Assessment Report of the Intergovernmental Panel on Climate Change.* Stocker, T.F., D. Qin, G.-K. Plattner, M. Tignor, S.K. Allen, J. Boschung, A. Nauels, Y. Xia, V. Bex, and P.M. Midgley, Eds. Cambridge University Press, Cambridge, United Kingdom and New York, NY, USA, 383–464. http://www.climatechange2013.org/report/full-report/

12. Andreadis, K.M. and D.P. Lettenmaier, 2006: Trends in 20th century drought over the continental United States. *Geophysical Research Letters*, **33**, L10403. http://dx.doi.org/10.1029/2006GL025711

13. Seager, R., M. Hoerling, S. Schubert, H. Wang, B. Lyon, A. Kumar, J. Nakamura, and N. Henderson, 2015: Causes of the 2011–14 California drought. *Journal of Climate*, **28**, 6997-7024. http://dx.doi.org/10.1175/JCLI-D-14-00860.1

14. Trenberth, K.E., A. Dai, G. van der Schrier, P.D. Jones, J. Barichivich, K.R. Briffa, and J. Sheffield, 2014: Global warming and changes in drought. *Nature Climate Change*, **4**, 17-22. http://dx.doi.org/10.1038/nclimate2067

15. Cheng, L., M. Hoerling, A. AghaKouchak, B. Livneh, X.-W. Quan, and J. Eischeid, 2016: How has human-induced climate change affected California drought risk? *Journal of Climate*, **29**, 111-120. http://dx.doi.org/10.1175/JCLI-D-15-0260.1

16. Hoerling, M., M. Chen, R. Dole, J. Eischeid, A. Kumar, J.W. Nielsen-Gammon, P. Pegion, J. Perlwitz, X.-W. Quan, and T. Zhang, 2013: Anatomy of an extreme event. *Journal of Climate,* **26,** 2811–2832. http://dx.doi.org/10.1175/JCLI-D-12-00270.1

17. Hoerling, M., J. Eischeid, A. Kumar, R. Leung, A. Mariotti, K. Mo, S. Schubert, and R. Seager, 2014: Causes and predictability of the 2012 Great Plains drought. *Bulletin of the American Meteorological Society,* **95,** 269-282. http://dx.doi.org/10.1175/BAMS-D-13-00055.1

18. Angélil, O., D. Stone, M. Wehner, C.J. Paciorek, H. Krishnan, and W. Collins, 2017: An independent assessment of anthropogenic attribution statements for recent extreme temperature and rainfall events. *Journal of Climate,* **30,** 5-16. http://dx.doi.org/10.1175/JCLI-D-16-0077.1

19. Rupp, D.E., P.W. Mote, N. Massey, F.E.L. Otto, and M.R. Allen, 2013: Human influence on the probability of low precipitation in the central United States in 2012 [in "Explaining Extreme Events of 2013 from a Climate Perspective"]. *Bulletin of the American Meteorological Society,* **94 (9),** S2-S6. http://dx.doi.org/10.1175/BAMS-D-13-00085.1

20. Mo, K.C. and D.P. Lettenmaier, 2016: Precipitation deficit flash droughts over the United States. *Journal of Hydrometeorology,* **17,** 1169-1184. http://dx.doi.org/10.1175/jhm-d-15-0158.1

21. Diffenbaugh, N.S. and M. Scherer, 2013: Likelihood of July 2012 U.S. temperatures in pre-industrial and current forcing regimes [in "Explaining Extreme Events of 2013 from a Climate Perspective"]. *Bulletin of the American Meteorological Society,* **94 (9),** S6-S9. http://dx.doi.org/10.1175/BAMS-D-13-00085.1

22. Livneh, B. and M.P. Hoerling, 2016: The physics of drought in the U.S. Central Great Plains. *Journal of Climate,* **29,** 6783-6804. http://dx.doi.org/10.1175/JCLI-D-15-0697.1

23. Seager, R., M. Hoerling, D.S. Siegfried, h. Wang, B. Lyon, A. Kumar, J. Nakamura, and N. Henderson, 2014: Causes and Predictability of the 2011-14 California Drought. National Oceanic and Atmospheric Administration, Drought Task Force Narrative Team, 40 pp. http://dx.doi.org/10.7289/V58K771F

24. Swain, D., M. Tsiang, M. Haughen, D. Singh, A. Charland, B. Rajarthan, and N.S. Diffenbaugh, 2014: The extraordinary California drought of 2013/14: Character, context and the role of climate change [in "Explaining Extreme Events of 2013 from a Climate Perspective"]. *Bulletin of the American Meteorological Society,* **95 (9),** S3-S6. http://dx.doi.org/10.1175/1520-0477-95.9.S1.1

25. Bond, N.A., M.F. Cronin, H. Freeland, and N. Mantua, 2015: Causes and impacts of the 2014 warm anomaly in the NE Pacific. *Geophysical Research Letters,* **42,** 3414-3420. http://dx.doi.org/10.1002/2015GL063306

26. Hartmann, D.L., 2015: Pacific sea surface temperature and the winter of 2014. *Geophysical Research Letters,* **42,** 1894-1902. http://dx.doi.org/10.1002/2015GL063083

27. Lee, M.-Y., C.-C. Hong, and H.-H. Hsu, 2015: Compounding effects of warm sea surface temperature and reduced sea ice on the extreme circulation over the extratropical North Pacific and North America during the 2013–2014 boreal winter. *Geophysical Research Letters,* **42,** 1612-1618. http://dx.doi.org/10.1002/2014GL062956

28. Diffenbaugh, N.S., D.L. Swain, and D. Touma, 2015: Anthropogenic warming has increased drought risk in California. *Proceedings of the National Academy of Sciences,* **112,** 3931-3936. http://dx.doi.org/10.1073/pnas.1422385112

29. Wang, H. and S. Schubert, 2014: Causes of the extreme dry conditions over California during early 2013 [in "Explaining Extreme Events of 2013 from a Climate Perspective"]. *Bulletin of the American Meteorological Society,* **95 (9),** S7-S11. http://dx.doi.org/10.1175/1520-0477-95.9.S1.1

30. Mao, Y., B. Nijssen, and D.P. Lettenmaier, 2015: Is climate change implicated in the 2013–2014 California drought? A hydrologic perspective. *Geophysical Research Letters,* **42,** 2805-2813. http://dx.doi.org/10.1002/2015GL063456

31. Mote, P.W., D.E. Rupp, S. Li, D.J. Sharp, F. Otto, P.F. Uhe, M. Xiao, D.P. Lettenmaier, H. Cullen, and M.R. Allen, 2016: Perspectives on the causes of exceptionally low 2015 snowpack in the western United States. *Geophysical Research Letters,* **43,** 10,980-10,988. http://dx.doi.org/10.1002/2016GL069965

32. Funk, C., A. Hoell, and D. Stone, 2014: Examining the contribution of the observed global warming trend to the California droughts of 2012/13 and 2013/14 [in "Explaining Extreme Events of 2013 from a Climate Perspective"]. *Bulletin of the American Meteorological Society,* **95 (9),** S11-S15. http://dx.doi.org/10.1175/1520-0477-95.9.S1.1

33. Williams, A.P., R. Seager, J.T. Abatzoglou, B.I. Cook, J.E. Smerdon, and E.R. Cook, 2015: Contribution of anthropogenic warming to California drought during 2012–2014. *Geophysical Research Letters,* **42,** 6819-6828. http://dx.doi.org/10.1002/2015GL064924

34. Walsh, J., D. Wuebbles, K. Hayhoe, J. Kossin, K. Kunkel, G. Stephens, P. Thorne, R. Vose, M. Wehner, J. Willis, D. Anderson, S. Doney, R. Feely, P. Hennon, V. Kharin, T. Knutson, F. Landerer, T. Lenton, J. Kennedy, and R. Somerville, 2014: Ch. 2: Our changing climate. *Climate Change Impacts in the United States: The Third National Climate Assessment.* Melillo, J.M., T.C. Richmond, and G.W. Yohe, Eds. U.S. Global Change Research Program, Washington, D.C., 19-67. http://dx.doi.org/10.7930/J0KW5CXT

35. Wehner, M., D.R. Easterling, J.H. Lawrimore, R.R. Heim, Jr., R.S. Vose, and B.D. Santer, 2011: Projections of future drought in the continental United States and Mexico. *Journal of Hydrometeorology*, **12**, 1359-1377. http://dx.doi.org/10.1175/2011JHM1351.1

36. Brown, P.M., E.K. Heyerdahl, S.G. Kitchen, and M.H. Weber, 2008: Climate effects on historical fires (1630–1900) in Utah. *International Journal of Wildland Fire*, **17**, 28-39. http://dx.doi.org/10.1071/WF07023

37. Milly, P.C.D. and K.A. Dunne, 2016: Potential evapotranspiration and continental drying. *Nature Climate Change*, **6**, 946-969. http://dx.doi.org/10.1038/nclimate3046

38. Cook, B.I., T.R. Ault, and J.E. Smerdon, 2015: Unprecedented 21st century drought risk in the American Southwest and Central Plains. *Science Advances*, **1**, e1400082. http://dx.doi.org/10.1126/sciadv.1400082

39. Berg, A., J. Sheffield, and P.C.D. Milly, 2017: Divergent surface and total soil moisture projections under global warming. *Geophysical Research Letters*, **44**, 236-244. http://dx.doi.org/10.1002/2016GL071921

40. Barnett, T.P. and D.W. Pierce, 2009: Sustainable water deliveries from the Colorado River in a changing climate. *Proceedings of the National Academy of Sciences*, **106**, 7334-7338. http://dx.doi.org/10.1073/pnas.0812762106

41. Barnett, T.P., D.W. Pierce, H.G. Hidalgo, C. Bonfils, B.D. Santer, T. Das, G. Bala, A.W. Wood, T. Nozawa, A.A. Mirin, D.R. Cayan, and M.D. Dettinger, 2008: Human-induced changes in the hydrology of the western United States. *Science*, **319**, 1080-1083. http://dx.doi.org/10.1126/science.1152538

42. Christensen, N.S. and D.P. Lettenmaier, 2007: A multimodel ensemble approach to assessment of climate change impacts on the hydrology and water resources of the Colorado River Basin. *Hydrology and Earth System Sciences*, **11**, 1417-1434. http://dx.doi.org/10.5194/hess-11-1417-2007

43. Hoerling, M., D. Lettenmaier, D. Cayan, and B. Udall, 2009: Reconciling future Colorado River flows. *Southwest Hydrology*, **8**, 20-21.31. http://www.swhydro.arizona.edu/archive/V8_N3/feature2.pdf

44. McCabe, G.J. and D.M. Wolock, 2014: Spatial and temporal patterns in conterminous United States streamflow characteristics. *Geophysical Research Letters*, **41**, 6889-6897. http://dx.doi.org/10.1002/2014GL061980

45. Stewart, I.T., D.R. Cayan, and M.D. Dettinger, 2005: Changes toward earlier streamflow timing across western North America. *Journal of Climate*, **18**, 1136-1155. http://dx.doi.org/10.1175/JCLI3321.1

46. Knowles, N., M.D. Dettinger, and D.R. Cayan, 2006: Trends in snowfall versus rainfall in the western United States. *Journal of Climate*, **19**, 4545-4559. http://dx.doi.org/10.1175/JCLI3850.1

47. Mote, P.W., 2003: Trends in snow water equivalent in the Pacific Northwest and their climatic causes. *Geophysical Research Letters*, **30**, 1601. http://dx.doi.org/10.1029/2003GL017258

48. Mote, P.W., A.F. Hamlet, M.P. Clark, and D.P. Lettenmaier, 2005: Declining mountain snowpack in western North America. *Bulletin of the American Meteorological Society*, **86**, 39-49. http://dx.doi.org/10.1175/BAMS-86-1-39

49. Hidalgo, H.G., T. Das, M.D. Dettinger, D.R. Cayan, D.W. Pierce, T.P. Barnett, G. Bala, A. Mirin, A.W. Wood, C. Bonfils, B.D. Santer, and T. Nozawa, 2009: Detection and attribution of streamflow timing changes to climate change in the western United States. *Journal of Climate*, **22**, 3838-3855. http://dx.doi.org/10.1175/2009jcli2470.1

50. Pierce, D.W., T.P. Barnett, H.G. Hidalgo, T. Das, C. Bonfils, B.D. Santer, G. Bala, M.D. Dettinger, D.R. Cayan, A. Mirin, A.W. Wood, and T. Nozawa, 2008: Attribution of declining western US snowpack to human effects. *Journal of Climate*, **21**, 6425-6444. http://dx.doi.org/10.1175/2008JCLI2405.1

51. Bonfils, C., B.D. Santer, D.W. Pierce, H.G. Hidalgo, G. Bala, T. Das, T.P. Barnett, D.R. Cayan, C. Doutriaux, A.W. Wood, A. Mirin, and T. Nozawa, 2008: Detection and attribution of temperature changes in the mountainous western United States. *Journal of Climate*, **21**, 6404-6424. http://dx.doi.org/10.1175/2008JCLI2397.1

52. Luce, C.H. and Z.A. Holden, 2009: Declining annual streamflow distributions in the Pacific Northwest United States, 1948–2006. *Geophysical Research Letters*, **36**, L16401. http://dx.doi.org/10.1029/2009GL039407

53. Luce, C.H., J.T. Abatzoglou, and Z.A. Holden, 2013: The missing mountain water: Slower westerlies decrease orographic enhancement in the Pacific Northwest USA. *Science*, **342**, 1360-1364. http://dx.doi.org/10.1126/science.1242335

54. Kormos, P.R., C.H. Luce, S.J. Wenger, and W.R. Berghuijs, 2016: Trends and sensitivities of low streamflow extremes to discharge timing and magnitude in Pacific Northwest mountain streams. *Water Resources Research*, **52**, 4990-5007. http://dx.doi.org/10.1002/2015WR018125

55. Dettinger, M., 2011: Climate change, atmospheric rivers, and floods in California–a multimodel analysis of storm frequency and magnitude changes. *Journal of the American Water Resources Association*, **47**, 514-523. http://dx.doi.org/10.1111/j.1752-1688.2011.00546.x

56. Guan, B., N.P. Molotch, D.E. Waliser, E.J. Fetzer, and P.J. Neiman, 2013: The 2010/2011 snow season in California's Sierra Nevada: Role of atmospheric rivers and modes of large-scale variability. *Water Resources Research*, **49**, 6731-6743. http://dx.doi.org/10.1002/wrcr.20537

57. Warner, M.D., C.F. Mass, and E.P. Salathé Jr., 2015: Changes in winter atmospheric rivers along the North American West Coast in CMIP5 climate models. *Journal of Hydrometeorology*, **16**, 118-128. http://dx.doi.org/10.1175/JHM-D-14-0080.1

58. Gao, Y., J. Lu, L.R. Leung, Q. Yang, S. Hagos, and Y. Qian, 2015: Dynamical and thermodynamical modulations on future changes of landfalling atmospheric rivers over western North America. *Geophysical Research Letters*, **42**, 7179-7186. http://dx.doi.org/10.1002/2015GL065435

59. Dettinger, M.D., 2013: Atmospheric rivers as drought busters on the U.S. West Coast. *Journal of Hydrometeorology*, **14**, 1721-1732. http://dx.doi.org/10.1175/JHM-D-13-02.1

60. Karl, T.R., J.T. Melillo, and T.C. Peterson, eds., 2009: *Global Climate Change Impacts in the United States.* Cambridge University Press: New York, NY, 189 pp. http://downloads.globalchange.gov/usimpacts/pdfs/climate-impacts-report.pdf

61. Williams, I.N. and M.S. Torn, 2015: Vegetation controls on surface heat flux partitioning, and land-atmosphere coupling. *Geophysical Research Letters*, **42**, 9416-9424. http://dx.doi.org/10.1002/2015GL066305

62. O'Gorman, P.A., 2014: Contrasting responses of mean and extreme snowfall to climate change. *Nature*, **512**, 416-418. http://dx.doi.org/10.1038/nature13625

63. Cayan, D.R., T. Das, D.W. Pierce, T.P. Barnett, M. Tyree, and A. Gershunov, 2010: Future dryness in the southwest US and the hydrology of the early 21st century drought. *Proceedings of the National Academy of Sciences*, **107**, 21271-21276. http://dx.doi.org/10.1073/pnas.0912391107

64. Das, T., D.W. Pierce, D.R. Cayan, J.A. Vano, and D.P. Lettenmaier, 2011: The importance of warm season warming to western U.S. streamflow changes. *Geophysical Research Letters*, **38**, L23403. http://dx.doi.org/10.1029/2011GL049660

65. Kapnick, S.B. and T.L. Delworth, 2013: Controls of global snow under a changed climate. *Journal of Climate*, **26**, 5537-5562. http://dx.doi.org/10.1175/JCLI-D-12-00528.1

66. Rhoades, A.M., P.A. Ullrich, and C.M. Zarzycki, 2017: Projecting 21st century snowpack trends in western USA mountains using variable-resolution CESM. *Climate Dynamics*, **Online First**, 1-28. http://dx.doi.org/10.1007/s00382-017-3606-0

67. Cayan, D., K. Kunkel, C. Castro, A. Gershunov, J. Barsugli, A. Ray, J. Overpeck, M. Anderson, J. Russell, B. Rajagopalan, I. Rangwala, and P. Duffy, 2013: Ch. 6: Future climate: Projected average. *Assessment of Climate Change in the Southwest United States: A Report Prepared for the National Climate Assessment.* Garfin, G., A. Jardine, R. Merideth, M. Black, and S. LeRoy, Eds. Island Press, Washington, D.C., 153-196. http://swccar.org/sites/all/themes/files/SW-NCA-color-FINALweb.pdf

68. Klos, P.Z., T.E. Link, and J.T. Abatzoglou, 2014: Extent of the rain–snow transition zone in the western U.S. under historic and projected climate. *Geophysical Research Letters*, **41**, 4560-4568. http://dx.doi.org/10.1002/2014GL060500

69. Luce, C.H., V. Lopez-Burgos, and Z. Holden, 2014: Sensitivity of snowpack storage to precipitation and temperature using spatial and temporal analog models. *Water Resources Research*, **50**, 9447-9462. http://dx.doi.org/10.1002/2013WR014844

70. Pierce, D.W. and D.R. Cayan, 2013: The uneven response of different snow measures to human-induced climate warming. *Journal of Climate*, **26**, 4148-4167. http://dx.doi.org/10.1175/jcli-d-12-00534.1

71. Fyfe, J.C., C. Derksen, L. Mudryk, G.M. Flato, B.D. Santer, N.C. Swart, N.P. Molotch, X. Zhang, H. Wan, V.K. Arora, J. Scinocca, and Y. Jiao, 2017: Large near-term projected snowpack loss over the western United States. *Nature Communications*, **8**, 14996. http://dx.doi.org/10.1038/ncomms14996

72. Neiman, P.J., L.J. Schick, F.M. Ralph, M. Hughes, and G.A. Wick, 2011: Flooding in western Washington: The connection to atmospheric rivers. *Journal of Hydrometeorology*, **12**, 1337-1358. http://dx.doi.org/10.1175/2011JHM1358.1

73. Ralph, F.M. and M.D. Dettinger, 2011: Storms, floods, and the science of atmospheric rivers. *Eos, Transactions, American Geophysical Union*, **92**, 265-266. http://dx.doi.org/10.1029/2011EO320001

74. Guan, B., D.E. Waliser, F.M. Ralph, E.J. Fetzer, and P.J. Neiman, 2016: Hydrometeorological characteristics of rain-on-snow events associated with atmospheric rivers. *Geophysical Research Letters*, **43**, 2964-2973. http://dx.doi.org/10.1002/2016GL067978

75. Archfield, S.A., R.M. Hirsch, A. Viglione, and G. Blöschl, 2016: Fragmented patterns of flood change across the United States. *Geophysical Research Letters*, **43**, 10,232-10,239. http://dx.doi.org/10.1002/2016GL070590

76. EPA, 2016: Climate Change Indicators in the United States, 2016. 4th edition. EPA 430-R-16-004. U.S. Environmental Protection Agency, Washington, D.C., 96 pp. https://www.epa.gov/sites/production/files/2016-08/documents/climate_indicators_2016.pdf

77. Hirsch, R.M. and K.R. Ryberg, 2012: Has the magnitude of floods across the USA changed with global $CO_2$ levels? *Hydrological Sciences Journal*, **57**, 1-9. http://dx.doi.org/10.1080/02626667.2011.621895

78. Mallakpour, I. and G. Villarini, 2015: The changing nature of flooding across the central United States. *Nature Climate Change*, **5**, 250-254. http://dx.doi.org/10.1038/nclimate2516

79. Wehner, M.F., 2013: Very extreme seasonal precipitation in the NARCCAP ensemble: Model performance and projections. *Climate Dynamics*, **40**, 59-80. http://dx.doi.org/10.1007/s00382-012-1393-1

80. Frei, A., K.E. Kunkel, and A. Matonse, 2015: The seasonal nature of extreme hydrological events in the northeastern United States. *Journal of Hydrometeorology*, **16**, 2065-2085. http://dx.doi.org/10.1175/JHM-D-14-0237.1

81. Berghuijs, W.R., R.A. Woods, C.J. Hutton, and M. Sivapalan, 2016: Dominant flood generating mechanisms across the United States. *Geophysical Research Letters*, **43**, 4382-4390. http://dx.doi.org/10.1002/2016GL068070

82. Groisman, P.Y., R.W. Knight, and T.R. Karl, 2001: Heavy precipitation and high streamflow in the contiguous United States: Trends in the twentieth century. *Bulletin of the American Meteorological Society*, **82**, 219-246. http://dx.doi.org/10.1175/1520-0477(2001)082<0219:hpahsi>2.3.co;2

83. Mallakpour, I. and G. Villarini, 2016: Investigating the relationship between the frequency of flooding over the central United States and large-scale climate. *Advances in Water Resources*, **92**, 159-171. http://dx.doi.org/10.1016/j.advwatres.2016.04.008

84. Novotny, E.V. and H.G. Stefan, 2007: Stream flow in Minnesota: Indicator of climate change. *Journal of Hydrology*, **334**, 319-333. http://dx.doi.org/10.1016/j.jhydrol.2006.10.011

85. Ryberg, K.R., W. Lin, and A.V. Vecchia, 2014: Impact of climate variability on runoff in the north-central United States. *Journal of Hydrologic Engineering*, **19**, 148-158. http://dx.doi.org/10.1061/(ASCE)HE.1943-5584.0000775

86. Slater, L.J., M.B. Singer, and J.W. Kirchner, 2015: Hydrologic versus geomorphic drivers of trends in flood hazard. *Geophysical Research Letters*, **42**, 370-376. http://dx.doi.org/10.1002/2014GL062482

87. Tomer, M.D. and K.E. Schilling, 2009: A simple approach to distinguish land-use and climate-change effects on watershed hydrology. *Journal of Hydrology*, **376**, 24-33. http://dx.doi.org/10.1016/j.jhydrol.2009.07.029

88. Villarini, G. and A. Strong, 2014: Roles of climate and agricultural practices in discharge changes in an agricultural watershed in Iowa. *Agriculture, Ecosystems & Environment*, **188**, 204-211. http://dx.doi.org/10.1016/j.agee.2014.02.036

89. Frans, C., E. Istanbulluoglu, V. Mishra, F. Munoz-Arriola, and D.P. Lettenmaier, 2013: Are climatic or land cover changes the dominant cause of runoff trends in the Upper Mississippi River Basin? *Geophysical Research Letters*, **40**, 1104-1110. http://dx.doi.org/10.1002/grl.50262

90. Wang, D. and M. Hejazi, 2011: Quantifying the relative contribution of the climate and direct human impacts on mean annual streamflow in the contiguous United States. *Water Resources Research*, **47**, W00J12. http://dx.doi.org/10.1029/2010WR010283

91. Armstrong, W.H., M.J. Collins, and N.P. Snyder, 2014: Hydroclimatic flood trends in the northeastern United States and linkages with large-scale atmospheric circulation patterns. *Hydrological Sciences Journal*, **59**, 1636-1655. http://dx.doi.org/10.1080/02626667.2013.862339

92. Woodhouse, C.A., S.T. Gray, and D.M. Meko, 2006: Updated streamflow reconstructions for the Upper Colorado River Basin. *Water Resources Research*, **42**. http://dx.doi.org/10.1029/2005WR004455

93. Munoz, S.E., K.E. Gruley, A. Massie, D.A. Fike, S. Schroeder, and J.W. Williams, 2015: Cahokia's emergence and decline coincided with shifts of flood frequency on the Mississippi River. *Proceedings of the National Academy of Sciences*, **112**, 6319-6324. http://dx.doi.org/10.1073/pnas.1501904112

94. Merz, B., S. Vorogushyn, S. Uhlemann, J. Delgado, and Y. Hundecha, 2012: HESS Opinions "More efforts and scientific rigour are needed to attribute trends in flood time series". *Hydrology and Earth System Sciences*, **16**, 1379-1387. http://dx.doi.org/10.5194/hess-16-1379-2012

95. Najafi, M.R. and H. Moradkhani, 2015: Multi-model ensemble analysis of runoff extremes for climate change impact assessments. *Journal of Hydrology*, **525**, 352-361. http://dx.doi.org/10.1016/j.jhydrol.2015.03.045

96. Dettinger, M.D., F.M. Ralph, T. Das, P.J. Neiman, and D.R. Cayan, 2011: Atmospheric rivers, floods and the water resources of California. *Water*, **3**, 445-478. http://dx.doi.org/10.3390/w3020445

97. Tebaldi, C., K. Hayhoe, J.M. Arblaster, and G.A. Meehl, 2006: Going to the extremes: An intercomparison of model-simulated historical and future changes in extreme events. *Climatic Change*, **79**, 185-211. http://dx.doi.org/10.1007/s10584-006-9051-4

98. AECOM, 2013: The Impact of Climate Change and Population Growth on the National Flood Insurance Program Through 2100. 257 pp. http://www.acclimatise.uk.com/login/uploaded/resources/FEMA_NFIP_report.pdf

99. SFPUC, 2016: Flood Resilience: Report, Task Order 57 (Draft: May 2016). San Francisco Public Utilities Commission, San Francisco, CA. 302 pp. http://sfwater.org/modules/showdocument.aspx?documentid=9176

100. Winters, B.A., J. Angel, C. Ballerine, J. Byard, A. Flegel, D. Gambill, E. Jenkins, S. McConkey, M. Markus, B.A. Bender, and M.J. O'Toole, 2015: Report for the Urban Flooding Awareness Act. Illinois Department of Natural Resources, Springfield, IL. 89 pp. https://www.dnr.illinois.gov/WaterResources/Documents/Final_UFAA_Report.pdf

101. IPCC, 2012: Summary for policymakers. *Managing the Risks of Extreme Events and Disasters to Advance Climate Change Adaptation. A Special Report of Working Groups I and II of the Intergovernmental Panel on Climate Change.* Field, C.B., V. Barros, T.F. Stocker, D. Qin, D.J. Dokken, K.L. Ebi, M.D. Mastrandrea, K.J. Mach, G.-K. Plattner, S.K. Allen, M. Tignor, and P.M. Midgley, Eds. Cambridge University Press, Cambridge, UK and New York, NY, 3-21. http://www.ipcc.ch/pdf/special-reports/srex/SREX_FD_SPM_final.pdf

102. Gochis, D., R. Schumacher, K. Friedrich, N. Doesken, M. Kelsch, J. Sun, K. Ikeda, D. Lindsey, A. Wood, B. Dolan, S. Matrosov, A. Newman, K. Mahoney, S. Rutledge, R. Johnson, P. Kucera, P. Kennedy, D. Sempere-Torres, M. Steiner, R. Roberts, J. Wilson, W. Yu, V. Chandrasekar, R. Rasmussen, A. Anderson, and B. Brown, 2015: The Great Colorado Flood of September 2013. *Bulletin of the American Meteorological Society,* **96 (12)**, 1461-1487. http://dx.doi.org/10.1175/BAMS-D-13-00241.1

103. Pall, P.C.M.P., M.F. Wehner, D.A. Stone, C.J. Paciorek, and W.D. Collins, 2017: Diagnosing anthropogenic contributions to heavy Colorado rainfall in September 2013. *Weather and Climate Extremes,* **17**, 1-6. http://dx.doi.org/10.1016/j.wace.2017.03.004

104. Abatzoglou, J.T. and A.P. Williams, 2016: Impact of anthropogenic climate change on wildfire across western US forests. *Proceedings of the National Academy of Sciences,* **113**, 11770-11775. http://dx.doi.org/10.1073/pnas.1607171113

105. Higuera, P.E., J.T. Abatzoglou, J.S. Littell, and P. Morgan, 2015: The changing strength and nature of fire-climate relationships in the northern Rocky Mountains, U.S.A., 1902-2008. *PLoS ONE,* **10**, e0127563. http://dx.doi.org/10.1371/journal.pone.0127563

106. Running, S.W., 2006: Is global warming causing more, larger wildfires? *Science,* **313**, 927-928. http://dx.doi.org/10.1126/science.1130370

107. Westerling, A.L., H.G. Hidalgo, D.R. Cayan, and T.W. Swetnam, 2006: Warming and earlier spring increase western U.S. forest wildfire activity. *Science,* **313**, 940-943. http://dx.doi.org/10.1126/science.1128834

108. Dennison, P.E., S.C. Brewer, J.D. Arnold, and M.A. Moritz, 2014: Large wildfire trends in the western United States, 1984–2011. *Geophysical Research Letters,* **41**, 2928-2933. http://dx.doi.org/10.1002/2014GL059576

109. Littell, J.S., D. McKenzie, D.L. Peterson, and A.L. Westerling, 2009: Climate and wildfire area burned in western U.S. ecoprovinces, 1916-2003. *Ecological Applications,* **19**, 1003-1021. http://dx.doi.org/10.1890/07-1183.1

110. Littell, J.S., E.E. Oneil, D. McKenzie, J.A. Hicke, J.A. Lutz, R.A. Norheim, and M.M. Elsner, 2010: Forest ecosystems, disturbance, and climatic change in Washington State, USA. *Climatic Change,* **102**, 129-158. http://dx.doi.org/10.1007/s10584-010-9858-x

111. Littell, J.S., D.L. Peterson, K.L. Riley, Y. Liu, and C.H. Luce, 2016: A review of the relationships between drought and forest fire in the United States. *Global Change Biology,* **22**, 2353-2369. http://dx.doi.org/10.1111/gcb.13275

112. Yoon, J.-H., B. Kravitz, P.J. Rasch, S.-Y.S. Wang, R.R. Gillies, and L. Hipps, 2015: Extreme fire season in California: A glimpse into the future? *Bulletin of the American Meteorological Society,* **96 (12)**, S5-S9. http://dx.doi.org/10.1175/bams-d-15-00114.1

113. Harvey, B.J., 2016: Human-caused climate change is now a key driver of forest fire activity in the western United States. *Proceedings of the National Academy of Sciences,* **113**, 11649-11650. http://dx.doi.org/10.1073/pnas.1612926113

114. Schoennagel, T., T.T. Veblen, and W.H. Romme, 2004: The interaction of fire, fuels, and climate across Rocky Mountain forests. *BioScience,* **54**, 661-676. http://dx.doi.org/10.1641/0006-3568(2004)054[0661:TIOFFA]2.0.CO;2

115. Stephens, S.L., J.K. Agee, P.Z. Fulé, M.P. North, W.H. Romme, T.W. Swetnam, and M.G. Turner, 2013: Managing forests and fire in changing climates. *Science,* **342**, 41-42. http://dx.doi.org/10.1126/science.1240294

116. Westerling, A.L., M.G. Turner, E.A.H. Smithwick, W.H. Romme, and M.G. Ryan, 2011: Continued warming could transform Greater Yellowstone fire regimes by mid-21st century. *Proceedings of the National Academy of Sciences,* **108**, 13165-13170. http://dx.doi.org/10.1073/pnas.1110199108

117. Stavros, E.N., J.T. Abatzoglou, D. McKenzie, and N.K. Larkin, 2014: Regional projections of the likelihood of very large wildland fires under a changing climate in the contiguous Western United States. *Climatic Change*, **126**, 455-468. http://dx.doi.org/10.1007/s10584-014-1229-6

118. Prestemon, J.P., U. Shankar, A. Xiu, K. Talgo, D. Yang, E. Dixon, D. McKenzie, and K.L. Abt, 2016: Projecting wildfire area burned in the south-eastern United States, 2011–60. *International Journal of Wildland Fire*, **25**, 715-729. http://dx.doi.org/10.1071/WF15124

119. Flannigan, M., B. Stocks, M. Turetsky, and M. Wotton, 2009: Impacts of climate change on fire activity and fire management in the circumboreal forest. *Global Change Biology*, **15**, 549-560. http://dx.doi.org/10.1111/j.1365-2486.2008.01660.x

120. Hu, F.S., P.E. Higuera, P. Duffy, M.L. Chipman, A.V. Rocha, A.M. Young, R. Kelly, and M.C. Dietze, 2015: Arctic tundra fires: Natural variability and responses to climate change. *Frontiers in Ecology and the Environment*, **13**, 369-377. http://dx.doi.org/10.1890/150063

121. Young, A.M., P.E. Higuera, P.A. Duffy, and F.S. Hu, 2017: Climatic thresholds shape northern high-latitude fire regimes and imply vulnerability to future climate change. *Ecography*, **40**, 606-617. http://dx.doi.org/10.1111/ecog.02205

122. Kasischke, E.S. and M.R. Turetsky, 2006: Recent changes in the fire regime across the North American boreal region—Spatial and temporal patterns of burning across Canada and Alaska. *Geophysical Research Letters*, **33**, L09703. http://dx.doi.org/10.1029/2006GL025677

123. Partain, J.L., Jr., S. Alden, U.S. Bhatt, P.A. Bieniek, B.R. Brettschneider, R. Lader, P.Q. Olsson, T.S. Rupp, H. Strader, R.L.T. Jr., J.E. Walsh, A.D. York, and R.H. Zieh, 2016: An assessment of the role of anthropogenic climate change in the Alaska fire season of 2015 [in "Explaining Extreme Events of 2015 from a Climate Perspective"]. *Bulletin of the American Meteorological Society*, **97 (12)**, S14-S18. http://dx.doi.org/10.1175/BAMS-D-16-0149.1

124. Sanford, T., R. Wang, and A. Kenwa, 2015: *The Age of Alaskan Wildfires*. Climate Central, Princeton, NJ, 32 pp. http://assets.climatecentral.org/pdfs/AgeofAlaskanWildfires.pdf

125. French, N.H.F., L.K. Jenkins, T.V. Loboda, M. Flannigan, R. Jandt, L.L. Bourgeau-Chavez, and M. Whitley, 2015: Fire in arctic tundra of Alaska: Past fire activity, future fire potential, and significance for land management and ecology. *International Journal of Wildland Fire*, **24**, 1045-1061. http://dx.doi.org/10.1071/WF14167

126. Kelly, R., M.L. Chipman, P.E. Higuera, I. Stefanova, L.B. Brubaker, and F.S. Hu, 2013: Recent burning of boreal forests exceeds fire regime limits of the past 10,000 years. *Proceedings of the National Academy of Sciences*, **110**, 13055-13060. http://dx.doi.org/10.1073/pnas.1305069110

127. Joly, K., P.A. Duffy, and T.S. Rupp, 2012: Simulating the effects of climate change on fire regimes in Arctic biomes: Implications for caribou and moose habitat. *Ecosphere*, **3**, 1-18. http://dx.doi.org/10.1890/ES12-00012.1

128. McGuire, A.D., L.G. Anderson, T.R. Christensen, S. Dallimore, L. Guo, D.J. Hayes, M. Heimann, T.D. Lorenson, R.W. MacDonald, and N. Roulet, 2009: Sensitivity of the carbon cycle in the Arctic to climate change. *Ecological Monographs*, **79**, 523-555. http://dx.doi.org/10.1890/08-2025.1

129. Meehl, G.A., C. Tebaldi, G. Walton, D. Easterling, and L. McDaniel, 2009: Relative increase of record high maximum temperatures compared to record low minimum temperatures in the US. *Geophysical Research Letters*, **36**, L23701. http://dx.doi.org/10.1029/2009GL040736

130. Rupp, D.E., P.W. Mote, N. Massey, C.J. Rye, R. Jones, and M.R. Allen, 2012: Did human influence on climate make the 2011 Texas drought more probable? [in "Explaining Extreme Events of 2011 from a Climate Perspective"]. *Bulletin of the American Meteorological Society*, **93**, 1052-1054. http://dx.doi.org/10.1175/BAMS-D-12-00021.1

131. Knutson, T.R., F. Zeng, and A.T. Wittenberg, 2014: Seasonal and annual mean precipitation extremes occurring during 2013: A U.S. focused analysis [in "Explaining Extreme Events of 2013 from a Climate Perspective"]. *Bulletin of the American Meteorological Society*, **95 (9)**, S19-S23. http://dx.doi.org/10.1175/1520-0477-95.9.S1.1

# 9

# Extreme Storms

## KEY FINDINGS

1. Human activities have contributed substantially to observed ocean–atmosphere variability in the Atlantic Ocean (*medium confidence*), and these changes have contributed to the observed upward trend in North Atlantic hurricane activity since the 1970s (*medium confidence*).

2. Both theory and numerical modeling simulations generally indicate an increase in tropical cyclone (TC) intensity in a warmer world, and the models generally show an increase in the number of very intense TCs. For Atlantic and eastern North Pacific hurricanes and western North Pacific typhoons, increases are projected in precipitation rates (*high confidence*) and intensity (*medium confidence*). The frequency of the most intense of these storms is projected to increase in the Atlantic and western North Pacific (*low confidence*) and in the eastern North Pacific (*medium confidence*).

3. Tornado activity in the United States has become more variable, particularly over the 2000s, with a decrease in the number of days per year with tornadoes and an increase in the number of tornadoes on these days (*medium confidence*). Confidence in past trends for hail and severe thunderstorm winds, however, is *low*. Climate models consistently project environmental changes that would putatively support an increase in the frequency and intensity of severe thunderstorms (a category that combines tornadoes, hail, and winds), especially over regions that are currently prone to these hazards, but confidence in the details of this projected increase is *low*.

4. There has been a trend toward earlier snowmelt and a decrease in snowstorm frequency on the southern margins of climatologically snowy areas (*medium confidence*). Winter storm tracks have shifted northward since 1950 over the Northern Hemisphere (*medium confidence*). Projections of winter storm frequency and intensity over the United States vary from increasing to decreasing depending on region, but model agreement is poor and confidence is *low*. Potential linkages between the frequency and intensity of severe winter storms in the United States and accelerated warming in the Arctic have been postulated, but they are complex, and, to some extent, contested, and confidence in the connection is currently *low*.

5. The frequency and severity of landfalling "atmospheric rivers" on the U.S. West Coast (narrow streams of moisture that account for 30%–40% of the typical snowpack and annual precipitation in the region and are associated with severe flooding events) will increase as a result of increasing evaporation and resulting higher atmospheric water vapor that occurs with increasing temperature. (*Medium confidence*)

**Recommended Citation for Chapter**

**Kossin**, J.P., T. Hall, T. Knutson, K.E. Kunkel, R.J. Trapp, D.E. Waliser, and M.F. Wehner, 2017: Extreme storms. In: *Climate Science Special Report: Fourth National Climate Assessment, Volume I* [Wuebbles, D.J., D.W. Fahey, K.A. Hibbard, D.J. Dokken, B.C. Stewart, and T.K. Maycock (eds.)]. U.S. Global Change Research Program, Washington, DC, USA, pp. 257-276, doi: 10.7930/J07S7KXX.

## 9.1 Introduction

Extreme storms have numerous impacts on lives and property. Quantifying how broad-scale average climate influences the behavior of extreme storms is particularly challenging, in part because extreme storms are comparatively rare short-lived events and occur within an environment of largely random variability. Additionally, because the physical mechanisms linking climate change and extreme storms can manifest in a variety of ways, even the sign of the changes in the extreme storms can vary in a warming climate. This makes detection and attribution of trends in extreme storm characteristics more difficult than detection and attribution of trends in the larger environment in which the storms evolve (e.g., Ch. 6: Temperature Change). Projecting changes in severe storms is also challenging because of model constraints in how they capture and represent small-scale, highly local physics. Despite the challenges, good progress is being made for a variety of storm types, such as tropical cyclones, severe convective storms (thunderstorms), winter storms, and atmospheric river events.

## 9.2 Tropical Cyclones (Hurricanes and Typhoons)

Detection and attribution (Ch. 3: Detection and Attribution) of past changes in tropical cyclone (TC) behavior remain a challenge due to the nature of the historical data, which are highly heterogeneous in both time and among the various regions that collect and analyze the data.[1, 2, 3] While there are ongoing efforts to reanalyze and homogenize the data (e.g., Landsea et al. 2015;[4] Kossin et al. 2013[2]), there is still low confidence that any reported long-term (multidecadal to centennial) increases in TC activity are robust, after accounting for past changes in observing capabilities [which is unchanged from the Intergovernmental Panel on Climate Change Fifth Assessment Report (IPCC AR5) assessment statement[5]]. This is not meant to

imply that no such increases have occurred, but rather that the data are not of a high enough quality to determine this with much confidence. Furthermore, it has been argued that within the period of highest data quality (since around 1980), the globally observed changes in the environment would not necessarily support a detectable trend in tropical cyclone intensity.[2] That is, the trend signal has not yet had time to rise above the background variability of natural processes.

Both theory and numerical modeling simulations (in general) indicate an increase in TC intensity in a warmer world, and the models generally show an increase in the number of very intense TCs.[6, 7, 8, 9, 10] In some cases, climate models can be used to make attribution statements about TCs without formal detection (see also Ch. 3: Detection and Attribution). For example, there is evidence that, in addition to the effects of El Niño, anthropogenic forcing made the extremely active 2014 Hawaiian hurricane season substantially more likely, although no significant rising trend in TC frequency near Hawai'i was detected.[11]

Changes in frequency and intensity are not the only measures of TC behavior that may be affected by climate variability and change, and there is evidence that the locations where TCs reach their peak intensity has migrated poleward over the past 30 years in the Northern and Southern Hemispheres, apparently in concert with environmental changes associated with the independently observed expansion of the tropics.[12] The poleward migration in the western North Pacific,[13] which includes a number of U.S. territories, appears particularly consistent among the various available TC datasets and remains significant over the past 60–70 years after accounting for the known modes of natural variability in the region (Figure 9.1). The migration, which can substantially change patterns of TC hazard exposure and mortality risk, is also evident in 21st century Coupled Model Intercomparison Project Phase 5 (CMIP5) projections following the RCP8.5 emissions trajectories, suggesting a possible link to human activities. Further analysis comparing observed past TC behavior with climate model historical forcing runs (and with model control runs simulating multidecadal internal climate variability alone) are needed to better understand this process, but it is expected that this will be an area of heightened future research.

In the Atlantic, observed multidecadal variability of the ocean and atmosphere, which TCs are shown to respond to, has been ascribed (Ch. 3: Detection and Attribution) to natural internal variability via meridional overturning ocean circulation changes,[14] natural external variability caused by volcanic eruptions[15, 16] and Saharan dust outbreaks,[17, 18] and anthropogenic external forcing via greenhouse gases and sulfate aerosols.[19, 20, 21] Determining the relative contributions of each mechanism to the observed multidecadal variability in the Atlantic, and even whether natural or anthropogenic factors have dominated, is presently a very active area of research and debate, and no consensus has yet been reached.[22, 23, 24, 25, 26, 27] Despite the level of disagreement about the relative magnitude of human influences, there is broad agreement that human factors have had an impact on the observed oceanic and atmospheric variability in the North Atlantic, and there is *medium confidence* that this has contributed to the observed increase in hurricane activity since the 1970s. This is essentially unchanged from the IPCC AR5 statement,[6] although the post-AR5 literature has only served to further support this statement.[28] This is expected to remain an active research topic in the foreseeable future.

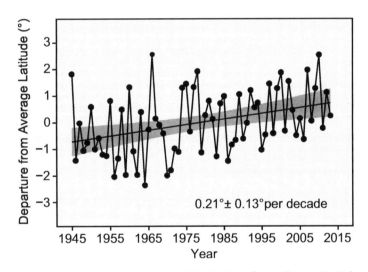

**Figure 9.1:** Poleward migration, in degrees of latitude, of the location of annual mean tropical cyclone (TC) peak lifetime intensity in the western North Pacific Ocean, after accounting for the known regional modes of interannual (El Niño–Southern Oscillation; ENSO) and interdecadal (Pacific Decadal Oscillation; PDO) variability. The time series shows residuals of the multivariate regression of annually averaged latitude of TC peak lifetime intensity onto the mean Niño-3.4 and PDO indices. Data are taken from the Joint Typhoon Warning Center (JTWC). Shading shows 95% confidence bounds for the trend. Annotated values at lower right show the mean migration rate and its 95% confidence interval in degrees per decade for the period 1945–2013. (Figure source: adapted from Kossin et al. 2016;[13] © American Meteorological Society. Used with permission.)

The IPCC AR5 consensus TC projections for the late 21st century (IPCC Figure 14.17)[8] include an increase in global mean TC intensity, precipitation rate, and frequency of very intense (Saffir-Simpson Category 4–5) TCs, and a decrease, or little change, in global TC frequency. Since the IPCC AR5, some studies have provided additional support for this consensus, and some have challenged an aspect of it. For example, a recent study[9] projects increased mean hurricane intensities in the Atlantic Ocean basin and in most, but not all, other TC-supporting basins (see Table 3 in Knutson et al. 2015[9]). In their study, the global occurrence of Saffir–Simpson Category 4–5 storms was projected to increase significantly, with the most significant basin-scale changes projected for the Northeast Pacific basin, potentially increasing intense hurricane risk to Hawai'i (Figure 9.2) over the coming century. However, another recent (post-AR5) study proposed that increased thermal stratification of the upper ocean in CMIP5 climate warming scenarios should substantially reduce the warming-induced intensification of TCs estimated in previous studies.[29] Follow-up studies, however, estimate that the effect of such increased stratification is relatively small, reducing the projected intensification of TCs by only about 10%–15%.[30, 31]

Another recent study challenged the IPCC AR5 consensus projection of a decrease, or little change, in global tropical cyclone frequency by simulating increased global TC frequency over the 21st century under the higher scenario (RCP8.5).[32] However, another modeling study has found that neither direct analysis of CMIP5-class simulations, nor indirect inferences from the simulations (such as those of Emanuel 2013[32]), could reproduce the decrease in TC frequency projected in a warmer world by high-resolution TC-permitting climate models,[33] which adds uncertainty to the results of Emanuel.[32]

In summary, despite new research that challenges one aspect of the AR5 consensus for late 21st century-projected TC activity, it remains *likely* that global mean tropical cyclone maximum wind speeds and precipitation rates will increase; and it is *more likely than not* that the global frequency of occurrence of TCs will either decrease or remain essentially the same. Confidence in projected global increases of intensity and tropical cyclone precipitation rates is *medium* and *high*, respectively, as there is better model consensus. Confidence is further heightened, particularly for projected increases in precipitation rates, by a robust physical understanding of the processes that lead to these increases. Confidence in projected increases in the frequency of very intense TCs is generally lower (*medium* in the eastern North Pacific and *low* in the western North Pacific and Atlantic) due to comparatively fewer studies available and due to the competing influences of projected reductions in overall storm frequency and increased mean intensity on the frequency of the most intense storms. Both the magnitude and sign of projected changes in individual ocean basins appears to depend on the large-scale pattern of changes to atmospheric circulation and ocean surface temperature (e.g., Knutson et al. 2015[9]). Projections of these regional patterns of change—apparently critical for TC projections—are uncertain, leading to uncertainty in regional TC projections.

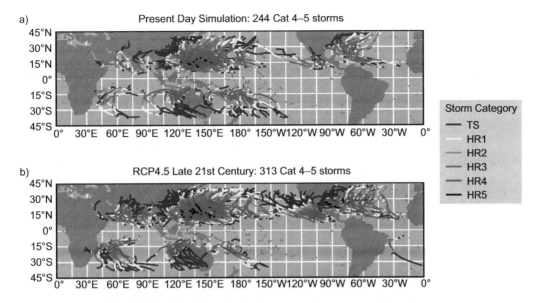

Figure 9.2: Tracks of simulated Saffir–Simpson Category 4–5 tropical cyclones for (a) present-day or (b) late-21st-century conditions, based on dynamical downscaling of climate conditions from the CMIP5 multimodel ensemble (lower scenario; RCP4.5). The tropical cyclones were initially simulated using a 50-km grid global atmospheric model, but each individual tropical cyclone was re-simulated at higher resolution using the GFDL hurricane model to provide more realistic storm intensities and structure. Storm categories or intensities are shown over the lifetime of each simulated storm, according to the Saffir–Simpson scale. The categories are depicted by the track colors, varying from tropical storm (blue) to Category 5 (black; see legend). (Figure source: Knutson et al. 2015;[9] © American Meteorological Society. Used with permission.)

## Box 9.1: U.S. Landfalling Major Hurricane "Drought"

Hurricane Harvey made landfall as a major hurricane (Saffir–Simpson Category 3 or higher) in Texas in 2017, breaking what has sometimes been colloquially referred to as the "hurricane drought." Prior to Harvey, the last major hurricane to make landfall in the continental United States was Wilma in 2005. The 11-year (2006–2016) absence of U.S. major hurricane landfall events is unprecedented in the historical records dating back to the mid-19th century and has occurred in tandem with average to above-average basin-wide major hurricane counts. Was the 11-year absence of U.S. landfalling major hurricanes due to random luck, or were there systematic changes in climate that drove this?

One recent study indicates that the absence of U.S. landfalling major hurricanes cannot readily be attributed to any sustained changes in the climate patterns that affect hurricanes.[34] Based on a statistical analysis of the historical North Atlantic hurricane database, the study found no evidence of a connection between the number of major U.S. landfalls from one year to the next and concluded that the 11-year absence of U.S. landfalling major hurricanes was random. A subsequent recent study did identify a systematic pattern of atmosphere/ocean conditions that vary in such a way that conditions conducive to hurricane intensification in the deep tropics occur in concert with conditions conducive to weakening near the U.S. coast.[35] This result suggests a possible relationship between climate and hurricanes; increasing basin-wide hurricane counts are associated with a decreasing fraction of major hurricanes making U.S. landfall, as major hurricanes approaching the U.S. coast are more likely to

## Box 9.1 (*continued*)

weaken during active North Atlantic hurricane periods (such as the present period). It is unclear to what degree this relationship has affected absolute hurricane landfall counts during the recent active hurricane period from the mid-1990s, as the basin-wide number and landfalling fraction are in opposition (that is, there are more major hurricanes but a smaller fraction make landfall as major hurricanes). It is also unclear how this relationship may change as the climate continues to warm. Other studies have identified systematic interdecadal hurricane track variability that may affect landfalling hurricane and major hurricane frequency.[36, 37, 38]

Another recent study[39] shows that the extent of the absence is sensitive to uncertainties in the historical data and even small variations in the definition of a major hurricane, which is somewhat arbitrary. It is also sensitive to the definition of U.S. landfall, which is a geopolitical-border-based constraint and has no physical meaning. In fact, many areas outside of the U.S. border have experienced major hurricane landfalls in the past 11 years. In this sense, the frequency of U.S. landfalling major hurricanes is not a particularly robust metric with which to study questions about hurricane activity and its relationship with climate variability. Furthermore, the 11-year absence of U.S. landfalling major hurricanes is not a particularly relevant metric in terms of coastal hazard exposure and risk. For example, Hurricanes Ike (2008), Irene (2011), Sandy (2012), and most recently Hurricane Matthew (2016) brought severe impacts to the U.S. coast despite not making landfall in the United States while classified as major hurricanes. In the case of Hurricane Sandy, extreme rainfall and storm surge (see also Ch. 12: Sea Level Rise) during landfall caused extensive destruction in and around the New York City area, despite Sandy's designation as a post-tropical cyclone at that time. In the case of Hurricane Matthew, the center came within about 40 miles of the Florida coast while Matthew was a major hurricane, which is close enough to significantly impact the coast but not close enough to break the "drought" as it is defined.

In summary, the absence of U.S. landfalling major hurricanes from Wilma in 2005 to Harvey in 2017 was anomalous. There is some evidence that systematic atmosphere/ocean variability has reduced the fraction of hurricanes making U.S. landfall since the mid-1990s, but this is at least partly countered by increased basin-wide numbers, and the net effect on landfall rates is unclear. Moreover, there is a large random element, and the metric itself suffers from lack of physical basis due to the arbitrary intensity threshold and geopolitically based constraints. Additionally, U.S. coastal risk, particularly from storm surge and freshwater flooding, depends strongly on storm size, propagation speed and direction, and rainfall rates. There is some danger, in the form of evoking complacency, in placing too much emphasis on an absence of a specific subset of hurricanes.

## 9.3 Severe Convective Storms (Thunderstorms)

Tornado and severe thunderstorm events cause significant loss of life and property: more than one-third of the $1 billion weather disasters in the United States during the past 25 years were due to such events, and, relative to other extreme weather, the damages from convective weather hazards have undergone the largest increase since 1980.[40] A particular challenge in quantifying the existence and intensity of these events arises from the data source: rather than measurements, the occurrence of tornadoes and severe thunderstorms is determined by visual sightings by eye-witnesses (such as "storm spotters" and law enforcement officials) or post-storm damage assessments. The reporting has been susceptible to changes in population density, modifications to reporting procedures and training, the introduction of video and social media, and so on. These have led to systematic, non-meteorological biases in the long-term data record.

Nonetheless, judicious use of the report database has revealed important information about tornado trends. Since the 1970s, the United States has experienced a decrease in the number of days per year on which tornadoes occur, but an increase in the number of tornadoes that form on such days.[41] One important implication is that the frequency of days with large numbers of tornadoes—tornado outbreaks—appears to be increasing (Figure 9.3). The extent of the season over which such tornado activity occurs is increasing as well: although tornadoes in the United States are observed in all months of the year, an earlier calendar-day start to the season of high activity is emerging. In general, there is more interannual variability, or volatility, in tornado occurrence (see also Elsner et al. 2015[42]).[43]

Evaluations of hail and (non-tornadic) thunderstorm wind reports have thus far been less revealing. Although there is evidence of an increase in the number of hail days per year, the inherent uncertainty in reported hail size reduces the confidence in such a conclusion.[44]

Thunderstorm wind reports have proven to be even less reliable, because, as compared to tornadoes and hail, there is less tangible visual evidence; thus, although the United States has lately experienced several significant thunderstorm wind events (sometimes referred to as "derechos"), the lack of studies that explore long-term trends in wind events and the uncertainties in the historical data preclude any robust assessment.

It is possible to bypass the use of reports by exploiting the fact that the temperature, humidity, and wind in the larger vicinity—or "environment"—of a developing thunderstorm ultimately control the intensity, morphology, and hazardous tendency of the storm. Thus, the premise is that quantifications of the vertical profiles of temperature, humidity, and wind can be used as a proxy for actual severe thunderstorm occurrence. In particular, a thresholded product of convective available potential energy (CAPE) and vertical wind shear over a surface-to-6 km layer (S06) constitutes one widely used means of

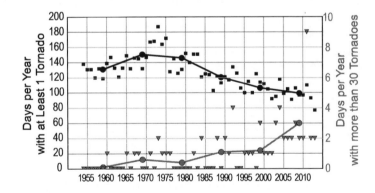

**Annual Tornado Activity in the U.S. (1955–2013)**

Figure 9.3: Annual tornado activity in the United States over the period 1955–2013. The black squares indicate the number of days per year with at least one tornado rated (E)F1 or greater, and the black circles and line show the decadal mean line of such *tornado days*. The red triangles indicate the number of days per year with more than 30 tornadoes rated (E)F1 or greater, and the red circles and line show the decadal mean of these *tornado outbreaks*. (Figure source: redrawn from Brooks et al. 2014[41]).

representing the frequency of severe thunderstorms.[45] This environmental-proxy approach avoids the biases and other issues with eyewitness storm reports and is readily evaluated using the relatively coarse global datasets and global climate models. It has the disadvantage of assuming that a thunderstorm will necessarily form and then realize its environmental potential.

Upon employing global climate models (GCMs) to evaluate CAPE and S06, a consistent finding among a growing number of proxy-based studies is a projected increase in the frequency of severe thunderstorm environments in the United States over the mid- to late 21st century.[46, 47, 48, 49, 50, 51] The most robust projected increases in frequency are over the U.S. Midwest and southern Great Plains, during March-April-May (MAM).[46] Based on the increased frequency of very high CAPE, increases in storm intensity are also projected over this same period (see also Del Genio et al. 2007[52]).

Key limitations of the environmental proxy approach are being addressed through the applications of high-resolution dynamical downscaling, wherein sufficiently fine model grids are used so that individual thunderstorms are explicitly resolved, rather than implicitly represented (as through environmental proxies). The individually modeled thunderstorms can then be quantified and assessed in terms of severity.[53, 54, 55] The dynamical-downscaling results have thus far supported the basic findings of the environmental proxy studies, particularly in terms of the seasons and geographical regions projected to experience the largest increases in severe thunderstorm occurrence.[46]

The computational expense of high-resolution dynamical downscaling makes it difficult to generate model ensembles over long time

periods, and thus to assess the uncertainty of the downscaled projections. Because these dynamical downscaling implementations focus on the statistics of storm occurrence rather than on faithful representations of individual events, they have generally been unconcerned with specific extreme convective events in history. So, for example, such downscaling does not address whether the intensity of an event like the Joplin, Missouri, tornado of May 22, 2011, would be amplified under projected future climates. Recently, the "pseudo-global warming" (PGW) methodology (see Schär et al. 1996[56]), which is a variant of dynamical downscaling, has been adapted to address these and related questions. As an example, when the parent "supercell" of select historical tornado events forms under the climate conditions projected during the late 21st century, it does not evolve into a benign, unorganized thunderstorm but instead maintains its supercellular structure.[57] As measured by updraft strength, the intensity of these supercells under PGW is relatively higher, although not in proportion to the theoretical intensity based on the projected higher levels of CAPE. The adverse effects of enhanced precipitation loading under PGW has been offered as one possible explanation for such shortfalls in projected updraft strength.

## 9.4 Winter Storms

The frequency of large snowfall years has decreased in the southern United States and Pacific Northwest and increased in the northern United States (see Ch. 7: Precipitation Change). The winters of 2013/2014 and 2014/2015 have contributed to this trend. They were characterized by frequent storms and heavier-than-normal snowfalls in the Midwest and Northeast and drought in the western United States. These were related to blocking (a large-scale pressure pattern with little or no movement) of the wintertime circulation in the Pacific sector of the Northern

Hemisphere (e.g., Marinaro et al. 2015[58]) that put the midwestern and northeastern United States in the primary winter storm track, while at the same time reducing the number of winter storms in California, causing severe drought conditions.[59] While some observational studies suggest a linkage between blocking affecting the U.S. climate and enhanced arctic warming (arctic amplification), specifically for an increase in highly amplified jet stream patterns in winter over the United States,[60] other studies show mixed results.[61, 62, 63] Therefore, a definitive understanding of the effects of arctic amplification on midlatitude winter weather remains elusive. Other explanations have been offered for the weather patterns of recent winters, such as anomalously strong Pacific trade winds,[64] but these have not been linked to anthropogenic forcing (e.g., Delworth et al. 2015[65]).

Analysis of storm tracks indicates that there has been an increase in winter storm frequency and intensity since 1950, with a slight shift in tracks toward the poles.[66, 67, 68] Current global climate models (CMIP5) do in fact predict an increase in extratropical cyclone (ETC) frequency over the eastern United States, including the most intense ETCs, under the higher scenario (RCP8.5).[69] However, there are large model-to-model differences in the realism of ETC simulations and in the projected changes. Moreover, projected ETC changes have large regional variations, including a decreased total frequency in the North Atlantic, further highlighting the complexity of the response to climate change.

## 9.5 Atmospheric Rivers

The term "atmospheric rivers" (ARs) refers to the relatively narrow streams of moisture transport that often occur within and across midlatitudes[70] (Figure 9.4), in part because they often transport as much water as in the Amazon River.[71] While ARs occupy less than 10% of the circumference of Earth at any given time, they account for 90% of the poleward moisture transport across midlatitudes (a more complete discussion of precipitation variability is found in Ch. 7: Precipitation Change). In many regions of the world, they account for a substantial fraction of the precipitation,[72] and thus water supply, often delivered in the form of an extreme weather and precipitation event (Figure 9.4). For example, ARs account for 30%–40% of the typical snowpack in the Sierra Nevada mountains and annual precipitation in the U.S. West Coast states[73, 74]—an essential summertime source of water for agriculture, consumption, and ecosystem health. However, this vital source of water is also associated with severe flooding—with observational evidence showing a close connection between historically high streamflow events and floods with landfalling AR events—in the west and other sectors of the United States.[75, 76, 77] More recently, research has also demonstrated that ARs are often found to be critical in ending droughts in the western United States.[78]

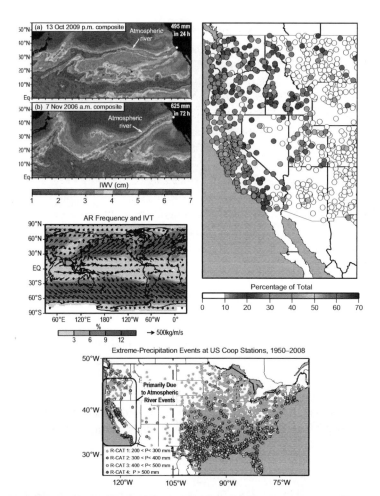

**Figure 9.4:** (upper left) Atmospheric rivers depicted in Special Sensor Microwave Imager (SSM/I) measurements of SSM/I total column water vapor leading to extreme precipitation events at landfall locations. (middle left) Annual mean frequency of atmospheric river occurrence (for example, 12% means about 1 every 8 days) and their integrated vapor transport (IVT).[72] (bottom) ARs are the dominant synoptic storms for the U.S. West Coast in terms of extreme precipitation[93] and (right) supply a large fraction of the annual precipitation in the U.S. West Coast states.[73] [Figure source: (upper and middle left) Ralph et al. 2011,[94] (upper right) Guan and Waliser 2015,[72] (lower left) Ralph and Dettinger 2012,[93] (lower right) Dettinger et al. 2011;[73] left panels, © American Meteorological Society. Used with permission.]

Given the important role that ARs play in the water supply of the western United States and their role in weather and water extremes in the west and occasionally other parts of the United States (e.g., Rutz et al. 2014[79]), it is critical to examine how climate change and the expected intensification of the global water cycle and atmospheric transports (e.g., Held and Soden 2006;[80] Lavers et al. 2015[81]) are projected to impact ARs (e.g., Dettinger and Ingram 2013[82]).

Under climate change conditions, ARs may be altered in a number of ways, namely their frequency, intensity, duration, and locations. In association with landfalling ARs, any of these would be expected to result in impacts on hazards and water supply given the discussion above. Assessments of ARs in climate change projections for the United States have been undertaken for central California from CMIP3,[73] and a number of studies have been done for the West Coast of North America,[83, 84, 85, 86, 87] and these studies have uniformly shown that ARs are likely to become more frequent and intense in the future. For example, one recent study reveals a large increase of AR days along the West Coast by the end of the 21st century under the higher scenario (RCP8.5), with fractional increases between 50% and 600%, depending on the seasons and landfall locations.[83] Results from these studies (and Lavers et al. 2013[88] for ARs impacting the United Kingdom) show that these AR changes were predominantly driven by increasing atmospheric specific humidity, with little discernible change in the low-level winds. The higher atmospheric water vapor content in a warmer climate is to be expected because of an increase in saturation water vapor pressure with air temperature (Ch. 2: Physical Drivers of Climate Change). While the thermodynamic effect appears to dominate the climate change impact on ARs, leading to projected increases in ARs, there is evidence for a dynamical effect (that is, location change) related to the projected poleward shift of the subtropical jet that diminished the thermodynamic effect in the southern portion of the West Coast of North America.[83]

Presently, there is no clear consensus on whether the consistently projected increases in AR frequency and intensity will translate to increased precipitation in California. This is mostly because previous studies did not examine this explicitly and because the model resolution is poor and thus the topography is poorly represented, and the topography is a key aspect of forcing the precipitation out of the systems.[89] The evidence for considerable increases in the number and intensity of ARs depends (as do all climate variability studies based on dynamical models) on the model fidelity in representing ARs and their interactions with the global climate/circulation. Additional confidence comes from studies that show qualitatively similar projected increases while also providing evidence that the models represent AR frequency, transports, and spatial distributions relatively well compared to observations.[84, 85] A caveat associated with drawing conclusions from any given study or differences between two is that they typically use different detection methodologies that are typically tailored to a regional setting (cf. Guan and Waliser 2015[72]). Additional research is warranted to examine these storms from a global perspective, with additional and more in-depth, process-oriented diagnostics/metrics. Stepping away from the sensitivities associated with defining atmospheric rivers, one study examined the intensification of the integrated vapor transport (IVT), which is easily and unambiguously defined.[81] That study found that for the higher scenario (RCP8.5), multimodel mean IVT and the IVT associated with extremes above 95% percentile increase by 30%–40% in the North Pacific. These results, along with the uniform findings of the studies above examining projected changes in ARs for western North America and the United Kingdom, give *high confidence* that the frequency of AR storms will increase in association with rising global temperatures.

# TRACEABLE ACCOUNTS

### Key Finding 1

Human activities have contributed substantially to observed ocean–atmosphere variability in the Atlantic Ocean (*medium confidence*), and these changes have contributed to the observed upward trend in North Atlantic hurricane activity since the 1970s (*medium confidence*).

### Description of evidence base

The Key Finding and supporting text summarizes extensive evidence documented in the climate science literature and is similar to statements made in previous national (NCA3)[90] and international[91] assessments. Data limitations are documented in Kossin et al. 2013[2] and references therein. Contributions of natural and anthropogenic factors in observed multidecadal variability are quantified in Carslaw et al. 2013;[22] Zhang et al. 2013;[27] Tung and Zhou 2013;[26] Mann et al. 2014;[23] Stevens 2015;[25] Sobel et al. 2016;[24] Walsh et al. 2015.[10]

### Major uncertainties

Key remaining uncertainties are due to known and substantial heterogeneities in the historical tropical cyclone data and lack of robust consensus in determining the precise relative contributions of natural and anthropogenic factors in past variability of the tropical environment.

### Assessment of confidence based on evidence and agreement, including short description of nature of evidence and level of agreement

Confidence in this finding is rated as *medium*. Although the range of estimates of natural versus anthropogenic contributions in the literature is fairly broad, virtually all studies identify a measurable, and generally substantial, anthropogenic influence. This does constitute a consensus for human contribution to the increases in tropical cyclone activity since 1970.

### Summary sentence or paragraph that integrates the above information

The key message and supporting text summarizes extensive evidence documented in the climate science peer-reviewed literature. The uncertainties and points of consensus that were described in the NCA3 and IPCC assessments have continued.

### Key Finding 2

Both theory and numerical modeling simulations generally indicate an increase in tropical cyclone (TC) intensity in a warmer world, and the models generally show an increase in the number of very intense TCs. For Atlantic and eastern North Pacific hurricanes and western North Pacific typhoons, increases are projected in precipitation rates (*high confidence*) and intensity (*medium confidence*). The frequency of the most intense of these storms is projected to increase in the Atlantic and western North Pacific (*low confidence*) and in the eastern North Pacific (*medium confidence*).

### Description of evidence base

The Key Finding and supporting text summarizes extensive evidence documented in the climate science literature and is similar to statements made in previous national (NCA3)[90] and international[91] assessments. Since these assessments, more recent downscaling studies have further supported these assessments (e.g., Knutson et al. 2015[9]), though pointing out that the changes (future increased intensity and tropical cyclone precipitation rates) may not occur in all ocean basins.

### Major uncertainties

A key uncertainty remains in the lack of a supporting detectable anthropogenic signal in the historical data to add further confidence to these projections. As such, confidence in the projections is based on agreement among different modeling studies and physical understanding (for example, potential intensity theory for tropical cyclone intensities and the expectation of stronger moisture convergence, and thus higher precipitation rates, in tropical cyclones in a warmer environment containing greater amounts of environmental atmospheric moisture). Additional uncertainty stems from uncertainty in both the projected pattern and magnitude of future sea surface temperatures.[9]

**Assessment of confidence based on evidence and agreement, including short description of nature of evidence and level of agreement**

Confidence is rated as *high* in tropical cyclone rainfall projections and *medium* in intensity projections since there are a number of publications supporting these overall conclusions, fairly well-established theory, general consistency among different studies, varying methods used in studies, and still a fairly strong consensus among studies. However, a limiting factor for confidence in the results is the lack of a supporting detectable anthropogenic contribution in observed tropical cyclone data.

There is *low* to *medium confidence* for increased occurrence of the most intense tropical cyclones for most ocean basins, as there are relatively few formal studies that focus on these changes, and the change in occurrence of such storms would be enhanced by increased intensities, but reduced by decreased overall frequency of tropical cyclones.

**Summary sentence or paragraph that integrates the above information**

Models are generally in agreement that tropical cyclones will be more intense and have higher precipitation rates, at least in most ocean basins. Given the agreement between models and support of theory and mechanistic understanding, there is *medium* to *high confidence* in the overall projection, although there is some limitation on confidence levels due to the lack of a supporting detectable anthropogenic contribution to tropical cyclone intensities or precipitation rates.

**Key Finding 3**

Tornado activity in the United States has become more variable, particularly over the 2000s, with a decrease in the number of days per year with tornadoes and an increase in the number of tornadoes on these days (*medium confidence*). Confidence in past trends for hail and severe thunderstorm winds, however, is *low*. Climate models consistently project environmental changes that would putatively support an increase in the frequency and intensity of severe thunderstorms (a category that combines tornadoes, hail, and winds), especially over

regions that are currently prone to these hazards, but confidence in the details of this projected increase is *low*.

**Description of evidence base**

Evidence for the first and second statement comes from the U.S. database of tornado reports. There are well known biases in this database, but application of an intensity threshold [greater than or equal to a rating of 1 on the (Enhanced) Fujita scale], and the quantification of tornado activity in terms of tornado days instead of raw numbers of reports are thought to reduce these biases. It is not known at this time whether the variability and trends are necessarily due to climate change.

The third statement is based on projections from a wide range of climate models, including GCMs and RCMs, run over the past 10 years (e.g., see the review by Brooks 2013[92]). The evidence is derived from an "environmental-proxy" approach, which herein means that severe thunderstorm occurrence is related to the occurrence of two key environmental parameters: CAPE and vertical wind shear. A limitation of this approach is the assumption that the thunderstorm will necessarily form and then realize its environmental potential. This assumption is indeed violated, albeit at levels that vary by region and season.

**Major uncertainties**

Regarding the first and second statements, there is still some uncertainty in the database, even when the data are filtered. The major uncertainty in the third statement equates to the aforementioned limitation (that is, the thunderstorm will necessarily form and then realize its environmental potential).

**Assessment of confidence based on evidence and agreement, including short description of nature of evidence and level of agreement**

*Medium*: That the variability in tornado activity has increased.

*Medium*: That the severe-thunderstorm environmental conditions will change with a changing climate, but

*Low*: on the precise (geographical and seasonal) realization of the environmental conditions as actual severe thunderstorms.

**Summary sentence or paragraph that integrates the above information**

With an established understanding of the data biases, careful analysis provides useful information about past changes in severe thunderstorm and tornado activity. This information suggests that tornado variability has increased in the 2000s, with a concurrent decrease in the number of days per year experiencing tornadoes and an increase in the number of tornadoes on these days. Similarly, the development of novel applications of climate models provides information about possible future severe storm and tornado activity, and although confidence in these projections is low, they do suggest that the projected environments are at least consistent with environments that would putatively support an increase in frequency and intensity of severe thunderstorms.

**Key Finding 4**

There has been a trend toward earlier snowmelt and a decrease in snowstorm frequency on the southern margins of climatologically snowy areas (*medium confidence*). Winter storm tracks have shifted northward since 1950 over the Northern Hemisphere (*medium confidence*). Projections of winter storm frequency and intensity over the United States vary from increasing to decreasing depending on region, but model agreement is poor and confidence is *low*. Potential linkages between the frequency and intensity of severe winter storms in the United States and accelerated warming in the Arctic have been postulated, but they are complex, and, to some extent, contested, and confidence in the connection is currently *low*.

**Description of evidence base**

The Key Finding and supporting text summarizes evidence documented in the climate science literature.

Evidence for changes in winter storm track changes are documented in a small number of studies.[67, 68] Future changes are documented in one study,[69] but there are large model-to-model differences. The effects of arctic amplification on U.S. winter storms have been studied, but the results are mixed,[60, 61, 62, 63] leading to considerable uncertainties.

**Major uncertainties**

Key remaining uncertainties relate to the sensitivity of observed snow changes to the spatial distribution of observing stations and to historical changes in station location and observing practices. There is conflicting evidence about the effects of arctic amplification on CONUS winter weather.

**Assessment of confidence based on evidence and agreement, including short description of nature of evidence and level of agreement**

There is *high confidence* that warming has resulted in earlier snowmelt and decreased snowfall on the warm margins of areas with consistent snowpack based on a number of observational studies. There is *medium confidence* that Northern Hemisphere storm tracks have shifted north based on a small number of studies. There is *low confidence* in future changes in winter storm frequency and intensity based on conflicting evidence from analysis of climate model simulations.

**Summary sentence or paragraph that integrates the above information**

Decreases in snowfall on southern and low elevation margins of currently climatologically snowy areas are likely but winter storm frequency and intensity changes are uncertain.

**Key Finding 5**

The frequency and severity of landfalling "atmospheric rivers" on the U.S. West Coast (narrow streams of moisture that account for 30%–40% of the typical snowpack and annual precipitation in the region and are associated with severe flooding events) will increase as a result of increasing evaporation and resulting higher atmospheric water vapor that occurs with increasing temperature (*medium confidence*).

**Description of evidence base**

The Key Finding and supporting text summarizes evidence documented in the climate science literature.

Evidence for the expectation of an increase in the frequency and severity of landfalling atmospheric rivers on the U.S. West Coast comes from the CMIP-based climate change projection studies of Dettinger et al. 2011;[73] Warner et al. 2015;[87] Payne and Magnusdottir 2015;[85] Gao et al. 2015;[83] Radić et al. 2015;[86] and Hagos et al. 2016.[84] The close connection between atmospheric rivers and water availability and flooding is based on the present-day observation studies of Guan et al. 2010;[74] Dettinger et al. 2011;[73] Ralph et al. 2006;[77] Neiman et al. 2011;[76] Moore et al. 2012;[75] and Dettinger 2013.[78]

**Major uncertainties**

A modest uncertainty remains in the lack of a supporting detectable anthropogenic signal in the historical data to add further confidence to these projections. However, the overall increase in atmospheric rivers projected/expected is based to a very large degree on the *very high confidence* that the atmospheric water vapor will increase. Thus, increasing water vapor coupled with little projected change in wind structure/intensity still indicates increases in the frequency/intensity of atmospheric rivers. A modest uncertainty arises in quantifying the expected change at a regional level (for example, northern Oregon vs. southern Oregon) given that there are some changes expected in the position of the jet stream that might influence the degree of increase for different locations along the West Coast. Uncertainty in the projections of the number and intensity of ARs is introduced by uncertainties in the models' ability to represent ARs and their interactions with climate.

**Assessment of confidence based on evidence and agreement, including short description of nature of evidence and level of agreement**

Confidence in this finding is rated as *medium* based on qualitatively similar projections among different studies.

**Summary sentence or paragraph that integrates the above information**

Increases in atmospheric river frequency and intensity are expected along the U.S. West Coast, leading to the likelihood of more frequent flooding conditions, with uncertainties remaining in the details of the spatial structure of theses along the coast (for example, northern vs. southern California).

# REFERENCES

1. Klotzbach, P.J. and C.W. Landsea, 2015: Extremely intense hurricanes: Revisiting Webster et al. (2005) after 10 years. *Journal of Climate*, **28**, 7621-7629. http://dx.doi.org/10.1175/JCLI-D-15-0188.1

2. Kossin, J.P., T.L. Olander, and K.R. Knapp, 2013: Trend analysis with a new global record of tropical cyclone intensity. *Journal of Climate*, **26**, 9960-9976. http://dx.doi.org/10.1175/JCLI-D-13-00262.1

3. Walsh, K.J.E., J.L. McBride, P.J. Klotzbach, S. Balachandran, S.J. Camargo, G. Holland, T.R. Knutson, J.P. Kossin, T.-c. Lee, A. Sobel, and M. Sugi, 2016: Tropical cyclones and climate change. *Wiley Interdisciplinary Reviews: Climate Change*, **7**, 65-89. http://dx.doi.org/10.1002/wcc.371

4. Landsea, C., J. Franklin, and J. Beven, 2015: The revised Atlantic hurricane database (HURDAT2). National Hurricane Center, Miami, FL.

5. Hartmann, D.L., A.M.G. Klein Tank, M. Rusticucci, L.V. Alexander, S. Brönnimann, Y. Charabi, F.J. Dentener, E.J. Dlugokencky, D.R. Easterling, A. Kaplan, B.J. Soden, P.W. Thorne, M. Wild, and P.M. Zhai, 2013: Observations: Atmosphere and surface. *Climate Change 2013: The Physical Science Basis. Contribution of Working Group I to the Fifth Assessment Report of the Intergovernmental Panel on Climate Change*. Stocker, T.F., D. Qin, G.-K. Plattner, M. Tignor, S.K. Allen, J. Boschung, A. Nauels, Y. Xia, V. Bex, and P.M. Midgley, Eds. Cambridge University Press, Cambridge, United Kingdom and New York, NY, USA, 159–254. http://www.climatechange2013.org/report/full-report/

6. **Bindoff, N.L., P.A. Stott, K.M. AchutaRao, M.R. Allen, N. Gillett, D. Gutzler, K. Hansingo, G. Hegerl, Y. Hu, S. Jain, I.I. Mokhov, J. Overland, J. Perlwitz, R. Sebbari, and X. Zhang, 2013: Detection and attribution of climate change: From global to regional.** *Climate Change 2013: The Physical Science Basis. Contribution of Working Group I to the Fifth Assessment Report of the Intergovernmental Panel on Climate Change*. Stocker, T.F., D. Qin, G.-K. Plattner, M. Tignor, S.K. Allen, J. Boschung, A. Nauels, Y. Xia, V. Bex, and P.M. Midgley, Eds. Cambridge University Press, Cambridge, United Kingdom and New York, NY, USA, 867–952. http://www.climatechange2013.org/report/full-report/

7. Camargo, S.J., 2013: Global and regional aspects of tropical cyclone activity in the CMIP5 models. *Journal of Climate*, **26**, 9880-9902. http://dx.doi.org/10.1175/jcli-d-12-00549.1

8. Christensen, J.H., K. Krishna Kumar, E. Aldrian, S.-I. An, I.F.A. Cavalcanti, M. de Castro, W. Dong, P. Goswami, A. Hall, J.K. Kanyanga, A. Kitoh, J. Kossin, N.-C. Lau, J. Renwick, D.B. Stephenson, S.-P. Xie, and T. Zhou, 2013: Climate phenomena and their relevance for future regional climate change. *Climate Change 2013: The Physical Science Basis. Contribution of Working Group I to the Fifth Assessment Report of the Intergovernmental Panel on Climate Change*. Stocker, T.F., D. Qin, G.-K. Plattner, M. Tignor, S.K. Allen, J. Boschung, A. Nauels, Y. Xia, V. Bex, and P.M. Midgley, Eds. Cambridge University Press, Cambridge, United Kingdom and New York, NY, USA, 1217–1308. http://www.climatechange2013.org/report/full-report/

9. Knutson, T.R., J.J. Sirutis, M. Zhao, R.E. Tuleya, M. Bender, G.A. Vecchi, G. Villarini, and D. Chavas, 2015: Global projections of intense tropical cyclone activity for the late twenty-first century from dynamical downscaling of CMIP5/RCP4.5 scenarios. *Journal of Climate*, **28**, 7203-7224. http://dx.doi.org/10.1175/JCLI-D-15-0129.1

10. Walsh, K.J.E., S.J. Camargo, G.A. Vecchi, A.S. Daloz, J. Elsner, K. Emanuel, M. Horn, Y.-K. Lim, M. Roberts, C. Patricola, E. Scoccimarro, A.H. Sobel, S. Strazzo, G. Villarini, M. Wehner, M. Zhao, J.P. Kossin, T. LaRow, K. Oouchi, S. Schubert, H. Wang, J. Bacmeister, P. Chang, F. Chauvin, C. Jablonowski, A. Kumar, H. Murakami, T. Ose, K.A. Reed, R. Saravanan, Y. Yamada, C.M. Zarzycki, P.L. Vidale, J.A. Jonas, and N. Henderson, 2015: Hurricanes and climate: The U.S. CLIVAR Working Group on Hurricanes. *Bulletin of the American Meteorological Society*, **96** (12), 997-1017. http://dx.doi.org/10.1175/BAMS-D-13-00242.1

11. **Murakami, H., G.A. Vecchi, T.L. Delworth, K. Paffendorf, L. Jia, R. Gudgel, and F. Zeng, 2015: Investigating the influence of anthropogenic forcing and natural variability on the 2014 Hawaiian hurricane season** [in "Explaining Extreme Events of 2014 from a Climate Perspective"]. *Bulletin of the American Meteorological Society*, **96** (12), S115-S119. http://dx.doi.org/10.1175/BAMS-D-15-00119.1

12. Kossin, J.P., K.A. Emanuel, and G.A. Vecchi, 2014: The poleward migration of the location of tropical cyclone maximum intensity. *Nature*, **509**, 349-352. http://dx.doi.org/10.1038/nature13278

13. Kossin, J.P., K.A. Emanuel, and S.J. Camargo, 2016: **Past and projected changes in western North Pacific** tropical cyclone exposure. *Journal of Climate*, **29**, 5725-5739. http://dx.doi.org/10.1175/JCLI-D-16-0076.1

14. Delworth, L.T. and E.M. Mann, 2000: Observed and simulated multidecadal variability in the Northern Hemisphere. *Climate Dynamics*, **16**, 661-676. http://dx.doi.org/10.1007/s003820000075

15. Evan, A.T., 2012: Atlantic hurricane activity following two major volcanic eruptions. *Journal of Geophysical Research*, **117**, D06101. http://dx.doi.org/10.1029/2011JD016716

16. Thompson, D.W.J. and S. Solomon, 2009: Understanding recent stratospheric climate change. *Journal of Climate*, 22, 1934-1943. http://dx.doi.org/10.1175/2008JCLI2482.1

17. Evan, A.T., G.R. Foltz, D. Zhang, and D.J. Vimont, 2011: Influence of African dust on ocean-atmosphere variability in the tropical Atlantic. *Nature Geoscience*, **4**, 762-765. http://dx.doi.org/10.1038/ngeo1276

18. Evan, A.T., D.J. Vimont, A.K. Heidinger, J.P. Kossin, and R. Bennartz, 2009: The role of aerosols in the evolution of tropical North Atlantic Ocean temperature anomalies. *Science*, **324**, 778-781. http://dx.doi.org/10.1126/science.1167404

19. Booth, B.B.B., N.J. Dunstone, P.R. Halloran, T. Andrews, and N. Bellouin, 2012: Aerosols implicated as a prime driver of twentieth-century North Atlantic climate variability. *Nature*, **484**, 228-232. http://dx.doi.org/10.1038/nature10946

20. Dunstone, N.J., D.M. Smith, B.B.B. Booth, L. Hermanson, and R. Eade, 2013: Anthropogenic aerosol forcing of Atlantic tropical storms. *Nature Geoscience*, **6**, 534-539. http://dx.doi.org/10.1038/ngeo1854

21. Mann, M.E. and K.A. Emanuel, 2006: Atlantic hurricane trends linked to climate change. *Eos, Transactions, American Geophysical Union*, **87**, 233-244. http://dx.doi.org/10.1029/2006EO240001

22. Carslaw, K.S., L.A. Lee, C.L. Reddington, K.J. Pringle, A. Rap, P.M. Forster, G.W. Mann, D.V. Spracklen, M.T. Woodhouse, L.A. Regayre, and J.R. Pierce, 2013: Large contribution of natural aerosols to uncertainty in indirect forcing. *Nature*, **503**, 67-71. http://dx.doi.org/10.1038/nature12674

23. Mann, M.E., B.A. Steinman, and S.K. Miller, 2014: On forced temperature changes, internal variability, and the AMO. *Geophysical Research Letters*, **41**, 3211-3219. http://dx.doi.org/10.1002/2014GL059233

24. Sobel, A.H., S.J. Camargo, T.M. Hall, C.-Y. Lee, M.K. Tippett, and A.A. Wing, 2016: Human influence on tropical cyclone intensity. *Science*, **353**, 242-246. http://dx.doi.org/10.1126/science.aaf6574

25. Stevens, B., 2015: Rethinking the lower bound on aerosol radiative forcing. *Journal of Climate*, **28**, 4794-4819. http://dx.doi.org/10.1175/JCLI-D-14-00656.1

26. Tung, K.-K. and J. Zhou, 2013: Using data to attribute episodes of warming and cooling in instrumental records. *Proceedings of the National Academy of Sciences*, **110**, 2058-2063. http://dx.doi.org/10.1073/pnas.1212471110

27. Zhang, R., T.L. Delworth, R. Sutton, D.L.R. Hodson, K.W. Dixon, I.M. Held, Y. Kushnir, J. Marshall, Y. Ming, R. Msadek, J. Robson, A.J. Rosati, M. Ting, and G.A. Vecchi, 2013: Have aerosols caused the observed Atlantic multidecadal variability? *Journal of the Atmospheric Sciences*, **70**, 1135-1144. http://dx.doi.org/10.1175/jas-d-12-0331.1

28. Kossin, J.P., T.R. Karl, T.R. Knutson, K.A. Emanuel, K.E. Kunkel, and J.J. O'Brien, 2015: Reply to "Comments on 'Monitoring and understanding trends in extreme storms: State of knowledge'". *Bulletin of the American Meteorological Society*, **96** (12), 1177-1179. http://dx.doi.org/10.1175/BAMS-D-14-00261.1

29. Huang, P., I.I. Lin, C. Chou, and R.-H. Huang, 2015: Change in ocean subsurface environment to suppress tropical cyclone intensification under global warming. *Nature Communications*, **6**, 7188. http://dx.doi.org/10.1038/ncomms8188

30. Emanuel, K., 2015: Effect of upper-ocean evolution on projected trends in tropical cyclone activity. *Journal of Climate*, **28**, 8165-8170. http://dx.doi.org/10.1175/JCLI-D-15-0401.1

31. Tuleya, R.E., M. Bender, T.R. Knutson, J.J. Sirutis, B. Thomas, and I. Ginis, 2016: Impact of upper-tropospheric temperature anomalies and vertical wind shear on tropical cyclone evolution using an idealized version of the operational GFDL hurricane model. *Journal of the Atmospheric Sciences*, **73**, 3803-3820. http://dx.doi.org/10.1175/JAS-D-16-0045.1

32. Emanuel, K.A., 2013: Downscaling CMIP5 climate models shows increased tropical cyclone activity over the 21st century. *Proceedings of the National Academy of Sciences*, **110**, 12219-12224. http://dx.doi.org/10.1073/pnas.1301293110

33. Wehner, M., Prabhat, K.A. Reed, D. Stone, W.D. Collins, and J. Bacmeister, 2015: Resolution dependence of future tropical cyclone projections of CAM5.1 in the U.S. CLIVAR Hurricane Working Group idealized configurations. *Journal of Climate*, **28**, 3905-3925. http://dx.doi.org/10.1175/JCLI-D-14-00311.1

34. Hall, T. and K. Hereid, 2015: The frequency and duration of U.S. hurricane droughts. *Geophysical Research Letters*, **42**, 3482-3485. http://dx.doi.org/10.1002/2015GL063652

35. Kossin, J.P., 2017: Hurricane intensification along U.S. coast suppressed during active hurricane periods. *Nature*, **541**, 390-393. http://dx.doi.org/10.1038/nature20783

36. Colbert, A.J. and B.J. Soden, 2012: Climatological variations in North Atlantic tropical cyclone tracks. *Journal of Climate*, **25**, 657-673. http://dx.doi.org/10.1175/jcli-d-11-00034.1

37. Kossin, J.P. and D.J. Vimont, 2007: A more general framework for understanding Atlantic hurricane variability and trends. *Bulletin of the American Meteorological Society*, **88**, 1767-1781. http://dx.doi.org/10.1175/bams-88-11-1767

38. Wang, C., H. Liu, S.-K. Lee, and R. Atlas, 2011: Impact of the Atlantic warm pool on United States landfalling hurricanes. *Geophysical Research Letters*, **38**, L19702. http://dx.doi.org/10.1029/2011gl049265

39. Hart, R.E., D.R. Chavas, and M.P. Guishard, 2016: The arbitrary definition of the current Atlantic major hurricane landfall drought. *Bulletin of the American Meteorological Society*, **97**, 713-722. http://dx.doi.org/10.1175/BAMS-D-15-00185.1

40. Smith, A.B. and R.W. Katz, 2013: U.S. billion-dollar weather and climate disasters: Data sources, trends, accuracy and biases. *Natural Hazards*, **67**, 387-410. http://dx.doi.org/10.1007/s11069-013-0566-5

41. Brooks, H.E., G.W. Carbin, and P.T. Marsh, 2014: Increased variability of tornado occurrence in the United States. *Science*, **346**, 349-352. http://dx.doi.org/10.1126/science.1257460

42. Elsner, J.B., S.C. Elsner, and T.H. Jagger, 2015: The increasing efficiency of tornado days in the United States. *Climate Dynamics*, **45**, 651-659. http://dx.doi.org/10.1007/s00382-014-2277-3

43. Tippett, M.K., 2014: Changing volatility of U.S. annual tornado reports. *Geophysical Research Letters*, **41**, 6956-6961. http://dx.doi.org/10.1002/2014GL061347

44. Allen, J.T. and M.K. Tippett, 2015: The Characteristics of United States Hail Reports: 1955-2014. *Electronic Journal of Severe Storms Meteorology*.

45. Brooks, H.E., J.W. Lee, and J.P. Craven, 2003: The spatial distribution of severe thunderstorm and tornado environments from global reanalysis data. *Atmospheric Research*, **67–68**, 73-94. http://dx.doi.org/10.1016/S0169-8095(03)00045-0

46. Diffenbaugh, N.S., M. Scherer, and R.J. Trapp, 2013: Robust increases in severe thunderstorm environments in response to greenhouse forcing. *Proceedings of the National Academy of Sciences*, **110**, 16361-16366. http://dx.doi.org/10.1073/pnas.1307758110

47. Gensini, V.A., C. Ramseyer, and T.L. Mote, 2014: Future convective environments using NARCCAP. *International Journal of Climatology*, **34**, 1699-1705. http://dx.doi.org/10.1002/joc.3769

48. Seeley, J.T. and D.M. Romps, 2015: The effect of global warming on severe thunderstorms in the United States. *Journal of Climate*, **28**, 2443-2458. http://dx.doi.org/10.1175/JCLI-D-14-00382.1

49. Trapp, R.J., N.S. Diffenbaugh, H.E. Brooks, M.E. Baldwin, E.D. Robinson, and J.S. Pal, 2007: Changes in severe thunderstorm environment frequency during the 21st century caused by anthropogenically enhanced global radiative forcing. *Proceedings of the National Academy of Sciences*, **104**, 19719-19723. http://dx.doi.org/10.1073/pnas.0705494104

50. Trapp, R.J., N.S. Diffenbaugh, and A. Gluhovsky, 2009: Transient response of severe thunderstorm forcing to elevated greenhouse gas concentrations. *Geophysical Research Letters*, **36**, L01703. http://dx.doi.org/10.1029/2008GL036203

51. Van Klooster, S.L. and P.J. Roebber, 2009: Surface-based convective potential in the contiguous United States in a business-as-usual future climate. *Journal of Climate*, **22**, 3317-3330. http://dx.doi.org/10.1175/2009JCLI2697.1

52. Del Genio, A.D., M.S. Yao, and J. Jonas, 2007: Will moist convection be stronger in a warmer climate? *Geophysical Research Letters*, **34**, L16703. http://dx.doi.org/10.1029/2007GL030525

53. Trapp, R.J., E.D. Robinson, M.E. Baldwin, N.S. Diffenbaugh, and B.R.J. Schwedler, 2011: Regional climate of hazardous convective weather through high-resolution dynamical downscaling. *Climate Dynamics*, **37**, 677-688. http://dx.doi.org/10.1007/s00382-010-0826-y

54. Robinson, E.D., R.J. Trapp, and M.E. Baldwin, 2013: The geospatial and temporal distributions of severe thunderstorms from high-resolution dynamical downscaling. *Journal of Applied Meteorology and Climatology*, **52**, 2147-2161. http://dx.doi.org/10.1175/JAMC-D-12-0131.1

55. Gensini, V.A. and T.L. Mote, 2014: Estimations of hazardous convective weather in the United States using dynamical downscaling. *Journal of Climate*, **27**, 6581-6589. http://dx.doi.org/10.1175/JCLI-D-13-00777.1

56. Schär, C., C. Frei, D. Lüthi, and H.C. Davies, 1996: Surrogate climate-change scenarios for regional climate models. *Geophysical Research Letters*, **23**, 669-672. http://dx.doi.org/10.1029/96GL00265

57. Trapp, R.J. and K.A. Hoogewind, 2016: The realization of extreme tornadic storm events under future anthropogenic climate change. *Journal of Climate*, **29**, 5251-5265. http://dx.doi.org/10.1175/JCLI-D-15-0623.1

58. Marinaro, A., S. Hilberg, D. Changnon, and J.R. Angel, 2015: The North Pacific–driven severe Midwest winter of 2013/14. *Journal of Applied Meteorology and Climatology*, **54**, 2141-2151. http://dx.doi.org/10.1175/JAMC-D-15-0084.1

59. Chang, E.K.M., C. Zheng, P. Lanigan, A.M.W. Yau, and J.D. Neelin, 2015: Significant modulation of variability and projected change in California winter precipitation by extratropical cyclone activity. *Geophysical Research Letters*, **42**, 5983-5991. http://dx.doi.org/10.1002/2015GL064424

60. Francis, J. and N. Skific, 2015: Evidence linking rapid Arctic warming to mid-latitude weather patterns. *Philosophical Transactions of the Royal Society A: Mathematical, Physical and Engineering Sciences*, **373**, 20140170. http://dx.doi.org/10.1098/rsta.2014.0170

61. Barnes, E.A. and L.M. Polvani, 2015: CMIP5 projections of Arctic amplification, of the North American/North Atlantic circulation, and of their relationship. *Journal of Climate*, **28**, 5254-5271. http://dx.doi.org/10.1175/JCLI-D-14-00589.1

62. Perlwitz, J., M. Hoerling, and R. Dole, 2015: Arctic tropospheric warming: Causes and linkages to lower latitudes. *Journal of Climate*, **28**, 2154-2167. http://dx.doi.org/10.1175/JCLI-D-14-00095.1

63. Screen, J.A., C. Deser, and L. Sun, 2015: Projected changes in regional climate extremes arising from Arctic sea ice loss. *Environmental Research Letters*, **10**, 084006. http://dx.doi.org/10.1088/1748-9326/10/8/084006

64. Yang, X., G.A. Vecchi, T.L. Delworth, K. Paffendorf, L. Jia, R. Gudgel, F. Zeng, and S.D. Underwood, 2015: Extreme North America winter storm season of 2013/14: Roles of radiative forcing and the global warming hiatus [in "Explaining Extreme Events of 2014 from a Climate Perspective"]. *Bulletin of the American Meteorological Society*, **96** (12), S25-S28. http://dx.doi.org/10.1175/BAMS-D-15-00133.1

65. Delworth, T.L., F. Zeng, A. Rosati, G.A. Vecchi, and A.T. Wittenberg, 2015: A link between the hiatus in global warming and North American drought. *Journal of Climate*, **28**, 3834-3845. http://dx.doi.org/10.1175/jcli-d-14-00616.1

66. Vose, R.S., S. Applequist, M.A. Bourassa, S.C. Pryor, R.J. Barthelmie, B. Blanton, P.D. Bromirski, H.E. Brooks, A.T. DeGaetano, R.M. Dole, D.R. Easterling, R.E. Jensen, T.R. Karl, R.W. Katz, K. Klink, M.C. Kruk, K.E. Kunkel, M.C. MacCracken, T.C. Peterson, K. Shein, B.R. Thomas, J.E. Walsh, X.L. Wang, M.F. Wehner, D.J. Wuebbles, and R.S. Young, 2014: Monitoring and understanding changes in extremes: Extratropical storms, winds, and waves. *Bulletin of the American Meteorological Society*, **95**, 377-386. http://dx.doi.org/10.1175/BAMS-D-12-00162.1

67. Wang, X.L., Y. Feng, G.P. Compo, V.R. Swail, F.W. Zwiers, R.J. Allan, and P.D. Sardeshmukh, 2012: Trends and low frequency variability of extra-tropical cyclone activity in the ensemble of twentieth century reanalysis. *Climate Dynamics*, **40**, 2775-2800. http://dx.doi.org/10.1007/s00382-012-1450-9

68. Wang, X.L., V.R. Swail, and F.W. Zwiers, 2006: Climatology and changes of extratropical cyclone activity: Comparison of ERA-40 with NCEP-NCAR reanalysis for 1958-2001. *Journal of Climate*, **19**, 3145-3166. http://dx.doi.org/10.1175/JCLI3781.1

69. Colle, B.A., Z. Zhang, K.A. Lombardo, E. Chang, P. Liu, and M. Zhang, 2013: Historical evaluation and future prediction of eastern North American and western Atlantic extratropical cyclones in the CMIP5 models during the cool season. *Journal of Climate*, **26**, 6882-6903. http://dx.doi.org/10.1175/JCLI-D-12-00498.1

70. Zhu, Y. and R.E. Newell, 1998: A proposed algorithm for moisture fluxes from atmospheric rivers. *Monthly Weather Review*, **126**, 725-735. http://dx.doi.org/10.1175/1520-0493(1998)126<0725:APAFMF>2.0.CO;2

71. Newell, R.E., N.E. Newell, Y. Zhu, and C. Scott, 1992: Tropospheric rivers? – A pilot study. *Geophysical Research Letters*, **19**, 2401-2404. http://dx.doi.org/10.1029/92GL02916

72. Guan, B. and D.E. Waliser, 2015: Detection of atmospheric rivers: Evaluation and application of an algorithm for global studies. *Journal of Geophysical Research Atmospheres*, **120**, 12514-12535. http://dx.doi.org/10.1002/2015JD024257

73. Dettinger, M.D., F.M. Ralph, T. Das, P.J. Neiman, and D.R. Cayan, 2011: Atmospheric rivers, floods and the water resources of California. *Water*, **3**, 445-478. http://dx.doi.org/10.3390/w3020445

74. Guan, B., N.P. Molotch, D.E. Waliser, E.J. Fetzer, and P.J. Neiman, 2010: Extreme snowfall events linked to atmospheric rivers and surface air temperature via satellite measurements. *Geophysical Research Letters*, **37**, L20401. http://dx.doi.org/10.1029/2010GL044696

75. Moore, B.J., P.J. Neiman, F.M. Ralph, and F.E. Barthold, 2012: Physical processes associated with heavy flooding rainfall in Nashville, Tennessee, and vicinity during 1–2 May 2010: The role of an atmospheric river and mesoscale convective systems. *Monthly Weather Review*, **140**, 358-378. http://dx.doi.org/10.1175/MWR-D-11-00126.1

76. Neiman, P.J., L.J. Schick, F.M. Ralph, M. Hughes, and G.A. Wick, 2011: Flooding in western Washington: The connection to atmospheric rivers. *Journal of Hydrometeorology*, **12**, 1337-1358. http://dx.doi.org/10.1175/2011JHM1358.1

77. Ralph, F.M., P.J. Neiman, G.A. Wick, S.I. Gutman, M.D. Dettinger, D.R. Cayan, and A.B. White, 2006: Flooding on California's Russian River: Role of atmospheric rivers. *Geophysical Research Letters*, **33**, L13801. http://dx.doi.org/10.1029/2006GL026689

78. Dettinger, M.D., 2013: Atmospheric rivers as drought busters on the U.S. West Coast. *Journal of Hydrometeorology*, **14**, 1721-1732. http://dx.doi.org/10.1175/JHM-D-13-02.1

79. Rutz, J.J., W.J. Steenburgh, and F.M. Ralph, 2014: Climatological characteristics of atmospheric rivers and their inland penetration over the western United States. *Monthly Weather Review*, **142**, 905-921. http://dx.doi.org/10.1175/MWR-D-13-00168.1

80. Held, I.M. and B.J. Soden, 2006: Robust responses of the hydrological cycle to global warming. *Journal of Climate*, **19**, 5686-5699. http://dx.doi.org/10.1175/jcli3990.1

81. Lavers, D.A., F.M. Ralph, D.E. Waliser, A. Gershunov, and M.D. Dettinger, 2015: Climate change intensification of horizontal water vapor transport in CMIP5. *Geophysical Research Letters*, **42**, 5617-5625. http://dx.doi.org/10.1002/2015GL064672

82. Dettinger, M.D. and B.L. Ingram, 2013: The coming megafloods. *Scientific American*, 308, 64-71. http://dx.doi.org/10.1038/scientificamerican0113-64

83. Gao, Y., J. Lu, L.R. Leung, Q. Yang, S. Hagos, and Y. Qian, 2015: Dynamical and thermodynamical modulations on future changes of landfalling atmospheric rivers over western North America. *Geophysical Research Letters*, **42**, 7179-7186. http://dx.doi.org/10.1002/2015GL065435

84. Hagos, S.M., L.R. Leung, J.-H. Yoon, J. Lu, and Y. Gao, 2016: A projection of changes in landfalling atmospheric river frequency and extreme precipitation over western North America from the Large Ensemble CESM simulations. *Geophysical Research Letters*, **43**, 1357-1363. http://dx.doi.org/10.1002/2015GL067392

85. Payne, A.E. and G. Magnusdottir, 2015: An evaluation of atmospheric rivers over the North Pacific in CMIP5 and their response to warming under RCP 8.5. *Journal of Geophysical Research Atmospheres*, **120**, 11,173-11,190. http://dx.doi.org/10.1002/2015JD023586

86. Radić, V., A.J. Cannon, B. Menounos, and N. Gi, 2015: Future changes in autumn atmospheric river events in British Columbia, Canada, as projected by CMIP5 global climate models. *Journal of Geophysical Research Atmospheres*, **120**, 9279-9302. http://dx.doi.org/10.1002/2015JD023279

87. Warner, M.D., C.F. Mass, and E.P. Salathé Jr., 2015: Changes in winter atmospheric rivers along the North American West Coast in CMIP5 climate models. *Journal of Hydrometeorology*, **16**, 118-128. http://dx.doi.org/10.1175/JHM-D-14-0080.1

88. Lavers, D.A., R.P. Allan, G. Villarini, B. Lloyd-Hughes, D.J. Brayshaw, and A.J. Wade, 2013: Future changes in atmospheric rivers and their implications for winter flooding in Britain. *Environmental Research Letters*, **8**, 034010. http://dx.doi.org/10.1088/1748-9326/8/3/034010

89. Pierce, D.W., D.R. Cayan, T. Das, E.P. Maurer, N.L. Miller, Y. Bao, M. Kanamitsu, K. Yoshimura, M.A. Snyder, L.C. Sloan, G. Franco, and M. Tyree, 2013: The key role of heavy precipitation events in climate model disagreements of future annual precipitation changes in California. *Journal of Climate*, **26**, 5879-5896. http://dx.doi.org/10.1175/jcli-d-12-00766.1

90. Melillo, J.M., T.C. Richmond, and G.W. Yohe, eds., 2014: *Climate Change Impacts in the United States: The Third National Climate Assessment*. U.S. Global Change Research Program: Washington, D.C., 841 pp. http://dx.doi.org/10.7930/J0Z31WJ2

91. IPCC, 2013: *Climate Change 2013: The Physical Science Basis. Contribution of Working Group I to the Fifth Assessment Report of the Intergovernmental Panel on Climate Change*. Cambridge University Press, Cambridge, UK and New York, NY, 1535 pp. http://www.climatechange2013.org/report/

92. Brooks, H.E., 2013: Severe thunderstorms and climate change. *Atmospheric Research*, **123**, 129-138. http://dx.doi.org/10.1016/j.atmosres.2012.04.002

93. Ralph, F.M. and M.D. Dettinger, 2012: Historical and national perspectives on extreme West Coast precipitation associated with atmospheric rivers during December 2010. *Bulletin of the American Meteorological Society*, **93**, 783-790. http://dx.doi.org/10.1175/BAMS-D-11-00188.1

94. Ralph, F.M., P.J. Neiman, G.N. Kiladis, K. Weickmann, and D.W. Reynolds, 2011: A multiscale observational case study of a Pacific atmospheric river exhibiting tropical–extratropical connections and a mesoscale frontal wave. *Monthly Weather Review*, **139**, 1169-1189. http://dx.doi.org/10.1175/2010mwr3596.1

# 10

# Changes in Land Cover and Terrestrial Biogeochemistry

**KEY FINDINGS**

1. Changes in land use and land cover due to human activities produce physical changes in land surface albedo, latent and sensible heat, and atmospheric aerosol and greenhouse gas concentrations. The combined effects of these changes have recently been estimated to account for 40% ± 16% of the human-caused global radiative forcing from 1850 to present day (*high confidence*). In recent decades, land use and land cover changes have turned the terrestrial biosphere (soil and plants) into a net "sink" for carbon (drawing down carbon from the atmosphere), and this sink has steadily increased since 1980 (*high confidence*). Because of the uncertainty in the trajectory of land cover, the possibility of the land becoming a net carbon source cannot be excluded (*very high confidence*).

2. Climate change and induced changes in the frequency and magnitude of extreme events (e.g., droughts, floods, and heat waves) have led to large changes in plant community structure with subsequent effects on the biogeochemistry of terrestrial ecosystems. Uncertainties about how climate change will affect land cover change make it difficult to project the magnitude and sign of future climate feedbacks from land cover changes (*high confidence*).

3. Since 1901, regional averages of both the consecutive number of frost-free days and the length of the corresponding growing season have increased for the seven contiguous U.S. regions used in this assessment. However, there is important variability at smaller scales, with some locations actually showing decreases of a few days to as much as one to two weeks. Plant productivity has not increased commensurate with the increased number of frost-free days or with the longer growing season due to plant-specific temperature thresholds, plant–pollinator dependence, and seasonal limitations in water and nutrient availability (*very high confidence*). Future consequences of changes to the growing season for plant productivity are uncertain.

4. Recent studies confirm and quantify that surface temperatures are higher in urban areas than in surrounding rural areas for a number of reasons, including the concentrated release of heat from buildings, vehicles, and industry. In the United States, this urban heat island effect results in daytime temperatures 0.9°–7.2°F (0.5°–4.0°C) higher and nighttime temperatures 1.8°– 4.5°F (1.0°–2.5°C) higher in urban areas, with larger temperature differences in humid regions (primarily in the eastern United States) and in cities with larger and denser populations. The urban heat island effect will strengthen in the future as the structure, spatial extent, and population density of urban areas change and grow (*high confidence*).

**Recommended Citation for Chapter**

**Hibbard**, K.A., F.M. Hoffman, D. Huntzinger, and T.O. West, 2017: Changes in land cover and terrestrial biogeochemistry. In: *Climate Science Special Report: Fourth National Climate Assessment, Volume I* [Wuebbles, D.J., D.W. Fahey, K.A. Hibbard, D.J. Dokken, B.C. Stewart, and T.K. Maycock (eds.)]. U.S. Global Change Research Program, Washington, DC, USA, pp. 277-302, doi: 10.7930/J0416V6X.

## 10.1 Introduction

Direct changes in land use by humans are contributing to radiative forcing by altering land cover and therefore albedo, contributing to climate change (Ch. 2: Physical Drivers of Climate Change). This forcing is spatially variable in both magnitude and sign; globally averaged, it is negative (climate cooling; Figure 2.3). Climate changes, in turn, are altering the biogeochemistry of land ecosystems through extended growing seasons, increased numbers of frost-free days, altered productivity in agricultural and forested systems, longer fire seasons, and urban-induced thunderstorms.[1,2] Changes in land use and land cover interact with local, regional, and global climate processes.[3] The resulting ecosystem responses alter Earth's albedo, the carbon cycle, and atmospheric aerosols, constituting a mix of positive and negative feedbacks to climate change (Figure 10.1 and Chapter 2, Section 2.6.2).[4,5] Thus, changes to terrestrial ecosystems or land cover are a direct driver of climate change and they are further altered by climate change in ways that affect both ecosystem productivity and, through feedbacks, the climate itself. The following sections describe advances since the Third National Climate Assessment (NCA3)[6] in scientific understanding of land cover and associated biogeochemistry and their impacts on the climate system.

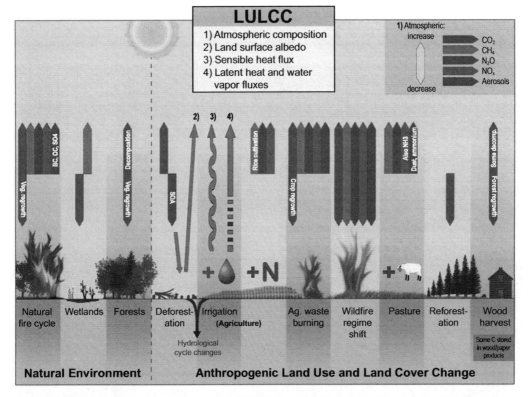

**Figure 10.1:** This graphical representation summarizes land–atmosphere interactions from natural and anthropogenic land-use and land-cover change (LULCC) contributions to radiative forcing. Emissions and sequestration of carbon and fluxes of nitrogen oxides, aerosols, and water shown here were used to calculate net radiative forcing from LULCC. (Figure source: Ward et al. 2014[5]).

## 10.2 Terrestrial Ecosystem Interactions with the Climate System

Other chapters of this report discuss changes in temperature (Ch. 6: Temperature Change), precipitation (Ch. 7: Precipitation Change), hydrology (Ch. 8: Droughts, Floods, and Wildfires), and extreme events (Ch. 9: Extreme Storms). Collectively, these processes affect the phenology, structure, productivity, and biogeochemical processes of all terrestrial ecosystems, and as such, climate change will alter land cover and ecosystem services.

### 10.2.1 Land Cover and Climate Forcing

Changes in land cover and land use have long been recognized as important contributors to global climate forcing (e.g., Feddema et al. 2005[7]). Historically, studies that account for the contribution of the land cover to radiative forcing have accounted for albedo forcings only and not those from changes in land surface geophysical properties (e.g., plant transpiration, evaporation from soils, plant community structure and function) or in aerosols. Physical climate effects from land-cover or land-use change do not lend themselves directly to quantification using the traditional radiative forcing concept. However, a framework to attribute the indirect contributions of land cover to radiative forcing and the climate system—including effects on seasonal and interannual soil moisture and latent/sensible heat, evapotranspiration, biogeochemical cycle ($CO_2$) fluxes from soils and plants, aerosol and aerosol precursor emissions, ozone precursor emissions, and snowpack—was reported in NRC.[8] Predicting future consequences of changes in land cover on the climate system will require not only the traditional calculations of surface albedo but also surface net radiation partitioning between latent and sensible heat exchange and the effects of resulting changes in biogeochemical trace gas and aerosol fluxes. Future trajectories of land use and land cover change are uncertain and

will depend on population growth, changes in agricultural yield driven by the competing demands for production of fuel (i.e., bioenergy crops), food, feed, and fiber as well as urban expansion. The diversity of future land cover and land use changes as implemented by the models that developed the Representative Concentration Pathways (RCPs) to attain target goals of radiative forcing by 2100 is discussed by Hurtt et al.[9] For example, the higher scenario (RCP8.5)[10] features an increase of cultivated land by about 185 million hectares from 2000 to 2050 and another 120 million hectares from 2050 to 2100. In the mid-high scenario (RCP6.0)—the Asia Pacific Integrated Model (AIM),[11] urban land use increases due to population and economic growth while cropland area expands due to increasing food demand. Grassland areas decline while total forested area extent remains constant throughout the century.[9] The Global Change Assessment Model (GCAM), under a lower scenario (RCP4.5), preserved and expanded forested areas throughout the 21st century. Agricultural land declined slightly due to this afforestation, yet food demand is met through crop yield improvements, dietary shifts, production efficiency, and international trade.[9, 12] As with the higher scenario (RCP8.5), the even lower scenario (RCP2.6)[13] reallocated agricultural production from developed to developing countries, with increased bioenergy production.[9] Continued land-use change is projected across all RCPs (2.6, 4.5, 6.0, and 8.5) and is expected to contribute between 0.9 and 1.9 W/m[2] to direct radiative forcing by 2100.[5] The RCPs demonstrate that land-use management and change combined with policy, demographic, energy technological innovations and change, and lifestyle changes all contribute to future climate (see Ch. 4: Projections for more detail on RCPs).[14]

Traditional calculations of radiative forcing by land-cover change yield small forcing values

(Ch. 2: Physical Drivers of Climate Change) because they account only for changes in surface albedo (e.g., Myhre and Myhre 2003;[15] Betts et al. 2007;[16] Jones et al. 2015[17]). Recent assessments (Myhre et al. 2013[4] and references therein) are beginning to calculate the relative contributions of land-use and land-cover change (LULCC) to radiative forcing in addition to albedo and/or aerosols.[5] Radiative forcing data reported in this chapter are largely from observations (see Table 8.2 in Myhre et al. 2013[4]). Ward et al.[5] performed an independent modeling study to partition radiative forcing from natural and anthropogenic land use and land cover change and related land management activities into contributions from carbon dioxide ($CO_2$), methane ($CH_4$), nitrous oxide ($N_2O$), aerosols, halocarbons, and ozone ($O_3$).

The more extended effects of land–atmosphere interactions from natural and anthropogenic land-use and land-cover change (LULCC; Figure 10.1) described above have recently been reviewed and estimated by atmospheric constituent (Figure 10.2).[4, 5] The combined albedo and greenhouse gas radiative forcing for land-cover change is estimated to account for 40% ± 16% of the human-caused global radiative forcing from 1850 to 2010 (Figure 10.2).[5] These calculations for total radiative forcing (from LULCC sources and all other sources) are consistent with Myhre et al. 2013[4] (2.23 W/m² and 2.22 W/m² for Ward et al. 2014[5] and Myhre et al. 2013[4], respectively). The contributions of $CO_2$, $CH_4$, $N_2O$, and aerosols/$O_3$/albedo effects to total LULCC radiative forcing are about 47%, 34%, 15%, and 4%, respectively, highlighting the importance of non-albedo contributions to LULCC and radiative forcing. The net radiative forcing due specifically to fire—after accounting for short-lived forcing agents ($O_3$ and aerosols), long-lived greenhouse gases, and land albedo change both now and in the future—is estimated to be near

zero due to regrowth of forests which offsets the release of $CO_2$ from fire.[18]

### 10.2.2 Land Cover and Climate Feedbacks

Earth system models differ significantly in projections of terrestrial carbon uptake,[19] with large uncertainties in the effects of increasing atmospheric $CO_2$ concentrations (i.e., $CO_2$ fertilization) and nutrient downregulation on plant productivity, as well as the strength of carbon cycle feedbacks (Ch. 2: Physical Drivers of Climate Change).[20, 21] When $CO_2$ effects on photosynthesis and transpiration are removed from global gridded crop models, simulated response to climate across the models is comparable, suggesting that model parameterizations representing these processes remain uncertain.[22]

A recent analysis shows large-scale greening in the Arctic and boreal regions of North America and browning in the boreal forests of eastern Alaska for the period 1984–2012.[23] Satellite observations and ecosystem models suggest that biogeochemical interactions of carbon dioxide ($CO_2$) fertilization, nitrogen (N) deposition, and land-cover change are responsible for 25%–50% of the global greening of the Earth and 4% of Earth's browning between 1982 and 2009.[24, 25] While several studies have documented significant increases in the rate of green-up periods, the lengthening of the growing season (Section 10.3.1) also alters the timing of green-up (onset of growth) and brown-down (senescence); however, where ecosystems become depleted of water resources as a result of a lengthening growing season, the actual period of productive growth can be truncated.[26]

Large-scale die-off and disturbances resulting from climate change have potential effects beyond the biogeochemical and carbon cycle effects. Biogeophysical feedbacks can strengthen or reduce climate forcing. The low albedo

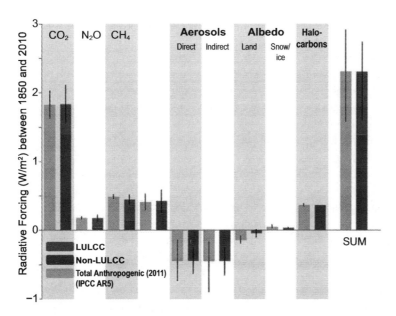

**Figure 10.2:** Anthropogenic radiative forcing (RF) contributions, separated by land-use and land-cover change (LUL-CC) and non-LULCC sources (green and maroon bars, respectively), are decomposed by atmospheric constituent to year 2010 in this diagram, using the year 1850 as the reference. Total anthropogenic RF contributions by atmospheric constituent[4] (see also Figure 2.3) are shown for comparison (yellow bars). Error bars represent uncertainties for total anthropogenic RF (yellow bars) and for the LULCC components (green bars).[5] The SUM bars indicate the net RF when all anthropogenic forcing agents are combined. (Figure source: Ward et al. 2014[5]).

of boreal forests provides a positive feedback, but those albedo effects are mitigated in tropical forests through evaporative cooling; for temperate forests, the evaporative effects are less clear.[27] Changes in surface albedo, evaporation, and surface roughness can have feedbacks to local temperatures that are larger than the feedback due to the change in carbon sequestration.[28] Forest management frameworks (e.g., afforestation, deforestation, and avoided deforestation) that account for biophysical (e.g., land surface albedo and surface roughness) properties can be used as climate protection or mitigation strategies.[29]

### 10.2.3 Temperature Change

Interactions between temperature changes, land cover, and biogeochemistry are more complex than commonly assumed. Previous research suggested a fairly direct relationship between increasing temperatures, longer growing seasons (see Section 10.3.1),

increasing plant productivity (e.g., Walsh et al. 2014[30]), and therefore also an increase in $CO_2$ uptake. Without water or nutrient limitations, increased $CO_2$ concentrations and warm temperatures have been shown to extend the growing season, which may contribute to longer periods of plant activity and carbon uptake, but do not affect reproduction rates.[31] However, a longer growing season can also increase plant water demand, affecting regional water availability, and result in conditions that exceed plant physiological thresholds for growth, producing subsequent feedbacks to radiative forcing and climate. These consequences could offset potential benefits of a longer growing season (e.g., Georgakakos et al. 2014[32]; Hibbard et al. 2014[33]). For instance, increased dry conditions can lead to wildfire (e.g., Hatfield et al. 2014;[34] Joyce et al. 2014;[35] Ch. 8: Droughts, Floods and Wildfires) and urban temperatures can contribute to urban-induced thunderstorms in the southeastern United States.[36] Temperature benefits of early onset of plant development in a longer growing season can be offset by 1) freeze damage caused by late-season frosts; 2) limits to growth because of shortening of the photoperiod later in the season; or 3) by shorter chilling periods required for leaf unfolding by many plants.[37, 38] MODIS data provided insight into the coterminous U.S. 2012 drought, when a warm spring reduced the carbon cycle impact of the drought by inducing earlier carbon uptake.[39] New evidence points to longer temperature-driven growing seasons for grasslands that may facilitate earlier onset of growth, but also that senescence is typically earlier.[40] In addition to changing $CO_2$ uptake, higher temperatures can also enhance soil decomposition rates, thereby adding more $CO_2$ to the atmosphere. Similarly, temperature, as well as changes in the seasonality and intensity of precipitation, can influence nutrient and water availability, leading to both shortages and excesses, thereby influencing rates and magnitudes of decomposition.[1]

### 10.2.4 Water Cycle Changes

The global hydrological cycle is expected to intensify under climate change as a consequence of increased temperatures in the troposphere. The consequences of the increased water-holding capacity of a warmer atmosphere include longer and more frequent droughts and less frequent but more severe precipitation events and cyclonic activity (see Ch. 9: Extreme Storms for an in-depth discussion of extreme storms). More intense rain events and storms can lead to flooding and ecosystem disturbances, thereby altering ecosystem function and carbon cycle dynamics. For an extensive review of precipitation changes and droughts, floods, and wildfires, see Chapters 7 and 8 in this report, respectively.

From the perspective of the land biosphere, drought has strong effects on ecosystem

productivity and carbon storage by reducing photosynthesis and increasing the risk of wildfire, pest infestation, and disease susceptibility. Thus, droughts of the future will affect carbon uptake and storage, leading to feedbacks to the climate system (Chapter 2, Section 2.6.2; also see Chapter 11 for Arctic/climate/wildfire feedbacks).[41] Reduced productivity as a result of extreme drought events can also extend for several years post-drought (i.e., drought legacy effects).[42, 43, 44] In 2011, the most severe drought on record in Texas led to statewide regional tree mortality of 6.2%, or nearly nine times greater than the average annual mortality in this region (approximately 0.7%).[45] The net effect on carbon storage was estimated to be a redistribution of 24–30 TgC from the live to dead tree carbon pool, which is equal to 6%–7% of pre-drought live tree carbon storage in Texas state forestlands.[45] Another way to think about this redistribution is that the single Texas drought event equals approximately 36% of annual global carbon losses due to deforestation and land-use change.[46] The projected increases in temperatures and in the magnitude and frequency of heavy precipitation events, changes to snowpack, and changes in the subsequent water availability for agriculture and forestry may lead to similar rates of mortality or changes in land cover. Increasing frequency and intensity of drought across northern ecosystems reduces total observed organic matter export, has led to oxidized wetland soils, and releases stored contaminants into streams after rain events.[47]

### 10.2.5 Biogeochemistry

Terrestrial biogeochemical cycles play a key role in Earth's climate system, including by affecting land–atmosphere fluxes of many aerosol precursors and greenhouse gases, including carbon dioxide ($CO_2$), methane ($CH_4$), and nitrous oxide ($N_2O$). As such, changes in the terrestrial ecosphere can drive climate change. At the same time, biogeochemical cycles are sensitive to changes in climate and atmospheric composition.

Increased atmospheric $CO_2$ concentrations are often assumed to lead to increased plant production (known as $CO_2$ fertilization) and longer-term storage of carbon in biomass and soils. Whether increased atmospheric $CO_2$ will continue to lead to long-term storage of carbon in terrestrial ecosystems depends on whether $CO_2$ fertilization simply intensifies the rate of short-term carbon cycling (for example, by stimulating respiration, root exudation, and high turnover root growth), how water and other nutrients constrain $CO_2$ fertilization, or whether the additional carbon is used by plants to build more wood or tissues that, once senesced, decompose into long-lived soil organic matter. Under increased $CO_2$ concentrations, plants have been observed to optimize water use due to reduced stomatal conductance, thereby increasing water-use efficiency.[48] This change in water-use efficiency can affect plants' tolerance to stress and specifically to drought.[49] Due to the complex interactions of the processes that govern terrestrial biogeochemical cycling, terrestrial ecosystem responses to increasing $CO_2$ levels remain one of the largest uncertainties in long-term climate feedbacks and therefore in predicting longer-term climate change (Ch. 2: Physical Drivers of Climate Change).

Nitrogen is a principal nutrient for plant growth and can limit or stimulate plant productivity (and carbon uptake), depending on availability. As a result, increased nitrogen deposition and natural nitrogen-cycle responses to climate change will influence the global carbon cycle. For example, nitrogen limitation can inhibit the $CO_2$ fertilization response of plants to elevated atmospheric $CO_2$ (e.g., Norby et al. 2005;[50] Zaehle et al. 2010[51]). Conversely, increased decomposition of soil organic matter in response to climate warm-ing increases nitrogen mineralization. This shift of nitrogen from soil to vegetation can increase ecosystem carbon storage.[46, 52] While the effects of increased nitrogen deposition may counteract some nitrogen limitation on $CO_2$ fertilization, the importance of nitrogen in future carbon–climate interactions is not clear. Nitrogen dynamics are being integrated into the simulation of land carbon cycle modeling, but only two of the models in CMIP5 included coupled carbon–nitrogen interactions.[53]

Many factors, including climate, atmospheric $CO_2$ concentrations, and nitrogen deposition rates influence the structure of the plant community and therefore the amount and biochemical quality of inputs into soils.[54, 55, 56] For example, though $CO_2$ losses from soils may decrease with greater nitrogen deposition, increased emissions of other greenhouse gases, such as methane ($CH_4$) and nitrous oxide ($N_2O$), can offset the reduction in $CO_2$.[57] The dynamics of soil organic carbon under the influence of climate change is poorly understood and therefore not well represented in models. As a result, there is high uncertainty in soil carbon stocks in model simulations.[58, 59]

Future emissions of many aerosol precursors are expected to be affected by a number of climate-related factors, in part because of changes in aerosol and aerosol precursors from the terrestrial biosphere. For example, volatile organic compounds (VOCs) are a significant source of secondary organic aerosols, and biogenic sources of VOCs exceed emissions from the industrial and transportation sectors.[60] Isoprene is one of the most important biogenic VOCs, and isoprene emissions are strongly dependent on temperature and light, as well as other factors like plant type and leaf age.[60] Higher temperatures are expected to lead to an increase in biogenic VOC emissions. Atmospheric $CO_2$ concentration can also affect isoprene emissions (e.g., Rosenstiel et al. 2003[61]).

Changes in biogenic VOC emissions can impact aerosol formation and feedbacks with climate (Ch. 2: Physical Drivers of Climate Change, Section 2.6.1; Feedbacks via changes in atmospheric composition). Increased biogenic VOC emissions can also impact ozone and the atmospheric oxidizing capacity.[62] Conversely, increases in nitrogen oxide ($NO_x$) pollution produce tropospheric ozone ($O_3$), which has damaging effects on vegetation. For example, a recent study estimated yield losses for maize and soybean production of up to 5% to 10% due to increases in $O_3$.[63]

### 10.2.6 Extreme Events and Disturbance

This section builds on the physical overview provided in earlier chapters to frame how the intersections of climate, extreme events, and disturbance affect regional land cover and biogeochemistry. In addition to overall trends in temperature (Ch. 6: Temperature Change) and precipitation (Ch. 7: Precipitation Change), changes in modes of variability such as the Pacific Decadal Oscillation (PDO) and the El Niño–Southern Oscillation (ENSO) (Ch. 5: Circulation and Variability) can contribute to drought in the United States, which leads to unanticipated changes in disturbance regimes in the terrestrial biosphere (e.g., Kam et al. 2014[64]). Extreme climatic events can increase the susceptibility of ecosystems to invasive plants and plant pests by promoting transport of propagules into affected regions, decreasing the resistance of native communities to establishment, and by putting existing native species at a competitive disadvantage.[65] For example, drought may exacerbate the rate of plant invasions by non-native species in rangelands and grasslands.[45] Land-cover changes such as encroachment and invasion of non-native species can in turn lead to increased frequency of disturbance such as fire. Disturbance events alter soil moisture, which, in addition to being affected by evapotranspiration and precipitation (Ch. 8: Droughts, Floods, and Wildfires),

is controlled by canopy and rooting architecture as well as soil physics. Invasive plants may be directly responsible for changes in fire regimes through increased biomass, changes in the distribution of flammable biomass, increased flammability, and altered timing of fuel drying, while others may be "fire followers" whose abundances increase as a result of shortening the fire return interval (e.g., Lambert et al. 2010[66]). Changes in land cover resulting from alteration of fire return intervals, fire severity, and historical disturbance regimes affect long-term carbon exchange between the atmosphere and biosphere (e.g., Moore et al. 2016[45]). Recent extensive diebacks and changes in plant cover due to drought have interacted with regional carbon cycle dynamics, including carbon release from biomass and reductions in carbon uptake from the atmosphere; however, plant regrowth may offset emissions.[67] The 2011–2015 meteorological drought in California (described in Ch. 8: Droughts, Floods, and Wildfires), combined with future warming, will lead to long-term changes in land cover, leading to increased probability of climate feedbacks (e.g., drought and wildfire) and in ecosystem shifts.[68] California's recent drought has also resulted in measureable canopy water losses, posing long-term hazards to forest health and biophysical feedbacks to regional climate.[44, 69, 70] Multiyear or severe meteorological and hydrological droughts (see Ch. 8: Droughts, Floods, and Wildfires for definitions) can also affect stream biogeochemistry and riparian ecosystems by concentrating sediments and nutrients.[67]

Changes in the variability of hurricanes and winter storm events (Ch. 9: Extreme Storms) also affect the terrestrial biosphere, as shown in studies comparing historic and future (projected) extreme events in the western United States and how these translate into changes in regional water balance, fire, and streamflow.

Composited across 10 global climate models (GCMs), summer (June–August) water-balance deficit in the future (2030–2059) increases compared to that under historical (1916–2006) conditions. Portions of the Southwest that have significant monsoon precipitation and some mountainous areas of the Pacific Northwest are exempt from this deficit.[71] Projections for 2030–2059 suggest that extremely low flows that have historically occurred (1916–2006) in the Columbia Basin, upper Snake River, southeastern California, and southwestern Oregon are less likely to occur. Given the historical relationships between fire occurrence and drought indicators such as water-balance deficit and streamflow, climate change can be expected to have significant effects on fire occurrence and area burned.[71, 72, 73]

Climate change in the northern high latitudes is directly contributing to increased fire occurrence (Ch. 11: Arctic Changes); in the coterminous United States, climate-induced changes in fires, changes in direct human ignitions, and land-management practices all significantly contribute to wildfire trends. Wildfires in the western United States are often ignited by lightning, but management practices such as fire suppression contribute to fuels and amplify the intensity and spread of wildfire. Fires initiated from unintentional ignition, such as by campfires, or intentional human-caused ignitions are also intensified by increasingly dry and vulnerable fuels, which build up with fire suppression or human settlements (See also Ch. 8: Droughts, Floods, and Wildfires).

## 10.3 Climate Indicators and Agricultural and Forest Responses

Recent studies indicate a correlation between the expansion of agriculture and the global amplitude of $CO_2$ uptake and emissions.[74, 75] Conversely, agricultural production is increasingly disrupted by climate and extreme weather events, and these effects are expected to be augmented by mid-century and beyond for most crops.[76, 77] Precipitation extremes put pressure on agricultural soil and water assets and lead to increased irrigation, shrinking aquifers, and ground subsidence.

### 10.3.1 Changes in the Frost-Free and Growing Seasons

The concept that longer growing seasons are increasing productivity in some agricultural and forested ecosystems was discussed in the Third National Climate Assessment (NCA3).[6] However, there are other consequences to a lengthened growing season that can offset gains in productivity. Here we discuss these emerging complexities as well as other aspects of how climate change is altering and interacting with terrestrial ecosystems. The growing season is the part of the year in which temperatures are favorable for plant growth. A basic metric by which this is measured is the frost-free period. The U.S. Department of Agriculture Natural Resources Conservation Service defines the frost-free period using a range of thresholds. They calculate the average date of the last day with temperature below 24°F (−4.4°C), 28°F (−2.2°C), and 32°F (0°C) in the spring and the average date of the first day with temperature below 24°F, 28°F, and 32°F in the fall, at various probabilities. They then define the frost-free period at three index temperatures (32°F, 28°F, and 24°F), also with a range of probabilities. A single temperature threshold (for example, temperature below 32°F) is often used when discussing growing season; however, different plant cover-types (e.g., forest, agricultural, shrub, and tundra) have different temperature thresholds for growth, and different requirements/thresholds for chilling.[34, 78] For the purposes of this report, we use the metric with a 32°F (0°C) threshold to define the change in the number of "frost-free" days, and a temperature threshold of 41°F (5°C) as a first-order measure of how the growing season length has changed over the observational record.[78]

The NCA3 reported an increase in the growing season length of as much as several weeks as a result of higher temperatures occurring earlier and later in the year (e.g., Walsh et al. 2014;[30] Hatfield et al. 2014;[34] Joyce et al. 2014[35]). NCA3 used a threshold of 32°F (0°C) (i.e., the frost-free period) to define the growing season. An update to this finding is presented in Figures 10.3 and 10.4, which show changes in the frost-free period and growing season, respectively, as defined above. Overall, the length of the frost-free period has increased in the contiguous United States during the past century (Figure 10.3). However, growing season changes are more variable: growing season length increased until the late 1930s, declined slightly until the early 1970s, increased again until about 1990, and remained quasi-stable thereafter (Figure 10.4). This contrasts somewhat with changes in the length of the frost-free period presented in NCA3, which showed a continuing increase after 1980. This difference is attributable to the temperature thresholds used in each indicator to define the start and end of these periods. Specifically, there are now more frost-free days (32°F threshold) in winter than the growing season (41°F threshold).

The lengthening of the growing season has been somewhat greater in the northern and western United States, which experienced increases of 1–2 weeks in many locations. In contrast, some areas in the Midwest, Southern Great Plains, and the Southeast had decreases of a week or more between the periods 1986–2015 and 1901–1960.[2] These differences reflect the more general pattern of warming and cooling nationwide (Ch. 6: Temperature Changes). Observations and models have verified that the growing season has generally increased plant productivity over most of the United States.[25]

Consistent with increases in growing season length and the coldest temperature of the year, plant hardiness zones have shifted northward in many areas.[79] The widespread increase in temperature has also impacted the distribution of other climate zones in parts of the United States. For instance, there have been moderate changes in the range of the temperate and continental climate zones of the eastern United States since 1950[80] as well as changes in the coverage of some extreme climate zones in the western United States. In particular, the spatial extent of the "alpine tundra" zone has decreased in high-elevation areas,[81] while the extent of the "hot arid" zone has increased in the Southwest.[82]

The period over which plants are actually productive, that is, their true growing season, is a function of multiple climate factors, including air temperature, number of frost-free days, and rainfall, as well as biophysical factors, including soil physics, daylight hours, and the biogeochemistry of ecosystems.[83] Temperature-induced changes in plant phenology, like flowering or spring leaf onset, could result in a timing mismatch (phenological asynchrony) with pollinator activity, affecting seasonal plant growth and reproduction and pollinator survival.[84, 85, 86, 87] Further, while growing season length is generally referred to in the context of agricultural productivity, the factors that govern which plant types will grow in a given location are common to all plants whether they are in agricultural, natural, or managed landscapes. Changes in both the length and the seasonality of the growing season, in concert with local environmental conditions, can have multiple effects on agricultural productivity and land cover.

In the context of agriculture, a longer growing season could allow for the diversification of cropping systems or allow multiple harvests within a growing season. For example, shifts in cold hardiness zones across the contiguous United States suggest widespread expansion of thermally suitable areas for the cultivation of cold-intolerant perennial crops[88] as well as for biological invasion of non-native plants and plant pests.[89] However, changes in available water, conversion from dry to irrigated farming, and changes in sensible and latent heat exchange associated with these shifts need to be considered. Increasingly dry conditions under a longer growing season can alter terrestrial organic matter export and catalyze oxidation of wetland soils, releasing stored contaminants (for example, copper and nickel) into streamflow after rainfall.[47] Similarly, a longer growing season, particularly in years where water is limited, is not due to warming alone, but is exacerbated by higher atmospheric $CO_2$ concentrations that extend the active period of growth by plants.[31] Longer growing seasons can also limit the types of crops that can be grown, encourage invasive species encroachment or weed growth, or in-

crease demand for irrigation, possibly beyond the limits of water availability. They could also disrupt the function and structure of a region's ecosystems and could, for example, alter the range and types of animal species in the area.

A longer and temporally shifted growing season also affects the role of terrestrial ecosystems in the carbon cycle. Neither seasonality of growing season (spring and summer) nor carbon, water, and energy fluxes should be interpreted separately when analyzing the impacts of climate extremes such as drought (Ch. 8: Droughts, Floods, and Wildfires).[39, 90] Observations and data-driven model studies suggest that losses in net terrestrial carbon uptake during record warm springs followed by severely hot and dry summers can be largely offset by carbon gains in record-exceeding warmth and early arrival of spring.[39] Depending on soil physics and land cover, a cool spring, however, can deplete soil water resources less rapidly, making the subsequent impacts of precipitation deficits less severe.[90] Depletion of soil moisture through early plant activity in a warm spring can potentially amplify summer heating, a typical lagged direct

(a)  Observed Increase in Frost-Free Season Length       (b)  Projected Changes in Frost-free Season Length

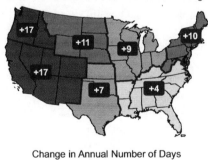
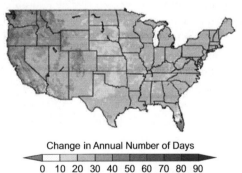

Change in Annual Number of Days

0–4   5–9   10–14   15+

Change in Annual Number of Days

0  10  20  30  40  50  60  70  80  90

Figure 10.3: (a) Observed changes in the length of the frost-free season by region, where the frost-free season is defined as the number of days between the last spring occurrence and the first fall occurrence of a minimum temperature at or below 32°F. This change is expressed as the change in the average number of frost-free days in 1986–2015 compared to 1901–1960. (b) Projected changes in the length of the frost-free season at mid-century (2036–2065 as compared to 1976–2005) under the higher scenario (RCP8.5). Gray indicates areas that are not projected to experience a freeze in more than 10 of the 30 years (Figure source: (a) updated from Walsh et al. 2014;[30] (b) NOAA NCEI and CICS-NC, data source: LOCA dataset).

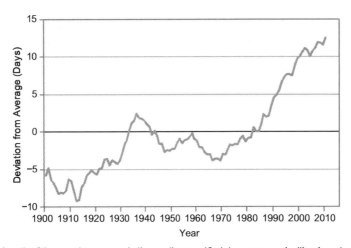

**Figure 10.4:** The length of the growing season in the contiguous 48 states compared with a long-term average (1895–2015), where "growing season" is defined by a daily minimum temperature threshold of 41°F. For each year, the line represents the number of days shorter or longer than the long-term average. The line was smoothed using an 11-year moving average. Choosing a different long-term average for comparison would not change the shape of the data over time. (Figure source: Kunkel 2016[2]).

effect of an extremely warm spring.[42] Ecosystem responses to the phenological changes of timing and extent of growing season and subsequent biophysical feedbacks are therefore strongly dependent on the timing of climate extremes (Ch. 8: Droughts, Floods, and Wildfires; Ch. 9: Extreme Storms).[90]

The global Coupled Model Intercomparison Project Phase 5 (CMIP5) analyses did not explicitly explore future changes to the growing season length. Many of the projected changes in North American climate are generally consistent across CMIP5 models, but there is substantial inter-model disagreement in projections of some metrics important to productivity in biophysical systems, including the sign of regional precipitation changes and extreme heat events across the northern United States.[91]

**10.3.2 Water Availability and Drought**
Drought is generally parameterized in most agricultural models as limited water availability and is an integrated response of both meteorological and agricultural drought, as described in Chapter 8: Droughts, Floods,

and Wildfires. However, physiological as well as biophysical processes that influence land cover and biogeochemistry interact with drought through stomatal closure induced by elevated atmospheric $CO_2$ levels.[48, 49] This has direct impacts on plant transpiration, atmospheric latent heat fluxes, and soil moisture, thereby influencing local and regional climate. Drought is often offset by management through groundwater withdrawals, with increasing pressure on these resources to maintain plant productivity. This results in indirect climate effects by altering land surface exchange of water and energy with the atmosphere.[92]

**10.3.3 Forestry Considerations**
Climate change and land-cover change in forested areas interact in many ways, such as through changes in mortality rates driven by changes in the frequency and magnitude of fire, insect infestations, and disease. In addition to the direct economic benefits of forestry, unquantified societal benefits include ecosystem services, like protection of watersheds and wildlife habitat, and recreation and human health value. United States forests and

related wood products also absorb and store the equivalent of 16% of all $CO_2$ emitted by fossil fuel burning in the United States each year.[6] Climate change is expected to reduce the carbon sink strength of forests overall.

Effective management of forests offers the opportunity to reduce future climate change—for example, as given in proposals for Reduced Emissions from Deforestation and forest Degradation (REDD+; https://www.forestcarbonpartnership.org/what-redd) in developing countries and tropical ecosystems (see Ch. 14: Mitigation)—by capturing and storing carbon in forest ecosystems and long-term wood products.[93] Afforestation in the United States has the potential to capture and store 225 million tons of additional carbon per year from 2010 to 2110.[94, 95] However, the projected maturation of United States forests[96] and land-cover change, driven in particular by the expansion of urban and suburban areas along with projected increased demands for food and bioenergy, threaten the extent of forests and their carbon storage potential.[97]

Changes in growing season length, combined with drought and accompanying wildfire are reshaping California's mountain ecosystems. The California drought led to the lowest snowpack in 500 years, the largest wildfires in post-settlement history, greater than 23% stress mortality in Sierra mid-elevation forests, and associated post-fire erosion.[69] It is anticipated that slow recovery, possibly to different ecosystem types, with numerous shifts to species' ranges will result in long-term changes to land surface biophysical as well as ecosystem structure and function in this region (http://www.fire.ca.gov/treetaskforce/).[69]

While changes in forest stocks, composition, and the ultimate use of forest products can influence net emissions and climate, the future net changes in forest stocks remain uncertain.[9, 27, 98, 99, 100] This

uncertainty is due to a combination of uncertainties in future population size, population distribution and subsequent land-use change, harvest trends, wildfire management practices (for example, large-scale thinning of forests), and the impact of maturing U.S. forests.

## 10.4 Urban Environments and Climate Change

Urban areas exhibit several characteristics that affect land-surface and geophysical attributes, including building infrastructure (rougher, more uneven surfaces compared to rural or natural systems), increased emissions and concentrations of aerosols and other greenhouse gasses, and increased anthropogenic heat sources.[101, 102] The understanding that urban areas modify their surrounding environment has been accepted for over a century, but the mechanisms through which this occurs have only begun to be understood and analyzed for more than 40 years.[102, 103] Prior to the 1970s, the majority of urban climate research was observational and descriptive,[104] but since that time, more importance has been given to physical dynamics that are a function of land surface (for example, built environment and change to surface roughness); hydrologic, aerosol, and other greenhouse gas emissions; thermal properties of the built environment; and heat generated from human activities (Seto et al. 2016[105] and references therein).

There is now strong evidence that urban environments modify local microclimates, with implications for regional and global climate change.[102, 104] Urban systems affect various climate attributes, including temperature, rainfall intensity and frequency, winter precipitation (snowfall), and flooding. New observational capabilities—including NASA's dual polarimetric radar, advanced satellite remote sensing (for example, the Global Precipitation Measurement Mission-GPM), and regionalized, coupled land–surface–atmospheric modeling systems for urban systems—are now available to evaluate aspects of daytime and nighttime temperature fluctuations; urban precipitation; contribution of aerosols; how the urban built environment impacts the seasonality and type of precipitation (rain or snow) as well as the amount and distribution of precipitation; and the significance of the extent of urban metropolitan areas.[101, 102, 106, 107]

The urban heat island (UHI) is characterized by increased surface and canopy temperatures as a result of heat-retaining asphalt and concrete, a lack of vegetation, and anthropogenic generation of heat and greenhouse gasses.[107] The heat gain due to the storage capacity of urban built structures, reductions in local evapotranspiration, and anthropogenically generated heat alter the spatio-temporal pattern of temperature and leads to the UHI phenomenon. The UHI physical processes that affect the climate system include generation of heat storage in buildings during the day, nighttime release of latent heat storage by buildings, and sensible heat generated by human activities, include heating of buildings, air conditioning, and traffic.[108]

The strength of the effect is correlated with the spatial extent and population density of urban areas; however, because of varying definitions of urban vs. non-urban, impervious surface area is a more objective metric for estimating the extent and intensity of urbanization.[109] Based on land surface temperature measurements, on average, the UHI effect increases urban temperature by 5.2°F (2.9°C), but it has been measured at 14.4°F (8°C) in cities built in areas dominated by temperate forests.[109] In arid regions, however, urban areas can be more than 3.6°F (2°C) cooler than surrounding shrublands.[110] Similarly, urban settings lose up to 12% of precipitation through impervious surface runoff, versus just over 3% loss to runoff in vegetated regions. Carbon losses

from the biosphere to the atmosphere through urbanization account for almost 2% of the continental terrestrial biosphere total, a significant proportion given that urban areas only account for around 1% of land in the United States.[110] Similarly, statistical analyses of the relationship between climate and urban land use suggest an empirical relationship between the patterns of urbanization and precipitation deficits during the dry season. Causal factors for this reduction may include changes to runoff (for example, impervious-surface versus natural-surface hydrology) that extend beyond the urban heat island effect and energy-related aerosol emissions.[111]

The urban heat island effect is more significant during the night and during winter than during the day, and it is affected by the shape, size, and geometry of buildings in urban centers as well as by infrastructure along gradients from urban to rural settlements.[101, 105, 106] Recent research points to mounting evidence that urbanization also affects cycling of water, carbon, aerosols, and nitrogen in the climate system.[106]

Coordinated modeling and observational studies have revealed other mechanisms by which the physical properties of urban areas can influence local weather and climate. It has been suggested that urban-induced wind convergence can determine storm initiation; aerosol concentrations and composition then influence the amount of cloud water and ice present in the clouds. Aerosols can also influence updraft and downdraft intensities, their life span, and surface precipitation totals.[107] A pair of studies investigated rainfall efficiency in sea-breeze thunderstorms and found that integrated moisture convergence in urban areas influenced storm initiation and mid-level moisture, thereby affecting precipitation dynamics.[112, 113]

According to the World Bank, over 81% of the United States population currently resides in urban settings.[114] Climate mitigation efforts to offset UHI are often stalled by the lack of quantitative data and understanding of the specific factors of urban systems that contribute to UHI. A recent study set out to quantitatively determine contributors to the intensity of UHI across North America.[115] The study found that population strongly influenced nighttime UHI, but that daytime UHI varied spatially following precipitation gradients. The model applied in this study indicated that the spatial variation in the UHI signal was controlled most strongly by impacts on the atmospheric convection efficiency. Because of the impracticality of managing convection efficiency, results from Zhao et al.[115] support albedo management as an efficient strategy to mitigate UHI on a large scale.

## TRACEABLE ACCOUNTS

### Key Finding 1

Changes in land use and land cover due to human activities produce physical changes in land surface albedo, latent and sensible heat, and atmospheric aerosol and greenhouse gas concentrations. The combined effects of these changes have recently been estimated to account for 40% ± 16% of the human-caused global radiative forcing from 1850 to present day (*high confidence*). In recent decades, land use and land cover changes have turned the terrestrial biosphere (soil and plants) into a net "sink" for carbon (drawing down carbon from the atmosphere), and this sink has steadily increased since 1980 (*high confidence*). Because of the uncertainty in the trajectory of land cover, the possibility of the land becoming a net carbon source cannot be excluded (*very high confidence*).

### Description of evidence base

Traditional methods that estimate albedo changes for calculating radiative forcing due to land-use change were identified by NRC.[8] That report recommended that indirect contributions of land-cover change to climate-relevant variables, such as soil moisture, greenhouse gas (e.g., $CO_2$ and water vapor) sources and sinks, snow cover, aerosols, and aerosol and ozone precursor emissions also be considered. Several studies have documented physical land surface processes such as albedo, surface roughness, sensible and latent heat exchange, and land-use and land-cover change that interact with regional atmospheric processes (e.g., Marotz et al. 1975;[116] Barnston and Schickendanz 1984;[117] Alpert and Mandel 1986;[118] Pielke and Zeng 1989;[119] Feddema et al. 2005;[7] Pielke et al. 2007[120]); however, traditional calculations of radiative forcing by land-cover change in global climate model simulations yield small forcing values (Ch. 2: Physical Drivers of Climate Change) because they account only for changes in surface albedo (e.g., Myhre and Myhre 2003;[15] Betts et al. 2007;[16] Jones et al. 2015[17]).

Recent studies that account for the physical as well as biogeochemical changes in land cover and land use radiative forcing estimated that these drivers contribute 40% of present radiative forcing due to land-use/land-cover change (0.9 W/m²).[4, 5] These studies utilized AR5 and follow-on model simulations to estimate changes in land-cover and land-use climate forcing and feedbacks for the greenhouse gases—carbon dioxide, methane, and nitrous oxide—that contribute to total anthropogenic radiative forcing from land-use and land-cover change.[4, 5] This research is grounded in long-term observations that have been documented for over 40 years and recently implemented into global Earth system models.[4, 20] For example, IPCC 2013: Summary for Policymakers states: "From 1750 to 2011, $CO_2$ emissions from fossil fuel combustion and cement production have released 375 [345 to 405] GtC to the atmosphere, while deforestation and other land-use changes are estimated to have released 180 [100 to 260] GtC. This results in cumulative anthropogenic emissions of 555 [470 to 640] GtC."[121] IPCC 2013, Working Group 1, Chapter 14 states for North America: "In summary, it is very likely that by mid-century the anthropogenic warming signal will be large compared to natural variability such as that stemming from the NAO, ENSO, PNA, PDO, and the NAMS in all North America regions throughout the year".[122]

### Major uncertainties

Uncertainty exists in the future land-cover and land-use change as well as uncertainties in regional calculations of land-cover change and associated radiative forcing. The role of the land as a current sink has *very high confidence;* however, future strength of the land sink is uncertain.[96, 97] The existing impact of land systems on climate forcing has *high confidence.*[4] Based on current RCP scenarios for future radiative forcing targets ranging from 2.6 to 8.5 W/m², the future forcing has lower confidence because it is difficult to estimate changes in land cover and land use into the future.[14] Compared to 2000, the $CO_2$-eq. emissions consistent with RCP8.5 more than double by 2050 and increase by three by 2100.[10] About one quarter of this increase is due to increasing use of fertilizers and intensification of agricultural production, giving rise to the primary source of $N_2O$ emissions. In addition, increases in livestock population, rice production, and enteric fermentation processes increase $CH_4$ emissions.[10] Therefore, if existing trends in land-use and land-cover change continue, the contribution of land cover to forcing will increase with *high confidence*. Overall, future scenarios from the RCPs suggest that land-cover change based on policy, bioenergy, and food demands could lead to significantly different distribution of land cover types (forest, agriculture, urban) by 2100.[9, 10, 11, 12, 13, 14]

### Summary sentence or paragraph that integrates the above information

The key finding is based on basic physics and biophysical models that have been well established for decades with regards to the contribution of land albedo to radiative forcing (NRC 2005). Recent assessments specifically address additional biogeochemical contributions of land-cover and land-use change to radiative forcing.[4, 8] The role of current sink strength of the land is also uncertain.[96, 97] The future distribution of land cover and contributions to total radiative forcing are uncertain and depend on policy, energy demand and food consumption, dietary demands.[14]

### Key Finding 2

Climate change and induced changes in the frequency and magnitude of extreme events (e.g., droughts, floods, and heat waves) have led to large changes in plant community structure with subsequent effects on the biogeochemistry of terrestrial ecosystems. Uncertainties about how climate change will affect land cover change make it difficult to project the magnitude and sign of future climate feedbacks from land cover changes (*high confidence*).

### Description of evidence base

From the perspective of the land biosphere, drought has strong effects on ecosystem productivity and carbon storage by reducing microbial activity and photosynthesis and by increasing the risk of wildfire, pest infestation, and disease susceptibility. Thus, future droughts will affect carbon uptake and storage, leading to feedbacks to the climate system.[41] Reduced productivity as a result of extreme drought events can also extend for several years post-drought (i.e., drought legacy effects).[42, 43, 44] Under increased $CO_2$ concentrations, plants have been observed to optimize water use due to reduced stomatal conductance, thereby increasing water-use efficiency.[48] This change in water-use efficiency can affect plants' tolerance to stress and specifically to drought.[49]

Recent severe droughts in the western United States (Texas and California) have led to significant mortality and carbon cycle dynamics (http://www.fire.ca.gov/treetaskforce/).[45, 69] Carbon redistribution through mortality in the Texas drought was around 36% of global carbon losses due to deforestation and land use change.[46]

### Major uncertainties

Major uncertainties include how future land-use/land-cover changes will occur as a result of policy and/or mitigation strategies in addition to climate change. Ecosystem responses to phenological changes are strongly dependent on the timing of climate extremes.[90] Due to the complex interactions of the processes that govern terrestrial biogeochemical cycling, terrestrial ecosystem response to increasing $CO_2$ levels remains one of the largest uncertainties in long-term climate feedbacks and therefore in predicting longer-term climate change effects on ecosystems (e.g., Swann et al. 2016[49]).

### Summary sentence or paragraph that integrates the above information

The timing, frequency, magnitude, and extent of climate extremes strongly influence plant community structure and function, with subsequent effects on terrestrial biogeochemistry and feedbacks to the climate system. Future interactions between land cover and the climate system are uncertain and depend on human land-use decisions, the evolution of the climate system, and the timing, frequency, magnitude, and extent of climate extremes.

### Key Finding 3

Since 1901, regional averages of both the consecutive number of frost-free days and the length of the corresponding growing season have increased for the seven contiguous U.S. regions used in this assessment. However, there is important variability at smaller scales, with some locations actually showing decreases of a few days to as much as one to two weeks. Plant productivity has not increased commensurate with the increased number of frost-free days or with the longer growing season due to plant-specific temperature thresholds, plant–pollinator dependence, and seasonal limitations in water and nutrient availability (*very high confidence*). Future consequences of changes to the growing season for plant productivity are uncertain.

### Description of evidence base

Data on the lengthening and regional variability of the growing season since 1901 were updated by Kunkel.[2] Many of these differences reflect the more general pattern of warming and cooling nationwide (Ch. 6: Temperature Changes). Without nutrient limitations, increased $CO_2$ concentrations and warm temperatures have been shown to extend the growing season, which may contribute to longer periods of plant activity and carbon uptake but do not affect reproduction rates.[31] However, other confounding variables that coincide with climate change (for example, drought, increased ozone, and reduced photosynthesis due to increased or extreme heat) can offset increased growth associated with longer growing seasons[26] as well as changes in water availability and demand for water (e.g., Georgakakos et al. 2014;[32] Hibbard et al. 2014[33]). Increased dry conditions can lead to wildfire (e.g., Hatfield et al. 2014;[34] Joyce et al. 2014;[35] Ch. 8: Droughts, Floods and Wildfires) and urban temperatures can contribute to urban-induced thunderstorms in the southeastern United States.[36] Temperature benefits of early onset of plant development in a longer growing season can be offset by 1) freeze damage caused by late-season frosts; 2) limits to growth because of shortening of the photoperiod later in the season; or 3) by shorter chilling periods required for leaf unfolding by many plants.[37, 38]

### Major uncertainties

Uncertainties exist in future response of the climate system to anthropogenic forcings (land use/land cover as well as fossil fuel emissions) and associated feedbacks among variables such as temperature and precipitation interactions with carbon and nitrogen cycles as well as land-cover change that impact the length of

the growing season (Ch. 6: Temperature Changes and Ch. 8: Droughts, Floods and Wildfires).[26, 31, 34]

### Summary sentence or paragraph that integrates the above information

Changes in growing season length and interactions with climate, biogeochemistry, and land cover were covered in 12 chapters of NCA3[6] but with sparse assessment of how changes in the growing season might offset plant productivity and subsequent feedbacks to the climate system. This key finding provides an assessment of the current state of the complex nature of the growing season.

### Key Finding 4

Recent studies confirm and quantify higher surface temperatures in urban areas than in surrounding rural areas for a number of reasons, including the concentrated release of heat from buildings, vehicles, and industry. In the United States, this urban heat island effect results in daytime temperatures 0.9°–7.2°F (0.5°–4.0°C) higher and nighttime temperatures 1.8°– 4.5°F (1.0°–2.5°C) higher in urban areas, with larger temperature differences in humid regions (primarily in the eastern United States) and in cities with larger and denser populations. The urban heat island effect will strengthen in the future as the structure, spatial extent, and population density of urban areas change and grow (*high confidence*).

### Description of evidence base

Urban interactions with the climate system have been investigated for more than 40 years.[102, 103] The heat gain due to the storage capacity of urban built structures, reduction in local evapotranspiration, and anthropogenically generated heat alter the spatio-temporal pattern of temperature and leads to the well-known urban heat island (UHI) phenomenon.[101, 105, 106] The urban heat island (UHI) effect is correlated with the extent of impervious surfaces, which alter albedo or the saturation of radiation.[109] The urban-rural difference that defines the UHI is greatest for cities built in temperate forest ecosystems.[109] The average temperature increase is 2.9°C, except for urban areas in biomes with arid and semiarid climates.[109, 110]

**Major uncertainties**

The largest uncertainties about urban forcings or feedbacks to the climate system are how urban settlements will evolve and how energy consumption and efficiencies, and their interactions with land cover and water, may change from present times.[10, 14, 33, 105]

**Summary sentence or paragraph that integrates the above information**

Key Finding 4 is based on simulated and satellite land surface measurements analyzed by Imhoff et al.[109] Bounoua et al.,[110] Shepherd,[107] Seto and Shepherd,[106] Grimmond et al.,[101] and Seto et al.[105] provide specific references with regard to how building materials and spatio-temporal patterns of urban settlements influence radiative forcing and feedbacks of urban areas to the climate system.

# REFERENCES

1. Galloway, J.N., W.H. Schlesinger, C.M. Clark, N.B. Grimm, R.B. Jackson, B.E. Law, P.E. Thornton, A.R. Townsend, and R. Martin, 2014: Ch. 15: Biogeochemical cycles. *Climate Change Impacts in the United States: The Third National Climate Assessment.* Melillo, J.M., Terese (T.C.) Richmond, and G.W. Yohe, Eds. U.S. Global Change Research Program, Washington, DC, 350-368. http://dx.doi.org/10.7930/J0X63JT0

2. Kunkel, K.E., 2016: Update to data orginally published in: Kunkel, K.E., D. R. Easterling, K. Hubbard, K. Redmond, 2004: Temporal variations in frost-free season in the United States: 1895 - 2000. *Geophysical Research Letters,* **31**, L03201. http://dx.doi.org/10.1029/2003gl018624

3. Brown, D.G., C. Polsky, P. Bolstad, S.D. Brody, D. Hulse, R. Kroh, T.R. Loveland, and A. Thomson, 2014: Ch. 13: Land use and land cover change. *Climate Change Impacts in the United States: The Third National Climate Assessment.* Melillo, J.M., Terese (T.C.) Richmond, and G.W. Yohe, Eds. U.S. Global Change Research Program, Washington, DC, 318-332. http://dx.doi.org/10.7930/J05Q4T1Q

4. Myhre, G., D. Shindell, F.-M. Bréon, W. Collins, J. Fuglestvedt, J. Huang, D. Koch, J.-F. Lamarque, D. Lee, B. Mendoza, T. Nakajima, A. Robock, G. Stephens, T. Takemura, and H. Zhang, 2013: Anthropogenic and natural radiative forcing. *Climate Change 2013: The Physical Science Basis. Contribution of Working Group I to the Fifth Assessment Report of the Intergovernmental Panel on Climate Change.* Stocker, T.F., D. Qin, G.-K. Plattner, M. Tignor, S.K. Allen, J. Boschung, A. Nauels, Y. Xia, V. Bex, and P.M. Midgley, Eds. Cambridge University Press, Cambridge, United Kingdom and New York, NY, USA, 659–740. http://www.climatechange2013.org/report/full-report/

5. Ward, D.S., N.M. Mahowald, and S. Kloster, 2014: Potential climate forcing of land use and land cover change. *Atmospheric Chemistry and Physics,* **14**, 12701-12724. http://dx.doi.org/10.5194/acp-14-12701-2014

6. Melillo, J.M., T.C. Richmond, and G.W. Yohe, eds., 2014: *Climate Change Impacts in the United States: The Third National Climate Assessment.* U.S. Global Change Research Program: Washington, D.C., 841 pp. http://dx.doi.org/10.7930/J0Z31WJ2

7. Feddema, J.J., K.W. Oleson, G.B. Bonan, L.O. Mearns, L.E. Buja, G.A. Meehl, and W.M. Washington, 2005: The importance of land-cover change in simulating future climates. *Science,* **310**, 1674-1678. http://dx.doi.org/10.1126/science.1118160

8. NRC, 2005: *Radiative Forcing of Climate Change: Expanding the Concept and Addressing Uncertainties.* National Academies Press, Washington, D.C., 222 pp. http://dx.doi.org/10.17226/11175

9. Hurtt, G.C., L.P. Chini, S. Frolking, R.A. Betts, J. Feddema, G. Fischer, J.P. Fisk, K. Hibbard, R.A. Houghton, A. Janetos, C.D. Jones, G. Kindermann, T. Kinoshita, K. Klein Goldewijk, K. Riahi, E. Shevliakova, S. Smith, E. Stehfest, A. Thomson, P. Thornton, D.P. van Vuuren, and Y.P. Wang, 2011: Harmonization of land-use scenarios for the period 1500–2100: 600 years of global gridded annual land-use transitions, wood harvest, and resulting secondary lands. *Climatic Change,* **109**, 117. http://dx.doi.org/10.1007/s10584-011-0153-2

10. Riahi, K., S. Rao, V. Krey, C. Cho, V. Chirkov, G. Fischer, G. Kindermann, N. Nakicenovic, and P. Rafaj, 2011: RCP 8.5—A scenario of comparatively high greenhouse gas emissions. *Climatic Change,* **109**, 33. http://dx.doi.org/10.1007/s10584-011-0149-y

11. Fujimori, S., T. Masui, and Y. Matsuoka, 2014: Development of a global computable general equilibrium model coupled with detailed energy end-use technology. *Applied Energy,* **128**, 296-306. http://dx.doi.org/10.1016/j.apenergy.2014.04.074

12. Thomson, A.M., K.V. Calvin, S.J. Smith, G.P. Kyle, A. Volke, P. Patel, S. Delgado-Arias, B. Bond-Lamberty, M.A. Wise, and L.E. Clarke, 2011: RCP4.5: A pathway for stabilization of radiative forcing by 2100. *Climatic Change,* **109**, 77-94. http://dx.doi.org/10.1007/s10584-011-0151-4

13. van Vuuren, D.P., S. Deetman, M.G.J. den Elzen, A. Hof, M. Isaac, K. Klein Goldewijk, T. Kram, A. Mendoza Beltran, E. Stehfest, and J. van Vliet, 2011: RCP2.6: Exploring the possibility to keep global mean temperature increase below 2°C. *Climatic Change,* **109**, 95-116. http://dx.doi.org/10.1007/s10584-011-0152-3

14. van Vuuren, D.P., J. Edmonds, M. Kainuma, K. Riahi, A. Thomson, K. Hibbard, G.C. Hurtt, T. Kram, V. Krey, and J.F. Lamarque, 2011: The representative concentration pathways: An overview. *Climatic Change,* **109**, 5-31. http://dx.doi.org/10.1007/s10584-011-0148-z

15. Myhre, G. and A. Myhre, 2003: Uncertainties in radiative forcing due to surface albedo changes caused by land-use changes. *Journal of Climate,* **16**, 1511-1524. http://dx.doi.org/10.1175/1520-0442(2003)016<1511:uirfdt>2.0.co;2

16. Betts, R.A., P.D. Falloon, K.K. Goldewijk, and N. Ramankutty, 2007: Biogeophysical effects of land use on climate: Model simulations of radiative forcing and large-scale temperature change. *Agricultural and Forest Meteorology,* **142**, 216-233. http://dx.doi.org/10.1016/j.agrformet.2006.08.021

17. Jones, A.D., K.V. Calvin, W.D. Collins, and J. Edmonds, 2015: Accounting for radiative forcing from albedo change in future global land-use scenarios. *Climatic Change,* **131**, 691-703. http://dx.doi.org/10.1007/s10584-015-1411-5

18. Ward, D.S. and N.M. Mahowald, 2015: Local sources of global climate forcing from different categories of land use activities. *Earth System Dynamics*, **6**, 175-194. http://dx.doi.org/10.5194/esd-6-175-2015

19. Lovenduski, N.S. and G.B. Bonan, 2017: Reducing uncertainty in projections of terrestrial carbon uptake. *Environmental Research Letters*, **12**, 044020. http://dx.doi.org/10.1088/1748-9326/aa66b8

20. Anav, A., P. Friedlingstein, M. Kidston, L. Bopp, P. Ciais, P. Cox, C. Jones, M. Jung, R. Myneni, and Z. Zhu, 2013: Evaluating the land and ocean components of the global carbon cycle in the CMIP5 earth system models. *Journal of Climate*, **26**, 6801-6843. http://dx.doi.org/10.1175/jcli-d-12-00417.1

21. Hoffman, F.M., J.T. Randerson, V.K. Arora, Q. Bao, P. Cadule, D. Ji, C.D. Jones, M. Kawamiya, S. Khatiwala, K. Lindsay, A. Obata, E. Shevliakova, K.D. Six, J.F. Tjiputra, E.M. Volodin, and T. Wu, 2014: Causes and implications of persistent atmospheric carbon dioxide biases in Earth System Models. *Journal of Geophysical Research Biogeosciences*, **119**, 141-162. http://dx.doi.org/10.1002/2013JG002381

22. Rosenzweig, C., J. Elliott, D. Deryng, A.C. Ruane, C. Müller, A. Arneth, K.J. Boote, C. Folberth, M. Glotter, N. Khabarov, K. Neumann, F. Piontek, T.A.M. Pugh, E. Schmid, E. Stehfest, H. Yang, and J.W. Jones, 2014: Assessing agricultural risks of climate change in the 21st century in a global gridded crop model intercomparison. *Proceedings of the National Academy of Sciences*, **111**, 3268-3273. http://dx.doi.org/10.1073/pnas.1222463110

23. Ju, J. and J.G. Masek, 2016: The vegetation greenness trend in Canada and US Alaska from 1984–2012 Landsat data. *Remote Sensing of Environment*, **176**, 1-16. http://dx.doi.org/10.1016/j.rse.2016.01.001

24. Zhu, Z., S. Piao, R.B. Myneni, M. Huang, Z. Zeng, J.G. Canadell, P. Ciais, S. Sitch, P. Friedlingstein, A. Arneth, C. Cao, L. Cheng, E. Kato, C. Koven, Y. Li, X. Lian, Y. Liu, R. Liu, J. Mao, Y. Pan, S. Peng, J. Penuelas, B. Poulter, T.A.M. Pugh, B.D. Stocker, N. Viovy, X. Wang, Y. Wang, Z. Xiao, H. Yang, S. Zaehle, and N. Zeng, 2016: Greening of the Earth and its drivers. *Nature Climate Change*, **6**, 791-795. http://dx.doi.org/10.1038/nclimate3004

25. Mao, J., A. Ribes, B. Yan, X. Shi, P.E. Thornton, R. Seferian, P. Ciais, R.B. Myneni, H. Douville, S. Piao, Z. Zhu, R.E. Dickinson, Y. Dai, D.M. Ricciuto, M. Jin, F.M. Hoffman, B. Wang, M. Huang, and X. Lian, 2016: Human-induced greening of the northern extratropical land surface. *Nature Climate Change*, **6**, 959-963. http://dx.doi.org/10.1038/nclimate3056

26. Adams, H.D., A.D. Collins, S.P. Briggs, M. Vennetier, L.T. Dickman, S.A. Sevanto, N. Garcia-Forner, H.H. Powers, and N.G. McDowell, 2015: Experimental drought and heat can delay phenological development and reduce foliar and shoot growth in semiarid trees. *Global Change Biology*, **21**, 4210-4220. http://dx.doi.org/10.1111/gcb.13030

27. Bonan, G.B., 2008: Forests and climate change: Forcings, feedbacks, and the climate benefits of forests. *Science*, **320**, 1444-1449. http://dx.doi.org/10.1126/science.1155121

28. Jackson, R.B., J.T. Randerson, J.G. Canadell, R.G. Anderson, R. Avissar, D.D. Baldocchi, G.B. Bonan, K. Caldeira, N.S. Diffenbaugh, C.B. Field, B.A. Hungate, E.G. Jobbágy, L.M. Kueppers, D.N. Marcelo, and D.E. Pataki, 2008: Protecting climate with forests. *Environmental Research Letters*, **3**, 044006. http://dx.doi.org/10.1088/1748-9326/3/4/044006

29. Anderson, R.G., J.G. Canadell, J.T. Randerson, R.B. Jackson, B.A. Hungate, D.D. Baldocchi, G.A. Ban-Weiss, G.B. Bonan, K. Caldeira, L. Cao, N.S. Diffenbaugh, K.R. Gurney, L.M. Kueppers, B.E. Law, S. Luyssaert, and T.L. O'Halloran, 2011: Biophysical considerations in forestry for climate protection. *Frontiers in Ecology and the Environment*, **9**, 174-182. http://dx.doi.org/10.1890/090179

30. Walsh, J., D. Wuebbles, K. Hayhoe, J. Kossin, K. Kunkel, G. Stephens, P. Thorne, R. Vose, M. Wehner, J. Willis, D. Anderson, S. Doney, R. Feely, P. Hennon, V. Kharin, T. Knutson, F. Landerer, T. Lenton, J. Kennedy, and R. Somerville, 2014: Ch. 2: Our changing climate. *Climate Change Impacts in the United States: The Third National Climate Assessment*. Melillo, J.M., T.C. Richmond, and G.W. Yohe, Eds. U.S. Global Change Research Program, Washington, D.C., 19-67. http://dx.doi.org/10.7930/J0KW5CXT

31. Reyes-Fox, M., H. Steltzer, M.J. Trlica, G.S. McMaster, A.A. Andales, D.R. LeCain, and J.A. Morgan, 2014: Elevated CO2 further lengthens growing season under warming conditions. *Nature*, **510**, 259-262. http://dx.doi.org/10.1038/nature13207

32. Georgakakos, A., P. Fleming, M. Dettinger, C. Peters-Lidard, T.C. Richmond, K. Reckhow, K. White, and D. Yates, 2014: Ch. 3: Water resources. *Climate Change Impacts in the United States: The Third National Climate Assessment*. Melillo, J.M., T.C. Richmond, and G.W. Yohe, Eds. U.S. Global Change Research Program, Washington, D.C., 69-112. http://dx.doi.org/10.7930/J0G44N6T

33. Hibbard, K., T. Wilson, K. Averyt, R. Harriss, R. Newmark, S. Rose, E. Shevliakova, and V. Tidwell, 2014: Ch. 10: Energy, water, and land use. *Climate Change Impacts in the United States: The Third National Climate Assessment*. Melillo, J.M., Terese (T.C.) Richmond, and G.W. Yohe, Eds. U.S. Global Change Research Program, Washington, DC, 257-281. http://dx.doi.org/10.7930/J0JW8BSF

34. Hatfield, J., G. Takle, R. Grotjahn, P. Holden, R.C. Izaurralde, T. Mader, E. Marshall, and D. Liverman, 2014: Ch. 6: Agriculture. *Climate Change Impacts in the United States: The Third National Climate Assessment.* Melillo, J.M., Terese (T.C.) Richmond, and G.W. Yohe, Eds. U.S. Global Change Research Program, Washington, DC, 150-174. http://dx.doi.org/10.7930/J02Z13FR

35. Joyce, L.A., S.W. Running, D.D. Breshears, V.H. Dale, R.W. Malmsheimer, R.N. Sampson, B. Sohngen, and C.W. Woodall, 2014: Ch. 7: Forests. *Climate Change Impacts in the United States: The Third National Climate Assessment.* Melillo, J.M., Terese (T.C.) Richmond, and G.W. Yohe, Eds. U.S. Global Change Research Program, Washington, DC, 175-194. http://dx.doi.org/10.7930/J0Z60KZC

36. Ashley, W.S., M.L. Bentley, and J.A. Stallins, 2012: Urban-induced thunderstorm modification in the southeast United States. *Climatic Change,* **113,** 481-498. http://dx.doi.org/10.1007/s10584-011-0324-1

37. Fu, Y.H., H. Zhao, S. Piao, M. Peaucelle, S. Peng, G. Zhou, P. Ciais, M. Huang, A. Menzel, J. Penuelas, Y. Song, Y. Vitasse, Z. Zeng, and I.A. Janssens, 2015: Declining global warming effects on the phenology of spring leaf unfolding. *Nature,* **526,** 104-107. http://dx.doi.org/10.1038/nature15402

38. Gu, L., P.J. Hanson, W. Mac Post, D.P. Kaiser, B. Yang, R. Nemani, S.G. Pallardy, and T. Meyers, 2008: The 2007 eastern US spring freezes: Increased cold damage in a warming world? *BioScience,* **58,** 253-262. http://dx.doi.org/10.1641/b580311

39. Wolf, S., T.F. Keenan, J.B. Fisher, D.D. Baldocchi, A.R. Desai, A.D. Richardson, R.L. Scott, B.E. Law, M.E. Litvak, N.A. Brunsell, W. Peters, and I.T. van der Laan-Luijkx, 2016: Warm spring reduced carbon cycle impact of the 2012 US summer drought. *Proceedings of the National Academy of Sciences,* **113,** 5880-5885. http://dx.doi.org/10.1073/pnas.1519620113

40. Fridley, J.D., J.S. Lynn, J.P. Grime, and A.P. Askew, 2016: Longer growing seasons shift grassland vegetation towards more-productive species. *Nature Climate Change,* **6,** 865-868. http://dx.doi.org/10.1038/nclimate3032

41. Schlesinger, W.H., M.C. Dietze, R.B. Jackson, R.P. Phillips, C.C. Rhoads, L.E. Rustad, and J.M. Vose, 2016: Forest biogeochemistry in response to drought. *Effects of Drought on Forests and Rangelands in the United States: A comprehensive science synthesis.* Vose, J., J.S. Clark, C. Luce, and T. Patel-Weynand, Eds. U.S. Department of Agriculture, Forest Service, Washington Office, Washington, DC, 97-106. http://www.treesearch.fs.fed.us/pubs/50261

42. Frank, D., M. Reichstein, M. Bahn, K. Thonicke, D. Frank, M.D. Mahecha, P. Smith, M. van der Velde, S. Vicca, F. Babst, C. Beer, N. Buchmann, J.G. Canadell, P. Ciais, W. Cramer, A. Ibrom, F. Miglietta, B. Poulter, A. Rammig, S.I. Seneviratne, A. Walz, M. Wattenbach, M.A. Zavala, and J. Zscheischler, 2015: Effects of climate extremes on the terrestrial carbon cycle: Concepts, processes and potential future impacts. *Global Change Biology,* **21,** 2861-2880. http://dx.doi.org/10.1111/gcb.12916

43. Reichstein, M., M. Bahn, P. Ciais, D. Frank, M.D. Mahecha, S.I. Seneviratne, J. Zscheischler, C. Beer, N. Buchmann, D.C. Frank, D. Papale, A. Rammig, P. Smith, K. Thonicke, M. van der Velde, S. Vicca, A. Walz, and M. Wattenbach, 2013: Climate extremes and the carbon cycle. *Nature,* **500,** 287-295. http://dx.doi.org/10.1038/nature12350

44. Anderegg, W.R.L., C. Schwalm, F. Biondi, J.J. Camarero, G. Koch, M. Litvak, K. Ogle, J.D. Shaw, E. Shevliakova, A.P. Williams, A. Wolf, E. Ziaco, and S. Pacala, 2015: Pervasive drought legacies in forest ecosystems and their implications for carbon cycle models. *Science,* **349,** 528-532. http://dx.doi.org/10.1126/science.aab1833

45. Moore, G.W., C.B. Edgar, J.G. Vogel, R.A. Washington-Allen, Rosaleen G. March, and R. Zehnder, 2016: Tree mortality from an exceptional drought spanning mesic to semiarid ecoregions. *Ecological Applications,* **26,** 602-611. http://dx.doi.org/10.1890/15-0330

46. Ciais, P., C. Sabine, G. Bala, L. Bopp, V. Brovkin, J. Canadell, A. Chhabra, R. DeFries, J. Galloway, M. Heimann, C. Jones, C. Le Quéré, R.B. Myneni, S. Piao, and P. Thornton, 2013: Carbon and other biogeochemical cycles. *Climate Change 2013: The Physical Science Basis. Contribution of Working Group I to the Fifth Assessment Report of the Intergovernmental Panel on Climate Change.* Stocker, T.F., D. Qin, G.-K. Plattner, M. Tignor, S.K. Allen, J. Boschung, A. Nauels, Y. Xia, V. Bex, and P.M. Midgley, Eds. Cambridge University Press, Cambridge, United Kingdom and New York, NY, USA, 465–570. http://www.climatechange2013.org/report/full-report/

47. Szkokan-Emilson, E.J., B.W. Kielstra, S.E. Arnott, S.A. Watmough, J.M. Gunn, and A.J. Tanentzap, 2017: Dry conditions disrupt terrestrial–aquatic linkages in northern catchments. *Global Change Biology,* **23,** 117-126. http://dx.doi.org/10.1111/gcb.13361

48. Keenan, T.F., D.Y. Hollinger, G. Bohrer, D. Dragoni, J.W. Munger, H.P. Schmid, and A.D. Richardson, 2013: Increase in forest water-use efficiency as atmospheric carbon dioxide concentrations rise. *Nature,* **499,** 324-327. http://dx.doi.org/10.1038/nature12291

49. Swann, A.L.S., F.M. Hoffman, C.D. Koven, and J.T. Randerson, 2016: Plant responses to increasing $CO_2$ reduce estimates of climate impacts on drought severity. *Proceedings of the National Academy of Sciences,* **113,** 10019-10024. http://dx.doi.org/10.1073/pnas.1604581113

50. Norby, R.J., E.H. DeLucia, B. Gielen, C. Calfapietra, C.P. Giardina, J.S. King, J. Ledford, H.R. McCarthy, D.J.P. Moore, R. Ceulemans, P. De Angelis, A.C. Finzi, D.F. Karnosky, M.E. Kubiske, M. Lukac, K.S. Pregitzer, G.E. Scarascia-Mugnozza, W.H. Schlesinger, and R. Oren, 2005: Forest response to elevated CO2 is conserved across a broad range of productivity. *Proceedings of the National Academy of Sciences of the United States of America*, **102**, 18052-18056. http://dx.doi.org/10.1073/pnas.0509478102

51. Zaehle, S., P. Friedlingstein, and A.D. Friend, 2010: Terrestrial nitrogen feedbacks may accelerate future climate change. *Geophysical Research Letters*, **37**, L01401. http://dx.doi.org/10.1029/2009GL041345

52. Melillo, J.M., S. Butler, J. Johnson, J. Mohan, P. Steudler, H. Lux, E. Burrows, F. Bowles, R. Smith, L. Scott, C. Vario, T. Hill, A. Burton, Y.M. Zhou, and J. Tang, 2011: Soil warming, carbon-nitrogen interactions, and forest carbon budgets. *Proceedings of the National Academy of Sciences*, **108**, 9508-9512. http://dx.doi.org/10.1073/pnas.1018189108

53. Knutti, R. and J. Sedláček, 2013: Robustness and uncertainties in the new CMIP5 climate model projections. *Nature Climate Change*, **3**, 369-373. http://dx.doi.org/10.1038/nclimate1716

54. Jandl, R., M. Lindner, L. Vesterdal, B. Bauwens, R. Baritz, F. Hagedorn, D.W. Johnson, K. Minkkinen, and K.A. Byrne, 2007: How strongly can forest management influence soil carbon sequestration? *Geoderma*, **137**, 253-268. http://dx.doi.org/10.1016/j.geoderma.2006.09.003

55. McLauchlan, K., 2006: The nature and longevity of agricultural impacts on soil carbon and nutrients: A review. *Ecosystems*, **9**, 1364-1382. http://dx.doi.org/10.1007/s10021-005-0135-1

56. Smith, P., S.J. Chapman, W.A. Scott, H.I.J. Black, M. Wattenbach, R. Milne, C.D. Campbell, A. Lilly, N. Ostle, P.E. Levy, D.G. Lumsdon, P. Millard, W. Towers, S. Zaehle, and J.U. Smith, 2007: Climate change cannot be entirely responsible for soil carbon loss observed in England and Wales, 1978–2003. *Global Change Biology*, **13**, 2605-2609. http://dx.doi.org/10.1111/j.1365-2486.2007.01458.x

57. Liu, L.L. and T.L. Greaver, 2009: A review of nitrogen enrichment effects on three biogenic GHGs: The $CO_2$ sink may be largely offset by stimulated $N_2O$ and $CH_4$ emission. *Ecology Letters*, **12**, 1103-1117. http://dx.doi.org/10.1111/j.1461-0248.2009.01351.x

58. Todd-Brown, K.E.O., J.T. Randerson, W.M. Post, F.M. Hoffman, C. Tarnocai, E.A.G. Schuur, and S.D. Allison, 2013: Causes of variation in soil carbon simulations from CMIP5 Earth system models and comparison with observations. *Biogeosciences*, **10**, 1717-1736. http://dx.doi.org/10.5194/bg-10-1717-2013

59. Tian, H., C. Lu, J. Yang, K. Banger, D.N. Huntzinger, C.R. Schwalm, A.M. Michalak, R. Cook, P. Ciais, D. Hayes, M. Huang, A. Ito, A.K. Jain, H. Lei, J. Mao, S. Pan, W.M. Post, S. Peng, B. Poulter, W. Ren, D. Ricciuto, K. Schaefer, X. Shi, B. Tao, W. Wang, Y. Wei, Q. Yang, B. Zhang, and N. Zeng, 2015: Global patterns and controls of soil organic carbon dynamics as simulated by multiple terrestrial biosphere models: Current status and future directions. *Global Biogeochemical Cycles*, **29**, 775-792. http://dx.doi.org/10.1002/2014GB005021

60. Guenther, A., T. Karl, P. Harley, C. Wiedinmyer, P.I. Palmer, and C. Geron, 2006: Estimates of global terrestrial isoprene emissions using MEGAN (Model of Emissions of Gases and Aerosols from Nature). *Atmospheric Chemistry and Physics*, **6**, 3181-3210. http://dx.doi.org/10.5194/acp-6-3181-2006

61. **Rosenstiel, T.N., M.J. Potosnak, K.L. Griffin, R. Fall,** and R.K. Monson, 2003: Increased CO2 uncouples growth from isoprene emission in an agriforest ecosystem. *Nature*, **421**, 256-259. http://dx.doi.org/10.1038/nature01312

62. Pyle, J.A., N. Warwick, X. Yang, P.J. Young, and G. Zeng, 2007: Climate/chemistry feedbacks and biogenic emissions. *Philosophical Transactions of the Royal Society A: Mathematical, Physical and Engineering Sciences*, **365**, 1727-40. http://dx.doi.org/10.1098/rsta.2007.2041

63. McGrath, J.M., A.M. Betzelberger, S. Wang, E. Shook, X.-G. Zhu, S.P. Long, and E.A. Ainsworth, 2015: An analysis of ozone damage to historical maize and soybean yields in the United States. *Proceedings of the National Academy of Sciences*, **112**, 14390-14395. http://dx.doi.org/10.1073/pnas.1509777112

64. **Kam, J., J. Sheffield, and E.F. Wood, 2014: Changes** in drought risk over the contiguous United States (1901–2012): The influence of the Pacific and Atlantic Oceans. *Geophysical Research Letters*, **41**, 5897-5903. http://dx.doi.org/10.1002/2014GL060973

65. Diez, J.M., C.M. D'Antonio, J.S. Dukes, E.D. Grosholz, J.D. Olden, C.J.B. Sorte, D.M. Blumenthal, B.A. Bradley, R. Early, I. Ibáñez, S.J. Jones, J.J. Lawler, and L.P. Miller, 2012: Will extreme climatic events facilitate biological invasions? *Frontiers in Ecology and the Environment*, **10**, 249-257. http://dx.doi.org/10.1890/110137

66. Lambert, A.M., C.M. D'Antonio, and T.L. Dudley, 2010: Invasive species and fire in California ecosystems. *Fremontia*, **38**, 29-36.

67. Vose, J., J.S. Clark, C. Luce, and T. Patel-Weynand, eds., 2016: *Effects of Drought on Forests and Rangelands in the United States: A Comprehensive Science Synthesis.* U.S. Department of Agriculture, Forest Service, **Washington Office: Washington, DC, 289 pp.** http://www.treesearch.fs.fed.us/pubs/50261

68. Diffenbaugh, N.S., D.L. Swain, and D. Touma, 2015: Anthropogenic warming has increased drought risk in California. *Proceedings of the National Academy of Sciences*, **112**, 3931-3936. http://dx.doi.org/10.1073/pnas.1422385112

69. Asner, G.P., P.G. Brodrick, C.B. Anderson, N. Vaughn, D.E. Knapp, and R.E. Martin, 2016: Progressive forest canopy water loss during the 2012–2015 California drought. *Proceedings of the National Academy of Sciences*, **113**, E249-E255. http://dx.doi.org/10.1073/pnas.1523397113

70. Mann, M.E. and P.H. Gleick, 2015: Climate change and California drought in the 21st century. *Proceedings of the National Academy of Sciences*, **112**, 3858-3859. http://dx.doi.org/10.1073/pnas.1503667112

71. Littell, J.S., D.L. Peterson, K.L. Riley, Y.-Q. Liu, and C.H. Luce, 2016: Fire and drought. *Effects of Drought on Forests and Rangelands in the United States: A comprehensive science synthesis*. Vose, J., J.S. Clark, C. Luce, and T. Patel-Weynand, Eds. U.S. Department of Agriculture, Forest Service, Washington Office, Washington, DC, 135-150. http://www.treesearch.fs.fed.us/pubs/50261

72. Littell, J.S., M.M. Elsner, G.S. Mauger, E.R. Lutz, A.F. Hamlet, and E.P. Salathé, 2011: Regional Climate and Hydrologic Change in the Northern U.S. Rockies and Pacific Northwest: Internally Consistent Projections of Future Climate for Resource Management. University of Washington, Seattle. https://cig.uw.edu/publications/regional-climate-and-hydrologic-change-in-the-northern-u-s-rockies-and-pacific-northwest-internally-consistent-projections-of-future-climate-for-resource-management/

73. Elsner, M.M., L. Cuo, N. Voisin, J.S. Deems, A.F. Hamlet, J.A. Vano, K.E.B. Mickelson, S.Y. Lee, and D.P. Lettenmaier, 2010: Implications of 21st century climate change for the hydrology of Washington State. *Climatic Change*, **102**, 225-260. http://dx.doi.org/10.1007/s10584-010-9855-0

74. Zeng, N., F. Zhao, G.J. Collatz, E. Kalnay, R.J. Salawitch, T.O. West, and L. Guanter, 2014: Agricultural green revolution as a driver of increasing atmospheric CO2 seasonal amplitude. *Nature*, **515**, 394-397. http://dx.doi.org/10.1038/nature13893

75. Gray, J.M., S. Frolking, E.A. Kort, D.K. Ray, C.J. Kucharik, N. Ramankutty, and M.A. Friedl, 2014: Direct human influence on atmospheric CO2 seasonality from increased cropland productivity. *Nature*, **515**, 398-401. http://dx.doi.org/10.1038/nature13957

76. Challinor, A.J., J. Watson, D.B. Lobell, S.M. Howden, D.R. Smith, and N. Chhetri, 2014: A meta-analysis of crop yield under climate change and adaptation. *Nature Climate Change*, **4**, 287-291. http://dx.doi.org/10.1038/nclimate2153

77. Lobell, D.B. and C. Tebaldi, 2014: Getting caught with our plants down: The risks of a global crop yield slowdown from climate trends in the next two decades. *Environmental Research Letters*, **9**, 074003. http://dx.doi.org/10.1088/1748-9326/9/7/074003

78. Zhang, X., L. Alexander, G.C. Hegerl, P. Jones, A.K. Tank, T.C. Peterson, B. Trewin, and F.W. Zwiers, 2011: Indices for monitoring changes in extremes based on daily temperature and precipitation data. *Wiley Interdisciplinary Reviews: Climate Change*, **2**, 851-870. http://dx.doi.org/10.1002/wcc.147

79. Daly, C., M.P. Widrlechner, M.D. Halbleib, J.I. Smith, and W.P. Gibson, 2012: Development of a new USDA plant hardiness zone map for the United States. *Journal of Applied Meteorology and Climatology*, **51**, 242-264. http://dx.doi.org/10.1175/2010JAMC2536.1

80. Chan, D. and Q. Wu, 2015: Significant anthropogenic-induced changes of climate classes since 1950. *Scientific Reports*, **5**, 13487. http://dx.doi.org/10.1038/srep13487

81. Diaz, H.F. and J.K. Eischeid, 2007: Disappearing "alpine tundra" Köppen climatic type in the western United States. *Geophysical Research Letters*, **34**, L18707. http://dx.doi.org/10.1029/2007GL031253

82. Grundstein, A., 2008: Assessing climate change in the contiguous United States using a modified Thornthwaite climate classification scheme. *The Professional Geographer*, **60**, 398-412. http://dx.doi.org/10.1080/00330120802046695

83. EPA, 2016: Climate Change Indicators in the United States, 2016. 4th edition. EPA 430-R-16-004. U.S. Environmental Protection Agency, Washington, D.C., 96 pp. https://www.epa.gov/sites/production/files/2016-08/documents/climate_indicators_2016.pdf

84. Yang, L.H. and V.H.W. Rudolf, 2010: Phenology, ontogeny and the effects of climate change on the timing of species interactions. *Ecology Letters*, **13**, 1-10. http://dx.doi.org/10.1111/j.1461-0248.2009.01402.x

85. Rafferty, N.E. and A.R. Ives, 2011: Effects of experimental shifts in flowering phenology on plant–pollinator interactions. *Ecology Letters*, **14**, 69-74. http://dx.doi.org/10.1111/j.1461-0248.2010.01557.x

86. Kudo, G. and T.Y. Ida, 2013: Early onset of spring increases the phenological mismatch between plants and pollinators. *Ecology*, **94**, 2311-2320. http://dx.doi.org/10.1890/12-2003.1

87. Forrest, J.R.K., 2015: Plant–pollinator interactions and phenological change: What can we learn about climate impacts from experiments and observations? *Oikos*, **124**, 4-13. http://dx.doi.org/10.1111/oik.01386

88. Parker, L.E. and J.T. Abatzoglou, 2016: Projected changes in cold hardiness zones and suitable overwinter ranges of perennial crops over the United States. *Environmental Research Letters*, **11**, 034001. http://dx.doi.org/10.1088/1748-9326/11/3/034001

89. Hellmann, J.J., J.E. Byers, B.G. Bierwagen, and J.S. Dukes, 2008: Five potential consequences of climate change for invasive species. *Conservation Biology*, **22**, 534-543. http://dx.doi.org/10.1111/j.1523-1739.2008.00951.x

90. Sippel, S., J. Zscheischler, and M. Reichstein, 2016: Ecosystem impacts of climate extremes crucially depend on the timing. *Proceedings of the National Academy of Sciences*, **113**, 5768-5770. http://dx.doi.org/10.1073/pnas.1605667113

91. Maloney, E.D., S.J. Camargo, E. Chang, B. Colle, R. Fu, K.L. Geil, Q. Hu, X. Jiang, N. Johnson, K.B. Karnauskas, J. Kinter, B. Kirtman, S. Kumar, B. Langenbrunner, K. Lombardo, L.N. Long, A. Mariotti, J.E. Meyerson, K.C. Mo, J.D. Neelin, Z. Pan, R. Seager, Y. Serra, A. Seth, J. Sheffield, J. Stroeve, J. Thibeault, S.-P. Xie, C. Wang, B. Wyman, and M. Zhao, 2014: North American climate in CMIP5 experiments: Part III: Assessment of twenty-first-century projections. *Journal of Climate*, **27**, 2230-2270. http://dx.doi.org/10.1175/JCLI-D-13-00273.1

92. Marston, L., M. Konar, X. Cai, and T.J. Troy, 2015: Virtual groundwater transfers from overexploited aquifers in the United States. *Proceedings of the National Academy of Sciences*, **112**, 8561-8566. http://dx.doi.org/10.1073/pnas.1500457112

93. Lippke, B., E. Oneil, R. Harrison, K. Skog, L. Gustavsson, and R. Sathre, 2011: Life cycle impacts of forest management and wood utilization on carbon mitigation: Knowns and unknowns. *Carbon Management*, **2**, 303-333. http://dx.doi.org/10.4155/CMT.11.24

94. EPA, 2005: Greenhouse Gas Mitigation Potential in U.S. Forestry and Agriculture. EPA 430-R-05-006. Environmental Protection Agency, Washington, D.C., 154 pp. https://www3.epa.gov/climatechange/Downloads/ccs/ghg_mitigation_forestry_ag_2005.pdf

95. King, S.L., D.J. Twedt, and R.R. Wilson, 2006: The role of the wetland reserve program in conservation efforts in the Mississippi River alluvial valley. *Wildlife Society Bulletin*, **34**, 914-920. http://dx.doi.org/10.2193/0091-7648(2006)34[914:TROTWR]2.0.CO;2

96. Wear, D.N. and J.W. Coulston, 2015: From sink to source: Regional variation in U.S. forest carbon futures. *Scientific Reports*, **5**, 16518. http://dx.doi.org/10.1038/srep16518

97. McKinley, D.C., M.G. Ryan, R.A. Birdsey, C.P. Giardina, M.E. Harmon, L.S. Heath, R.A. Houghton, R.B. Jackson, J.F. Morrison, B.C. Murray, D.E. Pataki, and K.E. Skog, 2011: A synthesis of current knowledge on forests and carbon storage in the United States. *Ecological Applications*, **21**, 1902-1924. http://dx.doi.org/10.1890/10-0697.1

98. Pan, Y., R.A. Birdsey, J. Fang, R. Houghton, P.E. Kauppi, W.A. Kurz, O.L. Phillips, A. Shvidenko, S.L. Lewis, J.G. Canadell, P. Ciais, R.B. Jackson, S.W. Pacala, A.D. McGuire, S. Piao, A. Rautiainen, S. Sitch, and D. Hayes, 2011: A large and persistent carbon sink in the world's forests. *Science*, **333**, 988-93. http://dx.doi.org/10.1126/science.1201609

99. Hansen, M.C., P.V. Potapov, R. Moore, M. Hancher, S.A. Turubanova, A. Tyukavina, D. Thau, S.V. Stehman, S.J. Goetz, T.R. Loveland, A. Kommareddy, A. Egorov, L. Chini, C.O. Justice, and J.R.G. Townshend, 2013: High-resolution global maps of 21st-century forest cover change. *Science*, **342**, 850-853. http://dx.doi.org/10.1126/science.1244693

100. Williams, A.P., C.D. Allen, A.K. Macalady, D. Griffin, C.A. Woodhouse, D.M. Meko, T.W. Swetnam, S.A. Rauscher, R. Seager, H.D. Grissino-Mayer, J.S. Dean, E.R. Cook, C. Gangodagamage, M. Cai, and N.G. McDowell, 2013: Temperature as a potent driver of regional forest drought stress and tree mortality. *Nature Climate Change*, **3**, 292-297. http://dx.doi.org/10.1038/nclimate1693

101. Grimmond, C.S.B., H.C. Ward, and S. Kotthaus, 2016: Effects of urbanization on local and regional climate. *Routledge Handbook of Urbanization and Global Environment Change*. Seto, K.C., W.D. Solecki, and C.A. Griffith, Eds. Routledge, London, 169-187.

102. Mitra, C. and J.M. Shepherd, 2016: Urban precipitation: A global perspective. *Routledge Handbook of Urbanization and Global Environment Change*. Seto, K.C., W.D. Solecki, and C.A. Griffith, Eds. Routledge, London, 152-168.

103. Landsberg, H.E., 1970: Man-made climatic changes: Man's activities have altered the climate of urbanized areas and may affect global climate in the future. *Science*, **170**, 1265-1274. http://dx.doi.org/10.1126/science.170.3964.1265

104. Mills, G., 2007: Cities as agents of global change. *International Journal of Climatology*, **27**, 1849-1857. http://dx.doi.org/10.1002/joc.1604

105. Seto, K.C., W.D. Solecki, and C.A. Griffith, eds., 2016: *Routledge Handbook on Urbanization and Global Environmental Change*. Routledge: London, 582 pp.

106. Seto, K.C. and J.M. Shepherd, 2009: Global urban land-use trends and climate impacts. *Current Opinion in Environmental Sustainability*, **1**, 89-95. http://dx.doi.org/10.1016/j.cosust.2009.07.012

107. Shepherd, J.M., 2013: Impacts of urbanization on precipitation and storms: Physical insights and vulnerabilities *Climate Vulnerability: Understanding and Addressing Threats to Essential Resources.* Academic Press, Oxford, 109-125. http://dx.doi.org/10.1016/B978-0-12-384703-4.00503-7

108. Hidalgo, J., V. Masson, A. Baklanov, G. Pigeon, and L. Gimeno, 2008: Advances in urban cimate mdeling. *Annals of the New York Academy of Sciences,* **1146,** 354-374. http://dx.doi.org/10.1196/annals.1446.015

109. Imhoff, M.L., P. Zhang, R.E. Wolfe, and L. Bounoua, 2010: Remote sensing of the urban heat island effect across biomes in the continental USA. *Remote Sensing of Environment,* **114,** 504-513. http://dx.doi.org/10.1016/j.rse.2009.10.008

110. Bounoua, L., P. Zhang, G. Mostovoy, K. Thome, J. Masek, M. Imhoff, M. Shepherd, D. Quattrochi, J. Santanello, J. Silva, R. Wolfe, and A.M. Toure, 2015: Impact of urbanization on US surface climate. *Environmental Research Letters,* **10,** 084010. http://dx.doi.org/10.1088/1748-9326/10/8/084010

111. Kaufmann, R.K., K.C. Seto, A. Schneider, Z. Liu, L. Zhou, and W. Wang, 2007: Climate response to rapid urban growth: Evidence of a human-induced precipitation deficit. *Journal of Climate,* **20,** 2299-2306. http://dx.doi.org/10.1175/jcli4109.1

112. Shepherd, J.M., B.S. Ferrier, and P.S. Ray, 2001: Rainfall morphology in Florida convergence zones: A numerical study. *Monthly Weather Review,* **129,** 177-197. http://dx.doi.org/10.1175/1520-0493(2001)129<0177:rmifcz>2.0.co;2

113. van den Heever, S.C. and W.R. Cotton, 2007: Urban aerosol impacts on downwind convective storms. *Journal of Applied Meteorology and Climatology,* **46,** 828-850. http://dx.doi.org/10.1175/jam2492.1

114. World Bank, 2017: Urban population (% of total): United States (1960-2015). World Bank Open Data. http://data.worldbank.org/indicator/SP.URB.TOTL.IN.ZS?locations=US

115. Zhao, L., X. Lee, R.B. Smith, and K. Oleson, 2014: Strong contributions of local background climate to urban heat islands. *Nature,* **511,** 216-219. http://dx.doi.org/10.1038/nature13462

116. Marotz, G.A., J. Clark, J. Henry, and R. Standfast, 1975: Cloud fields over irrigated areas in southwestern Kansas—Data and speculations. *The Professional Geographer,* **27,** 457-461. http://dx.doi.org/10.1111/j.0033-0124.1975.00457.x

117. Barnston, A.G. and P.T. Schickedanz, 1984: The effect of irrigation on warm season precipitation in the southern Great Plains. *Journal of Climate and Applied Meteorology,* **23,** 865-888. http://dx.doi.org/10.1175/1520-0450(1984)023<0865:TEOIOW>2.0.CO;2

118. Alpert, P. and M. Mandel, 1986: Wind variability—An indicator for a mesoclimatic change in Israel. *Journal of Climate and Applied Meteorology,* **25,** 1568-1576. http://dx.doi.org/10.1175/1520-0450(1986)025<1568:wvifam>2.0.co;2

119. Pielke, R.A., Sr. and X. Zeng, 1989: Influence on severe storm development of irrigated land. *National Weather Digest* **14,** 16-17.

120. Pielke, R.A., Sr., J. Adegoke, A. BeltráN-Przekurat, C.A. Hiemstra, J. Lin, U.S. Nair, D. Niyogi, and T.E. Nobis, 2007: An overview of regional land-use and land-cover impacts on rainfall. *Tellus B,* **59,** 587-601. http://dx.doi.org/10.1111/j.1600-0889.2007.00251.x

121. IPCC, 2013: Summary for policymakers. *Climate Change 2013: The Physical Science Basis. Contribution of Working Group I to the Fifth Assessment Report of the Intergovernmental Panel on Climate Change.* Stocker, T.F., D. Qin, G.-K. Plattner, M. Tignor, S.K. Allen, J. Boschung, A. Nauels, Y. Xia, V. Bex, and P.M. Midgley, Eds. Cambridge University Press, Cambridge, United Kingdom and New York, NY, USA, 1–30. http://www.climatechange2013.org/report/

122. Christensen, J.H., K. Krishna Kumar, E. Aldrian, S.-I. An, I.F.A. Cavalcanti, M. de Castro, W. Dong, P. Goswami, A. Hall, J.K. Kanyanga, A. Kitoh, J. Kossin, N.-C. Lau, J. Renwick, D.B. Stephenson, S.-P. Xie, and T. Zhou, 2013: Climate phenomena and their relevance for future regional climate change. *Climate Change 2013: The Physical Science Basis. Contribution of Working Group I to the Fifth Assessment Report of the Intergovernmental Panel on Climate Change.* Stocker, T.F., D. Qin, G.-K. Plattner, M. Tignor, S.K. Allen, J. Boschung, A. Nauels, Y. Xia, V. Bex, and P.M. Midgley, Eds. Cambridge University Press, Cambridge, United Kingdom and New York, NY, USA, 1217–1308. http://www.climatechange2013.org/report/full-report/

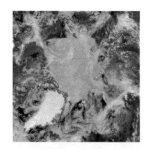

# 11

# Arctic Changes and their Effects on Alaska and the Rest of the United States

## KEY FINDINGS

1. Annual average near-surface air temperatures across Alaska and the Arctic have increased over the last 50 years at a rate more than twice as fast as the global average temperature (*very high confidence*).

2. Rising Alaskan permafrost temperatures are causing permafrost to thaw and become more discontinuous; this process releases additional carbon dioxide and methane, resulting in an amplifying feedback and additional warming (*high confidence*). The overall magnitude of the permafrost–carbon feedback is uncertain; however, it is clear that these emissions have the potential to compromise the ability to limit global temperature increases.

3. Arctic land and sea ice loss observed in the last three decades continues, in some cases accelerating (*very high confidence*). It is *virtually certain* that Alaska glaciers have lost mass over the last 50 years, with each year since 1984 showing an annual average ice mass less than the previous year. Based on gravitational data from satellites, average ice mass loss from Greenland was −269 Gt per year between April 2002 and April 2016, accelerating in recent years (*high confidence*). Since the early 1980s, annual average arctic sea ice has decreased in extent between 3.5% and 4.1% per decade, become thinner by between 4.3 and 7.5 feet, and began melting at least 15 more days each year. September sea ice extent has decreased between 10.7% and 15.9% per decade (*very high confidence*). Arctic-wide ice loss is expected to continue through the 21st century, *very likely* resulting in nearly sea ice-free late summers by the 2040s (*very high confidence*).

4. It is *very likely* that human activities have contributed to observed arctic surface temperature warming, sea ice loss, glacier mass loss, and Northern Hemisphere snow extent decline (*high confidence*).

5. Atmospheric circulation patterns connect the climates of the Arctic and the contiguous United States. Evidenced by recent record warm temperatures in the Arctic and emerging science, the midlatitude circulation has influenced observed arctic temperatures and sea ice (*high confidence*). However, confidence is *low* regarding whether or by what mechanisms observed arctic warming may have influenced the midlatitude circulation and weather patterns over the continental United States. The influence of arctic changes on U.S. weather over the coming decades remains an open question with the potential for significant impact.

### Recommended Citation for Chapter

**Taylor**, P.C., W. Maslowski, J. Perlwitz, and D.J. Wuebbles, 2017: Arctic changes and their effects on Alaska and the rest of the United States. In: *Climate Science Special Report: Fourth National Climate Assessment, Volume I* [Wuebbles, D.J., D.W. Fahey, K.A. Hibbard, D.J. Dokken, B.C. Stewart, and T.K. Maycock (eds.)]. U.S. Global Change Research Program, Washington, DC, USA, pp. 303-332, doi: 10.7930/J00863GK.

## 11.1 Introduction

Climate changes in Alaska and across the Arctic continue to outpace changes occurring across the globe. The Arctic, defined as the area north of the Arctic Circle, is a vulnerable and complex system integral to Earth's climate. The vulnerability stems in part from the extensive cover of ice and snow, where the freezing point marks a critical threshold that when crossed has the potential to transform the region. Because of its high sensitivity to radiative forcing and its role in amplifying warming,[1] the arctic cryosphere is a key indicator of the global climate state. Accelerated melting of multiyear sea ice, mass loss from the Greenland Ice Sheet (GrIS), reduction of terrestrial snow cover, and permafrost degradation are stark examples of the rapid Arctic-wide response to global warming. These local arctic changes influence global sea level, ocean salinity, the carbon cycle, and potentially atmospheric and oceanic circulation patterns. Arctic climate change has altered the global climate in the past[2] and will influence climate in the future.

As an arctic nation, United States' decisions regarding climate change adaptation and mitigation, resource development, trade, national security, transportation, etc., depend on projections of future Alaskan and arctic climate. Aside from uncertainties due to natural variability, scientific uncertainty, and human activities including greenhouse gas emissions (see Ch. 4: Projections), additional unique uncertainties in our understanding of arctic processes thwart projections, including mixed-phase cloud processes;[3] boundary layer processes;[4] sea ice mechanics;[4] and ocean currents, eddies, and tides that affect the advection of heat into and around the Arctic Ocean.[5, 6] The inaccessibility of the Arctic has made it difficult to sustain the high-quality observations of the atmosphere, ocean, land, and ice required to improve physically-based models.

Improved data quality and increased observational coverage would help address societally relevant arctic science questions.

Despite these challenges, our scientific knowledge is sufficiently advanced to effectively inform policy. This chapter documents significant scientific progress and knowledge about how the Alaskan and arctic climate has changed and will continue to change.

## 11.2 Arctic Changes

### 11.2.1 Alaska and Arctic Temperature

Surface temperature—an essential component of the arctic climate system—drives and signifies change, fundamentally controlling the melting of ice and snow. Further, the vertical profile of boundary layer temperature modulates the exchange of mass, energy, and momentum between the surface and atmosphere, influencing other components such as clouds.[7, 8] Arctic temperatures exhibit spatial and interannual variability due to interactions and feedbacks between sea ice, snow cover, atmospheric heat transports, vegetation, clouds, water vapor, and the surface energy budget.[9, 10, 11] Interannual variations in Alaskan temperatures are strongly influenced by decadal variability like the Pacific Decadal Oscillation (Ch. 5: Circulation and Variability).[12, 13] However, observed temperature trends exceed this variability.

Arctic surface and atmospheric temperatures have substantially increased in the observational record. Multiple observation sources, including land-based surface stations since at least 1950 and available meteorological reanalysis datasets, provide evidence that arctic near-surface air temperatures have increased more than twice as fast as the global average.[14, 15, 16, 17, 18] Showing enhanced arctic warming since 1981, satellite-observed arctic average surface skin temperatures have increased by 1.08° ± 0.13°F (+0.60° ± 0.07°C) per decade.[19]

As analyzed in Chapter 6: Temperature Change (Figure 6.1), strong near-surface air temperature warming has occurred across Alaska exceeding 1.5°F (0.8°C) over the last 30 years. Especially strong warming has occurred over Alaska's North Slope during autumn. For example, Utqiagvik's (formally Barrow) warming since 1979 exceeds 7°F (3.8°C) in September, 12°F (6.6°C) in October, and 10°F (5.5°C) in November.[20]

Enhanced arctic warming is a robust feature of the climate response to anthropogenic forcing.[21, 22] An anthropogenic contribution to arctic and Alaskan surface temperature warming over the past 50 years is *very likely*.[23, 24, 25, 26, 27] One study argues that the natural forcing has not contributed to the long-term arctic warming in a discernable way.[27] Also, other anthropogenic forcings (mostly aerosols) have *likely* offset up to 60% of the high-latitude greenhouse gas warming since 1913,[27] suggesting that arctic warming to date would have been larger without the offsetting influence of aerosols. Other studies argue for a more significant contribution of natural variability to observed arctic temperature trends[24, 28] and indicate that natural variability alone cannot explain observed warming. It is *very likely* that arctic surface temperatures will continue to increase faster than the global mean through the 21st century.[25, 26, 27, 29]

### 11.2.2 Arctic Sea Ice Change

Arctic sea ice strongly influences Alaskan, arctic, and global climate by modulating exchanges of mass, energy, and momentum between the ocean and the atmosphere. Variations in arctic sea ice cover also influence atmospheric temperature and humidity, wind patterns, clouds, ocean temperature, thermal stratification, and ecosystem productivity.[7, 10, 30, 31, 32, 33, 34, 35, 36, 37] Arctic sea ice exhibits significant interannual, spatial, and seasonal variability driven by atmospheric wind patterns and cyclones, atmospheric temperature and humidity structure, clouds, radiation, sea ice dynamics, and the ocean. [38, 39, 40, 41, 42, 43, 44]

Overwhelming evidence indicates that the character of arctic sea ice is rapidly changing. Observational evidence shows Arctic-wide sea ice decline since 1979, accelerating ice loss since 2000, and some of the fastest loss along the Alaskan coast.[19, 20, 45, 46] Although sea ice loss is found in all months, satellite observations show the fastest loss in late summer and autumn.[45] Since 1979, the annual average arctic sea ice extent has *very likely* decreased at a rate of 3.5%–4.1% per decade.[19, 37] Regional sea ice melt along the Alaskan coasts exceeds the arctic average rates with declines in the Beaufort and Chukchi Seas of −4.1% and −4.7% per decade, respectively.[20] The annual minimum and maximum sea ice extent have decreased over the last 35 years by −13.3% ± 2.6% and −2.7% ± 0.5% per decade, respectively.[47] The ten lowest September sea ice extents over the satellite period have all occurred in the last ten years, the lowest in 2012. The 2016 September sea ice minimum tied with 2007 for the second lowest on record, but rapid refreezing resulted in the 2016 September monthly average extent being the fifth lowest. Despite the rapid initial refreezing, sea ice extent was again in record low territory during fall–winter 2016/2017 due to anomalously warm temperatures in the marginal seas around Alaska,[47] contributing to a new record low in winter ice-volume (see http://psc.apl.uw.edu/research/projects/arctic-sea-ice-volume-anomaly).[48]

Other important characteristics of arctic sea ice have also changed, including thickness, age, and volume. Sea ice thickness is monitored using an array of satellite, aircraft, and vessel measurements.[37, 45] The mean thickness of the arctic sea ice during winter between 1980 and 2008 has decreased between 4.3 and 7.5 feet (1.3 and 2.3 meters).[37] The age distribution

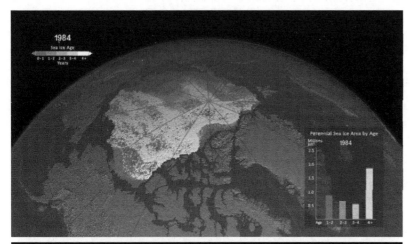

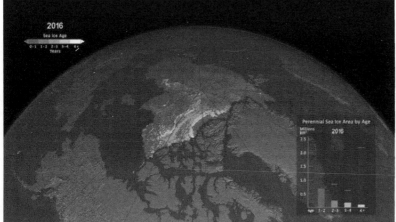

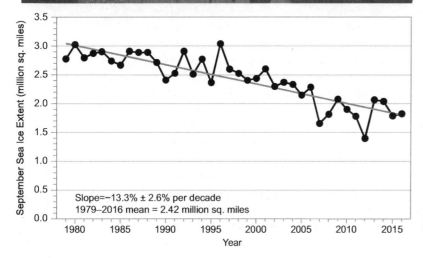

Slope=−13.3% ± 2.6% per decade
1979–2016 mean = 2.42 million sq. miles

Figure 11.1: September sea ice extent and age shown for (a) 1984 and (b) 2016, illustrating significant reductions in sea ice extent and age (thickness). Bar graph in the lower right of each panel illustrates the sea ice area (unit: million km$^2$) covered within each age category (>1 year), and the green bars represent the maximum value for each age range during the record. The year 1984 is representative of September sea ice characteristics during the 1980s. The years 1984 and 2016 are selected as endpoints in the time series; a movie of the complete time series is available at http://svs.gsfc.nasa.gov/cgi-bin/details.cgi?aid=4489. (c) Shows the satellite-era arctic sea ice areal extent trend from 1979 to 2016 for September (unit: million mi$^2$). [Figure source: Panels (a),(b): NASA Science Visualization Studio; data: Tschudi et al. 2016;[49] Panel (c) data: Fetterer et al. 2016[209]].

of sea ice has become younger since 1988. In March 2016, first-year (multi-year) sea ice accounted for 78% (22%) of the total extent, whereas in the 1980s first-year (multi-year) sea ice accounted for 55% (45%).[47] Moreover, ice older than four years accounted for 16% of the March 1985 icepack but accounted for only 1.2% of the icepack in March 2016, indicating significant changes in sea ice volume.[47] The top two panels in Figure 11.1 show the September sea ice extent and age in 1984 and 2016, illustrating significant reductions in sea ice age.[49] While these panels show only two years (beginning point and ending point) of the complete time series, these two years are representative of the overall trends discussed and shown in the September sea ice extent time series in the bottom panel of Fig 11.1. Younger, thinner sea ice is more susceptible to melt, therefore reductions in age and thickness imply a larger interannual variability of extent.

Sea ice melt season—defined as the number of days between spring melt onset and fall freeze-up—has lengthened Arctic-wide by at least five days per decade since 1979, with larger regional changes.[46, 50] Some of the largest observed changes in sea ice melt season (Figure 11.2) are found along Alaska's northern and western coasts, lengthening the melt season by 20–30 days per decade and increasing the annual number of ice-free days by more than 90.[50] Summer sea ice retreat along coastal Alaska has led to longer open water seasons, making the Alaskan coastline more vulnerable to erosion.[51, 52] Increased melt season length corresponds to increased absorption of solar radiation by the Arctic Ocean during summer and increases upper ocean temperature, delaying fall freeze-up. Overall, this process significantly contributes to reductions in arctic sea ice.[42, 46] Wind-driven sea ice export through the Fram Strait has not increased over the last 80 years;[37] however, one recent study suggests that it may have increased since 1979.[53]

It is *very likely* that there is an anthropogenic contribution to the observed arctic sea ice decline since 1979. A range of modeling studies analyzing the September sea ice extent trends in simulations with and without anthropogenic forcing conclude that these declines cannot be explained by natural variability alone.[54, 55, 56, 57, 58, 59] Further, observational-based analyses considering a range of anthropogenic and natural forcing mechanisms for September sea ice loss reach the same conclusion.[60] Considering the occurrence of individual September sea ice anomalies, internal climate variability alone *very likely* could not have caused recently observed record low arctic sea ice extents, such as in September 2012.[61, 62] The potential contribution of natural variability to arctic sea ice trends is significant.[55, 63, 64] One recent study[28] indicates that internal variability dominates arctic atmospheric circulation trends, accounting for 30%–50% of the sea ice reductions since 1979, and up to 60% in September. However, previous studies indicate that the contributions from internal variability are smaller than 50%.[54, 55] This apparent significant contribution of natural variability to sea ice decline indicates that natural variability alone cannot explain the observed sea ice decline and is consistent with the statement that it is *very likely* there is an anthropogenic contribution to the observed arctic sea ice decline since 1979.

Continued sea ice loss is expected across the Arctic, which is *very likely* to result in late summers becoming nearly ice-free (areal extent less than $10^6$ km² or approximately 3.9 × $10^5$ mi²) by the 2040s.[21, 65] Natural variability,[66] future scenarios, and model uncertainties[64, 67, 68] all influence sea ice projections. One study suggests that internal variability alone accounts for a 20-year prediction uncertainty in

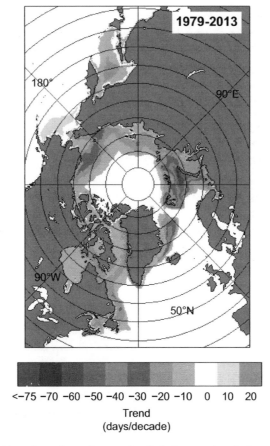

Figure 11.2: A 35-year trend in arctic sea ice melt season length, in days per decade, from passive microwave satellite observations, illustrating that the sea ice season has shortened by more than 60 days in coastal Alaska over the last 30 years. (Figure source: adapted from Parkinson 2014[50]).

the timing of the first occurrence of an ice-free summer, whereas differences between a higher scenario (RCP8.5) and a lower scenario (RCP4.5) add only 5 years.[63] Projected September sea ice reductions by 2081–2100 range from 43% for an even lower scenario (RCP2.6) to 94% for RCP8.5.[21] However, September sea ice projections over the next few decades are similar for the different anthropogenic forcing associated with these scenarios; scenario dependent sea ice loss only becomes apparent after 2050. Another study[69] indicates that the total sea ice loss scales roughly linearly with $CO_2$ emissions, such that an additional 1,000 GtC from present day levels corresponds to

ice-free conditions in September. A key message from the Third National Climate Assessment (NCA3)[70] was that arctic sea ice is disappearing. The fundamental conclusion of this assessment is unchanged; additional research corroborates the NCA3 statement.

### 11.2.3 Arctic Ocean and Marginal Seas
*Sea Surface Temperature*
Arctic Ocean sea surface temperatures (SSTs) have increased since comprehensive records became available in 1982. Satellite-observed Arctic Ocean SSTs, poleward of 60°N, exhibit a trend of 0.16° ± 0.02°F (0.09° ± 0.01°C) per decade.[19] Arctic Ocean SST is controlled by a

combination of factors, including solar radiation and energy transport from ocean currents and atmospheric winds. Summertime Arctic Ocean SST trends and patterns strongly couple with sea ice extent; however, clouds, ocean color, upper-ocean thermal structure, and atmospheric circulation also play a role.[40, 71] Along coastal Alaska, SSTs in the Chukchi Sea exhibit a statistically significant (95% confidence) trend of 0.9° ± 0.5°F (0.5° ± 0.3°C) per decade.[72]

Arctic Ocean temperatures also increased at depth.[71, 73] Since 1970, Arctic Ocean Intermediate Atlantic Water—located between 150 and 900 meters—has warmed by 0.86° ± 0.09°F (0.48° ± 0.05°C) per decade; the most recent decade being the warmest.[73] The observed temperature level is unprecedented in the last 1,150 years for which proxy indicators provide records.[74, 75] The influence of Intermediate Atlantic Water warming on future Alaska and arctic sea ice loss is unclear.[38, 76]

*Alaskan Sea Level Rise*
The Alaskan coastline is vulnerable to sea level rise (SLR); however, strong regional variability exists in current trends and future projections. Some regions are experiencing relative sea level fall, whereas others are experiencing relative sea level rise, as measured by tide gauges that are part of NOAA's National Water Level Observation Network. These tide gauge data show sea levels rising fastest along the northern coast of Alaska but still slower than the global average, due to isostatic rebound (Ch. 12: Sea Level Rise).[77] However, considerable uncertainty in relative sea level rise exists due to a lack of tide gauges; for example, no tide gauges are located between Bristol Bay and Norton Sound or between Cape Lisburne and Prudhoe Bay. Under almost all future scenarios, SLR along most of the Alaskan coastline is projected to be less than the global average (Ch. 12: Sea Level Rise).

*Salinity*
Arctic Ocean salinity influences the freezing temperature of sea ice (less salty water freezes more readily) and the density profile representing the integrated effects of freshwater transport, river runoff, evaporation, and sea ice processes. Arctic Ocean salinity exhibits multidecadal variability, hampering the assessment of long-term trends.[78] Emerging evidence suggests that the Arctic Ocean and marginal sea salinity has decreased in recent years despite short-lived regional salinity increases between 2000 and 2005.[71] Increased river runoff, rapid melting of sea and land ice, and changes in freshwater transport have influenced observed Arctic Ocean salinity.[71, 79]

*Ocean Acidification*
Arctic Ocean acidification is occurring at a faster rate than the rest of the globe (see also Ch. 13: Ocean Changes).[80] Coastal Alaska and its ecosystems are especially vulnerable to ocean acidification because of the high sensitivity of Arctic Ocean water chemistry to changes in sea ice, respiration of organic matter, upwelling, and increasing river runoff.[80] Sea ice loss and a longer melt season contribute to increased vulnerability of the Arctic Ocean to acidification by lowering total alkalinity, permitting greater upwelling, and influencing the primary production characteristics in coastal Alaska.[81, 82, 83, 84, 85, 86] Global-scale modeling studies suggest that the largest and most rapid changes in pH will continue along Alaska's coast, indicating that ocean acidification may increase enough by the 2030s to significantly influence coastal ecosystems.[80]

### 11.2.4 Boreal Wildfires
Alaskan wildfire activity has increased in recent decades. This increase has occurred both in the boreal forest[87] and in the arctic tundra,[88] where fires historically were smaller and less frequent. A shortened snow cover season and higher temperatures over the last 50 years[89] make the Arctic more vulnerable to wildfire.[87,

[88, 90] Total area burned and the number of large fires (those with area greater than 1,000 km² or 386 mi²) in Alaska exhibit significant interannual and decadal variability, from influences of atmospheric circulation patterns and controlled burns, but have *likely* increased since 1959.[91] The most recent decade has seen an unusually large number of years with anomalously large wildfires in Alaska.[92] Studies indicate that anthropogenic climate change has *likely* lengthened the wildfire season and increased the risk of severe fires.[93] Further, wildfire risks are expected to increase through the end of the century due to warmer, drier conditions.[90, 94] Using climate simulations to force an ecosystem model over Alaska (Alaska Frame-Based Ecosystem Code, ALFRESCO), the total area burned is projected to increase between 25% and 53% by 2100.[95] A transition into a regime of fire activity unprecedented in the last 10,000 years is possible.[96] We conclude that there is *medium confidence* for a human-caused climate change contribution to increased forest fire activity in Alaska in recent decades. See Chapter 8: Drought, Floods, and Wildfires for more details.

A significant amount of the total global soil carbon is found in the boreal forest and tundra ecosystems, including permafrost.[97, 98, 99] Increased fire activity could deplete these stores, releasing them to the atmosphere to serve as an additional source of atmospheric $CO_2$.[97, 100] Increased fires may also enhance the degradation of Alaska's permafrost by blackening the ground, reducing surface albedo, and removing protective vegetation.[101, 102, 103, 104]

### 11.2.5 Snow Cover

Snow cover extent has significantly decreased across the Northern Hemisphere and Alaska over the last decade (see also Ch. 7: Precipitation Change and Ch. 10: Land Cover).[105, 106] Northern Hemisphere June snow cover decreased by more than 65% between 1967 and 2012,[37, 107] at a trend of −17.2% per decade since 1979.[89] June snow cover dipped below 3 million square km (approximately 1.16 million square miles) for the fifth time in six years between 2010 and 2015, a threshold not crossed in the previous 43 years of record.[89] Early season snow cover in May, which affects the accumulation of solar insolation through the summer, has also declined at −7.3% per decade, due to reduced winter accumulation from warmer temperatures. Regional trends in snow cover duration vary, with some showing earlier onsets while others show later onsets.[89] In Alaska, the 2016 May statewide snow coverage of 595,000 square km (approximately 372,000 square miles) was the lowest on record dating back to 1967; the snow coverage of 2015 was the second lowest, and 2014 was the fourth lowest.

Human activities have *very likely* contributed to observed snow cover declines over the last 50 years. Attribution studies indicate that observed trends in Northern Hemisphere snow cover cannot be explained by natural forcing alone, but instead require anthropogenic forcing.[24, 106, 108] Declining snow cover is expected to continue and will be affected by both the anthropogenic forcing and evolution of arctic ecosystems. The observed tundra shrub expansion and greening[109, 110] affects melt by influencing snow depth, melt dynamics, and the local surface energy budget. Nevertheless, model simulations show that future reductions in snow cover influence biogeochemical feedbacks and warming more strongly than changes in vegetation cover and fire in the North American Arctic.[111]

### 11.2.6 Continental Ice Sheets and Mountain Glaciers

Mass loss from ice sheets and glaciers influences sea level rise, the oceanic thermohaline circulation, and the global energy budget. Moreover, the relative contribution of GrIS to global sea level rise continues to increase, exceeding the contribution from thermal expansion (see Ch. 12: Sea Level Rise). Observational and modeling studies indicate that GrIS and glaciers in Alaska are out of mass balance with current climate conditions and are rapidly losing mass.[37, 112] In recent years, mass loss has accelerated and is expected to continue.[112, 113]

Dramatic changes have occurred across GrIS, particularly at its margins. GrIS average annual mass loss from January 2003 to May 2013 was −244 ± 6 Gt per year (approximately 0.26 inches per decade sea level equivalent).[113] One study indicates that ice mass loss from Greenland was −269 Gt per year between April 2002 and April 2016.[47] Increased surface melt, runoff, and increased outlet glacier discharge from warmer air temperatures are primary contributing factors.[114, 115, 116, 117, 118] The effects of warmer air and ocean temperatures on GrIS can be amplified by ice dynamical feedbacks, such as faster sliding, greater calving, and increased submarine melting.[116, 119, 120, 121] Shallow ocean warming and regional ocean and atmospheric circulation changes also contribute to mass loss.[122, 123, 124] The underlying mechanisms of the recent discharge speed-up remain unclear;[125, 126] however, warmer subsurface ocean and atmospheric temperatures[118, 127, 128] and meltwater penetration to the glacier bed[125, 129] *very likely* contribute.

Annual average ice mass from Arctic-wide glaciers has decreased every year since 1984,[112, 130, 131] with significant losses in Alaska, especially over the past two decades (Figure 11.3).[37, 132] Figure 11.4 illustrates observed changes from U.S. Geological Survey repeat photography of Alaska's Muir Glacier, retreating more than 4 miles between 1941 and 2004, and its tributary the Riggs Glacier. Total glacial ice mass in the Gulf of Alaska region has declined steadily since 2003.[113] NASA's Gravity Recovery and Climate Experiment (GRACE) indicates mass loss from the northern and southern parts of the Gulf of Alaska region of −36 ± 4 Gt per year and −4 ± 3 Gt per year, respectively.[113] Studies suggest an anthropogenic imprint on imbalances in Alaskan glaciers, indicating that melt will continue through the 21st century.[112, 133, 134] Multiple datasets indicate that it is *virtually certain* that Alaskan glaciers have lost mass over the last 50 years and will continue to do so.[135]

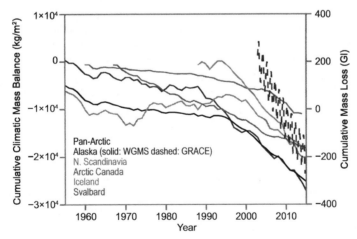

Figure 11.3: Time series of the cumulative climatic mass balance (units: kg/m²) in five arctic regions and for the Pan-Arctic from the World Glacier Monitoring Service (WGMS;[210] Wolken et al.;[211] solid lines, left y-axis), plus Alaskan glacial mass loss observed from NASA GRACE[113] (dashed blue line, right y-axis). (Figure source: Harig and Simons 2016[113] and Wolken et al. 2016;[211] © American Meteorological Society, used with permission.)

(a)

(b)

Figure 11.4: Two northeast-looking photographs of the Muir Glacier located in southeastern Alaska taken from a Glacier Bay Photo station in (a) 1941 and (b) 2004. U.S. Geological Survey repeat photography allows the tracking of glacier changes, illustrating that between 1941 and 2004 the Muir Glacier has retreated more than 4 miles to the northwest and out of view. Riggs Glacier (in view) is a tributary to Muir Glacier and has retreated by as much as 0.37 miles and thinned by more than 0.16 miles. The photographs also illustrate a significant change in the surface type between 1941 and 2004 as bare rock in the foreground has been replaced by dense vegetation (Figure source: USGS 2004[212]).

## 11.3 Arctic Feedbacks on the Lower 48 and Globally

### 11.3.1 Linkages between Arctic Warming and Lower Latitudes

Midlatitude circulation influences arctic climate and climate change.[11, 136, 137, 138, 139, 140, 141, 142, 143, 144, 145] Record warm arctic temperatures in winter 2016 resulted primarily from the transport of midlatitude air into the Arctic, demonstrating the significant midlatitude influence.[146] Emerging science demonstrates that warm, moist air intrusions from midlatitudes results in increased downwelling longwave radiation, warming the arctic surface and hindering wintertime sea ice growth.[139, 141, 147, 148]

The extent to which enhanced arctic surface warming and sea ice loss influence the large-scale atmospheric circulation and midlatitude weather and climate extremes has become an active research area.[137, 146] Several pathways have been proposed (see references in Cohen et al.[149] and Barnes and Screen[150]): reduced meridional temperature gradient, a more sinuous jet-stream, trapped atmospheric waves, modified storm tracks, weakened stratospheric polar vortex. While modeling studies link a reduced meridional temperature gradient to fewer cold temperature extremes in the continental United States,[151, 152, 153, 154] other studies hypothesize that a slower jet stream may amplify Rossby waves and increase the frequency of atmospheric blocking, causing more persistent and extreme weather in midlatitudes.[155]

Multiple observational studies suggest that the concurrent changes in the Arctic and Northern Hemisphere large-scale circulation since the 1990s did not occur by chance, but were caused by arctic amplification.[149, 150, 156] Reanalysis data suggest a relationship between arctic amplification and observed changes in persistent circulation phenomena like blocking and planetary wave amplitude.[155, 157, 158] The recent multi-year California drought serves as an example of an event caused by persistent circulation phenomena (see Ch. 5: Circulation and Variability and Ch. 8: Drought, Floods, and Wildfires).[159, 160, 161] Robust empirical evidence is lacking because the arctic sea ice observational record is too short[162] or because the atmospheric response to arctic amplification depends on the prior state of the atmospheric circulation, reducing detectability.[146] Furthermore, it is not possible to draw conclusions regarding the direction of the relationship between arctic warming and midlatitude circulation based on empirical correlation and covariance analyses alone. Observational analyses have been combined with modeling studies to test causality statements.

Studies with simple models and Atmospheric General Circulation Models (AGCMs) provide evidence that arctic warming can affect midlatitude jet streams and location of storm tracks.[137, 146, 150] In addition, analysis of CMIP5 models forced with increasing greenhouse gases suggests that the magnitude of arctic amplification affects the future midlatitude jet position, specifically during boreal winter.[163] However, the effect of arctic amplification on blocking is not clear (Ch. 5: Circulation and Variability).[164]

Regarding attribution, AGCM simulations forced with observed changes in arctic sea ice suggest that the sea ice loss effect on observed recent midlatitude circulation changes and winter climate in the continental United States is small compared to natural large-scale atmospheric variability.[142, 144, 154, 165] It is argued, however, that climate models do not properly

reproduce the linkages between arctic amplification and lower latitude climate due to model errors, including incorrect sea ice–atmosphere coupling and poor representation of stratospheric processes.[137, 166]

In summary, emerging science demonstrates a strong influence of the midlatitude circulation on the Arctic, affecting temperatures and sea ice (*high confidence*). The influence of arctic changes on the midlatitude circulation and weather patterns are an area of active research. Currently, confidence is *low* regarding whether or by what mechanisms observed arctic warming may have influenced midlatitude circulation and weather patterns over the continental United States. The nature and magnitude of arctic amplification's influence on U.S. weather over the coming decades remains an open question.

### 11.3.2 Freshwater Effects on Ocean Circulation

The addition of freshwater to the Arctic Ocean from melting sea ice and land ice can influence important arctic climate system characteristics, including ocean salinity, altering ocean circulation, density stratification, and sea ice characteristics. Observations indicate that river runoff is increasing, driven by land ice melt, adding freshwater to the Arctic Ocean.[167] Melting arctic sea and land ice combined with time-varying atmospheric forcing[79, 168] control Arctic Ocean freshwater export to the North Atlantic. Large-scale circulation variability in the central Arctic not only controls the redistribution and storage of freshwater in the Arctic[79] but also the export volume.[169] Increased freshwater fluxes can weaken open ocean convection and deep water formation in the Labrador and Irminger seas, weakening the Atlantic meridional overturning circulation (AMOC).[170, 171] AMOC-associated poleward heat transport substantially contributes to North American and continental European climate; any AMOC slowdown could have implications for global climate change as well (see Ch. 15: Potential Surprises).[172, 173] Connections to subarctic ocean variations and the Atlantic Meridional Overturning Circulation have not been conclusively established and require further investigation (see Ch. 13: Ocean Changes).

### 11.3.3 Permafrost–Carbon Feedback

Alaska and arctic permafrost characteristics have responded to increased temperatures and reduced snow cover in most regions since the 1980s.[130] The permafrost warming rate varies regionally; however, colder permafrost is warming faster than warmer permafrost.[37, 174] This feature is most evident across Alaska, where permafrost on the North Slope is warming more rapidly than in the interior. Permafrost temperatures across the North Slope at various depths ranging from 39 to 65 feet (12 to 20 meters) have warmed between 0.3° and 1.3°F (0.2° and 0.7°C) per decade over the observational period (Figure 11.5).[175] Permafrost active layer thickness increased across much of the Arctic while showing strong regional

variations.[37, 130, 176] Further, recent geologic survey data indicate significant permafrost thaw slumping in northwestern Canada and across the circumpolar Arctic that indicate significant ongoing permafrost thaw, potentially priming the region for more rapid thaw in the future.[177] Continued degradation of permafrost and a transition from continuous to discontinuous permafrost is expected over the 21st century.[37, 178, 179]

Permafrost contains large stores of carbon. Though the total contribution of these carbon stores to global methane emission is uncertain, Alaska's permafrost contains rich and vulnerable organic carbon soils.[99, 179, 180] Thus, warming Alaska permafrost is a concern for the global carbon cycle as it provides a possibility for a significant and potentially uncontrollable release of carbon, complicating the ability to limit global temperature increases. Current methane emissions from Alaskan arctic tundra and boreal forests contribute a small fraction of the global methane ($CH_4$) budget.[181] Howev-

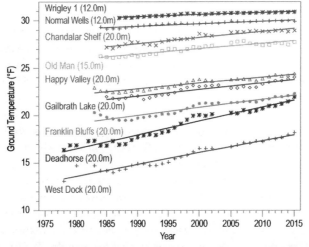

Figure 11.5: Time series of annual mean permafrost temperatures (units: °F) at various depths from 39 to 65 feet (12 to 20 meters) from 1977 through 2015 at several sites across Alaska, including the North Slope continuous permafrost region (purple/blue/green shades), and the discontinuous permafrost (orange/pink/red shades) in Alaska and northwestern Canada. Solid lines represent the linear trends drawn to highlight that permafrost temperatures are warming faster in the colder, coastal permafrost regions than the warmer interior regions. (Figure Source: adapted from Romanovsky et al. 2016;[175] © American Meteorological Society, used with permission.)

er, gas flux measurements have directly measured the release of $CO_2$ and $CH_4$ from arctic permafrost.[182] Recent measurements indicate that cold season methane emissions (after snowfall) are greater than summer emissions in Alaska, and methane emissions in upland tundra are greater than in wetland tundra.[183]

The permafrost–carbon feedback represents the additional release of $CO_2$ and $CH_4$ from thawing permafrost soils providing additional radiative forcing, a source of a potential surprise (Ch. 15: Potential Surprises).[184] Thawing permafrost makes previously frozen organic matter available for microbial decomposition, producing $CO_2$ and $CH_4$. The specific condition under which microbial decomposition occurs, aerobic or anaerobic, determines the proportion of $CO_2$ and $CH_4$ released. This distinction has potentially significant implications, as $CH_4$ has a 100-year global warming potential 35 times that of $CO_2$.[185] Emerging science indicates that 3.4 times more carbon is released under aerobic conditions than anaerobic conditions, and 2.3 times more carbon after accounting for the stronger greenhouse effect of $CH_4$.[186] Additionally, $CO_2$ and $CH_4$ production strongly depends on vegetation and soil properties.[184]

Combined data and modeling studies indicate a positive permafrost–carbon feedback with a global sensitivity between −14 and −19 GtC per °C (approximately −25 to −34 GtC per °F) soil carbon loss[187, 188] resulting in a total 120 ± 85 GtC release from permafrost by 2100 and an additional global temperature increase of 0.52° ± 0.38°F (0.29° ± 0.21°C) by the permafrost–carbon feedback.[189] More recently, Chadburn et al.[190] infer a −4 million km$^2$ per °C (or approximately 858,000 mi$^2$ per °F) reduction in permafrost area to globally averaged warming at stabilization by constraining climate models with the observed spatial distribution of permafrost; this sensitivity is 20% higher

than previous studies. In the coming decades, enhanced high-latitude plant growth and its associated $CO_2$ sink should partially offset the increased emissions from permafrost thaw;[179, 189, 191] thereafter, decomposition is expected to dominate uptake. Permafrost thaw is occurring faster than models predict due to poorly understood deep soil, ice wedge, and thermokarst processes.[188, 192, 193] Additionally, uncertainty stems from the surprising uptake of methane from mineral soils.[194] There is *high confidence* in the positive sign of the permafrost–carbon feedback, but *low confidence* in the feedback magnitude.

### 11.3.4 Methane Hydrate Instability

Significant stores of $CH_4$, in the form of methane hydrates (also called clathrates), lie within and below permafrost and under the global ocean on continental margins. The estimated total global inventory of methane hydrates ranges from 500 to 3,000 GtC[195, 196, 197] with a central estimate of 1,800 GtC.[198] Methane hydrates are solid compounds formed at high pressures and cold temperatures, trapping methane gas within the crystalline structure of water. Methane hydrates within upper continental slopes of the Pacific, Atlantic, and Gulf of Mexico margins and beneath the Alaskan arctic continental shelf may be vulnerable to small increases in ocean temperature.[197, 198, 199, 200, 201, 202, 203]

Rising sea levels and warming oceans have a competing influence on methane hydrate stability.[199, 204] Studies indicate that the temperature effect dominates and that the overall influence is *very likely* a destabilizing effect.[198] Projected warming rates for the 21st century Arctic Ocean are not expected to lead to sudden or catastrophic destabilization of seafloor methane hydrates.[205] Recent observations indicate increased $CH_4$ emission from the arctic seafloor near Svalbard; however, these emissions are not reaching the atmosphere.[198, 206]

# TRACEABLE ACCOUNTS

## Key Finding 1

Annual average near-surface air temperatures across Alaska and the Arctic have increased over the last 50 years at a rate more than twice as fast as the global average temperature. (*Very high confidence*)

### Description of evidence base

The Key Finding is supported by observational evidence from ground-based observing stations, satellites, and data-model temperature analyses from multiple sources and independent analysis techniques.[14, 15, 16, 17, 18, 19, 20] For more than 40 years, climate models have predicted enhanced arctic warming, indicating a solid grasp on the underlying physics and positive feedbacks driving the accelerated arctic warming.[1, 21, 22] Lastly, similar statements have been made in NCA3,[70] IPCC AR5,[17] and in other arctic-specific assessments such as the Arctic Climate Impacts Assessment[207] and Snow, Water, Ice and Permafrost in the Arctic.[130]

### Major Uncertainties

The lack of high quality and restricted spatial resolution of surface and ground temperature data over many arctic land regions and essentially no measurements over the Central Arctic Ocean hamper the ability to better refine the rate of arctic warming and completely restrict our ability to quantify and detect regional trends, especially over the sea ice. Climate models generally produce an arctic warming between two to three times the global mean warming. A key uncertainty is our quantitative knowledge of the contributions from individual feedback processes in driving the accelerated arctic warming. Reducing this uncertainty will help constrain projections of future arctic warming.

### Assessment of confidence based on evidence and agreement, including short description of nature of evidence and level of agreement

*Very high confidence* that the arctic surface and air temperatures have warmed across Alaska and the Arctic at a much faster rate than the global average is provided by the multiple datasets analyzed by multiple independent groups indicating the same conclusion. Additionally, climate models capture the enhanced warming in the Arctic, indicating a solid understanding of the underlying physical mechanisms.

### If appropriate, estimate likelihood of impact or consequence, including short description of basis of estimate

It is *very likely* that the accelerated rate of arctic warming will have a significant consequence for the United States due to accelerated land and sea ice melt driving changes in the ocean including sea level rise threatening our coastal communities and freshening of sea water that is influencing marine ecology.

### Summary sentence or paragraph that integrates the above information

Annual average near-surface air temperatures across Alaska and the Arctic have increased over the last 50 years at a rate more than twice the global average. Observational studies using ground-based observing stations and satellites analyzed by multiple independent groups support this finding. The enhanced sensitivity of the arctic climate system to anthropogenic forcing is also supported by climate modeling evidence, indicating a solid grasp on the underlying physics. These multiple lines of evidence provide *very high confidence* of enhanced arctic warming with potentially significant impacts on coastal communities and marine ecosystems.

## Key Finding 2

Rising Alaskan permafrost temperatures are causing permafrost to thaw and become more discontinuous; this process releases additional carbon dioxide and methane resulting in an amplifying feedback and additional warming (*high confidence*). The overall magnitude of the permafrost–carbon feedback is uncertain; however, it is clear that these emissions have the potential to compromise the ability to limit global temperature increases.

### Description of evidence base

The Key Finding is supported by observational evidence of warming permafrost temperatures and a deepening active layer, in situ gas measurements and laboratory incubation experiments of $CO_2$ and $CH_4$ release, and model studies.[37, 179, 186, 187, 188, 192, 193] Alaska and arctic permafrost characteristics have responded to increased temperatures and reduced snow cover in most regions since the 1980s, with colder permafrost warming faster than warmer permafrost.[37, 130, 175] Large carbon soil pools (more than 50% of the global below-ground organic carbon pool) are locked up in the permafrost soils,[180] with the potential to be released. Thawing permafrost makes previously frozen organic matter available for microbial decomposition. In situ gas flux measurements have directly measured the release of $CO_2$ and $CH_4$ from arctic permafrost.[182, 183] The specific conditions of microbial decomposition, aerobic or anaerobic, determines the relative production of $CO_2$ and $CH_4$. This distinction is significant as $CH_4$ is a much more powerful greenhouse gas than $CO_2$.[185] However, incubation studies indicate that 3.4 times more carbon is released under aerobic conditions than anaerobic conditions, leading to a 2.3 times the stronger radiative forcing under aerobic conditions.[186] Combined data and modeling studies suggest a global sensitivity of the permafrost–carbon feedback warming global temperatures in 2100 by 0.52° ± 0.38°F (0.29° ± 0.21°C) alone.[189] Chadburn et al.[190] infer the sensitivity of permafrost area to globally averaged warming to be 4 million km² by constraining a group of climate models with the observed spatial distribution of permafrost; this sensitivity is 20% higher than previous studies. Permafrost thaw is occurring faster than models predict due to poorly understood deep soil, ice wedge, and thermokarst processes.[188, 192, 193, 208] Additional uncertainty stems from the surprising uptake of methane from mineral soils[194] and dependence of emissions on vegetation and soil properties.[184] The observational and modeling evidence supports the Key Finding that the permafrost–carbon cycle is positive.

### Major uncertainties

A major limiting factor is the sparse observations of permafrost in Alaska and remote areas across the Arctic. Major uncertainties are related to deep soil, ice wedging, and thermokarst processes and the dependence of $CO_2$ and $CH_4$ uptake and production on vegetation and soil properties. Uncertainties also exist in relevant

soil processes during and after permafrost thaw, especially those that control unfrozen soil carbon storage and plant carbon uptake and net ecosystem exchange. Many processes with the potential to drive rapid permafrost thaw (such as thermokarst) are not included in current earth system models.

### Assessment of confidence based on evidence and agreement, including short description of nature of evidence and level of agreement

There is *high confidence* that permafrost is thawing, becoming discontinuous, and releasing $CO_2$ and $CH_4$. Physically-based arguments and observed increases in $CO_2$ and $CH_4$ emissions as permafrost thaws indicate that the feedback is positive. This confidence level is justified based on observations of rapidly changing permafrost characteristics.

### If appropriate, estimate likelihood of impact or consequence, including short description of basis of estimate

Thawing permafrost *very likely* has significant impacts to the global carbon cycle and serves as a source of $CO_2$ and $CH_4$ emission that complicates the ability to limit global temperature increases.

### Summary sentence or paragraph that integrates the above information

Permafrost is thawing, becoming more discontinuous, and releasing $CO_2$ and $CH_4$. Observational and modeling evidence indicates that permafrost has thawed and released additional $CO_2$ and $CH_4$ indicating that the permafrost–carbon cycle feedback is positive accounting for additional warming of approximately 0.08° to 0.50°C on top of climate model projections. Although the magnitude of the permafrost–carbon feedback is uncertain due to a range of poorly understood processes (deep soil and ice wedge processes, plant carbon uptake, dependence of uptake and emissions on vegetation and soil type, and the role of rapid permafrost thaw processes, such as thermokarst), emerging science and the newest estimates continue to indicate that this feedback is more likely on the larger side of the range. Impacts of permafrost thaw and the permafrost carbon feedback complicates our ability to limit global temperature increases by adding a currently unconstrained radiative forcing to the climate system.

### Key Finding 3

Arctic land and sea ice loss observed in the last three decades continues, in some cases accelerating (*very high confidence*). It is *virtually certain* that Alaska glaciers have lost mass over the last 50 years, with each year since 1984 showing an annual average ice mass less than the previous year. Based on gravitational data from satellites, average ice mass loss from Greenland was −269 Gt per year between April 2002 and April 2016, accelerating in recent years (*high confidence*). Since the early 1980s, annual average arctic sea ice has decreased in extent between 3.5% and 4.1% per decade, become thinner by between 4.3 and 7.5 feet, and began melting at least 15 more days each year. September sea ice extent has decreased between 10.7% and 15.9% per decade (*very high confidence*). Arctic-wide ice loss is expected to continue through the 21st century, *very likely* resulting in nearly sea ice-free late summers by the 2040s (*very high confidence*).

### Description of evidence base

The Key Finding is supported by observational evidence from multiple ground-based and satellite-based observational techniques (including passive microwave, laser and radar altimetry, and gravimetry) analyzed by independent groups using different techniques reaching similar conclusions.[19, 37, 45, 47, 112, 113, 134, 135] Additionally, the U.S. Geological Survey repeat photography database shows the glacier retreat for many Alaskan glaciers (Figure 11.4: Muir Glacier). Several independent model analysis studies using a wide array of climate models and different analysis techniques indicate that sea ice loss will continue across the Arctic, *very likely* resulting in late summers becoming nearly ice-free by the 2040s.[21, 59, 65]

### Major uncertainties

Key uncertainties remain in the quantification and modeling of key physical processes that contribute to the acceleration of land and sea ice melting. Climate models are unable to capture the rapid pace of ob-

served sea and land ice melt over the last 15 years; a major factor is our inability to quantify and accurately model the physical processes driving the accelerated melting. The interactions between atmospheric circulation, ice dynamics and thermodynamics, clouds, and specifically the influence on the surface energy budget are key uncertainties. Mechanisms controlling marine-terminating glacier dynamics—specifically the roles of atmospheric warming, seawater intrusions under floating ice shelves, and the penetration of surface meltwater to the glacier bed—are key uncertainties in projecting Greenland Ice Sheet melt.

### Assessment of confidence based on evidence and agreement, including short description of nature of evidence and level of agreement

There is *very high confidence* that arctic sea and land ice melt is accelerating and mountain glacier ice mass is declining given the multiple observational sources and analysis techniques documented in the peer-reviewed climate science literature.

### If appropriate, estimate likelihood of impact or consequence, including short description of basis of estimate

It is *very likely* that accelerating arctic land and sea ice melt impacts the United States. Accelerating Arctic Ocean sea ice melt increases coastal erosion in Alaska and makes Alaskan fisheries more susceptible to ocean acidification by changing Arctic Ocean chemistry. Greenland Ice Sheet and Alaska mountain glacier melt drives sea level rise threatening coastal communities in the United States and worldwide, influencing marine ecology, and potentially altering the thermohaline circulation.

### Summary sentence or paragraph that integrates the above information

Arctic land and sea ice loss observed in the last three decades continues, in some cases accelerating. A diverse range of observational evidence from multiple data sources and independent analysis techniques provide consistent evidence of substantial declines in arctic sea ice extent, thickness, and volume since at least 1979, mountain glacier melt over the last 50 years, and accelerating mass loss from Greenland. An array of different models and independent analyses indicate that future declines in ice across the Arctic are expected resulting in late summers in the Arctic becoming ice free by the 2040s.

### Key Finding 4

It is *very likely* that human activities have contributed to observed arctic surface temperature warming, sea ice loss, glacier mass loss, and Northern Hemisphere snow extent decline (*high confidence*).

### Description of evidence base

The Key Finding is supported by many attribution studies using a wide array of climate models documenting the anthropogenic influence on arctic temperature, sea ice, mountain glaciers, and snow extent.[23, 24, 25, 26, 27, 29, 54, 55, 56, 57, 58, 59, 61, 62, 106, 108, 133] Observation-based analyses also support an anthropogenic influence.[60, 69] Najafi et al.[27] show that the greenhouse warming signal in the Arctic could be even stronger, as a significant portion of greenhouse gas induced warming (approximately 60%) has been offset by anthropogenic aerosol emissions. The emerging science of extreme event attribution indicates that natural variability alone could not have caused the recently observed record low arctic sea ice extents, such as in September 2012.[61, 62] Natural variability in the Arctic is significant,[63, 64] however the majority of studies indicate that the contribution from individual sources of internal variability to observed trends in arctic temperature and sea ice are less than 50%[28, 54, 55] and alone cannot explain the observed trends over the satellite era. This Key Finding marks an increased confidence relative to the IPCC AR5[24] moving from *likely* to *very likely*. In our assessment, the new understanding of the anthropogenic forcing,[27] its relationship to arctic climate change,[69] arctic climate variability,[28, 63, 64] and especially extreme event attribution studies[61, 62] reaffirms previous studies and warrants the increased likelihood of an anthropogenic influence on arctic climate change. Multiple lines evidence, independent analysis techniques, models, and studies support the Key Finding.

### Major uncertainties

A major limiting factor in our ability to attribute arctic sea ice and glacier melt to human activities is the significant natural climate variability in the Arctic. Longer data records and a better understanding of the physical mechanisms that drive natural climate variability in the Arctic are required to reduce this uncertainty. Another major uncertainty is the ability of climate models to capture the relevant physical processes and climate changes at a fine spatial scale, especially those at the land and ocean surface in the Arctic.

### Assessment of confidence based on evidence and agreement, including short description of nature of evidence and level of agreement

There is *high confidence* that human activities have contributed to arctic surface temperature warming, sea ice loss since 1979, glacier mass loss, and Northern Hemisphere snow extent given multiple independent analysis techniques from independent groups using many different climate models indicate the same conclusion.

### If appropriate, estimate likelihood of impact or consequence, including short description of basis of estimate

Arctic sea ice and glacier mass loss impacts the United States by affecting coastal erosion in Alaska and key Alaskan fisheries through an increased vulnerability to ocean acidification. Glacier mass loss is a significant driver of sea level rise threatening coastal communities in the United States and worldwide, influencing marine ecology, and potentially altering the Atlantic Meridional Overturning Circulation.[172]

### Summary sentence or paragraph that integrates the above information

Evidenced by the multiple independent studies, analysis techniques, and the array of different climate models used over the last 20 years, it is *very likely* that human activities have contributed to arctic surface temperature warming, sea ice loss since 1979, glacier mass loss, and Northern Hemisphere snow extent decline observed across the Arctic. Key uncertainties remain in the understanding and modeling of arctic climate variability; however, many independent studies indicate that internal variability alone cannot explain the trends or extreme events observed in arctic temperature and sea ice over the satellite era.

**Key Finding 5**

Atmospheric circulation patterns connect the climates of the Arctic and the contiguous United States. Evidenced by recent record warm temperatures in the Arctic and emerging science, the midlatitude circulation has influenced observed arctic temperatures and sea ice (*high confidence*). However, confidence is *low* regarding whether or by what mechanisms observed arctic warming may have influenced the midlatitude circulation and weather patterns over the continental United States. The influence of arctic changes on U.S. weather over the coming decades remains an open question with the potential for significant impact.

**Description of evidence base**

The midlatitude circulation influences the Arctic through the transport of warm, moist air, altering the Arctic surface energy budget.[138, 142, 143, 144] The intrusion of warm, moist air from midlatitudes increases downwelling longwave radiation, warming the arctic surface and hindering wintertime sea ice growth.[139, 147] Emerging research provides a new understanding of the importance of synoptic time scales and the episodic nature of midlatitude air intrusions.[139, 141, 148] The combination of recent observational and model-based evidence as well as the physical understanding of the mechanisms of midlatitude circulation effects on arctic climate supports this Key Finding.

In addition, research on the impact of arctic climate on midlatitude circulation is rapidly evolving, including observational analysis and modeling studies. Multiple observational studies provide evidence for concurrent changes in the Arctic and Northern Hemisphere large-scale circulation changes.[149, 150, 156] Further, modeling studies demonstrate that arctic warming can influence the midlatitude jet stream and storm track.[137, 146, 150, 163] However, attribution studies indicate that the observed midlatitude circulation changes over the continental United States are smaller than natural variability and are therefore not detectable in the observational re-

cord.[142, 144, 154, 165] This disagreement between independent studies using different analysis techniques and the lack of understanding of the physical mechanism(s) supports this Key Finding.

**Major uncertainties**

A major limiting factor is our understanding and modeling of natural climate variability in the Arctic. Longer data records and a better understanding of the physical mechanisms that drive natural climate variability in the Arctic are required to reduce this uncertainty. The inability of climate models to accurately capture interactions between sea ice and the atmospheric circulation and polar stratospheric processes limits our current understanding.

**Assessment of confidence based on evidence and agreement, including short description of nature of evidence and level of agreement**

*High confidence* in the impact of midlatitude circulation on arctic changes from the consistency between observations and models as well as a solid physical understanding.

*Low confidence* on the detection of an impact of arctic warming on midlatitude climate is based on short observational data record, model uncertainty, and lack of physical understanding.

**Summary sentence or paragraph that integrates the above information**

The midlatitude circulation has influenced observed arctic temperatures, supported by recent observational and model-based evidence as well as the physical understanding from emerging science. In turn, confidence is low regarding the mechanisms by which observed arctic warming has influenced the midlatitude circulation and weather patterns over the continental United States, due to the disagreement between numerous studies and a lack of understanding of the physical mechanism(s). Resolving the remaining questions requires longer data records and improved understanding and modeling of physics in the Arctic. The influence of arctic changes on U.S. weather over the coming decades remains an open question with the potential for significant impact.

# REFERENCES

1. Manabe, S. and R.T. Wetherald, 1975: The effects of doubling the CO2 concentration on the climate of a General Circulation Model. *Journal of the Atmospheric Sciences,* **32,** 3-15. http://dx.doi.org/10.1175/1520-0469(1975)032<0003:teodtc>2.0.co;2

2. Knies, J., P. Cabedo-Sanz, S.T. Belt, S. Baranwal, S. Fietz, and A. Rosell-Melé, 2014: The emergence of modern sea ice cover in the Arctic Ocean. *Nature Communications,* **5,** 5608. http://dx.doi.org/10.1038/ncomms6608

3. Wyser, K., C.G. Jones, P. Du, E. Girard, U. Willén, J. Cassano, J.H. Christensen, J.A. Curry, K. Dethloff, J.-E. Haugen, D. Jacob, M. Køltzow, R. Laprise, A. Lynch, S. Pfeifer, A. Rinke, M. Serreze, M.J. Shaw, M. Tjernström, and M. Zagar, 2008: An evaluation of Arctic cloud and radiation processes during the SHEBA year: Simulation results from eight Arctic regional climate models. *Climate Dynamics,* **30,** 203-223. http://dx.doi.org/10.1007/s00382-007-0286-1

4. Bourassa, M.A., S.T. Gille, C. Bitz, D. Carlson, I. Cerovecki, C.A. Clayson, M.F. Cronin, W.M. Drennan, C.W. Fairall, R.N. Hoffman, G. Magnusdottir, R.T. Pinker, I.A. Renfrew, M. Serreze, K. Speer, L.D. Talley, and G.A. Wick, 2013: High-latitude ocean and sea ice surface fluxes: Challenges for climate research. *Bulletin of the American Meteorological Society,* **94,** 403-423. http://dx.doi.org/10.1175/BAMS-D-11-00244.1

5. Maslowski, W., J. Clement Kinney, M. Higgins, and A. Roberts, 2012: The future of Arctic sea ice. *Annual Review of Earth and Planetary Sciences,* **40,** 625-654. http://dx.doi.org/10.1146/annurev-earth-042711-105345

6. Maslowski, W., J. Clement Kinney, S.R. Okkonen, R. Osinski, A.F. Roberts, and W.J. Williams, 2014: The large scale ocean circulation and physical processes controlling Pacific-Arctic interactions. *The Pacific Arctic Region: Ecosystem Status and Trends in a Rapidly Changing Environment.* Grebmeier, M.J. and W. Maslowski, Eds. Springer Netherlands, Dordrecht, 101-132. http://dx.doi.org/10.1007/978-94-017-8863-2_5

7. Kay, J.E. and A. Gettelman, 2009: Cloud influence on and response to seasonal Arctic sea ice loss. *Journal of Geophysical Research,* **114,** D18204. http://dx.doi.org/10.1029/2009JD011773

8. Taylor, P.C., S. Kato, K.-M. Xu, and M. Cai, 2015: Covariance between Arctic sea ice and clouds within atmospheric state regimes at the satellite footprint level. *Journal of Geophysical Research Atmospheres,* **120,** 12656-12678. http://dx.doi.org/10.1002/2015JD023520

9. Overland, J., E. Hanna, I. Hanssen-Bauer, S.-J. Kim, J. Wlash, M. Wang, and U.S. Bhatt, 2015: [The Arctic] Arctic air temperature [in "State of the Climate in 2014"]. *Bulletin of the American Meteorological Society,* **96 (12),** S128-S129. http://dx.doi.org/10.1175/2015BAMSStateoftheClimate.1

10. Johannessen, O.M., S.I. Kuzmina, L.P. Bobylev, and M.W. Miles, 2016: Surface air temperature variability and trends in the Arctic: New amplification assessment and regionalisation. *Tellus A,* **68,** 28234. http://dx.doi.org/10.3402/tellusa.v68.28234

11. Overland, J.E. and M. Wang, 2016: Recent extreme Arctic temperatures are due to a split polar vortex. *Journal of Climate,* **29,** 5609-5616. http://dx.doi.org/10.1175/JCLI-D-16-0320.1

12. Hartmann, B. and G. Wendler, 2005: The significance of the 1976 Pacific climate shift in the climatology of Alaska. *Journal of Climate,* **18,** 4824-4839. http://dx.doi.org/10.1175/JCLI3532.1

13. McAfee, S.A., 2014: Consistency and the lack thereof in Pacific Decadal Oscillation impacts on North American winter climate. *Journal of Climate,* **27,** 7410-7431. http://dx.doi.org/10.1175/JCLI-D-14-00143.1

14. Serreze, M.C., A.P. Barrett, J.C. Stroeve, D.N. Kindig, and M.M. Holland, 2009: The emergence of surface-based Arctic amplification. *The Cryosphere,* **3,** 11-19. http://dx.doi.org/10.5194/tc-3-11-2009

15. Bekryaev, R.V., I.V. Polyakov, and V.A. Alexeev, 2010: Role of polar amplification in long-term surface air temperature variations and modern Arctic warming. *Journal of Climate,* **23,** 3888-3906. http://dx.doi.org/10.1175/2010jcli3297.1

16. Screen, J.A. and I. Simmonds, 2010: The central role of diminishing sea ice in recent Arctic temperature amplification. *Nature,* **464,** 1334-1337. http://dx.doi.org/10.1038/nature09051

17. Hartmann, D.L., A.M.G. Klein Tank, M. Rusticucci, L.V. Alexander, S. Brönnimann, Y. Charabi, F.J. Dentener, E.J. Dlugokencky, D.R. Easterling, A. Kaplan, B.J. Soden, P.W. Thorne, M. Wild, and P.M. Zhai, 2013: Observations: Atmosphere and surface. *Climate Change 2013: The Physical Science Basis. Contribution of Working Group I to the Fifth Assessment Report of the Intergovernmental Panel on Climate Change.* Stocker, T.F., D. Qin, G.-K. Plattner, M. Tignor, S.K. Allen, J. Boschung, A. Nauels, Y. Xia, V. Bex, and P.M. Midgley, Eds. Cambridge University Press, Cambridge, United Kingdom and New York, NY, USA, 159–254. http://www.climatechange2013.org/report/full-report/

18. Overland, J., E. Hanna, I. Hanssen-Bauer, S.-J. Kim, J. Walsh, M. Wang, and U. Bhatt, 2014: Air temperature [in Arctic Report Card 2014]. ftp://ftp.oar.noaa.gov/arctic/documents/ArcticReportCard_full_report2014.pdf

19. Comiso, J.C. and D.K. Hall, 2014: Climate trends in the Arctic as observed from space. *Wiley Interdisciplinary Reviews: Climate Change*, **5**, 389-409. http://dx.doi.org/10.1002/wcc.277

20. Wendler, G., B. Moore, and K. Galloway, 2014: Strong temperature increase and shrinking sea ice in Arctic Alaska. *The Open Atmospheric Science Journal*, **8**, 7-15. http://dx.doi.org/10.2174/1874282301408010007

21. Collins, M., R. Knutti, J. Arblaster, J.-L. Dufresne, T. Fichefet, P. Friedlingstein, X. Gao, W.J. Gutowski, T. Johns, G. Krinner, M. Shongwe, C. Tebaldi, A.J. Weaver, and M. Wehner, 2013: Long-term climate change: Projections, commitments and irreversibility. *Climate Change 2013: The Physical Science Basis. Contribution of Working Group I to the Fifth Assessment Report of the Intergovernmental Panel on Climate Change*. Stocker, T.F., D. Qin, G.-K. Plattner, M. Tignor, S.K. Allen, J. Boschung, A. Nauels, Y. Xia, V. Bex, and P.M. Midgley, Eds. Cambridge University Press, Cambridge, United Kingdom and New York, NY, USA, 1029–1136. http://www.climatechange2013.org/report/full-report/

22. Taylor, P.C., M. Cai, A. Hu, J. Meehl, W. Washington, and G.J. Zhang, 2013: A decomposition of feedback contributions to polar warming amplification. *Journal of Climate*, **26**, 7023-7043. http://dx.doi.org/10.1175/JCLI-D-12-00696.1

23. Gillett, N.P., D.A. Stone, P.A. Stott, T. Nozawa, A.Y. Karpechko, G.C. Hegerl, M.F. Wehner, and P.D. Jones, 2008: Attribution of polar warming to human influence. *Nature Geoscience*, **1**, 750-754. http://dx.doi.org/10.1038/ngeo338

24. Bindoff, N.L., P.A. Stott, K.M. AchutaRao, M.R. Allen, N. Gillett, D. Gutzler, K. Hansingo, G. Hegerl, Y. Hu, S. Jain, I.I. Mokhov, J. Overland, J. Perlwitz, R. Sebbari, and X. Zhang, 2013: Detection and attribution of climate change: From global to regional. *Climate Change 2013: The Physical Science Basis. Contribution of Working Group I to the Fifth Assessment Report of the Intergovernmental Panel on Climate Change*. Stocker, T.F., D. Qin, G.-K. Plattner, M. Tignor, S.K. Allen, J. Boschung, A. Nauels, Y. Xia, V. Bex, and P.M. Midgley, Eds. Cambridge University Press, Cambridge, United Kingdom and New York, NY, USA, 867–952. http://www.climatechange2013.org/report/full-report/

25. Fyfe, J.C., K. von Salzen, N.P. Gillett, V.K. Arora, G.M. Flato, and J.R. McConnell, 2013: One hundred years of Arctic surface temperature variation due to anthropogenic influence. *Scientific Reports*, **3**, 2645. http://dx.doi.org/10.1038/srep02645

26. Chylek, P., N. Hengartner, G. Lesins, J.D. Klett, O. Humlum, M. Wyatt, and M.K. Dubey, 2014: Isolating the anthropogenic component of Arctic warming. *Geophysical Research Letters*, **41**, 3569-3576. http://dx.doi.org/10.1002/2014GL060184

27. Najafi, M.R., F.W. Zwiers, and N.P. Gillett, 2015: Attribution of Arctic temperature change to greenhouse-gas and aerosol influences. *Nature Climate Change*, **5**, 246-249. http://dx.doi.org/10.1038/nclimate2524

28. Ding, Q., A. Schweiger, M. Lheureux, D.S. Battisti, S. Po-Chedley, N.C. Johnson, E. Blanchard-Wrigglesworth, K. Harnos, Q. Zhang, R. Eastman, and E.J. Steig, 2017: Influence of high-latitude atmospheric circulation changes on summertime Arctic sea ice. *Nature Climate Change*, **7**, 289-295. http://dx.doi.org/10.1038/nclimate3241

29. Christensen, J.H., K. Krishna Kumar, E. Aldrian, S.-I. An, I.F.A. Cavalcanti, M. de Castro, W. Dong, P. Goswami, A. Hall, J.K. Kanyanga, A. Kitoh, J. Kossin, N.-C. Lau, J. Renwick, D.B. Stephenson, S.-P. Xie, and T. Zhou, 2013: Climate phenomena and their relevance for future regional climate change. *Climate Change 2013: The Physical Science Basis. Contribution of Working Group I to the Fifth Assessment Report of the Intergovernmental Panel on Climate Change*. Stocker, T.F., D. Qin, G.-K. Plattner, M. Tignor, S.K. Allen, J. Boschung, A. Nauels, Y. Xia, V. Bex, and P.M. Midgley, Eds. Cambridge University Press, Cambridge, United Kingdom and New York, NY, USA, 1217–1308. http://www.climatechange2013.org/report/full-report/

30. Boisvert, L.N., T. Markus, and T. Vihma, 2013: Moisture flux changes and trends for the entire Arctic in 2003–2011 derived from EOS Aqua data. *Journal of Geophysical Research Oceans*, **118**, 5829-5843. http://dx.doi.org/10.1002/jgrc.20414

31. Boisvert, L.N., D.L. Wu, and C.L. Shie, 2015: Increasing evaporation amounts seen in the Arctic between 2003 and 2013 from AIRS data. *Journal of Geophysical Research Atmospheres*, **120**, 6865-6881. http://dx.doi.org/10.1002/2015JD023258

32. Boisvert, L.N., D.L. Wu, T. Vihma, and J. Susskind, 2015: Verification of air/surface humidity differences from AIRS and ERA-Interim in support of turbulent flux estimation in the Arctic. *Journal of Geophysical Research Atmospheres*, **120**, 945-963. http://dx.doi.org/10.1002/2014JD021666

33. Kay, J.E., K. Raeder, A. Gettelman, and J. Anderson, 2011: The boundary layer response to recent Arctic sea ice loss and implications for high-latitude climate feedbacks. *Journal of Climate*, **24**, 428-447. http://dx.doi.org/10.1175/2010JCLI3651.1

34. Pavelsky, T.M., J. Boé, A. Hall, and E.J. Fetzer, 2011: Atmospheric inversion strength over polar oceans in winter regulated by sea ice. *Climate Dynamics*, **36**, 945-955. http://dx.doi.org/10.1007/s00382-010-0756-8

35. Solomon, A., M.D. Shupe, O. Persson, H. Morrison, T. Yamaguchi, P.M. Caldwell, and G.d. Boer, 2014: The sensitivity of springtime Arctic mixed-phase stratocumulus clouds to surface-layer and cloud-top inversion-layer moisture sources. *Journal of the Atmospheric Sciences*, **71**, 574-595. http://dx.doi.org/10.1175/JAS-D-13-0179.1

36. Taylor, P.C., R.G. Ellingson, and M. Cai, 2011: Geographical distribution of climate feedbacks in the NCAR CCSM3.0. *Journal of Climate*, **24**, 2737-2753. http://dx.doi.org/10.1175/2010JCLI3788.1

37. Vaughan, D.G., J.C. Comiso, I. Allison, J. Carrasco, G. Kaser, R. Kwok, P. Mote, T. Murray, F. Paul, J. Ren, E. Rignot, O. Solomina, K. Steffen, and T. Zhang, 2013: Observations: Cryosphere. *Climate Change 2013: The Physical Science Basis. Contribution of Working Group I to the Fifth Assessment Report of the Intergovernmental Panel on Climate Change.* Stocker, T.F., D. Qin, G.-K. Plattner, M. Tignor, S.K. Allen, J. Boschung, A. Nauels, Y. Xia, V. Bex, and P.M. Midgley, Eds. Cambridge University Press, Cambridge, United Kingdom and New York, NY, USA, 317–382. http://www.climatechange2013.org/report/full-report/

38. Carmack, E., I. Polyakov, L. Padman, I. Fer, E. Hunke, J. Hutchings, J. Jackson, D. Kelley, R. Kwok, C. Layton, H. Melling, D. Perovich, O. Persson, B. Ruddick, M.-L. Timmermans, J. Toole, T. Ross, S. Vavrus, and P. Winsor, 2015: Toward quantifying the increasing role of oceanic heat in sea ice loss in the new Arctic. *Bulletin of the American Meteorological Society*, **96 (12)**, 2079-2105. http://dx.doi.org/10.1175/BAMS-D-13-00177.1

39. Kwok, R. and N. Untersteiner, 2011: The thinning of Arctic sea ice. *Physics Today*, **64**, 36-41. http://dx.doi.org/10.1063/1.3580491

40. Ogi, M. and I.G. Rigor, 2013: Trends in Arctic sea ice and the role of atmospheric circulation. *Atmospheric Science Letters*, **14**, 97-101. http://dx.doi.org/10.1002/asl2.423

41. Ogi, M. and J.M. Wallace, 2007: Summer minimum Arctic sea ice extent and the associated summer atmospheric circulation. *Geophysical Research Letters*, **34**, L12705. http://dx.doi.org/10.1029/2007GL029897

42. Stroeve, J.C., V. Kattsov, A. Barrett, M. Serreze, T. Pavlova, M. Holland, and W.N. Meier, 2012: Trends in Arctic sea ice extent from CMIP5, CMIP3 and observations. *Geophysical Research Letters*, **39**, L16502. http://dx.doi.org/10.1029/2012GL052676

43. Stroeve, J.C., M.C. Serreze, M.M. Holland, J.E. Kay, J. Malanik, and A.P. Barrett, 2012: The Arctic's rapidly shrinking sea ice cover: A research synthesis. *Climatic Change*, **110**, 1005-1027. http://dx.doi.org/10.1007/s10584-011-0101-1

44. Taylor, P.C., R.G. Ellingson, and M. Cai, 2011: Seasonal variations of climate feedbacks in the NCAR CCSM3. *Journal of Climate*, **24**, 3433-3444. http://dx.doi.org/10.1175/2011jcli3862.1

45. Stroeve, J., A. Barrett, M. Serreze, and A. Schweiger, 2014: Using records from submarine, aircraft and satellites to evaluate climate model simulations of Arctic sea ice thickness. *The Cryosphere*, **8**, 1839-1854. http://dx.doi.org/10.5194/tc-8-1839-2014

46. Stroeve, J.C., T. Markus, L. Boisvert, J. Miller, and A. Barrett, 2014: Changes in Arctic melt season and implications for sea ice loss. *Geophysical Research Letters*, **41**, 1216-1225. http://dx.doi.org/10.1002/2013GL058951

47. Perovich, D., W. Meier, M. Tschudi, S. Farrell, S. Gerland, S. Hendricks, T. Krumpen, and C. Hass, 2016: Sea ice [in Arctic Report Cart 2016]. http://www.arctic.noaa.gov/Report-Card/Report-Card-2016/ArtMID/5022/ArticleID/286/Sea-Ice

48. Schweiger, A., R. Lindsay, J. Zhang, M. Steele, H. Stern, and R. Kwok, 2011: Uncertainty in modeled Arctic sea ice volume. *Journal of Geophysical Research*, **116**, C00D06. http://dx.doi.org/10.1029/2011JC007084

49. Tschudi, M., C. Fowler, J. Maslanik, J.S. Stewart, and W. Meier, 2016: EASE-Grid Sea Ice Age, Version 3. In: NASA (ed.). National Snow and Ice Data Center Distributed Active Archive Center, Boulder, CO.

50. Parkinson, C.L., 2014: Spatially mapped reductions in the length of the Arctic sea ice season. *Geophysical Research Letters*, **41**, 4316-4322. http://dx.doi.org/10.1002/2014GL060434

51. Chapin III, F.S., S.F. Trainor, P. Cochran, H. Huntington, C. Markon, M. McCammon, A.D. McGuire, and M. Serreze, 2014: Ch. 22: Alaska. *Climate Change Impacts in the United States: The Third National Climate Assessment.* Melillo, J.M., Terese (T.C.) Richmond, and G.W. Yohe, Eds. U.S. Global Change Research Program, Washington, DC, 514-536. http://dx.doi.org/10.7930/J00Z7150

52. Gibbs, A.E. and B.M. Richmond, 2015: National Assessment of Shoreline Change: Historical Shoreline Change Along the North Coast of Alaska, U.S.–Canadian Border to Icy Cape. U.S. Geological Survey, 96 pp. http://dx.doi.org/10.3133/ofr20151048

53. Smedsrud, L.H., M.H. Halvorsen, J.C. Stroeve, R. Zhang, and K. Kloster, 2017: Fram Strait sea ice export variability and September Arctic sea ice extent over the last 80 years. *The Cryosphere*, **11**, 65-79. http://dx.doi.org/10.5194/tc-11-65-2017

54. Day, J.J., J.C. Hargreaves, J.D. Annan, and A. Abe-Ouchi, 2012: Sources of multi-decadal variability in Arctic sea ice extent. *Environmental Research Letters*, **7**, 034011. http://dx.doi.org/10.1088/1748-9326/7/3/034011

55. Kay, J.E., M.M. Holland, and A. Jahn, 2011: Inter-annual to multi-decadal Arctic sea ice extent trends in a warming world. *Geophysical Research Letters*, **38**, L15708. http://dx.doi.org/10.1029/2011GL048008

56. Min, S.-K., X. Zhang, F.W. Zwiers, and T. Agnew, 2008: Human influence on Arctic sea ice detectable from early 1990s onwards. *Geophysical Research Letters*, **35**, L21701. http://dx.doi.org/10.1029/2008GL035725

57. Stroeve, J., M.M. Holland, W. Meier, T. Scambos, and M. Serreze, 2007: Arctic sea ice decline: Faster than forecast. *Geophysical Research Letters*, **34**, L09501. http://dx.doi.org/10.1029/2007GL029703

58. Vinnikov, K.Y., A. Robock, R.J. Stouffer, J.E. Walsh, C.L. Parkinson, D.J. Cavalieri, J.F.B. Mitchell, D. Garrett, and V.F. Zakharov, 1999: Global warming and Northern Hemisphere sea ice extent. *Science*, **286**, 1934-1937. http://dx.doi.org/10.1126/science.286.5446.1934

59. Wang, M. and J.E. Overland, 2012: A sea ice free summer Arctic within 30 years: An update from CMIP5 models. *Geophysical Research Letters*, **39**, L18501. http://dx.doi.org/10.1029/2012GL052868

60. Notz, D. and J. Marotzke, 2012: Observations reveal external driver for Arctic sea-ice retreat. *Geophysical Research Letters*, **39**, L08502. http://dx.doi.org/10.1029/2012GL051094

61. Kirchmeier-Young, M.C., F.W. Zwiers, and N.P. Gillett, 2017: Attribution of extreme events in Arctic sea ice extent. *Journal of Climate*, **30**, 553-571. http://dx.doi.org/10.1175/jcli-d-16-0412.1

62. Zhang, R. and T.R. Knutson, 2013: The role of global climate change in the extreme low summer Arctic sea ice extent in 2012 [in "Explaining Extreme Events of 2012 from a Climate Perspective"]. *Bulletin of the American Meteorological Society*, **94 (9)**, S23-S26. http://dx.doi.org/10.1175/BAMS-D-13-00085.1

63. Jahn, A., J.E. Kay, M.M. Holland, and D.M. Hall, 2016: How predictable is the timing of a summer ice-free Arctic? *Geophysical Research Letters*, **43**, 9113-9120. http://dx.doi.org/10.1002/2016GL070067

64. Swart, N.C., J.C. Fyfe, E. Hawkins, J.E. Kay, and A. Jahn, 2015: Influence of internal variability on Arctic sea-ice trends. *Nature Climate Change*, **5**, 86-89. http://dx.doi.org/10.1038/nclimate2483

65. Snape, T.J. and P.M. Forster, 2014: Decline of Arctic sea ice: Evaluation and weighting of CMIP5 projections. *Journal of Geophysical Research Atmospheres*, **119**, 546-554. http://dx.doi.org/10.1002/2013JD020593

66. Wettstein, J.J. and C. Deser, 2014: Internal variability in projections of twenty-first-century Arctic sea ice loss: Role of the large-scale atmospheric circulation. *Journal of Climate*, **27**, 527-550. http://dx.doi.org/10.1175/JCLI-D-12-00839.1

67. Gagné, M.È., N.P. Gillett, and J.C. Fyfe, 2015: Impact of aerosol emission controls on future Arctic sea ice cover. *Geophysical Research Letters*, **42**, 8481-8488. http://dx.doi.org/10.1002/2015GL065504

68. Stroeve, J. and D. Notz, 2015: Insights on past and future sea-ice evolution from combining observations and models. *Global and Planetary Change*, **135**, 119-132. http://dx.doi.org/10.1016/j.gloplacha.2015.10.011

69. Notz, D. and J. Stroeve, 2016: Observed Arctic sea-ice loss directly follows anthropogenic $CO_2$ emission. *Science*, **354**, 747-750. http://dx.doi.org/10.1126/science.aag2345

70. Melillo, J.M., T.C. Richmond, and G.W. Yohe, eds., 2014: *Climate Change Impacts in the United States: The Third National Climate Assessment*. U.S. Global Change Research Program: Washington, D.C., 841 pp. http://dx.doi.org/10.7930/J0Z31WJ2

71. Rhein, M., S.R. Rintoul, S. Aoki, E. Campos, D. Chambers, R.A. Feely, S. Gulev, G.C. Johnson, S.A. Josey, A. Kostianoy, C. Mauritzen, D. Roemmich, L.D. Talley, and F. Wang, 2013: Observations: Ocean. *Climate Change 2013: The Physical Science Basis. Contribution of Working Group I to the Fifth Assessment Report of the Intergovernmental Panel on Climate Change*. Stocker, T.F., D. Qin, G.-K. Plattner, M. Tignor, S.K. Allen, J. Boschung, A. Nauels, Y. Xia, V. Bex, and P.M. Midgley, Eds. Cambridge University Press, Cambridge, United Kingdom and New York, NY, USA, 255–316. http://www.climatechange2013.org/report/full-report/

72. Timmermans, M.-L. and A. Proshutinsky, 2015: [The Arctic] Sea surface temperature [in "State of the Climate in 2014"]. *Bulletin of the American Meteorological Society*, **96 (12)**, S147-S148. http://dx.doi.org/10.1175/2015BAMSStateoftheClimate.1

73. Polyakov, I.V., A.V. Pnyushkov, and L.A. Timokhov, 2012: Warming of the intermediate Atlantic water of the Arctic Ocean in the 2000s. *Journal of Climate*, **25**, 8362-8370. http://dx.doi.org/10.1175/JCLI-D-12-00266.1

74. Jungclaus, J.H., K. Lohmann, and D. Zanchettin, 2014: Enhanced 20th-century heat transfer to the Arctic simulated in the context of climate variations over the last millennium. *Climate of the Past*, **10**, 2201-2213. http://dx.doi.org/10.5194/cp-10-2201-2014

75. Spielhagen, R.F., K. Werner, S.A. Sørensen, K. Zamelczyk, E. Kandiano, G. Budeus, K. Husum, T.M. Marchitto, and M. Hald, 2011: Enhanced modern heat transfer to the Arctic by warm Atlantic water. *Science*, **331**, 450-453. http://dx.doi.org/10.1126/science.1197397

76. Döscher, R., T. Vihma, and E. Maksimovich, 2014: Recent advances in understanding the Arctic climate system state and change from a sea ice perspective: A review. *Atmospheric Chemistry and Physics*, **14**, 13571-13600. http://dx.doi.org/10.5194/acp-14-13571-2014

77. Church, J.A., P.U. Clark, A. Cazenave, J.M. Gregory, S. Jevrejeva, A. Levermann, M.A. Merrifield, G.A. Milne, R.S. Nerem, P.D. Nunn, A.J. Payne, W.T. Pfeffer, D. Stammer, and A.S. Unnikrishnan, 2013: Sea level change. *Climate Change 2013: The Physical Science Basis. Contribution of Working Group I to the Fifth Assessment Report of the Intergovernmental Panel on Climate Change*. Stocker, T.F., D. Qin, G.-K. Plattner, M. Tignor, S.K. Allen, J. Boschung, A. Nauels, Y. Xia, V. Bex, and P.M. Midgley, Eds. Cambridge University Press, Cambridge, United Kingdom and New York, NY, USA, 1137–1216. http://www.climatechange2013.org/report/full-report/

78. Rawlins, M.A., M. Steele, M.M. Holland, J.C. Adam, J.E. Cherry, J.A. Francis, P.Y. Groisman, L.D. Hinzman, T.G. Huntington, D.L. Kane, J.S. Kimball, R. Kwok, R.B. Lammers, C.M. Lee, D.P. Lettenmaier, K.C. McDonald, E. Podest, J.W. Pundsack, B. Rudels, M.C. Serreze, A. Shiklomanov, Ø. Skagseth, T.J. Troy, C.J. Vörösmarty, M. Wensnahan, E.F. Wood, R. Woodgate, D. Yang, K. Zhang, and T. Zhang, 2010: Analysis of the Arctic system for freshwater cycle intensification: Observations and expectations. *Journal of Climate*, **23**, 5715-5737. http://dx.doi.org/10.1175/2010JCLI3421.1

79. Köhl, A. and N. Serra, 2014: Causes of decadal changes of the freshwater content in the Arctic Ocean. *Journal of Climate*, **27**, 3461-3475. http://dx.doi.org/10.1175/JCLI-D-13-00389.1

80. Mathis, J.T., J.N. Cross, W. Evans, and S.C. Doney, 2015: Ocean acidification in the surface waters of the Pacific–Arctic boundary regions. *Oceanography*, **28**, 122-135. http://dx.doi.org/10.5670/oceanog.2015.36

81. Arrigo, K.R., G. van Dijken, and S. Pabi, 2008: Impact of a shrinking Arctic ice cover on marine primary production. *Geophysical Research Letters*, **35**, L19603. http://dx.doi.org/10.1029/2008GL035028

82. Bates, N.R., R. Garley, K.E. Frey, K.L. Shake, and J.T. Mathis, 2014: Sea-ice melt $CO_2$–carbonate chemistry in the western Arctic Ocean: Meltwater contributions to air–sea $CO_2$ gas exchange, mixed-layer properties and rates of net community production under sea ice. *Biogeosciences*, **11**, 6769-6789. http://dx.doi.org/10.5194/bg-11-6769-2014

83. Cai, W.-J., L. Chen, B. Chen, Z. Gao, S.H. Lee, J. Chen, D. Pierrot, K. Sullivan, Y. Wang, X. Hu, W.-J. Huang, Y. Zhang, S. Xu, A. Murata, J.M. Grebmeier, E.P. Jones, and H. Zhang, 2010: Decrease in the $CO_2$ uptake capacity in an ice-free Arctic Ocean basin. *Science*, **329**, 556-559. http://dx.doi.org/10.1126/science.1189338

84. Hunt, G.L., Jr., K.O. Coyle, L.B. Eisner, E.V. Farley, R.A. Heintz, F. Mueter, J.M. Napp, J.E. Overland, P.H. Ressler, S. Salo, and P.J. Stabeno, 2011: Climate impacts on eastern Bering Sea foodwebs: A synthesis of new data and an assessment of the Oscillating Control Hypothesis. *ICES Journal of Marine Science*, **68**, 1230-1243. http://dx.doi.org/10.1093/icesjms/fsr036

85. Mathis, J.T., R.S. Pickart, R.H. Byrne, C.L. McNeil, G.W.K. Moore, L.W. Juranek, X. Liu, J. Ma, R.A. Easley, M.M. Elliot, J.N. Cross, S.C. Reisdorph, F. Bahr, J. Morison, T. Lichendorf, and R.A. Feely, 2012: Storm-induced upwelling of high pCO2 waters onto the continental shelf of the western Arctic Ocean and implications for carbonate mineral saturation states. *Geophysical Research Letters*, **39**, L16703. http://dx.doi.org/10.1029/2012GL051574

86. Stabeno, P.J., E.V. Farley, Jr., N.B. Kachel, S. Moore, C.W. Mordy, J.M. Napp, J.E. Overland, A.I. Pinchuk, and M.F. Sigler, 2012: A comparison of the physics of the northern and southern shelves of the eastern Bering Sea and some implications for the ecosystem. *Deep Sea Research Part II: Topical Studies in Oceanography*, **65-70**, 14-30. http://dx.doi.org/10.1016/j.dsr2.2012.02.019

87. Flannigan, M., B. Stocks, M. Turetsky, and M. Wotton, 2009: Impacts of climate change on fire activity and fire management in the circumboreal forest. *Global Change Biology*, **15**, 549-560. http://dx.doi.org/10.1111/j.1365-2486.2008.01660.x

88. Hu, F.S., P.E. Higuera, P. Duffy, M.L. Chipman, A.V. Rocha, A.M. Young, R. Kelly, and M.C. Dietze, 2015: Arctic tundra fires: Natural variability and responses to climate change. *Frontiers in Ecology and the Environment*, **13**, 369-377. http://dx.doi.org/10.1890/150063

89. Derksen, C., R. Brown, L. Mudryk, and K. Luojus, 2015: Terrestrial snow cover [in Arctic Report Card 2015]. ftp://ftp.oar.noaa.gov/arctic/documents/ArcticReportCard_full_report2015.pdf

90. Young, A.M., P.E. Higuera, P.A. Duffy, and F.S. Hu, 2017: Climatic thresholds shape northern high-latitude fire regimes and imply vulnerability to future climate change. *Ecography*, **40**, 606-617. http://dx.doi.org/10.1111/ecog.02205

91. Kasischke, E.S. and M.R. Turetsky, 2006: Recent changes in the fire regime across the North American boreal region—Spatial and temporal patterns of burning across Canada and Alaska. *Geophysical Research Letters*, **33**, L09703. http://dx.doi.org/10.1029/2006GL025677

92. Sanford, T., R. Wang, and A. Kenwa, 2015: *The Age of Alaskan Wildfires*. Climate Central, Princeton, NJ, 32 pp. http://assets.climatecentral.org/pdfs/AgeofAlaskanWildfires.pdf

93. Partain, J.L., Jr., S. Alden, U.S. Bhatt, P.A. Bieniek, B.R. Brettschneider, R. Lader, P.Q. Olsson, T.S. Rupp, H. Strader, R.L.T. Jr., J.E. Walsh, A.D. York, and R.H. Zieh, 2016: An assessment of the role of anthropogenic climate change in the Alaska fire season of 2015 [in "Explaining Extreme Events of 2015 from a Climate Perspective"]. *Bulletin of the American Meteorological Society*, **97 (12)**, S14-S18. http://dx.doi.org/10.1175/BAMS-D-16-0149.1

94. French, N.H.F., L.K. Jenkins, T.V. Loboda, M. Flannigan, R. Jandt, L.L. Bourgeau-Chavez, and M. Whitley, 2015: Fire in arctic tundra of Alaska: Past fire activity, future fire potential, and significance for land management and ecology. *International Journal of Wildland Fire*, **24**, 1045-1061. http://dx.doi.org/10.1071/WF14167

95. Joly, K., P.A. Duffy, and T.S. Rupp, 2012: Simulating the effects of climate change on fire regimes in Arctic biomes: Implications for caribou and moose habitat. *Ecosphere*, **3**, 1-18. http://dx.doi.org/10.1890/ES12-00012.1

96. Kelly, R., M.L. Chipman, P.E. Higuera, I. Stefanova, L.B. Brubaker, and F.S. Hu, 2013: Recent burning of boreal forests exceeds fire regime limits of the past 10,000 years. *Proceedings of the National Academy of Sciences*, **110**, 13055-13060. http://dx.doi.org/10.1073/pnas.1305069110

97. McGuire, A.D., L.G. Anderson, T.R. Christensen, S. Dallimore, L. Guo, D.J. Hayes, M. Heimann, T.D. Lorenson, R.W. MacDonald, and N. Roulet, 2009: Sensitivity of the carbon cycle in the Arctic to climate change. *Ecological Monographs*, **79**, 523-555. http://dx.doi.org/10.1890/08-2025.1

98. Mishra, U., J.D. Jastrow, R. Matamala, G. Hugelius, C.D. Koven, J.W. Harden, C.L. Ping, G.J. Michaelson, Z. Fan, R.M. Miller, A.D. McGuire, C. Tarnocai, P. Kuhry, W.J. Riley, K. Schaefer, E.A.G. Schuur, M.T. Jorgenson, and L.D. Hinzman, 2013: Empirical estimates to reduce modeling uncertainties of soil organic carbon in permafrost regions: A review of recent progress and remaining challenges. *Environmental Research Letters*, **8**, 035020. http://dx.doi.org/10.1088/1748-9326/8/3/035020

99. Mishra, U. and W.J. Riley, 2012: Alaskan soil carbon stocks: Spatial variability and dependence on environmental factors. *Biogeosciences*, **9**, 3637-3645. http://dx.doi.org/10.5194/bg-9-3637-2012

100. Kelly, R., H. Genet, A.D. McGuire, and F.S. Hu, 2016: Palaeodata-informed modelling of large carbon losses from recent burning of boreal forests. *Nature Climate Change*, **6**, 79-82. http://dx.doi.org/10.1038/nclimate2832

101. Brown, D.R.N., M.T. Jorgenson, T.A. Douglas, V.E. Romanovsky, K. Kielland, C. Hiemstra, E.S. Euskirchen, and R.W. Ruess, 2015: Interactive effects of wildfire and climate on permafrost degradation in Alaskan lowland forests. *Journal of Geophysical Research Biogeosciences*, **120**, 1619-1637. http://dx.doi.org/10.1002/2015JG003033

102. Myers-Smith, I.H., J.W. Harden, M. Wilmking, C.C. Fuller, A.D. McGuire, and F.S. Chapin Iii, 2008: Wetland succession in a permafrost collapse: interactions between fire and thermokarst. *Biogeosciences*, **5**, 1273-1286. http://dx.doi.org/10.5194/bg-5-1273-2008

103. Swanson, D.K., 1996: Susceptibility of permafrost soils to deep thaw after forest fires in interior Alaska, U.S.A., and some ecologic implications. *Arctic and Alpine Research*, **28**, 217-227. http://dx.doi.org/10.2307/1551763

104. Yoshikawa, K., W.R. Bolton, V.E. Romanovsky, M. Fukuda, and L.D. Hinzman, 2002: Impacts of wildfire on the permafrost in the boreal forests of Interior Alaska. *Journal of Geophysical Research*, **107**, 8148. http://dx.doi.org/10.1029/2001JD000438

105. Derksen, C. and R. Brown, 2012: Snow [in Arctic Report Card 2012]. ftp://ftp.oar.noaa.gov/arctic/documents/ArcticReportCard_full_report2012.pdf

106. Kunkel, K.E., D.A. Robinson, S. Champion, X. Yin, T. Estilow, and R.M. Frankson, 2016: Trends and extremes in Northern Hemisphere snow characteristics. *Current Climate Change Reports*, **2**, 65-73. http://dx.doi.org/10.1007/s40641-016-0036-8

107. Brown, R.D. and D.A. Robinson, 2011: Northern Hemisphere spring snow cover variability and change over 1922–2010 including an assessment of uncertainty. *The Cryosphere*, **5**, 219-229. http://dx.doi.org/10.5194/tc-5-219-2011

108. Rupp, D.E., P.W. Mote, N.L. Bindoff, P.A. Stott, and D.A. Robinson, 2013: Detection and attribution of observed changes in Northern Hemisphere spring snow cover. *Journal of Climate*, **26**, 6904-6914. http://dx.doi.org/10.1175/JCLI-D-12-00563.1

109. Mao, J., A. Ribes, B. Yan, X. Shi, P.E. Thornton, R. Seferian, P. Ciais, R.B. Myneni, H. Douville, S. Piao, Z. Zhu, R.E. Dickinson, Y. Dai, D.M. Ricciuto, M. Jin, F.M. Hoffman, B. Wang, M. Huang, and X. Lian, 2016: Human-induced greening of the northern extratropical land surface. *Nature Climate Change*, **6**, 959-963. http://dx.doi.org/10.1038/nclimate3056

110. Myers-Smith, I.H., B.C. Forbes, M. Wilmking, M. Hallinger, T. Lantz, D. Blok, K.D. Tape, M. Macias-Fauria, U. Sass-Klaassen, E. Lévesque, S. Boudreau, P. Ropars, L. Hermanutz, A. Trant, L.S. Collier, S. Weijers, J. Rozema, S.A. Rayback, N.M. Schmidt, G. Schaepman-Strub, S. Wipf, C. Rixen, C.B. Ménard, S. Venn, S. Goetz, L. Andreu-Hayles, S. Elmendorf, V. Ravolainen, J. Welker, P. Grogan, H.E. Epstein, and D.S. Hik, 2011: Shrub expansion in tundra ecosystems: Dynamics, impacts and research priorities. *Environmental Research Letters*, **6**, 045509. http://dx.doi.org/10.1088/1748-9326/6/4/045509

111. Euskirchen, E.S., A.P. Bennett, A.L. Breen, H. Genet, M.A. Lindgren, T.A. Kurkowski, A.D. McGuire, and T.S. Rupp, 2016: Consequences of changes in vegetation and snow cover for climate feedbacks in Alaska and northwest Canada. *Environmental Research Letters*, **11**, 105003. http://dx.doi.org/10.1088/1748-9326/11/10/105003

112. Zemp, M., H. Frey, I. Gärtner-Roer, S.U. Nussbaumer, M. Hoelzle, F. Paul, W. Haeberli, F. Denzinger, A.P. Ahlstrøm, B. Anderson, S. Bajracharya, C. Baroni, L.N. Braun, B.E. Cáceres, G. Casassa, G. Cobos, L.R. Dávila, H. Delgado Granados, M.N. Demuth, L. Espizua, A. Fischer, K. Fujita, B. Gadek, A. Ghazanfar, J.O. Hagen, P. Holmlund, N. Karimi, Z. Li, M. Pelto, P. Pitte, V.V. Popovnin, C.A. Portocarrero, R. Prinz, C.V. Sangewar, I. Severskiy, O. Sigurðsson, A. Soruco, R. Usubaliev, and C. Vincent, 2015: Historically unprecedented global glacier decline in the early 21st century. *Journal of Glaciology*, **61**, 745-762. http://dx.doi.org/10.3189/2015JoG15J017

113. Harig, C. and F.J. Simons, 2016: Ice mass loss in Greenland, the Gulf of Alaska, and the Canadian Archipelago: Seasonal cycles and decadal trends. *Geophysical Research Letters*, **43**, 3150-3159. http://dx.doi.org/10.1002/2016GL067759

114. Howat, I.M., I. Joughin, M. Fahnestock, B.E. Smith, and T.A. Scambos, 2008: Synchronous retreat and acceleration of southeast Greenland outlet glaciers 2000–06: Ice dynamics and coupling to climate. *Journal of Glaciology*, **54**, 646-660. http://dx.doi.org/10.3189/002214308786570908

115. Khan, S.A., K.H. Kjaer, M. Bevis, J.L. Bamber, J. Wahr, K.K. Kjeldsen, A.A. Bjork, N.J. Korsgaard, L.A. Stearns, M.R. van den Broeke, L. Liu, N.K. Larsen, and I.S. Muresan, 2014: Sustained mass loss of the northeast Greenland ice sheet triggered by regional warming. *Nature Climate Change*, **4**, 292-299. http://dx.doi.org/10.1038/nclimate2161

116. Rignot, E., M. Koppes, and I. Velicogna, 2010: Rapid submarine melting of the calving faces of West Greenland glaciers. *Nature Geoscience*, **3**, 187-191. http://dx.doi.org/10.1038/ngeo765

117. Straneo, F., R.G. Curry, D.A. Sutherland, G.S. Hamilton, C. Cenedese, K. Vage, and L.A. Stearns, 2011: Impact of fjord dynamics and glacial runoff on the circulation near Helheim Glacier. *Nature Geoscience*, **4**, 322-327. http://dx.doi.org/10.1038/ngeo1109

118. van den Broeke, M., J. Bamber, J. Ettema, E. Rignot, E. Schrama, W.J. van de Berg, E. van Meijgaard, I. Velicogna, and B. Wouters, 2009: Partitioning recent Greenland mass loss. *Science*, **326**, 984-986. http://dx.doi.org/10.1126/science.1178176

119. Bartholomew, I.D., P. Nienow, A. Sole, D. Mair, T. Cowton, M.A. King, and S. Palmer, 2011: Seasonal variations in Greenland Ice Sheet motion: Inland extent and behaviour at higher elevations. *Earth and Planetary Science Letters*, **307**, 271-278. http://dx.doi.org/10.1016/j.epsl.2011.04.014

120. Holland, D.M., R.H. Thomas, B. de Young, M.H. Ribergaard, and B. Lyberth, 2008: Acceleration of Jakobshavn Isbrae triggered by warm subsurface ocean waters. *Nature Geoscience*, **1**, 659-664. http://dx.doi.org/10.1038/ngeo316

121. Joughin, I., S.B. Das, M.A. King, B.E. Smith, I.M. Howat, and T. Moon, 2008: Seasonal speedup along the western flank of the Greenland Ice Sheet. *Science*, **320**, 781-783. http://dx.doi.org/10.1126/science.1153288

122. Dupont, T.K. and R.B. Alley, 2005: Assessment of the importance of ice-shelf buttressing to ice-sheet flow. *Geophysical Research Letters*, **32**, L04503. http://dx.doi.org/10.1029/2004GL022024

123. Lim, Y.-K., D.S. Siegfried, M.J.N. Sophie, N.L. Jae, M.M. Andrea, I.C. Richard, Z. Bin, and V. Isabella, 2016: Atmospheric summer teleconnections and Greenland Ice Sheet surface mass variations: Insights from MERRA-2. *Environmental Research Letters*, **11**, 024002. http://dx.doi.org/10.1088/1748-9326/11/2/024002

124. Tedesco, M., T. Mote, X. Fettweis, E. Hanna, J. Jeyaratnam, J.F. Booth, R. Datta, and K. Briggs, 2016: Arctic cut-off high drives the poleward shift of a new Greenland melting record. *Nature Communications*, **7**, 11723. http://dx.doi.org/10.1038/ncomms11723

125. Johannessen, O.M., A. Korablev, V. Miles, M.W. Miles, and K.E. Solberg, 2011: Interaction between the warm subsurface Atlantic water in the Sermilik Fjord and Helheim Glacier in southeast Greenland. *Surveys in Geophysics*, **32**, 387-396. http://dx.doi.org/10.1007/s10712-011-9130-6

126. Straneo, F., G.S. Hamilton, D.A. Sutherland, L.A. Stearns, F. Davidson, M.O. Hammill, G.B. Stenson, and A. Rosing-Asvid, 2010: Rapid circulation of warm subtropical waters in a major glacial fjord in East Greenland. *Nature Geoscience*, **3**, 182-186. http://dx.doi.org/10.1038/ngeo764

127. Andresen, C.S., F. Straneo, M.H. Ribergaard, A.A. Bjork, T.J. Andersen, A. Kuijpers, N. Norgaard-Pedersen, K.H. Kjaer, F. Schjoth, K. Weckstrom, and A.P. Ahlstrom, 2012: Rapid response of Helheim Glacier in Greenland to climate variability over the past century. *Nature Geoscience*, **5**, 37-41. http://dx.doi.org/10.1038/ngeo1349

128. Velicogna, I., 2009: Increasing rates of ice mass loss from the Greenland and Antarctic ice sheets revealed by GRACE. *Geophysical Research Letters*, **36**, L19503. http://dx.doi.org/10.1029/2009GL040222

129. Mernild, S.H., J.K. Malmros, J.C. Yde, and N.T. Knudsen, 2012: Multi-decadal marine- and land-terminating glacier recession in the Ammassalik region, southeast Greenland. *The Cryosphere*, **6**, 625-639. http://dx.doi.org/10.5194/tc-6-625-2012

130. AMAP, 2011: Snow, Water, Ice and Permafrost in the Arctic (SWIPA): Climate Change and the Cryosphere. Oslo, Norway. 538 pp. http://www.amap.no/documents/download/1448

131. Pelto, M.S., 2015: [Global Climate] Alpine glaciers [in "State of the Climate in 2014"]. *Bulletin of the American Meteorological Society*, **96 (12)**, S19-S20. http://dx.doi.org/10.1175/2015BAMSStateoftheClimate.1

132. Sharp, M., G. Wolken, D. Burgess, J.G. Cogley, L. Copland, L. Thomson, A. Arendt, B. Wouters, J. Kohler, L.M. Andreassen, S. O'Neel, and M. Pelto, 2015: [Global Climate] Glaciers and ice caps outside Greenland [in "State of the Climate in 2014"]. *Bulletin of the American Meteorological Society*, **96 (12)**, S135-S137. http://dx.doi.org/10.1175/2015BAMSStateoftheClimate.1

133. Marzeion, B., J.G. Cogley, K. Richter, and D. Parkes, 2014: Attribution of global glacier mass loss to anthropogenic and natural causes. *Science*, **345**, 919-921. http://dx.doi.org/10.1126/science.1254702

134. Mengel, M., A. Levermann, K. Frieler, A. Robinson, B. Marzeion, and R. Winkelmann, 2016: Future sea level rise constrained by observations and long-term commitment. *Proceedings of the National Academy of Sciences*, **113**, 2597-2602. http://dx.doi.org/10.1073/pnas.1500515113

135. Larsen, C.F., E. Burgess, A.A. Arendt, S. O'Neel, A.J. Johnson, and C. Kienholz, 2015: Surface melt dominates Alaska glacier mass balance. *Geophysical Research Letters*, **42**, 5902-5908. http://dx.doi.org/10.1002/2015GL064349

136. Ding, Q., J.M. Wallace, D.S. Battisti, E.J. Steig, A.J.E. Gallant, H.-J. Kim, and L. Geng, 2014: Tropical forcing of the recent rapid Arctic warming in northeastern Canada and Greenland. *Nature*, **509**, 209-212. http://dx.doi.org/10.1038/nature13260

137. Francis, J.A., S.J. Vavrus, and J. Cohen, 2017: Amplified Arctic warming and mid-latitude weather: Emerging connections. *Wiley Interdisciplinary Review: Climate Change*, **8**, e474. http://dx.doi.org/10.1002/wcc.474

138. Graversen, R.G., 2006: Do changes in the midlatitude circulation have any impact on the Arctic surface air temperature trend? *Journal of Climate*, **19**, 5422-5438. http://dx.doi.org/10.1175/JCLI3906.1

139. Lee, S., 2014: A theory for polar amplification from a general circulation perspective. *Asia-Pacific Journal of Atmospheric Sciences*, **50**, 31-43. http://dx.doi.org/10.1007/s13143-014-0024-7

140. Lee, S., T. Gong, N. Johnson, S.B. Feldstein, and D. Pollard, 2011: On the possible link between tropical convection and the Northern Hemisphere Arctic surface air temperature change between 1958 and 2001. *Journal of Climate*, **24**, 4350-4367. http://dx.doi.org/10.1175/2011JCLI4003.1

141. Park, H.-S., S. Lee, S.-W. Son, S.B. Feldstein, and Y. Kosaka, 2015: The impact of poleward moisture and sensible heat flux on Arctic winter sea ice variability. *Journal of Climate*, **28**, 5030-5040. http://dx.doi.org/10.1175/JCLI-D-15-0074.1

142. Perlwitz, J., M. Hoerling, and R. Dole, 2015: Arctic tropospheric warming: Causes and linkages to lower latitudes. *Journal of Climate*, **28**, 2154-2167. http://dx.doi.org/10.1175/JCLI-D-14-00095.1

143. Rigor, I.G., J.M. Wallace, and R.L. Colony, 2002: Response of sea ice to the Arctic oscillation. *Journal of Climate*, **15**, 2648-2663. http://dx.doi.org/10.1175/1520-0442(2002)015<2648:ROSITT>2.0.CO;2

144. Screen, J.A., C. Deser, and I. Simmonds, 2012: Local and remote controls on observed Arctic warming. *Geophysical Research Letters*, **39**, L10709. http://dx.doi.org/10.1029/2012GL051598

145. Screen, J.A. and J.A. Francis, 2016: Contribution of sea-ice loss to Arctic amplification is regulated by Pacific Ocean decadal variability. *Nature Climate Change*, **6**, 856-860. http://dx.doi.org/10.1038/nclimate3011

146. Overland, J., E. Hanna, I. Hanssen-Bauer, S.-J. Kim, J. Walsh, M. Wang, U. Bhatt, and R.L. Thoman, 2016: Surface air temperature [in Arctic Report Card 2016]. http://arctic.noaa.gov/Report-Card/Report-Card-2016/ArtMID/5022/ArticleID/271/Surface-Air-Temperature

147. Liu, Y. and J.R. Key, 2014: Less winter cloud aids summer 2013 Arctic sea ice return from 2012 minimum. *Environmental Research Letters*, **9**, 044002. http://dx.doi.org/10.1088/1748-9326/9/4/044002

148. Woods, C. and R. Caballero, 2016: The role of moist intrusions in winter Arctic warming and sea ice decline. *Journal of Climate*, **29**, 4473-4485. http://dx.doi.org/10.1175/jcli-d-15-0773.1

149. Cohen, J., J.A. Screen, J.C. Furtado, M. Barlow, D. Whittleston, D. Coumou, J. Francis, K. Dethloff, D. Entekhabi, J. Overland, and J. Jones, 2014: Recent Arctic amplification and extreme mid-latitude weather. *Nature Geoscience*, **7**, 627-637. http://dx.doi.org/10.1038/ngeo2234

150. Barnes, E.A. and J.A. Screen, 2015: The impact of Arctic warming on the midlatitude jet-stream: Can it? Has it? Will it? *Wiley Interdisciplinary Reviews: Climate Change*, **6**, 277-286. http://dx.doi.org/10.1002/wcc.337

151. Ayarzagüena, B. and J.A. Screen, 2016: Future Arctic sea ice loss reduces severity of cold air outbreaks in midlatitudes. *Geophysical Research Letters*, **43**, 2801-2809. http://dx.doi.org/10.1002/2016GL068092

152. Screen, J.A., C. Deser, and L. Sun, 2015: Reduced risk of North American cold extremes due to continued Arctic sea ice loss. *Bulletin of the American Meteorological Society*, **96 (12)**, 1489-1503. http://dx.doi.org/10.1175/BAMS-D-14-00185.1

153. Screen, J.A., C. Deser, and L. Sun, 2015: Projected changes in regional climate extremes arising from Arctic sea ice loss. *Environmental Research Letters*, **10**, 084006. http://dx.doi.org/10.1088/1748-9326/10/8/084006

154. Sun, L., J. Perlwitz, and M. Hoerling, 2016: What caused the recent "Warm Arctic, Cold Continents" trend pattern in winter temperatures? *Geophysical Research Letters*, **43**, 5345-5352. http://dx.doi.org/10.1002/2016GL069024

155. Francis, J.A. and S.J. Vavrus, 2012: Evidence linking Arctic amplification to extreme weather in mid-latitudes. *Geophysical Research Letters*, **39**, L06801. http://dx.doi.org/10.1029/2012GL051000

156. Vihma, T., 2014: Effects of Arctic sea ice decline on weather and climate: A review. *Surveys in Geophysics*, **35**, 1175-1214. http://dx.doi.org/10.1007/s10712-014-9284-0

157. Francis, J. and N. Skific, 2015: Evidence linking rapid Arctic warming to mid-latitude weather patterns. *Philosophical Transactions of the Royal Society A: Mathematical, Physical and Engineering Sciences*, **373**, 20140170. http://dx.doi.org/10.1098/rsta.2014.0170

158. Francis, J.A. and S.J. Vavrus, 2015: Evidence for a wavier jet stream in response to rapid Arctic warming. *Environmental Research Letters*, **10**, 014005. http://dx.doi.org/10.1088/1748-9326/10/1/014005

159. Seager, R., M. Hoerling, S. Schubert, H. Wang, B. Lyon, A. Kumar, J. Nakamura, and N. Henderson, 2015: Causes of the 2011–14 California drought. *Journal of Climate*, **28**, 6997-7024. http://dx.doi.org/10.1175/JCLI-D-14-00860.1

160. Swain, D., M. Tsiang, M. Haughen, D. Singh, A. Charland, B. Rajarthan, and N.S. Diffenbaugh, 2014: The extraordinary California drought of 2013/14: Character, context and the role of climate change [in "Explaining Extreme Events of 2013 from a Climate Perspective"]. *Bulletin of the American Meteorological Society*, **95 (9)**, S3-S6. http://dx.doi.org/10.1175/1520-0477-95.9.S1.1

161. Teng, H. and G. Branstator, 2017: Causes of extreme ridges that induce California droughts. *Journal of Climate*, **30**, 1477-1492. http://dx.doi.org/10.1175/jcli-d-16-0524.1

162. Overland, J., J.A. Francis, R. Hall, E. Hanna, S.-J. Kim, and T. Vihma, 2015: The melting Arctic and midlatitude weather patterns: Are they connected? *Journal of Climate*, **28**, 7917-7932. http://dx.doi.org/10.1175/JCLI-D-14-00822.1

163. Barnes, E.A. and L.M. Polvani, 2015: CMIP5 projections of Arctic amplification, of the North American/ North Atlantic circulation, and of their relationship. *Journal of Climate*, **28**, 5254-5271. http://dx.doi.org/10.1175/JCLI-D-14-00589.1

164. Hoskins, B. and T. Woollings, 2015: Persistent extratropical regimes and climate extremes. *Current Climate Change Reports*, **1**, 115-124. http://dx.doi.org/10.1007/s40641-015-0020-8

165. Sigmond, M. and J.C. Fyfe, 2016: Tropical Pacific impacts on cooling North American winters. *Nature Climate Change*, **6**, 970-974. http://dx.doi.org/10.1038/nclimate3069

166. Cohen, J., J. Jones, J.C. Furtado, and E. Tzipermam, 2013: Warm Arctic, cold continents: A common pattern related to Arctic sea ice melt, snow advance, and extreme winter weather. . *Oceanography*, **26**, 150-160. http://dx.doi.org/10.5670/oceanog.2013.70

167. Nummelin, A., M. Ilicak, C. Li, and L.H. Smedsrud, 2016: Consequences of future increased Arctic runoff on Arctic Ocean stratification, circulation, and sea ice cover. *Journal of Geophysical Research Oceans*, **121**, 617-637. http://dx.doi.org/10.1002/2015JC011156

168. Giles, K.A., S.W. Laxon, A.L. Ridout, D.J. Wingham, and S. Bacon, 2012: Western Arctic Ocean freshwater storage increased by wind-driven spin-up of the Beaufort Gyre. *Nature Geoscience*, **5**, 194-197. http://dx.doi.org/10.1038/ngeo1379

169. Morison, J., R. Kwok, C. Peralta-Ferriz, M. Alkire, I. Rigor, R. Andersen, and M. Steele, 2012: Changing Arctic Ocean freshwater pathways. *Nature*, **481**, 66-70. http://dx.doi.org/10.1038/nature10705

170. Rahmstorf, S., J.E. Box, G. Feulner, M.E. Mann, A. Robinson, S. Rutherford, and E.J. Schaffernicht, 2015: Exceptional twentieth-century slowdown in Atlantic Ocean overturning circulation. *Nature Climate Change*, **5**, 475-480. http://dx.doi.org/10.1038/nclimate2554

171. Yang, Q., T.H. Dixon, P.G. Myers, J. Bonin, D. Chambers, and M.R. van den Broeke, 2016: Recent increases in Arctic freshwater flux affects Labrador Sea convection and Atlantic overturning circulation. *Nature Communications*, **7**, 10525. http://dx.doi.org/10.1038/ncomms10525

172. Liu, W., S.-P. Xie, Z. Liu, and J. Zhu, 2017: Overlooked possibility of a collapsed Atlantic Meridional Overturning Circulation in warming climate. *Science Advances*, **3**, e1601666. http://dx.doi.org/10.1126/sciadv.1601666

173. Smeed, D.A., G.D. McCarthy, S.A. Cunningham, E. Frajka-Williams, D. Rayner, W.E. Johns, C.S. Meinen, M.O. Baringer, B.I. Moat, A. Duchez, and H.L. Bryden, 2014: Observed decline of the Atlantic meridional overturning circulation 2004–2012. *Ocean Science*, **10**, 29-38. http://dx.doi.org/10.5194/os-10-29-2014

174. Romanovsky, V.E., S.L. Smith, H.H. Christiansen, N.I. Shiklomanov, D.A. Streletskiy, D.S. Drozdov, G.V. Malkova, N.G. Oberman, A.L. Kholodov, and S.S. Marchenko, 2015: [The Arctic] Terrestrial permafrost [in "State of the Climate in 2014"]. *Bulletin of the American Meteorological Society*, **96 (12)**, S139-S141. http://dx.doi.org/10.1175/2015BAMSStateoftheClimate.1

175. Romanovsky, V.E., S.L. Smith, K. Isaksen, N.I. Shiklomanov, D.A. Streletskiy, A.L. Kholodov, H.H. Christiansen, D.S. Drozdov, G.V. Malkova, and S.S. Marchenko, 2016: [The Arctic] Terrestrial permafrost [in "State of the Climate in 2015"]. *Bulletin of the American Meteorological Society*, **97**, S149-S152. http://dx.doi.org/10.1175/2016BAMSStateoftheClimate.1

176. Shiklomanov, N.E., D.A. Streletskiy, and F.E. Nelson, 2012: Northern Hemisphere component of the global Circumpolar Active Layer Monitory (CALM) program. In *Proceedings of the 10th International Conference on Permafrost*, Salekhard, Russia. Kane, D.L. and K.M. Hinkel, Eds., 377-382. http://research.iarc.uaf.edu/NICOP/proceedings/10th/TICOP_vol1.pdf

177. Kokelj, S.V., T.C. Lantz, J. Tunnicliffe, R. Segal, and D. Lacelle, 2017: Climate-driven thaw of permafrost preserved glacial landscapes, northwestern Canada. *Geology*, **45**, 371-374. http://dx.doi.org/10.1130/g38626.1

178. Grosse, G., S. Goetz, A.D. McGuire, V.E. Romanovsky, and E.A.G. Schuur, 2016: Changing permafrost in a warming world and feedbacks to the Earth system. *Environmental Research Letters*, **11**, 040201. http://dx.doi.org/10.1088/1748-9326/11/4/040201

179. Schuur, E.A.G., A.D. McGuire, C. Schadel, G. Grosse, J.W. Harden, D.J. Hayes, G. Hugelius, C.D. Koven, P. Kuhry, D.M. Lawrence, S.M. Natali, D. Olefeldt, V.E. Romanovsky, K. Schaefer, M.R. Turetsky, C.C. Treat, and J.E. Vonk, 2015: Climate change and the permafrost carbon feedback. *Nature*, **520**, 171-179. http://dx.doi.org/10.1038/nature14338

180. Tarnocai, C., J.G. Canadell, E.A.G. Schuur, P. Kuhry, G. Mazhitova, and S. Zimov, 2009: Soil organic carbon pools in the northern circumpolar permafrost region. *Global Biogeochemical Cycles*, **23**, GB2023. http://dx.doi.org/10.1029/2008GB003327

181. Chang, R.Y.-W., C.E. Miller, S.J. Dinardo, A. Karion, C. Sweeney, B.C. Daube, J.M. Henderson, M.E. Mountain, J. Eluszkiewicz, J.B. Miller, L.M.P. Bruhwiler, and S.C. Wofsy, 2014: Methane emissions from Alaska in 2012 from CARVE airborne observations. *Proceedings of the National Academy of Sciences*, **111**, 16694-16699. http://dx.doi.org/10.1073/pnas.1412953111

182. Schuur, E.A.G., J.G. Vogel, K.G. Crummer, H. Lee, J.O. Sickman, and T.E. Osterkamp, 2009: The effect of permafrost thaw on old carbon release and net carbon exchange from tundra. *Nature*, **459**, 556-559. http://dx.doi.org/10.1038/nature08031

183. Zona, D., B. Gioli, R. Commane, J. Lindaas, S.C. Wofsy, C.E. Miller, S.J. Dinardo, S. Dengel, C. Sweeney, A. Karion, R.Y.-W. Chang, J.M. Henderson, P.C. Murphy, J.P. Goodrich, V. Moreaux, A. Liljedahl, J.D. Watts, J.S. Kimball, D.A. Lipson, and W.C. Oechel, 2016: Cold season emissions dominate the Arctic tundra methane budget. *Proceedings of the National Academy of Sciences*, **113**, 40-45. http://dx.doi.org/10.1073/pnas.1516017113

184. Treat, C.C., S.M. Natali, J. Ernakovich, C.M. Iversen, M. Lupascu, A.D. McGuire, R.J. Norby, T. Roy Chowdhury, A. Richter, H. Šantrůčková, C. Schädel, E.A.G. Schuur, V.L. Sloan, M.R. Turetsky, and M.P. Waldrop, 2015: A pan-Arctic synthesis of $CH_4$ and $CO_2$ production from anoxic soil incubations. *Global Change Biology*, **21**, 2787-2803. http://dx.doi.org/10.1111/gcb.12875

185. Myhre, G., D. Shindell, F.-M. Bréon, W. Collins, J. Fuglestvedt, J. Huang, D. Koch, J.-F. Lamarque, D. Lee, B. Mendoza, T. Nakajima, A. Robock, G. Stephens, T. Takemura, and H. Zhang, 2013: Anthropogenic and natural radiative forcing. *Climate Change 2013: The Physical Science Basis. Contribution of Working Group I to the Fifth Assessment Report of the Intergovernmental Panel on Climate Change*. Stocker, T.F., D. Qin, G.-K. Plattner, M. Tignor, S.K. Allen, J. Boschung, A. Nauels, Y. Xia, V. Bex, and P.M. Midgley, Eds. Cambridge University Press, Cambridge, United Kingdom and New York, NY, USA, 659–740. http://www.climatechange2013.org/report/full-report/

186. Schädel, C., M.K.F. Bader, E.A.G. Schuur, C. Biasi, R. Bracho, P. Capek, S. De Baets, K. Diakova, J. Ernakovich, C. Estop-Aragones, D.E. Graham, I.P. Hartley, C.M. Iversen, E. Kane, C. Knoblauch, M. Lupascu, P.J. Martikainen, S.M. Natali, R.J. Norby, J.A. O'Donnell, T.R. Chowdhury, H. Santruckova, G. Shaver, V.L. Sloan, C.C. Treat, M.R. Turetsky, M.P. Waldrop, and K.P. Wickland, 2016: Potential carbon emissions dominated by carbon dioxide from thawed permafrost soils. *Nature Climate Change*, **6**, 950-953. http://dx.doi.org/10.1038/nclimate3054

187. Koven, C.D., D.M. Lawrence, and W.J. Riley, 2015: Permafrost carbon–climate feedback is sensitive to deep soil carbon decomposability but not deep soil nitrogen dynamics. *Proceedings of the National Academy of Sciences*, **112**, 3752-3757. http://dx.doi.org/10.1073/pnas.1415123112

188. Koven, C.D., E.A.G. Schuur, C. Schädel, T.J. Bohn, E.J. Burke, G. Chen, X. Chen, P. Ciais, G. Grosse, J.W. Harden, D.J. Hayes, G. Hugelius, E.E. Jafarov, G. Krinner, P. Kuhry, D.M. Lawrence, A.H. MacDougall, S.S. Marchenko, A.D. McGuire, S.M. Natali, D.J. Nicolsky, D. Olefeldt, S. Peng, V.E. Romanovsky, K.M. Schaefer, J. Strauss, C.C. Treat, and M. Turetsky, 2015: A simplified, data-constrained approach to estimate the permafrost carbon–climate feedback. *Philosophical Transactions of the Royal Society A: Mathematical, Physical and Engineering Sciences*, **373**, 20140423. http://dx.doi.org/10.1098/rsta.2014.0423

189. Schaefer, K., H. Lantuit, E.R. Vladimir, E.A.G. Schuur, and R. Witt, 2014: The impact of the permafrost carbon feedback on global climate. *Environmental Research Letters*, **9**, 085003. http://dx.doi.org/10.1088/1748-9326/9/8/085003

190. Chadburn, S.E., E.J. Burke, P.M. Cox, P. Friedlingstein, G. Hugelius, and S. Westermann, 2017: An observation-based constraint on permafrost loss as a function of global warming. *Nature Climate Change*, **7**, 340-344. http://dx.doi.org/10.1038/nclimate3262

191. Friedlingstein, P., P. Cox, R. Betts, L. Bopp, W.v. Bloh, V. Brovkin, P. Cadule, S. Doney, M. Eby, I. Fung, G. Bala, J. John, C. Jones, F. Joos, T. Kato, M. Kawamiya, W. Knorr, K. Lindsay, H.D. Matthews, T. Raddatz, P. Rayner, C. Reick, E. Roeckner, K.-G. Schnitzler, R. Schnur, K. Strassmann, A.J. Weaver, C. Yoshikawa, and N. Zeng, 2006: Climate–carbon cycle feedback analysis: Results from the C⁴MIP model intercomparison. *Journal of Climate*, **19**, 3337-3353. http://dx.doi.org/10.1175/JCLI3800.1

192. Fisher, J.B., M. Sikka, W.C. Oechel, D.N. Huntzinger, J.R. Melton, C.D. Koven, A. Ahlström, M.A. Arain, I. Baker, J.M. Chen, P. Ciais, C. Davidson, M. Dietze, B. El-Masri, D. Hayes, C. Huntingford, A.K. Jain, P.E. Levy, M.R. Lomas, B. Poulter, D. Price, A.K. Sahoo, K. Schaefer, H. Tian, E. Tomelleri, H. Verbeeck, N. Viovy, R. Wania, N. Zeng, and C.E. Miller, 2014: Carbon cycle uncertainty in the Alaskan Arctic. *Biogeosciences*, **11**, 4271-4288. http://dx.doi.org/10.5194/bg-11-4271-2014

193. Liljedahl, A.K., J. Boike, R.P. Daanen, A.N. Fedorov, G.V. Frost, G. Grosse, L.D. Hinzman, Y. Iijma, J.C. Jorgenson, N. Matveyeva, M. Necsoiu, M.K. Raynolds, V.E. Romanovsky, J. Schulla, K.D. Tape, D.A. Walker, C.J. Wilson, H. Yabuki, and D. Zona, 2016: Pan-Arctic ice-wedge degradation in warming permafrost and its influence on tundra hydrology. *Nature Geoscience*, **9**, 312-318. http://dx.doi.org/10.1038/ngeo2674

194. Oh, Y., B. Stackhouse, M.C.Y. Lau, X. Xu, A.T. Trugman, J. Moch, T.C. Onstott, C.J. Jørgensen, L. D'Imperio, B. Elberling, C.A. Emmerton, V.L. St. Louis, and D. Medvigy, 2016: A scalable model for methane consumption in Arctic mineral soils. *Geophysical Research Letters*, **43**, 5143-5150. http://dx.doi.org/10.1002/2016GL069049

195. Archer, D., 2007: Methane hydrate stability and anthropogenic climate change. *Biogeosciences*, **4**, 521-544. http://dx.doi.org/10.5194/bg-4-521-2007

196. Piñero, E., M. Marquardt, C. Hensen, M. Haeckel, and K. Wallmann, 2013: Estimation of the global inventory of methane hydrates in marine sediments using transfer functions. *Biogeosciences*, **10**, 959-975. http://dx.doi.org/10.5194/bg-10-959-2013

197. Ruppel, C.D. *Methane hydrates and contemporary climate change.* Nature Education Knowledge, 2011. **3**.

198. Ruppel, C.D. and J.D. Kessler, 2017: The interaction of climate change and methane hydrates. *Reviews of Geophysics*, **55**, 126-168. http://dx.doi.org/10.1002/2016RG000534

199. Bollmann, M., T. Bosch, F. Colijn, R. Ebinghaus, R. Froese, K. Güssow, S. Khalilian, S. Krastel, A. Körtzinger, M. Langenbuch, M. Latif, B. Matthiessen, F. Melzner, A. Oschlies, S. Petersen, A. Proelß, M. Quaas, J. Reichenbach, T. Requate, T. Reusch, P. Rosenstiel, J.O. Schmidt, K. Schrottke, H. Sichelschmidt, U. Siebert, R. Soltwedel, U. Sommer, K. Stattegger, H. Sterr, R. Sturm, T. Treude, A. Vafeidis, C.v. Bernem, J.v. Beusekom, R. Voss, M. Visbeck, M. Wahl, K. Wallmann, and F. Weinberger, 2010: *World Ocean Review: Living With the Oceans.* maribus gGmbH, 232 pp. http://worldoceanreview.com/wp-content/downloads/wor1/WOR1_english.pdf

200. Brothers, L.L., B.M. Herman, P.E. Hart, and C.D. Ruppel, 2016: Subsea ice-bearing permafrost on the U.S. Beaufort Margin: 1. Minimum seaward extent defined from multichannel seismic reflection data. *Geochemistry, Geophysics, Geosystems*, **17**, 4354-4365. http://dx.doi.org/10.1002/2016GC006584

201. Johnson, H.P., U.K. Miller, M.S. Salmi, and E.A. Solomon, 2015: Analysis of bubble plume distributions to evaluate methane hydrate decomposition on the continental slope. *Geochemistry, Geophysics, Geosystems*, **16**, 3825-3839. http://dx.doi.org/10.1002/2015GC005955

202. Ruppel, C.D., B.M. Herman, L.L. Brothers, and P.E. Hart, 2016: Subsea ice-bearing permafrost on the U.S. Beaufort Margin: 2. Borehole constraints. *Geochemistry, Geophysics, Geosystems*, **17**, 4333-4353. http://dx.doi.org/10.1002/2016GC006582

203. Skarke, A., C. Ruppel, M. Kodis, D. Brothers, and E. Lobecker, 2014: Widespread methane leakage from the sea floor on the northern US Atlantic margin. *Nature Geoscience*, **7**, 657-661. http://dx.doi.org/10.1038/ngeo2232

204. Hunter, S.J., D.S. Goldobin, A.M. Haywood, A. Ridgwell, and J.G. Rees, 2013: Sensitivity of the global submarine hydrate inventory to scenarios of future climate change. *Earth and Planetary Science Letters*, **367**, 105-115. http://dx.doi.org/10.1016/j.epsl.2013.02.017

205. Kretschmer, K., A. Biastoch, L. Rüpke, and E. Burwicz, 2015: Modeling the fate of methane hydrates under global warming. *Global Biogeochemical Cycles*, **29**, 610-625. http://dx.doi.org/10.1002/2014GB005011

206. Graves, C.A., L. Steinle, G. Rehder, H. Niemann, D.P. Connelly, D. Lowry, R.E. Fisher, A.W. Stott, H. Sahling, and R.H. James, 2015: Fluxes and fate of dissolved methane released at the seafloor at the landward limit of the gas hydrate stability zone offshore western Svalbard. *Journal of Geophysical Research Oceans*, **120**, 6185-6201. http://dx.doi.org/10.1002/2015JC011084

207. ACIA, 2005: Arctic Climate Impact Assessment. ACIA Secretariat and Cooperative Institute for Arctic Research, 1042 pp. http://www.acia.uaf.edu/pages/scientific.html

208. Hollesen, J., H. Matthiesen, A.B. Møller, and B. Elberling, 2015: Permafrost thawing in organic Arctic soils accelerated by ground heat production. *Nature Climate Change*, **5**, 574-578. http://dx.doi.org/10.1038/nclimate2590

209. Fetterer, F., K. Knowles, W. Meier, and M. Savoie, 2016, updated daily: Sea Ice Index, Version 2. National Snow and Ice Data Center, Boulder, CO.

210. WGMS, 2016: Fluctuations of Glaciers Database. World Glacier Monitoring Service, Zurich, Switzerland.

211. Wolken, G., M. Sharp, L.M. Andreassen, A. Arendt, D. Burgess, J.G. Cogley, L. Copland, J. Kohler, S. O'Neel, M. Pelto, L. Thomson, and B. Wouters, 2016: [The Arctic] Glaciers and ice caps outside Greenland [in "State of the Climate in 2015"]. *Bulletin of the American Meteorological Society*, **97**, S142-S145. http://dx.doi.org/10.1175/2016BAMSStateoftheClimate.1

212. USGS, 2004: Repeat Photography of Alaskan Glaciers: Muir Glacier (USGS Photograph by Bruce F. Molnia). Department of the Interior, U.S. Geological Survey. https://www2.usgs.gov/climate_landuse/glaciers/repeat_photography.asp

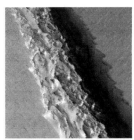

# 12

# Sea Level Rise

**KEY FINDINGS**

1. Global mean sea level (GMSL) has risen by about 7–8 inches (about 16–21 cm) since 1900, with about 3 of those inches (about 7 cm) occurring since 1993 (*very high confidence*). Human-caused climate change has made a substantial contribution to GMSL rise since 1900 (*high confidence*), contributing to a rate of rise that is greater than during any preceding century in at least 2,800 years (*medium confidence*).

2. Relative to the year 2000, GMSL is *very likely* to rise by 0.3–0.6 feet (9–18 cm) by 2030, 0.5–1.2 feet (15–38 cm) by 2050, and 1.0–4.3 feet (30–130 cm) by 2100 (*very high confidence in lower bounds; medium confidence in upper bounds for 2030 and 2050; low confidence in upper bounds for 2100*). Future pathways have little effect on projected GMSL rise in the first half of the century, but significantly affect projections for the second half of the century (*high confidence*). Emerging science regarding Antarctic ice sheet stability suggests that, for high emission scenarios, a GMSL rise exceeding 8 feet (2.4 m) by 2100 is physically possible, although the probability of such an extreme outcome cannot currently be assessed. Regardless of pathway, it is *extremely likely* that GMSL rise will continue beyond 2100 (*high confidence*).

3. Relative sea level (RSL) rise in this century will vary along U.S. coastlines due, in part, to changes in Earth's gravitational field and rotation from melting of land ice, changes in ocean circulation, and vertical land motion (*very high confidence*). For almost all future GMSL rise scenarios, RSL rise is *likely* to be greater than the global average in the U.S. Northeast and the western Gulf of Mexico. In intermediate and low GMSL rise scenarios, RSL rise is *likely* to be less than the global average in much of the Pacific Northwest and Alaska. For high GMSL rise scenarios, RSL rise is *likely* to be higher than the global average along all U.S. coastlines outside Alaska. Almost all U.S. coastlines experience more than global mean sea level rise in response to Antarctic ice loss, and thus would be particularly affected under extreme GMSL rise scenarios involving substantial Antarctic mass loss (*high confidence*).

4. As sea levels have risen, the number of tidal floods each year that cause minor impacts (also called "nuisance floods") have increased 5- to 10-fold since the 1960s in several U.S. coastal cities (*very high confidence*). Rates of increase are accelerating in over 25 Atlantic and Gulf Coast cities (*very high confidence*). Tidal flooding will continue increasing in depth, frequency, and extent this century (*very high confidence*).

## KEY FINDINGS *(continued)*

5. Assuming storm characteristics do not change, sea level rise will increase the frequency and extent of extreme flooding associated with coastal storms, such as hurricanes and nor'easters (*very high confidence*). A projected increase in the intensity of hurricanes in the North Atlantic (*medium confidence*) could increase the probability of extreme flooding along most of the U.S. Atlantic and Gulf Coast states beyond what would be projected based solely on RSL rise. However, there is *low confidence* in the projected increase in frequency of intense Atlantic hurricanes, and the associated flood risk amplification and flood effects could be offset or amplified by such factors as changes in overall storm frequency or tracks.

### Recommended Citation for Chapter

**Sweet**, W.V., R. Horton, R.E. Kopp, A.N. LeGrande, and A. Romanou, 2017: Sea level rise. In: *Climate Science Special Report: Fourth National Climate Assessment, Volume I* [Wuebbles, D.J., D.W. Fahey, K.A. Hibbard, D.J. Dokken, B.C. Stewart, and T.K. Maycock (eds.)]. U.S. Global Change Research Program, Washington, DC, USA, pp. 333-363, doi: 10.7930/J0VM49F2.

## 12.1 Introduction

Sea level rise is closely linked to increasing global temperatures. Thus, even as uncertainties remain about just how much sea level may rise this century, it is virtually certain that sea level rise this century and beyond will pose a growing challenge to coastal communities, infrastructure, and ecosystems from increased (permanent) inundation, more frequent and extreme coastal flooding, erosion of coastal landforms, and saltwater intrusion within coastal rivers and aquifers. Assessment of vulnerability to rising sea levels requires consideration of physical causes, historical evidence, and projections. A risk-based perspective on sea level rise points to the need for emphasis on how changing sea levels alter the coastal zone and interact with coastal flood risk at local scales.

This chapter reviews the physical factors driving changes in global mean sea level (GMSL) and those causing additional regional variations in relative sea level (RSL). It presents geological and instrumental observations of historical sea level changes and an assessment of the human contribution to sea level change. It then describes a range of scenarios for future levels and rates of sea level change, and the relationship of these scenarios to the Representative Concentration Pathways (RCPs). Finally, it assesses the impact of changes in sea level on extreme water levels.

While outside the scope of this chapter, it is important to note the myriad of other potential impacts associated with RSL rise, wave action, and increases in coastal flooding. These impacts include loss of life, damage to infrastructure and the built environment, salinization of coastal aquifers, mobilization of pollutants, changing sediment budgets, coastal erosion, and ecosystem changes such as marsh loss and threats to endangered flora and fauna.[1] While all of these impacts are inherently important, some also have the potential to influence local rates of RSL rise and the extent of wave-driven and coastal flooding impacts. For example, there is evidence that wave action and flooding of beaches and marshes can induce changes in coastal geomorphology, such as sediment build up, that may iteratively modify the future flood risk profile of communities and ecosystems.[2]

## 12.2 Physical Factors Contributing to Sea Level Rise

Sea level change is driven by a variety of mechanisms operating at different spatial and temporal scales (see Kopp et al. 2015[3] for a review). GMSL rise is primarily driven by two factors: 1) increased volume of seawater due to thermal expansion of the ocean as it warms, and 2) increased mass of water in the ocean due to melting ice from mountain glaciers and the Antarctic and Greenland ice sheets.[4] The overall amount (mass) of ocean water, and thus sea level, is also affected to a lesser extent by changes in global land-water storage, which reflects changes in the impoundment of water in dams and reservoirs and river runoff from groundwater extraction, inland sea and wetland drainage, and global precipitation patterns, such as occur during phases of the El Niño–Southern Oscillation (ENSO).[4, 5, 6, 7, 8]

Sea level and its changes are not uniform globally for several reasons. First, atmosphere–ocean dynamics—driven by ocean circulation, winds, and other factors—are associated with differences in the height of the sea surface, as are differences in density arising from the distribution of heat and salinity in the ocean. Changes in any of these factors will affect sea surface height. For example, a weakening of the Gulf Stream transport in the mid-to-late 2000s may have contributed to enhanced sea level rise in the ocean environment extending to the northeastern U.S. coast,[9, 10, 11] a trend that many models project will continue into the future.[12]

Second, the locations of land ice melting and land water reservoir changes impart distinct regional "static-equilibrium fingerprints" on sea level, based on gravitational, rotational, and crustal deformation effects (Figure 12.1a–d).[13] For example, sea level falls near a melting ice sheet because of the reduced gravitational attraction of the ocean toward the ice sheet;

reciprocally, it rises by greater than the global average far from the melting ice sheet.

Third, the Earth's mantle is still moving in response to the loss of the great North American (Laurentide) and European ice sheets of the Last Glacial Maximum; the associated changes in the height of the land, the shape of the ocean basin, and the Earth's gravitational field give rise to glacial-isostatic adjustment (Figure 12.1e). For example, in areas once covered by the thickest parts of the great ice sheets of the Last Glacial Maximum, such as in Hudson Bay and in Scandinavia, post-glacial rebound of the land is causing RSL to fall. Along the flanks of the ice sheets, such as along most of the east coast of the United States, subsidence of the bulge that flanked the ice sheet is causing RSL to rise.

Finally, a variety of other factors can cause local vertical land movement. These include natural sediment compaction, compaction caused by local extraction of groundwater and fossil fuels, and processes related to plate tectonics, such as earthquakes and more gradual seismic creep (Figure 12.1f).[14, 15]

Compared to many climate variables, the trend signal for sea level change tends to be large relative to natural variability. However, at inter-annual timescales, changes in ocean dynamics, density, and wind can cause substantial sea level variability in some regions. For example, there has been a multidecadal suppression of sea level rise off the Pacific coast[16] and large year-to-year variations in sea level along the Northeast U.S. coast.[17] Local rates of land height change have also varied dramatically on decadal timescales in some locations, such as along the western Gulf Coast, where rates of subsurface extraction of fossil fuels and groundwater have varied over time.[18]

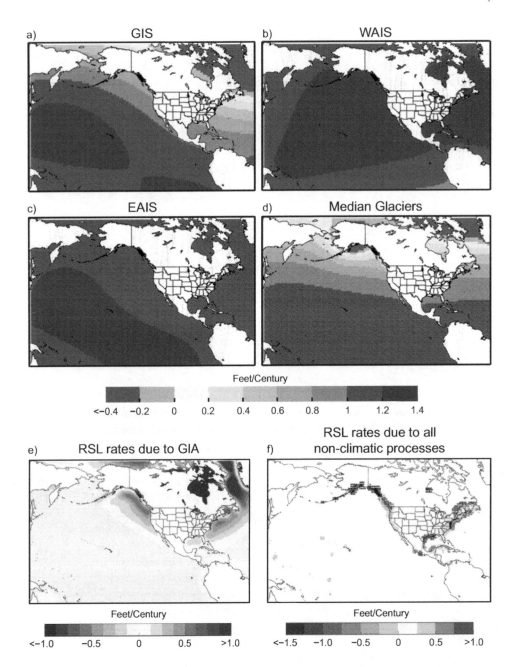

Figure 12.1: (a–d) Static-equilibrium fingerprints of the relative sea level (RSL) effect of land ice melt, in units of feet of RSL change per feet of global mean sea level (GMSL) change, for mass loss from (a) Greenland, (b) West Antarctica, (c) East Antarctica, and (d) the median projected combination of melting glaciers, after Kopp et al.[3, 76] (e) Model projections of the rate of RSL rise due to glacial-isostatic adjustment (units of feet/century), after Kopp et al.[3] (f) Tide gauge-based estimates of the non-climatic, long term contribution to RSL rise, including the effects of glacial isostatic adjustment, tectonics, and sediment compaction (units of feet/century).[76] (Figure source: (a)–(d) Kopp et al. 2015,[3] (e) adapted from Kopp et al. 2015;[3] (f) adapted from Sweet et al. 2017[71]).

## 12.3 Paleo Sea Level

Geological records of temperature and sea level indicate that during past warm periods over the last several millions of years, GMSL was higher than it is today.[19, 20] During the Last Interglacial stage, about 125,000 years ago, global average sea surface temperature was about 0.5° ± 0.3°C (0.9° ± 0.5°F) above the preindustrial level [that is, comparable to the average over 1995–2014, when global mean temperature was about 0.8°C (1.4°F) above the preindustrial levels].[21] Polar temperatures were comparable to those projected for 1°–2°C (1.8°–3.6°F) of global mean warming above the preindustrial level. At this time, GMSL was about 6–9 meters (about 20–30 feet) higher than today (Figure 12.2a).[22, 23] This geological benchmark may indicate the probable long-term response of GMSL to the minimum magnitude of temperature change projected for the current century.

Similarly, during the mid-Pliocene warm period, about 3 million years ago, global mean temperature was about 1.8°–3.6°C (3.2°–6.5°F) above the preindustrial level.[24] Estimates of GMSL are less well constrained than during the Last Interglacial, due to the smaller number of local geological sea level reconstruction and the possibility of significant vertical land motion over millions of years.[20] Some reconstructions place mid-Pliocene GMSL at about 10–30 meters (about 30–100 feet) higher than today.[25] Sea levels this high would require a significantly reduced Antarctic ice sheet, highlighting the risk of significant Antarctic ice sheet loss under such levels of warming (Figure 12.2a).

For the period since the Last Glacial Maximum, about 26,000 to 19,000 years ago,[26] geologists can produce detailed reconstructions of sea levels as well as rates of sea level change. To do this, they use proxies such as the heights of fossil coral reefs and the populations of different salinity-sensitive microfossils within salt marsh sediments.[27] During the main portion of the deglaciation, from about 17,000 to 8,000 years ago, GMSL rose at an average rate of about 12 mm/year (0.5 inches/year).[28] However, there were periods of faster rise. For example, during Meltwater Pulse 1a, lasting from about 14,600 to 14,300 years ago, GMSL may have risen at an average rate about 50 mm/year (2 inches/year).[29]

Since the disappearance of the last remnants of the North American (Laurentide) Ice Sheet about 7,000 years ago[30] to about the start of the 20th century, however, GMSL has been relatively stable. During this period, total GMSL rise is estimated to have been about 4 meters (about 13 feet), most of which occurred between 7,000 and 4,000 years ago.[28] The Third National Climate Assessment (NCA3) noted, based on a geological data set from North Carolina,[31] that the 20th century GMSL rise was much faster than at any time over the past 2,000 years. Since NCA3, high-resolution sea level reconstructions have been developed for multiple locations, and a new global analysis of such reconstructions strengthens this finding.[32] Over the last 2,000 years, prior to the industrial era, GMSL exhibited small fluctuations of about ±8 cm (3 inches), with a significant decline of about 8 cm (3 inches) between the years 1000 and 1400 CE coinciding with about 0.2°C (0.4°F) of global mean cooling.[32] The rate of rise in the last century, about 14 cm/century (5.5 inches/century), was greater than during any preceding century in at least 2,800 years (Figure 12.2b).[32]

(a)

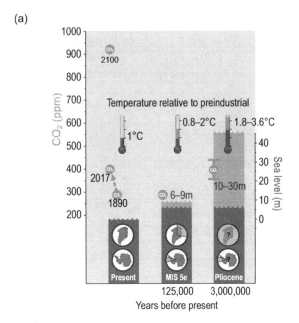

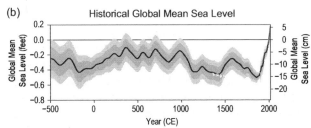

**Figure 12.2: (a)** The relationship between peak global mean temperature, atmospheric $CO_2$, maximum global mean sea level (GMSL), and source(s) of meltwater for two periods in the past with global mean temperature comparable to or warmer than present. Light blue shading indicates uncertainty of GMSL maximum. Red pie charts over Greenland and Antarctica denote fraction, not location, of ice retreat. Atmospheric $CO_2$ levels in 2100 are shown under RCP8.5. **(b)** GMSL rise from −500 to 1900 CE, from Kopp et al.'s[32] geological and tide gauge-based reconstruction (blue), from 1900 to 2010 from Hay et al.'s[33] tide gauge-based reconstruction (black), and from 1992 to 2015 from the satellite-based reconstruction updated from Nerem et al.[35] (magenta). (Figure source: (a) adapted from Dutton et al. 2015[20] and (b) Sweet et al. 2017[71]).

## 12.4 Recent Past Trends (20th and 21st Centuries)

### 12.4.1 Global Tide Gauge Network and Satellite Observations

A global tide gauge network provides the century-long observations of local RSL, whereas satellite altimetry provides broader coverage of sea surface heights outside the polar regions starting in 1993. GMSL can be estimated through statistical analyses of either data set. GMSL trends over the 1901–1990 period vary slightly (Hay et al. 2015:[33] 1.2 ± 0.2 mm/year [0.05 inches/year]; Church and White 2011:[34] 1.5 ± 0.2 mm/year [0.06 inches/year]) with differences amounting to about 1 inch over 90 years. Thus, these results indicate about 11–14 cm (4–5 inches) of GMSL rise from 1901 to 1990.

Tide gauge analyses indicate that GMSL rose at a considerably faster rate of about 3 mm/year (0.12 inches/year) since 1993,[33, 34] a result supported by satellite data indicating a trend of 3.4 ± 0.4 mm/year (0.13 ± 0.02 inches/year) over 1993–2015 (update to Nerem et al. 2010[35]). These results indicate an additional GMSL rise of about 7 cm (about 3 inches) since 1990 (Figure 12.2b, Figure 12.3a) and about 16–21 cm (about 7–8 inches) since 1900. Satellite (altimetry and gravity) and in situ water column (Argo floats) measurements show that, since 2005, about one third of GMSL rise has been from steric changes (primarily thermal expansion) and about two thirds from the addition of mass to the ocean, which represents a growing land-ice contribution (compared to steric) and a departure from the relative contributions earlier in the 20th century (Figure 12.3a).[4, 36, 37, 38, 39, 40]

In addition to land ice, the mass-addition contribution also includes net changes in global land-water storage. This term varied in sign over the course of the last century, with human-induced changes in land-water storage being negative (perhaps as much as about −0.6 mm/year [−0.02 inches/year]) during the period of heavy dam construction in the middle of the last century, and turning positive in the 1990s as groundwater withdrawal came to dominate.[8] On decadal timescales, precipitation variability can dominate human-induced changes in land water storage; recent satellite-gravity estimates suggest that, over 2002–2014, a human-caused land-water contribution to GMSL of 0.4 mm/year (0.02 inches/year) **was more than offset by −0.7 mm/year (−0.03 inches/year) due to natural variability.**[5]

Comparison of results from a variety of approaches supports the conclusion that a substantial fraction of GMSL rise since 1900 is attributable to human-caused climate change.[32, 41, 42, 43, 44, 45, 46, 47, 48] For example, based on the long term historical relationship between temperature and rate of GMSL change, Kopp et al.[32] found that GMSL rise would *extremely likely* have been less than 59% of observed in the absence of 20th century global warming, and that it is *very likely* that GMSL has been higher since 1960 than it would have been without 20th century global warming (Figure 12.3b). Similarly, using a variety of models for individual components, Slangen et al.[41] found that about 80% of the GMSL rise they simulated for 1970–2005 and about half of that which they simulated for 1900–2005 was attributable to anthropogenic forcing.

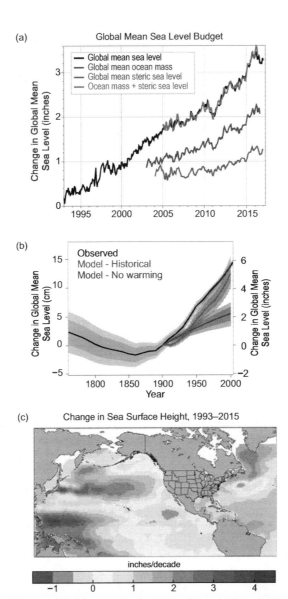

**Figure 12.3: (a)** Contributions of ocean mass changes from land ice and land water storage (measured by satellite gravimetry) and ocean volume changes (or steric, primarily from thermal expansion measured by in situ ocean profilers) and their comparison to global mean sea level (GMSL) change (measured by satellite altimetry) since 1993. **(b)** An estimate of modeled GMSL rise in the absence of 20th century warming (blue), from the same model with observed warming (red), and compared to observed GMSL change (black). Heavy/light shading indicates the 17th–83rd and 5th–95th percentiles. **(c)** Rates of change from 1993 to 2015 in sea surface height from satellite altimetry data; updated from Kopp et al.[3] using data updated from Church and White.[34] (Figure source: (a) adapted and updated from Leuliette and Nerem 2016,[40] (b) adapted from Kopp et al. 2016[32] and (c) adapted and updated from Kopp et al. 2015[3]).

Over timescales of a few decades, ocean–atmosphere dynamics drive significant variability in sea surface height, as can be observed by satellite (Figure 12.3c) and in tide gauge records that have been adjusted to account for background rates of rise due to long term factors like glacio-isostatic adjustments. For example, the U.S. Pacific Coast experienced a slower-than-global increase between about 1980 and 2011, while the western tropical Pacific experienced a faster-than-global increase in the 1990s and 2000s. This pattern was associated with changes in average winds linked to the Pacific Decadal Oscillation (PDO)[16, 49, 50] and appears to have reversed since about 2012.[51] Along the Atlantic coast, the U.S. Northeast has experienced a faster-than-global increase since the 1970s, while the U.S. Southeast has experienced a slower-than-global increase since the 1970s. This pattern appears to be tied to changes in the Gulf Stream,[10, 12, 52, 53] although whether these changes represent natural variability or a long-term trend remains uncertain.[54]

### 12.4.2 Ice Sheet Gravity and Altimetry and Visual Observations

Since NCA3, Antarctica and Greenland have continued to lose ice mass, with mounting evidence accumulating that mass loss is accelerating. Studies using repeat gravimetry (GRACE satellites), repeat altimetry, GPS monitoring, and mass balance calculations generally agree on accelerating mass loss in Antarctica.[55, 56, 57, 58] Together, these indicate a mass loss of roughly 100 Gt/year (gigatonnes/year) over the last decade (a contribution to GMSL of about 0.3 mm/year [0.01 inches/year]). Positive accumulation rate anomalies in East Antarctica, especially in Dronning Maud Land,[59] have contributed to the trend of slight growth there (e.g., Seo et al. 2015;[57] Martín-Español et al. 2016[58]), but this is more than offset by mass loss elsewhere, especially in West Antarctica along the coast facing the Amundsen Sea,[60, 61]

Totten Glacier in East Antarctica,[62, 63] and along the Antarctic Peninsula.[57, 58, 64] Floating ice shelves around Antarctica are losing mass at an accelerating rate.[65] Mass loss from floating ice shelves does not directly affect GMSL, but does allow faster flow of ice from the ice sheet into the ocean.

Estimates of mass loss in Greenland based on mass balance from input-output, repeat gravimetry, repeat altimetry, and aerial imagery as discussed in Chapter 11: Arctic Changes reveal a recent acceleration.[66] Mass loss averaged approximately 75 Gt/year (about 0.2 mm/year [0.01 inches/year] GMSL rise) from 1900 to 1983, continuing at a similar rate of approximately 74 Gt/year through 2003 before accelerating to 186 Gt/year (0.5 mm/year [0.02 inches/year] GMSL rise) from 2003 to 2010.[67] Strong interannual variability does exist (see Ch. 11: Arctic Changes), such as during the exceptional melt year from April 2012 to April 2013, which resulted in mass loss of approximately 560 Gt (1.6 mm/year [0.06 inches/year]).[68] More recently (April 2014–April 2015), annual mass losses have resumed the accelerated rate of 186 Gt/year.[67, 69] Mass loss over the last century has reversed the long-term trend of slow thickening linked to the continuing evolution of the ice sheet from the end of the last ice age.[70]

## 12.5 Projected Sea Level Rise

### 12.5.1 Scenarios of Global Mean Sea Level Rise

No single physical model is capable of accurately representing all of the major processes contributing to GMSL and regional/local RSL rise. Accordingly, the U.S. Interagency Sea Level Rise Task Force (henceforth referred to as "Interagency")[71] has revised the GMSL rise scenarios for the United States and now provides six scenarios that can be used for assessment and risk-framing purposes (Figure 12.4a; Table 12.1). The low scenario of 30 cm (about 1 foot) GMSL rise by 2100 is consistent with a continuation of the recent approximately 3 mm/year (0.12 inches/year) rate of rise through to 2100 (Table 12.2), while the five other scenarios span a range of GMSL rise be-

tween 50 and 250 cm (1.6 and 8.2 feet) in 2100, with corresponding rise rates between 5 mm/year (0.2 inches/year) to 44 mm/year (1.7 inches/year) towards the end of this century (Table 12.2). The highest scenario of 250 cm is consistent with several literature estimates of the maximum physically plausible level of 21st century sea level rise (e.g., Pfeffer et al. 2008,[72] updated with Sriver et al. 2012[73] estimates of thermal expansion and Bamber and Aspinall 2013[74] estimates of Antarctic contribution, and incorporating land water storage, as discussed in Miller et al. 2013[75] and Kopp et al. 2014[76]). It is It is also consistent with the high end of recent projections of Antarctic ice sheet melt discussed below.[77] The Interagency

(a)

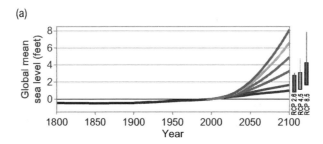

(b)

Projected Relative Sea Level Change for 2100 under the Intermediate Scenario

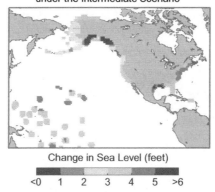

Change in Sea Level (feet)

<0   1   2   3   4   5   >6

Figure 12.4: (a) Global mean sea level (GMSL) rise from 1800 to 2100, based on Figure 12.2b from 1800 to 2015, the six Interagency[71] GMSL scenarios (navy blue, royal blue, cyan, green, orange, and red curves), the *very likely* ranges in 2100 for different RCPs (colored boxes), and lines augmenting the *very likely* ranges by the difference between the median Antarctic contribution of Kopp et al.[76] and the various median Antarctic projections of DeConto and Pollard.[77] (b) Relative sea level (RSL) rise (feet) in 2100 projected for the Interagency Intermediate Scenario (1-meter [3.3 feet] GMSL rise by 2100) (Figure source: Sweet et al. 2017[71]).

Table 12.1. The Interagency GMSL rise scenarios in meters (feet) relative to 2000. All values are 19-year averages of GMSL centered at the identified year. To convert from a 1991–2009 tidal datum to the 1983–2001 tidal datum, add 2.4 cm (0.9 inches).

| Scenario | 2020 | 2030 | 2050 | 2100 |
|---|---|---|---|---|
| Low | 0.06 (0.2) | 0.09 (0.3) | 0.16 (0.5) | 0.30 (1.0) |
| Intermediate-Low | 0.08 (0.3) | 0.13 (0.4) | 0.24 (0.8) | 0.50 (1.6) |
| Intermediate | 0.10 (0.3) | 0.16 (0.5) | 0.34 (1.1) | 1.0 (3.3) |
| Intermediate-High | 0.10 (0.3) | 0.19 (0.6) | 0.44 (1.4) | 1.5 (4.9) |
| High | 0.11 (0.4) | 0.21 (0.7) | 0.54 (1.8) | 2.0 (6.6) |
| Extreme | 0.11 (0.4) | 0.24 (0.8) | 0.63 (2.1) | 2.5 (8.2) |

Table 12.2. Rates of GMSL rise in the Interagency scenarios in mm/year (inches/year). All values represent 19-year average rates of change, centered at the identified year.

| Scenario | 2020 | 2030 | 2050 | 2090 |
|---|---|---|---|---|
| Low | 3 (0.1) | 3 (0.1) | 3 (0.1) | 3 (0.1) |
| Intermediate-Low | 5 (0.2) | 5 (0.2) | 5 (0.2) | 5 (0.2) |
| Intermediate | 6 (0.2) | 7 (0.3) | 10 (0.4) | 15 (0.6) |
| Intermediate-High | 7 (0.3) | 10 (0.4) | 15 (0.6) | 24 (0.9) |
| High | 8 (0.3) | 13 (0.5) | 20 (0.8) | 35 (1.4) |
| Extreme | 10 (0.4) | 15 (0.6) | 25 (1.0) | 44 (1.7) |

Table 12.3. Interpretations of the Interagency GMSL rise scenarios

| Scenario | Interpretation |
|---|---|
| Low | Continuing current rate of GMSL rise, as calculated since 1993<br>Low end of *very likely* range under RCP2.6 |
| Intermediate-Low | Modest increase in rate<br>Middle of *likely* range under RCP2.6<br>Low end of *likely* range under RCP4.5<br>Low end of *very likely* range under RCP8.5 |
| Intermediate | High end of *very likely* range under RCP4.5<br>High end of *likely* range under RCP8.5<br>Middle of *likely* range under RCP4.5 when accounting for possible ice cliff instabilities |
| Intermediate-High | Slightly above high end of *very likely* range under RCP8.5<br>Middle of *likely* range under RCP8.5 when accounting for possible ice cliff instabilities |
| High | High end of *very likely* range under RCP8.5 when accounting for possible ice cliff instabilities |
| Extreme | Consistent with estimates of physically possible "worst case" |

GMSL scenario interpretations are shown in Table 12.3.

The Interagency scenario approach is similar to local RSL rise scenarios of Hall et al.[78] used for all coastal U.S. Department of Defense installations worldwide. The Interagency approach starts with a probabilistic projection framework to generate time series and regional projections consistent with each GMSL rise scenario for 2100.[76] That framework combines probabilistic estimates of contributions to GMSL and regional RSL rise from ocean processes, cryospheric processes, geological processes, and anthropogenic land-water storage. Pooling the Kopp et al.[76] projections across even lower, lower, and higher scenarios (RCP2.6, 4.5, and 8.5), the probabilistic projections are filtered to identify pathways consistent with each of these 2100 levels, with the median (and 17th and 83rd percentiles) picked from each of the filtered subsets.

### 12.5.2 Probabilities of Different Sea Level Rise Scenarios

Several studies have estimated the probabilities of different amounts of GMSL rise under different pathways (e.g., Church et al. 2013;[4] Kopp et al. 2014;[76] Slangen et al. 2014;[79] Jevrejeva et al. 2014;[80] Grinsted et al. 2015;[81] Kopp et al. 2016;[32] Mengel et al. 2016;[82] Jackson and

Jevrejeva 2016[83]) using a variety of methods, including both statistical and physical models. Most of these studies are in general agreement that GMSL rise by 2100 is *very likely* to be between about 25–80 cm (0.8–2.6 feet) under an even lower scenario (RCP2.6), 35–95 cm (1.1–3.1 feet) under a lower scenario (RCP4.5), and 50–130 cm (1.6–4.3 feet) under a higher scenario (RCP8.5), although some projections extend the *very likely* range for RCP8.5 as high as 160–180 cm (5–6 feet) (Kopp et al. 2014,[76] sensitivity study).[80, 83] Based on Kopp et al.,[76] the probability of exceeding the amount of GMSL in 2100 under the Interagency scenarios is shown in Table 12.4.

The Antarctic projections of Kopp et al.,[76] the GMSL projections of which underlie Table 12.4, are consistent with a statistical-physical model of the onset of marine ice sheet instability calibrated to observations of ongoing retreat in the Amundsen Embayment sector of West Antarctica.[84] Ritz et al.'s[84] 95th percentile Antarctic contribution to GMSL of 30 cm by 2100 is comparable to Kopp et al.'s[76] 95th percentile projection of 33 cm under the higher scenario (RCP8.5). However, emerging science suggests that these projections may understate the probability of faster-than-expected ice sheet melt, particularly for high-end warming scenarios. While these probability estimates

**Table 12.4.** Probability of exceeding the Interagency GMSL scenarios in 2100 per Kopp et al.[76] New evidence regarding the Antarctic ice sheet, if sustained, may significantly increase the probability of the intermediate-high, high, and extreme scenarios, particularly under the higher scenario (RCP8.5), but these results have not yet been incorporated into a probabilistic analysis.

| Scenario | RCP2.6 | RCP4.5 | RCP8.5 |
|---|---|---|---|
| Low | 94% | 98% | 100% |
| Intermediate-Low | 49% | 73% | 96% |
| Intermediate | 2% | 3% | 17% |
| Intermediate-High | 0.4% | 0.5% | 1.3% |
| High | 0.1% | 0.1% | 0.3% |
| Extreme | 0.05% | 0.05% | 0.1% |

are consistent with the assumption that the relationship between global temperature and GMSL in the coming century will be similar to that observed over the last two millennia,[32, 85] emerging positive feedbacks (self-amplifying cycles) in the Antarctic Ice Sheet especially[86, 87] may invalidate that assumption. Physical feedbacks that until recently were not incorporated into ice sheet models[88] could add about 0–10 cm (0–0.3 feet), 20–50 cm (0.7–1.6 feet) and 60–110 cm (2.0–3.6 feet) to central estimates of current century sea level rise under even lower, lower, and higher scenarios (RCP2.6, RCP4.5 and RCP8.5, respectively).[77] In addition to marine ice sheet instability, examples of these interrelated processes include ice cliff instability and ice shelf hydrofracturing. Processes underway in Greenland may also be leading to accelerating high-end melt risk. Much of the research has focused on changes in surface albedo driven by the melt-associated unmasking and concentration of impurities in snow and ice.[69] However, ice dynamics at the bottom of the ice sheet may be important as well, through interactions with surface runoff or a warming ocean. As an example of the latter, Jakobshavn Isbræ, Kangerdlugssuaq Glacier, and the Northeast Greenland ice stream may be vulnerable to marine ice sheet instability.[66]

### 12.5.3 Sea Level Rise after 2100

GMSL rise will not stop in 2100, and so it is useful to consider extensions of GMSL rise projections beyond this point. By 2200, the 0.3–

2.5 meter (1.0–8.2 feet) range spanned by the six Interagency GMSL scenarios in year 2100 increases to about 0.4–9.7 meters (1.3–31.8 feet), as shown in Table 12.5. These six scenarios imply average rates of GMSL rise over the first half of the next century of 1.4 mm/year (0.06 inch/year), 4.6 mm/yr (0.2 inch/year), 16 mm/year (0.6 inch/year), 32 mm/year (1.3 inches/year), 46 mm/yr (1.8 inches/year) and 60 mm/year (2.4 inches/year), respectively. Excluding the possible effects of still emerging science regarding ice cliffs and ice shelves, it is very likely that by 2200 GMSL will have risen by 0.3–2.4 meters (1.0–7.9 feet) under an even lower scenario (RCP2.6), 0.4–2.7 meters (1.3–8.9 feet) under a lower scenario (RCP4.5), and 1.0–3.7 meters (3.3–12 feet) under the higher scenario (RCP8.5).[76]

Under most projections, GMSL rise will also not stop in 2200. The concept of a "sea level rise commitment" refers to the long-term projected sea level rise were the planet's temperature to be stabilized at a given level (e.g., Levermann et al. 2013;[89] Golledge et al. 2015[90]). The paleo sea level record suggests that even 2°C (3.6°F) of global average warming above the preindustrial temperature may represent a commitment to several meters of rise. One modeling study suggesting a 2,000-year commitment of 2.3 m/°C (4.2 feet/°F)[89] indicates that emissions through 2100 would lock in a likely 2,000-year GMSL rise commitment of about 0.7–4.2 meters (2.3–14 feet) under an even lower scenario (RCP2.6), about 1.7–5.6

**Table 12.5.** Post-2100 extensions of the Interagency GMSL rise scenarios in meters (feet)

| Scenario | 2100 | 2120 | 2150 | 2200 |
|---|---|---|---|---|
| Low | 0.30 (1.0) | 0.34 (1.1) | 0.37 (1.2) | 0.39 (1.3) |
| Intermediate-Low | 0.50 (1.6) | 0.60 (2.0) | 0.73 (2.4) | 0.95 (3.1) |
| Intermediate | 1.0 (3.3) | 1.3 (4.3) | 1.8 (5.9) | 2.8 (9.2) |
| Intermediate-High | 1.5 (4.9) | 2.0 (6.6) | 3.1 (10) | 5.1 (17) |
| High | 2.0 (6.6) | 2.8 (9.2) | 4.3 (14) | 7.5 (25) |
| Extreme | 2.5 (8.2) | 3.6 (12) | 5.5 (18) | 9.7 (32) |

meters (5.6–19 feet) under a lower scenario (RCP4.5), and about 4.3–9.9 meters (14–33 feet) under the higher scenario (RCP8.5).[91] However, as with the 21st century projections, emerging science regarding the sensitivity of the Antarctic Ice Sheet may increase the estimated sea level rise over the next millennium, especially for a higher scenario.[77] Large-scale climate geoengineering might reduce these commitments,[92, 93] but may not be able to avoid lock-in of significant change.[94, 95, 96, 97] Once changes are realized, they will be effectively irreversible for many millennia, even if humans artificially accelerate the removal of $CO_2$ from the atmosphere.[77]

The 2,000-year commitment understates the full sea level rise commitment, due to the long response time of the polar ice sheets. Paleo sea level records (Figure 12.2a) suggest that 1°C of warming may already represent a long-term commitment to more than 6 meters (20 feet) of GMSL rise.[20, 22, 23] A 10,000-year modeling study[98] suggests that 2°C warming represents a 10,000-year commitment to about 25 meters (80 feet) of GMSL rise, driven primarily by a loss of about one-third of the Antarctic ice sheet and three-fifths of the Greenland ice sheet, while 21st century emissions consistent with a higher scenario (RCP8.5) represent a 10,000-year commitment to about 38 meters (125 feet) of GMSL rise, including a complete loss of the Greenland ice sheet over about 6,000 years.

### 12.5.4 Regional Projections of Sea Level Change

Because the different factors contributing to sea level change give rise to different spatial patterns, projecting future RSL change at specific locations requires not just an estimate of GMSL change but estimates of the different processes contributing to GMSL change—each of which has a different associated spatial pattern—as well as of the processes contributing exclusively to regional or local change. Based

on the process-level projections of the Interagency GMSL scenarios, several key regional patterns are apparent in future U.S. RSL rise as shown for the Intermediate (1 meter [3.3 feet] GMSL rise by 2100 scenario) in Figure 12.4b.

1. RSL rise due to Antarctic Ice Sheet melt is greater than GMSL rise along all U.S. coastlines due to static-equilibrium effects.

2. RSL rise due to Greenland Ice Sheet melt is less than GMSL rise along the coastline of the continental United States due to static-equilibrium effects. This effect is especially strong in the Northeast.

3. RSL rise is additionally augmented in the Northeast by the effects of glacial isostatic adjustment.

4. The Northeast is also exposed to rise due to changes in the Gulf Stream and reductions in the Atlantic meridional overturning circulation (AMOC). Were the AMOC to collapse entirely—an outcome viewed as unlikely in the 21st century—it could result in as much as approximately 0.5 meters (1.6 feet) of additional regional sea level rise (see Ch. 15: Potential Surprises for further discussion).[99, 100]

5. The western Gulf of Mexico and parts of the U.S. Atlantic Coast south of New York are currently experiencing significant RSL rise caused by the withdrawal of groundwater (along the Atlantic Coast) and of both fossil fuels and groundwater (along the Gulf Coast). Continuation of these practices will further amplify RSL rise.

6. The presence of glaciers in Alaska and their proximity to the Pacific Northwest reduces RSL rise in these regions, due to both the ongoing glacial isostatic adjustment to past glacier shrinkage and to

the static-equilibrium effects of projected future losses.

7. Because they are far from all glaciers and ice sheets, RSL rise in Hawai'i and other Pacific islands due to any source of melting land ice is amplified by the static-equilibrium effects.

## 12.6 Extreme Water Levels

### 12.6.1 Observations

Coastal flooding during extreme high-water events has become deeper due to local RSL rise and more frequent from a fixed-elevation perspective.[78, 101, 102, 103] Trends in annual frequencies surpassing local emergency preparedness thresholds for minor tidal flooding (i.e., "nuisance" levels of about 30–60 cm [1–2 feet]) that begin to flood infrastructure and trigger coastal flood "advisories" by NOAA's National Weather Service have increased 5- to 10-fold or more since the 1960s along the U.S. coastline,[104] as shown in Figure 12.5a. Locations experiencing such trend changes (based upon fits of flood days per year of Sweet and Park 2014[105]) include Atlantic City and Sandy Hook, NJ; Philadelphia, PA; Baltimore and Annapolis, MD; Norfolk, VA; Wilmington, NC; Charleston, SC; Savannah, GA; Mayport and Key West, FL; Port Isabel, TX, La Jolla, CA; and Honolulu, HI. In fact, over the last several decades, minor tidal flood rates have been accelerating within several (more than 25) East and Gulf Coast cities with established elevation thresholds for minor (nuisance) flood impacts, fastest where elevation thresholds are lower, local RSL rise is higher, and extreme variability less.[104, 105, 106]

Trends in extreme water levels (for example, monthly maxima) in excess of mean sea levels (for example, monthly means) exist, but are not commonplace.[48, 101, 107, 108, 109] More common are regional time dependencies in high-water probabilities, which can co-vary on an interan-

nual basis with climatic and other patterns.[101, 110, 111, 112, 113, 114, 115] These patterns are often associated with anomalous oceanic and atmospheric conditions.[116, 117] For instance, the probability of experiencing minor tidal flooding is compounded during El Niño periods along portions of the West and Mid-Atlantic Coasts[105] from a combination of higher sea levels and enhanced synoptic forcing and storm surge frequency.[112, 118, 119, 120]

### 12.6.2 Influence of Projected Sea Level Rise on Coastal Flood Frequencies

The extent and depth of minor-to-major coastal flooding during high-water events will continue to increase in the future as local RSL rises.[71, 76, 78, 105, 121, 122, 123, 124, 125] Relative to fixed elevations, the frequency of high-water events will increase the fastest where extreme variability is less and the rate of local RSL rise is higher.[71, 76, 105, 121, 124, 126] Under the RCP-based probabilistic RSL projections of Kopp et al. 2014,[76] at tide gauge locations along the contiguous U.S. coastline, a median 8-fold increase (range of 1.1- to 430-fold increase) is expected by 2050 in the annual number of floods exceeding the elevation of the current 100-year flood event (measured with respect to a 1991–2009 baseline sea level).[124] Under the same forcing, the frequency of minor tidal flooding (with contemporary recurrence intervals generally <1 year[104]) will increase even more so in the coming decades[105, 127] and eventually occur on a daily basis (Figure 12.5b). With only about 0.35 m (<14 inches) of additional local RSL rise (with respect to the year 2000), annual frequencies of moderate level flooding—those locally with a 5-year recurrence interval (Figure 12.5c) and associated with a NOAA coastal flood warning of serious risk to life and property—will increase 25-fold at the majority of NOAA tide gauge locations along the U.S. coastline (outside of Alaska) by or about (±5 years) 2080, 2060, 2040, and 2030 under the Interagency Low, Intermediate-Low,

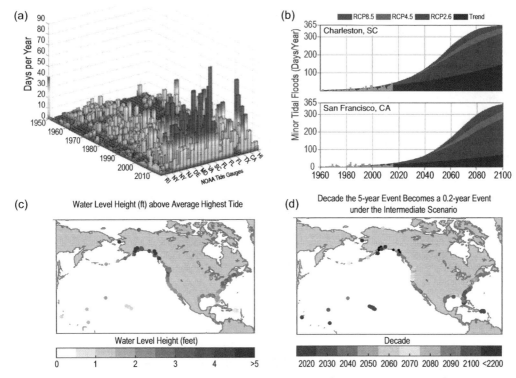

(a)

(b)

(c) Water Level Height (ft) above Average Highest Tide

(d) Decade the 5-year Event Becomes a 0.2-year Event under the Intermediate Scenario

**Figure 12.5:** (a) Tidal floods (days per year) exceeding NOAA thresholds for minor impacts at 28 NOAA tide gauges through 2015. (b) Historical exceedances (orange), future projections through 2100 based upon the continuation of the historical trend (blue), and future projections under median RCP2.6, 4.5 and 8.5 conditions, for two of the locations—Charleston, SC and San Francisco, CA. (c) Water level heights above average highest tide associated with a local 5-year recurrence probability, and (d) the future decade when the 5-year event becomes a 0.2-year (5 or more times per year) event under the Interagency Intermediate scenario; black dots imply that a 5-year to 0.2-year frequency change does not unfold by 2200 under the Intermediate scenario. (Figure source: (a) adapted from Sweet and Marra 2016,[165] (b) adapted from Sweet and Park 2014,[105] (c) and (d) Sweet et al. 2017[71]).

Intermediate, and Intermediate-High GMSL scenarios, respectively.[71] Figure 12.5d, which shows the decade in which the frequency of such moderate level flooding will increase 25-fold under the Interagency Intermediate Scenario, highlights that the mid- and Southeast Atlantic, western Gulf of Mexico, California, and the Island States and Territories are most susceptible to rapid changes in potentially damaging flood frequencies.

### 12.6.3 Waves and Impacts

The combination of a storm surge at high tide with additional dynamic effects from waves[128, 129] creates the most damaging coastal hydraulic conditions.[130] Simply with higher-than-nor-

mal sea levels, wave action increases the likelihood for extensive coastal erosion[131, 132, 133] and low-island overwash.[134] Wave runup is often the largest water level component during extreme events, especially along island coastlines where storm surge is constrained by bathymetry.[78, 121, 123] On an interannual basis, wave impacts are correlated across the Pacific Ocean with phases of ENSO.[135, 136] Over the last half century, there has been an increasing trend in wave height and power within the North Pacific Ocean[137, 138] that is modulated by the PDO.[137, 139] Resultant increases in wave run-up have been more of a factor than RSL rise in terms of impacts along the U.S. Northwest Pacific Coast over the last several decades.[140] In the Northwest Atlantic Ocean, no long-term trends in wave power have been observed over the last half century,[141] though hurricane activity drives interannual variability.[142] In terms of future conditions this century, increases in mean and maximum seasonal wave heights are projected within parts of the northeast Pacific, northwest Atlantic, and Gulf of Mexico.[138, 143, 144, 145]

### 12.6.4 Sea Level Rise, Changing Storm Characteristics, and Their Interdependencies

Future probabilities of extreme coastal floods will depend upon the amount of local RSL rise, changes in coastal storm characteristics, and their interdependencies. For instance, there have been more storms producing concurrent locally extreme storm surge and rainfall (not captured in tide gauge data) along the U.S. East and Gulf Coasts over the last 65 years, with flooding further compounded by local RSL rise.[166] Hemispheric-scale extratropical cyclones may experience a northward shift this century, with some studies projecting an overall decrease in storm number (Colle et al. 2015[117] and references therein). The research is mixed about strong extratropical storms; studies find potential increases in frequency and intensity in some regions, like within the Northeast,[146] whereas others project decreases in strong extratropical storms in some regions (e.g., Zappa et al. 2013[147]).

For tropical cyclones, model projections for the North Atlantic mostly agree that intensities and precipitation rates will increase this century (see Ch. 9: Extreme Storms), although some model evidence suggests that track changes could dampen the effect in the U.S. Mid-Atlantic and Northeast.[148] Assuming other storm characteristics do not change, sea level rise will increase the frequency and extent of extreme flooding associated with coastal storms, such as hurricanes and nor'easters. A projected increase in the intensity of hurricanes in the North Atlantic could increase the probability of extreme flooding along most of the U.S. Atlantic and Gulf Coast states beyond what would be projected based solely on RSL rise.[110, 149, 150, 151] In addition, RSL increases are projected to cause a nonlinear increase in storm surge heights in shallow bathymetry environments[152, 153, 154, 155, 156] and extend wave propagation and impacts landward.[152, 153] However, there is low confidence in the magnitude of the increase in intensity and the associated flood risk amplification, and it could be offset or amplified by other factors, such as changes in storm frequency or tracks (e.g., Knutson et al. 2013,[157] 2015[158]).

# TRACEABLE ACCOUNTS

### Key Finding 1

Global mean sea level (GMSL) has risen by about 7–8 inches (about 16–21 cm) since 1900, with about 3 of those inches (about 7 cm) occurring since 1993 (*very high confidence*). Human-caused climate change has made a substantial contribution to GMSL rise since 1900 (*high confidence*), contributing to a rate of rise that is greater than during any preceding century in at least 2,800 years (*medium confidence*).

### Description of evidence base

Multiple researchers, using different statistical approaches, have integrated tide gauge records to estimate GMSL rise since the late nineteenth century (e.g., Church and White 2006,[159] 2011;[34] Hay et al. 2015;[33] Jevrejeva et al. 2009).[42] The most recent published rate estimates are 1.2 ± 0.2[33] or 1.5 ± 0.2[34] mm/year over 1901–1990. Thus, these results indicate about 11–14 cm (4–5 inches) of GMSL rise from 1901 to 1990. Tide gauge analyses indicate that GMSL rose at a considerably faster rate of about 3 mm/year (0.12 inches/year) since 1993,[33, 34] a result supported by satellite data indicating a trend of 3.4 ± 0.4 mm/year (0.13 inches/year) over 1993–2015 (update to Nerem et al. 2010[35]) (Figure 12.3a). These results indicate an additional GMSL rise of about 7 cm (about 3 inches) rise since 1990. Thus, total GMSL rise since 1900 is about 16–21 cm (about 7–8 inches).

The finding regarding the historical context of the 20th century change is based upon Kopp et al.[32], who conducted a meta-analysis of geological RSL reconstructions spanning the last 3,000 years from 24 locations around the world as well as tide gauge data from 66 sites and the tide gauge based GMSL reconstruction of Hay et al.[33] By constructing a spatio-temporal statistical model of these data sets, they identified the common global sea level signal over the last three millennia and its uncertainties. They found a 95% probability that the average rate of GMSL change over 1900–2000 was greater than during any preceding century in at least 2,800 years.

The finding regarding the substantial human contribution is based upon several lines of evidence. Kopp et al.,[32] based on the long term historical relationship between temperature and the rate of sea level change, found that it is *extremely likely* that GMSL rise would have been <59% of observed in the absence of 20th century global warming, and that it is *very likely* that GMSL has been higher since 1960 than it would have been without 20th century global warming. Using a variety of models for individual components, Slangen et al.[41] found that 69% ± 31% out of the 87% ± 20% of GMSL rise over 1970–2005 that their models simulated was attributable to anthropogenic forcing, and that 37% ± 38% out of 74% ± 22% simulated was attributable over 1900–2005. Jevrejeva et al.,[42] using the relationship between forcing and GMSL over 1850 and 2001 and CMIP3 models, found that ~75% of GMSL rise in the 20th century is attributable to anthropogenic forcing. Marcos and Amores,[45] using CMIP5 models, found that ~87% of ocean heat uptake since 1970 in the top 700 m of the ocean has been due to anthropogenic forcing. Slangen et al.,[46] using CMIP5, found that anthropogenic forcing was required to explain observed thermosteric SLR over 1957–2005. Marzeion et al.[47] found that 25% ± 35% of glacial loss over 1851–2010, and 69% ± 24% over 1991–2010, was attributable to anthropogenic forcing. Dangendorf et al.,[43] based on time series analysis, found that >45% of observed GMSL trend since 1900 cannot (with 99% probability) be explained by multi-decadal natural variability. Becker et al.,[44] based on time series analysis, found a 99% probability that at least 1.0 or 1.3 mm/year of GMSL rise over 1880–2010 is anthropogenic.

### Major uncertainties

Uncertainties in reconstructed GMSL change relate to the sparsity of tide gauge records, particularly before **the middle of the twentieth century**, and to different statistical approaches for estimating GMSL change from these sparse records. Uncertainties in reconstructed GMSL change before the twentieth century also relate to the sparsity of geological proxies for sea level change, the interpretation of these proxies, and the dating of these proxies. Uncertainty in attribution relates to the reconstruction of past changes and the magnitude of unforced variability.

**Assessment of confidence based on evidence and agreement, including short description of nature of evidence and level of agreement**

Confidence is *very high* in the rate of GMSL rise since 1900, based on multiple different approaches to estimating GMSL rise from tide gauges and satellite altimetry. Confidence is *high* in the substantial human contribution to GMSL rise since 1900, based on both statistical and physical modeling evidence. It is *medium* that the magnitude of the observed rise since 1900 is unprecedented in the context of the previous 2,800 years, based on meta-analysis of geological proxy records.

**Summary sentence or paragraph that integrates the above information**

This key finding is based upon multiple analyses of tide gauge and satellite altimetry records, on a meta-analysis of multiple geological proxies for pre-instrumental sea level change, and on both statistical and physical analyses of the human contribution to GMSL rise since 1900.

**Key Finding 2**

Relative to the year 2000, GMSL is *very likely* to rise by 0.3–0.6 feet (9–18 cm) by 2030, 0.5–1.2 feet (15–38 cm) by 2050, and 1.0–4.3 feet (30–130 cm) by 2100 (*very high confidence in lower bounds; medium confidence in upper bounds for 2030 and 2050; low confidence in upper bounds for 2100*). Future pathways have little effect on projected GMSL rise in the first half of the century, but significantly affect projections for the second half of the century (*high confidence*). Emerging science regarding Antarctic ice sheet stability suggests that, for high emission scenarios, a GMSL rise exceeding 8 feet (2.4 m) by 2100 is physically possible, although the probability of such an extreme outcome cannot currently be assessed. Regardless of pathway, it is *extremely likely* that GMSL rise will continue beyond 2100 (*high confidence*).

**Description of evidence base**

The lower bound of the *very likely* range is based on a continuation of the observed approximately 3 mm/year rate of GMSL rise. The upper end of the *very likely* range is based upon estimates for the higher scenario (RCP8.5) from three studies producing fully probabilistic projections across multiple RCPs. Kopp et al. 2014[76] fused multiple sources of information accounting for the different individual process contributing to GMSL rise. Kopp et al. 2016[32] constructed a semi-empirical sea level model calibrated to the Common Era sea level reconstruction. Mengel et al.[82] constructed a set of semi-empirical models of the different contributing processes. All three studies show negligible RCP dependence in the first half of this century, becoming more prominent in the second half of the century. A sensitivity study by Kopp et al. 2014,[76] as well as studies by Jevrejeva et al.[80] and by Jackson and Jevrejeva,[83] used frameworks similar to Kopp et al. 2016[32] but incorporated directly an expert elicitation study on ice sheet stability.[74] (This study was incorporated in Kopp et al. 2014's[76] main results with adjustments for consistency with Church et al. 2013[4]). These studies extend the *very likely* range for the higher scenario (RCP8.5) as high as 160–180 cm (5–6 feet) (Kopp et al. 2014,[76] sensitivity study).[80, 83]

To estimate the effect of incorporating the DeConto and Pollard[77] projections of Antarctic ice sheet melt, we note that Kopp et al. (2014)'s[76] median projection of Antarctic melt in 2100 is 4 cm (1.6 inches) (RCP2.6), 5 cm (2 inches) (RCP4.5), or 6 cm (2.4 inches) (RCP8.5). By contrast, DeConto and Pollard's[77] ensemble mean projections are (varying the assumptions for the size of Pliocene mass loss and the bias correction in the Amundsen Sea) 2–14 cm (0.1–0.5 foot) for an even lower scenario (RCP2.6), 26–58 cm (0.9–1.9 feet) for a lower scenario (RCP4.5), and 64–114 cm (2.1–3.7 ft) for the higher scenario (RCP8.5). Thus, we conclude that DeConto and Pollard's[77] projection would lead to a –10 cm (–0.1–0.3 ft) increase in median RCP2.6 projections, a 21–53 cm (0.7–1.7 feet) increase in median RCP4.5 projections, and a 58–108 cm (1.9–3.5 feet) increase in median RCP8.5 projections.

Very likely ranges, 2030 relative to 2000 in cm (feet)

|  | Kopp et al. (2014)[76] | Kopp et al. (2016)[32] | Mengel et al. (2016)[8] |
|---|---|---|---|
| RCP8.5 | 11–18 (0.4–0.6) | 8–15 (0.3–0.5) | 7–12 (0.2–0.4) |
| RCP4.5 | 10–18 (0.3–0.6) | 8–15 (0.3–0.5) | 7–12 (0.2–0.4) |
| RCP2.6 | 10–18 (0.3–0.6) | 8–15 (0.3–0.5) | 7–12 (0.2–0.4) |

Very likely ranges, 2050 relative to 2000 in cm (feet)

|  | Kopp et al. (2014)[76] | Kopp et al. (2016)[32] | Mengel et al. (2016)[8] |
|---|---|---|---|
| RCP8.5 | 21–38 (0.7–1.2) | 16–34 (0.5–1.1) | 15–28 (0.5–0.9) |
| RCP4.5 | 18–35 (0.6–1.1) | 15–31 (0.5–1.0) | 14–25 (0.5–0.8) |
| RCP2.6 | 18–33 (0.6–1.1) | 14–29 (0.5–1.0) | 13–23 (0.4–0.8) |

Very likely ranges, 2100 relative to 2000 in cm (feet)

|  | Kopp et al. (2014)[76] | Kopp et al. (2016)[32] | Mengel et al. (2016)[8] |
|---|---|---|---|
| RCP8.5 | 55–121 (1.8–4.0) | 52–131 (1.7–4.3) | 57–131 (1.9–4.3) |
| RCP4.5 | 36–93 (1.2–3.1) | 33–85 (1.1–2.8) | 37–77 (1.2–2.5) |
| RCP2.6 | 29–82 (1.0–2.7) | 24–61 (0.8–2.0) | 28–56 (0.9–1.8) |

**Major uncertainties**

Since NCA3, multiple different approaches have been used to generate probabilistic projections of GMSL rise, conditional upon the RCPs. These approaches are in general agreement. However, emerging results indicate that marine-based sectors of the Antarctic Ice Sheet are more unstable than previous modeling indicated. The rate of ice sheet mass changes remains challenging to project.

**Assessment of confidence based on evidence and agreement, including short description of nature of evidence and level of agreement**

There is *very high* confidence that future GMSL rise over the next several decades will be at least as fast as a con-

tinuation of the historical trend over the last quarter century would indicate. There is *medium* confidence in the upper end of very likely ranges for 2030 and 2050. Due to possibly large ice sheet contributions, there is *low* confidence in the upper end of very likely ranges for 2100. Based on multiple projection methods, there is *high confidence* that differences between emission scenarios are small before 2050 but significant beyond 2050.

**Summary sentence or paragraph that integrates the above information**

This key finding is based upon multiple methods for estimating the probability of future sea level change and on new modeling results regarding the stability of marine based ice in Antarctica.

**Key Finding 3**

Relative sea level (RSL) rise in this century will vary along U.S. coastlines due, in part, to changes in Earth's gravitational field and rotation from melting of land ice, changes in ocean circulation, and vertical land motion (*very high confidence*). For almost all future GMSL rise scenarios, RSL rise is *likely* to be greater than the global average in the U.S. Northeast and the western Gulf of Mexico. In intermediate and low GMSL rise scenarios, RSL rise is *likely* to be less than the global average in much of the Pacific Northwest and Alaska. For high GMSL rise scenarios, RSL rise is *likely* to be higher than the global average along all U.S. coastlines outside Alaska. Almost all U.S. coastlines experience more than global-mean sea-level rise in response to Antarctic ice loss, and thus would be particularly affected under extreme GMSL rise scenarios involving substantial Antarctic mass loss (*high confidence*).

**Description of evidence base**

The processes that cause geographic variability in RSL change are reviewed by Kopp et al.[3] Long tide gauge data sets show the RSL rise caused by vertical land motion due to glacio-isostatic adjustment and fluid withdrawal along many U.S. coastlines.[160, 161] These observations are corroborated by glacio-isostatic adjustment models, by GPS observations, and by geological data (e.g., Engelhart and Horton 2012[162]). The physics of the gravitational, rotational and flexural "static-equilibrium fingerprint" response of sea level to redistribution of mass from land ice to the oceans is well established.[13, 163] GCM studies indicate the potential for a Gulf Stream contribution to sea level rise in the U.S. Northeast.[12, 164] Kopp et al.[76] and Slangen et al.[46] accounted for land motion (only glacial isostatic adjustment for Slangen et al.), fingerprint, and ocean dynamic responses. Comparing projections of local RSL change and GMSL change in these studies indicate that local rise is likely to be greater than the global average along the U.S. Atlantic and Gulf Coasts and less than the global average in most of the Pacific Northwest. Sea level rise projections in this report are developed by an Interagency Sea Level Rise Task Force.[71]

### Major uncertainties

Since NCA3, multiple authors have produced global or regional studies synthesizing the major process that causes global and local sea level change to diverge. The largest sources of uncertainty in the geographic variability of sea level change are ocean dynamic sea level change and, for those regions where sea level fingerprints for Greenland and Antarctica differ from the global mean in different directions, the relative contributions of these two sources to projected sea level change.

### Assessment of confidence based on evidence and agreement, including short description of nature of evidence and level of agreement

Because of the enumerated physical processes, there is *very high* confidence that RSL change will vary across U.S. coastlines. There is *high* confidence in the likely differences of RSL change from GMSL change under different levels of GMSL change, based on projections incorporating the different relevant processes.

### Summary sentence or paragraph that integrates the above information

The part of the key finding regarding the existence of geographic variability is based upon a broader observational, modeling, and theoretical literature. The specific differences are based upon the scenarios described by the Interagency Sea Level Rise Task Force.[71]

### Key Finding 4

As sea levels have risen, the number of tidal floods each year that cause minor impacts (also called "nuisance floods") have increased 5- to 10-fold since the 1960s in several U.S. coastal cities (*very high confidence*). Rates of increase are accelerating in over 25 Atlantic and Gulf Coast cities (*very high confidence*). Tidal flooding will continue increasing in depth, frequency, and extent this century (*very high confidence*).

### Description of evidence base

Sweet et al.[104] examined 45 NOAA tide gauge locations with hourly data since 1980 and Sweet and Park[105] examined a subset of these (27 locations) with hourly data prior to 1950, all with a National Weather Service elevation threshold established for minor "nuisance" flood impacts. Using linear or quadratic fits of annual number of days exceeding the minor thresholds, Sweet and Park[105] find increases in trend-derived values between 1960 and 2010 greater than 10-fold at 8 locations, greater than 5-fold at 6 locations, and greater than 3-fold at 7 locations. Sweet et al.,[104] Sweet and Park,[105] and Ezer and Atkinson[106] find that annual minor tidal flood frequencies since 1980 are accelerating along locations on the East and Gulf Coasts (>25 locations[104]) due to continued exceedance of a typical high-water distribution above elevation thresholds for minor impacts.

Historical changes over the last 60 years in flood probabilities have occurred most rapidly where RSL rates were highest and where tide ranges and extreme variability is less (Sweet and Park 2014). In terms of future rates of changes in extreme event probabilities relative to fixed elevations, Hunter,[126] Tebaldi et al.,[121] Kopp et al.,[76] Sweet and Park[105] and Sweet et al.[71] all find that locations with less extreme variability and higher RSL rise rates are most prone.

### Major uncertainties

Minor flooding probabilities have been only assessed where a tide gauge is present with >30 years of data and where a NOAA National Weather Service elevation threshold for impacts has been established. There are likely many other locations experiencing similar flooding patterns, but an expanded assessment is not possible at this time.

**Assessment of confidence based on evidence and agreement, including short description of nature of evidence and level of agreement**

There is *very high* confidence that exceedance probabilities of high tide flooding at dozens of local-specific elevation thresholds have significantly increased over the last half century, often in an accelerated fashion, and that exceedance probabilities will continue to increase this century.

**Summary sentence or paragraph that integrates the above information**

This key finding is based upon several studies finding historic and projecting future changes in high-water probabilities for local-specific elevation thresholds for flooding.

**Key Finding 5**

Assuming storm characteristics do not change, sea level rise will increase the frequency and extent of extreme flooding associated with coastal storms, such as hurricanes and nor'easters (*very high confidence*). A projected increase in the intensity of hurricanes in the North Atlantic (*medium confidence*) could increase the probability of extreme flooding along most of the U.S. Atlantic and Gulf Coast states beyond what would be projected based solely on RSL rise. However, there is *low confidence* in the projected increase in frequency of intense Atlantic hurricanes, and the associated flood risk amplification and flood effects could be offset or amplified by such factors as changes in overall storm frequency or tracks.

**Description of evidence base**

The frequency, extent, and depth of extreme event-driven (for example, 5- to 100-year event probabilities) coastal flooding relative to existing infrastructure will continue to increase in the future as local RSL rises.[71, 76, 78, 103, 121, 122, 123, 124] Extreme flood probabilities will increase regardless of change in storm characteristics, which may exacerbate such changes. Model-based projections of tropical storms and related major storm

surges within the North Atlantic mostly agree that intensities and frequencies of the most intense storms will increase this century.[110, 149, 150, 151, 157] However, the projection of increased hurricane intensity is more robust across models than the projection of increased frequency of the most intense storms, since a number of models project a substantial decrease in the overall number of tropical storms and hurricanes in the North Atlantic. Changes in the frequency of intense hurricanes depends on changes in both the overall frequency of tropical cyclones storms and their intensities. High-resolution models generally project an increase in mean hurricane intensity in the Atlantic (e.g., Knutson et al. 2013[157]). In addition, there is model evidence for a change in tropical cyclone tracks in warm years that minimizes the increase in landfalling hurricanes in the U.S. Mid-Atlantic or Northeast.[148]

**Major uncertainties**

Uncertainties remain large with respect to the precise change in future risk of a major coastal impact at a specific location from changes in the most intense tropical cyclone characteristics and tracks beyond changes imposed from local sea level rise.

**Assessment of confidence based on evidence and agreement, including short description of nature of evidence and level of agreement**

There is *low confidence* that the flood risk at specific locations will be amplified from a major tropical storm this century.

**Summary sentence or paragraph that integrates the above information**

This key finding is based upon several modeling studies of future hurricane characteristics and associated increases in major storm surge risk amplification.

# REFERENCES

1. Wong, P.P., I.J. Losada, J.-P. Gattuso, J. Hinkel, A. Khattabi, K.L. McInnes, Y. Saito, and A. Sallenger, 2014: Coastal systems and low-lying areas. *Climate Change 2014: Impacts,Adaptation, and Vulnerability. Part A: Global and Sectoral Aspects. Contribution of Working Group II to the Fifth Assessment Report of the Intergovernmental Panel on Climate Change.* Field, C.B., V.R. Barros, D.J. Dokken, K.J. Mach, M.D. Mastrandrea, T.E. Bilir, M. Chatterjee, K.L. Ebi, Y.O. Estrada, R.C. Genova, B. Girma, E.S. Kissel, A.N. Levy, S. MacCracken, P.R. Mastrandrea, and L.L.White, Eds. Cambridge University Press, Cambridge, United Kingdom and New York, NY, USA, 361-409. http://www.ipcc.ch/report/ar5/wg2/

2. Lentz, E.E., E.R. Thieler, N.G. Plant, S.R. Stippa, R.M. Horton, and D.B. Gesch, 2016: Evaluation of dynamic coastal response to sea-level rise modifies inundation likelihood. *Nature Climate Change,* **6**, 696-700. http://dx.doi.org/10.1038/nclimate2957

3. Kopp, R.E., C.C. Hay, C.M. Little, and J.X. Mitrovica, 2015: Geographic variability of sea-level change. *Current Climate Change Reports,* **1**, 192-204. http://dx.doi.org/10.7282/T37W6F4P

4. Church, J.A., P.U. Clark, A. Cazenave, J.M. Gregory, S. Jevrejeva, A. Levermann, M.A. Merrifield, G.A. Milne, R.S. Nerem, P.D. Nunn, A.J. Payne, W.T. Pfeffer, D. Stammer, and A.S. Unnikrishnan, 2013: Sea level change. *Climate Change 2013: The Physical Science Basis. Contribution of Working Group I to the Fifth Assessment Report of the Intergovernmental Panel on Climate Change.* Stocker, T.F., D. Qin, G.-K. Plattner, M. Tignor, S.K. Allen, J. Boschung, A. Nauels, Y. Xia, V. Bex, and P.M. Midgley, Eds. Cambridge University Press, Cambridge, United Kingdom and New York, NY, USA, 1137–1216. http://www.climatechange2013.org/report/full-report/

5. Reager, J.T., A.S. Gardner, J.S. Famiglietti, D.N. Wiese, A. Eicker, and M.-H. Lo, 2016: A decade of sea level rise slowed by climate-driven hydrology. *Science,* **351**, 699-703. http://dx.doi.org/10.1126/science.aad8386

6. Rietbroek, R., S.-E. Brunnabend, J. Kusche, J. Schröter, and C. Dahle, 2016: Revisiting the contemporary sea-level budget on global and regional scales. *Proceedings of the National Academy of Sciences,* **113**, 1504-1509. http://dx.doi.org/10.1073/pnas.1519132113

7. Wada, Y., M.-H. Lo, P.J.F. Yeh, J.T. Reager, J.S. Famiglietti, R.-J. Wu, and Y.-H. Tseng, 2016: Fate of water pumped from underground and contributions to sea-level rise. *Nature Climate Change,* **6**, 777-780. http://dx.doi.org/10.1038/nclimate3001

8. Wada, Y., J.T. Reager, B.F. Chao, J. Wang, M.-H. Lo, C. Song, Y. Li, and A.S. Gardner, 2017: Recent changes in land water storage and its contribution to sea level variations. *Surveys in Geophysics,* **38**, 131-152. http://dx.doi.org/10.1007/s10712-016-9399-6

9. Boon, J.D., 2012: Evidence of sea level acceleration at U.S. and Canadian tide stations, Atlantic Coast, North America. *Journal of Coastal Research,* 1437-1445. http://dx.doi.org/10.2112/JCOASTRES-D-12-00102.1

10. Ezer, T., 2013: Sea level rise, spatially uneven and temporally unsteady: Why the U.S. East Coast, the global tide gauge record, and the global altimeter data show different trends. *Geophysical Research Letters,* **40**, 5439-5444. http://dx.doi.org/10.1002/2013GL057952

11. Sallenger, A.H., K.S. Doran, and P.A. Howd, 2012: Hotspot of accelerated sea-level rise on the Atlantic coast of North America. *Nature Climate Change,* **2**, 884-888. http://dx.doi.org/10.1038/nclimate1597

12. Yin, J. and P.B. Goddard, 2013: Oceanic control of sea level rise patterns along the East Coast of the United States. *Geophysical Research Letters,* **40**, 5514-5520. http://dx.doi.org/10.1002/2013GL057992

13. Mitrovica, J.X., N. Gomez, E. Morrow, C. Hay, K. Latychev, and M.E. Tamisiea, 2011: On the robustness of predictions of sea level fingerprints. *Geophysical Journal International,* **187**, 729-742. http://dx.doi.org/10.1111/j.1365-246X.2011.05090.x

14. Zervas, C., S. Gill, and W.V. Sweet, 2013: Estimating Vertical Land Motion From Long-term Tide Gauge Records. National Oceanic and Atmospheric Administration, National Ocean Service, 22 pp. https://tidesandcurrents.noaa.gov/publications/Technical_Report_NOS_CO-OPS_065.pdf

15. Wöppelmann, G. and M. Marcos, 2016: Vertical land motion as a key to understanding sea level change and variability. *Reviews of Geophysics,* **54**, 64-92. http://dx.doi.org/10.1002/2015RG000502

16. Bromirski, P.D., A.J. Miller, R.E. Flick, and G. Auad, 2011: Dynamical suppression of sea level rise along the Pacific coast of North America: Indications for imminent acceleration. *Journal of Geophysical Research,* **116**, C07005. http://dx.doi.org/10.1029/2010JC006759

17. Goddard, P.B., J. Yin, S.M. Griffies, and S. Zhang, 2015: An extreme event of sea-level rise along the Northeast coast of North America in 2009–2010. *Nature Communications,* **6**, 6346. http://dx.doi.org/10.1038/ncomms7346

18. Galloway, D., D.R. Jones, and S.E. Ingebritsen, 1999: Land Subsidence in the United States. U.S. Geological Survey, Reston, VA. 6 pp. https://pubs.usgs.gov/circ/circ1182/

19. Miller, K.G., M.A. Kominz, J.V. Browning, J.D. Wright, G.S. Mountain, M.E. Katz, P.J. Sugarman, B.S. Cramer, N. Christie-Blick, and S.F. Pekar, 2005: The Phanerozoic record of global sea-level change. *Science*, **310**, 1293-1298. http://dx.doi.org/10.1126/science.1116412

20. Dutton, A., A.E. Carlson, A.J. Long, G.A. Milne, P.U. Clark, R. DeConto, B.P. Horton, S. Rahmstorf, and M.E. Raymo, 2015: Sea-level rise due to polar ice-sheet mass loss during past warm periods. *Science*, **349**, aaa4019. http://dx.doi.org/10.1126/science.aaa4019

21. Hoffman, J.S., P.U. Clark, A.C. Parnell, and F. He, 2017: Regional and global sea-surface temperatures during the last interglaciation. *Science*, **355**, 276-279. http://dx.doi.org/10.1126/science.aai8464

22. Dutton, A. and K. Lambeck, 2012: Ice volume and sea level during the Last Interglacial. *Science*, **337**, 216-219. http://dx.doi.org/10.1126/science.1205749

23. Kopp, R.E., F.J. Simons, J.X. Mitrovica, A.C. Maloof, and M. Oppenheimer, 2009: Probabilistic assessment of sea level during the last interglacial stage. *Nature*, **462**, 863-867. http://dx.doi.org/10.1038/nature08686

24. Haywood, A.M., D.J. Hill, A.M. Dolan, B.L. Otto-Bliesner, F. Bragg, W.L. Chan, M.A. Chandler, C. Contoux, H.J. Dowsett, A. Jost, Y. Kamae, G. Lohmann, D.J. Lunt, A. Abe-Ouchi, S.J. Pickering, G. Ramstein, N.A. Rosenbloom, U. Salzmann, L. Sohl, C. Stepanek, H. Ueda, Q. Yan, and Z. Zhang, 2013: Large-scale features of Pliocene climate: Results from the Pliocene Model Intercomparison Project. *Climate of the Past*, **9**, 191-209. http://dx.doi.org/10.5194/cp-9-191-2013

25. Miller, K.G., J.D. Wright, J.V. Browning, A. Kulpecz, M. Kominz, T.R. Naish, B.S. Cramer, Y. Rosenthal, W.R. Peltier, and S. Sosdian, 2012: High tide of the warm Pliocene: Implications of global sea level for Antarctic deglaciation. *Geology*, **40**, 407-410. http://dx.doi.org/10.1130/g32869.1

26. Clark, P.U., A.S. Dyke, J.D. Shakun, A.E. Carlson, J. Clark, B. Wohlfarth, J.X. Mitrovica, S.W. Hostetler, and A.M. McCabe, 2009: The last glacial maximum. *Science*, **325**, 710-714. http://dx.doi.org/10.1126/science.1172873

27. Shennan, I., A.J. Long, and B.P. Horton, eds., 2015: *Handbook of Sea-Level Research*. John Wiley & Sons, Ltd, 581 pp. http://dx.doi.org/10.1002/9781118452547

28. Lambeck, K., H. Rouby, A. Purcell, Y. Sun, and M. Sambridge, 2014: Sea level and global ice volumes from the Last Glacial Maximum to the Holocene. *Proceedings of the National Academy of Sciences*, **111**, 15296-15303. http://dx.doi.org/10.1073/pnas.1411762111

29. Deschamps, P., N. Durand, E. Bard, B. Hamelin, G. Camoin, A.L. Thomas, G.M. Henderson, J.i. Okuno, and Y. Yokoyama, 2012: Ice-sheet collapse and sea-level rise at the Bolling warming 14,600 years ago. *Nature*, **483**, 559-564. http://dx.doi.org/10.1038/nature10902

30. Carlson, A.E., A.N. LeGrande, D.W. Oppo, R.E. Came, G.A. Schmidt, F.S. Anslow, J.M. Licciardi, and E.A. Obbink, 2008: Rapid early Holocene deglaciation of the Laurentide ice sheet. *Nature Geoscience*, **1**, 620-624. http://dx.doi.org/10.1038/ngeo285

31. Kemp, A.C., B.P. Horton, J.P. Donnelly, M.E. Mann, M. Vermeer, and S. Rahmstorf, 2011: Climate related sea-level variations over the past two millennia. *Proceedings of the National Academy of Sciences*, **108**, 11017-11022. http://dx.doi.org/10.1073/pnas.1015619108

32. Kopp, R.E., A.C. Kemp, K. Bittermann, B.P. Horton, J.P. Donnelly, W.R. Gehrels, C.C. Hay, J.X. Mitrovica, E.D. Morrow, and S. Rahmstorf, 2016: Temperature-driven global sea-level variability in the Common Era. *Proceedings of the National Academy of Sciences*, **113**, E1434-E1441. http://dx.doi.org/10.1073/pnas.1517056113

33. Hay, C.C., E. Morrow, R.E. Kopp, and J.X. Mitrovica, 2015: Probabilistic reanalysis of twentieth-century sea-level rise. *Nature*, **517**, 481-484. http://dx.doi.org/10.1038/nature14093

34. Church, J.A. and N.J. White, 2011: Sea-level rise from the late 19th to the early 21st century. *Surveys in Geophysics*, **32**, 585-602. http://dx.doi.org/10.1007/s10712-011-9119-1

35. Nerem, R.S., D.P. Chambers, C. Choe, and G.T. Mitchum, 2010: Estimating mean sea level change from the TOPEX and Jason altimeter missions. *Marine Geodesy*, **33**, 435-446. http://dx.doi.org/10.1080/01490419.2010.491031

36. Llovel, W., J.K. Willis, F.W. Landerer, and I. Fukumori, 2014: Deep-ocean contribution to sea level and energy budget not detectable over the past decade. *Nature Climate Change*, **4**, 1031-1035. http://dx.doi.org/10.1038/nclimate2387

37. Leuliette, E.W., 2015: The balancing of the sea-level budget. *Current Climate Change Reports*, **1**, 185-191. http://dx.doi.org/10.1007/s40641-015-0012-8

38. Merrifield, M.A., P. Thompson, E. Leuliette, G.T. Mitchum, D.P. Chambers, S. Jevrejeva, R.S. Nerem, M. Menéndez, W. Sweet, B. Hamlington, and J.J. Marra, 2015: [Global Oceans] Sea level variability and change [in "State of the Climate in 2014"]. *Bulletin of the American Meteorological Society*, **96 (12)**, S82-S85. http://dx.doi.org/10.1175/2015BAMSStateoftheClimate.1

39. Chambers, D.P., A. Cazenave, N. Champollion, H. Dieng, W. Llovel, R. Forsberg, K. von Schuckmann, and Y. Wada, 2017: Evaluation of the global mean sea level budget between 1993 and 2014. *Surveys in Geophysics*, **38**, 309–327. http://dx.doi.org/10.1007/s10712-016-9381-3

40. Leuliette, E.W. and R.S. Nerem, 2016: Contributions of Greenland and Antarctica to global and regional sea level change. *Oceanography*, **29**, 154-159. http://dx.doi.org/10.5670/oceanog.2016.107

41. Slangen, A.B.A., J.A. Church, C. Agosta, X. Fettweis, B. Marzeion, and K. Richter, 2016: Anthropogenic forcing dominates global mean sea-level rise since 1970. *Nature Climate Change*, **6**, 701-705. http://dx.doi.org/10.1038/nclimate2991

42. Jevrejeva, S., A. Grinsted, and J.C. Moore, 2009: Anthropogenic forcing dominates sea level rise since 1850. *Geophysical Research Letters*, **36**, L20706. http://dx.doi.org/10.1029/2009GL040216

43. Dangendorf, S., M. Marcos, A. Müller, E. Zorita, R. Riva, K. Berk, and J. Jensen, 2015: Detecting anthropogenic footprints in sea level rise. *Nature Communications*, **6**, 7849. http://dx.doi.org/10.1038/ncomms8849

44. Becker, M., M. Karpytchev, and S. Lennartz-Sassinek, 2014: Long-term sea level trends: Natural or anthropogenic? *Geophysical Research Letters*, **41**, 5571-5580. http://dx.doi.org/10.1002/2014GL061027

45. Marcos, M. and A. Amores, 2014: Quantifying anthropogenic and natural contributions to thermosteric sea level rise. *Geophysical Research Letters*, **41**, 2502-2507. http://dx.doi.org/10.1002/2014GL059766

46. Slangen, A.B.A., J.A. Church, X. Zhang, and D. Monselesan, 2014: Detection and attribution of global mean thermosteric sea level change. *Geophysical Research Letters*, **41**, 5951-5959. http://dx.doi.org/10.1002/2014GL061356

47. Marzeion, B., J.G. Cogley, K. Richter, and D. Parkes, 2014: Attribution of global glacier mass loss to anthropogenic and natural causes. *Science*, **345**, 919-921. http://dx.doi.org/10.1126/science.1254702

48. Marcos, M., B. Marzeion, S. Dangendorf, A.B.A. Slangen, H. Palanisamy, and L. Fenoglio-Marc, 2017: Internal variability versus anthropogenic forcing on sea level and its components. *Surveys in Geophysics*, **38**, 329–348. http://dx.doi.org/10.1007/s10712-016-9373-3

49. Zhang, X. and J.A. Church, 2012: Sea level trends, interannual and decadal variability in the Pacific Ocean. *Geophysical Research Letters*, **39**, L21701. http://dx.doi.org/10.1029/2012GL053240

50. Merrifield, M.A., 2011: A shift in western tropical Pacific sea level trends during the 1990s. *Journal of Climate*, **24**, 4126-4138. http://dx.doi.org/10.1175/2011JCLI3932.1

51. Hamlington, B.D., S.H. Cheon, P.R. Thompson, M.A. Merrifield, R.S. Nerem, R.R. Leben, and K.Y. Kim, 2016: An ongoing shift in Pacific Ocean sea level. *Journal of Geophysical Research Oceans*, **121**, 5084-5097. http://dx.doi.org/10.1002/2016JC011815

52. Kopp, R.E., 2013: Does the mid-Atlantic United States sea level acceleration hot spot reflect ocean dynamic variability? *Geophysical Research Letters*, **40**, 3981-3985. http://dx.doi.org/10.1002/grl.50781

53. Kopp, R.E., B.P. Horton, A.C. Kemp, and C. Tebaldi, 2015: Past and future sea-level rise along the coast of North Carolina, USA. *Climatic Change*, **132**, 693-707. http://dx.doi.org/10.1007/s10584-015-1451-x

54. Rahmstorf, S., J.E. Box, G. Feulner, M.E. Mann, A. Robinson, S. Rutherford, and E.J. Schaffernicht, 2015: Exceptional twentieth-century slowdown in Atlantic Ocean overturning circulation. *Nature Climate Change*, **5**, 475-480. http://dx.doi.org/10.1038/nclimate2554

55. Shepherd, A., E.R. Ivins, A. Geruo, V.R. Barletta, M.J. Bentley, S. Bettadpur, K.H. Briggs, D.H. Bromwich, R. Forsberg, N. Galin, M. Horwath, S. Jacobs, I. Joughin, M.A. King, J.T.M. Lenaerts, J. Li, S.R.M. Ligtenberg, A. Luckman, S.B. Luthcke, M. McMillan, R. Meister, G. Milne, J. Mouginot, A. Muir, J.P. Nicolas, J. Paden, A.J. Payne, H. Pritchard, E. Rignot, H. Rott, L. Sandberg Sørensen, T.A. Scambos, B. Scheuchl, E.J.O. Schrama, B. Smith, A.V. Sundal, J.H. van Angelen, W.J. van de Berg, M.R. van den Broeke, D.G. Vaughan, I. Velicogna, J. Wahr, P.L. Whitehouse, D.J. Wingham, D. Yi, D. Young, and H.J. Zwally, 2012: A reconciled estimate of ice-sheet mass balance. *Science*, **338**, 1183-1189. http://dx.doi.org/10.1126/science.1228102

56. Scambos, T. and C. Shuman, 2016: Comment on 'Mass gains of the Antarctic ice sheet exceed losses' by H. J. Zwally and others. *Journal of Glaciology*, **62**, 599-603. http://dx.doi.org/10.1017/jog.2016.59

57. Seo, K.-W., C.R. Wilson, T. Scambos, B.-M. Kim, D.E. Waliser, B. Tian, B.-H. Kim, and J. Eom, 2015: Surface mass balance contributions to acceleration of Antarctic ice mass loss during 2003–2013. *Journal of Geophysical Research Solid Earth*, **120**, 3617-3627. http://dx.doi.org/10.1002/2014JB011755

58. Martín-Español, A., A. Zammit-Mangion, P.J. Clarke, T. Flament, V. Helm, M.A. King, S.B. Luthcke, E. Petrie, F. Rémy, N. Schön, B. Wouters, and J.L. Bamber, 2016: Spatial and temporal Antarctic Ice Sheet mass trends, glacio-isostatic adjustment, and surface processes from a joint inversion of satellite altimeter, gravity, and GPS data. *Journal of Geophysical Research Earth Surface*, **121**, 182-200. http://dx.doi.org/10.1002/2015JF003550

59. Helm, V., A. Humbert, and H. Miller, 2014: Elevation and elevation change of Greenland and Antarctica derived from CryoSat-2. *The Cryosphere*, **8**, 1539-1559. http://dx.doi.org/10.5194/tc-8-1539-2014

60. Sutterley, T.C., I. Velicogna, E. Rignot, J. Mouginot, T. Flament, M.R. van den Broeke, J.M. van Wessem, and C.H. Reijmer, 2014: Mass loss of the Amundsen Sea embayment of West Antarctica from four independent techniques. *Geophysical Research Letters*, **41**, 8421-8428. http://dx.doi.org/10.1002/2014GL061940

61. Mouginot, J., E. Rignot, and B. Scheuchl, 2014: Sustained increase in ice discharge from the Amundsen Sea Embayment, West Antarctica, from 1973 to 2013. *Geophysical Research Letters*, **41**, 1576-1584. http://dx.doi.org/10.1002/2013GL059069

62. Khazendar, A., M.P. Schodlok, I. Fenty, S.R.M. Ligtenberg, E. Rignot, and M.R. van den Broeke, 2013: Observed thinning of Totten Glacier is linked to coastal polynya variability. *Nature Communications*, **4**, 2857. http://dx.doi.org/10.1038/ncomms3857

63. Li, X., E. Rignot, M. Morlighem, J. Mouginot, and B. Scheuchl, 2015: Grounding line retreat of Totten Glacier, East Antarctica, 1996 to 2013. *Geophysical Research Letters*, **42**, 8049-8056. http://dx.doi.org/10.1002/2015GL065701

64. Wouters, B., A. Martin-Español, V. Helm, T. Flament, J.M. van Wessem, S.R.M. Ligtenberg, M.R. van den Broeke, and J.L. Bamber, 2015: Dynamic thinning of glaciers on the Southern Antarctic Peninsula. *Science*, **348**, 899-903. http://dx.doi.org/10.1126/science.aaa5727

65. Paolo, F.S., H.A. Fricker, and L. Padman, 2015: Volume loss from Antarctic ice shelves is accelerating. *Science*, **348**, 327-331. http://dx.doi.org/10.1126/science.aaa0940

66. Khan, S.A., K.H. Kjaer, M. Bevis, J.L. Bamber, J. Wahr, K.K. Kjeldsen, A.A. Bjork, N.J. Korsgaard, L.A. Stearns, M.R. van den Broeke, L. Liu, N.K. Larsen, and I.S. Muresan, 2014: Sustained mass loss of the northeast Greenland ice sheet triggered by regional warming. *Nature Climate Change*, **4**, 292-299. http://dx.doi.org/10.1038/nclimate2161

67. Kjeldsen, K.K., N.J. Korsgaard, A.A. Bjørk, S.A. Khan, J.E. Box, S. Funder, N.K. Larsen, J.L. Bamber, W. Colgan, M. van den Broeke, M.-L. Siggaard-Andersen, C. Nuth, A. Schomacker, C.S. Andresen, E. Willerslev, and K.H. Kjær, 2015: Spatial and temporal distribution of mass loss from the Greenland Ice Sheet since AD 1900. *Nature*, **528**, 396-400. http://dx.doi.org/10.1038/nature16183

68. Tedesco, M., X. Fettweis, T. Mote, J. Wahr, P. Alexander, J.E. Box, and B. Wouters, 2013: Evidence and analysis of 2012 Greenland records from spaceborne observations, a regional climate model and reanalysis data. *The Cryosphere*, **7**, 615-630. http://dx.doi.org/10.5194/tc-7-615-2013

69. Tedesco, M., S. Doherty, X. Fettweis, P. Alexander, J. Jeyaratnam, and J. Stroeve, 2016: The darkening of the Greenland ice sheet: Trends, drivers, and projections (1981–2100). *The Cryosphere*, **10**, 477-496. http://dx.doi.org/10.5194/tc-10-477-2016

70. MacGregor, J.A., W.T. Colgan, M.A. Fahnestock, M. Morlighem, G.A. Catania, J.D. Paden, and S.P. Gogineni, 2016: Holocene deceleration of the Greenland Ice Sheet. *Science*, **351**, 590-593. http://dx.doi.org/10.1126/science.aab1702

71. Sweet, W.V., R.E. Kopp, C.P. Weaver, J. Obeysekera, R.M. Horton, E.R. Thieler, and C. Zervas, 2017: Global and Regional Sea Level Rise Scenarios for the United States. National Oceanic and Atmospheric Administration, National Ocean Service, Silver Spring, MD. 75 pp. https://tidesandcurrents.noaa.gov/publications/techrpt83_Global_and_Regional_SLR_Scenarios_for_the_US_final.pdf

72. Pfeffer, W.T., J.T. Harper, and S. O'Neel, 2008: Kinematic constraints on glacier contributions to 21st-century sea-level rise. *Science*, **321**, 1340-1343. http://dx.doi.org/10.1126/science.1159099

73. Sriver, R.L., N.M. Urban, R. Olson, and K. Keller, 2012: Toward a physically plausible upper bound of sea-level rise projections. *Climatic Change*, **115**, 893-902. http://dx.doi.org/10.1007/s10584-012-0610-6

74. Bamber, J.L. and W.P. Aspinall, 2013: An expert judgement assessment of future sea level rise from the ice sheets. *Nature Climate Change*, **3**, 424-427. http://dx.doi.org/10.1038/nclimate1778

75. Miller, K.G., R.E. Kopp, B.P. Horton, J.V. Browning, and A.C. Kemp, 2013: A geological perspective on sea-level rise and its impacts along the U.S. mid-Atlantic coast. *Earth's Future*, **1**, 3-18. http://dx.doi.org/10.1002/2013EF000135

76. Kopp, R.E., R.M. Horton, C.M. Little, J.X. Mitrovica, M. Oppenheimer, D.J. Rasmussen, B.H. Strauss, and C. Tebaldi, 2014: Probabilistic 21st and 22nd century sea-level projections at a global network of tide-gauge sites. *Earth's Future*, **2**, 383-406. http://dx.doi.org/10.1002/2014EF000239

77. DeConto, R.M. and D. Pollard, 2016: Contribution of Antarctica to past and future sea-level rise. *Nature*, **531**, 591-597. http://dx.doi.org/10.1038/nature17145

78. Hall, J.A., S. Gill, J. Obeysekera, W. Sweet, K. Knuti, and J. Marburger, 2016: Regional Sea Level Scenarios for Coastal Risk Management: Managing the Uncertainty of Future Sea Level Change and Extreme Water Levels for Department of Defense Coastal Sites Worldwide. U.S. Department of Defense, Strategic Environmental Research and Development Program, Alexandria VA. 224 pp. https://www.usfsp.edu/icar/files/2015/08/CARSWG-SLR-FINAL-April-2016.pdf

79. Slangen, A.B.A., M. Carson, C.A. Katsman, R.S.W. van de Wal, A. Köhl, L.L.A. Vermeersen, and D. Stammer, 2014: Projecting twenty-first century regional sea-level changes. *Climatic Change*, **124**, 317-332. http://dx.doi.org/10.1007/s10584-014-1080-9

80. Jevrejeva, S., A. Grinsted, and J.C. Moore, 2014: Upper limit for sea level projections by 2100. *Environmental Research Letters*, **9**, 104008. http://dx.doi.org/10.1088/1748-9326/9/10/104008

81. Grinsted, A., S. Jevrejeva, R.E.M. Riva, and D. Dahl-Jensen, 2015: Sea level rise projections for northern Europe under RCP8.5. *Climate Research*, **64**, 15-23. http://dx.doi.org/10.3354/cr01309

82. Mengel, M., A. Levermann, K. Frieler, A. Robinson, B. Marzeion, and R. Winkelmann, 2016: Future sea level rise constrained by observations and long-term commitment. *Proceedings of the National Academy of Sciences*, **113**, 2597-2602. http://dx.doi.org/10.1073/pnas.1500515113

83. Jackson, L.P. and S. Jevrejeva, 2016: A probabilistic approach to 21st century regional sea-level projections using RCP and High-end scenarios. *Global and Planetary Change*, **146**, 179-189. http://dx.doi.org/10.1016/j.gloplacha.2016.10.006

84. Ritz, C., T.L. Edwards, G. Durand, A.J. Payne, V. Peyaud, and R.C.A. Hindmarsh, 2015: Potential sea-level rise from Antarctic ice-sheet instability constrained by observations. *Nature*, **528**, 115-118. http://dx.doi.org/10.1038/nature16147

85. Rahmstorf, S., 2007: A semi-empirical approach to projecting future sea-level rise. *Science*, **315**, 368-370. http://dx.doi.org/10.1126/science.1135456

86. Rignot, E., J. Mouginot, M. Morlighem, H. Seroussi, and B. Scheuchl, 2014: Widespread, rapid grounding line retreat of Pine Island, Thwaites, Smith, and Kohler Glaciers, West Antarctica, from 1992 to 2011. *Geophysical Research Letters*, **41**, 3502-3509. http://dx.doi.org/10.1002/2014GL060140

87. Joughin, I., B.E. Smith, and B. Medley, 2014: Marine ice sheet collapse potentially under way for the Thwaites Glacier Basin, West Antarctica. *Science*, **344**, 735-738. http://dx.doi.org/10.1126/science.1249055

88. Pollard, D., R.M. DeConto, and R.B. Alley, 2015: Potential Antarctic Ice Sheet retreat driven by hydrofracturing and ice cliff failure. *Earth and Planetary Science Letters*, **412**, 112-121. http://dx.doi.org/10.1016/j.epsl.2014.12.035

89. Levermann, A., P.U. Clark, B. Marzeion, G.A. Milne, D. Pollard, V. Radic, and A. Robinson, 2013: The multimillennial sea-level commitment of global warming. *Proceedings of the National Academy of Sciences*, **110**, 13745-13750. http://dx.doi.org/10.1073/pnas.1219414110

90. Golledge, N.R., D.E. Kowalewski, T.R. Naish, R.H. Levy, C.J. Fogwill, and E.G.W. Gasson, 2015: The multi-millennial Antarctic commitment to future sea-level rise. *Nature*, **526**, 421-425. http://dx.doi.org/10.1038/nature15706

91. Strauss, B.H., S. Kulp, and A. Levermann, 2015: Carbon choices determine US cities committed to futures below sea level. *Proceedings of the National Academy of Sciences*, **112**, 13508-13513. http://dx.doi.org/10.1073/pnas.1511186112

92. Irvine, P.J., D.J. Lunt, E.J. Stone, and A. Ridgwell, 2009: The fate of the Greenland Ice Sheet in a geo-engineered, high $CO_2$ world. *Environmental Research Letters*, **4**, 045109. http://dx.doi.org/10.1088/1748-9326/4/4/045109

93. Applegate, P.J. and K. Keller, 2015: How effective is albedo modification (solar radiation management geoengineering) in preventing sea-level rise from the Greenland Ice Sheet? *Environmental Research Letters*, **10**, 084018. http://dx.doi.org/10.1088/1748-9326/10/8/084018

94. Lenton, T.M., 2011: Early warning of climate tipping points. *Nature Climate Change*, **1**, 201-209. http://dx.doi.org/10.1038/nclimate1143

95. Barrett, S., T.M. Lenton, A. Millner, A. Tavoni, S. Carpenter, J.M. Anderies, F.S. Chapin, III, A.-S. Crepin, G. Daily, P. Ehrlich, C. Folke, V. Galaz, T. Hughes, N. Kautsky, E.F. Lambin, R. Naylor, K. Nyborg, S. Polasky, M. Scheffer, J. Wilen, A. Xepapadeas, and A. de Zeeuw, 2014: Climate engineering reconsidered. *Nature Climate Change*, **4**, 527-529. http://dx.doi.org/10.1038/nclimate2278

96. Markusson, N., F. Ginn, N. Singh Ghaleigh, and V. Scott, 2014: 'In case of emergency press here': Framing geoengineering as a response to dangerous climate change. *Wiley Interdisciplinary Reviews: Climate Change*, **5**, 281-290. http://dx.doi.org/10.1002/wcc.263

97. Sillmann, J., T.M. Lenton, A. Levermann, K. Ott, M. Hulme, F. Benduhn, and J.B. Horton, 2015: Climate emergencies do not justify engineering the climate. *Nature Climate Change*, **5**, 290-292. http://dx.doi.org/10.1038/nclimate2539

98. Clark, P.U., J.D. Shakun, S.A. Marcott, A.C. Mix, M. Eby, S. Kulp, A. Levermann, G.A. Milne, P.L. Pfister, B.D. Santer, D.P. Schrag, S. Solomon, T.F. Stocker, B.H. Strauss, A.J. Weaver, R. Winkelmann, D. Archer, E. Bard, A. Goldner, K. Lambeck, R.T. Pierrehumbert, and G.-K. Plattner, 2016: Consequences of twenty-first-century policy for multi-millennial climate and sea-level change. *Nature Climate Change*, **6**, 360-369. http://dx.doi.org/10.1038/nclimate2923

99. Gregory, J.M. and J.A. Lowe, 2000: Predictions of global and regional sea-level rise using AOG-CMs with and without flux adjustment. *Geophysical Research Letters*, **27**, 3069-3072. http://dx.doi.org/10.1029/1999GL011228

100. Levermann, A., A. Griesel, M. Hofmann, M. Montoya, and S. Rahmstorf, 2005: Dynamic sea level changes following changes in the thermohaline circulation. *Climate Dynamics*, **24**, 347-354. http://dx.doi.org/10.1007/s00382-004-0505-y

101. Menéndez, M. and P.L. Woodworth, 2010: Changes in extreme high water levels based on a quasi-global tide-gauge data set. *Journal of Geophysical Research*, **115**, C10011. http://dx.doi.org/10.1029/2009JC005997

102. Kemp, A.C. and B.P. Horton, 2013: Contribution of relative sea-level rise to historical hurricane flooding in New York City. *Journal of Quaternary Science*, **28**, 537-541. http://dx.doi.org/10.1002/jqs.2653

103. Sweet, W.V., C. Zervas, S. Gill, and J. Park, 2013: Hurricane Sandy inundation probabilities of today and tomorrow [in "Explaining Extreme Events of 2012 from a Climate Perspective"]. *Bulletin of the American Meteorological Society*, **94 (9)**, S17-S20. http://dx.doi.org/10.1175/BAMS-D-13-00085.1

104. Sweet, W., J. Park, J. Marra, C. Zervas, and S. Gill, 2014: Sea Level Rise and Nuisance Flood Frequency Changes around the United States. NOAA Technical Report NOS CO-OPS 073. National Oceanic and Atmospheric Administration, National Ocean Service, Silver Spring, MD. 58 pp. http://tidesandcurrents.noaa.gov/publications/NOAA_Technical_Report_NOS_COOPS_073.pdf

105. Sweet, W.V. and J. Park, 2014: From the extreme to the mean: Acceleration and tipping points of coastal inundation from sea level rise. *Earth's Future*, **2**, 579-600. http://dx.doi.org/10.1002/2014EF000272

106. Ezer, T. and L.P. Atkinson, 2014: Accelerated flooding along the U.S. East Coast: On the impact of sea-level rise, tides, storms, the Gulf Stream, and the North Atlantic Oscillations. *Earth's Future*, **2**, 362-382. http://dx.doi.org/10.1002/2014EF000252

107. Talke, S.A., P. Orton, and D.A. Jay, 2014: Increasing storm tides in New York Harbor, 1844–2013. *Geophysical Research Letters*, **41**, 3149-3155. http://dx.doi.org/10.1002/2014GL059574

108. Wahl, T. and D.P. Chambers, 2015: Evidence for multidecadal variability in US extreme sea level records. *Journal of Geophysical Research Oceans*, **120**, 1527-1544. http://dx.doi.org/10.1002/2014JC010443

109. Reed, A.J., M.E. Mann, K.A. Emanuel, N. Lin, B.P. Horton, A.C. Kemp, and J.P. Donnelly, 2015: Increased threat of tropical cyclones and coastal flooding to New York City during the anthropogenic era. *Proceedings of the National Academy of Sciences*, **112**, 12610-12615. http://dx.doi.org/10.1073/pnas.1513127112

110. Grinsted, A., J.C. Moore, and S. Jevrejeva, 2013: Projected Atlantic hurricane surge threat from rising temperatures. *Proceedings of the National Academy of Sciences*, **110**, 5369-5373. http://dx.doi.org/10.1073/pnas.1209980110

111. Marcos, M., F.M. Calafat, Á. Berihuete, and S. Dangendorf, 2015: Long-term variations in global sea level extremes. *Journal of Geophysical Research Oceans*, **120**, 8115-8134. http://dx.doi.org/10.1002/2015JC011173

112. Woodworth, P.L. and M. Menéndez, 2015: Changes in the mesoscale variability and in extreme sea levels over two decades as observed by satellite altimetry. *Journal of Geophysical Research Oceans*, **120**, 64-77. http://dx.doi.org/10.1002/2014JC010363

113. Wahl, T. and D.P. Chambers, 2016: Climate controls multidecadal variability in U. S. extreme sea level records. *Journal of Geophysical Research Oceans*, **121**, 1274-1290. http://dx.doi.org/10.1002/2015JC011057

114. Mawdsley, R.J. and I.D. Haigh, 2016: Spatial and temporal variability and long-term trends in skew surges globally. *Frontiers in Marine Science*, **3**, Art. 26. http://dx.doi.org/10.3389/fmars.2016.00029

115. Sweet, W., M. Menendez, A. Genz, J. Obeysekera, J. Park, and J. Marra, 2016: In tide's way: Southeast Florida's September 2015 sunny-day flood [in "Explaining Extreme Events of 2015 from a Climate Perspective"]. *Bulletin of the American Meteorological Society*, **97 (12)**, S25-S30. http://dx.doi.org/10.1175/BAMS-D-16-0117.1

116. Feser, F., M. Barcikowska, O. Krueger, F. Schenk, R. Weisse, and L. Xia, 2015: Storminess over the North Atlantic and northwestern Europe—A review. *Quarterly Journal of the Royal Meteorological Society*, **141**, 350-382. http://dx.doi.org/10.1002/qj.2364

117. Colle, B.A., J.F. Booth, and E.K.M. Chang, 2015: A review of historical and future changes of extratropical cyclones and associated impacts along the US East Coast. *Current Climate Change Reports*, **1**, 125-143. http://dx.doi.org/10.1007/s40641-015-0013-7

118. Sweet, W.V. and C. Zervas, 2011: Cool-season sea level anomalies and storm surges along the U.S. East Coast: Climatology and comparison with the 2009/10 El Niño. *Monthly Weather Review*, **139**, 2290-2299. http://dx.doi.org/10.1175/MWR-D-10-05043.1

119. Thompson, P.R., G.T. Mitchum, C. Vonesch, and J. Li, 2013: Variability of winter storminess in the eastern United States during the twentieth century from tide gauges. *Journal of Climate*, **26**, 9713-9726. http://dx.doi.org/10.1175/JCLI-D-12-00561.1

120. Hamlington, B.D., R.R. Leben, K.Y. Kim, R.S. Nerem, L.P. Atkinson, and P.R. Thompson, 2015: The effect of the El Niño–Southern Oscillation on U.S. regional and coastal sea level. *Journal of Geophysical Research Oceans*, **120**, 3970-3986. http://dx.doi.org/10.1002/2014JC010602

121. Tebaldi, C., B.H. Strauss, and C.E. Zervas, 2012: Modelling sea level rise impacts on storm surges along US coasts. *Environmental Research Letters*, **7**, 014032. http://dx.doi.org/10.1088/1748-9326/7/1/014032

122. Horton, R.M., V. Gornitz, D.A. Bader, A.C. Ruane, R. Goldberg, and C. Rosenzweig, 2011: Climate hazard assessment for stakeholder adaptation planning in New York City. *Journal of Applied Meteorology and Climatology*, **50**, 2247-2266. http://dx.doi.org/10.1175/2011JAMC2521.1

123. Woodruff, J.D., J.L. Irish, and S.J. Camargo, 2013: Coastal flooding by tropical cyclones and sea-level rise. *Nature*, **504**, 44-52. http://dx.doi.org/10.1038/nature12855

124. Buchanan, M.K., R.E. Kopp, M. Oppenheimer, and C. Tebaldi, 2016: Allowances for evolving coastal flood risk under uncertain local sea-level rise. *Climatic Change*, **137**, 347-362. http://dx.doi.org/10.1007/s10584-016-1664-7

125. Dahl, K.A., M.F. Fitzpatrick, and E. Spanger-Siegfried, 2017: Sea level rise drives increased tidal flooding frequency at tide gauges along the U.S. East and Gulf Coasts: Projections for 2030 and 2045. *PLoS ONE*, **12**, e0170949. http://dx.doi.org/10.1371/journal.pone.0170949

126. Hunter, J., 2012: A simple technique for estimating an allowance for uncertain sea-level rise. *Climatic Change*, **113**, 239-252. http://dx.doi.org/10.1007/s10584-011-0332-1

127. Moftakhari, H.R., A. AghaKouchak, B.F. Sanders, D.L. Feldman, W. Sweet, R.A. Matthew, and A. Luke, 2015: Increased nuisance flooding along the coasts of the United States due to sea level rise: Past and future. *Geophysical Research Letters*, **42**, 9846-9852. http://dx.doi.org/10.1002/2015GL066072

128. Stockdon, H.F., R.A. Holman, P.A. Howd, and A.H. Sallenger, Jr., 2006: Empirical parameterization of setup, swash, and runup. *Coastal Engineering*, **53**, 573-588. http://dx.doi.org/10.1016/j.coastaleng.2005.12.005

129. Sweet, W.V., J. Park, S. Gill, and J. Marra, 2015: New ways to measure waves and their effects at NOAA tide gauges: A Hawaiian-network perspective. *Geophysical Research Letters*, **42**, 9355-9361. http://dx.doi.org/10.1002/2015GL066030

130. Moritz, H., K. White, B. Gouldby, W. Sweet, P. Ruggiero, M. Gravens, P. O'Brien, H. Moritz, T. Wahl, N.C. Nadal-Caraballo, and W. Veatch, 2015: USACE adaptation approach for future coastal climate conditions. *Proceedings of the Institution of Civil Engineers - Maritime Engineering*, **168**, 111-117. http://dx.doi.org/10.1680/jmaen.15.00015

131. Barnard, P.L., J. Allan, J.E. Hansen, G.M. Kaminsky, P. Ruggiero, and A. Doria, 2011: The impact of the 2009–10 El Niño Modoki on U.S. West Coast beaches. *Geophysical Research Letters*, **38**, L13604. http://dx.doi.org/10.1029/2011GL047707

132. Theuerkauf, E.J., A.B. Rodriguez, S.R. Fegley, and R.A. Luettich, 2014: Sea level anomalies exacerbate beach erosion. *Geophysical Research Letters*, **41**, 5139-5147. http://dx.doi.org/10.1002/2014GL060544

133. Serafin, K.A. and P. Ruggiero, 2014: Simulating extreme total water levels using a time-dependent, extreme value approach. *Journal of Geophysical Research Oceans*, **119**, 6305-6329. http://dx.doi.org/10.1002/2014JC010093

134. Hoeke, R.K., K.L. McInnes, J.C. Kruger, R.J. McNaught, J.R. Hunter, and S.G. Smithers, 2013: Widespread inundation of Pacific islands triggered by distant-source wind-waves. *Global and Planetary Change*, **108**, 128-138. http://dx.doi.org/10.1016/j.gloplacha.2013.06.006

135. Stopa, J.E. and K.F. Cheung, 2014: Periodicity and patterns of ocean wind and wave climate. *Journal of Geophysical Research Oceans*, **119**, 5563-5584. http://dx.doi.org/10.1002/2013JC009729

136. Barnard, P.L., A.D. Short, M.D. Harley, K.D. Splinter, S. Vitousek, I.L. Turner, J. Allan, M. Banno, K.R. Bryan, A. Doria, J.E. Hansen, S. Kato, Y. Kuriyama, E. Randall-Goodwin, P. Ruggiero, I.J. Walker, and D.K. Heathfield, 2015: Coastal vulnerability across the Pacific dominated by El Niño/Southern Oscillation. *Nature Geoscience*, **8**, 801-807. http://dx.doi.org/10.1038/ngeo2539

137. Bromirski, P.D., D.R. Cayan, J. Helly, and P. Wittmann, 2013: Wave power variability and trends across the North Pacific. *Journal of Geophysical Research Oceans*, **118**, 6329-6348. http://dx.doi.org/10.1002/2013JC009189

138. Erikson, L.H., C.A. Hegermiller, P.L. Barnard, P. Ruggiero, and M. van Ormondt, 2015: Projected wave conditions in the Eastern North Pacific under the influence of two CMIP5 climate scenarios. *Ocean Modelling*, **96 (12)**, Part 1, 171-185. http://dx.doi.org/10.1016/j.ocemod.2015.07.004

139. Aucan, J., R. Hoeke, and M.A. Merrifield, 2012: Wave-driven sea level anomalies at the Midway tide gauge as an index of North Pacific storminess over the past 60 years. *Geophysical Research Letters*, **39**, L17603. http://dx.doi.org/10.1029/2012GL052993

140. Ruggiero, P., 2013: Is the intensifying wave climate of the U.S. Pacific Northwest increasing flooding and erosion risk faster than sea-level rise? *Journal of Waterway, Port, Coastal, and Ocean Engineering*, **139**, 88-97. http://dx.doi.org/10.1061/(ASCE)WW.1943-5460.0000172

141. Bromirski, P.D. and D.R. Cayan, 2015: Wave power variability and trends across the North Atlantic influenced by decadal climate patterns. *Journal of Geophysical Research Oceans*, **120**, 3419-3443. http://dx.doi.org/10.1002/2014JC010440

142. Bromirski, P.D. and J.P. Kossin, 2008: Increasing hurricane wave power along the U.S. Atlantic and Gulf coasts. *Journal of Geophysical Research*, **113**, C07012. http://dx.doi.org/10.1029/2007JC004706

143. Graham, N.E., D.R. Cayan, P.D. Bromirski, and R.E. Flick, 2013: Multi-model projections of twenty-first century North Pacific winter wave climate under the IPCC A2 scenario. *Climate Dynamics*, **40**, 1335-1360. http://dx.doi.org/10.1007/s00382-012-1435-8

144. Wang, X.L., Y. Feng, and V.R. Swail, 2014: Changes in global ocean wave heights as projected using multimodel CMIP5 simulations. *Geophysical Research Letters*, **41**, 1026-1034. http://dx.doi.org/10.1002/2013GL058650

145. Shope, J.B., C.D. Storlazzi, L.H. Erikson, and C.A. Hegermiller, 2016: Changes to extreme wave climates of islands within the western tropical Pacific throughout the 21st century under RCP 4.5 and RCP 8.5, with implications for island vulnerability and sustainability. *Global and Planetary Change*, **141**, 25-38. http://dx.doi.org/10.1016/j.gloplacha.2016.03.009

146. Colle, B.A., Z. Zhang, K.A. Lombardo, E. Chang, P. Liu, and M. Zhang, 2013: Historical evaluation and future prediction of eastern North American and western Atlantic extratropical cyclones in the CMIP5 models during the cool season. *Journal of Climate*, **26**, 6882-6903. http://dx.doi.org/10.1175/JCLI-D-12-00498.1

147. Zappa, G., L.C. Shaffrey, K.I. Hodges, P.G. Sansom, and D.B. Stephenson, 2013: A multimodel assessment of future projections of North Atlantic and European extratropical cyclones in the CMIP5 climate models. *Journal of Climate*, **26**, 5846-5862. http://dx.doi.org/10.1175/jcli-d-12-00573.1

148. Hall, T. and E. Yonekura, 2013: North American tropical cyclone landfall and SST: A statistical model study. *Journal of Climate*, **26**, 8422-8439. http://dx.doi.org/10.1175/jcli-d-12-00756.1

149. Lin, N., K. Emanuel, M. Oppenheimer, and E. Vanmarcke, 2012: Physically based assessment of hurricane surge threat under climate change. *Nature Climate Change*, **2**, 462-467. http://dx.doi.org/10.1038/nclimate1389

150. Little, C.M., R.M. Horton, R.E. Kopp, M. Oppenheimer, and S. Yip, 2015: Uncertainty in twenty-first-century CMIP5 sea level projections. *Journal of Climate*, **28**, 838-852. http://dx.doi.org/10.1175/JCLI-D-14-00453.1

151. Lin, N., R.E. Kopp, B.P. Horton, and J.P. Donnelly, 2016: Hurricane Sandy's flood frequency increasing from year 1800 to 2100. *Proceedings of the National Academy of Sciences*, **113**, 12071-12075. http://dx.doi.org/10.1073/pnas.1604386113

152. Smith, J.M., M.A. Cialone, T.V. Wamsley, and T.O. McAlpin, 2010: Potential impact of sea level rise on coastal surges in southeast Louisiana. *Ocean Engineering*, **37**, 37-47. http://dx.doi.org/10.1016/j.oceaneng.2009.07.008

153. Atkinson, J., J.M. Smith, and C. Bender, 2013: Sea-level rise effects on storm surge and nearshore waves on the Texas coast: Influence of landscape and storm characteristics. *Journal of Waterway, Port, Coastal, and Ocean Engineering*, **139**, 98-117. http://dx.doi.org/10.1061/(ASCE)WW.1943-5460.0000187

154. Bilskie, M.V., S.C. Hagen, S.C. Medeiros, and D.L. Passeri, 2014: Dynamics of sea level rise and coastal flooding on a changing landscape. *Geophysical Research Letters*, **41**, 927-934. http://dx.doi.org/10.1002/2013GL058759

155. Passeri, D.L., S.C. Hagen, S.C. Medeiros, M.V. Bilskie, K. Alizad, and D. Wang, 2015: The dynamic effects of sea level rise on low-gradient coastal landscapes: A review. *Earth's Future*, **3**, 159-181. http://dx.doi.org/10.1002/2015EF000298

156. Bilskie, M.V., S.C. Hagen, K. Alizad, S.C. Medeiros, D.L. Passeri, H.F. Needham, and A. Cox, 2016: Dynamic simulation and numerical analysis of hurricane storm surge under sea level rise with geomorphologic changes along the northern Gulf of Mexico. *Earth's Future*, **4**, 177-193. http://dx.doi.org/10.1002/2015EF000347

157. Knutson, T.R., J.J. Sirutis, G.A. Vecchi, S. Garner, M. Zhao, H.-S. Kim, M. Bender, R.E. Tuleya, I.M. Held, and G. Villarini, 2013: Dynamical downscaling projections of twenty-first-century Atlantic hurricane activity: CMIP3 and CMIP5 model-based scenarios. *Journal of Climate*, **27**, 6591-6617. http://dx.doi.org/10.1175/jcli-d-12-00539.1

158. Knutson, T.R., J.J. Sirutis, M. Zhao, R.E. Tuleya, M. Bender, G.A. Vecchi, G. Villarini, and D. Chavas, 2015: Global projections of intense tropical cyclone activity for the late twenty-first century from dynamical downscaling of CMIP5/RCP4.5 scenarios. *Journal of Climate*, **28**, 7203-7224. http://dx.doi.org/10.1175/JCLI-D-15-0129.1

159. Church, J.A. and N.J. White, 2006: A 20th century acceleration in global sea-level rise. *Geophysical Research Letters*, **33**, L01602. http://dx.doi.org/10.1029/2005GL024826

160. PSMSL, 2016: Obtaining tide guage data. Permanent Service for Mean Sea Level. http://www.psmsl.org/data/obtaining/

161. Holgate, S.J., A. Matthews, P.L. Woodworth, L.J. Rickards, M.E. Tamisiea, E. Bradshaw, P.R. Foden, K.M. Gordon, S. Jevrejeva, and J. Pugh, 2013: New data systems and products at the Permanent Service for Mean Sea Level. *Journal of Coastal Research*, **29**, 493-504. http://dx.doi.org/10.2112/JCOAS-TRES-D-12-00175.1

162. Engelhart, S.E. and B.P. Horton, 2012: Holocene sea level database for the Atlantic coast of the United States. *Quaternary Science Reviews*, **54**, 12-25. http://dx.doi.org/10.1016/j.quascirev.2011.09.013

163. Farrell, W.E. and J.A. Clark, 1976: On postglacial sea level. *Geophysical Journal International*, **46**, 647-667. http://dx.doi.org/10.1111/j.1365-246X.1976.tb01252.x

164. Yin, J., M.E. Schlesinger, and R.J. Stouffer, 2009: Model projections of rapid sea-level rise on the northeast coast of the United States. *Nature Geoscience*, **2**, 262-266. http://dx.doi.org/10.1038/ngeo462

165. Sweet, W.V. and J.J. Marra, 2016: State of U.S. Nuisance Tidal Flooding. Supplement to State of the Climate: National Overview for May 2016. National Oceanic and Atmospheric Administration, National Centers for Environmental Information, 5 pp. http://www.ncdc.noaa.gov/monitoring-content/sotc/national/2016/may/sweet-marra-nuisance-flooding-2015.pdf

166. Wahl, T., S. Jain, J. Bender, S.D. Meyers, and M.E. Luther, 2015: Increasing risk of compound flooding from storm surge and rainfall for major US cities. *Nature Climate Change*, **5**, 1093-1097. http://dx.doi.org/10.1038/nclimate2736

# 13

# Ocean Acidification and Other Ocean Changes

## KEY FINDINGS

1. The world's oceans have absorbed about 93% of the excess heat caused by greenhouse gas warming since the mid-20th century, making them warmer and altering global and regional climate feedbacks. Ocean heat content has increased at all depths since the 1960s and surface waters have warmed by about $1.3° ± 0.1°F$ ($0.7° ± 0.08°C$) per century globally since 1900 to 2016. Under a higher scenario, a global increase in average sea surface temperature of $4.9° ± 1.3°F$ ($2.7° ± 0.7°C$) by 2100 is projected, with even higher changes in some U.S. coastal regions. (*Very high confidence*)

2. The potential slowing of the Atlantic meridional overturning circulation (AMOC; of which the Gulf Stream is one component)—as a result of increasing ocean heat content and freshwater driven buoyancy changes—could have dramatic climate feedbacks as the ocean absorbs less heat and $CO_2$ from the atmosphere. This slowing would also affect the climates of North America and Europe. Any slowing documented to date cannot be directly tied to anthropogenic forcing primarily due to lack of adequate observational data and to challenges in modeling ocean circulation changes. Under a higher scenario (RCP8.5) in CMIP5 simulations, the AMOC weakens over the 21st century by 12% to 54% (*low confidence*).

3. The world's oceans are currently absorbing more than a quarter of the $CO_2$ emitted to the atmosphere annually from human activities, making them more acidic (*very high confidence*), with potential detrimental impacts to marine ecosystems. In particular, higher-latitude systems typically have a lower buffering capacity against pH change, exhibiting seasonally corrosive conditions sooner than low-latitude systems. Acidification is regionally increasing along U.S. coastal systems as a result of upwelling (for example, in the Pacific Northwest) (*high confidence*), changes in freshwater inputs (for example, in the Gulf of Maine) (*medium confidence*), and nutrient input (for example, in agricultural watersheds and urbanized estuaries) (*high confidence*). The rate of acidification is unparalleled in at least the past 66 million years (*medium confidence*). Under the higher scenario (RCP8.5), the global average surface ocean acidity is projected to increase by 100% to 150% (*high confidence*).

4. Increasing sea surface temperatures, rising sea levels, and changing patterns of precipitation, winds, nutrients, and ocean circulation are contributing to overall declining oxygen concentrations at intermediate depths in various ocean locations and in many coastal areas. Over the last half century, major oxygen losses have occurred in inland seas, estuaries, and in the coastal and open ocean (*high confidence*). Ocean oxygen levels are projected to decrease by as much as 3.5% under the higher scenario (RCP8.5) by 2100 relative to preindustrial values (*high confidence*).

**Recommended Citation for Chapter**

**Jewett**, L. and A. Romanou, 2017: Ocean acidification and other ocean changes. In: *Climate Science Special Report: Fourth National Climate Assessment, Volume I* [Wuebbles, D.J., D.W. Fahey, K.A. Hibbard, D.J. Dokken, B.C. Stewart, and T.K. Maycock (eds.)]. U.S. Global Change Research Program, Washington, DC, USA, pp. 364-392, doi: 10.7930/J0QV3JQB.

## 13.0 A Changing Ocean

Anthropogenic perturbations to the global Earth system have included important alterations in the chemical composition, temperature, and circulation of the oceans. Some of these changes will be distinguishable from the background natural variability in nearly half of the global open ocean within a decade, with important consequences for marine ecosystems and their services.[1] However, the timeframe for detection will vary depending on the parameter featured.[2, 3]

## 13.1 Ocean Warming

### 13.1.1 General Background

Approximately 93% of excess heat energy trapped since the 1970s has been absorbed into the oceans, lessening atmospheric warming and leading to a variety of changes in ocean conditions, including sea level rise and ocean circulation (see Ch. 2: Physical Drivers of Climate Change, Ch. 6: Temperature Change, and Ch. 12: Sea Level Rise in this report).[1, 4] This is the result of the high heat capacity of seawater relative to the atmosphere, the relative area of the ocean compared to the land, and the ocean circulation that enables the transport of heat into deep waters. This large heat absorption by the oceans moderates the effects of increased anthropogenic greenhouse emissions on terrestrial climates while altering the fundamental physical properties of the ocean and indirectly impacting chemical properties such as the biological pump through increased stratification.[1, 5] Although upper ocean temperature varies over short- and medium timescales (for example, seasonal and regional patterns), there are clear long-term increases in surface temperature and ocean heat content over the past 65 years.[4, 6, 7]

### 13.1.2 Ocean Heat Content

Ocean heat content (OHC) is an ideal variable to monitor changing climate as it is calculated using the entire water column, so ocean warming can be documented and compared between particular regions, ocean basins, and depths. However, for years prior to the 1970s, estimates of ocean uptake are confined to the upper ocean (up to 700 m) due to sparse spatial and temporal coverage and limited vertical capabilities of many of the instruments in use. OHC estimates are improved for time periods after 1970 with increased sampling coverage and depth.[4, 8] Estimates of OHC have been calculated going back to the 1950s using averages over longer time intervals (i.e., decadal or 5-year intervals) to compensate for sparse data distributions, allowing for clear long-term trends to emerge (e.g., Levitus et al. 2012[7]).

From 1960 to 2015, OHC significantly increased for both 0–700 and 700–2,000 m depths, for a total ocean warming of about $33.5 \pm 7.0 \times 10^{22}$ J (a net heating of $0.37 \pm 0.08$ W/m²; Figure 13.1).[6] During this period, there is evidence of an acceleration of ocean warming beginning in 1998,[9] with a total heat increase of about $15.2 \times 10^{22}$ J.[6] Robust ocean warming occurs in the upper 700 m and is slow to penetrate into the deep ocean. However, the 700–2,000 m depths constitute an increasing portion of the total ocean energy budget as compared to the surface ocean (Figure 13.1).[6] The role of the deep ocean (below 2,000 m [6,600 ft]) in ocean heat uptake remains uncertain, both in the magnitude but also the sign of the uptake.[10, 11] Penetration of surface waters to the deep ocean is a slow process, which means that while it takes only about a decade for near-surface temperatures to respond to increased heat energy, the deep ocean will continue to warm, and as a result sea levels will rise for centuries to millennia even if all further emissions cease.[4]

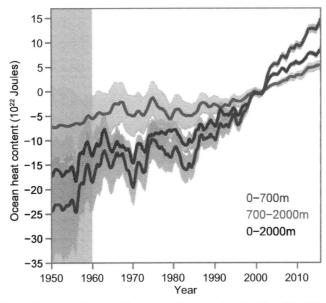

**Figure 13.1:** Global Ocean heat content change time series. Ocean heat content from 0 to 700 m (blue), 700 to 2,000 m (red), and 0 to 2,000 m (dark gray) from 1955 to 2015 with an uncertainty interval of ±2 standard deviations shown in shading. All time series of the analysis performed by Cheng et al.[6] are smoothed by a 12-month running mean filter, relative to the 1997–2005 base period. (Figure source: Cheng et al. 2017[6]).

Several sources have documented warming in all ocean basins from 0–2,000 m depths over the past 50 years (Figure 13.2).[6, 7, 12] Annual fluctuations in surface temperatures and OHC are attributed to the combination of a long-term secular trend and decadal and smaller time scale variations, such as the Pacific Decadal Oscillation (PDO) and the Atlantic Multidecadal Oscillation (AMO) (Ch. 5: Circulation & Variability; Ch. 12: Sea Level Rise).[13, 14] The transport of heat to the deep ocean is likely linked to the strength of the Atlantic Meridional Overturning Circulation (see Section 13.2.1), where the Atlantic and Southern Ocean accounts for the dominant portion of total OHC change at the 700–2,000 m depth.[6, 8, 9, 15] Decadal variabil-

ity in ocean heat uptake is mostly attributed to ENSO phases (with El Niños warming and La Niñas cooling). For instance, La Niña conditions over the past decade have led to colder ocean temperatures in the eastern tropical Pacific.[6, 8, 9, 16] For the Pacific and Indian Oceans, the decadal shifts are primarily observed in the upper 350 m depth, likely due to shallow subtropical circulation, leading to an abrupt increase of OHC in the Indian Ocean carried by the Indonesian throughflow from the Pacific Ocean over the last decade.[9] Although there is natural variability in ocean temperature, there remain clear increasing trends due to anthropogenic influences.

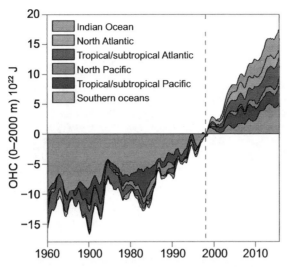

Figure 13.2: Ocean heat content changes from 1960 to 2015 for different ocean basins for 0 to 2,000 m depths. Time series is relative to the 1997–1999 base period and smoothed by a 12-month running filter by Cheng et al.[6] The curves are additive, and the ocean heat content changes in different ocean basins are shaded in different colors (Figure source: Cheng et al. 2017[6]).

### 13.1.3 Sea Surface Temperature and U.S. Regional Warming

In addition to OHC, sea surface temperature (SST) measurements are widely available. SST measurements are useful because 1) the measurements have been taken over 150 years (albeit using different platforms, instruments, and depths through time); 2) SST reflects the lower boundary condition of the atmosphere; and 3) SST can be used to predict specific regional impacts of global warming on terrestrial and coastal systems.[15, 17, 18] Globally, surface ocean temperatures have increased by 1.3° ± 0.1°F (0.70° ± 0.08°C) per century from 1900 to 2016 for the Extended Reconstructed Sea Surface Temperature version 4 (ERSST v4) record.[19] All U.S. coastal waters have warmed by more than 0.7°F (0.4°C) over this period as shown in both Table 13.1 and Chapter 6: Temperature Change, Figure 6.6. During the past 60 years, the rates of increase of SSTs for the coastal waters of three U.S. regions were above the global average rate. These included the waters around Alaska, the Northeast, and the Southwest (Table 13.1). Over the last decade, some regions have experienced

increased high ocean temperature anomalies. SST in the Northeast has warmed faster than 99% of the global ocean since 2004, and a peak temperature for the region in 2012 was part of a large "ocean heat wave" in the Northwest Atlantic that persisted for nearly 18 months.[20, 21] Projections indicate that the Northeast will continue to warm more quickly than other ocean regions through the end of the century.[22] In the Northwest, a resilient ridge of high pressure over the North American West Coast suppressed storm activity and mixing, which intensified heat in the upper ocean in a phenomenon known as "The Blob".[23] Anomalously warm waters persisted in the coastal waters of the Alaskan and Pacific Northwest from 2013 until 2015. Under a higher scenario (RCP8.5), SSTs are projected to increase by an additional 4.9°F (2.7°C) by 2100 (Figure 13.3), whereas for a lower scenario (RCP4.5) the SST increase would be 2.3°F (1.3°C).[24] In all U.S. coastal regions, the warming since 1901 is detectable compared to natural variability and attributable to anthropogenic forcing, according to an analysis of the CMIP5 models (Ch. 6: Temperature Change, Figure 6.5).

**Table 13.1.** Historical sea surface temperature trends (°C per century) and projected trends by 2080 (°C) for eight U.S. coastal regions and globally. Historical temperature trends are presented for the 1900–2016 and 1950–2016 periods with 95% confidence level, observed using the Extended Reconstructed Sea Surface Temperature version 4 (ERSSTv4).[19] Global and regional predictions are calculated for lower and higher scenarios (RCP4.5 and RCP8.5, respectively) with 80% spread of all the CMIP5 members compared to the 1976–2005 period.[151] The historical trends were analyzed for the latitude and longitude in the table, while the projected trends were analyzed for the California Current instead of the Northwest and Southwest separately and for the Bering Sea in Alaska (NOAA).

| Region | Latitude and Longitude | Historical Trend (°C/100 years) | | Projected Trend 2080 (relative to 1976–2005 climate) (°C) | |
|---|---|---|---|---|---|
| | | 1900–2016 | 1950–2016 | RCP4.5 | RCP8.5 |
| Global | | 0.70 ± 0.08 | 1.00 ± 0.11 | 1.3 ± 0.6 | 2.7 ± 0.7 |
| Alaska | 50°–66°N, 150°–170°W | 0.82 ± 0.26 | 1.22 ± 0.59 | 2.5 ± 0.6 | 3.7 ± 1.0 |
| Northwest (NW) | 40°–50°N, 120°–132°W | 0.64 ± 0.30 | 0.68 ± 0.70 | 1.7 ± 0.4 | 2.8 ± 0.6 |
| Southwest (SW) | 30°–40°N, 116°–126°W | 0.73 ± 0.33 | 1.02 ± 0.79 | | |
| Hawaii (HI) | 18°–24°N, 152°–162°W | 0.58 ± 0.19 | 0.46 ± 0.39 | 1.6 ± 0.4 | 2.8 ± 0.6 |
| Northeast (NE) | 36°–46°N, 64°–76°W | 0.63 ± 0.31 | 1.10 ± 0.71 | 2.0 ± 0.3 | 3.2 ± 0.6 |
| Southeast (SE) | 24°–34°N, 64°–80°W | 0.40 ± 0.18 | 0.13 ± 0.34 | 1.6 ± 0.3 | 2.7 ± 0.4 |
| Gulf of Mexico (GOM) | 20°–30°N, 80°–96°W | 0.52 ± 0.14 | 0.37 ± 0.27 | 1.6 ± 0.3 | 2.8 ± 0.3 |
| Caribbean | 10°–20°N, 66°–86°W | 0.76 ± 0.15 | 0.77 ± 0.32 | 1.5 ± 0.4 | 2.6 ± 0.3 |

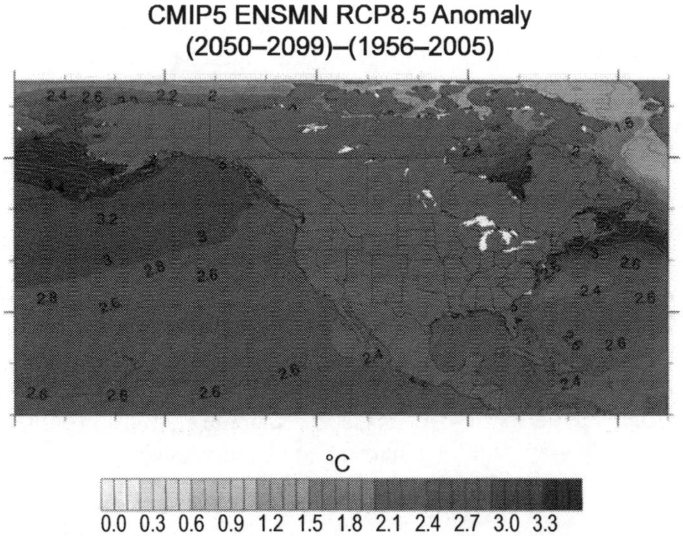

## CMIP5 ENSMN RCP8.5 Anomaly
## (2050–2099)–(1956–2005)

°C

0.0  0.3  0.6  0.9  1.2  1.5  1.8  2.1  2.4  2.7  3.0  3.3

**Figure 13.3:** Projected changes in sea surface temperature (°C) for the coastal United States under the higher scenario (RCP8.5). Projected anomalies for the 2050–2099 period are calculated using a comparison from the average sea surface temperatures over 1956–2005. Projected changes are examined using the Coupled Model Intercomparison Project Phase 5 (CMIP5) suite of model simulations. (Figure source: NOAA).

### 13.1.4 Ocean Heat Feedback

The residual heat not taken up by the oceans increases land surface temperatures (approximately 3%) and atmospheric temperatures (approximately 1%), and melts both land and sea ice (approximately 3%), leading to sea level rise (see Ch. 12: Sea Level Rise).[4, 6, 25] The meltwater from land and sea ice amplifies further subsurface ocean warming and ice shelf melting, primarily due to increased thermal stratification, which reduces the ocean's efficiency in transporting heat to deep waters.[4] Surface ocean stratification has increased by about 4% during the period 1971–2010[26] due to thermal heating and freshening from increased freshwater inputs (precipitation and evaporation changes and land and sea ice melting). The increase of ocean stratification will contribute to further feedback of ocean warming and, indirectly, mean sea level. In addition, increases in stratification are associated with suppression of tropical cyclone intensification,[27] retreat of the polar ice sheets,[28] and reductions of the convective mixing at higher latitudes that transports heat to the

deep ocean through the Atlantic Meridional Overturning Circulation.[29] Ocean heat uptake therefore represents an important feedback that will have a significant influence on future shifts in climate (see Ch. 2: Physical Drivers of Climate Change).

### 13.2 Ocean Circulation

#### 13.2.1 Atlantic Meridional Overturning Circulation

The Atlantic Meridional Overturning Circulation (AMOC) refers to the three-dimensional, time-dependent circulation of the Atlantic Ocean, which has been a high priority topic of study in recent decades. The AMOC plays an important role in climate through its transport of heat, freshwater, and carbon (e.g., Johns et al. 2011;[30] McDonagh et al. 2015;[31] Talley et al. 2016[32]). AMOC-associated poleward heat transport substantially contributes to North American and continental European climate (see Ch. 5: Circulation and Variability). The Gulf Stream, in contrast to other western boundary currents, is expected to slow down because of the weakening of the AMOC, which would impact the European climate.[33] Variability in the AMOC has been attributed to wind forcing on intra-annual time scales and to geostrophic forces on interannual to decadal timescales.[34] Increased freshwater fluxes from melting Arctic Sea and land ice can weaken open ocean convection and deep-water formation in the Labrador and Irminger Seas, which could weaken the AMOC (Ch. 11: Arctic Changes; also see Ch. 5, Section 5.2.3: North Atlantic Oscillation and Northern Annular Mode).[29, 33]

While one recent study has suggested that the AMOC has slowed since preindustrial times[29] and another suggested slowing on faster time scales,[35] there is at present insufficient observational evidence to support a finding of long term slowdown of AMOC strength over the 20th century[4] or within the last 50 years[34] as decadal ocean variability can obscure long-term trends. Some studies show long-term trends,[36, 37] but the combination of sparse data and large seasonal variability may also lead to incorrect interpretations (e.g., Kanzow et al. 2010[38]). Several recent high resolution modeling studies constrained with the limited existing observational data[39] and/or with reconstructed freshwater fluxes[40] suggest that the recently observed AMOC slowdown at 26°N (off the Florida coast) since 2004 (e.g., as described in Smeed et al. 2014[35]) is mainly due to natural variability, and that anthropogenic forcing has not yet caused a significant AMOC slowdown. In addition, direct observations of the AMOC in the South Atlantic fail to unambiguously demonstrate anthropogenic trends (e.g., Dong et al. 2015;[41] Garzoli et al. 2013[42]).

Under a higher scenario (RCP8.5) in CMIP5 simulations, it is very likely that the AMOC will weaken over the 21st century. The projected decline ranges from 12% to 54%,[43] with the range width reflecting substantial uncertainty in quantitative projections of AMOC behavior. In lower scenarios (like RCP4.5), CMIP5 mod-

els predict a 20% weakening of the AMOC during the first half of the 21st century and a stabilization and slight recovery after that.[44] The projected slowdown of the AMOC will be counteracted by the warming of the deep ocean (below 700 m [2,300 ft]), which will tend to strengthen the AMOC.[45] The situation is further complicated due to the known bias in coupled climate models related to the direction of the salinity transport in models versus observations, which is an indicator of AMOC stability (e.g., Drijhout et al. 2011;[46] Bryden et al. 2011;[47] Garzoli et al. 2013[42]). Some argue that coupled climate models should be corrected for this known bias and that AMOC variations could be even larger than the gradual decrease most models predict if the AMOC were to shut down completely and "flip states".[48] Any AMOC slowdown could result in less heat and $CO_2$ absorbed by the ocean from the atmosphere, which is a positive feedback to climate change (also see Ch. 2: Physical Drivers of Climate Change).[49, 50, 51]

### 13.2.2 Changes in Salinity Structure

As a response to warming, increased atmospheric moisture leads to stronger evaporation or precipitation in terrestrial and oceanic environments and melting of land and sea ice. Approximately 80% of precipitation/evaporation events occur over the ocean, leading to patterns of higher salt content or freshwater anomalies and changes in ocean circulation (see Ch. 2: Physical Drivers of Climate Change and Ch. 6: Temperature Change).[52] Over 1950–2010, average global amplification of the surface salinity pattern amounted to 5.3%; where fresh regions in the ocean became fresher and salty regions became saltier.[53] However, the long-term trends of these physical and chemical changes to the ocean are difficult to isolate from natural large-scale variability. In particular, ENSO displays particular salinity and precipitation/evaporation patterns that skew the trends. More research and data are necessary to better model changes to ocean salinity. Several models have shown a similar spatial structure of surface salinity changes, including general salinity increases in the subtropical gyres, a strong basin-wide salinity increase in the Atlantic Ocean, and reduced salinity in the western Pacific warm pools and the North Pacific subpolar regions.[52, 53] There is also a stronger distinction between the upper salty thermocline and fresh intermediate depth through the century. The regional changes in salinity to ocean basins will have an overall impact on ocean circulation and net primary production, leading to corresponding carbon export (see Ch. 2: Physical Drivers of Climate Change). In particular, the freshening of the Arctic Ocean due to melting of land and sea ice can lead to buoyancy changes which could slow down the AMOC (see Section 13.2.1).

### 13.2.3 Changes in Upwelling

Significant changes to ocean stratification and circulation can also be observed regionally, along the eastern ocean boundaries and at the equator. In these areas, wind-driven upwelling brings colder, nutrient- and carbon-rich water to the surface; this upwelled water is more efficient in heat and anthropogenic $CO_2$ uptake. There is some evidence that coastal upwelling in mid- to high-latitude eastern boundary regions has increased in intensity and/or frequency,[54] but in more tropical areas of the western Atlantic, such as in the Caribbean Sea, it has decreased between 1990 and 2010.[55, 56] This has led to a decrease in primary productivity in the southern Caribbean Sea.[55] Within the continental United States, the California Current is experiencing fewer (by about 23%–40%) but stronger upwelling events.[57, 58, 59] Stronger offshore upwelling combined with cross-shelf advection brings nutrients from the deeper ocean but also increased offshore transport.[60] The net nutrient load in the coastal regions is responsible for increased productivity and ecosystem function.

IPCC 2013 concluded that there is low confidence in the current understanding of how eastern upwelling systems will be altered under future climate change because of the obscuring role of multidecadal climate variability.[26] However, subsequent studies show that by 2100, upwelling is predicted to start earlier in the year, end later, and intensify in three of the four major eastern boundary upwelling systems (not in the California Current).[61] In the California Current, upwelling is projected to intensify in spring but weaken in summer, with changes emerging from the envelope of natural variability primarily in the second half of the 21st century.[62] Southern Ocean upwelling will intensify while the Atlantic equatorial upwelling systems will weaken.[57, 61] The intensification is attributed to the strengthening of regional coastal winds as observations already show,[58] and model projections under the higher scenario (RCP8.5) estimate wind intensifying near poleward boundaries (including northern California Current) and weakening near equatorward boundaries (including southern California Current) for the 21st century.[61, 63]

## 13.3 Ocean Acidification

### 13.3.1 General Background

In addition to causing changes in climate, increasing atmospheric levels of carbon dioxide ($CO_2$) from the burning of fossil fuels and other human activities, including changes in land use, have a direct effect on ocean carbonate chemistry that is termed ocean acidification.[64, 65] Surface ocean waters absorb part of the increasing $CO_2$ in the atmosphere, which causes a variety of chemical changes in seawater: an increase in the partial pressure of $CO_2$ (p$CO_2$,sw), dissolved inorganic carbon (DIC), and the concentration of hydrogen and bicarbonate ions and a decrease in the concentration of carbonate ions (Figure 13.4). In brief, $CO_2$ is an acid gas that combines with water to form carbonic acid, which then dissociates to hydrogen and bicarbonate ions. Increasing concentrations of seawater hydrogen ions result in a decrease of carbonate ions through their conversion to bicarbonate ions. The concentration of carbonate ions in seawater affects saturation states for calcium carbonate compounds, which many marine species use to build their shells and skeletons. Ocean acidity refers to the concentration of hydrogen ions in ocean seawater regardless of ocean pH, which is fundamentally basic (e.g., pH > 7). Ocean

surface waters have become 30% more acidic over the last 150 years as they have absorbed large amounts of $CO_2$ from the atmosphere,[66] and anthropogenically sourced $CO_2$ is gradually invading into oceanic deep waters. Since the preindustrial period, the oceans have absorbed approximately 29% of all $CO_2$ emitted to the atmosphere.[67] Oceans currently absorb about 26% of the human-caused $CO_2$ anthropogenically emitted into the atmosphere.[67]

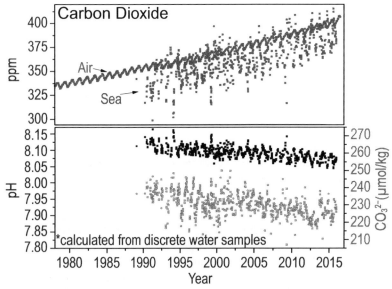

Figure 13.4: Trends in surface (< 50 m) ocean carbonate chemistry calculated from observations obtained at the Hawai'i Ocean Time-series (HOT) Program in the North Pacific over 1988–2015. The upper panel shows the linked increase in atmospheric (red points) and seawater (blue points) $CO_2$ concentrations. The bottom panel shows a decline in seawater pH (black points, primary y-axis) and carbonate ion concentration (green points, secondary y-axis). Ocean chemistry data were obtained from the Hawai'i Ocean Time-series Data Organization & Graphical System (HOT-DOGS, http://hahana.soest.hawaii.edu/hot/hot-dogs/index.html). (Figure source: NOAA).

### 13.3.2 Open Ocean Acidification

Surface waters in the open ocean experience changes in carbonate chemistry reflective of large-scale physical oceanic processes (see Ch. 2: Physical Drivers of Climate Change). These processes include both the global uptake of atmospheric $CO_2$ and the shoaling of naturally acidified subsurface waters due to vertical mixing and upwelling. In general, the rate of ocean acidification in open ocean surface waters at a decadal time-scale closely approximates the rate of atmospheric $CO_2$ increase.[68] Large, multidecadal phenomena such as the Atlantic Multidecadal Oscillation and Pacific Decadal Oscillation can add variability to the observed rate of change.[68]

### 13.3.3 Coastal Acidification

Coastal shelf and nearshore waters are influenced by the same processes as open ocean surface waters such as absorption of atmospheric $CO_2$ and upwelling, as well as a number of additional, local-level processes, including freshwater, nutrient, sulfur, and nitrogen inputs.[69, 70] Coastal acidification generally exhibits higher-frequency variability and short-term episodic events relative to open-ocean acidification.[71, 72, 73, 74] Upwelling is of particular importance in coastal waters, especially along the U.S. West Coast. Deep waters that shoal with upwelling are enriched in $CO_2$ due to uptake of anthropogenic atmospheric $CO_2$ when last in contact with the atmosphere, coupled with deep water respiration processes and lack of gas exchange with the atmosphere.[65, 75] Freshwater inputs to coastal waters change seawater chemistry in ways that make it more susceptible to acidification, largely by freshening ocean waters and contributing varying amounts of dissolved inorganic carbon (DIC), total alkalinity (TA), dissolved and particulate organic carbon, and nutrients from riverine and estuarine sources. Coastal waters of the East Coast and mid-Atlantic are far more influenced by freshwater inputs than are Pacific Coast waters.[76] Coastal waters can episodically experience riverine and glacial melt plumes that create conditions in which seawater can dissolve calcium carbonate structures.[77, 78] While these processes have persisted historically, climate-induced increases in glacial melt and high-intensity precipitation events can yield larger freshwater plumes than have occurred in the past. Nutrient runoff can increase coastal acidification by creating conditions that enhance biological respiration. In brief, nutrient loading typically promotes phytoplankton blooms, which, when they die, are consumed by bacteria. Bacteria respire $CO_2$ and thus bacterial blooms can result in acidification events whose intensity depends on local hydrographic conditions, including water column stratification and residence time.[72] Long-term changes in nutrient loading, precipitation, and/or ice melt may also impart long-term, secular changes in the magnitude of coastal acidification.

### 13.3.4 Latitudinal Variation

Ocean carbon chemistry is highly influenced by water temperature, largely because the solubility of $CO_2$ in seawater increases as water temperature declines. Thus, cold, high-latitude surface waters can retain more $CO_2$ than warm, lower-latitude surface waters.[76, 79] Because carbonate minerals also more readily dissolve in colder waters, these waters can more regularly become undersaturated with respect to calcium carbonate whereby mineral dissolution is energetically favored. This chemical state, often referred to as seawater being "corrosive" to calcium carbonate, is important when considering the ecological implications of ocean acidification as many species make structures such as shells and skeletons from calcium carbonate. Seawater conditions undersaturated with respect to calcium carbonate are common at depth, but currently and historically rare at the surface and near-surface.[80] Some high-latitude surface and near-surface waters now experience such corrosive conditions, which are rarely documented in low-latitude surface or near-surface systems. For example, corrosive conditions at a range of ocean depths have been documented in the Arctic and northeastern Pacific Oceans.[74, 79, 81, 82] Storm-induced upwelling could cause undersaturation in tropical areas in the future.[83] It is important to note that low-latitude waters are experiencing a greater absolute rate of change in calcium carbonate saturation state than higher latitudes, though these low-latitude waters are not approaching the undersaturated state except within nearshore or some benthic habitats.[84]

### 13.3.5 Paleo Evidence
Evidence suggests that the current rate of ocean acidification is the fastest in the last 66 million years (the K-Pg boundary) and possibly even the last 300 million years (when the first pelagic calcifiers evolved providing proxy information and also a strong carbonate buffer, characteristic of the modern ocean).[85, 86] The Paleo-Eocene Thermal Maximum (PETM; around 56 million years ago) is often referenced as the closest analogue to the present, although the overall rate of change in $CO_2$ conditions during that event (estimated between 0.6 and 1.1 GtC/year) was much lower than the current increase in atmospheric $CO_2$ of 10 GtC/year.[86, 87] The relatively slower rate of atmospheric $CO_2$ increase at the PETM likely led to relatively small changes in carbonate ion concentration in seawater compared with the contemporary acidification rate, due to the ability of rock weathering to buffer the change over the longer time period.[86] Some of the presumed acidification events in Earth's history have been linked to selective extinction events suggestive of how guilds of species may respond to the current acidification event.[85]

### 13.3.6 Projected Changes
Projections indicate that by the end of the century under higher scenarios, such as SRES A1FI or RCP8.5, open-ocean surface pH will decline from the current average level of 8.1 to a possible average of 7.8 (Figure 13.5).[1] When the entire ocean volume is considered under the same scenario, the volume of waters undersaturated with respect to calcium carbonate could expand from 76% in the 1990s to 91% in 2100, resulting in a shallowing of the saturation horizons—depths below which undersaturation occurs.[1, 88] Saturation horizons, which naturally vary among ocean basins, influence ocean carbon cycles and organisms with calcium carbonate structures, especially as they shoal into the zones where most biota lives.[81, 89] As discussed above, for a variety of reasons, not all ocean and coastal regions will experience acidification in the same way depending on other compounding factors. For instance, recent observational data from the Arctic Basin show that the Beaufort Sea became undersaturated, for part of the year, with respect to aragonite in 2001, while other continental shelf seas in the Arctic Basin are projected to do so closer to the middle of the century (e.g., the Chukchi Sea in about 2033 and Bering Sea in about 2062).[90] Deviation from the global average rate of acidification will be especially true in coastal and estuarine areas where the rate of acidification is influenced by other drivers than atmospheric $CO_2$, some of which are under the control of local management decisions (for example, nutrient pollution loads).

Surface pH in 2090s (RCP8.5, changes from 1990s)

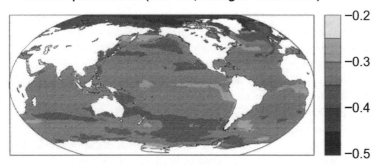

Figure 13.5: Predicted change in sea surface pH in 2090–2099 relative to 1990–1999 under the higher scenario (RCP8.5), based on the Community Earth System Models–Large Ensemble Experiments CMIP5 (Figure source: adapted from Bopp et al. 2013[24]).

## 13.4 Ocean Deoxygenation

### 13.4.1 General Background

Oxygen is essential to most life in the ocean, governing a host of biogeochemical and biological processes. Oxygen influences metabolic, physiological, reproductive, behavioral, and ecological processes, ultimately shaping the composition, diversity, abundance, and distribution of organisms from microbes to whales. Increasingly, climate-induced oxygen loss (deoxygenation) associated with ocean warming and reduced ventilation to deep waters has become evident locally, regionally, and globally. Deoxygenation can also be attributed to anthropogenic nutrient input, especially in the coastal regions, where the nutrients can lead to the proliferation of primary production and, consequently, enhanced drawdown of dissolved oxygen by microbes.[91] In addition, acidification (Section 13.2) can co-occur with deoxygenation as a result of warming-enhanced biological respiration.[92] As aerobic organisms respire, $O_2$ is consumed and $CO_2$ is produced. Understanding the combined effect of both low $O_2$ and low pH on marine ecosystems is an area of active research.[93] Warming also raises biological metabolic rates which, in combination with intensified coastal and estuarine stratification, exacerbates eutrophication-induced hypoxia. We now see earlier onset and longer periods of seasonal hypoxia in many eutrophic sites, most of which occur in areas that are also warming.[91]

### 13.4.2 Climate Drivers of Ocean Deoxygenation

Global ocean deoxygenation is a direct effect of warming. Ocean warming reduces the solubility of oxygen (that is, warmer water can hold less oxygen) and changes physical mixing (for example, upwelling and circulation) of oxygen in the oceans. The increased temperature of global oceans accounts for about 15% of current global oxygen loss,[94] although changes in temperature and oxygen are not uniform throughout the ocean.[15] Warming also exerts direct influence on thermal stratification and enhances salinity stratification through ice melt and climate change-associated precipitation effects. Intensified stratification leads to reduced ventilation (mixing of oxygen into the ocean interior) and accounts for up to 85% of global ocean oxygen loss.[94] Effects of ocean temperature change and stratification on oxygen loss are strongest in intermediate or mode waters at bathyal depths (in general, 200–3,000 m) and also nearshore and in the open ocean; these changes are especially evident in tropical and subtropical waters globally, in the Eastern Pacific,[95] and in the Southern Ocean.[94]

There are also other, less direct effects of global temperature increase. Warming on land reduces terrestrial plant water efficiency (through effects on stomata; see Ch. 8: Drought, Floods, and Wildfires, Key Message 3), leading to greater runoff, on average, into coastal zones (see Ch. 8: Drought, Floods, and Wildfires for other hydrological effects of warming) and further enhancing hypoxia potential because greater runoff can mean more nutrient transport (See Ch. 2: Physical Drivers of Climate Change).[96, 97] Estuaries, especially ones with minimal tidal mixing, are particularly vulnerable to oxygen-depleted dead zones from the enhanced runoff and stratification. Warming can induce dissociation of frozen methane in gas hydrates buried on continental margins, leading to further drawdown of oxygen through aerobic methane oxidation in the water column.[98] On eastern ocean boundaries, warming can enhance the land–sea temperature differential, causing increased upwelling due to higher winds with (a) greater nutrient input leading to production, sinking, decay, and biochemical drawdown of oxygen and (b) upwelling of naturally low-oxygen, high-$CO_2$ waters onto the upper slope and shelf environments.[58, 65] However, in the California Current, upwelling intensification has occurred only in the poleward regions (north of San Francisco), and the drivers may not be associated with land–sea temperature differences.[63] Taken together, the effects of warming are manifested as low-oxygen water in open oceans are being transported to and upwelled along coastal regions. These low-oxygen upwelled waters are then coupled with eutrophication-induced hypoxia, further reducing oxygen content in coastal areas.

Changes in precipitation, winds, circulation, airborne nutrients, and sea level can also contribute to ocean deoxygenation. Projected increases in precipitation in some regions will intensify stratification, reducing vertical mixing and ventilation, and intensify nutrient input to coastal waters through excess runoff, which leads to increased algal biomass and concurrent dissolved oxygen consumption via community respiration.[99] Coastal wetlands that might remove these nutrients before they reach the ocean may be lost through rising sea level, further exacerbating hypoxia.[97] Some observations of oxygen decline are linked to regional changes in circulation involving low-oxygen water masses. Enhanced fluxes of airborne iron and nitrogen are interacting with natural climate variability and contributing to fertilization, enhanced respiration, and oxygen loss in the tropical Pacific.[100]

### 13.4.3 Biogeochemical Feedbacks of Deoxygenation to Climate and Elemental Cycles

Climate patterns and ocean circulation have a large effect on global nitrogen and oxygen cycles, which in turn affect phosphorus and trace metal availability and generate feedbacks to the atmosphere and oceanic production. Global ocean productivity may be affected by climate-driven changes below the tropical and subtropical thermocline which control the volume of suboxic waters (< 5 micromolar $O_2$), and consequently the loss of fixed nitrogen through denitrification.[101, 102] The extent of suboxia in the open ocean also regulates the production of the greenhouse gas nitrous oxide ($N_2O$); as oxygen declines, greater $N_2O$ production may intensify global warming, as $N_2O$ is about 310 times more effective at trapping heat than $CO_2$ (see Ch. 2: Physical Drivers of Climate Change, Section 2.3.2).[103, 104] Production of hydrogen sulfide ($H_2S$, which is highly toxic) and intensified phosphorus recycling can occur at low oxygen levels.[105] Other feedbacks may emerge as oxygen minimum zone (OMZ) shoaling diminishes the depths of diurnal vertical migrations by fish and invertebrates, and as their huge biomass and associated oxygen consumption deplete oxygen.[106]

### 13.4.4 Past Trends

Over hundreds of millions of years, oxygen has varied dramatically in the atmosphere and ocean and has been linked to biodiversity gains and losses.[107, 108] Variation in oxygenation in the paleo record is very sensitive to climate—with clear links to temperature and often $CO_2$ variation.[109] OMZs expand and contract in synchrony with warming and cooling events, respectively.[110] Episodic climate events that involve rapid temperature increases over decades, followed by a cool period lasting a few hundred years, lead to major fluctuations in the intensity of Pacific and Indian Ocean OMZs (i.e., DO of < 20 $\mu$M). These events are associated with rapid variations in North Atlantic deep water formation.[111] Ocean oxygen fluctuates on glacial-interglacial timescales of thousands of years in the Eastern Pacific.[112, 113]

### 13.4.5 Modern Observations (last 50+ years)

Long-term oxygen records made over the last 50 years reflect oxygen declines in inland seas,[114, 115, 116] in estuaries,[117, 118] and in coastal waters.[119, 120, 121, 122] The number of coastal, eutrophication-induced hypoxic sites in the United States has grown dramatically over the past 40 years.[123] Over larger scales, global syntheses show hypoxic waters have expanded by 4.5 million km² at a depth of 200 m,[95] with widespread loss of oxygen in the Southern Ocean,[94] Western Pacific,[124] and North Atlantic.[125] Overall oxygen declines have been greater in coastal oceans than in the open ocean[126] and often greater inshore than offshore.[127] The emergence of a deoxygenation signal in regions with naturally high oxygen variability will unfold over longer time periods (20–50 years from now).[128]

### 13.4.6 Projected Changes

*Global Models*

Global models generally agree that ocean deoxygenation is occurring; this finding is also reflected in in situ observations from past 50 years. Compilations of 10 Earth System models predict a global average loss of **oxygen of −3.5% (higher scenario, RCP8.5) to −2.4% (lower scenario, RCP4.5) by 2100,** but much stronger losses regionally, and in intermediate and mode waters (Figure 13.6).[24] The North Pacific, North Atlantic, Southern Ocean, subtropical South Pacific, and South Indian Oceans all are expected to experience deoxygenation, with $O_2$ decreases of as much as 17% in the North Pacific by 2100 for the RCP8.5 pathway. However, the tropical Atlantic and tropical Indian Oceans show increasing $O_2$ concentrations. In the many areas where oxygen is declining, high natural variability makes it difficult to identify anthropogenically forced trends.[128]

## Projected Change in Dissolved Oxygen

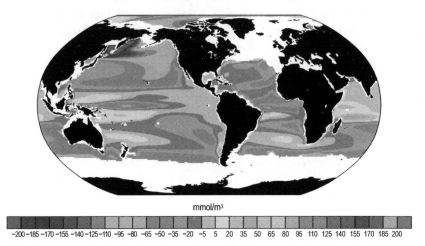

mmol/m³

-200 -185 -170 -155 -140 -125 -110 -95 -80 -65 -50 -35 -20 -5  5  20  35  50  65  80  95  110 125 140 155 170 185 200

Figure 13.6: Predicted change in dissolved oxygen on the $\sigma_\theta$ = 26.5 (average depth of approximately 290 m) potential density surface, between the 1981–2000 and 2081–2100, based on the Community Earth System Models–Large Ensemble Experiments (Figure source: redrawn from Long et al. 2016[128]).

*Regional Models*

Regional models are critical because many oxygen drivers are local, influenced by bathymetry, winds, circulation, and fresh water and nutrient inputs. Most eastern boundary upwelling areas are predicted to experience intensified upwelling to 2100,[61] although on the West Coast projections for increasing upwelling for the northern California Current occur only north of San Francisco (see Section 13.2.3).

Particularly notable for the western United States, variation in trade winds in the eastern Pacific Ocean can affect nutrient inputs, leading to centennial periods of oxygen decline or oxygen increase distinct from global oxygen decline.[129] Oxygen dynamics in the Eastern Tropical Pacific are highly sensitive to equatorial circulation changes.[130]

Regional modeling also shows that year-to-year variability in precipitation in the central United States affects the nitrate–N flux by the Mississippi River and the extent of hypox-

ia in the Gulf of Mexico.[131] A host of climate influences linked to warming and increased precipitation are predicted to lower dissolved oxygen in Chesapeake Bay.[132]

### 13.5 Other Coastal Changes

#### 13.5.1 Sea Level Rise

Sea level is an important variable that affects coastal ecosystems. Global sea level rose rapidly at the end of the last glaciation, as glaciers and the polar ice sheets thinned and melted at their fringes. On average around the globe, sea level is estimated to have risen at rates exceeding 2.5 mm/year between about 8,000 and 6,000 years before present. These rates steadily decreased to less than 2.0 mm/year through about 4,000 years ago and stabilized at less than 0.4 mm/year through the late 1800s. Global sea level rise has accelerated again within the last 100 years, and now averages about 1 to 2 mm/year.[133] See Chapter 12: Sea Level Rise for more thorough analysis of how sea level rise has already and will affect the U.S. coasts.

### 13.5.2 Wet and Dry Deposition

Dust transported from continental desert regions to the marine environment deposits nutrients such as iron, nitrogen, phosphorus, and trace metals that stimulate growth of phytoplankton and increase marine productivity.[134] U.S. continental and coastal regions experience large dust deposition fluxes originating from the Saharan desert to the East and from Central Asia and China to the Northwest.[135] Changes in drought frequency or intensity resulting from anthropogenically forced climate change, as well as other anthropogenic activities such as agricultural practices and land-use changes may play an important role in the future viability and strength of these dust sources (e.g., Mulitza et al. 2010[136]).

Additionally, oxidized nitrogen, released during high-temperature combustion over land, and reduced nitrogen, released from intensive agriculture, are emitted in high population areas in North America and are carried away and deposited through wet or dry deposition over coastal and open ocean ecosystems via local wind circulation. Wet deposition of pollutants produced in urban areas is known to play an important role in changes of ecosystem structure in coastal and open ocean systems through intermediate changes in the biogeochemistry, for instance in dissolved oxygen or various forms of carbon.[137]

### 13.5.3 Primary Productivity

Marine phytoplankton represent about half of the global net primary production (NPP) (approximately $50 \pm 28$ GtC /year), fixing atmospheric $CO_2$ into a bioavailable form for utilization by higher trophic levels (see also Ch. 2: Physical Drivers of Climate Change).[138, 139] As such, NPP represents a critical component in the role of the oceans in climate feedback. The effect of climate change on primary productivity varies across the coasts depending on local conditions. For instance, nutrients that stimulate phytoplankton growth are impacted by various climate conditions, such as increased stratification which limits the transport of nutrient-rich deep water to the surface, changes in circulation leading to variability in dry and wet deposition of nutrients to coasts, and altered precipitation/evaporation which changes runoff of nutrients from coastal communities. The effect of the multiple physical factors on NPP is complex and leads to model uncertainties.[140] There is considerable variation in model projections for NPP, from estimated decreases or no changes, to the potential increases by 2100.[141, 142, 143] Simulations from nine Earth system models projected total NPP in 2090 to decrease by 2%–16% and export production (that is, particulate flux to the deep ocean) to drop by 7%–18% as compared to 1990 (RCP8.5).[142] More information on phytoplankton species response and associated ecosystem dynamics is needed as any reduction of NPP and the associated export production would have an impact on carbon cycling and marine ecosystems.

### 13.5.4 Estuaries

Estuaries are critical ecosystems of biological, economic, and social importance in the United States. They are highly dynamic, influenced by the interactions of atmospheric, freshwater, terrestrial, oceanic, and benthic components. Of the 28 national estuarine research reserves in the United States and Puerto Rico, all are being impacted by climate change to varying levels.[144] In particular, sea level rise, saltwater intrusion, and the degree of freshwater discharge influence the forces and processes within these estuaries.[145] Sea level rise and subsidence are leading to drowning of existing salt marshes and/or subsequent changes in the relative area of the marsh plain, if adaptive upslope movement is impeded due to urbanization along shorelines. Several model scenarios indicate a decline in salt marsh habitat quality and an accelerated degradation as the rate of sea level rise increases in the latter half of the century.[146, 147] The increase in sea level as well as alterations to oceanic and atmospheric circulation can result in extreme wave conditions and storm surges, impacting coastal communities.[144] Additional climate change impacts to the physical and chemical estuarine processes include more extreme sea surface temperatures (higher highs and lower lows compared to the open ocean due to shallower depths and influence from land temperatures), changes in flow rates due to changes in precipitation, and potentially greater extents of salinity intrusion.

## TRACEABLE ACCOUNTS

### Key Finding 1

The world's oceans have absorbed about 93% of the excess heat caused by greenhouse gas warming since the mid-20th century, making them warmer and altering global and regional climate feedbacks. Ocean heat content has increased at all depths since the 1960s and surface waters have warmed by about 1.3° ± 0.1°F (0.7° ± 0.08°C) per century globally since 1900 to 2016. Under a higher scenario, a global increase in average sea surface temperature of 4.9° ± 1.3°F (2.7° ± 0.7°C) by 2100 is projected, with even higher changes in some U.S. coastal regions. (*Very high confidence*)

### Description of evidence base

The key finding and supporting text summarizes the evidence documented in climate science literature, including Rhein et al. 2013.[4] Oceanic warming has been documented in a variety of data sources, most notably the World Ocean Circulation Experiment (WOCE) (http://www.nodc.noaa.gov/woce/wdiu/) and Argo databases (https://www.nodc.noaa.gov/argo/) and Extended Reconstructed Sea Surface Temperature (ERSST) v4 (https://www.ncdc.noaa.gov/data-access/marineocean-data/extended-reconstructed-sea-surface-temperature-ersst-v4). There is particular confidence in calculated warming for the time period since 1971 due to increased spatial and depth coverage and the level of agreement among independent SST observations from satellites, surface drifters and ships, and independent studies using differing analyses, bias corrections, and data sources.[6, 7, 11] Other observations such as the increase in mean sea level rise (see Ch. 12: Sea Level Rise) and reduced Arctic/Antarctic ice sheets (see Ch. 11: Arctic Changes) further confirm the increase in thermal expansion. For the purpose of extending the selected time periods back from 1900 to 2016 and analyzing U.S. regional SSTs, the ERSST version 4 (ERSSTv4)[19] is used. For the centennial time scale changes over 1900–2016, warming trends in all regions are statistically significant with the 95% confidence level. U.S. regional SST warming is similar between calculations using ERSSTv4 in this report and those published by Belkin,[148] suggesting confidence in these findings. The projected increase in SST is based on evidence from the latest generation of Earth System Models (CMIP5).

### Major uncertainties

Uncertainties in the magnitude of ocean warming stem from the disparate measurements of ocean temperature over the last century. There is low uncertainty in warming trends of the upper ocean temperature from 0–700 m depth, whereas there is more uncertainty for deeper ocean depths of 700–2,000 m due to the short record of measurements from those areas. Data on warming trends at depths greater than 2,000 m are even more sparse. There are also uncertainties in the timing and reasons for particular decadal and interannual variations in ocean heat content and the contributions that different ocean basins play in the overall ocean heat uptake.

### Summary sentence or paragraph that integrates the above information

There is *very high confidence* in measurements that show increases in the ocean heat content and warming of the ocean, based on the agreement of different methods. However, long-term data in total ocean heat uptake in the deep ocean are sparse leading to limited knowledge of the transport of heat between and within ocean basins.

### Key Finding 2

The potential slowing of the Atlantic Meridional Overturning Circulation (AMOC; of which the Gulf Stream is one component)—as a result of increasing ocean heat content and freshwater driven buoyancy changes—could have dramatic climate feedbacks as the ocean absorbs less heat and $CO_2$ from the atmosphere.[51] This slowing would also affect the climates of North America and Europe. Any slowing documented to date cannot be directly tied to anthropogenic forcing primarily due to lack of adequate observational data and to challenges in modeling ocean circulation changes. Under a higher scenario (RCP8.5) in CMIP5 simulations, the AMOC weakens over the 21st century by 12% to 54% (*low confidence*).

## Description of evidence base

Investigations both through direct observations and models since 2013[4] have raised significant concerns about whether there is enough evidence to determine the existence of an overall slowdown in the AMOC. As a result, more robust international observational campaigns are underway currently to measure AMOC circulation. Direct observations have determined a statistically significant slowdown at the 95% confidence level at 26°N (off Florida; see Baringer et al. 2016[149]) but modeling studies constrained with observations cannot attribute this to anthropogenic forcing.[39] The study[29] which seemed to indicate broad-scale slowing has since been discounted due to its heavy reliance on sea surface temperature cooling as proxy for slowdown rather than actual direct observations. Since Rhein et al. 2013,[4] more observations have led to increased statistical confidence in the measurement of the AMOC. Current observation trends indicate the AMOC slowing down at the 95% confidence level at 26°N and 41°N but a more limited in situ estimate at 35°S, shows an increase in the AMOC.[35, 149] There is no one collection spot for AMOC-related data, but the U.S. Climate Variability and Predictability Program (US CLIVAR) has a U.S. AMOC priority focus area and a webpage with relevant data sites (https://usclivar.org/amoc/amoc-time-series).

The IPCC 2013 WG1 projections indicate a high likelihood of AMOC slowdown in the next 100 years, however overall understanding is limited by both a lack of direct observations (which is being remedied) and a lack of model skill to resolve deep ocean dynamics. As a result, this key finding was given an overall assessment of *low confidence*.

## Major uncertainties

As noted, uncertainty about the overall trend of the AMOC is high given opposing trends in northern and southern ocean time series observations. Although earth system models do indicate a high likelihood of AMOC slowdown as a result of a warming, climate projections are subject to high uncertainty. This uncertainty stems from intermodel differences, internal variability that is different in each model, uncertainty in stratification changes, and most importantly uncer-

tainty in both future freshwater input at high latitudes as well as the strength of the subpolar gyre circulation.

## Summary sentence or paragraph that integrates the above information

The increased focus on direct measurements of the AMOC should lead to a better understanding of 1) how it is changing and its variability by region, and 2) whether those changes are attributable to climate drivers through both model improvements and incorporation of those expanded observations into the models.

## Key Finding 3

The world's oceans are currently absorbing more than a quarter of the $CO_2$ emitted to the atmosphere annually from human activities, making them more acidic (*very high confidence*), with potential detrimental impacts to marine ecosystems. In particular, higher-latitude systems typically have a lower buffering capacity against pH change, exhibiting seasonally corrosive conditions sooner than low-latitude systems. Acidification is regionally increasing along U.S. coastal systems as a result of upwelling (for example, in the Pacific Northwest) (*high confidence*), changes in freshwater inputs (for example, in the Gulf of Maine) (*medium confidence*), and nutrient input (for example, in agricultural watersheds and urbanized estuaries) (*high confidence*). The rate of acidification is unparalleled in at least the past 66 million years (*medium confidence*). Under the higher scenario (RCP8.5), the global average surface ocean acidity is projected to increase by 100% to 150% (*high confidence*).

## Description of evidence base

Evidence on the magnitude of the ocean sink is obtained from multiple biogeochemical and transport ocean models and two observation-based estimates from the 1990s for the uptake of the anthropogenic $CO_2$. Estimates of the carbonate system (DIC and alkalinity) were based on multiple survey cruises in the global ocean in the 1990s (WOCE – now GO-SHIP, JGOFS). Coastal carbon and acidification surveys have been executed along the U.S. coastal large marine ecosystem since at least 2007, documenting significantly elevated pCO$_2$ and low pH conditions relative to oceanic waters. The data are available from the National Centers for Environmental Information (https://www. ncei.noaa.gov/). Other sources of biogeochemical bottle data can be found from HOT-DOGS ALOHA (http:// hahana.soest.hawaii.edu/hot/hot-dogs) or CCHDO (https://cchdo.ucsd.edu/). Rates of change associated with the Palaeocene-Eocene Thermal Maximum (PETM, 56 million years ago) were derived using stable carbon and oxygen isotope records preserved in the sedimentary record from the New Jersey shelf using time series analysis and carbon cycle–climate modelling. This evidence supports a carbon release during the onset of the PETM over no less than 4,000 years, yielding a maximum sustained carbon release rate of less than 1.1 GtC per year.[86] The projected increase in global surface ocean acidity is based on evidence from ten of the latest generation earth system models which include six distinct biogeochemical models that were included in the latest IPCC AR5 2013.

**Major uncertainties**

In 2014 the ocean sink was $2.6 \pm 0.5$ GtC (9.5 GtCO$_2$), equivalent to 26% of the total emissions attributed to fossil fuel use and land use changes.[67] Estimates of the PETM ocean acidification event evidenced in the geological record remains a matter of some debate within the community. Evidence for the 1.1 GtC per year cited by Zeebe et al.,[86] could be biased as a result of brief pulses of carbon input above average rates of emissions were they to transpire over timescales $\lesssim$ 40 years.

**Summary sentence or paragraph that integrates the above information**

There is *very high confidence* in evidence that the oceans absorb about a quarter of the carbon dioxide emitted in the atmosphere and hence become more acidic. The magnitude of the ocean carbon sink is known at a *high confidence* level because it is estimated using a series of disparate data sources and analysis methods, while the magnitude of the interannual variability is based only on model studies. There is *medium confidence* that the current rate of climate acidification is unprecedented in the past 66 million years. There is also *high confidence* that oceanic pH will continue to decrease.

**Key Finding 4**

Increasing sea surface temperatures, rising sea levels, and changing patterns of precipitation, winds, nutrients, and ocean circulation are contributing to overall declining oxygen concentrations at intermediate depths in various ocean locations and in many coastal areas. Over the last half century, major oxygen losses have occurred in inland seas, estuaries, and in the coastal and open ocean (*high confidence*). Ocean oxygen levels are projected to decrease by as much as 3.5% under the higher scenario (RCP8.5) by 2100 relative to preindustrial values (*high confidence*).

**Description of evidence base**

The key finding and supporting text summarizes the evidence documented in climate science literature including Rhein et al. 2013,[4] Bopp et al. 2013,[24] and Schmidtko et al. 2017.[150] Evidence arises from extensive global measurements of the WOCE after 1989 and individual profiles before that.[94] The first basin-wide dissolved oxygen surveys were performed in the 1920s.[150] The confidence level is based on globally integrated O$_2$ distributions in a variety of ocean models. Although the global mean exhibits low interannual variability, regional contrasts are large.

**Major uncertainties**

Uncertainties (as estimated from the intermodel spread) in the global mean are moderate mainly because ocean oxygen content exhibits low interannual variability when globally averaged. Uncertainties in long-term decreases of the global averaged oxygen concentration amount to 25% in the upper 1,000 m for the 1970–1992 period and 28% for the 1993–2003 period. Remaining uncertainties relate to regional variability driven by mesoscale eddies and intrinsic climate variability such as ENSO.

**Summary sentence or paragraph that integrates the above information**

Major ocean deoxygenation is taking place in bodies of water inland, at estuaries, and in the coastal and the open ocean (*high confidence*). Regionally, the phenomenon is exacerbated by local changes in weather, ocean circulation, and continental inputs to the oceans.

# REFERENCES

1. Gattuso, J.-P., A. Magnan, R. Billé, W.W.L. Cheung, E.L. Howes, F. Joos, D. Allemand, L. Bopp, S.R. Cooley, C.M. Eakin, O. Hoegh-Guldberg, R.P. Kelly, H.-O. Pörtner, A.D. Rogers, J.M. Baxter, D. Laffoley, D. Osborn, A. Rankovic, J. Rochette, U.R. Sumaila, S. Treyer, and C. Turley, 2015: Contrasting futures for ocean and society from different anthropogenic CO$_2$ emissions scenarios. *Science*, **349**, aac4722. http://dx.doi.org/10.1126/science.aac4722

2. Henson, S.A., J.L. Sarmiento, J.P. Dunne, L. Bopp, I. Lima, S.C. Doney, J. John, and C. Beaulieu, 2010: Detection of anthropogenic climate change in satellite records of ocean chlorophyll and productivity. *Biogeosciences*, **7**, 621-640. http://dx.doi.org/10.5194/bg-7-621-2010

3. Henson, S.A., C. Beaulieu, and R. Lampitt, 2016: Observing climate change trends in ocean biogeochemistry: When and where. *Global Change Biology*, **22**, 1561-1571. http://dx.doi.org/10.1111/gcb.13152

4. Rhein, M., S.R. Rintoul, S. Aoki, E. Campos, D. Chambers, R.A. Feely, S. Gulev, G.C. Johnson, S.A. Josey, A. Kostianoy, C. Mauritzen, D. Roemmich, L.D. Talley, and F. Wang, 2013: Observations: Ocean. *Climate Change 2013: The Physical Science Basis. Contribution of Working Group I to the Fifth Assessment Report of the Intergovernmental Panel on Climate Change*. Stocker, T.F., D. Qin, G.-K. Plattner, M. Tignor, S.K. Allen, J. Boschung, A. Nauels, Y. Xia, V. Bex, and P.M. Midgley, Eds. Cambridge University Press, Cambridge, United Kingdom and New York, NY, USA, 255–316. http://www.climatechange2013.org/report/full-report/

5. Rossby, C.-G., 1959: Current problems in meteorology. *The Atmosphere and the Sea in Motion*. Bolin, B., Ed. Rockefeller Institute Press, New York, 9-50.

6. Cheng, L., K.E. Trenberth, J. Fasullo, T. Boyer, J. Abraham, and J. Zhu, 2017: Improved estimates of ocean heat content from 1960 to 2015. *Science Advances*, **3**, e1601545. http://dx.doi.org/10.1126/sciadv.1601545

7. Levitus, S., J.I. Antonov, T.P. Boyer, O.K. Baranova, H.E. Garcia, R.A. Locarnini, A.V. Mishonov, J.R. Reagan, D. Seidov, E.S. Yarosh, and M.M. Zweng, 2012: World ocean heat content and thermosteric sea level change (0–2000 m), 1955–2010. *Geophysical Research Letters*, **39**, L10603. http://dx.doi.org/10.1029/2012GL051106

8. Abraham, J.P., M. Baringer, N.L. Bindoff, T. Boyer, L.J. Cheng, J.A. Church, J.L. Conroy, C.M. Domingues, J.T. Fasullo, J. Gilson, G. Goni, S.A. Good, J.M. Gorman, V. Gouretski, M. Ishii, G.C. Johnson, S. Kizu, J.M. Lyman, A.M. Macdonald, W.J. Minkowycz, S.E. Moffitt, M.D. Palmer, A.R. Piola, F. Reseghetti, K. Schuckmann, K.E. Trenberth, I. Velicogna, and J.K. Willis, 2013: A review of global ocean temperature observations: Implications for ocean heat content estimates and climate change. *Reviews of Geophysics*, **51**, 450-483. http://dx.doi.org/10.1002/rog.20022

9. Lee, S.-K., W. Park, M.O. Baringer, A.L. Gordon, B. Huber, and Y. Liu, 2015: Pacific origin of the abrupt increase in Indian Ocean heat content during the warming hiatus. *Nature Geoscience*, **8**, 445-449. http://dx.doi.org/10.1038/ngeo2438

10. Purkey, S.G. and G.C. Johnson, 2010: Warming of global abyssal and deep Southern Ocean waters between the 1990s and 2000s: Contributions to global heat and sea level rise budgets. *Journal of Climate*, **23**, 6336-6351. http://dx.doi.org/10.1175/2010JCLI3682.1

11. Llovel, W., J.K. Willis, F.W. Landerer, and I. Fukumori, 2014: Deep-ocean contribution to sea level and energy budget not detectable over the past decade. *Nature Climate Change*, **4**, 1031-1035. http://dx.doi.org/10.1038/nclimate2387

12. Boyer, T., C.M. Domingues, S.A. Good, G.C. Johnson, J.M. Lyman, M. Ishii, V. Gouretski, J.K. Willis, J. Antonov, S. Wijffels, J.A. Church, R. Cowley, and N.L. Bindoff, 2016: Sensitivity of global upper-ocean heat content estimates to mapping methods, XBT bias corrections, and baseline climatologies. *Journal of Climate*, **29**, 4817-4842. http://dx.doi.org/10.1175/jcli-d-15-0801.1

13. Trenberth, K.E., J.T. Fasullo, and M.A. Balmaseda, 2014: Earth's energy imbalance. *Journal of Climate*, **27**, 3129-3144. http://dx.doi.org/10.1175/jcli-d-13-00294.1

14. Steinman, B.A., M.E. Mann, and S.K. Miller, 2015: Atlantic and Pacific Multidecadal Oscillations and Northern Hemisphere temperatures. *Science*, **347**, 988-991. http://dx.doi.org/10.1126/science.1257856

15. Roemmich, D., J. Church, J. Gilson, D. Monselesan, P. Sutton, and S. Wijffels, 2015: Unabated planetary warming and its ocean structure since 2006. *Nature Climate Change*, **5**, 240-245. http://dx.doi.org/10.1038/nclimate2513

16. Kosaka, Y. and S.-P. Xie, 2013: Recent global-warming hiatus tied to equatorial Pacific surface cooling. *Nature*, **501**, 403-407. http://dx.doi.org/10.1038/nature12534

17. Yan, X.-H., T. Boyer, K. Trenberth, T.R. Karl, S.-P. Xie, V. Nieves, K.-K. Tung, and D. Roemmich, 2016: The global warming hiatus: Slowdown or redistribution? *Earth's Future*, **4**, 472-482. http://dx.doi.org/10.1002/2016EF000417

18. Matthews, J.B.R., 2013: Comparing historical and modern methods of sea surface temperature measurement – Part 1: Review of methods, field comparisons and dataset adjustments. *Ocean Science*, **9**, 683-694. http://dx.doi.org/10.5194/os-9-683-2013

19. Huang, B., V.F. Banzon, E. Freeman, J. Lawrimore, W. Liu, T.C. Peterson, T.M. Smith, P.W. Thorne, S.D. Woodruff, and H.-M. Zhang, 2015: Extended Reconstructed Sea Surface Temperature Version 4 (ERSST. v4). Part I: Upgrades and intercomparisons. *Journal of Climate*, **28**, 911-930. http://dx.doi.org/10.1175/JCLI-D-14-00006.1

20. Pershing, A.J., M.A. Alexander, C.M. Hernandez, L.A. Kerr, A. Le Bris, K.E. Mills, J.A. Nye, N.R. Record, H.A. Scannell, J.D. Scott, G.D. Sherwood, and A.C. Thomas, 2015: Slow adaptation in the face of rapid warming leads to collapse of the Gulf of Maine cod fishery. *Science*, **350**, 809-812. http://dx.doi.org/10.1126/science.aac9819

21. Mills, K.E., A.J. Pershing, C.J. Brown, Y. Chen, F.-S. Chiang, D.S. Holland, S. Lehuta, J.A. Nye, J.C. Sun, A.C. Thomas, and R.A. Wahle, 2013: Fisheries management in a changing climate: Lessons from the 2012 ocean heat wave in the Northwest Atlantic. *Oceanography*, **26 (2)**, 191–195. http://dx.doi.org/10.5670/oceanog.2013.27

22. Saba, V.S., S.M. Griffies, W.G. Anderson, M. Winton, M.A. Alexander, T.L. Delworth, J.A. Hare, M.J. Harrison, A. Rosati, G.A. Vecchi, and R. Zhang, 2016: Enhanced warming of the Northwest Atlantic Ocean under climate change. *Journal of Geophysical Research Oceans*, **121**, 118-132. http://dx.doi.org/10.1002/2015JC011346

23. Bond, N.A., M.F. Cronin, H. Freeland, and N. Mantua, 2015: Causes and impacts of the 2014 warm anomaly in the NE Pacific. *Geophysical Research Letters*, **42**, 3414-3420. http://dx.doi.org/10.1002/2015GL063306

24. Bopp, L., L. Resplandy, J.C. Orr, S.C. Doney, J.P. Dunne, M. Gehlen, P. Halloran, C. Heinze, T. Ilyina, R. Séférian, J. Tjiputra, and M. Vichi, 2013: Multiple stressors of ocean ecosystems in the 21st century: Projections with CMIP5 models. *Biogeosciences*, **10**, 6225-6245. http://dx.doi.org/10.5194/bg-10-6225-2013

25. Nieves, V., J.K. Willis, and W.C. Patzert, 2015: Recent hiatus caused by decadal shift in Indo-Pacific heating. *Science*, **349**, 532-535. http://dx.doi.org/10.1126/science.aaa4521

26. Ciais, P., C. Sabine, G. Bala, L. Bopp, V. Brovkin, J. Canadell, A. Chhabra, R. DeFries, J. Galloway, M. Heimann, C. Jones, C. Le Quéré, R.B. Myneni, S. Piao, and P. Thornton, 2013: Carbon and other biogeochemical cycles. *Climate Change 2013: The Physical Science Basis. Contribution of Working Group I to the Fifth Assessment Report of the Intergovernmental Panel on Climate Change*. Stocker, T.F., D. Qin, G.-K. Plattner, M. Tignor, S.K. Allen, J. Boschung, A. Nauels, Y. Xia, V. Bex, and P.M. Midgley, Eds. Cambridge University Press, Cambridge, United Kingdom and New York, NY, USA, 465–570. http://www.climatechange2013.org/report/full-report/

27. Mei, W., S.-P. Xie, F. Primeau, J.C. McWilliams, and C. Pasquero, 2015: Northwestern Pacific typhoon intensity controlled by changes in ocean temperatures. *Science Advances*, **1**, e1500014. http://dx.doi.org/10.1126/sciadv.1500014

28. Straneo, F. and P. Heimbach, 2013: North Atlantic warming and the retreat of Greenland's outlet glaciers. *Nature*, **504**, 36-43. http://dx.doi.org/10.1038/nature12854

29. Rahmstorf, S., J.E. Box, G. Feulner, M.E. Mann, A. Robinson, S. Rutherford, and E.J. Schaffernicht, 2015: Exceptional twentieth-century slowdown in Atlantic Ocean overturning circulation. *Nature Climate Change*, **5**, 475-480. http://dx.doi.org/10.1038/nclimate2554

30. Johns, W.E., M.O. Baringer, L.M. Beal, S.A. Cunningham, T. Kanzow, H.L. Bryden, J.J.M. Hirschi, J. Marotzke, C.S. Meinen, B. Shaw, and R. Curry, 2011: Continuous, array-based estimates of Atlantic ocean heat transport at 26.5°N. *Journal of Climate*, **24**, 2429-2449. http://dx.doi.org/10.1175/2010jcli3997.1

31. McDonagh, E.L., B.A. King, H.L. Bryden, P. Courtois, Z. Szuts, M. Baringer, S.A. Cunningham, C. Atkinson, and G. McCarthy, 2015: Continuous estimate of Atlantic oceanic freshwater flux at 26.5°N. *Journal of Climate*, **28**, 8888-8906. http://dx.doi.org/10.1175/jcli-d-14-00519.1

32. Talley, L.D., R.A. Feely, B.M. Sloyan, R. Wanninkhof, M.O. Baringer, J.L. Bullister, C.A. Carlson, S.C. Doney, R.A. Fine, E. Firing, N. Gruber, D.A. Hansell, M. Ishii, G.C. Johnson, K. Katsumata, R.M. Key, M. Kramp, C. Langdon, A.M. Macdonald, J.T. Mathis, E.L. McDonagh, S. Mecking, F.J. Millero, C.W. Mordy, T. Nakano, C.L. Sabine, W.M. Smethie, J.H. Swift, T. Tanhua, A.M. Thurnherr, M.J. Warner, and J.-Z. Zhang, 2016: Changes in ocean heat, carbon content, and ventilation: A review of the first decade of GO-SHIP global repeat hydrography. *Annual Review of Marine Science*, **8**, 185-215. http://dx.doi.org/10.1146/annurev-marine-052915-100829

33. Yang, H., G. Lohmann, W. Wei, M. Dima, M. Ionita, and J. Liu, 2016: Intensification and poleward shift of subtropical western boundary currents in a warming climate. *Journal of Geophysical Research Oceans*, **121**, 4928-4945. http://dx.doi.org/10.1002/2015JC011513

34. Buckley, M.W. and J. Marshall, 2016: Observations, inferences, and mechanisms of the Atlantic Meridional Overturning Circulation: A review. *Reviews of Geophysics*, **54**, 5-63. http://dx.doi.org/10.1002/2015RG000493

35. Smeed, D.A., G.D. McCarthy, S.A. Cunningham, E. Frajka-Williams, D. Rayner, W.E. Johns, C.S. Meinen, M.O. Baringer, B.I. Moat, A. Duchez, and H.L. Bryden, 2014: Observed decline of the Atlantic meridional overturning circulation 2004–2012. *Ocean Science*, **10**, 29-38. http://dx.doi.org/10.5194/os-10-29-2014

36. Longworth, H.R., H.L. Bryden, and M.O. Baringer, 2011: Historical variability in Atlantic meridional baroclinic transport at 26.5°N from boundary dynamic height observations. *Deep Sea Research Part II: Topical Studies in Oceanography*, **58**, 1754-1767. http://dx.doi.org/10.1016/j.dsr2.2010.10.057

37. Bryden, H.L., H.R. Longworth, and S.A. Cunningham, 2005: Slowing of the Atlantic meridional overturning circulation at 25°N. *Nature*, **438**, 655-657. http://dx.doi.org/10.1038/nature04385

38. Kanzow, T., S.A. Cunningham, W.E. Johns, J.J.-M. Hirschi, J. Marotzke, M.O. Baringer, C.S. Meinen, M.P. Chidichimo, C. Atkinson, L.M. Beal, H.L. Bryden, and J. Collins, 2010: Seasonal variability of the Atlantic meridional overturning circulation at 26.5°N. *Journal of Climate*, **23**, 5678-5698. http://dx.doi.org/10.1175/2010jcli3389.1

39. Jackson, L.C., K.A. Peterson, C.D. Roberts, and R.A. Wood, 2016: Recent slowing of Atlantic overturning circulation as a recovery from earlier strengthening. *Nature Geoscience*, **9**, 518-522. http://dx.doi.org/10.1038/ngeo2715

40. Böning, C.W., E. Behrens, A. Biastoch, K. Getzlaff, and J.L. Bamber, 2016: Emerging impact of Greenland meltwater on deepwater formation in the North Atlantic Ocean. *Nature Geoscience*, **9**, 523-527. http://dx.doi.org/10.1038/ngeo2740

41. Dong, S., G. Goni, and F. Bringas, 2015: Temporal variability of the South Atlantic Meridional Overturning Circulation between 20°S and 35°S. *Geophysical Research Letters*, **42**, 7655-7662. http://dx.doi.org/10.1002/2015GL065603

42. Garzoli, S.L., M.O. Baringer, S. Dong, R.C. Perez, and Q. Yao, 2013: South Atlantic meridional fluxes. *Deep Sea Research Part I: Oceanographic Research Papers*, **71**, 21-32. http://dx.doi.org/10.1016/j.dsr.2012.09.003

43. Collins, M., R. Knutti, J. Arblaster, J.-L. Dufresne, T. Fichefet, P. Friedlingstein, X. Gao, W.J. Gutowski, T. Johns, G. Krinner, M. Shongwe, C. Tebaldi, A.J. Weaver, and M. Wehner, 2013: Long-term climate change: Projections, commitments and irreversibility. *Climate Change 2013: The Physical Science Basis. Contribution of Working Group I to the Fifth Assessment Report of the Intergovernmental Panel on Climate Change*. Stocker, T.F., D. Qin, G.-K. Plattner, M. Tignor, S.K. Allen, J. Boschung, A. Nauels, Y. Xia, V. Bex, and P.M. Midgley, Eds. Cambridge University Press, Cambridge, United Kingdom and New York, NY, USA, 1029–1136. http://www.climatechange2013.org/report/full-report/

44. Cheng, W., J.C.H. Chiang, and D. Zhang, 2013: Atlantic Meridional Overturning Circulation (AMOC) in CMIP5 models: RCP and historical simulations. *Journal of Climate*, **26**, 7187-7197. http://dx.doi.org/10.1175/jcli-d-12-00496.1

45. Patara, L. and C.W. Böning, 2014: Abyssal ocean warming around Antarctica strengthens the Atlantic overturning circulation. *Geophysical Research Letters*, **41**, 3972-3978. http://dx.doi.org/10.1002/2014GL059923

46. Drijfhout, S.S., S.L. Weber, and E. van der Swaluw, 2011: The stability of the MOC as diagnosed from model projections for pre-industrial, present and future climates. *Climate Dynamics*, **37**, 1575-1586. http://dx.doi.org/10.1007/s00382-010-0930-z

47. Bryden, H.L., B.A. King, and G.D. McCarthy, 2011: South Atlantic overturning circulation at 24°S. *Journal of Marine Research*, **69**, 38-55. http://dx.doi.org/10.1357/002224011798147633

48. Liu, W., S.-P. Xie, Z. Liu, and J. Zhu, 2017: Overlooked possibility of a collapsed Atlantic Meridional Overturning Circulation in warming climate. *Science Advances*, **3**, e1601666. http://dx.doi.org/10.1126/sciadv.1601666

49. Zickfeld, K., M. Eby, and A.J. Weaver, 2008: Carbon-cycle feedbacks of changes in the Atlantic meridional overturning circulation under future atmospheric CO2. *Global Biogeochemical Cycles*, **22**, GB3024. http://dx.doi.org/10.1029/2007GB003118

50. Halloran, P.R., B.B.B. Booth, C.D. Jones, F.H. Lambert, D.J. McNeall, I.J. Totterdell, and C. Völker, 2015: The mechanisms of North Atlantic $CO_2$ uptake in a large Earth System Model ensemble. *Biogeosciences*, **12**, 4497-4508. http://dx.doi.org/10.5194/bg-12-4497-2015

51. Romanou, A., J. Marshall, M. Kelley, and J. Scott, 2017: Role of the ocean's AMOC in setting the uptake efficiency of transient tracers. *Geophysical Research Letters*, **44**, 5590-5598. http://dx.doi.org/10.1002/2017GL072972

52. Durack, P.J. and S.E. Wijffels, 2010: Fifty-year trends in global ocean salinities and their relationship to broad-scale warming. *Journal of Climate*, **23**, 4342-4362. http://dx.doi.org/10.1175/2010jcli3377.1

53. Skliris, N., R. Marsh, S.A. Josey, S.A. Good, C. Liu, and R.P. Allan, 2014: Salinity changes in the World Ocean since 1950 in relation to changing surface freshwater fluxes. *Climate Dynamics*, **43**, 709-736. http://dx.doi.org/10.1007/s00382-014-2131-7

54. García-Reyes, M., W.J. Sydeman, D.S. Schoeman, R.R. Rykaczewski, B.A. Black, A.J. Smit, and S.J. Bograd, 2015: Under pressure: Climate change, upwelling, and eastern boundary upwelling ecosystems. *Frontiers in Marine Science*, **2**, Art. 109. http://dx.doi.org/10.3389/fmars.2015.00109

55. Taylor, G.T., F.E. Muller-Karger, R.C. Thunell, M.I. Scranton, Y. Astor, R. Varela, L.T. Ghinaglia, L. Lorenzoni, K.A. Fanning, S. Hameed, and O. Doherty, 2012: Ecosystem responses in the southern Caribbean Sea to global climate change. *Proceedings of the National Academy of Sciences*, **109**, 19315-19320. http://dx.doi.org/10.1073/pnas.1207514109

56. Astor, Y.M., L. Lorenzoni, R. Thunell, R. Varela, F. Muller-Karger, L. Troccoli, G.T. Taylor, M.I. Scranton, E. Tappa, and D. Rueda, 2013: Interannual variability in sea surface temperature and $fCO_2$ changes in the Cariaco Basin. *Deep Sea Research Part II: Topical Studies in Oceanography*, **93**, 33-43. http://dx.doi.org/10.1016/j.dsr2.2013.01.002

57. Hoegh-Guldberg, O., R. Cai, E.S. Poloczanska, P.G. Brewer, S. Sundby, K. Hilmi, V.J. Fabry, and S. Jung, 2014: The Ocean—Supplementary material. *Climate Change 2014: Impacts, Adaptation, and Vulnerability. Part B: Regional Aspects. Contribution of Working Group II to the Fifth Assessment Report of the Intergovernmental Panel of Climate Change*. Barros, V.R., C.B. Field, D.J. Dokken, M.D. Mastrandrea, K.J. Mach, T.E. Bilir, M. Chatterjee, K.L. Ebi, Y.O. Estrada, R.C. Genova, B. Girma, E.S. Kissel, A.N. Levy, S. MacCracken, P.R. Mastrandrea, and L.L. White, Eds. Cambridge University Press, Cambridge, United Kingdom and New York, NY, USA, 1655-1731. http://ipcc.ch/pdf/assessment-report/ar5/wg2/supplementary/WGIIAR5-Chap30_OLSM.pdf

58. Sydeman, W.J., M. García-Reyes, D.S. Schoeman, R.R. Rykaczewski, S.A. Thompson, B.A. Black, and S.J. Bograd, 2014: Climate change and wind intensification in coastal upwelling ecosystems. *Science*, **345**, 77-80. http://dx.doi.org/10.1126/science.1251635

59. Jacox, M.G., A.M. Moore, C.A. Edwards, and J. Fiechter, 2014: Spatially resolved upwelling in the California Current System and its connections to climate variability. *Geophysical Research Letters*, **41**, 3189-3196. http://dx.doi.org/10.1002/2014GL059589

60. Bakun, A., B.A. Black, S.J. Bograd, M. García-Reyes, A.J. Miller, R.R. Rykaczewski, and W.J. Sydeman, 2015: Anticipated effects of climate change on coastal upwelling ecosystems. *Current Climate Change Reports*, **1**, 85-93. http://dx.doi.org/10.1007/s40641-015-0008-4

61. Wang, D., T.C. Gouhier, B.A. Menge, and A.R. Ganguly, 2015: Intensification and spatial homogenization of coastal upwelling under climate change. *Nature*, **518**, 390-394. http://dx.doi.org/10.1038/nature14235

62. Brady, R.X., M.A. Alexander, N.S. Lovenduski, and R.R. Rykaczewski, 2017: Emergent anthropogenic trends in California Current upwelling. *Geophysical Research Letters*, **44**, 5044-5052. http://dx.doi.org/10.1002/2017GL072945

63. Rykaczewski, R.R., J.P. Dunne, W.J. Sydeman, M. García-Reyes, B.A. Black, and S.J. Bograd, 2015: Poleward displacement of coastal upwelling-favorable winds in the ocean's eastern boundary currents through the 21st century. *Geophysical Research Letters*, **42**, 6424-6431. http://dx.doi.org/10.1002/2015GL064694

64. Orr, J.C., V.J. Fabry, O. Aumont, L. Bopp, S.C. Doney, R.A. Feely, A. Gnanadesikan, N. Gruber, A. Ishida, F. Joos, R.M. Key, K. Lindsay, E. Maier-Reimer, R. Matear, P. Monfray, A. Mouchet, R.G. Najjar, G.-K. Plattner, K.B. Rodgers, C.L. Sabine, J.L. Sarmiento, R. Schlitzer, R.D. Slater, I.J. Totterdell, M.-F. Weirig, Y. Yamanaka, and A. Yool, 2005: Anthropogenic ocean acidification over the twenty-first century and its impact on calcifying organisms. *Nature*, **437**, 681-686. http://dx.doi.org/10.1038/nature04095

65. Feely, R.A., S.C. Doney, and S.R. Cooley, 2009: Ocean acidification: Present conditions and future changes in a high-$CO_2$ world. *Oceanography*, **22**, 36-47. http://dx.doi.org/10.5670/oceanog.2009.95

66. Feely, R.A., C.L. Sabine, K. Lee, W. Berelson, J. Kleypas, V.J. Fabry, and F.J. Millero, 2004: Impact of anthropogenic $CO_2$ on the $CaCO_3$ system in the oceans. *Science*, **305**, 362-366. http://dx.doi.org/10.1126/science.1097329

67. Le Quéré, C., R.M. Andrew, J.G. Canadell, S. Sitch, J.I. Korsbakken, G.P. Peters, A.C. Manning, T.A. Boden, P.P. Tans, R.A. Houghton, R.F. Keeling, S. Alin, O.D. Andrews, P. Anthoni, L. Barbero, L. Bopp, F. Chevallier, L.P. Chini, P. Ciais, K. Currie, C. Delire, S.C. Doney, P. Friedlingstein, T. Gkritzalis, I. Harris, J. Hauck, V. Haverd, M. Hoppema, K. Klein Goldewijk, A.K. Jain, E. Kato, A. Körtzinger, P. Landschützer, N. Lefèvre, A. Lenton, S. Lienert, D. Lombardozzi, J.R. Melton, N. Metzl, F. Millero, P.M.S. Monteiro, D.R. Munro, J.E.M.S. Nabel, S.I. Nakaoka, K. O'Brien, A. Olsen, A.M. Omar, T. Ono, D. Pierrot, B. Poulter, C. Rödenbeck, J. Salisbury, U. Schuster, J. Schwinger, R. Séférian, I. Skjelvan, B.D. Stocker, A.J. Sutton, T. Takahashi, H. Tian, B. Tilbrook, I.T. van der Laan-Luijkx, G.R. van der Werf, N. Viovy, A.P. Walker, A.J. Wiltshire, and S. Zaehle, 2016: Global carbon budget 2016. *Earth System Science Data*, **8**, 605-649. http://dx.doi.org/10.5194/essd-8-605-2016

68. Bates, N.R., Y.M. Astor, M.J. Church, K. Currie, J.E. Dore, M. González-Dávila, L. Lorenzoni, F. Muller-Karger, J. Olafsson, and J.M. Santana-Casiano, 2014: A time-series view of changing ocean chemistry due to ocean uptake of anthropogenic CO2 and ocean acidification. *Oceanography*, **27**, 126–141. http://dx.doi.org/10.5670/oceanog.2014.16

69. Duarte, C.M., I.E. Hendriks, T.S. Moore, Y.S. Olsen, A. Steckbauer, L. Ramajo, J. Carstensen, J.A. Trotter, and M. McCulloch, 2013: Is ocean acidification an open-ocean syndrome? Understanding anthropogenic impacts on seawater pH. *Estuaries and Coasts*, **36**, 221-236. http://dx.doi.org/10.1007/s12237-013-9594-3

70. Doney, S.C., N. Mahowald, I. Lima, R.A. Feely, F.T. Mackenzie, J.F. Lamarque, and P.J. Rasch, 2007: Impact of anthropogenic atmospheric nitrogen and sulfur deposition on ocean acidification and the inorganic carbon system. *Proc Natl Acad Sci U S A*, **104**, 14580-5. http://dx.doi.org/10.1073/pnas.0702218104

71. Borges, A.V. and N. Gypens, 2010: Carbonate chemistry in the coastal zone responds more strongly to eutrophication than ocean acidification. *Limnology and Oceanography*, **55**, 346-353. http://dx.doi.org/10.4319/lo.2010.55.1.0346

72. Waldbusser, G.G. and J.E. Salisbury, 2014: Ocean acidification in the coastal zone from an organism's perspective: Multiple system parameters, frequency domains, and habitats. *Annual Review of Marine Science*, **6**, 221-247. http://dx.doi.org/10.1146/annurev-marine-121211-172238

73. Hendriks, I.E., C.M. Duarte, Y.S. Olsen, A. Steckbauer, L. Ramajo, T.S. Moore, J.A. Trotter, and M. McCulloch, 2015: Biological mechanisms supporting adaptation to ocean acidification in coastal ecosystems. *Estuarine, Coastal and Shelf Science*, **152**, A1-A8. http://dx.doi.org/10.1016/j.ecss.2014.07.019

74. Sutton, A.J., C.L. Sabine, R.A. Feely, W.J. Cai, M.F. Cronin, M.J. McPhaden, J.M. Morell, J.A. Newton, J.H. Noh, S.R. Ólafsdóttir, J.E. Salisbury, U. Send, D.C. Vandemark, and R.A. Weller, 2016: Using present-day observations to detect when anthropogenic change forces surface ocean carbonate chemistry outside preindustrial bounds. *Biogeosciences*, **13**, 5065-5083. http://dx.doi.org/10.5194/bg-13-5065-2016

75. Harris, K.E., M.D. DeGrandpre, and B. Hales, 2013: Aragonite saturation state dynamics in a coastal upwelling zone. *Geophysical Research Letters*, **40**, 2720-2725. http://dx.doi.org/10.1002/grl.50460

76. Gledhill, D.K., M.M. White, J. Salisbury, H. Thomas, I. Mlsna, M. Liebman, B. Mook, J. Grear, A.C. Candelmo, R.C. Chambers, C.J. Gobler, C.W. Hunt, A.L. King, N.N. Price, S.R. Signorini, E. Stancioff, C. Stymiest, R.A. Wahle, J.D. Waller, N.D. Rebuck, Z.A. Wang, T.L. Capson, J.R. Morrison, S.R. Cooley, and S.C. Doney, 2015: Ocean and coastal acidification off New England and Nova Scotia. *Oceanography*, **28**, 182-197. http://dx.doi.org/10.5670/oceanog.2015.41

77. Evans, W., J.T. Mathis, and J.N. Cross, 2014: Calcium carbonate corrosivity in an Alaskan inland sea. *Biogeosciences*, **11**, 365-379. http://dx.doi.org/10.5194/bg-11-365-2014

78. Salisbury, J., M. Green, C. Hunt, and J. Campbell, 2008: Coastal acidification by rivers: A threat to shellfish? *Eos, Transactions, American Geophysical Union*, **89**, 513-513. http://dx.doi.org/10.1029/2008EO500001

79. Bates, N.R. and J.T. Mathis, 2009: The Arctic Ocean marine carbon cycle: Evaluation of air-sea CO₂ exchanges, ocean acidification impacts and potential feedbacks. *Biogeosciences*, **6**, 2433-2459. http://dx.doi.org/10.5194/bg-6-2433-2009

80. Jiang, L.-Q., R.A. Feely, B.R. Carter, D.J. Greeley, D.K. Gledhill, and K.M. Arzayus, 2015: Climatological distribution of aragonite saturation state in the global oceans. *Global Biogeochemical Cycles*, **29**, 1656-1673. http://dx.doi.org/10.1002/2015GB005198

81. Feely, R.A., C.L. Sabine, J.M. Hernandez-Ayon, D. Ianson, and B. Hales, 2008: Evidence for upwelling of corrosive "acidified" water onto the continental shelf. *Science*, **320**, 1490-1492. http://dx.doi.org/10.1126/science.1155676

82. Qi, D., L. Chen, B. Chen, Z. Gao, W. Zhong, R.A. Feely, L.G. Anderson, H. Sun, J. Chen, M. Chen, L. Zhan, Y. Zhang, and W.-J. Cai, 2017: Increase in acidifying water in the western Arctic Ocean. *Nature Climate Change*, **7**, 195-199. http://dx.doi.org/10.1038/nclimate3228

83. Manzello, D., I. Enochs, S. Musielewicz, R. Carlton, and D. Gledhill, 2013: Tropical cyclones cause CaCO3 undersaturation of coral reef seawater in a high-CO2 world. *Journal of Geophysical Research Oceans*, **118**, 5312-5321. http://dx.doi.org/10.1002/jgrc.20378

84. Friedrich, T., A. Timmermann, A. Abe-Ouchi, N.R. Bates, M.O. Chikamoto, M.J. Church, J.E. Dore, D.K. Gledhill, M. Gonzalez-Davila, M. Heinemann, T. Ilyina, J.H. Jungclaus, E. McLeod, A. Mouchet, and J.M. Santana-Casiano, 2012: Detecting regional anthropogenic trends in ocean acidification against natural variability. *Nature Climate Change,* **2,** 167-171. http://dx.doi.org/10.1038/nclimate1372

85. Hönisch, B., A. Ridgwell, D.N. Schmidt, E. Thomas, S.J. Gibbs, A. Sluijs, R. Zeebe, L. Kump, R.C. Martindale, S.E. Greene, W. Kiessling, J. Ries, J.C. Zachos, D.L. Royer, S. Barker, T.M. Marchitto, Jr., R. Moyer, C. Pelejero, P. Ziveri, G.L. Foster, and B. Williams, 2012: The geological record of ocean acidification. *Science,* **335,** 1058-1063. http://dx.doi.org/10.1126/science.1208277

86. Zeebe, R.E., A. Ridgwell, and J.C. Zachos, 2016: Anthropogenic carbon release rate unprecedented during the past 66 million years. *Nature Geoscience,* **9,** 325-329. http://dx.doi.org/10.1038/ngeo2681

87. Wright, J.D. and M.F. Schaller, 2013: Evidence for a rapid release of carbon at the Paleocene-Eocene thermal maximum. *Proceedings of the National Academy of Sciences,* **110,** 15908-15913. http://dx.doi.org/10.1073/pnas.1309188110

88. Caldeira, K. and M.E. Wickett, 2005: Ocean model predictions of chemistry changes from carbon dioxide emissions to the atmosphere and ocean. *Journal of Geophysical Research: Oceans,* **110,** C09S04. http://dx.doi.org/10.1029/2004JC002671

89. Feely, R.A., S.R. Alin, B. Carter, N. Bednaršek, B. Hales, F. Chan, T.M. Hill, B. Gaylord, E. Sanford, R.H. Byrne, C.L. Sabine, D. Greeley, and L. Juranek, 2016: Chemical and biological impacts of ocean acidification along the west coast of North America. *Estuarine, Coastal and Shelf Science,* **183, Part A,** 260-270. http://dx.doi.org/10.1016/j.ecss.2016.08.043

90. Mathis, J.T., S.R. Cooley, N. Lucey, S. Colt, J. Ekstrom, T. Hurst, C. Hauri, W. Evans, J.N. Cross, and R.A. Feely, 2015: Ocean acidification risk assessment for Alaska's fishery sector. *Progress in Oceanography,* **136,** 71-91. http://dx.doi.org/10.1016/j.pocean.2014.07.001

91. Altieri, A.H. and K.B. Gedan, 2015: Climate change and dead zones. *Global Change Biology,* **21,** 1395-1406. http://dx.doi.org/10.1111/gcb.12754

92. Breitburg, D.L., J. Salisbury, J.M. Bernhard, W.-J. Cai, S. Dupont, S.C. Doney, K.J. Kroeker, L.A. Levin, W.C. Long, L.M. Milke, S.H. Miller, B. Phelan, U. Passow, B.A. Seibel, A.E. Todgham, and A.M. Tarrant, 2015: And on top of all that… Coping with ocean acidification in the midst of many stressors. *Oceanography,* **28,** 48-61. http://dx.doi.org/10.5670/oceanog.2015.31

93. Gobler, C.J., E.L. DePasquale, A.W. Griffith, and H. Baumann, 2014: Hypoxia and acidification have additive and synergistic negative effects on the growth, survival, and metamorphosis of early life stage bivalves. *PLoS ONE,* **9,** e83648. http://dx.doi.org/10.1371/journal.pone.0083648

94. Helm, K.P., N.L. Bindoff, and J.A. Church, 2011: Observed decreases in oxygen content of the global ocean. *Geophysical Research Letters,* **38,** L23602. http://dx.doi.org/10.1029/2011GL049513

95. Stramma, L., S. Schmidtko, L.A. Levin, and G.C. Johnson, 2010: Ocean oxygen minima expansions and their biological impacts. *Deep Sea Research Part I: Oceanographic Research Papers,* **57,** 587-595. http://dx.doi.org/10.1016/j.dsr.2010.01.005

96. Reay, D.S., F. Dentener, P. Smith, J. Grace, and R.A. Feely, 2008: Global nitrogen deposition and carbon sinks. *Nature Geoscience,* **1,** 430-437. http://dx.doi.org/10.1038/ngeo230

97. Rabalais, N.N., R.E. Turner, R.J. Díaz, and D. Justić, 2009: Global change and eutrophication of coastal waters. *ICES Journal of Marine Science,* **66,** 1528-1537. http://dx.doi.org/10.1093/icesjms/fsp047

98. Boetius, A. and F. Wenzhofer, 2013: Seafloor oxygen consumption fuelled by methane from cold seeps. *Nature Geoscience,* **6,** 725-734. http://dx.doi.org/10.1038/ngeo1926

99. Lee, M., E. Shevliakova, S. Malyshev, P.C.D. Milly, and P.R. Jaffé, 2016: Climate variability and extremes, interacting with nitrogen storage, amplify eutrophication risk. *Geophysical Research Letters,* **43,** 7520-7528. http://dx.doi.org/10.1002/2016GL069254

100. Ito, T., A. Nenes, M.S. Johnson, N. Meskhidze, and C. Deutsch, 2016: Acceleration of oxygen decline in the tropical Pacific over the past decades by aerosol pollutants. *Nature Geoscience,* **9,** 443-447. http://dx.doi.org/10.1038/ngeo2717

101. Codispoti, L.A., J.A. Brandes, J.P. Christensen, A.H. Devol, S.W.A. Naqvi, H.W. Paerl, and T. Yoshinari, 2001: The oceanic fixed nitrogen and nitrous oxide budgets: Moving targets as we enter the anthropocene? *Scientia Marina,* **65,** 85-105. http://dx.doi.org/10.3989/scimar.2001.65s285

102. Deutsch, C., H. Brix, T. Ito, H. Frenzel, and L. Thompson, 2011: Climate-forced variability of ocean hypoxia. *Science,* **333,** 336-339. http://dx.doi.org/10.1126/science.1202422

103. Gruber, N., 2008: Chapter 1 - The marine nitrogen cycle: Overview and challenges. *Nitrogen in the Marine Environment (2nd Edition).* Academic Press, San Diego, 1-50. http://dx.doi.org/10.1016/B978-0-12-372522-6.00001-3

104. EPA, 2017: Inventory of U.S. Greenhouse Gas Emissions and Sinks: 1990-2015. EPA 430-P-17-001. U.S. Environmental Protection Agency, Washington, D.C., 633 pp. https://www.epa.gov/sites/production/files/2017-02/documents/2017_complete_report.pdf

105. Wallmann, K., 2003: Feedbacks between oceanic redox states and marine productivity: A model perspective focused on benthic phosphorus cycling. *Global Biogeochemical Cycles*, **17**, 1084. http://dx.doi.org/10.1029/2002GB001968

106. Bianchi, D., E.D. Galbraith, D.A. Carozza, K.A.S. Mislan, and C.A. Stock, 2013: Intensification of open-ocean oxygen depletion by vertically migrating animals. *Nature Geoscience*, **6**, 545-548. http://dx.doi.org/10.1038/ngeo1837

107. Knoll, A.H. and S.B. Carroll, 1999: Early animal evolution: Emerging views from comparative biology and geology. *Science*, **284**, 2129-2137. http://dx.doi.org/10.1126/science.284.5423.2129

108. McFall-Ngai, M., M.G. Hadfield, T.C.G. Bosch, H.V. Carey, T. Domazet-Lošo, A.E. Douglas, N. Dubilier, G. Eberl, T. Fukami, S.F. Gilbert, U. Hentschel, N. King, S. Kjelleberg, A.H. Knoll, N. Kremer, S.K. Mazmanian, J.L. Metcalf, K. Nealson, N.E. Pierce, J.F. Rawls, A. Reid, E.G. Ruby, M. Rumpho, J.G. Sanders, D. Tautz, and J.J. Wernegreen, 2013: Animals in a bacterial world, a new imperative for the life sciences. *Proceedings of the National Academy of Sciences*, **110**, 3229-3236. http://dx.doi.org/10.1073/pnas.1218525110

109. Falkowski, P.G., T. Algeo, L. Codispoti, C. Deutsch, S. Emerson, B. Hales, R.B. Huey, W.J. Jenkins, L.R. Kump, L.A. Levin, T.W. Lyons, N.B. Nelson, O.S. Schofield, R. Summons, L.D. Talley, E. Thomas, F. Whitney, and C.B. Pilcher, 2011: Ocean deoxygenation: Past, present, and future. *Eos, Transactions, American Geophysical Union*, **92**, 409-410. http://dx.doi.org/10.1029/2011EO460001

110. Robinson, R.S., A. Mix, and P. Martinez, 2007: Southern Ocean control on the extent of denitrification in the southeast Pacific over the last 70 ka. *Quaternary Science Reviews*, **26**, 201-212. http://dx.doi.org/10.1016/j.quascirev.2006.08.005

111. Schmittner, A., E.D. Galbraith, S.W. Hostetler, T.F. Pedersen, and R. Zhang, 2007: Large fluctuations of dissolved oxygen in the Indian and Pacific oceans during Dansgaard–Oeschger oscillations caused by variations of North Atlantic Deep Water subduction. *Paleoceanography*, **22**, PA3207. http://dx.doi.org/10.1029/2006PA001384

112. Galbraith, E.D., M. Kienast, T.F. Pedersen, and S.E. Calvert, 2004: Glacial-interglacial modulation of the marine nitrogen cycle by high-latitude O2 supply to the global thermocline. *Paleoceanography*, **19**, PA4007. http://dx.doi.org/10.1029/2003PA001000

113. Moffitt, S.E., R.A. Moffitt, W. Sauthoff, C.V. Davis, K. Hewett, and T.M. Hill, 2015: Paleoceanographic insights on recent oxygen minimum zone expansion: Lessons for modern oceanography. *PLoS ONE*, **10**, e0115246. http://dx.doi.org/10.1371/journal.pone.0115246

114. Justić, D., T. Legović, and L. Rottini-Sandrini, 1987: Trends in oxygen content 1911–1984 and occurrence of benthic mortality in the northern Adriatic Sea. *Estuarine, Coastal and Shelf Science*, **25**, 435-445. http://dx.doi.org/10.1016/0272-7714(87)90035-7

115. Zaitsev, Y.P., 1992: Recent changes in the trophic structure of the Black Sea. *Fisheries Oceanography*, **1**, 180-189. http://dx.doi.org/10.1111/j.1365-2419.1992.tb00036.x

116. Conley, D.J., J. Carstensen, J. Aigars, P. Axe, E. Bonsdorff, T. Eremina, B.-M. Haahti, C. Humborg, P. Jonsson, J. Kotta, C. Lännegren, U. Larsson, A. Maximov, M.R. Medina, E. Lysiak-Pastuszak, N. Remeikaitè-Nikienė, J. Walve, S. Wilhelms, and L. Zillén, 2011: Hypoxia is increasing in the coastal zone of the Baltic Sea. *Environmental Science & Technology*, **45**, 6777-6783. http://dx.doi.org/10.1021/es201212r

117. Brush, G.S., 2009: Historical land use, nitrogen, and coastal eutrophication: A paleoecological perspective. *Estuaries and Coasts*, **32**, 18-28. http://dx.doi.org/10.1007/s12237-008-9106-z

118. Gilbert, D., B. Sundby, C. Gobeil, A. Mucci, and G.-H. Tremblay, 2005: A seventy-two-year record of diminishing deep-water oxygen in the St. Lawrence estuary: The northwest Atlantic connection. *Limnology and Oceanography*, **50**, 1654-1666. http://dx.doi.org/10.4319/lo.2005.50.5.1654

119. Rabalais, N.N., R.E. Turner, B.K. Sen Gupta, D.F. Boesch, P. Chapman, and M.C. Murrell, 2007: Hypoxia in the northern Gulf of Mexico: Does the science support the plan to reduce, mitigate, and control hypoxia? *Estuaries and Coasts*, **30**, 753-772. http://dx.doi.org/10.1007/bf02841332

120. Rabalais, N.N., R.J. Díaz, L.A. Levin, R.E. Turner, D. Gilbert, and J. Zhang, 2010: Dynamics and distribution of natural and human-caused hypoxia. *Biogeosciences*, **7**, 585-619. http://dx.doi.org/10.5194/bg-7-585-2010

121. Booth, J.A.T., E.E. McPhee-Shaw, P. Chua, E. Kingsley, M. Denny, R. Phillips, S.J. Bograd, L.D. Zeidberg, and W.F. Gilly, 2012: Natural intrusions of hypoxic, low pH water into nearshore marine environments on the California coast. *Continental Shelf Research*, **45**, 108-115. http://dx.doi.org/10.1016/j.csr.2012.06.009

122. Baden, S.P., L.O. Loo, L. Pihl, and R. Rosenberg, 1990: Effects of eutrophication on benthic communities including fish — Swedish west coast. *Ambio*, **19**, 113-122. http://www.jstor.org/stable/4313676

123. Diaz, R.J. and R. Rosenberg, 2008: Spreading dead zones and consequences for marine ecosystems. *Science*, **321**, 926-929. http://dx.doi.org/10.1126/science.1156401

124. Takatani, Y., D. Sasano, T. Nakano, T. Midorikawa, and M. Ishii, 2012: Decrease of dissolved oxygen after the mid-1980s in the western North Pacific subtropical gyre along the 137°E repeat section. *Global Biogeochemical Cycles*, **26**, GB2013. http://dx.doi.org/10.1029/2011GB004227

125. Stendardo, I. and N. Gruber, 2012: Oxygen trends over five decades in the North Atlantic. *Journal of Geophysical Research*, **117**, C11004. http://dx.doi.org/10.1029/2012JC007909

126. Gilbert, D., N.N. Rabalais, R.J. Díaz, and J. Zhang, 2010: Evidence for greater oxygen decline rates in the coastal ocean than in the open ocean. *Biogeosciences*, **7**, 2283-2296. http://dx.doi.org/10.5194/bg-7-2283-2010

127. Bograd, S.J., M.P. Buil, E.D. Lorenzo, C.G. Castro, I.D. Schroeder, R. Goericke, C.R. Anderson, C. Benitez-Nelson, and F.A. Whitney, 2015: Changes in source waters to the Southern California Bight. *Deep Sea Research Part II: Topical Studies in Oceanography*, **112**, 42-52. http://dx.doi.org/10.1016/j.dsr2.2014.04.009

128. Long, M.C., C. Deutsch, and T. Ito, 2016: Finding forced trends in oceanic oxygen. *Global Biogeochemical Cycles*, **30**, 381-397. http://dx.doi.org/10.1002/2015GB005310

129. Deutsch, C., W. Berelson, R. Thunell, T. Weber, C. Tems, J. McManus, J. Crusius, T. Ito, T. Baumgartner, V. Ferreira, J. Mey, and A. van Geen, 2014: Centennial changes in North Pacific anoxia linked to tropical trade winds. *Science*, **345**, 665-668. http://dx.doi.org/10.1126/science.1252332

130. Montes, I., B. Dewitte, E. Gutknecht, A. Paulmier, I. Dadou, A. Oschlies, and V. Garçon, 2014: High-resolution modeling of the eastern tropical Pacific oxygen minimum zone: Sensitivity to the tropical oceanic circulation. *Journal of Geophysical Research Oceans*, **119**, 5515-5532. http://dx.doi.org/10.1002/2014JC009858

131. Donner, S.D. and D. Scavia, 2007: How climate controls the flux of nitrogen by the Mississippi River and the development of hypoxia in the Gulf of Mexico. *Limnology and Oceanography*, **52**, 856-861. http://dx.doi.org/10.4319/lo.2007.52.2.0856

132. Najjar, R.G., C.R. Pyke, M.B. Adams, D. Breitburg, C. Hershner, M. Kemp, R. Howarth, M.R. Mulholland, M. Paolisso, D. Secor, K. Sellner, D. Wardrop, and R. Wood, 2010: Potential climate-change impacts on the Chesapeake Bay. *Estuarine, Coastal and Shelf Science*, **86**, 1-20. http://dx.doi.org/10.1016/j.ecss.2009.09.026

133. Thompson, P.R., B.D. Hamlington, F.W. Landerer, and S. Adhikari, 2016: Are long tide gauge records in the wrong place to measure global mean sea level rise? *Geophysical Research Letters*, **43**, 10,403-10,411. http://dx.doi.org/10.1002/2016GL070552

134. Jickells, T. and C.M. Moore, 2015: The importance of atmospheric deposition for ocean productivity. *Annual Review of Ecology, Evolution, and Systematics*, **46**, 481-501. http://dx.doi.org/10.1146/annurev-ecolsys-112414-054118

135. Chiapello, I., 2014: Dust observations and climatology. *Mineral Dust: A Key Player in the Earth System*. Knippertz, P. and J.-B.W. Stuut, Eds. Springer Netherlands, Dordrecht, 149-177. http://dx.doi.org/10.1007/978-94-017-8978-3_7

136. Mulitza, S., D. Heslop, D. Pittauerova, H.W. Fischer, I. Meyer, J.-B. Stuut, M. Zabel, G. Mollenhauer, J.A. Collins, H. Kuhnert, and M. Schulz, 2010: Increase in African dust flux at the onset of commercial agriculture in the Sahel region. *Nature*, **466**, 226-228. http://dx.doi.org/10.1038/nature09213

137. Paerl, H.W., R.L. Dennis, and D.R. Whitall, 2002: Atmospheric deposition of nitrogen: Implications for nutrient over-enrichment of coastal waters. *Estuaries*, **25**, 677-693. http://dx.doi.org/10.1007/bf02804899

138. Carr, M.-E., M.A.M. Friedrichs, M. Schmeltz, M. Noguchi Aita, D. Antoine, K.R. Arrigo, I. Asanuma, O. Aumont, R. Barber, M. Behrenfeld, R. Bidigare, E.T. Buitenhuis, J. Campbell, A. Ciotti, H. Dierssen, M. Dowell, J. Dunne, W. Esaias, B. Gentili, W. Gregg, S. Groom, N. Hoepffner, J. Ishizaka, T. Kameda, C. Le Quéré, S. Lohrenz, J. Marra, F. Mélin, K. Moore, A. Morel, T.E. Reddy, J. Ryan, M. Scardi, T. Smyth, K. Turpie, G. Tilstone, K. Waters, and Y. Yamanaka, 2006: A comparison of global estimates of marine primary production from ocean color. *Deep Sea Research Part II: Topical Studies in Oceanography*, **53**, 741-770. http://dx.doi.org/10.1016/j.dsr2.2006.01.028

139. Franz, B.A., M.J. Behrenfeld, D.A. Siegel, and S.R. Signorini, 2016: Global ocean phytoplankton [in "State of the Climate in 2015"]. *Bulletin of the American Meteorological Society*, **97**, S87–S89. http://dx.doi.org/10.1175/2016BAMSStateoftheClimate.1

140. Chavez, F.P., M. Messié, and J.T. Pennington, 2011: Marine primary production in relation to climate variability and change. *Annual Review of Marine Science*, **3**, 227-260. http://dx.doi.org/10.1146/annurev.marine.010908.163917

141. Frölicher, T.L., K.B. Rodgers, C.A. Stock, and W.W.L. Cheung, 2016: Sources of uncertainties in 21st century projections of potential ocean ecosystem stressors. *Global Biogeochemical Cycles*, **30**, 1224-1243. http://dx.doi.org/10.1002/2015GB005338

142. Fu, W., J.T. Randerson, and J.K. Moore, 2016: Climate change impacts on net primary production (NPP) and export production (EP) regulated by increasing **stratification and phytoplankton community structure** in the CMIP5 models. *Biogeosciences*, **13**, 5151-5170. http://dx.doi.org/10.5194/bg-13-5151-2016

143. Laufkötter, C., M. Vogt, N. Gruber, M. Aita-Noguchi, O. Aumont, L. Bopp, E. Buitenhuis, S.C. Doney, J. Dunne, T. Hashioka, J. Hauck, T. Hirata, J. John, C. Le Quéré, I.D. Lima, H. Nakano, R. Seferian, I. Totterdell, M. Vichi, and C. Völker, 2015: Drivers and uncertainties of future global marine primary production in marine ecosystem models. *Biogeosciences*, **12**, 6955-6984. http://dx.doi.org/10.5194/bg-12-6955-2015

144. Robinson, P., A.K. Leight, D.D. Trueblood, and B. Wood, 2013: Climate sensitivity of the National Estuarine Research Reserve System. NERRS, NOAA National Ocean Service, Silver Spring, Maryland. 79 pp. https://coast.noaa.gov/data/docs/nerrs/Research_DataSyntheses_130725_climate%20sensitivity%20of%20nerrs_Final-Rpt-in-Layout_FINAL.pdf

145. Monbaliu, J., Z. Chen, D. Felts, J. Ge, F. Hissel, J. Kappenberg, S. Narayan, R.J. Nicholls, N. Ohle, D. Schuster, J. Sothmann, and P. Willems, 2014: Risk assessment of estuaries under climate change: Lessons from Western Europe. *Coastal Engineering*, **87**, 32-49. http://dx.doi.org/10.1016/j.coastaleng.2014.01.001

146. Schile, L.M., J.C. Callaway, J.T. Morris, D. Stralberg, V.T. Parker, and M. Kelly, 2014: Modeling tidal marsh distribution with sea-level rise: Evaluating the role of vegetation, sediment, and upland habitat in marsh resiliency. *PLoS ONE*, **9**, e88760. http://dx.doi.org/10.1371/journal.pone.0088760

147. Swanson, K.M., J.Z. Drexler, C.C. Fuller, and D.H. Schoellhamer, 2015: Modeling tidal freshwater marsh sustainability in the Sacramento–San Joaquin delta under a broad suite of potential future scenarios. *San Francisco Estuary and Watershed Science*, **13**, 21. http://dx.doi.org/10.15447/sfews.2015v13iss1art3

148. Belkin, I., 2016: Chapter 5.2: Sea surface temperature trends in large marine ecosystems. *Large Marine Ecosystems: Status and Trends*. United Nations Environment Programme, Nairobi, 101-109. http://wedocs.unep.org/bitstream/handle/20.500.11822/13456/UNEP_DEWA_TWAP%20VOLUME%204%20REPORT_FINAL_4_MAY.pdf?sequence=1&isAllowed=y,%20English%20-%20Summary

149. Baringer, M.O., M. Lankhorst, D. Volkov, S. Garzoli, S. Dong, U. Send, and C. Meinen, 2016: Meridional oceanic overturning circulation and heat transport in the Atlantic Ocean [in "State of the Climate in 2015"]. *Bulletin of the American Meteorological Society*, **97**, S84–S87. http://dx.doi.org/10.1175/2015BAMSStateoftheClimate.1

150. Schmidtko, S., L. Stramma, and M. Visbeck, 2017: Decline in global oceanic oxygen content during the **past five decades**. *Nature*, **542**, 335-339. http://dx.doi.org/10.1038/nature21399

151. Scott, J.D., M.A. Alexander, D.R. Murray, D. Swales, and J. Eischeid, 2016: The climate change web portal: A system to access and display climate and earth system model output from the CMIP5 archive. *Bulletin of the American Meteorological Society*, **97**, 523-530. http://dx.doi.org/10.1175/bams-d-15-00035.1

# 14

# Perspectives on Climate Change Mitigation

## KEY FINDINGS

1. Reducing net emissions of $CO_2$ is necessary to limit near-term climate change and long-term warming. Other greenhouse gases (for example, methane) and black carbon aerosols exert stronger warming effects than $CO_2$ on a per ton basis, but they do not persist as long in the atmosphere; therefore, mitigation of non-$CO_2$ species contributes substantially to near-term cooling benefits but cannot be relied upon for ultimate stabilization goals. (*Very high confidence*)

2. Stabilizing global mean temperature to less than 3.6°F (2°C) above preindustrial levels requires substantial reductions in net global $CO_2$ emissions prior to 2040 relative to present-day values and likely requires net emissions to become zero or possibly negative later in the century. After accounting for the temperature effects of non-$CO_2$ species, cumulative global $CO_2$ emissions must stay below about 800 GtC in order to provide a two-thirds likelihood of preventing 3.6°F (2°C) of warming. Given estimated cumulative emissions since 1870, no more than approximately 230 GtC may be emitted in the future to remain under this temperature threshold. Assuming global emissions are equal to or greater than those consistent with the RCP4.5 scenario, this cumulative carbon threshold would be exceeded in approximately two decades. (*High confidence*)

3. Achieving global greenhouse gas emissions reductions before 2030 consistent with targets and actions announced by governments in the lead up to the 2015 Paris climate conference would hold open the possibility of meeting the long-term temperature goal of limiting global warming to 3.6°F (2°C) above preindustrial levels, whereas there would be virtually no chance if net global emissions followed a pathway well above those implied by country announcements. Actions in the announcements are, by themselves, insufficient to meet a 3.6°F (2°C) goal; the likelihood of achieving that goal depends strongly on the magnitude of global emissions reductions after 2030. (*High confidence*)

4. Further assessments of the technical feasibilities, costs, risks, co-benefits, and governance challenges of climate intervention or geoengineering strategies, which are as yet unproven at scale, are a necessary step before judgments about the benefits and risks of these approaches can be made with high confidence. (*High confidence*)

**Recommended Citation for Chapter**

**DeAngelo**, B., J. Edmonds, D.W. Fahey, and B.M. Sanderson, 2017: Perspectives on climate change mitigation. In: *Climate Science Special Report: Fourth National Climate Assessment, Volume I* [Wuebbles, D.J., D.W. Fahey, K.A. Hibbard, D.J. Dokken, B.C. Stewart, and T.K. Maycock (eds.)]. U.S. Global Change Research Program, Washington, DC, USA, pp. 393-410, doi: 10.7930/J0M32SZG.

## Introduction

This chapter provides scientific context for key issues regarding the long-term mitigation of climate change. As such, this chapter first addresses the science underlying the timing of when and how $CO_2$ and other greenhouse gas (GHG) mitigation activities that occur in the present affect the climate of the future. When do we see the benefits of a GHG emission reduction activity? Chapter 4: Projections provides further context for this topic. Relatedly, the present chapter discusses the significance of the relationship between net cumulative $CO_2$ emissions and eventual global warming levels. The chapter reviews recent analyses of global emissions pathways associated with preventing 3.6°F (2°C) or 2.7°F (1.5°C) of warming relative to preindustrial times. And finally, this chapter briefly reviews the status of climate intervention proposals and how these types of mitigation actions could possibly play a role in avoiding future climate change.

## 14.1 The Timing of Benefits from Mitigation Actions

### 14.1.1 Lifetime of Greenhouse Gases and Inherent Delays in the Climate System

Carbon dioxide ($CO_2$) concentrations in the atmosphere are directly affected by human activities in the form of $CO_2$ emissions. Atmospheric $CO_2$ concentrations adjust to human emissions of $CO_2$ over long time scales, spanning from decades to millennia.[1, 2] The IPCC estimated that 15% to 40% of $CO_2$ emitted until 2100 will remain in the atmosphere longer than 1,000 years.[1] The persistence of warming is longer than the atmospheric lifetime of $CO_2$ and other GHGs, owing in large part to the thermal inertia of the ocean.[3] Climate change resulting from anthropogenic $CO_2$ emissions, and any associated risks to the environment, human health and society, are thus essentially irreversible on human timescales.[4] The world is committed to some degree of irreversible

warming and associated climate change resulting from emissions to date.

The long lifetime in the atmosphere of $CO_2$[2] and some other key GHGs, coupled with the time lag in the response of the climate system to atmospheric forcing,[5] has timing implications for the benefits (i.e., avoided warming or risk) of mitigation actions. Large reductions in emissions of the long-lived GHGs are estimated to have modest temperature effects in the *near term* (e.g., over one to two decades) because total atmospheric concentration levels require long periods to adjust,[6] but are necessary in the *long term* to achieve any objective of preventing warming of any desired magnitude. Near-term projections of global mean surface temperature are therefore not strongly influenced by changes in near-term emissions but rather dominated by natural variability, the Earth system response to past and current GHG emissions, and by model spread (i.e., the different climate outcomes associated with different models using the same emissions pathway).[7] Long-term projections of global surface temperature (after mid-century), on the other hand, show that the choice of global emissions pathway, and thus the long-term mitigation pathway the world chooses, is the dominant source of future uncertainty in climate outcomes.[3, 8]

Some studies have nevertheless shown the potential for some near-term benefits of mitigation. For example, one study found that, even at the regional scale, heat waves would already be significantly more severe by the 2030s in a non-mitigation scenario compared to a moderate mitigation scenario.[9] The mitigation of non-$CO_2$ GHGs with short atmospheric lifetimes (such as methane, some hydrofluorocarbons [HFCs], and ozone) and black carbon (an aerosol that absorbs solar radiation; see Ch. 2: Physical Drivers of Climate Change), collectively referred to as short-lived climate pollutants

(SLCPs), has been highlighted as a particular way to achieve more rapid climate benefits (e.g., Zaelke and Borgford-Parnell 2015[10]). SLCPs are substances that not only have an atmospheric lifetime shorter (for example, weeks to a decade) than $CO_2$ but also exert a stronger radiative forcing (and hence temperature effect) compared to $CO_2$ on a per ton basis.[11] For these reasons, mitigation of SLCP emissions produces more rapid radiative responses. In the case of black carbon, with an atmospheric lifetime of a few days to weeks,[12] emissions (and therefore reductions of those emissions) produce strong regional effects. Mitigation of black carbon and methane also generate direct health co-benefits.[13, 14] Reductions and/or avoidances of SLCP emissions could be a significant contribution to staying at or below a 3.6°F (2°C) increase or any other chosen global mean temperature increase.[15, 16, 17, 18] The recent Kigali Amendment to the Montreal Protocol seeks to phase down global HFC production and consumption in order to avoid substantial GHG emissions in coming decades. Stringent and continuous SLCP mitigation could potentially increase allowable $CO_2$ budgets for avoiding warming beyond any desired future level, by up to 25% under certain scenarios.[18] However, given that economic and technological factors tend to couple $CO_2$ and many SLCP emissions to varying degrees, significant SLCP emissions reductions would be a co-benefit of $CO_2$ mitigation.

### 14.1.2 Stock and Stabilization: Cumulative $CO_2$ and the Role of Other Greenhouse Gases

Net cumulative $CO_2$ emissions in the industrial era will largely determine long-term, global mean temperature change. A robust feature of model climate change simulations is a nearly linear relationship between cumulative $CO_2$ emissions and global mean temperature increases, irrespective of the details and exact timing of the emissions pathway (see Figure 14.1; see also Ch. 4: Projections). Limiting and stabilizing warming to any level implies that there is a physical upper limit to the cumulative amount of $CO_2$ that can be added to the atmosphere.[3] Eventually stabilizing the global temperature requires $CO_2$ emissions to approach zero.[19] Thus, for a 3.6°F (2°C) or any desired global mean warming goal, an estimated range of cumulative $CO_2$ emissions from the current period onward can be calculated. The key sources of uncertainty for any compatible, forward looking $CO_2$ budget associated with a given future warming objective include the climate sensitivity, the response of the carbon cycle including feedbacks (for example, the release of GHGs from permafrost thaw), the amount of past $CO_2$ emissions, and the influence of past and future non-$CO_2$ species.[3, 19] Increasing the probability that any given temperature goal will be reached therefore implies tighter constraints on cumulative $CO_2$ emissions. Relatedly, for any given cumulative $CO_2$ budget, higher emissions in the near term imply the need for steeper reductions in the long term.

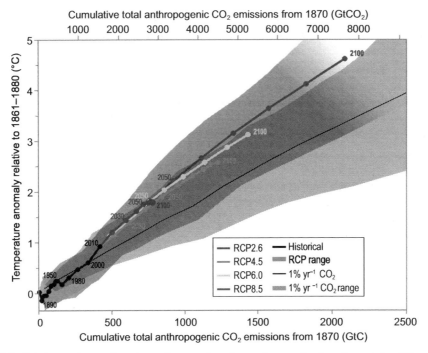

Cumulative total anthropogenic CO₂ emissions from 1870 (GtCO₂)

**Figure 14.1:** Global mean temperature change for a number of scenarios as a function of cumulative CO₂ emissions from preindustrial conditions, with time progressing along each individual line for each scenario. (Figure source: IPCC 2013;[42] ©IPCC. Used with permission).

Between 1870 and 2015, human activities, primarily the burning of fossil fuels and defor- estation, emitted about 560 GtC in the form of CO₂ into the atmosphere.[20] According to best estimates in the literature, 1,000 GtC is the total cumulative amount of CO₂ that could be emitted yet still provide a two-thirds likeli- hood of preventing 3.6°F (2°C) of global mean warming since preindustrial times.[3, 21] That estimate, however, ignores the additional ra- diative forcing effects of non-CO₂ species (that is, the net positive forcing resulting from the forcing of other well-mixed GHGs, including halocarbons, plus the other ozone precursor gases and aerosols). Considering both histori- cal and projected non-CO₂ effects reduces the estimated cumulative CO₂ budget compatible with any future warming goal,[18] and in the case of 3.6°F (2°C) it reduces the aforemen-

tioned estimate to 790 GtC.[3] Given this more comprehensive estimate, limiting the global average temperature increase to below 3.6°F (2°C) means approximately 230 GtC more CO₂ could be emitted globally. To illustrate, if one assumes future global emissions follow a pathway consistent with the lower scenario (RCP4.5), this cumulative carbon threshold is exceeded by around 2037, while under the higher scenario (RCP8.5) this occurs by around 2033. To limit the global average tem- perature increase to 2.7°F (1.5°C), the estimat- ed cumulative CO₂ budget is about 590 GtC (assuming linear scaling with the compatible 3.6°F (2°C) budget that also considers non-CO₂ effects), meaning only about 30 GtC more of CO₂ could be emitted. Further emissions of 30 GtC (in the form of CO₂) are projected to occur in the next few years (Table 14.1).

Table 14.1: Dates illustrating when cumulative CO$_2$ emissions thresholds associated with eventual warming of 3.6°F or 2.7°F above preindustrial levels might be reached. RCP4.5 and RCP8.5 refer, respectively, to emissions consistent with the lower and higher scenarios used throughout this report. The estimated cumulative CO$_2$ emissions (measured in Gigatons (Gt) of carbon) associated with different probabilities (e.g., 66%) of preventing 3.6°F (2°C) of warming are from the IPCC.[3] The cumulative emissions compatible with 2.7°F (1.5°C) are linearly derived from the estimates associated with 3.6°F (2°C). The cumulative CO$_2$ estimates take into account the additional net warming effects associated with past and future non-CO$_2$ emissions consistent with the RCP scenarios. Historical CO$_2$ emissions from 1870–2015 (including fossil fuel combustion, land use change, and cement manufacturing) are from Le Quéré et al.[20] See Traceable Accounts for further details.

| | Dates by when cumulative carbon emissions (GtC) since 1870 reach amount commensurate with 3.6°F (2°C), when accounting for non-CO$_2$ forcings | | |
|---|---|---|---|
| | 66% = 790 GtC | 50% = 820 GtC | 33% = 900 GtC |
| RCP4.5 | 2037 | 2040 | 2047 |
| RCP8.5 | 2033 | 2035 | 2040 |
| | Dates by when cumulative carbon emissions (GtC) since 1870 reach amount commensurate with 2.7°F (1.5°C), when accounting for non-CO$_2$ forcings | | |
| | 66% = 593 GtC | 50% = 615 GtC | 33% = 675 GtC |
| RCP4.5 | 2019 | 2021 | 2027 |
| RCP8.5 | 2019 | 2021 | 2025 |

## 14.2 Pathways Centered Around 3.6°F (2°C)

The idea of a 3.6°F (2°C) goal can be found in the scientific literature as early as 1975. Nordhaus[22] justified it by simply stating, "If there were global temperatures more than 2 or 3°C above the current average temperature, this would take the climate outside of the range of observations which have been made over the last several hundred thousand years." Since that time, the concept of a 3.6°F (2°C) goal gained attention in both scientific and policy discourse. For example, the Stockholm Environment Institute[23] published a report stating that 3.6°F (2°C) "can be viewed as an upper limit beyond which the risks of grave damage to ecosystems, and of non-linear responses, are expected to increase rapidly." And in 2007, the IPCC Fourth Assessment Report stated, among other things: "Confidence has increased that a 1 to 2°C increase in global mean temperature above 1990 levels (about 1.5 to 2.5°C above pre-industrial) poses

significant risks to many unique and threatened systems including many biodiversity hotspots." Most recently, the Paris Agreement of 2015 took on the long-term goal of "holding the increase in the global average temperature to well below 2°C above pre-industrial levels and pursuing efforts to limit the temperature increase to 1.5°C above pre-industrial levels." Many countries announced GHG emissions reduction targets and related actions (formally called Intended Nationally Determined Contributions [INDCs]) in the lead up to the Paris meeting; these announcements addressed emissions through 2025 or 2030 and take a wide range of forms. A number of studies have generated projections of future GHG emissions based on these announcements and evaluated whether, if implemented, the resulting emissions reductions would limit the increase in global average temperatures to 3.6°F (2°C) above preindustrial levels. In June 2017, the United States announced its intent to withdraw from the Paris Agreement. The scenarios assessed below were published prior to this announcement and therefore do not reflect the implications of this announcement.

Estimates of global emissions and temperature implications from emissions pathways consistent with targets and actions announced by governments in the lead up to the 2015 Paris climate conference[24, 25, 26, 27, 28] generally find that 1) these targets and actions would reduce GHG emissions growth by 2030 relative to a situation where these goals did not exist, though emissions are still not expected to be lower in 2030 than in 2015; and 2) the targets and actions would be a step towards limiting global mean temperature increase to 3.6°F (2°C), but by themselves, would be insufficient for this goal. According to one study, emissions pathways consistent with governments' announcements imply a median warming of 4.7°–5.6°F (2.6°–3.1°C) by 2100, though year 2100 temperature estimates depend on assumed emissions between 2030 and 2100.[24] For example, Climate Action Tracker,[26] using alternative post-2030 assumptions, put the range at 5.9°–7.0°F (3.3°–3.9°C).

Emissions pathways consistent with the targets and actions announced by governments in the lead up to the 2015 Paris conference have been evaluated in the context of the likelihood of global mean surface temperature change (Figure 14.2). It was found that the likelihood of limiting the global mean temperature increase to 3.6°F (2°C) or less was enhanced by these announced actions, but depended strongly on assumptions about subsequent policies and measures. Under a scenario in which countries maintain the same pace of decarbonization past 2030 as they announced in their first actions (leading up to 2025 or 2030) there is some likelihood (less than 10%) of preventing a global mean surface temperature change of 3.6°F (2°C) relative to preindustrial levels; this scenario thus holds open the possibility of achieving this goal, whereas there would be virtually no chance if emissions climbed to levels above those implied by country announcements (Figure 14.2).[27] Greater emissions reductions beyond 2030 (based on higher decarbonization rates past 2030) increase the likelihood of limiting warming to 3.6°F (2°C) or lower to about 30%, and almost eliminate the likelihood of a global mean temperature increase greater than 7°F (4°C). Scenarios that assume even greater emissions reductions past 2030 would be necessary to have at least a 50% probability of limiting warming to 3.6°F (2°C)[27] as discussed and illustrated further below.

(a)  Emissions pathways

(b)  Temperature probabilities

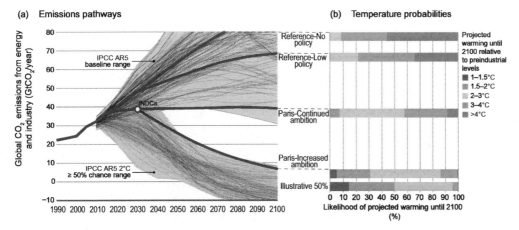

**Figure 14.2:** Global CO$_2$ emissions and probabilistic temperature outcomes of government announcements associated with the lead up to the Paris climate conference. (a) Global CO$_2$ emissions from energy and industry (includes CO$_2$ emissions from all fossil fuel production and use and industrial processes such as cement manufacture that also produce CO$_2$ as a byproduct) for emissions pathways following no policy, current policy, meeting the governments' announcements with constant country decarbonization rates past 2030, and meeting the governments' announcements with higher rates of decarbonization past 2030. **INDCs** refer to Intended Nationally Determined Contributions which is the term used for the governments' announced actions in the lead up to Paris. (b) Likelihoods of different levels of increase in global mean surface temperature during the 21st century relative to preindustrial levels for the four scenarios. Although (a) shows only CO$_2$ emissions from energy and industry, temperature outcomes are based on the full suite of GHG, aerosol, and short-lived species emissions across the full set of human activities and physical Earth systems. (Figure source: Fawcett et al. 2015[27]).

There is a limited range of pathways which enable the world to remain below 3.6°F (2°C) of warming (see Figure 14.3), and almost all but the most rapid near-term mitigation pathways are heavily reliant on the implementation of CO$_2$ removal from the atmosphere later in the century or other climate intervention, discussed below. If global emissions are in line with the first round of announced government actions by 2030, then the world likely needs to reduce effective GHG emissions to zero by 2080 and be significantly net negative by the end of the century (relying on as yet unproven technologies to remove GHGs from the atmosphere) in order to stay below 3.6°F (2°C) of warming. Avoiding 2.7°F (1.5°C) of warming requires more aggressive action still, with net zero emissions achieved by 2050 and net negative emissions thereafter. In either case, faster near-term emissions reductions significantly decrease the requirements for net negative emissions in the future.

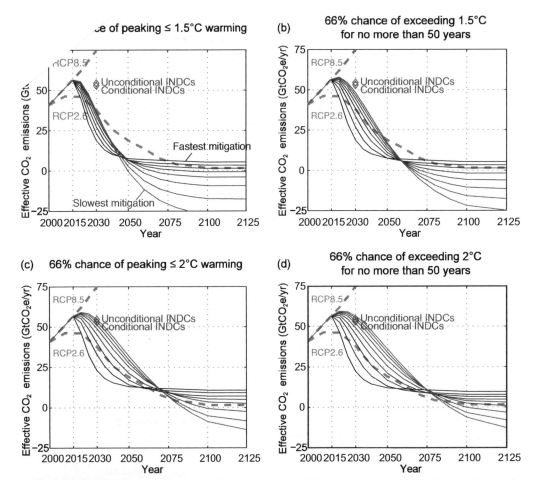

Figure 14.3: Global emissions pathways for GHGs, expressed as CO₂-equivalent emissions, which would be consistent with different temperature goals (relative to preindustrial temperatures). INDCs refer to Intended Nationally Determined Contributions which is the term used for the governments' announced actions in the lead up to Paris. (a) shows a set of pathways where global mean temperatures would likely (66%) not exceed 2.7°F (1.5°C). A number of pathways are consistent with the goal, ranging from the red curve (slowest near-term mitigation with large negative emissions requirements in the future) to the black curve with rapid near-term mitigation and less future negative emissions. (b) shows similar pathways with a 66% chance of exceeding 2.7°F (1.5°C) for only 50 years, where (c) and (d) show similar emission pathways for 3.6°F (2°C). (Figure source: Sanderson et al. 2016[25]).

## 14.3 The Potential Role of Climate Intervention in Mitigation Strategies

Limiting the global mean temperature increase through emissions reductions or adapting to the impacts of a greater-than-3.6°F (2°C) warmer world have been acknowledged as severely challenging tasks by the international science and policy communities. Consequently, there is increased interest by some scientists and policy makers in exploring additional measures designed to reduce net radiative forcing through other, as yet untested actions, which are often referred to as geoengineering or climate intervention (CI) actions. CI approaches are generally divided into two categories: carbon dioxide removal (CDR)[29] and solar radiation management (SRM).[30] CDR and SRM methods may have future roles in helping meet global temperature goals. Both methods would reduce global average temperature by reducing net global radiative forcing: CDR through reducing atmospheric $CO_2$ concentrations and SRM through increasing Earth's albedo.

The evaluation of the suitability and advisability of potential CI actions requires a decision framework that includes important dimensions beyond scientific and technical considerations. Among these dimensions to be considered are the potential development of global and national governance and oversight procedures, geopolitical relations, legal considerations, environmental, economic and societal impacts, ethical considerations, and the relationships to global climate policy and current GHG mitigation and adaptation actions. It is clear that these social science and other non-physical science dimensions are likely to be a major part of the decision framework and ultimately control the adoption and effectiveness of CI actions. This report only acknowledges these mostly non-physical scientific dimensions and must forego a detailed discussion.

By removing $CO_2$ from the atmosphere, CDR directly addresses the principal cause of climate change. Potential CDR approaches include direct air capture, currently well-understood biological methods on land (for example, afforestation), less well-understood and potentially risky methods in the ocean (for example, ocean fertilization), and accelerated weathering (for example, forming calcium carbonate on land or in the oceans).[29] While CDR is technically possible, the primary challenge is achieving the required scale of removal in a cost-effective manner, which in part presumes a comparison to the costs of other, more traditional GHG mitigation options.[31, 32] In principle, at large scale, CDR could measurably reduce $CO_2$ concentrations (that is, cause negative emissions). Point-source capture (as opposed to $CO_2$ capture from ambient air) and removal of $CO_2$ is a particularly effective CDR method. The climate value of avoided $CO_2$ emissions is essentially equivalent to that of the atmospheric removal of the same amount. To realize sustained climate benefits from CDR, however, the removal of $CO_2$ from the atmosphere must be essentially permanent— at least several centuries to millennia. In addition to high costs, CDR has the additional limitation of long implementation times.

By contrast, SRM approaches offer the only known CI methods of cooling Earth within a few years after inception. An important limitation of SRM is that it would not address damage to ocean ecosystems from increasing ocean acidification due to continued $CO_2$ uptake. SRM could theoretically have a significant global impact even if implemented by a small number of nations, and by nations that are not also the major emitters of GHGs; this could be viewed either as a benefit or risk of SRM.[30]

Proposed SRM concepts increase Earth's albedo through injection of sulfur gases or aerosols into the stratosphere (thereby simulating the effects of explosive volcanic eruptions) or marine cloud brightening through aerosol injection near the ocean surface. Injection of solid particles is an alternative to sulfur and yet other SRM methods could be deployed in space. Studies have evaluated the expected effort and effectiveness of various SRM methods.[30, 33] For example, model runs were performed in the GeoMIP project using the full CMIP5 model suite to illustrate the effect of reducing top-of-the-atmosphere insolation to offset climate warming from $CO_2$.[34] The idealized runs, which assumed an abrupt, globally-uniform insolation reduction in a $4 \times CO_2$ atmosphere, show that temperature increases are largely offset, most sea ice loss is avoided, average precipitation changes are small, and net primary productivity increases. However, important regional changes in climate variables are likely in SRM scenarios as discussed below.

As global ambitions increase to avoid or remove $CO_2$ emissions, probabilities of large increases in global temperatures by 2100 are proportionately reduced.[27] Scenarios in which large-scale CDR is used to meet a 3.6°F (2°C) limit while allowing business-as-usual consumption of fossil fuels are likely not feasible with present technologies. Model SRM scenarios have been developed that show reductions in radiative forcing up to 1 W/m² with annual stratospheric injections of 1 Mt of sulfur from aircraft or other platforms.[35, 36] Preliminary studies suggest that this could be accomplished at an implementation cost as low as a few billion dollars per year using current technology, enabling an individual country or subnational entity to conduct activities having significant global climate impacts.

SRM scenarios could in principle be designed to follow a particular radiative forcing trajectory, with adjustments made in response to monitoring of the climate effects.[37] SRM

could be used as an interim measure to avoid peaks in global average temperature and other climate parameters. The assumption is often made that SRM measures, once implemented, must continue indefinitely in order to avoid the rapid climate change that would occur if the measures were abruptly stopped. SRM could be used, however, as an interim measure to buy time for the implementation of emissions reductions and/or CDR, and SRM could be phased out as emissions reductions and CDR are phased in, to avoid abrupt changes in radiative forcing.[37]

SRM via marine cloud brightening derives from changes in cloud albedo from injection of aerosols into low-level clouds, primarily over the oceans. Clouds with smaller and more numerous droplets reflect more sunlight than clouds with fewer and larger droplets. Current models provide more confidence in the effects of stratospheric injection than in marine cloud brightening and in achieving scales large enough to reduce global forcing.[30]

CDR and SRM have substantial uncertainties regarding their effectiveness and unintended consequences. For example, CDR on a large scale may disturb natural systems and have important implications for land-use changes. For SRM actions, even if the reduction in global average radiative forcing from SRM was exactly equal to the radiative forcing from GHGs, the regional and temporal patterns of these forcings would have important differences. While SRM could rapidly lower global mean temperatures, the effects on precipitation patterns, light availability, crop yields, acid rain, pollution levels, temperature gradients, and atmospheric circulation in response to such actions are less well understood. Also, the reduction in sunlight from SRM may have effects on agriculture and ecosystems. In general, restoring regional preindustrial temperature and precipitation conditions through

SRM actions is not expected to be possible based on ensemble modeling studies.[38] As a consequence, optimizing the climate and geopolitical value of SRM actions would likely involve tradeoffs between regional temperature and precipitation changes.[39] Alternatively, intervention options have been proposed to address particular regional impacts.[40]

GHG forcing has the potential to push the climate farther into unprecedented states for human civilization and increase the likelihood of "surprises" (see Ch. 15: Potential Surprises). CI could prevent climate change from reaching a state with more unpredictable consequences. The potential for rapid changes upon initiation (or ceasing) of a CI action would require adaptation on timescales significantly more rapid than what would otherwise be necessary. The NAS[29, 30] and the Royal Society[41] recognized that research on the feasibilities and consequences of CI actions is incomplete and call for continued research to improve knowledge of the feasibility, risks, and benefits of CI techniques.

# TRACEABLE ACCOUNTS

### Key Finding 1

Reducing net emissions of $CO_2$ is necessary to limit near-term climate change and long-term warming. Other greenhouse gases (for example, methane) and black carbon aerosols exert stronger warming effects than $CO_2$ on a per ton basis, but they do not persist as long in the atmosphere; therefore, mitigation of non-$CO_2$ species contributes substantially to near-term cooling benefits but cannot be relied upon for ultimate stabilization goals. (*Very high confidence*)

### Description of evidence base

Joos et al.[2] and Ciais et al. (see Box 6.1 in particular)[1] describe the climate response of $CO_2$ pulse emissions, and Solomon et al.,[4] NRC,[19] and Collins et al.[3] describe the long-term warming and other climate effects associated with $CO_2$ emissions. Paltsev et al.[8] and Collins et al.[3] describe the near-term vs. long-term nature of climate outcomes resulting from GHG mitigation. Myhre et al.[11] synthesize numerous studies detailing information about the radiative forcing effects and atmospheric lifetimes of all GHGs and aerosols (see in particular Appendix 8A therein). A recent body of literature has emerged highlighting the particular role that non-$CO_2$ mitigation can play in providing near-term cooling benefits (e.g., Shindell et al. 2012;[17] Zaelke and Borgford-Parnell 2015;[10] Rogelj et al. 2015[18]). For each of the individual statements made in Key Finding 1, there are numerous literature sources that provide consistent grounds on which to make these statements with *very high confidence*.

### Major uncertainties

The Key Finding is comprised of qualitative statements that are traceable to the literature described above and in this chapter. Uncertainties affecting estimates of the exact timing and magnitude of the climate response following emissions (or avoidance of those emissions) of $CO_2$ and other GHGs involve the quantity of emissions, climate sensitivity, some uncertainty about the removal time or atmospheric lifetime of $CO_2$ and other GHGs, and the choice of model carrying out future simulations. The role of black carbon in climate change is

more uncertain compared to the role of the well-mixed GHGs (see Bond et al. 2013[12]).

### Assessment of confidence based on evidence and agreement, including short description of nature of evidence and level of agreement

Key Finding 1 is comprised of qualitative statements based on a body of literature for which there is a high level of agreement. There is a well-established understanding, based in the literature, of the atmospheric lifetime and warming effects of $CO_2$ vs. other GHGs after emission, and in turn how atmospheric concentration levels respond following the emission of $CO_2$ and other GHGs.

### Summary sentence or paragraph that integrates the above information

The qualitative statements contained in Key Finding 1 reflect aspects of fundamental scientific understanding, well grounded in the literature, that provide a relevant framework for considering the role of $CO_2$ and non-$CO_2$ species in mitigating climate change.

### Key Finding 2

Stabilizing global mean temperature to less than 3.6°F (2°C) above preindustrial levels requires substantial reductions in net global $CO_2$ emissions prior to 2040 relative to present-day values and likely requires net emissions to become zero or possibly negative later in the century. After accounting for the temperature effects of non-$CO_2$ species, cumulative global $CO_2$ emissions must stay below about 800 GtC in order to provide a two-thirds likelihood of preventing 3.6°F (2°C) of warming. Given estimated cumulative emissions since 1870, no more than approximately 230 GtC may be emitted in the future to remain under this temperature threshold. Assuming global emissions are equal to or greater than those consistent with the RCP4.5 scenario, this cumulative carbon threshold would be exceeded in approximately two decades. (*High confidence*)

### Description of evidence base

Key Finding 2 is a case study, focused on a pathway associated with 3.6°F (2°C) of warming, based on the more general concepts described in the chapter. As such, the evidence for the relationship between cumulative $CO_2$ emissions and global mean temperature response[3, 19, 21] also supports Key Finding 3.

Numerous studies have provided best estimates of cumulative $CO_2$ compatible with 3.6°F (2°C) of warming above preindustrial levels, including a synthesis by the IPCC.[3] Sanderson et al.[25] provide further recent evidence to support the statement that net $CO_2$ emissions would need to approach zero or become negative later in the century in order to avoid this level of warming. Rogelj et al. 2015[18] and the IPCC[3] demonstrate that the consideration of non-$CO_2$ species has the effect of further constraining the amount of cumulative $CO_2$ emissions compatible with 3.6°F (2°C) of warming.

Table 14.1 shows the IPCC estimates associated with different probabilities (66% [the one highlighted in Key Finding 2], 50%, and 33%) of cumulative $CO_2$ emissions compatible with warming of 3.6°F (2°C) above preindustrial levels, and the cumulative $CO_2$ emissions compatible with 2.7°F (1.5°C) are in turn linearly derived from those, based on the understanding that cumulative emissions scale linearly with global mean temperature response. The IPCC estimates take into account the additional radiative forcing effects—past and future—of non-$CO_2$ species based on the emissions pathways consistent with the RCP scenarios (available here: https://tntcat.iiasa.ac.at/RcpDb/dsd?Action=htmlpage&page=about#descript).

The authors calculated the dates shown in Table 14.1, which supports the last statement in Key Finding 2, based on Le Quéré et al.[20] and the publicly available RCP database. Le Quéré et al.[20] provide the widely used reference for historical global, annual $CO_2$ emissions from 1870 to 2015 (land-use change emissions were estimated up to year 2010 so are assumed to be constant between 2010 and 2015). Future $CO_2$ emissions are based on the lower and higher scenarios (RCP4.5 and RCP8.5, respectively); annual numbers between model-projected years (2020, 2030, 2040, etc.) are linearly interpolated.

**Major uncertainties**

There are large uncertainties about the course of future $CO_2$ and non-$CO_2$ emissions, but the fundamental point that $CO_2$ emissions need to eventually approach zero or possibly become net negative to stabilize warming below 3.6°F (2°C) holds regardless of future emissions scenario. There are also large uncertainties about the magnitude of past (since 1870 in this case) $CO_2$ and non-$CO_2$ emissions, which in turn influence the uncertainty about compatible cumulative emissions from the present day forward. Further uncertainties regarding non-$CO_2$ species, including aerosols, include their radiative forcing effects. The uncertainty in achieving the temperature targets for a given emissions pathway is, in large part, reflected by the range of probabilities shown in Table 14.1.

**Assessment of confidence based on evidence and agreement, including short description of nature of evidence and level of agreement**

There is *very high* confidence in the first statement of Key Finding 2 because it is based on a number of sources with a high level of agreement. The role of non-$CO_2$ species in particular introduces uncertainty in the second statement of Key Finding 2 regarding compatible cumulative $CO_2$ emissions that take into account past and future radiative forcing effects of non-$CO_2$ species; though this estimate is based on a synthesis of numerous studies by the IPCC. The last statement of Key Finding 2 is straightforward based on the best available estimates of historical emissions in combination with the widely used future projections of the RCP scenarios.

**Summary sentence or paragraph that integrates the above information**

Fundamental scientific understanding of the climate system provides a framework for considering potential pathways for achieving a target of preventing 3.6°F (2°C) of warming. There are uncertainties about cumulative $CO_2$ emissions compatible with this goal, in large part because of uncertainties about the role of non-$CO_2$ species, but it appears, based on past emissions and future projections, that the cumulative carbon threshold for this goal could be reached or exceeded in about two decades.

### Key Finding 3

Achieving global greenhouse gas emissions reductions before 2030 consistent with targets and actions announced by governments in the lead up to the 2015 Paris climate conference would hold open the possibility of meeting the long-term temperature goal of limiting global warming to 3.6°F (2°C) above preindustrial levels, whereas there would be virtually no chance if global net emissions followed a pathway well above those implied by country announcements. Actions in the announcements are, by themselves, insufficient to meet a 3.6°F (2°C) goal; the likelihood of achieving that goal depends strongly on the magnitude of global emissions reductions after 2030. (*High confidence*)

### Description of evidence base

The primary source supporting this key finding is Fawcett et al.;[27] it is also supported by Rogelj et al.,[24] Sanderson et al.,[25] and the Climate Action Tracker.[26] Each of these analyses evaluated the global climate implications of the aggregation of the individual country contributions thus far put forward under the Paris Agreement.

### Major uncertainties

The largest uncertainty lies in the assumption of achieving emissions reductions consistent with the announcements prior to December 2015; these reductions are assumed to be achieved but could either be over- or underachieved. This in turn creates uncertainty about the extent of emissions reductions that would be needed after the first round of government announcements in order to achieve the 2°C or any other target. The response of the climate system, the climate sensitivity, is also a source of uncertainty; the Fawcett et al. analysis used the IPCC AR5 range, 1.5° to 4.5°C.

### Assessment of confidence based on evidence and agreement, including short description of nature of evidence and level of agreement

There is *high* confidence in this key finding because a number of analyses have examined the implications of these announcements and have come to similar conclusions, as captured in this key finding.

### Summary sentence or paragraph that integrates the above information

Different analyses have estimated the implications for global mean temperature of the emissions reductions consistent with the actions announced by governments in the lead up to the 2015 Paris climate conference and have reached similar conclusions. Assuming emissions reductions indicated in these announcements are achieved, along with a range of climate sensitivities, these contributions provide some likelihood of meeting the long-term goal of limiting global warming to well below 3.6°F (2°C) above preindustrial levels, but much depends on assumptions about what happens after 2030.

### Key Finding 4

Further assessments of the technical feasibilities, costs, risks, co-benefits, and governance challenges of climate intervention or geoengineering strategies, which are as yet unproven at scale, are a necessary step before judgments about the benefits and risks of these approaches can be made with high confidence. (*High confidence*)

### Description of evidence base

Key Finding 4 contains qualitative statements based on the growing literature addressing this topic, including from such bodies as the National Academy of Sciences and the Royal Society, coupled with judgment by the authors about the future interest level in this topic.

### Major uncertainties

The major uncertainty is how public perception and interest among policymakers in climate intervention may change over time, even independently from the perceived level of progress made towards reducing $CO_2$ and other GHG emissions over time.

### Assessment of confidence based on evidence and agreement, including short description of nature of evidence and level of agreement

There is *high* confidence that climate intervention strategies may gain greater attention, especially if efforts to slow the buildup of atmospheric $CO_2$ and other GHGs are considered inadequate by many in the scientific and policy communities.

**Summary sentence or paragraph that integrates the above information**

The key finding is a qualitative statement based on the growing literature on this topic. The uncertainty moving forward is the comfort level and desire among numerous stakeholders to research and potentially carry out these climate intervention strategies, particularly in light of how progress by the global community to reduce GHG emissions is perceived.

# REFERENCES

1. Ciais, P., C. Sabine, G. Bala, L. Bopp, V. Brovkin, J. Canadell, A. Chhabra, R. DeFries, J. Galloway, M. Heimann, C. Jones, C. Le Quéré, R.B. Myneni, S. Piao, and P. Thornton, 2013: Carbon and other bio-geochemical cycles. *Climate Change 2013: The Physical Science Basis. Contribution of Working Group I to the Fifth Assessment Report of the Intergovernmental Panel on Climate Change.* Stocker, T.F., D. Qin, G.-K. Plattner, M. Tignor, S.K. Allen, J. Boschung, A. Nauels, Y. Xia, V. Bex, and P.M. Midgley, Eds. Cambridge University Press, Cambridge, United Kingdom and New York, NY, USA, 465–570. http://www.climatechange2013.org/report/full-report/

2. Joos, F., R. Roth, J.S. Fuglestvedt, G.P. Peters, I.G. Enting, W. von Bloh, V. Brovkin, E.J. Burke, M. Eby, N.R. Edwards, T. Friedrich, T.L. Frölicher, P.R. Halloran, P.B. Holden, C. Jones, T. Kleinen, F.T. Mackenzie, K. Matsumoto, M. Meinshausen, G.K. Plattner, A. Reisinger, J. Segschneider, G. Shaffer, M. Steinacher, K. Strassmann, K. Tanaka, A. Timmermann, and A.J. Weaver, 2013: Carbon dioxide and climate impulse response functions for the computation of greenhouse gas metrics: A multi-model analysis. *Atmospheric Chemistry and Physics*, **13**, 2793-2825. http://dx.doi.org/10.5194/acp-13-2793-2013

3. Collins, M., R. Knutti, J. Arblaster, J.-L. Dufresne, T. Fichefet, P. Friedlingstein, X. Gao, W.J. Gutowski, T. Johns, G. Krinner, M. Shongwe, C. Tebaldi, A.J. Weaver, and M. Wehner, 2013: Long-term climate change: Projections, commitments and irreversibility. *Climate Change 2013: The Physical Science Basis. Contribution of Working Group I to the Fifth Assessment Report of the Intergovernmental Panel on Climate Change.* Stocker, T.F., D. Qin, G.-K. Plattner, M. Tignor, S.K. Allen, J. Boschung, A. Nauels, Y. Xia, V. Bex, and P.M. Midgley, Eds. Cambridge University Press, Cambridge, United Kingdom and New York, NY, USA, 1029–1136. http://www.climatechange2013.org/report/full-report/

4. Solomon, S., G.K. Plattner, R. Knutti, and P. Friedlingstein, 2009: Irreversible climate change due to carbon dioxide emissions. *Proceedings of the National Academy of Sciences of the United States of America*, **106**, 1704-1709. http://dx.doi.org/10.1073/pnas.0812721106

5. Tebaldi, C. and P. Friedlingstein, 2013: Delayed detection of climate mitigation benefits due to climate inertia and variability. *Proceedings of the National Academy of Sciences*, **110**, 17229-17234. http://dx.doi.org/10.1073/pnas.1300005110

6. Prather, M.J., J.E. Penner, J.S. Fuglestvedt, A. Kurosawa, J.A. Lowe, N. Höhne, A.K. Jain, N. Andronova, L. Pinguelli, C. Pires de Campos, S.C.B. Raper, R.B. Skeie, P.A. Stott, J. van Aardenne, and F. Wagner, 2009: Tracking uncertainties in the causal chain from human activities to climate. *Geophysical Research Letters*, **36**, L05707. http://dx.doi.org/10.1029/2008GL036474

7. Kirtman, B., S.B. Power, J.A. Adedoyin, G.J. Boer, R. Bojariu, I. Camilloni, F.J. Doblas-Reyes, A.M. Fiore, M. Kimoto, G.A. Meehl, M. Prather, A. Sarr, C. Schär, R. Sutton, G.J. van Oldenborgh, G. Vecchi, and H.J. Wang, 2013: Near-term climate change: Projections and predictability. *Climate Change 2013: The Physical Science Basis. Contribution of Working Group I to the Fifth Assessment Report of the Intergovernmental Panel on Climate Change.* Stocker, T.F., D. Qin, G.-K. Plattner, M. Tignor, S.K. Allen, J. Boschung, A. Nauels, Y. Xia, V. Bex, and P.M. Midgley, Eds. Cambridge University Press, Cambridge, UK and New York, NY, USA, 953–1028. http://www.climatechange2013.org/report/full-report/

8. Paltsev, S., E. Monier, J. Scott, A. Sokolov, and J. Reilly, 2015: Integrated economic and climate projections for impact assessment. *Climatic Change*, **131**, 21-33. http://dx.doi.org/10.1007/s10584-013-0892-3

9. Tebaldi, C. and M.F. Wehner, 2016: Benefits of mitigation for future heat extremes under RCP4.5 compared to RCP8.5. *Climatic Change*, **First online**, 1-13. http://dx.doi.org/10.1007/s10584-016-1605-5

10. Zaelke, D. and N. Borgford-Parnell, 2015: The importance of phasing down hydrofluorocarbons and other short-lived climate pollutants. *Journal of Environmental Studies and Sciences*, **5**, 169-175. http://dx.doi.org/10.1007/s13412-014-0215-7

11. Myhre, G., D. Shindell, F.-M. Bréon, W. Collins, J. Fuglestvedt, J. Huang, D. Koch, J.-F. Lamarque, D. Lee, B. Mendoza, T. Nakajima, A. Robock, G. Stephens, T. Takemura, and H. Zhang, 2013: Anthropogenic and natural radiative forcing. *Climate Change 2013: The Physical Science Basis. Contribution of Working Group I to the Fifth Assessment Report of the Intergovernmental Panel on Climate Change.* Stocker, T.F., D. Qin, G.-K. Plattner, M. Tignor, S.K. Allen, J. Boschung, A. Nauels, Y. Xia, V. Bex, and P.M. Midgley, Eds. Cambridge University Press, Cambridge, United Kingdom and New York, NY, USA, 659–740. http://www.climatechange2013.org/report/full-report/

12. Bond, T.C., S.J. Doherty, D.W. Fahey, P.M. Forster, T. Berntsen, B.J. DeAngelo, M.G. Flanner, S. Ghan, B. Kärcher, D. Koch, S. Kinne, Y. Kondo, P.K. Quinn, M.C. Sarofim, M.G. Schultz, M. Schulz, C. Venkataraman, H. Zhang, S. Zhang, N. Bellouin, S.K. Guttikunda, P.K. Hopke, M.Z. Jacobson, J.W. Kaiser, Z. Klimont, U. Lohmann, J.P. Schwarz, D. Shindell, T. Storelvmo, S.G. Warren, and C.S. Zender, 2013: Bounding the role of black carbon in the climate system: A scientific assessment. *Journal of Geophysical Research Atmospheres*, **118**, 5380-5552. http://dx.doi.org/10.1002/jgrd.50171

13. Anenberg, S.C., J. Schwartz, D. Shindell, M. Amann, G. Faluvegi, Z. Klimont, G. Janssens-Maenhout, L. Pozzoli, R. Van Dingenen, E. Vignati, L. Emberson, N.Z. Muller, J.J. West, M. Williams, V. Demkine, W.K. Hicks, J. Kuylenstierna, F. Raes, and V. Ramanathan, 2012: Global air quality and health co-benefits of mitigating near-term climate change through methane and black carbon emission controls. *Environmental Health Perspectives*, **120**, 831-839. http://dx.doi.org/10.1289/ehp.1104301

14. Rao, S., Z. Klimont, J. Leitao, K. Riahi, R. van Dingenen, L.A. Reis, K. Calvin, F. Dentener, L. Drouet, S. Fujimori, M. Harmsen, G. Luderer, C. Heyes, J. Strefler, M. Tavoni, and D.P. van Vuuren, 2016: A multi-model assessment of the co-benefits of climate mitigation for global air quality. *Environmental Research Letters*, **11**, 124013. http://dx.doi.org/10.1088/1748-9326/11/12/124013

15. Hayhoe, K.A.S., H.S. Kheshgi, A.K. Jain, and D.J. Wuebbles, 1998: Tradeoffs in fossil fuel use: The effects of CO2, CH4, and SO2 aerosol emissions on climate. *World Resources Review*, **10**.

16. Shah, N., M. Wei, E.L. Virginie, and A.A. Phadke, 2015: Benefits of Leapfrogging to Superefficiency and Low Global Warming Potential Refrigerants in Room Air Conditioning. Lawrence Berkeley National Laboratory, Energy Technology Area, Berkeley, CA. 39 pp. https://eetd.lbl.gov/publications/benefits-of-leapfrogging-to-superef-0

17. Shindell, D., J.C.I. Kuylenstierna, E. Vignati, R. van Dingenen, M. Amann, Z. Klimont, S.C. Anenberg, N. Muller, G. Janssens-Maenhout, F. Raes, J. Schwartz, G. Faluvegi, L. Pozzoli, K. Kupiainen, L. Hoglund-Isaksson, L. Emberson, D. Streets, V. Ramanathan, K. Hicks, N.T.K. Oanh, G. Milly, M. Williams, V. Demkine, and D. Fowler, 2012: Simultaneously mitigating near-term climate change and improving human health and food security. *Science*, **335**, 183-189. http://dx.doi.org/10.1126/science.1210026

18. Rogelj, J., M. Meinshausen, M. Schaeffer, R. Knutti, and K. Riahi, 2015: Impact of short-lived non-CO$_2$ mitigation on carbon budgets for stabilizing global warming. *Environmental Research Letters*, **10**, 075001. http://dx.doi.org/10.1088/1748-9326/10/7/075001

19. NRC, 2011: *Climate Stabilization Targets: Emissions, Concentrations, and Impacts over Decades to Millennia*. National Research Council. The National Academies Press, Washington, D.C., 298 pp. http://dx.doi.org/10.17226/12877

20. Le Quéré, C., R.M. Andrew, J.G. Canadell, S. Sitch, J.I. Korsbakken, G.P. Peters, A.C. Manning, T.A. Boden, P.P. Tans, R.A. Houghton, R.F. Keeling, S. Alin, O.D. Andrews, P. Anthoni, L. Barbero, L. Bopp, F. Chevallier, L.P. Chini, P. Ciais, K. Currie, C. Delire, S.C. Doney, P. Friedlingstein, T. Gkritzalis, I. Harris, J. Hauck, V. Haverd, M. Hoppema, K. Klein Goldewijk, A.K. Jain, E. Kato, A. Körtzinger, P. Landschützer, N. Lefèvre, A. Lenton, S. Lienert, D. Lombardozzi, J.R. Melton, N. Metzl, F. Millero, P.M.S. Monteiro, D.R. Munro, J.E.M.S. Nabel, S.I. Nakaoka, K. O'Brien, A. Olsen, A.M. Omar, T. Ono, D. Pierrot, B. Poulter, C. Rödenbeck, J. Salisbury, U. Schuster, J. Schwinger, R. Séférian, I. Skjelvan, B.D. Stocker, A.J. Sutton, T. Takahashi, H. Tian, B. Tilbrook, I.T. van der Laan-Luijkx, G.R. van der Werf, N. Viovy, A.P. Walker, A.J. Wiltshire, and S. Zaehle, 2016: Global carbon budget 2016. *Earth System Science Data*, **8**, 605-649. http://dx.doi.org/10.5194/essd-8-605-2016

21. Allen, M.R., D.J. Frame, C. Huntingford, C.D. Jones, J.A. Lowe, M. Meinshausen, and N. Meinshausen, 2009: Warming caused by cumulative carbon emissions towards the trillionth tonne. *Nature*, **458**, 1163-1166. http://dx.doi.org/10.1038/nature08019

22. Nordhaus, W.D., 1975: Can We Control Carbon Dioxide? International Institute for Applied Systems Analysis (IIASA), Laxenburg, Austria. 47 pp. http://pure.iiasa.ac.at/365/

23. Stockholm Environment Institute, 1990: Targets and Indicators of Climatic Change. Rijsberman, F.R. and R.J. Swart (Eds.). Stockholm Environment Institute, Stockholm, Sweden. 166 pp. https://www.sei-international.org/mediamanager/documents/Publications/SEI-Report-TargetsAndIndicatorsOfClimaticChange-1990.pdf

24. Rogelj, J., M. den Elzen, N. Höhne, T. Fransen, H. Fekete, H. Winkler, R. Schaeffer, F. Sha, K. Riahi, and M. Meinshausen, 2016: Paris Agreement climate proposals need a boost to keep warming well below 2°C. *Nature*, **534**, 631-639. http://dx.doi.org/10.1038/nature18307

25. Sanderson, B.M., B.C. O'Neill, and C. Tebaldi, 2016: What would it take to achieve the Paris temperature targets? *Geophysical Research Letters*, **43**, 7133-7142. http://dx.doi.org/10.1002/2016GL069563

26. Climate Action Tracker, 2016: Climate Action Tracker. http://climateactiontracker.org/global.html

27. Fawcett, A.A., G.C. Iyer, L.E. Clarke, J.A. Edmonds, N.E. Hultman, H.C. McJeon, J. Rogelj, R. Schuler, J. Alsalam, G.R. Asrar, J. Creason, M. Jeong, J. McFarland, A. Mundra, and W. Shi, 2015: Can Paris pledges avert severe climate change? *Science*, **350**, 1168-1169. http://dx.doi.org/10.1126/science.aad5761

28. UNFCCC, 2015: Paris Agreement. United Nations Framework Convention on Climate Change, [Bonn, Germany]. 25 pp. http://unfccc.int/files/essential_background/convention/application/pdf/english_paris_agreement.pdf

29. NAS, 2015: *Climate Intervention: Carbon Dioxide Removal and Reliable Sequestration*. The National Academies Press, Washington, DC, 154 pp. http://dx.doi.org/10.17226/18805

30. NAS, 2015: *Climate Intervention: Reflecting Sunlight to Cool Earth*. The National Academies Press, Washington, DC, 260 pp. http://dx.doi.org/10.17226/18988

31. Fuss, S., J.G. Canadell, G.P. Peters, M. Tavoni, R.M. Andrew, P. Ciais, R.B. Jackson, C.D. Jones, F. Kraxner, N. Nakicenovic, C. Le Quere, M.R. Raupach, A. Sharifi, P. Smith, and Y. Yamagata, 2014: Betting on negative emissions. *Nature Climate Change*, **4**, 850-853. http://dx.doi.org/10.1038/nclimate2392

32. Smith, P., S.J. Davis, F. Creutzig, S. Fuss, J. Minx, B. Gabrielle, E. Kato, R.B. Jackson, A. Cowie, E. Kriegler, D.P. van Vuuren, J. Rogelj, P. Ciais, J. Milne, J.G. Canadell, D. McCollum, G. Peters, R. Andrew, V. Krey, G. Shrestha, P. Friedlingstein, T. Gasser, A. Grubler, W.K. Heidug, M. Jonas, C.D. Jones, F. Kraxner, E. Littleton, J. Lowe, J.R. Moreira, N. Nakicenovic, M. Obersteiner, A. Patwardhan, M. Rogner, E. Rubin, A. Sharifi, A. Torvanger, Y. Yamagata, J. Edmonds, and C. Yongsung, 2016: Biophysical and economic limits to negative CO2 emissions. *Nature Climate Change*, **6**, 42-50. http://dx.doi.org/10.1038/nclimate2870

33. Keith, D.W., R. Duren, and D.G. MacMartin, 2014: Field experiments on solar geoengineering: Report of a workshop exploring a representative research portfolio. *Philosophical Transactions of the Royal Society A: Mathematical, Physical and Engineering Sciences*, **372**, 20140175. http://dx.doi.org/10.1098/rsta.2014.0175

34. Kravitz, B., K. Caldeira, O. Boucher, A. Robock, P.J. Rasch, K. Alterskjær, D.B. Karam, J.N.S. Cole, C.L. Curry, J.M. Haywood, P.J. Irvine, D. Ji, A. Jones, J.E. Kristjánsson, D.J. Lunt, J.C. Moore, U. Niemeier, H. Schmidt, M. Schulz, B. Singh, S. Tilmes, S. Watanabe, S. Yang, and J.-H. Yoon, 2013: Climate model response from the Geoengineering Model Intercomparison Project (GeoMIP). *Journal of Geophysical Research Atmospheres*, **118**, 8320-8332. http://dx.doi.org/10.1002/jgrd.50646

35. Pierce, J.R., D.K. Weisenstein, P. Heckendorn, T. Peter, and D.W. Keith, 2010: Efficient formation of stratospheric aerosol for climate engineering by emission of condensible vapor from aircraft. *Geophysical Research Letters*, **37**, L18805. http://dx.doi.org/10.1029/2010GL043975

36. Tilmes, S., B.M. Sanderson, and B.C. O'Neill, 2016: Climate impacts of geoengineering in a delayed mitigation scenario. *Geophysical Research Letters*, **43**, 8222-8229. http://dx.doi.org/10.1002/2016GL070122

37. Keith, D.W. and D.G. MacMartin, 2015: A temporary, moderate and responsive scenario for solar geoengineering. *Nature Climate Change*, **5**, 201-206. http://dx.doi.org/10.1038/nclimate2493

38. Ricke, K.L., M.G. Morgan, and M.R. Allen, 2010: Regional climate response to solar-radiation management. *Nature Geoscience*, **3**, 537-541. http://dx.doi.org/10.1038/ngeo915

39. MacMartin, D.G., D.W. Keith, B. Kravitz, and K. Caldeira, 2013: Management of trade-offs in geoengineering through optimal choice of non-uniform radiative forcing. *Nature Climate Change*, **3**, 365-368. http://dx.doi.org/10.1038/nclimate1722

40. MacCracken, M.C., 2016: The rationale for accelerating regionally focused climate intervention research. *Earth's Future*, **4**, 649-657. http://dx.doi.org/10.1002/2016EF000450

41. Shepherd, J.G., K. Caldeira, P. Cox, J. Haigh, D. Keith, B. Launder, G. Mace, G. MacKerron, J. Pyle, S. Rayner, C. Redgwell, and A. Watson, 2009: *Geoengineering the Climate: Science, Governance and Uncertainty*. Royal Society, 82 pp. http://eprints.soton.ac.uk/156647/1/Geoengineering_the_climate.pdf

42. IPCC, 2013: Summary for policymakers. *Climate Change 2013: The Physical Science Basis. Contribution of Working Group I to the Fifth Assessment Report of the Intergovernmental Panel on Climate Change*. Stocker, T.F., D. Qin, G.-K. Plattner, M. Tignor, S.K. Allen, J. Boschung, A. Nauels, Y. Xia, V. Bex, and P.M. Midgley, Eds. Cambridge University Press, Cambridge, United Kingdom and New York, NY, USA, 1–30. http://www.climatechange2013.org/report/

# 15

# Potential Surprises: Compound Extremes and Tipping Elements

## KEY FINDINGS

1. Positive feedbacks (self-reinforcing cycles) within the climate system have the potential to accelerate human-induced climate change and even shift the Earth's climate system, in part or in whole, into new states that are very different from those experienced in the recent past (for example, ones with greatly diminished ice sheets or different large-scale patterns of atmosphere or ocean circulation). Some feedbacks and potential state shifts can be modeled and quantified; others can be modeled or identified but not quantified; and some are probably still unknown. (*Very high confidence* in the potential for state shifts and in the incompleteness of knowledge about feedbacks and potential state shifts).

2. The physical and socioeconomic impacts of compound extreme events (such as simultaneous heat and drought, wildfires associated with hot and dry conditions, or flooding associated with high precipitation on top of snow or waterlogged ground) can be greater than the sum of the parts (*very high confidence*). Few analyses consider the spatial or temporal correlation between extreme events.

3. While climate models incorporate important climate processes that can be well quantified, they do not include all of the processes that can contribute to feedbacks, compound extreme events, and abrupt and/or irreversible changes. For this reason, future changes outside the range projected by climate models cannot be ruled out (*very high confidence*). Moreover, the systematic tendency of climate models to underestimate temperature change during warm paleoclimates suggests that climate models are more likely to underestimate than to overestimate the amount of long-term future change (*medium confidence*).

**Recommended Citation for Chapter**

**Kopp**, R.E., K. Hayhoe, D.R. Easterling, T. Hall, R. Horton, K.E. Kunkel, and A.N. LeGrande, 2017: Potential surprises – compound extremes and tipping elements. In: *Climate Science Special Report: Fourth National Climate Assessment, Volume I* [Wuebbles, D.J., D.W. Fahey, K.A. Hibbard, D.J. Dokken, B.C. Stewart, and T.K. Maycock (eds.)]. U.S. Global Change Research Program, Washington, DC, USA, pp. 411-429, doi: 10.7930/J0GB227J.

## 15.1 Introduction

The Earth system is made up of many components that interact in complex ways across a broad range of temporal and spatial scales. As a result of these interactions the behavior of the system cannot be predicted by looking at individual components in isolation. Negative feedbacks, or self-stabilizing cycles, within and between components of the Earth system can dampen changes (Ch. 2: Physical Drivers of Climate Change). However, their stabilizing effects render such feedbacks of less concern from a risk perspective than positive feedbacks, or self-reinforcing cycles. Positive feedbacks magnify both natural and anthropogenic changes. Some Earth system components, such as arctic sea ice and the polar ice sheets, may exhibit thresholds beyond which these self-reinforcing cycles can drive the component, or the entire system, into a radically different state. Although the probabilities of these state shifts may be difficult to assess, their consequences could be high, potentially exceeding anything anticipated by climate model projections for the coming century.

Humanity's effect on the Earth system, through the large-scale combustion of fossil fuels and widespread deforestation and the resulting release of carbon dioxide ($CO_2$) into the atmosphere, as well as through emissions of other greenhouse gases and radiatively active substances from human activities, is unprecedented (Ch. 2: Physical Drivers of Climate Change). These forcings are driving changes in temperature and other climate variables. Previous chapters have covered a variety of observed and projected changes in such variables, including averages and extremes of temperature, precipitation, sea level, and storm events (see Chapters 1, 4–13).

While the distribution of climate model projections provides insight into the range of possible future changes, this range is limited by the fact that models do not include or fully represent all of the known processes and components of the Earth system (e.g., ice sheets or arctic carbon reservoirs),[1] nor do they include all of the interactions between these components that contribute to the self-stabilizing and self-reinforcing cycles mentioned above (e.g., the dynamics of the interactions between ice sheets, the ocean, and the atmosphere). They also do not include currently unknown processes that may become increasingly relevant under increasingly large climate forcings. This limitation is emphasized by the systematic tendency of climate models to underestimate temperature change during warm paleoclimates (Section 15.5). Therefore, there is significant potential for humanity's effect on the planet to result in unanticipated surprises and a broad consensus that the further and faster the Earth system is pushed towards warming, the greater the risk of such surprises.

Scientists have been surprised by the Earth system many times in the past. The discovery of the ozone hole is a clear example. Prior to groundbreaking work by Molina and Rowland[2], chlorofluorocarbons (CFCs) were viewed as chemically inert; the chemistry by which they catalyzed stratospheric ozone depletion was unknown. Within eleven years of Molina and Rowland's work, British Antarctic Survey scientists reported ground observations showing that spring ozone concentrations in the Antarctic, driven by chlorine from human-emitted CFCs, had fallen by about one-third since the late 1960s.[3] The problem quickly moved from being an "unknown unknown" to a "known known," and by 1987, the Montreal Protocol was adopted to phase out these ozone-depleting substances.

Another surprise has come from arctic sea ice. While the potential for powerful positive ice-albedo feedbacks has been understood since the late 19th century, climate models have struggled to capture the magnitude of these feedbacks and to include all the relevant dynamics that affect sea ice extent. As of 2007, the observed decline in arctic sea ice from the start of the satellite era in 1979 outpaced the declines projected by almost all the models used by the Intergovernmental Panel on Climate Change's Fourth Assessment Report (AR4),[4] and it was not until AR4 that the IPCC first raised the prospect of an ice-free summer Arctic during this century.[5] More recent studies are more consistent with observations and have moved the date of an ice-free summer Arctic up to approximately mid-century (see Ch. 11: Arctic Changes).[6] But continued rapid declines—2016 featured the lowest annually averaged arctic sea ice extent on record, and the 2017 winter maximum was also the lowest on record—suggest that climate models may still be underestimating or missing relevant feedback processes. These processes could include, for example, effects of melt ponds, changes in storminess and ocean wave impacts, and warming of near surface waters.[7, 8, 9]

This chapter focuses primarily on two types of potential surprises. The first arises from potential changes in correlations between extreme events that may not be surprising on their own but together can increase the likelihood of compound extremes, in which multiple events occur simultaneously or in rapid sequence. Increasingly frequent compound extremes—either of multiple types of events (such as paired extremes of droughts and intense rainfall) or over greater spatial or temporal scales (such as a drought occurring in multiple major agricultural regions around the world or lasting for multiple decades)— are often not captured by analyses that focus solely on one type of extreme.

The second type of surprise arises from self-reinforcing cycles, which can give rise to "tipping elements"—subcomponents of the Earth

system that can be stable in multiple different states and can be "tipped" between these states by small changes in forcing, amplified by positive feedbacks. Examples of potential tipping elements include ice sheets, modes of atmosphere–ocean circulation like the El Niño–Southern Oscillation, patterns of ocean circulation like the Atlantic meridional overturning circulation, and large-scale ecosystems like the Amazon rainforest.[10, 11] While compound extremes and tipping elements constitute at least partially "known unknowns," the paleoclimate record also suggests the possibility of "unknown unknowns." These possibilities arise in part from the tendency of current climate models to underestimate past responses to forcing, for reasons that may or may not be explained by current hypotheses (e.g., hypotheses related to positive feedbacks that are unrepresented or poorly represented in existing models).

## 15.2 Risk Quantification and Its Limits

Quantifying the risk of low-probability, high-impact events, based on models or observations, usually involves examining the tails of a probability distribution function (PDF). Robust detection, attribution, and projection of such events into the future is challenged by multiple factors, including an observational record that often does not represent the full range of physical possibilities in the climate system, as well as the limitations of the statistical tools, scientific understanding, and models used to describe these processes.[12]

The 2013 Boulder, Colorado, floods and the Dust Bowl of the 1930s in the central United States are two examples of extreme events whose magnitude and/or extent are unprecedented in the observational record. Statistical approaches such as Extreme Value Theory can be used to model and estimate the magnitude of rare events that may not have occurred in the observational record, such as the "1,000-

year flood event" (i.e., a flood event with a 0.1% chance of occurrence in any given year) (e.g., Smith 1987[13]). While useful for many applications, these are not physical models: they are statistical models that are typically based on the assumption that observed patterns of natural variability (that is, the sample from which the models derive their statistics) are both valid and stationary beyond the observational period. Extremely rare events can also be assessed based upon paleoclimate records and physical modeling. In the paleoclimatic record, numerous abrupt changes have occurred since the last deglaciation, many larger than those recorded in the instrumental record. For example, tree ring records of drought in the western United States show abrupt, long-lasting megadroughts that were similar to but more intense and longer-lasting than the 1930s Dust Bowl.[14]

Since models are based on physics rather than observational data, they are not inherently constrained to any given time period or set of physical conditions. They have been used to study the Earth in the distant past and even the climate of other planets (e.g., Lunt et al. 2012;[15] Navarro et al. 2014[16]). Looking to the future, thousands of years' worth of simulations can be generated and explored to characterize small-probability, high-risk extreme events, as well as correlated extremes (see Section 15.3). However, the likelihood that such model events represent real risks is limited by well-known uncertainties in climate modeling related to parameterizations, model resolution, and limits to scientific understanding (Ch. 4: Projections). For example, conventional convective parameterizations in global climate models systematically underestimate extreme precipitation.[17] In addition, models often do not accurately capture or even include the processes, such as permafrost feedbacks, by which abrupt, non-reversible change may occur (see Section 15.4). An analysis focusing on physical climate predictions over the last 20 years found a tendency for scientific assessments such as those of the IPCC to under-predict rather than over-predict changes that were subsequently observed.[18]

## 15.3 Compound Extremes

An important aspect of surprise is the potential for compound extreme events. These can be events that occur at the same time or in sequence (such as consecutive floods in the same region) and in the same geographic location or at multiple locations within a given country or around the world (such as the 2009 Australian floods and wildfires). They may consist of multiple extreme events or of events that by themselves may not be extreme but together produce a multi-event occurrence (such as a heat wave accompanied by drought[19]). It is possible for the net impact of these events to be less than the sum of the individual events if their effects cancel each other out. For example, increasing $CO_2$ concentrations and acceleration of the hydrological cycle may mitigate the future impact of extremes in gross primary productivity that currently impact the carbon cycle.[20] However, from a risk perspective, the primary concern relates to compound extremes with additive or even multiplicative effects.

Some areas are susceptible to multiple types of extreme events that can occur simultaneously. For example, certain regions are susceptible to both flooding from coastal storms and riverine flooding from snow melt, and a compound event would be the occurrence of both simultaneously. Compound events can also result from shared forcing factors, including natural cycles like the El Niño–Southern Oscillation (ENSO); large-scale circulation patterns, such as the ridge observed during the 2011–2017 California drought (e.g., Swain et al. 2016[21]; see also Ch. 8: Droughts, Floods, and Wildfires); or relatively greater regional sensitivity to global change, as may occur in "hot spots" such as the western United States.[22] Finally, compound events can result from mutually reinforcing cycles between individual events, such as the relationship between drought and heat, linked through soil moisture and evaporation, in water-limited areas.[23]

In a changing climate, the probability of compound events can be altered if there is an underlying trend in conditions such as mean temperature, precipitation, or sea level that alters the baseline conditions or vulnerability of a region. It can also be altered if there is a change in the frequency or intensity of individual extreme events relative to the changing mean (for example, stronger storm surges, more frequent heat waves, or heavier precipitation events).

The occurrence of warm/dry and warm/ wet conditions is discussed extensively in the literature; at the global scale, these conditions have increased since the 1950s,[24] and analysis of NOAA's billion-dollar disasters illustrates the correlation between temperature and precipitation extremes during the costliest climate and weather events since 1980 (Figure 15.1, right). In the future, hot summers will become more frequent, and although it is not always clear for every region whether drought frequency will change, droughts in already dry regions, such as the southwestern United States, are likely to be more intense in a warmer world due to faster evaporation and associated surface drying.[25, 26, 27] For other regions, however, the picture is not as clear. Recent examples of heat/drought events (in the southern Great Plains in 2011 or in California, 2012–2016) have highlighted the inadequacy of traditional univariate risk assessment methods.[28] Yet a bivariate analysis for the contiguous United States of precipitation deficits and positive temperature anomalies finds no significant trend in the last 30 years.[29]

Another compound event frequently discussed in the literature is the increase in wildfire risk resulting from the combined effects of high precipitation variability (wet seasons followed by dry), elevated temperature, and low humidity. If followed by heavy rain, wildfires can in turn increase the risk of landslides and erosion. They can also radically increase emissions of greenhouse gases, as demonstrated by the amount of carbon dioxide produced by the Fort McMurray fires of May 2016—more than 10% of Canada's annual emissions.

A third example of a compound event involves flooding arising from wet conditions due to precipitation or to snowmelt, which could be exacerbated by warm temperatures. These wet conditions lead to high groundwater levels, saturated soils, and/or elevated river flows, which can increase the risk of flooding associated with a given storm days or even months later.[23]

Compound events may surprise in two ways. The first is if known types of compound events recur, but are stronger, longer-lasting, and/or more widespread than those experienced in the observational record or projected by model simulations for the future. One example would be simultaneous drought events in different agricultural regions across the country, or even around the world, that challenge the ability of human systems to provide adequate affordable food. Regions that lack the ability to adapt would be most vulnerable to this risk (e.g., Fraser et al. 2013[30]). Another example would be the concurrent and more severe heavy precipitation events that have occurred in the U.S. Midwest in recent years. After record insurance payouts following the events, in 2014 several insurance companies, led by Farmers Insurance, sued the city of Chicago and surrounding counties for failing to adequately prepare for the impacts of a changing climate. Although the suit was dropped later that same year, their point was made: in some regions of the United States, the insurance industry is not able to cope with the increasing frequency and/or concurrence of certain types of extreme events.

The second way in which compound events could surprise would be the emergence of new types of compound events not observed in the historical record or predicted by model simulations, due to model limitations (in terms of both their spatial resolution as well as their ability to explicitly resolve the physical processes that would result in such compound events), an increase in the frequency of such events from human-induced climate change, or both. An example is Hurricane Sandy, where sea level rise, anomalously high ocean temperatures, and high tides combined to strengthen both the storm and the magnitude of the associated storm surge.[31] At the same time, a blocking ridge over Greenland—a feature whose strength and frequency may be related to both Greenland surface melt and reduced summer sea ice in the Arctic (see also Ch. 11: Arctic Changes)[32]—redirected the storm inland to what was, coincidentally, an exceptionally high-exposure location.

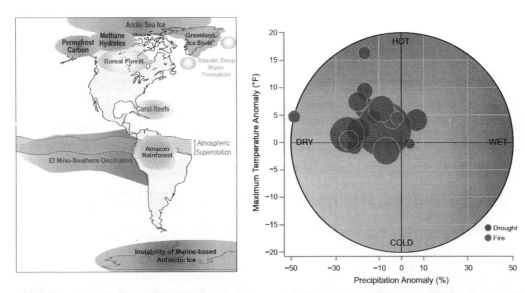

**Figure 15.1:** (left) Potential climatic tipping elements affecting the Americas (Figure source: adapted from Lenton et al. 2008[10]). (right) Wildfire and drought events from the NOAA Billion Dollar Weather Events list (1980–2016), and associated temperature and precipitation anomalies. Dot size scales with the magnitude of impact, as reflected by the cost of the event. These high-impact events occur preferentially under hot, dry conditions.

## 15.4 Climatic Tipping Elements

Different parts of the Earth system exhibit *critical thresholds*, sometimes called "tipping points" (e.g., Lenton et al. 2008;[10] Collins et al. 2013;[25] NRC 2013;[33] Kopp et al. 2016[11]). These parts, known as *tipping elements*, have the potential to enter into self-amplifying cycles that commit them to shifting from their current state into a new state: for example, from one in which the summer Arctic Ocean is covered by ice, to one in which it is ice-free. In some potential tipping elements, these state shifts occur abruptly; in others, the commitment to a state shift may occur rapidly, but the state shift itself may take decades, centuries, or even millennia to play out. Often the forcing that commits a tipping element to a shift in state is unknown. Sometimes, it is even unclear whether a proposed tipping element actually exhibits tipping behavior. Through a combination of physical modeling, paleoclimate observations, and expert elicitations, scientists have identified a number of possible tipping elements in atmosphere–ocean circulation, the cryosphere, the carbon cycle, and ecosystems (Figure 15.1, left; Table 15.1).

**Table 15.1: Potential tipping elements (adapted from Kopp et al. 2016[11]).**

| Candidate Climatic Tipping Element | State Shift | Main Impact Pathways |
|---|---|---|
| ***Atmosphere–ocean circulation*** | | |
| Atlantic meridional overturning circulation | Major reduction in strength | Regional temperature and precipitation; global mean temperature; regional sea level |
| El Niño–Southern Oscillation | Increase in amplitude | Regional temperature and precipitation |
| Equatorial atmospheric superrotation | Initiation | Cloud cover; climate sensitivity |
| Regional North Atlantic Ocean convection | Major reduction in strength | Regional temperature and precipitation |
| ***Cryosphere*** | | |
| Antarctic Ice Sheet | Major decrease in ice volume | Sea level; albedo; freshwater forcing on ocean circulation |
| Arctic sea ice | Major decrease in summertime and/or perennial area | Regional temperature and precipitation; albedo |
| Greenland Ice Sheet | Major decrease in ice volume | Sea level; albedo; freshwater forcing on ocean circulation |
| ***Carbon cycle*** | | |
| Methane hydrates | Massive release of carbon | Greenhouse gas emissions |
| Permafrost carbon | Massive release of carbon | Greenhouse gas emissions |
| ***Ecosystem*** | | |
| Amazon rainforest | Dieback, transition to grasslands | Greenhouse gas emissions; biodiversity |
| Boreal forest | Dieback, transition to grasslands | Greenhouse gas emissions; albedo; biodiversity |
| Coral reefs | Die-off | Biodiversity |

One important tipping element is the Atlantic meridional overturning circulation (AMOC), a major component of global ocean circulation. Driven by the sinking of cold, dense water in the North Atlantic near Greenland, its strength is projected to decrease with warming due to freshwater input from increased precipitation, glacial melt, and melt of the Greenland Ice Sheet (see also discussion in Ch. 11: Arctic Changes).[34] A decrease in AMOC strength is probable and may already be culpable for the "warming hole" observed in the North Atlantic,[34, 35] although it is still unclear whether this decrease represents a forced change or internal variability.[36] Given sufficient freshwater input, there is even the possibility of complete AMOC collapse. Most models do not predict such a collapse in the 21st century,[33] although one study that used observations to bias-correct climate model simulations found that $CO_2$ concentrations of 700 ppm led to a AMOC collapse within 300 years.[37]

A slowing or collapse of the AMOC would have several consequences for the United States. A decrease in AMOC strength would accelerate sea level rise off the northeastern United States,[38] while a full collapse could result in as much as approximately 1.6 feet (0.5 m) of regional sea level rise,[39, 40] as well as a cooling of approximately 0°–4°F (0°–2°C) over the country.[37, 41] These changes would occur in addition to preexisting global and regional sea level and temperature change. A slowdown of the AMOC would also lead to a reduction of ocean carbon dioxide uptake, and thus an acceleration of global-scale warming.[42]

Another tipping element is the atmospheric–oceanic circulation of the equatorial Pacific that, through a set of feedbacks, drives the state shifts of the El Niño–Southern Oscillation. This is an example of a tipping element that already shifts on a sub-decadal, interannual timescale, primarily in response to internal noise. Climate model experiments suggest that warming will reduce the threshold needed to trigger extremely strong El Niño and La Niña events.[43, 44] As evident from recent El Niño and La Niña events, such a shift would negatively impact many regions and sectors across the United States (for more on ENSO impacts, see Ch. 5: Circulation and Variability).

A third potential tipping element is arctic sea ice, which may exhibit abrupt state shifts into summer ice-free or year-round ice-free states.[45, 46] As discussed above, climate models have historically underestimated the rate of arctic sea ice loss. This is likely due to insufficient representation of critical positive feedbacks in models. Such feedbacks could include: greater high-latitude storminess and ocean wave penetration as sea ice declines; more northerly incursions of warm air and water; melting associated with increasing water vapor; loss of multiyear ice; and albedo decreases on the sea ice surface (e.g., Schröder et al. 2014;[7] Asplin et al. 2012;[8] Perovich et al. 2008[9]). At the same time, however, the point at which the threshold for an abrupt shift would be crossed also depends on the role of natural variability in a changing system; the relative importance of potential stabilizing negative feedbacks, such as more efficient heat transfer from the ocean to the atmosphere in fall and winter as sea declines; and how sea ice in other seasons, as well as the climate system more generally, responds once the first "ice-free" summer occurs (e.g., Ding et al. 2017[47]). It is also possible that summer sea ice may not abruptly collapse, but instead respond in a manner proportional to the increase in temperature.[48, 49, 50, 51] Moreover, an abrupt decrease in winter sea ice may result simply as the gradual warming of Arctic Ocean causes it to cross a critical temperature for ice formation, rather than from self-reinforcing cycles.[52]

Two possible tipping elements in the carbon cycle also lie in the Arctic. The first is buried in the permafrost, which contains an estimated 1,300–1,600 GtC (see also Ch. 11: Arctic Changes).[53] As the Arctic warms, about 5–15% is estimated to be vulnerable to release in this century.[53] Locally, the heat produced by the decomposition of organic carbon could serve as a positive feedback, accelerating carbon release.[54] However, the release of permafrost carbon, as well as whether that carbon is initially released as $CO_2$ or as the more potent greenhouse gas $CH_4$, is limited by many factors, including the freeze–thaw cycle, the rate with which heat diffuses into the permafrost, the potential for organisms to cycle permafrost carbon into new biomass, and oxygen availability. Though the release of permafrost carbon would probably not be fast enough to trigger a runaway self-amplifying cycle leading to a permafrost-free Arctic,[53] it still has the potential to significantly amplify both local and global warming, reduce the budget of human-caused $CO_2$ emissions consistent with global temperature targets, and drive continued warming even if human-caused emissions stopped altogether.[55, 56]

The second possible arctic carbon cycle tipping element is the reservoir of methane hydrates frozen into the sediments of continental shelves of the Arctic Ocean (see also Ch. 11: Arctic Changes). There is an estimated 500 to 3,000 GtC in methane hydrates,[57, 58, 59] with a most recent estimate of 1,800 GtC (equivalently, 2,400 Gt $CH_4$).[60] If released as methane rather than $CO_2$, this would be equivalent to about 82,000 Gt $CO_2$ using a global warming potential of 34.[61] While the existence of this reservoir has been known and discussed for several decades (e.g., Kvenvolden 1988[62]), only recently has it been hypothesized that warming bottom water temperatures may destabilize the hydrates over timescales shorter than millennia, leading to their release into the water column and eventually the atmosphere (e.g., Archer 2007;[57] Kretschmer et al. 2015[63]). Recent measurements of the release of methane from these sediments in summer find that, while methane hydrates on the continental shelf and upper slope are undergoing dissociation, the resulting emissions are not reaching the ocean surface in sufficient quantity to affect the atmospheric methane budget significantly, if at all.[60, 64] Estimates of plausible hydrate releases to the atmosphere over the next century are only a fraction of present-day anthropogenic methane emissions.[60, 63, 65]

These estimates of future emissions from permafrost and hydrates, however, neglect the possibility that humans may insert themselves into the physical feedback systems. With an estimated 53% of global fossil fuel reserves in the Arctic becoming increasingly accessible in a warmer world,[66] the risks associated with this carbon being extracted and burned, further exacerbating the influence of humans on global climate, are evident.[67, 68] Of less concern but still relevant, arctic ocean waters themselves are a source of methane, which could increase as sea ice decreases.[69]

The Antarctic and Greenland Ice Sheets are clear tipping elements. The Greenland Ice Sheet exhibits multiple stable states as a result of feedbacks involving the elevation of the ice sheet, atmosphere-ocean-sea ice dynamics, and albedo.[70, 71, 72, 73] At least one study suggests that warming of 2.9°F (1.6°C) above a preindustrial baseline could commit Greenland to an 85% reduction in ice volume and a 20 foot (6 m) contribution to global mean sea level over millennia.[71] One 10,000-year modeling study[74] suggests that following the higher RCP8.5 scenario (see Ch. 4: Projections) over the 21st century would lead to complete loss of the Greenland Ice Sheet over 6,000 years.

In Antarctica, the amount of ice that sits on bedrock below sea level is enough to raise global mean sea level by 75.5 feet (23 m).[75] This ice is vulnerable to collapse over centuries to millennia due to a range of feedbacks involving ocean-ice sheet-bedrock interactions.[74, 76, 77, 78, 79, 80] Observational evidence suggests that ice dynamics already in progress have committed the planet to as much as 3.9 feet (1.2 m) worth of sea level rise from the West Antarctic Ice Sheet alone, although that amount is projected to occur over the course of many centuries.[81, 82] Plausible physical modeling indicates that, under the higher RCP8.5 scenario, Antarctic ice could contribute 3.3 feet (1 m) or more to global mean sea level over the remainder of this century,[83] with some authors arguing that rates of change could be even faster.[84] Over 10,000 years, one modeling study suggests that 3.6°F (2°C) of sustained warming could lead to about 70 feet (25 m) of global mean sea level rise from Antarctica alone.[74]

Finally, tipping elements also exist in large-scale ecosystems. For example, boreal forests such as those in southern Alaska may expand northward in response to arctic warming. Because forests are darker than the tundra they replace, their expansion amplifies regional warming, which in turn accelerates their expansion.[85] As another example, coral reef ecosystems, such as those in Florida, are maintained by stabilizing ecological feedbacks among corals, coralline red algae, and grazing fish and invertebrates. However, these stabilizing feedbacks can be undermined by warming, increased risk of bleaching events, spread of disease, and ocean acidification, leading to abrupt reef collapse.[86] More generally, many ecosystems can undergo rapid regime shifts in response to a range of stressors, including climate change (e.g., Scheffer et al. 2001;[87] Folke et al. 2004[88]).

## 15.5 Paleoclimatic Hints of Additional Potential Surprises

The paleoclimatic record provides evidence for additional state shifts whose driving mechanisms are as yet poorly understood. As mentioned, global climate models tend to underestimate both the magnitude of global mean warming in response to higher $CO_2$ levels as well as its amplification at high latitudes, compared to reconstructions of temperature and $CO_2$ from the geological record. Three case studies—all periods well predating the first appearance of *Homo sapiens* around 200,000 years ago[89]—illustrate the limitations of current scientific understanding in capturing the full range of self-reinforcing cycles that operate within the Earth system, particularly over millennial time scales.

The first of these, the late Pliocene, occurred about 3.6 to 2.6 million years ago. Climate model simulations for this period systematically underestimate warming north of 30°N.[90] During the second of these, the middle Miocene (about 17–14.5 million years ago), models also fail to simultaneously replicate global mean temperature—estimated from proxies to be approximately 14° ± 4°F (8° ± 2°C) warmer than preindustrial—and the approximately 40% reduction in the pole-to-equator temperature gradient relative to today.[91] Although about one-third of the global mean temperature increase during the Miocene can be attributed to changes in geography and vegetation, geological proxies indicate $CO_2$ concentrations of around 400 ppm,[91, 92] similar to today. This suggests the possibility of as yet unmodeled feedbacks, perhaps related to a significant change in the vertical distribution of heat in the tropical ocean.[93]

The last of these case studies, the early Eocene, occurred about 56–48 million years ago. This period is characterized by the absence of permanent land ice, $CO_2$ concentrations peaking around $1,400 \pm 470$ ppm,[94] and global temperatures about $25°F \pm 5°F$ ($14°C \pm 3°C$) warmer than the preindustrial.[95] Like the late Pliocene and the middle Miocene, this period also exhibits about half the pole-to-equator temperature gradient of today.[15, 96] About one-third of the temperature difference is attributable to changes in geography, vegetation, and ice sheet coverage.[95] However, to reproduce both the elevated global mean temperature and the reduced pole-to-equator temperature gradient, climate models would require $CO_2$ concentrations that exceed those indicated by the proxy record by two to five times[15]—suggesting once again the presence of as yet poorly understood processes and feedbacks.

One possible explanation for this discrepancy is a planetary state shift that, above a particular $CO_2$ threshold, leads to a significant increase in the sensitivity of the climate to $CO_2$. Paleo-data for the last 800,000 years suggest a gradual increase in climate sensitivity with global mean temperature over glacial-interglacial cycles,[97, 98] although these results are based on a time period with $CO_2$ concentra-

tions lower than today. At higher $CO_2$ levels, one modeling study[95] suggests that an abrupt change in atmospheric circulation (the onset of equatorial atmospheric superrotation) between 1,120 and 2,240 ppm $CO_2$ could lead to a reduction in cloudiness and an approximate doubling of climate sensitivity. However, the critical threshold for such a transition is poorly constrained. If it occurred in the past at a lower $CO_2$ level, it might explain the Eocene discrepancy and potentially also the Miocene discrepancy: but in that case, it could also pose a plausible threat within the 21st century under the higher RCP8.5 scenario.

Regardless of the particular mechanism, the systematic paleoclimatic model-data mismatch for past warm climates suggests that climate models are omitting at least one, and probably more, processes crucial to future warming, especially in polar regions. For this reason, future changes outside the range projected by climate models cannot be ruled out, and climate models are more likely to underestimate than to overestimate the amount of long-term future change.

## TRACEABLE ACCOUNTS

### Key Finding 1

Positive feedbacks (self-reinforcing cycles) within the climate system have the potential to accelerate human-induced climate change and even shift the Earth's climate system, in part or in whole, into new states that are very different from those experienced in the recent past (for example, ones with greatly diminished ice sheets or different large-scale patterns of atmosphere or ocean circulation). Some feedbacks and potential state shifts can be modeled and quantified; others can be modeled or identified but not quantified; and some are probably still unknown. (*Very high confidence* in the potential for state shifts and in the incompleteness of knowledge about feedbacks and potential state shifts).

### Description of evidence base

This key finding is based on a large body of scientific literature recently summarized by Lenton et al.,[10] NRC,[33] and Kopp et al.[11] As NRC[33] (page vii) states, "A study of Earth's climate history suggests the inevitability of 'tipping points'—thresholds beyond which major and rapid changes occur when crossed—that lead to abrupt changes in the climate system" and (page xi), "Can all tipping points be foreseen? Probably not. Some will have no precursors, or may be triggered by naturally occurring variability in the climate system. Some will be difficult to detect, clearly visible only after they have been crossed and an abrupt change becomes inevitable." As IPCC AR5 WG1 Chapter 12, section 12.5.5[25] further states, "A number of components or phenomena within the Earth system have been proposed as potentially possessing critical thresholds (sometimes referred to as tipping points) beyond which abrupt or nonlinear transitions to a different state ensues." Collins et al.[25] further summarizes critical thresholds that can be modeled and others that can only be identified.

### Major uncertainties

The largest uncertainties are 1) whether proposed tipping elements actually undergo critical transitions; 2) the magnitude and timing of forcing that will be required to initiate critical transitions in tipping elements; 3) the speed of the transition once it has been triggered; 4) the characteristics of the new state that re-

sults from such transition; and 5) the potential for new tipping elements to exist that are yet unknown.

### Assessment of confidence based on evidence and agreement, including short description of nature of evidence and level of agreement

There is *very high confidence* in the likelihood of the existence of positive feedbacks, and the tipping elements statement is based on a large body of literature published over the last 25 years that draws from basic physics, observations, paleoclimate data, and modeling.

There is *very high confidence* that some feedbacks can be quantified, others are known but cannot be quantified, and others may yet exist that are currently unknown.

### Summary sentence or paragraph that integrates the above information

The key finding is based on NRC[33] and IPCC AR5 WG1 Chapter 12 section 12.5.5,[25] which made a thorough assessment of the relevant literature.

### Key Finding 2

The physical and socioeconomic impacts of compound extreme events (such as simultaneous heat and drought, wildfires associated with hot and dry conditions, or flooding associated with high precipitation on top of snow or waterlogged ground) can be greater than the sum of the parts (*very high confidence*). Few analyses consider the spatial or temporal correlation between extreme events.

### Description of evidence base

This key finding is based on a large body of scientific literature summarized in the 2012 IPCC Special Report on Extremes.[23] The report's Summary for Policymakers (page 6) states, "exposure and vulnerability are key determinants of disaster risk and of impacts when risk is realized... extreme impacts on human, ecological, or physical systems can result from individual extreme weather or climate events. Extreme impacts can also result from non-extreme events where exposure and vulnerability are high or from a compounding of events or their impacts. For example, drought, coupled with extreme heat and low humidity, can increase the risk of wildfire."

**Major uncertainties**

The largest uncertainties are in the temporal congruence of the events and the compounding nature of their impacts.

**Assessment of confidence based on evidence and agreement, including short description of nature of evidence and level of agreement**

There is *very high confidence* that the impacts of multiple events could exceed the sum of the impacts of events occurring individually.

**Summary sentence or paragraph that integrates the above information**

The key finding is based on the 2012 IPCC SREX report, particularly section 3.1.3 on compound or multiple events, which presents a thorough assessment of the relevant literature.

**Key Finding 3**

While climate models incorporate important climate processes that can be well quantified, they do not include all of the processes that can contribute to feedbacks, compound extreme events, and abrupt and/or irreversible changes. For this reason, future changes outside the range projected by climate models cannot be ruled out (*very high confidence*). Moreover, the systematic tendency of climate models to underestimate temperature change during warm paleoclimates suggests that climate models are more likely to underestimate than to overestimate the amount of long-term future change (*medium confidence*).

**Description of evidence base**

This key finding is based on the conclusions of IPCC AR5 WG1,[99] specifically Chapter 9;[1] the state of the art of global models is briefly summarized in Chapter 4: Projections of this report. The second half of this key finding is based upon the tendency of global climate models to underestimate, relative to geological reconstructions, the magnitude of both long-term global mean warming and the amplification of warming at high latitudes in past warm climates (e.g., Salzmann et al. 2013;[90] Goldner et al. 2014;[91] Caballeo and Huber 2013;[95] Lunt et al. 2012[15]).

**Major uncertainties**

The largest uncertainties are structural: are the models including all the important components and relationships necessary to model the feedbacks and if so, are these correctly represented in the models?

**Assessment of confidence based on evidence and agreement, including short description of nature of evidence and level of agreement**

There is *very high confidence* that the models are incomplete representations of the real world; and there is *medium confidence* that their tendency is to under- rather than over-estimate the amount of long-term future change.

**Summary sentence or paragraph that integrates the above information**

The key finding is based on the IPCC AR5 WG1 Chapter 9,[1] as well as systematic paleoclimatic model/data comparisons.

# REFERENCES

1. Flato, G., J. Marotzke, B. Abiodun, P. Braconnot, S.C. Chou, W. Collins, P. Cox, F. Driouech, S. Emori, V. Eyring, C. Forest, P. Gleckler, E. Guilyardi, C. Jakob, V. Kattsov, C. Reason, and M. Rummukainen, 2013: Evaluation of climate models. *Climate Change 2013: The Physical Science Basis. Contribution of Working Group I to the Fifth Assessment Report of the Intergovernmental Panel on Climate Change.* Stocker, T.F., D. Qin, G.-K. Plattner, M. Tignor, S.K. Allen, J. Boschung, A. Nauels, Y. Xia, V. Bex, and P.M. Midgley, Eds. Cambridge University Press, Cambridge, United Kingdom and New York, NY, USA, 741–866. http://www.climatechange2013.org/report/full-report/

2. Molina, M.J. and F.S. Rowland, 1974: Stratospheric sink for chlorofluoromethanes: Chlorine atomc-atalysed destruction of ozone. *Nature,* **249,** 810-812. http://dx.doi.org/10.1038/249810a0

3. Farman, J.C., B.G. Gardiner, and J.D. Shanklin, 1985: Large losses of total ozone in Antarctica reveal seasonal ClOx/NOx interaction. *Nature,* **315,** 207-210. http://dx.doi.org/10.1038/315207a0

4. Stroeve, J., M.M. Holland, W. Meier, T. Scambos, and M. Serreze, 2007: Arctic sea ice decline: Faster than forecast. *Geophysical Research Letters,* **34,** L09501. http://dx.doi.org/10.1029/2007GL029703

5. Meehl, G.A., T.F. Stocker, W.D. Collins, P. Friedlingstein, A.T. Gaye, J.M. Gregory, A. Kitoh, R. Knutti, J.M. Murphy, A. Noda, S.C.B. Raper, I.G. Watterson, A.J. Weaver, and Z.-C. Zhao, 2007: Global Climate Projections. *Climate Change 2007: The Physical Science Basis. Contribution of Working Group I to the Fourth Assessment Report of the Intergovernmental Panel on Climate Change.* Solomon, S., D. Qin, M. Manning, Z. Chen, M. Marquis, K.B. Averyt, M. Tignor, and H.L. Miller, Eds. Cambridge University Press, Cambridge, United Kingdom and New York, NY, USA, 747-845.

6. Stroeve, J.C., V. Kattsov, A. Barrett, M. Serreze, T. Pavlova, M. Holland, and W.N. Meier, 2012: Trends in Arctic sea ice extent from CMIP5, CMIP3 and observations. *Geophysical Research Letters,* **39,** L16502. http://dx.doi.org/10.1029/2012GL052676

7. Schröder, D., D.L. Feltham, D. Flocco, and M. Tsamados, 2014: September Arctic sea-ice minimum predicted by spring melt-pond fraction. *Nature Climate Change,* **4,** 353-357. http://dx.doi.org/10.1038/nclimate2203

8. Asplin, M.G., R. Galley, D.G. Barber, and S. Prinsenberg, 2012: Fracture of summer perennial sea ice by ocean swell as a result of Arctic storms. *Journal of Geophysical Research,* **117,** C06025. http://dx.doi.org/10.1029/2011JC007221

9. Perovich, D.K., J.A. Richter-Menge, K.F. Jones, and B. Light, 2008: Sunlight, water, and ice: Extreme Arctic sea ice melt during the summer of 2007. *Geophysical Research Letters,* **35,** L11501. http://dx.doi.org/10.1029/2008GL034007

10. Lenton, T.M., H. Held, E. Kriegler, J.W. Hall, W. Lucht, S. Rahmstorf, and H.J. Schellnhuber, 2008: Tipping elements in the Earth's climate system. *Proceedings of the National Academy of Sciences,* **105,** 1786-1793. http://dx.doi.org/10.1073/pnas.0705414105

11. Kopp, R.E., R.L. Shwom, G. Wagner, and J. Yuan, 2016: Tipping elements and climate–economic shocks: Pathways toward integrated assessment. *Earth's Future,* **4,** 346-372. http://dx.doi.org/10.1002/2016EF000362

12. Zwiers, F.W., L.V. Alexander, G.C. Hegerl, T.R. Knutson, J.P. Kossin, P. Naveau, N. Nicholls, C. Schär, S.I. Seneviratne, and X. Zhang, 2013: Climate extremes: Challenges in estimating and understanding recent changes in the frequency and intensity of extreme climate and weather events. *Climate Science for Serving Society: Research, Modeling and Prediction Priorities.* Asrar, G.R. and J.W. Hurrell, Eds. Springer Netherlands, Dordrecht, 339-389. http://dx.doi.org/10.1007/978-94-007-6692-1_13

13. Smith, J.A., 1987: Estimating the upper tail of flood frequency distributions. *Water Resources Research,* **23,** 1657-1666. http://dx.doi.org/10.1029/WR023i008p01657

14. Woodhouse, C.A. and J.T. Overpeck, 1998: 2000 years of drought variability in the central United States. *Bulletin of the American Meteorological Society,* **79,** 2693-2714. http://dx.doi.org/10.1175/1520-0477(1998)079<2693:YODVIT>2.0.CO;2

15. Lunt, D.J., T. Dunkley Jones, M. Heinemann, M. Huber, A. LeGrande, A. Winguth, C. Loptson, J. Marotzke, C.D. Roberts, J. Tindall, P. Valdes, and C. Winguth, 2012: A model–data comparison for a multi-model ensemble of early Eocene atmosphere–ocean simulations: EoMIP. *Climate of the Past,* **8,** 1717-1736. http://dx.doi.org/10.5194/cp-8-1717-2012

16. Navarro, T., J.B. Madeleine, F. Forget, A. Spiga, E. Millour, F. Montmessin, and A. Määttänen, 2014: Global climate modeling of the Martian water cycle with improved microphysics and radiatively active water ice clouds. *Journal of Geophysical Research Planets,* **119,** 1479-1495. http://dx.doi.org/10.1002/2013JE004550

17. Kang, I.-S., Y.-M. Yang, and W.-K. Tao, 2015: GCMs with implicit and explicit representation of cloud microphysics for simulation of extreme precipitation frequency. *Climate Dynamics,* **45,** 325-335. http://dx.doi.org/10.1007/s00382-014-2376-1

18. Brysse, K., N. Oreskes, J. O'Reilly, and M. Oppenheimer, 2013: Climate change prediction: Erring on the side of least drama? *Global Environmental Change*, **23**, 327-337. http://dx.doi.org/10.1016/j.gloenvcha.2012.10.008

19. Quarantelli, E.L., 1986: Disaster Crisis Management. University of Delaware, Newark, DE. 10 pp. http://udspace.udel.edu/handle/19716/487

20. Zscheischler, J., M. Reichstein, J. von Buttlar, M. Mu, J.T. Randerson, and M.D. Mahecha, 2014: Carbon cycle extremes during the 21st century in CMIP5 models: Future evolution and attribution to climatic drivers. *Geophysical Research Letters*, **41**, 8853-8861. http://dx.doi.org/10.1002/2014GL062409

21. Swain, D.L., D.E. Horton, D. Singh, and N.S. Diffenbaugh, 2016: Trends in atmospheric patterns conducive to seasonal precipitation and temperature extremes in California. *Science Advances*, **2**, e1501344. http://dx.doi.org/10.1126/sciadv.1501344

22. Diffenbaugh, N.S. and F. Giorgi, 2012: Climate change hotspots in the CMIP5 global climate model ensemble. *Climatic Change*, **114**, 813-822. http://dx.doi.org/10.1007/s10584-012-0570-x

23. IPCC, 2012: Managing the Risks of Extreme Events and Disasters to Advance Climate Change Adaptation. A Special Report of Working Groups I and II of the Intergovernmental Panel on Climate Change. Field, C.B., V. Barros, T.F. Stocker, D. Qin, D.J. Dokken, K.L. Ebi, M.D. Mastrandrea, K.J. Mach, G.-K. Plattner, S.K. Allen, M. Tignor, and P.M. Midgley (Eds.). Cambridge University Press, Cambridge, UK and New York, NY. 582 pp. https://www.ipcc.ch/pdf/special-reports/srex/SREX_Full_Report.pdf

24. Hao, Z., A. AghaKouchak, and T.J. Phillips, 2013: Changes in concurrent monthly precipitation and temperature extremes. *Environmental Research Letters*, **8**, 034014. http://dx.doi.org/10.1088/1748-9326/8/3/034014

25. Collins, M., R. Knutti, J. Arblaster, J.-L. Dufresne, T. Fichefet, P. Friedlingstein, X. Gao, W.J. Gutowski, T. Johns, G. Krinner, M. Shongwe, C. Tebaldi, A.J. Weaver, and M. Wehner, 2013: Long-term climate change: Projections, commitments and irreversibility. *Climate Change 2013: The Physical Science Basis. Contribution of Working Group I to the Fifth Assessment Report of the Intergovernmental Panel on Climate Change.* Stocker, T.F., D. Qin, G.-K. Plattner, M. Tignor, S.K. Allen, J. Boschung, A. Nauels, Y. Xia, V. Bex, and P.M. Midgley, Eds. Cambridge University Press, Cambridge, United Kingdom and New York, NY, USA, 1029–1136. http://www.climatechange2013.org/report/full-report/

26. Trenberth, K.E., A. Dai, G. van der Schrier, P.D. Jones, J. Barichivich, K.R. Briffa, and J. Sheffield, 2014: Global warming and changes in drought. *Nature Climate Change*, **4**, 17-22. http://dx.doi.org/10.1038/nclimate2067

27. Cook, B.I., T.R. Ault, and J.E. Smerdon, 2015: Unprecedented 21st century drought risk in the American Southwest and Central Plains. *Science Advances*, **1**, e1400082. http://dx.doi.org/10.1126/sciadv.1400082

28. AghaKouchak, A., L. Cheng, O. Mazdiyasni, and A. Farahmand, 2014: Global warming and changes in risk of concurrent climate extremes: Insights from the 2014 California drought. *Geophysical Research Letters*, **41**, 8847-8852. http://dx.doi.org/10.1002/2014GL062308

29. Serinaldi, F., 2016: Can we tell more than we can know? The limits of bivariate drought analyses in the United States. *Stochastic Environmental Research and Risk Assessment*, **30**, 1691-1704. http://dx.doi.org/10.1007/s00477-015-1124-3

30. Fraser, E.D.G., E. Simelton, M. Termansen, S.N. Gosling, and A. South, 2013: "Vulnerability hotspots": Integrating socio-economic and hydrological models to identify where cereal production may decline in the future due to climate change induced drought. *Agricultural and Forest Meteorology*, **170**, 195-205. http://dx.doi.org/10.1016/j.agrformet.2012.04.008

31. Reed, A.J., M.E. Mann, K.A. Emanuel, N. Lin, B.P. Horton, A.C. Kemp, and J.P. Donnelly, 2015: Increased threat of tropical cyclones and coastal flooding to New York City during the anthropogenic era. *Proceedings of the National Academy of Sciences*, **112**, 12610-12615. http://dx.doi.org/10.1073/pnas.1513127112

32. Liu, J., Z. Chen, J. Francis, M. Song, T. Mote, and Y. Hu, 2016: Has Arctic sea ice loss contributed to increased surface melting of the Greenland Ice Sheet? *Journal of Climate*, **29**, 3373-3386. http://dx.doi.org/10.1175/JCLI-D-15-0391.1

33. NRC, 2013: *Abrupt Impacts of Climate Change: Anticipating Surprises.* The National Academies Press, Washington, DC, 222 pp. http://dx.doi.org/10.17226/18373

34. Rahmstorf, S., J.E. Box, G. Feulner, M.E. Mann, A. Robinson, S. Rutherford, and E.J. Schaffernicht, 2015: Exceptional twentieth-century slowdown in Atlantic Ocean overturning circulation. *Nature Climate Change*, **5**, 475-480. http://dx.doi.org/10.1038/nclimate2554

35. Drijfhout, S., G.J.v. Oldenborgh, and A. Cimatoribus, 2012: Is a decline of AMOC causing the warming hole above the North Atlantic in observed and modeled warming patterns? *Journal of Climate*, **25**, 8373-8379. http://dx.doi.org/10.1175/jcli-d-12-00490.1

36. Cheng, J., Z. Liu, S. Zhang, W. Liu, L. Dong, P. Liu, and H. Li, 2016: Reduced interdecadal variability of Atlantic Meridional Overturning Circulation under global warming. *Proceedings of the National Academy of Sciences*, **113**, 3175-3178. http://dx.doi.org/10.1073/pnas.1519827113

37. Liu, W., S.-P. Xie, Z. Liu, and J. Zhu, 2017: Overlooked possibility of a collapsed Atlantic Meridional Overturning Circulation in warming climate. *Science Advances*, **3**, e1601666. http://dx.doi.org/10.1126/sciadv.1601666

38. Yin, J. and P.B. Goddard, 2013: Oceanic control of sea level rise patterns along the East Coast of the United States. *Geophysical Research Letters*, **40**, 5514-5520. http://dx.doi.org/10.1002/2013GL057992

39. Gregory, J.M. and J.A. Lowe, 2000: Predictions of global and regional sea-level rise using AOGCMs with and without flux adjustment. *Geophysical Research Letters*, **27**, 3069-3072. http://dx.doi.org/10.1029/1999GL011228

40. Levermann, A., A. Griesel, M. Hofmann, M. Montoya, and S. Rahmstorf, 2005: Dynamic sea level changes following changes in the thermohaline circulation. *Climate Dynamics*, **24**, 347-354. http://dx.doi.org/10.1007/s00382-004-0505-y

41. Jackson, L.C., R. Kahana, T. Graham, M.A. Ringer, T. Woollings, J.V. Mecking, and R.A. Wood, 2015: Global and European climate impacts of a slowdown of the AMOC in a high resolution GCM. *Climate Dynamics*, **45**, 3299-3316. http://dx.doi.org/10.1007/s00382-015-2540-2

42. Pérez, F.F., H. Mercier, M. Vazquez-Rodriguez, P. Lherminier, A. Velo, P.C. Pardo, G. Roson, and A.F. Rios, 2013: Atlantic Ocean $CO_2$ uptake reduced by weakening of the meridional overturning circulation. *Nature Geoscience*, **6**, 146-152. http://dx.doi.org/10.1038/ngeo1680

43. Cai, W., S. Borlace, M. Lengaigne, P. van Rensch, M. Collins, G. Vecchi, A. Timmermann, A. Santoso, M.J. McPhaden, L. Wu, M.H. England, G. Wang, E. Guilyardi, and F.-F. Jin, 2014: Increasing frequency of extreme El Niño events due to greenhouse warming. *Nature Climate Change*, **4**, 111-116. http://dx.doi.org/10.1038/nclimate2100

44. Cai, W., G. Wang, A. Santoso, M.J. McPhaden, L. Wu, F.-F. Jin, A. Timmermann, M. Collins, G. Vecchi, M. Lengaigne, M.H. England, D. Dommenget, K. Takahashi, and E. Guilyardi, 2015: Increased frequency of extreme La Niña events under greenhouse warming. *Nature Climate Change*, **5**, 132-137. http://dx.doi.org/10.1038/nclimate2492

45. Lindsay, R.W. and J. Zhang, 2005: The thinning of Arctic sea ice, 1988–2003: Have we passed a tipping point? *Journal of Climate*, **18**, 4879-4894. http://dx.doi.org/10.1175/jcli3587.1

46. Eisenman, I. and J.S. Wettlaufer, 2009: Nonlinear threshold behavior during the loss of Arctic sea ice. *Proceedings of the National Academy of Sciences*, **106**, 28-32. http://dx.doi.org/10.1073/pnas.0806887106

47. Ding, Q., A. Schweiger, M. Lheureux, D.S. Battisti, S. Po-Chedley, N.C. Johnson, E. Blanchard-Wrigglesworth, K. Harnos, Q. Zhang, R. Eastman, and E.J. Steig, 2017: Influence of high-latitude atmospheric circulation changes on summertime Arctic sea ice. *Nature Climate Change*, **7**, 289-295. http://dx.doi.org/10.1038/nclimate3241

48. Armour, K.C., I. Eisenman, E. Blanchard-Wrigglesworth, K.E. McCusker, and C.M. Bitz, 2011: The reversibility of sea ice loss in a state-of-the-art climate model. *Geophysical Research Letters*, **38**, L16705. http://dx.doi.org/10.1029/2011GL048739

49. Ridley, J.K., J.A. Lowe, and H.T. Hewitt, 2012: How reversible is sea ice loss? *The Cryosphere*, **6**, 193-198. http://dx.doi.org/10.5194/tc-6-193-2012

50. Li, C., D. Notz, S. Tietsche, and J. Marotzke, 2013: The transient versus the equilibrium response of sea ice to global warming. *Journal of Climate*, **26**, 5624-5636. http://dx.doi.org/10.1175/JCLI-D-12-00492.1

51. Wagner, T.J.W. and I. Eisenman, 2015: How climate model complexity influences sea ice stability. *Journal of Climate*, **28**, 3998-4014. http://dx.doi.org/10.1175/JCLI-D-14-00654.1

52. Bathiany, S., D. Notz, T. Mauritsen, G. Raedel, and V. Brovkin, 2016: On the potential for abrupt Arctic winter sea ice loss. *Journal of Climate*, **29**, 2703-2719. http://dx.doi.org/10.1175/JCLI-D-15-0466.1

53. Schuur, E.A.G., A.D. McGuire, C. Schadel, G. Grosse, J.W. Harden, D.J. Hayes, G. Hugelius, C.D. Koven, P. Kuhry, D.M. Lawrence, S.M. Natali, D. Olefeldt, V.E. Romanovsky, K. Schaefer, M.R. Turetsky, C.C. Treat, and J.E. Vonk, 2015: Climate change and the permafrost carbon feedback. *Nature*, **520**, 171-179. http://dx.doi.org/10.1038/nature14338

54. Hollesen, J., H. Matthiesen, A.B. Møller, and B. Elberling, 2015: Permafrost thawing in organic Arctic soils accelerated by ground heat production. *Nature Climate Change*, **5**, 574-578. http://dx.doi.org/10.1038/nclimate2590

55. MacDougall, A.H., C.A. Avis, and A.J. Weaver, 2012: Significant contribution to climate warming from the permafrost carbon feedback. *Nature Geoscience*, **5**, 719-721. http://dx.doi.org/10.1038/ngeo1573

56. MacDougall, A.H., K. Zickfeld, R. Knutti, and H.D. Matthews, 2015: Sensitivity of carbon budgets to permafrost carbon feedbacks and non-$CO_2$ forcings. *Environmental Research Letters*, **10**, 125003. http://dx.doi.org/10.1088/1748-9326/10/12/125003

57. Archer, D., 2007: Methane hydrate stability and anthropogenic climate change. *Biogeosciences*, **4**, 521-544. http://dx.doi.org/10.5194/bg-4-521-2007

58. Ruppel, C.D. *Methane hydrates and contemporary climate change*. Nature Education Knowledge, 2011. **3**.

59. Piñero, E., M. Marquardt, C. Hensen, M. Haeckel, and K. Wallmann, 2013: Estimation of the global inventory of methane hydrates in marine sediments using transfer functions. *Biogeosciences*, **10**, 959-975. http://dx.doi.org/10.5194/bg-10-959-2013

60. Ruppel, C.D. and J.D. Kessler, 2017: The interaction of climate change and methane hydrates. *Reviews of Geophysics*, **55**, 126-168. http://dx.doi.org/10.1002/2016RG000534

61. Myhre, G., D. Shindell, F.-M. Bréon, W. Collins, J. Fuglestvedt, J. Huang, D. Koch, J.-F. Lamarque, D. Lee, B. Mendoza, T. Nakajima, A. Robock, G. Stephens, T. Takemura, and H. Zhang, 2013: Anthropogenic and natural radiative forcing. *Climate Change 2013: The Physical Science Basis. Contribution of Working Group I to the Fifth Assessment Report of the Intergovernmental Panel on Climate Change*. Stocker, T.F., D. Qin, G.-K. Plattner, M. Tignor, S.K. Allen, J. Boschung, A. Nauels, Y. Xia, V. Bex, and P.M. Midgley, Eds. Cambridge University Press, Cambridge, United Kingdom and New York, NY, USA, 659–740. http://www.climatechange2013.org/report/full-report/

62. Kvenvolden, K.A., 1988: Methane hydrate — A major reservoir of carbon in the shallow geosphere? *Chemical Geology*, **71**, 41-51. http://dx.doi.org/10.1016/0009-2541(88)90104-0

63. Kretschmer, K., A. Biastoch, L. Rüpke, and E. Burwicz, 2015: Modeling the fate of methane hydrates under global warming. *Global Biogeochemical Cycles*, **29**, 610-625. http://dx.doi.org/10.1002/2014GB005011

64. Myhre, C.L., B. Ferré, S.M. Platt, A. Silyakova, O. Hermansen, G. Allen, I. Pisso, N. Schmidbauer, A. Stohl, J. Pitt, P. Jansson, J. Greinert, C. Percival, A.M. Fjaeraa, S.J. O'Shea, M. Gallagher, M. Le Breton, K.N. Bower, S.J.B. Bauguitte, S. Dalsøren, S. Vadakkepuliyambatta, R.E. Fisher, E.G. Nisbet, D. Lowry, G. Myhre, J.A. Pyle, M. Cain, and J. Mienert, 2016: Extensive release of methane from Arctic seabed west of Svalbard during summer 2014 does not influence the atmosphere. *Geophysical Research Letters*, **43**, 4624-4631. http://dx.doi.org/10.1002/2016GL068999

65. Stranne, C., M. O'Regan, G.R. Dickens, P. Crill, C. Miller, P. Preto, and M. Jakobsson, 2016: Dynamic simulations of potential methane release from East Siberian continental slope sediments. *Geochemistry, Geophysics, Geosystems*, **17**, 872-886. http://dx.doi.org/10.1002/2015GC006119

66. Lee, S.-Y. and G.D. Holder, 2001: Methane hydrates potential as a future energy source. *Fuel Processing Technology*, **71**, 181-186. http://dx.doi.org/10.1016/S0378-3820(01)00145-X

67. Jakob, M. and J. Hilaire, 2015: Climate science: Unburnable fossil-fuel reserves. *Nature*, **517**, 150-152. http://dx.doi.org/10.1038/517150a

68. McGlade, C. and P. Ekins, 2015: The geographical distribution of fossil fuels unused when limiting global warming to 2°C. *Nature*, **517**, 187-190. http://dx.doi.org/10.1038/nature14016

69. Kort, E.A., S.C. Wofsy, B.C. Daube, M. Diao, J.W. Elkins, R.S. Gao, E.J. Hintsa, D.F. Hurst, R. Jimenez, F.L. Moore, J.R. Spackman, and M.A. Zondlo, 2012: Atmospheric observations of Arctic Ocean methane emissions up to 82° north. *Nature Geoscience*, **5**, 318-321. http://dx.doi.org/10.1038/ngeo1452

70. Ridley, J., J.M. Gregory, P. Huybrechts, and J. Lowe, 2010: Thresholds for irreversible decline of the Greenland ice sheet. *Climate Dynamics*, **35**, 1049-1057. http://dx.doi.org/10.1007/s00382-009-0646-0

71. Robinson, A., R. Calov, and A. Ganopolski, 2012: Multistability and critical thresholds of the Greenland ice sheet. *Nature Climate Change*, **2**, 429-432. http://dx.doi.org/10.1038/nclimate1449

72. Levermann, A., P.U. Clark, B. Marzeion, G.A. Milne, D. Pollard, V. Radic, and A. Robinson, 2013: The multimillennial sea-level commitment of global warming. *Proceedings of the National Academy of Sciences*, **110**, 13745-13750. http://dx.doi.org/10.1073/pnas.1219414110

73. Koenig, S.J., R.M. DeConto, and D. Pollard, 2014: Impact of reduced Arctic sea ice on Greenland ice sheet variability in a warmer than present climate. *Geophysical Research Letters*, **41**, 3933-3942. http://dx.doi.org/10.1002/2014GL059770

74. Clark, P.U., J.D. Shakun, S.A. Marcott, A.C. Mix, M. Eby, S. Kulp, A. Levermann, G.A. Milne, P.L. Pfister, B.D. Santer, D.P. Schrag, S. Solomon, T.F. Stocker, B.H. Strauss, A.J. Weaver, R. Winkelmann, D. Archer, E. Bard, A. Goldner, K. Lambeck, R.T. Pierrehumbert, and G.-K. Plattner, 2016: Consequences of twenty-first-century policy for multi-millennial climate and sea-level change. *Nature Climate Change*, **6**, 360-369. http://dx.doi.org/10.1038/nclimate2923

75. Fretwell, P., H.D. Pritchard, D.G. Vaughan, J.L. Bamber, N.E. Barrand, R. Bell, C. Bianchi, R.G. Bingham, D.D. Blankenship, G. Casassa, G. Catania, D. Callens, H. Conway, A.J. Cook, H.F.J. Corr, D. Damaske, V. Damm, F. Ferraccioli, R. Forsberg, S. Fujita, Y. Gim, P. Gogineni, J.A. Griggs, R.C.A. Hindmarsh, P. Holmlund, J.W. Holt, R.W. Jacobel, A. Jenkins, W. Jokat, T. Jordan, E.C. King, J. Kohler, W. Krabill, M. Riger-Kusk, K.A. Langley, G. Leitchenkov, C. Leuschen, B.P. Luyendyk, K. Matsuoka, J. Mouginot, F.O. Nitsche, Y. Nogi, O.A. Nost, S.V. Popov, E. Rignot, D.M. Rippin, A. Rivera, J. Roberts, N. Ross, M.J. Siegert, A.M. Smith, D. Steinhage, M. Studinger, B. Sun, B.K. Tinto, B.C. Welch, D. Wilson, D.A. Young, C. Xiangbin, and A. Zirizzotti, 2013: Bedmap2: Improved ice bed, surface and thickness datasets for Antarctica. *The Cryosphere*, **7**, 375-393. http://dx.doi.org/10.5194/tc-7-375-2013

76. Schoof, C., 2007: Ice sheet grounding line dynamics: Steady states, stability, and hysteresis. *Journal of Geophysical Research*, **112**, F03S28. http://dx.doi.org/10.1029/2006JF000664

77. Gomez, N., J.X. Mitrovica, P. Huybers, and P.U. Clark, 2010: Sea level as a stabilizing factor for marine-ice-sheet grounding lines. *Nature Geoscience*, **3**, 850-853. http://dx.doi.org/10.1038/ngeo1012

78. Ritz, C., T.L. Edwards, G. Durand, A.J. Payne, V. Peyaud, and R.C.A. Hindmarsh, 2015: Potential sea-level rise from Antarctic ice-sheet instability constrained by observations. *Nature*, **528**, 115-118. http://dx.doi.org/10.1038/nature16147

79. Mengel, M. and A. Levermann, 2014: Ice plug prevents irreversible discharge from East Antarctica. *Nature Climate Change*, **4**, 451-455. http://dx.doi.org/10.1038/nclimate2226

80. Pollard, D., R.M. DeConto, and R.B. Alley, 2015: Potential Antarctic Ice Sheet retreat driven by hydrofracturing and ice cliff failure. *Earth and Planetary Science Letters*, **412**, 112-121. http://dx.doi.org/10.1016/j.epsl.2014.12.035

81. Joughin, I., B.E. Smith, and B. Medley, 2014: Marine ice sheet collapse potentially under way for the Thwaites Glacier Basin, West Antarctica. *Science*, **344**, 735-738. http://dx.doi.org/10.1126/science.1249055

82. Rignot, E., J. Mouginot, M. Morlighem, H. Seroussi, and B. Scheuchl, 2014: Widespread, rapid grounding line retreat of Pine Island, Thwaites, Smith, and Kohler Glaciers, West Antarctica, from 1992 to 2011. *Geophysical Research Letters*, **41**, 3502-3509. http://dx.doi.org/10.1002/2014GL060140

83. DeConto, R.M. and D. Pollard, 2016: Contribution of Antarctica to past and future sea-level rise. *Nature*, **531**, 591-597. http://dx.doi.org/10.1038/nature17145

84. Hansen, J., M. Sato, P. Hearty, R. Ruedy, M. Kelley, V. Masson-Delmotte, G. Russell, G. Tselioudis, J. Cao, E. Rignot, I. Velicogna, B. Tormey, B. Donovan, E. Kandiano, K. von Schuckmann, P. Kharecha, A.N. Legrande, M. Bauer, and K.W. Lo, 2016: Ice melt, sea level rise and superstorms: Evidence from paleoclimate data, climate modeling, and modern observations that 2°C global warming could be dangerous. *Atmospheric Chemistry and Physics*, **16**, 3761-3812. http://dx.doi.org/10.5194/acp-16-3761-2016

85. Jones, C., J. Lowe, S. Liddicoat, and R. Betts, 2009: Committed terrestrial ecosystem changes due to climate change. *Nature Geoscience*, **2**, 484-487. http://dx.doi.org/10.1038/ngeo555

86. Hoegh-Guldberg, O., P.J. Mumby, A.J. Hooten, R.S. Steneck, P. Greenfield, E. Gomez, C.D. Harvell, P.F. Sale, A.J. Edwards, K. Caldeira, N. Knowlton, C.M. Eakin, R. Iglesias-Prieto, N. Muthiga, R.H. Bradbury, A. Dubi, and M.E. Hatziolos, 2007: Coral reefs under rapid climate change and ocean acidification. *Science*, **318**, 1737-1742. http://dx.doi.org/10.1126/science.1152509

87. Scheffer, M., S. Carpenter, J.A. Foley, C. Folke, and B. Walker, 2001: Catastrophic shifts in ecosystems. *Nature*, **413**, 591-596. http://dx.doi.org/10.1038/35098000

88. Folke, C., S. Carpenter, B. Walker, M. Scheffer, T. Elmqvist, L. Gunderson, and C.S. Holling, 2004: Regime shifts, resilience, and biodiversity in ecosystem management. *Annual Review of Ecology, Evolution, and Systematics*, **35**, 557-581. http://dx.doi.org/10.1146/annurev.ecolsys.35.021103.105711

89. Tattersall, I., 2009: Human origins: Out of Africa. *Proceedings of the National Academy of Sciences*, **106**, 16018-16021. http://dx.doi.org/10.1073/pnas.0903207106

90. Salzmann, U., A.M. Dolan, A.M. Haywood, W.-L. Chan, J. Voss, D.J. Hill, A. Abe-Ouchi, B. Otto-Bliesner, F.J. Bragg, M.A. Chandler, C. Contoux, H.J. Dowsett, A. Jost, Y. Kamae, G. Lohmann, D.J. Lunt, S.J. Pickering, M.J. Pound, G. Ramstein, N.A. Rosenbloom, L. Sohl, C. Stepanek, H. Ueda, and Z. Zhang, 2013: Challenges in quantifying Pliocene terrestrial warming revealed by data-model discord. *Nature Climate Change*, **3**, 969-974. http://dx.doi.org/10.1038/nclimate2008

91. Goldner, A., N. Herold, and M. Huber, 2014: The challenge of simulating the warmth of the mid-Miocene climatic optimum in CESM1. *Climate of the Past*, **10**, 523-536. http://dx.doi.org/10.5194/cp-10-523-2014

92. Foster, G.L., C.H. Lear, and J.W.B. Rae, 2012: The evolution of $pCO_2$, ice volume and climate during the middle Miocene. *Earth and Planetary Science Letters*, **341–344**, 243-254. http://dx.doi.org/10.1016/j.epsl.2012.06.007

93. LaRiviere, J.P., A.C. Ravelo, A. Crimmins, P.S. Dekens, H.L. Ford, M. Lyle, and M.W. Wara, 2012: Late Miocene decoupling of oceanic warmth and atmospheric carbon dioxide forcing. *Nature*, **486**, 97-100. http://dx.doi.org/10.1038/nature11200

94. Anagnostou, E., E.H. John, K.M. Edgar, G.L. Foster, A. Ridgwell, G.N. Inglis, R.D. Pancost, D.J. Lunt, and P.N. Pearson, 2016: Changing atmospheric $CO_2$ concentration was the primary driver of early Cenozoic climate. *Nature*, **533**, 380-384. http://dx.doi.org/10.1038/nature17423

95. Caballero, R. and M. Huber, 2013: State-dependent climate sensitivity in past warm climates and its implications for future climate projections. *Proceedings of the National Academy of Sciences*, **110**, 14162-14167. http://dx.doi.org/10.1073/pnas.1303365110

96. Huber, M. and R. Caballero, 2011: The early Eocene equable climate problem revisited. *Climate of the Past*, **7**, 603-633. http://dx.doi.org/10.5194/cp-7-603-2011

97. von der Heydt, A.S., P. Köhler, R.S.W. van de Wal, and H.A. Dijkstra, 2014: On the state dependency of fast feedback processes in (paleo) climate sensitivity. *Geophysical Research Letters*, **41**, 6484-6492. http://dx-.doi.org/10.1002/2014GL061121

98. Friedrich, T., A. Timmermann, M. Tigchelaar, O. Elison Timm, and A. Ganopolski, 2016: Nonlinear climate sensitivity and its implications for future greenhouse warming. *Science Advances*, **2**, e1501923. http://dx.doi.org/10.1126/sciadv.1501923

99. IPCC, 2013: *Climate Change 2013: The Physical Science Basis. Contribution of Working Group I to the Fifth Assessment Report of the Intergovernmental Panel on Climate Change*. Cambridge University Press, Cambridge, UK and New York, NY, 1535 pp. http://www.climatechange2013.org/report/

# Appendix A
# Observational Datasets
# Used in Climate Studies

**Recommended Citation for Chapter**

**Wuebbles**, D.J., 2017: Observational datasets used in climate studies. In: *Climate Science Special Report: Fourth National Climate Assessment, Volume I* [Wuebbles, D.J., D.W. Fahey, K.A. Hibbard, D.J. Dokken, B.C. Stewart, and T.K. Maycock (eds.)]. U.S. Global Change Research Program, Washington, DC, USA, pp. 430-435, doi: 10.7930/J0BK19HT.

## Climate Datasets

Observations, including those from satellites, mobile platforms, field campaigns, and ground-based networks, provide the basis of knowledge on many temporal and spatial scales for understanding the changes occurring in Earth's climate system. These observations also inform the development, calibration, and evaluation of numerical models of the physics, chemistry, and biology being used in analyzing past changes in climate and for making future projections. As all observational data collected by support from Federal agencies are required to be made available free of charge with machine readable metadata, everyone can access these products for their personal analysis and research and for informing decisions. Many of these datasets are accessible through web services.

Many long-running observations worldwide have provided us with long-term records necessary for investigating climate change and its impacts. These include important climate variables such as surface temperature, sea ice extent, sea level rise, and streamflow. Perhaps one of the most iconic climatic datasets, that of atmospheric carbon dioxide measured at Mauna Loa, Hawai'i, has been recorded since the 1950s. The U.S. and Global Historical Climatology Networks have been used as authoritative sources of recorded surface temperature increases, with some stations having continuous records going back many decades. Satellite radar altimetry data (for example, TOPEX/JASON1 & 2 satellite data) have informed the development of the University of Colorado's 20+ year record of global sea level changes. In the United States, the USGS (U.S. Geological Survey) National Water Information System contains, in some instances, decades of daily streamflow records which inform not only climate but land-use studies as well. The U.S. Bureau of Reclamation and U.S. Army Corp of Engineers have maintained data about reservoir levels for decades where applicable. Of course, datasets based on shorter-term observations are used in conjunction with longer-term records for climate study, and the U.S. programs are aimed at providing continuous data records. Methods have been developed and applied to process these data so as to account for biases, collection method, earth surface geometry, the urban heat island effect, station relocations, and uncertainty (e.g., see Vose et al. 2012;[1] Rennie et al. 2014;[2] Karl et al. 2015[3]).

Even observations not designed for climate have informed climate research. These include ship logs containing descriptions of ice extent, readings of temperature and precipitation provided in newspapers, and harvest records. Today, observations recorded both manually and in automated fashions inform research and are used in climate studies.

The U.S Global Change Research Program (USGCRP) has established the Global Change Information System (GCIS) to better coordinate and integrate the use of federal information products on changes in the global environment and the implications of those changes for society. The GCIS is an open-source, web-based resource for traceable global change data, information, and products. Designed for use by scientists, decision makers, and the public, the GCIS provides coordinated links to a select group of information products produced, maintained, and disseminated by government agencies and organizations. Currently the GCIS is aimed at the datasets used in Third National Climate Assessment (NCA3) and the USGCRP Climate and Health Assessment. It will be updated for the datasets used in this report (The Climate Science Special Report, CSSR).

**Temperature and Precipitation Observational Datasets**

For analyses of surface temperature or precipitation, including determining changes over the globe or the United States, the starting point is accumulating observations of surface air temperature or precipitation taken at observing stations all over the world, and, in the case of temperature, sea surface temperatures (SSTs) taken by ships and buoys. These are direct measurements of the air temperature, sea surface temperature, and precipitation. The observations are quality assured to exclude clearly erroneous values. For tempera-

ture, additional analyses are performed on the data to correct for known biases in the way the temperatures were measured. These biases include the change to the observations that result from changes in observing practices or changes in the location or local environment of an observing station. One example is with SSTs where there was a change in practice from throwing a bucket over the side of the ship, pulling up seawater and measuring the temperature of the water in the bucket to measuring the temperature of the water in the engine intake. The bucket temperatures are systematically cooler than engine intake water and must be corrected.

For evaluating the globally averaged temperature, data are then compared to a long-term average for the location where the observations were taken (e.g., a 30-year average for an individual observing station) to create a deviation from that average, commonly referred to as an anomaly. Using anomalies allows the spatial averaging of stations in different climates and elevations to produce robust estimates of the spatially averaged temperature or precipitation for a given area.

To calculate the temperature or precipitation for a large area, like the globe or the United States, the area is divided into "grid boxes" usually in latitude/longitude space. For example, one common grid size has 5° x 5° latitude/longitude boxes, where each side of a grid box is 5° of longitude and 5° of latitude in length. All data anomalies in a given grid box are averaged together to produce a gridbox average. Some grid boxes contain no observations, but nearby grid boxes do contain observations, so temperatures or precipitation for the grid boxes with no observations are estimated as a function of the nearby grid boxes with observations for that date.

Calculating the temperature or precipitation value for the larger area, either the globe or the United States, is done by averaging the values for all the grid boxes to produce one number for each day, month, season, or year resulting in a time series. The time series in each of the grid boxes are also used to calculate long-term trends in the temperature or precipitation for each grid box. This provides a picture of how temperatures and precipitation are changing in different locations.

Evidence for changes in the climate of the United States arises from multiple analyses of data from *in situ*, satellite, and other records undertaken by many groups over several decades. The primary dataset for surface temperatures and precipitation in the United States is nClimGrid,[4, 5] though trends are similar in the U.S. Historical Climatology Network, the Global Historical Climatology Network, and other datasets. For temperature, several atmospheric reanalyses (e.g., 20th Century Reanalysis, Climate Forecast System Reanalysis, ERA-Interim, and Modern Era Reanalysis for Research and Applications) confirm rapid warming at the surface since 1979, with observed trends closely tracking

the ensemble mean of the reanalyses.[1] Several recently improved satellite datasets document changes in middle tropospheric temperatures.[6, 7, 8] Longer-term changes are depicted using multiple paleo analyses (e.g., Wahl and Smerdon 2012;[9] Trouet et al. 2013[10]).

**Satellite Temperature Datasets**

A special look is given to the satellite temperature datasets because of controversies associated with these datasets. Satellite-borne microwave sounders such as the Microwave Sounding Unit (MSU) and Advanced Microwave Sounding Unit (AMSU) instruments operating on NOAA polar-orbiting platforms take measurements of the temperature of thick layers of the atmosphere with near global coverage. Because the long-term data record requires the piecing together of measurements made by 16 different satellites, accurate instrument intercalibration is of critical importance. Over the mission lifetime of most satellites, the instruments drift in both calibration and local measurement time. Adjustments to counter the effects of these drifts need to be developed and applied before a long-term record can be assembled. For tropospheric measurements,

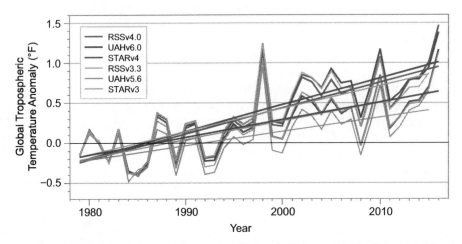

Figure A.1: Annual global (80°S–80°N) mean time series of tropospheric temperature for five recent datasets (see below). Each time series is adjusted so the mean value for the first three years is zero. This accentuates the differences in the long-term changes between the datasets. (Figure source: Remote Sensing Systems).

Table A.1.: Global Trends in Temperature Total Troposphere (TTT) since 1979 and 2000 (in °F per decade).

| Dataset | Trend (1979–2015) (°F/Decade) | Trend (2000–2015) (°F/Decade) |
|---|---|---|
| RSS V4.0 | 0.301 | 0.198 |
| UAH V6Beta5 | 0.196 | 0.141 |
| STAR V4.0 | 0.316 | 0.157 |
| RSS V3.3 | 0.208 | 0.105 |
| UAH V5.6 | 0.176 | 0.211 |
| STAR V3.0 | 0.286 | 0.061 |

the most challenging of these adjustments is the adjustment for drifting measurement time, which requires knowledge of the diurnal cycle in both atmospheric and surface temperature. Current versions of the sounder-based datasets account for the diurnal cycle by either using diurnal cycles deduced from model output[11, 12] or by attempting to derive the diurnal cycle from the satellite measurements themselves (an approach plagued by sampling issues and possible calibration drifts).[13, 14] Recently a hybrid approach has been developed, RSS Version 4.0,[6] that results in an increased warming signal relative to the other approach-es, particularly since 2000. Each of these methods has strengths and weaknesses, but none has sufficient accuracy to construct an unassailable long-term record of atmospheric temperature change. The resulting datasets show a greater spread in decadal-scale trends than do the surface temperature datasets for the same period, suggesting that they may be less reliable. Figure A.1 shows annual time series for the global mean tropospheric temperature for some recent versions of the satellite datasets. These data have been adjusted to remove the influence of stratospheric cooling.[15] Linear trend values are shown in Table A.1.

## DATA SOURCES

**All Satellite Data are "Temperature Total Troposphere" time series calculated from TMT and TLS**

(1.1*TMT) - (0.1*TLS). This combination reduces the effect of the lower stratosphere on the tropospheric temperature. (Fu, Qiang et al. "Contribution of stratospheric cooling to satellite-inferred tropospheric temperature trends." *Nature* 429.6987 (2004): 55-58.)

**UAH. UAH Version 6.0Beta5. Yearly (yyyy) text files of TMT and TLS are available from**

https://www.nsstc.uah.edu/data/msu/v6.0beta/tmt/

https://www.nsstc.uah.edu/data/msu/v6.0beta/tls/

Downloaded 5/15/2016.

**UAH. UAH Version 5.6. Yearly (yyyy) text files of TMT and TLS are available from**

http://vortex.nsstc.uah.edu/data/msu/t2/

http://vortex.nsstc.uah.edu/data/msu/t4/

Downloaded 5/15/2016.

**RSS. RSS Version 4.0.**

ftp://ftp.remss.com/msu/data/netcdf/RSS_Tb_Anom_Maps_ch_TTT_V4_0.nc

Downloaded 5/15/2016

**RSS. RSS Version 3.3.**

ftp://ftp.remss.com/msu/data/netcdf/RSS_Tb_Anom_Maps_ch_TTT_V3.3.nc

Downloaded 5/15/2016

**NOAA STAR. Star Version 3.0.**

ftp://ftp.star.nesdis.noaa.gov/pub/smcd/emb/mscat/data/MSU_AMSU_v3.0/Monthly_Atmospheric_Layer_Mean_Temperature/Merged_Deep-Layer_Temperature/NESDIS-STAR_TCDR_MSU-AMSUA_V03R00_TMT_S197811_E201709_C20171002.nc

ftp://ftp.star.nesdis.noaa.gov/pub/smcd/emb/mscat/data/MSU_AMSU_v3.0/Monthly_Atmospheric_Layer_Mean_Temperature/Merged_Deep-Layer_Temperature/NESDIS-STAR_TCDR_MSU-AMSUA_V03R00_TLS_S197811_E201709_C20171002.nc

Downloaded 5/18/2016.

# REFERENCES

1. Vose, R.S., D. Arndt, V.F. Banzon, D.R. Easterling, B. Gleason, B. Huang, E. Kearns, J.H. Lawrimore, M.J. Menne, T.C. Peterson, R.W. Reynolds, T.M. Smith, C.N. Williams, and D.L. Wuertz, 2012: NOAA's merged land-ocean surface temperature analysis. *Bulletin of the American Meteorological Society*, **93**, 1677-1685. http://dx.doi.org/10.1175/BAMS-D-11-00241.1

2. Rennie, J.J., J.H. Lawrimore, B.E. Gleason, P.W. Thorne, C.P. Morice, M.J. Menne, C.N. Williams, Jr., W.G. de Almeida, J.R. Christy, M. Flannery, M. Ishihara, K. Kamiguchi, A.M.G. Klein-Tank, A. Mhanda, D.H. Lister, V. Razuvaev, M. Renom, M. Rusticucci, J. Tandy, S.J. Worley, V. Venema, W. Angel, M. Brunet, B. Dattore, H. Diamond, M.A. Lazzara, F. Le Blancq, J. Luterbacher, H. Mächel, J. Revadekar, R.S. Vose, and X. Yin, 2014: The international surface temperature initiative global land surface databank: Monthly temperature data release description and methods. *Geoscience Data Journal*, **1**, 75-102. http://dx.doi.org/10.1002/gdj3.8

3. Karl, T.R., A. Arguez, B. Huang, J.H. Lawrimore, J.R. McMahon, M.J. Menne, T.C. Peterson, R.S. Vose, and H.-M. Zhang, 2015: Possible artifacts of data biases in the recent global surface warming hiatus. *Science*, **348**, 1469-1472. http://dx.doi.org/10.1126/science.aaa5632

4. Vose, R.S., S. Applequist, M. Squires, I. Durre, M.J. Menne, C.N. Williams, Jr., C. Fenimore, K. Gleason, and D. Arndt, 2014: Improved historical temperature and precipitation time series for U.S. climate divisions. *Journal of Applied Meteorology and Climatology*, **53**, 1232-1251. http://dx.doi.org/10.1175/JAMC-D-13-0248.1

5. Vose, R.S., M. Squires, D. Arndt, I. Durre, C. Fenimore, K. Gleason, M.J. Menne, J. Partain, C.N. Williams Jr., P.A. Bieniek, and R.L. Thoman, 2017: Deriving historical temperature and precipitation time series for Alaska climate divisions via climatologically aided interpolation. *Journal of Service Climatology* **10**, 20. https://www.stateclimate.org/sites/default/files/upload/pdf/journal-articles/2017-Ross-etal.pdf

6. Mears, C.A. and F.J. Wentz, 2016: Sensitivity of satellite-derived tropospheric temperature trends to the diurnal cycle adjustment. *Journal of Climate*, **29**, 3629-3646. http://dx.doi.org/10.1175/JCLI-D-15-0744.1

7. Spencer, R.W., J.R. Christy, and W.D. Braswell, 2017: UAH Version 6 global satellite temperature products: Methodology and results. *Asia-Pacific Journal of Atmospheric Sciences*, **53**, 121-130. http://dx.doi.org/10.1007/s13143-017-0010-y

8. Zou, C.-Z. and J. Li, 2014: NOAA MSU Mean Layer Temperature. National Oceanic and Atmospheric Administration, Center for Satellite Applications and Research, 35 pp. http://www.star.nesdis.noaa.gov/smcd/emb/mscat/documents/MSU_TCDR_CATBD_Zou_Li.pdf

9. Wahl, E.R. and J.E. Smerdon, 2012: Comparative performance of paleoclimate field and index reconstructions derived from climate proxies and noise-only predictors. *Geophysical Research Letters*, **39**, L06703. http://dx.doi.org/10.1029/2012GL051086

10. Trouet, V., H.F. Diaz, E.R. Wahl, A.E. Viau, R. Graham, N. Graham, and E.R. Cook, 2013: A 1500-year reconstruction of annual mean temperature for temperate North America on decadal-to-multidecadal time scales. *Environmental Research Letters*, **8**, 024008. http://dx.doi.org/10.1088/1748-9326/8/2/024008

11. Mears, C.A. and F.J. Wentz, 2009: Construction of the Remote Sensing Systems V3.2 atmospheric temperature records from the MSU and AMSU microwave sounders. *Journal of Atmospheric and Oceanic Technology*, **26**, 1040-1056. http://dx.doi.org/10.1175/2008JTECHA1176.1

12. Zou, C.-Z., M. Gao, and M.D. Goldberg, 2009: Error structure and atmospheric temperature trends in observations from the microwave sounding unit. *Journal of Climate*, **22**, 1661-1681. http://dx.doi.org/10.1175/2008JCLI2233.1

13. Christy, J.R., R.W. Spencer, W.B. Norris, W.D. Braswell, and D.E. Parker, 2003: Error estimates of version 5.0 of MSU–AMSU bulk atmospheric temperatures. *Journal of Atmospheric and Oceanic Technology*, **20**, 613-629. http://dx.doi.org/10.1175/1520-0426(2003)20<613:EEOVOM>2.0.CO;2

14. Po-Chedley, S., T.J. Thorsen, and Q. Fu, 2015: Removing diurnal cycle contamination in satellite-derived tropospheric temperatures: Understanding tropical tropospheric trend discrepancies. *Journal of Climate*, **28**, 2274-2290. http://dx.doi.org/10.1175/JCLI-D-13-00767.1

15. Fu, Q. and C.M. Johanson, 2005: Satellite-derived vertical dependence of tropical tropospheric temperature trends. *Geophysical Research Letters*, **32**, L10703. http://dx.doi.org/10.1029/2004GL022266

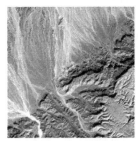

# Appendix B
# Model Weighting Strategy

**Recommended Citation for Chapter**

**Sanderson,** B.M. and M.F. Wehner, 2017: Model weighting strategy. In: *Climate Science Special Report: Fourth National Climate Assessment, Volume I* [Wuebbles, D.J., D.W. Fahey, K.A. Hibbard, D.J. Dokken, B.C. Stewart, and T.K. Maycock (eds.)]. U.S. Global Change Research Program, Washington, DC, USA, pp. 436-442, doi: 10.7930/J06T0JS3.

## Introduction

This document briefly describes a weighting strategy for use with the Climate Model Intercomparison Project, Phase 5 (CMIP5) multimodel archive in the Fourth National Climate Assessment (NCA4). This approach considers both skill in the climatological performance of models over North America and the interdependency of models arising from common parameterizations or tuning practices. The method exploits information relating to the climatological mean state of a number of projection-relevant variables as well as long-term metrics representing long-term statistics of weather extremes. The weights, once computed, can be used to simply compute weighted mean and significance information from an ensemble containing multiple initial condition members from co-dependent models of varying skill.

Our methodology is based on the concepts outlined in Sanderson et al. 2015,[1] and the specific application to the NCA4 is also described in that paper. The approach produces a single set of model weights that can be used to combine projections into a weighted mean result, with significance estimates which also treat the weighting appropriately.

The method, ideally, would seek to have two fundamental characteristics:

- If a duplicate of one ensemble member is added to the archive, the resulting mean and significance estimate for future change computed from the ensemble should not change.

- If a demonstrably unphysical model is added to the archive, the resulting mean and significance estimates should also not change.

## Method

The analysis requires an assessment of both model skill and an estimate of intermodel relationships—for which intermodel root mean square difference is taken as a proxy. The model and observational data used here is for the contiguous United States (CONUS), and most of Canada, using high-resolution data where available. Intermodel distances are computed as simple root mean square differences. Data is derived from a number of mean state fields and a number of fields that represent extreme behavior—these are listed in Table B.1. All fields are masked to only include information from CONUS/Canada.

The root mean square error (RMSE) between observations and each model can be used to produce an overall ranking for model simu-lations of the North American climate. Figure B.1 shows how this metric is influenced by different component variables.

Table B.1: Observational datasets used as observations.

| Field | Description | Source | Reference | Years |
|---|---|---|---|---|
| TS | Surface Temperature (seasonal) | Livneh, Hutchinson | (Hopkinson et al. 2012;[3] Hutchinson et al. 2009;[4] Livneh et al. 2013[5]) | 1950–2011 |
| PR | Mean Precipitation (seasonal) | Livneh, Hutchinson | (Hopkinson et al. 2012;[3] Hutchinson et al. 2009;[4] Livneh et al. 2013[5]) | 1950–2011 |
| RSUT | TOA Shortwave Flux (seasonal) | CERES-EBAF | (Wielicki et al. 1996[6]) | 2000–2005 |
| RLUT | TOA Longwave Flux (seasonal) | CERES-EBAF | (Wielicki et al. 1996[6]) | 2000–2005 |
| T | Vertical Temperature Profile (seasonal) | AIRS* | (Aumann et al. 2003[7]) | 2002–2010 |
| RH | Vertical Humidity Pro-file (seasonal) | AIRS | (Aumann et al. 2003[7]) | 2002–2010 |
| PSL | Surface Pressure (seasonal) | ERA-40 | (Uppala et al. 2005[8]) | 1970–2000 |
| Tnn | Coldest Night | Livneh, Hutchinson | (Hopkinson et al. 2012;[3] Hutchinson et al. 2009;[4] Livneh et al. 2013[5]) | 1950–2011 |
| Txn | Coldest Day | Livneh, Hutchinson | (Hopkinson et al. 2012;[3] Hutchinson et al. 2009;[4] Livneh et al. 2013[5]) | 1950–2011 |
| Tnx | Warmest Night | Livneh, Hutchinson | (Hopkinson et al. 2012;[3] Hutchinson et al. 2009;[4] Livneh et al. 2013[5]) | 1950–2011 |
| Txx | Warmest day | Livneh, Hutchinson | (Hopkinson et al. 2012;[3] Hutchinson et al. 2009;[4] Livneh et al. 2013[5]) | 1950–2011 |
| rx5day | seasonal max. 5-day total precip. | Livneh, Hutchinson | (Hopkinson et al. 2012;[3] Hutchinson et al. 2009;[4] Livneh et al. 2013[5]) | 1950–2011 |

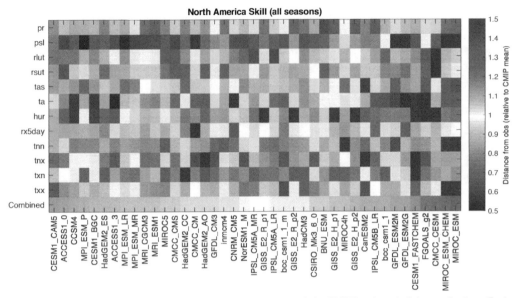

**Figure B.1:** A graphical representation of the intermodel distance matrix for CMIP5 and a set of observed values. Each row and column represents a single climate model (or observation). All scores are aggregated over seasons (individual seasons are not shown). Each box represents a pairwise distance, where warm (red) colors indicate a greater distance. Distances are measured as a fraction of the mean intermodel distance in the CMIP5 ensemble. (Figure source: Sanderson et al. 2017[2]).

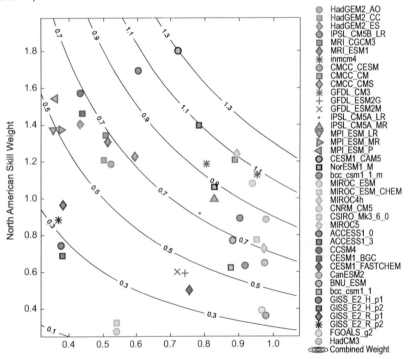

**Figure B.2:** Model skill and independence weights for the CMIP5 archive evaluated over the North American domain. Contours show the overall weighting, which is the product of the two individual weights. (Figure source: Sanderson et al. 2017[2]).

Models are downweighted for poor skill if their multivariate combined error is significantly greater than a "skill radius" term, which is a free parameter of the approach. The calibration of this parameter is determined through a perfect model study.[2] A pairwise distance matrix is computed to assess intermodel RMSE values for each model pair in the archive, and a model is downweighted for dependency if there exists another model with a pairwise distance to the original model significantly smaller than a "similarity radius." This is the second parameter of the approach, which is calibrated by considering known relationships within the archive. The resulting skill and independence weights are multiplied to give an overall "combined" weight—illustrated in Figure B.2 for the CMIP5 ensemble and listed in Table B.2.

The weights are used in the Climate Science Special Report (CSSR) to produce weighted mean and significance maps of future change, where the following protocol is used:

- Stippling—large changes, where the weighted multimodel average change is greater than double the standard deviation of the 20-year mean from control simulations runs, and 90% of the weight corresponds to changes of the same sign.

- Hatching—No significant change, where the weighted multimodel average change is less than the standard deviation of the 20-year means from control simulations runs.

- Whited out—Inconclusive, where the weighted multimodel average change is greater than double the standard deviation of the 20-year mean from control runs and less than 90% of the weight corresponds to changes of the same sign.

We illustrate the application of this method to future projections of precipitation change under the higher scenario (RCP8.5) in Figure B.3. The weights used in the report are chosen to be conservative, minimizing the risk of overconfidence and maximizing out-of-sample predictive skill for future projections. This results (as in Figure B.3) in only modest differences in the weighted and unweighted maps. It is shown in Sanderson et al. 2017[2] that a more aggressive weighting strategy, or one focused on a particular variable, tends to exhibit a stronger constraint on future change relative to the unweighted case. It is also notable that tradeoffs exist between skill and replication in the archive (evident in Figure B.2), such that the weighting for both skill and uniqueness has a compensating effect. As such, mean projections using the CMIP5 ensemble are not strongly influenced by the weighting. However, the establishment of the weighting strategy used in the CSSR provides some insurance against a potential case in future assessments where there is a highly replicated, but poorly performing model.

**Table B.2:** Uniqueness, skill, and combined weights for CMIP5.

| | Uniqueness Weight | Skill Weight | Combined |
|---|---|---|---|
| ACCESS1-0 | 0.60 | 1.69 | 1.02 |
| ACCESS1-3 | 0.78 | 1.40 | 1.09 |
| BNU-ESM | 0.88 | 0.77 | 0.68 |
| CCSM4 | 0.43 | 1.57 | 0.68 |
| CESM1-BGC | 0.44 | 1.46 | 0.64 |
| CESM1-CAM5 | 0.72 | 1.80 | 1.30 |
| CESM1-FASTCHEM | 0.76 | 0.50 | 0.38 |
| CMCC-CESM | 0.98 | 0.36 | 0.35 |
| CMCC-CM | 0.89 | 1.21 | 1.07 |
| CMCC-CMS | 0.59 | 1.23 | 0.73 |
| CNRM-CM5 | 0.94 | 1.08 | 1.01 |
| CSIRO-Mk3-6-0 | 0.95 | 0.77 | 0.74 |
| CanESM2 | 0.97 | 0.65 | 0.63 |
| FGOALS-g2 | 0.97 | 0.39 | 0.38 |
| GFDL-CM3 | 0.81 | 1.18 | 0.95 |
| GFDL-ESM2G | 0.74 | 0.59 | 0.44 |
| GFDL-ESM2M | 0.72 | 0.60 | 0.43 |
| GISS-E2-H-p1 | 0.38 | 0.74 | 0.28 |
| GISS-E2-H-p2 | 0.38 | 0.69 | 0.26 |
| GISS-E2-R-p1 | 0.38 | 0.97 | 0.37 |
| GISS-E2-R-p2 | 0.37 | 0.89 | 0.33 |
| HadCM3 | 0.98 | 0.89 | 0.87 |
| HadGEM2-AO | 0.52 | 1.19 | 0.62 |
| HadGEM2-CC | 0.50 | 1.21 | 0.60 |
| HadGEM2-ES | 0.43 | 1.40 | 0.61 |
| IPSL-CM5A-LR | 0.79 | 0.92 | 0.72 |
| IPSL-CM5A-MR | 0.83 | 0.99 | 0.82 |
| IPSL-CM5B-LR | 0.92 | 0.63 | 0.58 |
| MIROC-ESM | 0.54 | 0.28 | 0.15 |
| MIROC-ESM-CHEM | 0.54 | 0.32 | 0.17 |
| MIROC4h | 0.97 | 0.73 | 0.71 |
| MIROC5 | 0.89 | 1.24 | 1.11 |
| MPI-ESM-LR | 0.35 | 1.38 | 0.49 |
| MPI-ESM-MR | 0.38 | 1.37 | 0.52 |
| MPI-ESM-P | 0.36 | 1.54 | 0.56 |
| MRI-CGCM3 | 0.51 | 1.35 | 0.68 |
| MRI-ESM1 | 0.51 | 1.31 | 0.67 |
| NorESM1-M | 0.83 | 1.06 | 0.88 |
| bcc-csm1-1 | 0.88 | 0.62 | 0.55 |
| bcc-csm1-1-m | 0.90 | 0.89 | 0.80 |
| inmcm4 | 0.95 | 1.13 | 1.08 |

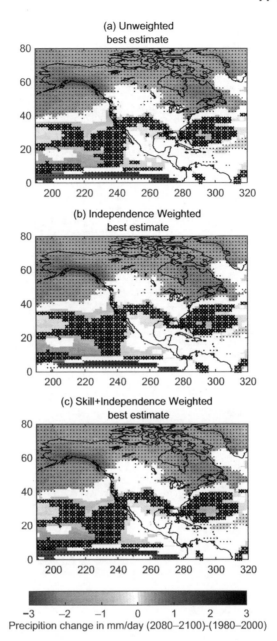

(a) Unweighted
best estimate

(b) Independence Weighted
best estimate

(c) Skill+Independence Weighted
best estimate

-3    -2    -1    0    1    2    3
Precipition change in mm/day (2080–2100)-(1980–2000)

Figure B.3: Projections of precipitation change over North America in 2080–2100, relative to 1980–2000 under the higher scenario (RCP8.5). (a) Shows the simple unweighted CMIP5 multimodel average, using the significance methodology from IPCC[9]; (b) shows the weighted results as outlined in Section 3 for models weighted by uniqueness only; and (c) shows weighted results for models weighted by both uniqueness and skill. (Figure source: Sanderson et al. 2017[2]).

# REFERENCES

1.  Sanderson, B.M., R. Knutti, and P. Caldwell, 2015: A representative democracy to reduce interdependency in a multimodel ensemble. *Journal of Climate*, **28**, 5171-5194. http://dx.doi.org/10.1175/jcli-d-14-00362.1

2.  Sanderson, B.M., M. Wehner, and R. Knutti, 2017: Skill and independence weighting for multi-model assessment. *Geoscientific Model Development*, **10**, 2379-2395. http://dx.doi.org/10.5194/gmd-10-2379-2017

3.  Hopkinson, R.F., M.F. Hutchinson, D.W. McKenney, E.J. Milewska, and P. Papadopol, 2012: Optimizing input data for gridding climate normals for Canada. *Journal of Applied Meteorology and Climatology*, **51**, 1508-1518. http://dx.doi.org/10.1175/jamc-d-12-018.1

4.  Hutchinson, M.F., D.W. McKenney, K. Lawrence, J.H. Pedlar, R.F. Hopkinson, E. Milewska, and P. Papadopol, 2009: Development and testing of Canada-wide interpolated spatial models of daily minimum–maximum temperature and precipitation for 1961–2003. *Journal of Applied Meteorology and Climatology*, **48**, 725-741. http://dx.doi.org/10.1175/2008jamc1979.1

5.  Livneh, B., E.A. Rosenberg, C. Lin, B. Nijssen, V. Mishra, K.M. Andreadis, E.P. Maurer, and D.P. Lettenmaier, 2013: A long-term hydrologically based dataset of land surface fluxes and states for the conterminous United States: Update and extensions. *Journal of Climate*, **26**, 9384-9392. http://dx.doi.org/10.1175/jcli-d-12-00508.1

6.  Wielicki, B.A., B.R. Barkstrom, E.F. Harrison, R.B. Lee III, G.L. Smith, and J.E. Cooper, 1996: Clouds and the Earth's Radiant Energy System (CERES): An Earth observing system experiment. *Bulletin of the American Meteorological Society*, **77**, 853-868. http://dx.doi.org/10.1175/1520-0477(1996)077<0853:catere>2.0.co;2

7.  Aumann, H.H., M.T. Chahine, C. Gautier, M.D. Goldberg, E. Kalnay, L.M. McMillin, H. Revercomb, P.W. Rosenkranz, W.L. Smith, D.H. Staelin, L.L. Strow, and J. Susskind, 2003: AIRS/AMSU/HSB on the Aqua mission: Design, science objectives, data products, and processing systems. *IEEE Transactions on Geoscience and Remote Sensing*, **41**, 253-264. http://dx.doi.org/10.1109/tgrs.2002.808356

8.  Uppala, S.M., P.W. KÅllberg, A.J. Simmons, U. Andrae, V.D.C. Bechtold, M. Fiorino, J.K. Gibson, J. Haseler, A. Hernandez, G.A. Kelly, X. Li, K. Onogi, S. Saarinen, N. Sokka, R.P. Allan, E. Andersson, K. Arpe, M.A. Balmaseda, A.C.M. Beljaars, L.V.D. Berg, J. Bidlot, N. Bormann, S. Caires, F. Chevallier, A. Dethof, M. Dragosavac, M. Fisher, M. Fuentes, S. Hagemann, E. Hólm, B.J. Hoskins, L. Isaksen, P.A.E.M. Janssen, R. Jenne, A.P. McNally, J.F. Mahfouf, J.J. Morcrette, N.A. Rayner, R.W. Saunders, P. Simon, A. Sterl, K.E. Trenberth, A. Untch, D. Vasiljevic, P. Viterbo, and J. Woollen, 2005: The ERA-40 re-analysis. *Quarterly Journal of the Royal Meteorological Society*, **131**, 2961-3012. http://dx.doi.org/10.1256/qj.04.176

9.  IPCC, 2013: *Climate Change 2013: The Physical Science Basis. Contribution of Working Group I to the Fifth Assessment Report of the Intergovernmental Panel on Climate Change*. Cambridge University Press, Cambridge, UK and New York, NY, 1535 pp. http://www.climatechange2013.org/report/

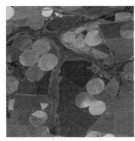

# Appendix C
# Detection and Attribution Methodologies Overview

**Recommended Citation for Chapter**

**Knutson**, T., 2017: Detection and attribution methodologies overview. In: *Climate Science Special Report: Fourth National Climate Assessment, Volume I* [Wuebbles, D.J., D.W. Fahey, K.A. Hibbard, D.J. Dokken, B.C. Stewart, and T.K. Maycock (eds.)]. U.S. Global Change Research Program, Washington, DC, USA, pp. 443-451, doi: 10.7930/J0319T2J.

## C.1 Introduction and Conceptual Framework

In this appendix, we present a brief overview of the methodologies and methodological issues for detection and attribution of climate change. Attributing an observed change or an event partly to a causal factor (such as anthropogenic climate forcing) normally requires that the change first be detectable.[1] A *detectable* observed change is one which is determined to be highly unlikely to occur (less than about a 10% chance) due to internal variability alone, without necessarily being ascribed to a causal factor. An *attributable* change refers to a change in which the relative contribution of causal factors has been evaluated along with an assignment of statistical confidence (e.g., Bindoff et al. 2013;[2] Hegerl et al. 2010[1]).

As outlined in Bindoff et al.,[2] the conceptual framework for most detection and attribution studies consists of four elements: 1) relevant observations; 2) the estimated time history of relevant climate forcings (such as greenhouse gas concentrations or volcanic activity); 3) a modeled estimate of the impact of the climate forcings on the climate variables of interest; and 4) an estimate of the internal (unforced) variability of the climate variables of inter-

est—that is, the changes that can occur due to natural unforced variations of the ocean, atmosphere, land, cryosphere, and other elements of the climate system in the absence of external forcings. The four elements above can be used together with a detection and attribution framework to assess possible causes of observed changes.

## C.2 Fingerprint-Based Methods

A key methodological approach for detection and attribution is the regression-based "fingerprint" method (e.g., Hasselmann 1997;[3] Allen and Stott 2003;[4] Hegerl et al. 2007;[5] Hegerl and Zwiers 2011;[6] Bindoff et al. 2013[2]), where observed changes are regressed onto a model-generated response pattern to a particular forcing (or set of forcings), and regression scaling factors are obtained. When a scaling factor for a forcing pattern is determined to be significantly different from zero, a detectable change has been identified. If the uncertainty bars on the scaling factor encompass unity, the observed change is consistent with the modeled response, and the observed change can be attributed, at least in part, to the associated forcing agent, according to this methodology. Zwiers et al.[7] showed how detection and attribution methods could be applied to the problem of changes in daily temperature extremes at the regional scale by using a generalized extreme value (GEV) approach. In their approach, a time-evolving pattern of GEV location parameters (i.e., "fingerprint") from models is fit to the observed extremes as a means of detecting and attributing changes in the extremes to certain forcing sets (for example, anthropogenic forcings).

A recent development in detection/attribution methodology[8] uses hypothesis testing and an additive decomposition approach rather than linear regression of patterns. The new approach makes use of the magnitudes of responses from the models rather than using the model patterns and deriving the scaling factors (magnitudes of responses) from regression. The new method, in a first application, gives very similar attributable anthropogenic warming estimates to the earlier methods as reported in Bindoff et al.[2] and shown in Figure 3.2. Some further methodological developments for performing optimal fingerprint detection and attribution studies are proposed in Hannart,[9] who, for example, focuses on the possible use of raw data in analyses without the use of dimensional reductions, such as projecting the data onto a limited number of basis functions, such as spherical harmonics, before analysis.

## C.3 Non-Fingerprint Based Methods

A simpler detection/attribution/consistency calculation, which does not involve regression and pattern scaling, compares observed and simulated time series to assess whether observations are consistent with natural variability simulations or with simulations forced by both natural and anthropogenic forcing agents.[10, 11] Cases where observations are inconsistent with model simulations using natural forcing only (a detectable change), while also being consistent with models that incorporate both anthropogenic and natural

forcings, are interpreted as having an attributable anthropogenic contribution, subject to caveats regarding uncertainties in observations, climate forcings, modeled responses, and simulated internal climate variability. This simpler method is useful for assessing trends over smaller regions such as sub-regions of the United States (see the example given in Figure 6.5 for regional surface temperature trends).

Delsole et al.[12] introduced a method of identifying internal (unforced) variability in climate data by decomposing variables by time scale, using a measure of their predictability. They found that while such internal variability could contribute to surface temperature trends of 30-years' duration or less, and could be responsible for the accelerated global warming during 1977–2008 compared to earlier decades, the strong (approximately 0.8°C, or 1.4°F) warming trend seen in observations over the past century was not explainable by such internal variability. Constructed circulation analogs[13, 14] is a method used to identify the part of observed surface temperature changes that is due to atmospheric circulation changes alone.

The time scale by which climate change signals will become detectable in various regions is a question of interest in detection and attribution studies, and methods of estimating this have been developed and applied (e.g., Mahlstein et al. 2011;[15] Deser et al. 2012[16]). These studies illustrate how natural variability can obscure forced climate signals for decades, particularly for smaller (less than continental) space scales.

Other examples of detection and attribution methods include the use of multiple linear regression with energy balance models (e.g., Canty et al. 2013[17]) and Granger causality tests (e.g., Stern and Kaufmann 2014[18]). These are typically attempting to relate forcing time series, such as the historical record of atmospheric $CO_2$ since 1860, to a climate response measure, such as global mean temperature or ocean heat content, but without using a full coupled climate model to explicitly estimate the response of the climate system to forcing (or the spatial pattern of the response to forcing). Granger causality, for example, explores the lead–lag relationships between different variables to infer causal relationships between them and attempts to control for any influence of a third variable that may be linked to the other two variables in question.

## C.4 Multistep Attribution and Attribution without Detection

A growing number of climate change and extreme event attribution studies use a *multistep attribution* approach,[1] based on attribution of a change in climate conditions that are closely related to the variable or event of interest. In the multistep approach, an observed change in the variable of interest is attributed to a change in climate or other environmental conditions, and then the changes in the climate or environmental conditions are separately attributed to an external forcing, such as anthropogenic emissions of greenhouse gases. As an example, some attribution statements for phenomena such as droughts or hurricane activity—where there are not necessarily detectable trends in occurrence of the phenomenon itself—are based on models and on detected changes in related variables such as surface temperature, as well as an understanding of the relevant physical processes linking surface temperatures to hurricanes or drought. For example, some studies of the recent California drought (e.g., Mao et al. 2015;[19] Williams et al. 2015[20]) attribute a fraction of the event to anthropogenic warming or to long-term warming based on modeling or statistical analysis, although without claiming that there was a detectable change in the drought frequency or magnitude.

The multistep approach and model simulations are both methods that, in principle, can allow for attribution of a climate change or a change in the likelihood of occurrence of an event to a causal factor without necessarily detecting a significant change in the occurrence rate of the phenomenon or event itself (though in some cases, there may also be a detectable change in the variable of interest). For example, Murakami et al.[21] used model simulations to conclude that the very active hurricane season observed near Hawai'i in 2014 was at least partially attributable to anthropogenic influence; they also show that there is no clear long-term detectable trend in historical hurricane occurrence near Hawai'i in available observations. If an attribution statement is made where there is not a detectable change in the phenomenon itself (for example, hurricane frequency or drought frequency) then this statement is an example of *attribution without detection*. Such an attribution without detection can be distinguished from a conventional single-step attribution (for example, global mean surface temperature) where in the latter case there is a detectable change in the variable of interest (or the scaling factor for a forcing pattern is significantly different from zero in observations) and attribution of the changes in that variable to specific external forcing agents. Regardless of whether a single-step or multistep attribution approach is used, or whether there is a detectable change in the variable of interest, attribution statements with relatively higher levels of confidence are underpinned by a thorough understanding of the physical processes involved.

There are reasons why attribution without detection statements can be appropriate, despite the lower confidence typically associated with such statements as compared to attribution statements that are supported by detection of a change in the phenomenon itself. For example, an event of interest may be so rare that a trend analysis for similar events is not practical. Including attribution without detection events in the analysis of climate change impacts reduces the chances of a false negative, that is, incorrectly concluding that climate change had no influence on a given extreme events[22] in a case where it did have an influence. However, avoiding this type of error through attribution without detection comes at the risk of increasing the rate of false positives, where one incorrectly concludes that anthropogenic climate change had a certain type of influence on an extreme event when in fact it did not have such an influence (see Box C.1).

## C.5 Extreme Event Attribution Methodologies

Since the release of the Intergovernmental Panel on Climate Change's Fifth Assessment Report (IPCC AR5) and the Third National Climate Assessment (NCA3),[23] there have been further advances in the science of detection and attribution of climate change. An emerging area in the science of detection and attribution is the attribution of extreme weather and climate events.[24, 25, 26] According to Hulme,[27] there are four general types of attribution methods that are applied in practice: physical reasoning, statistical analysis of time series, fraction of attributable risk (FAR) estimation, and the philosophical argument that there are no longer any purely natural weather events. As discussed in a recent National Academy of Sciences report,[24] possible anthropogenic influence on an extreme event can be assessed using a risk-based approach, which examines whether the odds of occurrence of a type of extreme event have changed, or through an ingredients-based or conditional attribution approach.

In the risk-based approach,[24, 27, 28] one typically uses a model to estimate the probability (p) of occurrence of a weather or climate event with-

in two climate states: one state with anthropogenic influence (where the probability is $p_1$) and the other state without anthropogenic influence (where the probability is $p_0$). Then the ratio ($p_1/p_0$) describes how much more or less likely the event is in the modeled climate with anthropogenic influence compared to a modeled hypothetical climate without anthropogenic influences. Another common metric used with this approach is the fraction **of attributable risk (FAR)**, defined as FAR = $1 - (p_0/p_1)$. Further refinements on such an approach using causal theory are discussed in Hannart et al.[29]

In the conditional or ingredients-based approach,[24, 30, 31, 32] an investigator may look for changes in occurrence of atmospheric circulation and weather patterns relevant to the extreme event, or at the impact of certain environmental changes (for example, greater atmospheric moisture) on the character of an extreme event. Conditional or ingredients-based attribution can be applied to extreme events or to climate changes in general. An example of the ingredients-based approach and more discussion of this type of attribution method is given in Box C.2.

Hannart et al.[29] have discussed how causal theory can also be applied to attribution studies in order to distinguish between necessary and sufficient causation. Hannart et al.[33] further propose methodologies to use data assimilation systems, which are now used operationally to update short-term numerical weather prediction models, for detection and attribution. They envision how such systems could be used in the future to implement near-real time systematic causal attribution of weather and climate-related events.

## Box C.1. On the Use of Significance Levels and Significance Tests in Attribution Studies

In detection/attribution studies, a detectable observed change is one which is determined to be highly unlikely to occur (less than about a 10% chance) due to internal variability alone. Some frequently asked questions concern the use of such a high statistical threshold (significance level) in attribution studies. In this box, we respond to several such questions received in the public review period.

*Why is such a high degree of confidence (for example, statistical significance at p level of 0.05) typically required before concluding that an attributable anthropogenic component to a climate change or event has been detected? For example, could attribution studies be reframed to ask whether there is a 5% or more chance that anthropogenic climate change contributed to the event?*

This question is partly related to the issue of risk avoidance. For example, if there is a particular climate change outcome that we wish to avoid (for example, global warming of 3°C, or 10°C, or a runaway greenhouse) then one can use the upper ranges of confidence intervals of climate model projections as guidance, based on available science, for avoiding such outcomes. Detection/attribution studies typically deal with smaller changes than climate projections over the next century or more. For detection/attribution studies, researchers are confronting models with historical data to explore whether or not observed climate change signals are emerging from the background of natural variability. Typically, the emergent signal is just a small fraction of what is predicted by the models for the coming century under continued strong greenhouse gas emission scenarios. Detecting that a change has emerged from natural variability is not the same as approaching a threshold to be avoided, unless the goal is to ensure no detectable anthropogenic influence on climate. Consequently, use of a relative strong confidence level (or *p*-value of 0.05) for determining climate change detection seems justified for the particular case of climate change detection, since one can also separately use risk-avoidance strategies or probability criteria to avoid reaching certain defined thresholds (for example, a 2°C global warming threshold).

A related question concerns ascribing blame for causing an extreme event. For example, if a damaging hurricane or typhoon strikes an area and causes much damage, affected residents may ask whether human-caused climate change was at least partially to blame for the event. In this case, climate scientists sometimes use the "Fraction of Attributable Risk" framework, where they examine whether the odds of some threshold event occurring have been increased due to anthropogenic climate change. This is typically a model-based calculation, where the probability distribution related to the event in question is modeled under preindustrial and present-day climate conditions, and the occurrence rates are compared for the two modeled distributions. Note that such an analysis can be done with or without the detection of a climate change signal for the occurrence of the event in question. In general, cases where there has been a detection and attribution of changes in the event in question to human causes, then the attribution of increased risk to anthropogenic forcing will be relatively more confident.

The question of whether it is more appropriate to use approaches that incorporate a high burden of statistical evidence before concluding that anthropogenic forcings contributed significantly (as in traditional detection/attribution studies) versus using models to estimate anthropogenic contributions when there may not even be a detectable signal present in the observations (as in some Fraction of Attributable Risk studies) may depend on what type of error or scenario one most wants to avoid. In the former case, one is attempting to avoid the error of concluding that anthropogenic forcing has contributed to some observed climate change, when in fact, it later turns out that anthropogenic forcing has not contributed to the change. In the second case, one is attempting to avoid the "error" of concluding that anthropogenic forcing has not contributed significantly to an observed climate change or event when (as it later comes to be known) anthropogenic forcing had evidently contributed to the change, just not at a level that was detectable at the time compared to natural variability.

*What is the tradeoff between false positives and false negatives in attribution statistical testing, and how is it decided which type of error one should focus on avoiding?*

As discussed above, there are different types of errors or scenarios that we would ideally like to avoid. However, the decision of what type of analysis to do may involve a tradeoff where one decides that it is more important to avoid either falsely concluding that anthropogenic forcing *has* contributed, or to avoid falsely concluding that anthropogenic forcing had *not* made a detectable contribution to the event. Since there is no correct answer that can apply in all cases, it would be helpful if, in requesting scientific assessments, policymakers provide some guidance about which type of error or scenario they would most desire be avoided in the analyses and assessments in question.

*Since substantial anthropogenic climate change (increased surface temperatures, increased atmospheric water vapor, etc.) has already occurred, aren't all extreme events affected to some degree by anthropogenic climate change?*

Climate scientists are aware from modeling experiments that very tiny changes to initial conditions in model simulations lead to very different realizations of internal climate variability "noise" in the model simulations. Comparing large samples of this random background noise from models against observed changes is one way to test whether the observed changes are statistically distinguishable from internal climate variability. In any case, this experience also teaches us that any anthropogenic influence on climate, no matter how tiny, has some effect on the future trajectory of climate variability, and thus could affect the timing and occurrence of extreme events. More meaningful questions are: 1) Has anthropogenic forcing produced a statistically significant change in the probability of occurrence of some class of extreme event? 2) Can we determine with confidence the net sign of influence of anthropogenic climate change on the frequency, intensity, etc., of a type of extreme event? 3) Can climate scientists quantify (with credible confidence intervals) the effect of climate change on the occurrence frequency, the intensity, or some other aspect of an observed extreme event?

## Box C.2 Illustration of Ingredients-based Event Attribution: The Case of Hurricane Sandy

To illustrate some aspects of the conditional or ingredients-based attribution approach, the case of Hurricane Sandy can be considered. If one considers Hurricane Sandy's surge event, there is strong evidence that sea level rise, at least partly anthropogenic in origin (see Ch. 12: Sea Level Rise), made Sandy's surge event worse, all other factors being equal.[34] The related question of whether anthropogenic climate change increased the risk of an event like Sandy involves not just the sea level ingredient to surge risk but also whether the frequency and/or intensity of Sandy-like storms has increased or decreased as a result of anthropogenic climate change. This latter question is more difficult and is briefly reviewed here.

A conditional or ingredients-based attribution approach, as applied to a hurricane event such as Sandy, may assume that the weather patterns in which the storm was embedded—and the storm itself—could have occurred in a preindustrial climate, and the event is re-simulated while changing only some aspects of the large-scale environment (for example, sea surface temperatures, atmospheric temperatures, and moisture) by an estimated anthropogenic climate change signal. Such an approach thus explores whether anthropogenic climate change to date has, for example, altered the intensity of a Hurricane Sandy-like storm, assuming the occurrence of a Sandy-like storm in both preindustrial and present-day climates. Modeling studies show, as expected, that the anomalously warm sea surface temperatures off the U.S. East Coast during Sandy led to a substantially more intense simulated storm than under present-day climatological conditions.[35] However, these anomalous sea surface temperatures and other environmental changes are a mixture of anthropogenic and natural influences, and so it is not generally possible to infer the anthropogenic component from such experiments. Another study[36] modeled the influence of just the anthropogenic changes to the thermodynamic environment (including sea surface temperatures, atmospheric temperatures, and moisture perturbations) and concluded that anthropogenic climate change to date had caused Hurricane Sandy to be about 5 hPa more intense, but that this modeled change was not statistically significant at the 95% confidence level. A third study used a statistical–dynamical model to compare simulated New York City-area tropical cyclones in pre-anthropogenic and anthropogenic time periods.[34] It concluded that there have been anthropogenically induced increases in the types of tropical cyclones that cause extreme surge events in the region, apart from the effects of sea level rise, such as increased radius of maximum winds in the anthropogenic era. However, the statistical–dynamical model used in the study simulates an unusually large increase in global tropical cyclone activity in 21st century projections[37] compared to other tropical cyclone modeling studies using dynamical models—a number of which simulate future decreases in late 21st century tropical storm frequency in the Atlantic basin (e.g., Christensen et al. 2013[38]). This range of uncertainty among various model simulations of Atlantic tropical cyclone activity under climate change imply that there is low confidence in determining the net impact to date of anthropogenic climate change on the risk of Sandy-like events, though anthropogenic sea level rise, all other things equal, has increased the surge risk.

In summary, while there is agreement that sea level rise alone has caused greater storm surge risk in the New York City area, there is low confidence on whether a number of other important determinants of storm surge climate risk, such as the frequency, size, or intensity of Sandy-like storms in the New York region, have increased or decreased due to anthropogenic warming to date.

# REFERENCES

1. Hegerl, G.C., O. Hoegh-Guldberg, G. Casassa, M.P. Hoerling, R.S. Kovats, C. Parmesan, D.W. Pierce, and P.A. Stott, 2010: Good practice guidance paper on detection and attribution related to anthropogenic climate change. *Meeting Report of the Intergovernmental Panel on Climate Change Expert Meeting on Detection and Attribution of Anthropogenic Climate Change.* Stocker, T.F., C.B. Field, D. Qin, V. Barros, G.-K. Plattner, M. Tignor, P.M. Midgley, and K.L. Ebi, Eds. IPCC Working Group I Technical Support Unit, University of Bern, Bern, Switzerland, 1-8. http://www.ipcc.ch/pdf/supporting-material/ipcc_good_practice_guidance_paper_anthropogenic.pdf

2. **Bindoff, N.L., P.A. Stott, K.M. AchutaRao, M.R. Allen, N. Gillett, D. Gutzler, K. Hansingo, G. Hegerl, Y. Hu, S. Jain, I.I. Mokhov, J. Overland, J. Perlwitz, R. Sebbari, and X. Zhang, 2013:** Detection and attribution of climate change: From global to regional. *Climate Change 2013: The Physical Science Basis. Contribution of Working Group I to the Fifth Assessment Report of the Intergovernmental Panel on Climate Change.* Stocker, T.F., D. Qin, G.-K. Plattner, M. Tignor, S.K. Allen, J. Boschung, A. Nauels, Y. Xia, V. Bex, and P.M. Midgley, Eds. Cambridge University Press, Cambridge, United Kingdom and New York, NY, USA, 867–952. http://www.climatechange2013.org/report/full-report/

3. **Hasselmann, K., 1997: Multi-pattern fingerprint** method for detection and attribution of climate change. *Climate Dynamics,* **13,** 601-611. http://dx.doi.org/10.1007/s003820050185

4. Allen, M.R. and P.A. Stott, 2003: Estimating signal **amplitudes in optimal fingerprinting, Part I:** Theory. *Climate Dynamics,* **21,** 477-491. http://dx.doi.org/10.1007/s00382-003-0313-9

5. Hegerl, G.C., F.W. Zwiers, P. Braconnot, N.P. Gillett, Y. Luo, J.A.M. Orsini, N. Nicholls, J.E. Penner, and P.A. Stott, 2007: Understanding and attributing climate change. *Climate Change 2007: The Physical Science Basis. Contribution of Working Group I to the Fourth Assessment Report of the Intergovernmental Panel on Climate Change.* Solomon, S., D. Qin, M. Manning, Z. Chen, M. Marquis, K.B. Averyt, M. Tignor, and H.L. Miller, Eds. Cambridge University Press, Cambridge, United Kingdom and New York, NY, USA, 663-745. http://www.ipcc.ch/publications_and_data/ar4/wg1/en/ch9.html

6. Hegerl, G. and F. Zwiers, 2011: Use of models in detection and attribution of climate change. *Wiley Interdisciplinary Reviews: Climate Change,* **2,** 570-591. http://dx.doi.org/10.1002/wcc.121

7. Zwiers, F.W., X.B. Zhang, and Y. Feng, 2011: **Anthropogenic influence on long return period daily temperature extremes at regional scales.** *Journal of Climate,* **24,** 881-892. http://dx.doi.org/10.1175/2010jcli3908.1

8. Ribes, A., F.W. Zwiers, J.-M. Azaïs, and P. Naveau, 2017: A new statistical approach to climate change detection and attribution. *Climate Dynamics,* **48,** 367-386. http://dx.doi.org/10.1007/s00382-016-3079-6

9. **Hannart, A., 2016: Integrated optimal fingerprinting:** Method description and illustration. *Journal of Climate,* **29,** 1977-1998. http://dx.doi.org/10.1175/jcli-d-14-00124.1

10. Knutson, T.R., F. Zeng, and A.T. Wittenberg, 2013: Multimodel assessment of regional surface temperature trends: CMIP3 and CMIP5 twentieth-century simulations. *Journal of Climate,* **26,** 8709-8743. http://dx.doi.org/10.1175/JCLI-D-12-00567.1

11. van Oldenborgh, G.J., F.J. Doblas Reyes, S.S. Drijfhout, and E. Hawkins, 2013: Reliability of regional climate model trends. *Environmental Research Letters,* **8,** 014055. http://dx.doi.org/10.1088/1748-9326/8/1/014055

12. DelSole, T., M.K. Tippett, and J. Shukla, 2011: A sig-**nificant component of unforced multidecadal variability in the recent acceleration of global warming.** *Journal of Climate,* **24,** 909-926. http://dx.doi.org/10.1175/2010jcli3659.1

13. van den Dool, H., J. Huang, and Y. Fan, 2003: Performance and analysis of the constructed analogue method applied to U.S. soil moisture over 1981–2001. *Journal of Geophysical Research,* **108,** 8617. http://dx.doi.org/10.1029/2002JD003114

14. Deser, C., L. Terray, and A.S. Phillips, 2016: Forced and internal components of winter air temperature trends over North America during the past 50 years: Mechanisms and implications. *Journal of Climate,* **29,** 2237-2258. http://dx.doi.org/10.1175/JCLI-D-15-0304.1

15. Mahlstein, I., R. Knutti, S. Solomon, and R.W. Portmann, 2011: **Early onset of significant local warming** in low latitude countries. *Environmental Research Letters,* **6,** 034009. http://dx.doi.org/10.1088/1748-9326/6/3/034009

16. Deser, C., R. Knutti, S. Solomon, and A.S. Phillips, 2012: Communication of the role of natural variability in future North American climate. *Nature Climate Change,* **2,** 775-779. http://dx.doi.org/10.1038/nclimate1562

17. Canty, T., N.R. Mascioli, M.D. Smarte, and R.J. Salawitch, 2013: An empirical model of global climate – Part 1: A critical evaluation of volcanic cooling. *Atmospheric Chemistry and Physics,* **13,** 3997-4031. http://dx.doi.org/10.5194/acp-13-3997-2013

18. Stern, D.I. and R.K. Kaufmann, 2014: Anthropogenic and natural causes of climate change. *Climatic Change,* **122,** 257-269. http://dx.doi.org/10.1007/s10584-013-1007-x

19. Mao, Y., B. Nijssen, and D.P. Lettenmaier, 2015: Is climate change implicated in the 2013–2014 California drought? A hydrologic perspective. *Geophysical Research Letters*, **42**, 2805-2813. http://dx.doi.org/10.1002/2015GL063456

20. Williams, A.P., R. Seager, J.T. Abatzoglou, B.I. Cook, J.E. Smerdon, and E.R. Cook, 2015: Contribution of anthropogenic warming to California drought during 2012–2014. *Geophysical Research Letters*, **42**, 6819-6828. http://dx.doi.org/10.1002/2015GL064924

21. Murakami, H., G.A. Vecchi, T.L. Delworth, K. Paffendorf, L. Jia, R. Gudgel, and F. Zeng, 2015: Investigating the influence of anthropogenic forcing and natural variability on the 2014 Hawaiian hurricane season [in "Explaining Extreme Events of 2014 from a Climate Perspective"]. *Bulletin of the American Meteorological Society*, **96 (12)**, S115-S119. http://dx.doi.org/10.1175/BAMS-D-15-00119.1

22. Anderegg, W.R.L., E.S. Callaway, M.T. Boykoff, G. Yohe, and T.y.L. Root, 2014: Awareness of both type 1 and 2 errors in climate science and assessment. *Bulletin of the American Meteorological Society*, **95**, 1445-1451. http://dx.doi.org/10.1175/BAMS-D-13-00115.1

23. Melillo, J.M., T.C. Richmond, and G.W. Yohe, eds., 2014: *Climate Change Impacts in the United States: The Third National Climate Assessment*. U.S. Global Change Research Program: Washington, D.C., 841 pp. http://dx.doi.org/10.7930/J0Z31WJ2

24. NAS, 2016: *Attribution of Extreme Weather Events in the Context of Climate Change*. The National Academies Press, Washington, DC, 186 pp. http://dx.doi.org/10.17226/21852

25. Stott, P., 2016: How climate change affects extreme weather events. *Science*, **352**, 1517-1518. http://dx.doi.org/10.1126/science.aaf7271

26. Easterling, D.R., K.E. Kunkel, M.F. Wehner, and L. Sun, 2016: Detection and attribution of climate extremes in the observed record. *Weather and Climate Extremes*, **11**, 17-27. http://dx.doi.org/10.1016/j.wace.2016.01.001

27. Hulme, M., 2014: Attributing weather extremes to 'climate change'. *Progress in Physical Geography*, **38**, 499-511. http://dx.doi.org/10.1177/0309133314538644

28. Stott, P.A., D.A. Stone, and M.R. Allen, 2004: Human contribution to the European heatwave of 2003. *Nature*, **432**, 610-614. http://dx.doi.org/10.1038/nature03089

29. Hannart, A., J. Pearl, F.E.L. Otto, P. Naveau, and M. Ghil, 2016: Causal counterfactual theory for the attribution of weather and climate-related events. *Bulletin of the American Meteorological Society*, **97**, 99-110. http://dx.doi.org/10.1175/bams-d-14-00034.1

30. Horton, R.M., J.S. Mankin, C. Lesk, E. Coffel, and C. Raymond, 2016: A review of recent advances in research on extreme heat events. *Current Climate Change Reports*, **2**, 242-259. http://dx.doi.org/10.1007/s40641-016-0042-x

31. Shepherd, T.G., 2016: A common framework for approaches to extreme event attribution. *Current Climate Change Reports*, **2**, 28-38. http://dx.doi.org/10.1007/s40641-016-0033-y

32. Trenberth, K.E., J.T. Fasullo, and T.G. Shepherd, 2015: Attribution of climate extreme events. *Nature Climate Change*, **5**, 725-730. http://dx.doi.org/10.1038/nclimate2657

33. Hannart, A., A. Carrassi, M. Bocquet, M. Ghil, P. Naveau, M. Pulido, J. Ruiz, and P. Tandeo, 2016: DADA: Data assimilation for the detection and attribution of weather and climate-related events. *Climatic Change*, **136**, 155-174. http://dx.doi.org/10.1007/s10584-016-1595-3

34. Reed, A.J., M.E. Mann, K.A. Emanuel, N. Lin, B.P. Horton, A.C. Kemp, and J.P. Donnelly, 2015: Increased threat of tropical cyclones and coastal flooding to New York City during the anthropogenic era. *Proceedings of the National Academy of Sciences*, **112**, 12610-12615. http://dx.doi.org/10.1073/pnas.1513127112

35. Magnusson, L., J.-R. Bidlot, S.T.K. Lang, A. Thorpe, N. Wedi, and M. Yamaguchi, 2014: Evaluation of medium-range forecasts for Hurricane Sandy. *Monthly Weather Review*, **142**, 1962-1981. http://dx.doi.org/10.1175/mwr-d-13-00228.1

36. Lackmann, G.M., 2015: Hurricane Sandy before 1900 and after 2100. *Bulletin of the American Meteorological Society*, **96 (12)**, 547-560. http://dx.doi.org/10.1175/BAMS-D-14-00123.1

37. Emanuel, K.A., 2013: Downscaling CMIP5 climate models shows increased tropical cyclone activity over the 21st century. *Proceedings of the National Academy of Sciences*, **110**, 12219-12224. http://dx.doi.org/10.1073/pnas.1301293110

38. Christensen, J.H., K. Krishna Kumar, E. Aldrian, S.-I. An, I.F.A. Cavalcanti, M. de Castro, W. Dong, P. Goswami, A. Hall, J.K. Kanyanga, A. Kitoh, J. Kossin, N.-C. Lau, J. Renwick, D.B. Stephenson, S.-P. Xie, and T. Zhou, 2013: Climate phenomena and their relevance for future regional climate change. *Climate Change 2013: The Physical Science Basis. Contribution of Working Group I to the Fifth Assessment Report of the Intergovernmental Panel on Climate Change*. Stocker, T.F., D. Qin, G.-K. Plattner, M. Tignor, S.K. Allen, J. Boschung, A. Nauels, Y. Xia, V. Bex, and P.M. Midgley, Eds. Cambridge University Press, Cambridge, United Kingdom and New York, NY, USA, 1217–1308. http://www.climatechange2013.org/report/full-report/

# Appendix D
# Acronyms and Units

doi: 10.7930/J0ZC811D

| | |
|---|---|
| **AGCM** | atmospheric general circulation model |
| **AIS** | Antarctic Ice Sheet |
| **AMO** | Atlantic Multidecadal Oscillation |
| **AMOC** | Atlantic meridional overturning circulation |
| **AMSU** | Advanced Microwave Sounding Unit |
| **AO** | Arctic Oscillation |
| **AOD** | aerosol optical depth |
| **AR** | atmospheric river |
| **AW** | Atlantic Water |
| **BAMS** | Bulletin of the American Meteorological Society |
| **BC** | black carbon |
| **BCE** | Before Common Era |
| **CAM5** | Community Atmospheric Model, Version 5 |
| **CAPE** | convective available potential energy |
| **CCN** | cloud condensation nuclei |
| **CCSM3** | Community Climate System Model, Version 3 |
| **CDR** | carbon dioxide removal |
| **CE** | Common Era |
| **CENRS** | Committee on Environment, Natural Resources, and Sustainability (National Science and Technology Council, White House) |

| | |
|---|---|
| **CESM-LE** | Community Earth System Model Large Ensemble Project |
| **CFCs** | chlorofluorocarbons |
| **CI** | climate intervention |
| **CMIP5** | Coupled Model Intercomparison Project, Fifth Phase (also CMIP3 and CMIP6) |
| **CONUS** | contiguous United States |
| **CP** | Central Pacific |
| **CSSR** | Climate Science Special Report |
| **DIC** | dissolved inorganic carbon |
| **DJF** | December-January-February |
| **DoD SERDP** | U.S. Department of Defense, Strategic Environmental Research and Development Program |
| **DOE** | U.S. Department of Energy |
| **EAIS** | East Antarctic Ice Sheet |
| **ECS** | equilibrium climate sensitivity |
| **ENSO** | El Niño–Southern Oscillation |
| **EOF analysis** | empirical orthogonal function analysis |
| **EP** | Eastern Pacific |
| **ERF** | effective radiative forcing |
| **ESD** | empirical statistical downscaling |
| **ESDM** | empirical statistical downscaling model |
| **ESM** | Earth System Model |
| **ESS** | Earth system sensitivity |
| **ETC** | extratropical cyclone |
| **ETCCDI** | Expert Team on Climate Change Detection Indices |
| **GBI** | Greenland Blocking Index |

| | |
|---|---|
| **GCIS** | Global Change Information System |
| **GCM** | global climate model |
| **GeoMIP** | Geoengineering Model Intercomparison Project |
| **GFDL HiRAM** | Geophysical Fluid Dynamics Laboratory, global HIgh Resolution Atmospheric Model (NOAA) |
| **GHCN** | Global Historical Climatology Network (National Centers for Environmental Information, NOAA) |
| **GHG** | greenhouse gas |
| **GMSL** | global mean sea level |
| **GMT** | global mean temperature |
| **GPS** | global positioning system |
| **GRACE** | Gravity Recovery and Climate Experiment |
| **GrIS** | Greenland Ice Sheet |
| **GWP** | global warming potential |
| **HadCM3** | Hadley Centre Coupled Model, Version 3 |
| **HadCRUT4** | Hadley Centre Climatic Research Unit Gridded Surface Temperature Dataset 4 |
| **HCFCs** | hydrochlorofluorocarbons |
| **HFCs** | hydrofluorocarbons |
| **HOT** | Hawai'i Ocean Time-series |
| **HOT-DOGS** | Hawai'i Ocean Time-series Data Organization & Graphical System |
| **HURDAT2** | revised Atlantic Hurricane Database (National Hurricane Center, NOAA) |
| **IAM** | integrated assessment model |
| **IAV** | impacts, adaptation, and vulnerability |
| **INMCM** | Institute for Numerical Mathematics Climate Model |
| **IPCC** | Intergovernmental Panel on Climate Change |

| | |
|---|---|
| **IPCC AR5** | Fifth Assessment Report of the IPCC; also SPM—Summary for Policymakers, and WG1, WG2, WG3—Working Groups 1–3 |
| **IPO** | Interdecadal Pacific Oscillation |
| **IVT** | integrated vapor transport |
| **JGOFS** | U.S. Joint Global Ocean Flux Study |
| **JJA** | June-July-August |
| **JTWC** | Joint Typhoon Warning Center |
| **LCC** | land-cover changes |
| **LULCC** | land-use and land-cover change |
| **MAM** | March-April-May |
| **MSU** | Microwave Sounding Unit |
| **NAM** | Northern Annular Mode |
| **NAO** | North Atlantic Oscillation |
| **NARCCAP** | North American Regional Climate Change Assessment Program (World Meteorological Organization) |
| **NAS** | National Academy of Sciences |
| **NASA** | National Aeronautics and Space Administration |
| **NCA** | National Climate Assessment |
| **NCA3** | Third National Climate Assessment |
| **NCA4** | Fourth National Climate Assessment |
| **NCEI** | National Centers for Environmental Information (NOAA) |
| **NDC** | nationally determined contribution |
| **NOAA** | National Oceanic and Atmospheric Administration |
| **NPI** | North Pacific Index |
| **NPO** | North Pacific oscillation |

| | |
|---|---|
| **NPP** | net primary production |
| **OMZs** | oxygen minimum zones |
| **OSTP** | Office of Science and Technology Policy (White House) |
| **PCA** | principle component analysis |
| **PDO** | Pacific Decadal Oscillation |
| **PDSI** | Palmer Drought Severity Index |
| **PETM** | Paleo-Eocene Thermal Maximum |
| **PFCs** | perfluorocarbons |
| **PGW** | pseudo-global warming |
| **PNA** | Pacific North American Pattern |
| **RCM** | regional climate models |
| **RCP** | Representative Concentration Pathway |
| **RF** | radiative forcing |
| **RFaci** | aerosol–cloud interaction (effect on RF) |
| **RFari** | aerosol–radiation interaction (effect on RF) |
| **RMSE** | root mean square error |
| **RSL** | relative sea level |
| **RSS** | remote sensing systems |
| **S06** | surface-to-6 km layer |
| **SCE** | snow cover extent |
| **SGCR** | Subcommittee on Global Change Research (National Science and Technology Council, White House) |
| **SLCF** | short-lived climate forcer |
| **SLCP** | short-lived climate pollutant |
| **SLR** | sea level rise |

| | |
|---|---|
| **SOC** | soil organic carbon |
| **SRES** | IPCC Special Report on Emissions Scenarios |
| **SREX** | IPCC Special Report on Managing the Risks of Extreme Events and Disasters to Advance Climate Change Adaptation |
| **SRM** | solar radiation management |
| **SSC** | Science Steering Committee |
| **SSI** | solar spectral irradiance |
| **SSP** | Shared Socioeconomic Pathway |
| **SST** | sea surface temperature |
| **STAR** | Center for Satellite Applications and Research (NOAA) |
| **SWCRE** | shortwave cloud radiative effect (on radiative fluxes) |
| **LWCRE** | longwave cloud radiative effect (on radiative fluxes) |
| **TA** | total alkalinity |
| **TC** | tropical cyclone |
| **TCR** | transient climate response |
| **TCRE** | transient climate response to cumulative carbon emissions |
| **TOPEX/JASON1,2** | Topography Experiment/Joint Altimetry Satellite Oceanography Network satellites (NASA) |
| **TSI** | total solar irradiance |
| **TTT** | temperature total troposphere |
| **UAH** | University of Alabama, Huntsville |
| **UHI** | urban heat island (effect) |
| **UNFCCC** | United Nations Framework Convention on Climate Change |
| **USGCRP** | U.S. Global Change Research Program |
| **USGS** | U.S. Geological Survey |

| | |
|---|---|
| **UV** | ultraviolet |
| **VOCs** | volatile organic compounds |
| **WAIS** | West Antarctic Ice Sheet |
| **WCRP** | World Climate Research Programme |
| **WMGHG** | well-mixed greenhouse gas |
| **WOCE** | World Ocean Circulation Experiment (JGOFS) |

## Abbreviations and Units

| | |
|---|---|
| **C** | carbon |
| **CO** | carbon monoxide |
| **CH$_4$** | methane |
| **cm** | centimeters |
| **CO$_2$** | carbon dioxide |
| **°C** | degrees Celsius |
| **°F** | degrees Fahrenheit |
| **GtC** | gigatonnes of carbon |
| **hPA** | hectopascal |
| **H$_2$S** | hydrogen sulfide |
| **H$_2$SO$_4$** | sulfuric acid |
| **km** | kilometers |
| **m** | meters |
| **mm** | millimeters |
| **Mt** | megaton |
| **μatm** | microatmosphere |
| **N** | nitrogen |

| | |
|---|---|
| **N₂O** | nitrous oxide |
| **NOₓ** | nitrogen oxides |
| **O₂** | molecular oxygen |
| **O₃** | ozone |
| **OH** | hydroxyl radical |
| **PgC** | petagrams of carbon |
| **ppb** | parts per billion |
| **ppm** | parts per million |
| **SF₆** | sulfur hexafluoride |
| **SO₂** | sulfur dioxide |
| **TgC** | teragrams of carbon |
| **W/m²** | Watts per meter squared |

# Appendix E
# Glossary Terms

doi: 10.7930/J0TM789P

**Abrupt climate change**

Change in the climate system on a timescale shorter than the timescale of the responsible forcing. In the case of anthropogenic forcing over the past century, abrupt change occurs over decades or less. Abrupt change need not be externally forced. (*CSSR, Ch. 15*)

**Aerosol–cloud interaction**

A process by which a perturbation to aerosol affects the microphysical properties and evolution of clouds through the aerosol role as cloud condensation nuclei or ice nuclei, particularly in ways that affect radiation or precipitation; such processes can also include the effect of clouds and precipitation on aerosol. The aerosol perturbation can be anthropogenic or come from some natural source. The radiative forcing from such interactions has traditionally been attributed to numerous indirect aerosol effects, but in this report, only two levels of radiative forcing (or effect) are distinguished:

The radiative forcing (or effect) due to aerosol–cloud interactions (**RFaci**) is the radiative forcing (or radiative effect, if the perturbation is internally generated) due to the change in number or size distribution of cloud droplets or ice crystals that is the proximate result of an aerosol perturbation, with other variables (in particular total cloud water content) remaining equal. In liquid clouds, an increase in cloud droplet concentration and surface area would increase the cloud albedo. This effect is also known as the cloud albedo effect, first indirect effect, or Twomey effect. It is a largely theoretical concept that cannot readily be isolated in observations or comprehensive process models due to the rapidity and ubiquity of rapid adjustments. This is contrasted with the effective radiative forcing (or effect) due to aerosol–cloud interactions (ERFaci)

The total effective radiative forcing due to both aerosol–cloud and aerosol–radiation interactions is denoted aerosol effective radiative forcing (ERFari+aci). See also **aerosol–radiation interaction**. (condensed from IPCC AR5 WGI Annex III: Glossary)

**Aerosol–radiation interaction (RFari)**

The radiative forcing (or radiative effect, if the perturbation is internally generated) of an aerosol perturbation due directly to aerosol–radiation interactions, with all environmental variables remaining unaffected. It is traditionally known in the literature as the *direct aerosol forcing* (or *effect*).

The total effective radiative forcing due to both aerosol–cloud and aerosol–radiation interactions is denoted aerosol effective radiative forcing (ERFari+aci). See also **aerosol-cloud interaction**. (condensed from IPCC AR5 WGI Annex III: Glossary)

**Agricultural drought**

See **drought**.

**Albedo**

The fraction of solar radiation reflected by a surface or object, often expressed as a percentage. Snow-covered surfaces have a high albedo, the albedo of soils ranges from high to low, and vegetation-covered surfaces and oceans have a low albedo. The Earth's planetary albedo varies mainly through varying cloudiness, snow, ice, leaf area, and land-cover changes. (IPCC AR5 WGI Annex III: Glossary)

**Altimetry**

A technique for measuring the height of the Earth's surface with respect to the geocenter of the Earth within a defined terrestrial reference frame (geocentric sea level). (IPCC AR5 WGI Annex III: Glossary)

### Anticyclonic circulation

Fluid motion having a sense of rotation about the local vertical opposite to that of the earth's rotation; that is, clockwise in the Northern Hemisphere, counterclockwise in the Southern Hemisphere, and undefined at the equator. It is the opposite of **cyclonic circulation**. (AMS glossary).

### Atlantic meridional overturning circulation (AMOC)

See **Meridional overturning circulation (MOC)**.

### Atmospheric blocking

See **Blocking**.

### Atmospheric river

A long, narrow, and transient corridor of strong horizontal water vapor transport that is typically associated with a low-level jet stream ahead of the cold front of an extratropical cyclone. The water vapor in atmospheric rivers is supplied by tropical and/or extratropical moisture sources. Atmospheric rivers frequently lead to heavy precipitation where they are forced upward—for example, by mountains or by ascent in the warm conveyor belt. Horizontal water vapor transport in the midlatitudes occurs primarily in atmospheric rivers and is focused in the lower troposphere. (AMS glossary).

### Baroclinicity

The state of stratification in a fluid in which surfaces of constant pressure (isobaric) intersect surfaces of constant density (isosteric). (AMS glossary).

### Bias correction method

One of two main statistical approaches used to alleviate the limitations of global and regional climate models, in which the statistics of the simulated model outputs are adjusted to those of the observation data. (The other approach is **empirical/stochastic downscaling**, described under **downscaling**). The rescaled variables can remove the effects of systematic errors in climate model outputs. (derived from Kim et al., 2015)

### Biological pump

The suite of biologically mediated processes responsible for transporting carbon against a concentration gradient from the upper ocean to the deep ocean. (Passow and Carlson, 2012)

### Blocking

Associated with persistent, slow-moving high pressure systems that obstruct the prevailing westerly winds in the middle and high latitudes and the normal eastward progress of extratropical transient storm systems. It is an important component of the intraseasonal climate variability in the extratropics and can cause long-lived weather conditions such as cold spells in winter and heat waves in summer. (IPCC AR5 WGI Annex III: Glossary)

### Carbon dioxide fertilization

The enhancement of the growth of plants as a result of increased atmospheric $CO_2$ concentration. (IPCC AR5 WGI Annex III: Glossary)

### Carbon dioxide removal

A set of techniques that aim to remove $CO_2$ directly from the atmosphere by either (1) increasing natural sinks for carbon or (2) using chemical engineering to remove the $CO_2$, with the intent of reducing the atmospheric $CO_2$ concentration. CDR methods involve the ocean, land and technical systems, including such methods as iron fertilization, large-scale afforestation and direct capture of $CO_2$ from the atmosphere using engineered chemical means. (truncated version from IPCC AR5 WGI Annex III: Glossary)

### Climate engineering

See **geoengineering**.

### Climate intervention

See **geoengineering**.

### Climate sensitivity

In Intergovernmental Panel on Climate Change (IPCC) reports, **equilibrium climate sensitivity** (units: °C) refers to the equilibrium (steady state) change in the annual global mean surface temperature following a doubling of the atmospheric equivalent carbon dioxide concentration. The **effective climate sensitivity** (units: °C) is an estimate of the global mean surface temperature response to doubled carbon dioxide concentration that is evaluated from model output or observations for evolving non-equilibrium conditions. It is a measure of the strengths of the climate feedbacks at a particular time and may vary with forcing history and climate state, and therefore may differ from equilibrium climate sensitivity. The **transient climate response** (units: °C) is the change in the global mean surface temperature, averaged over a 20-year period centered at the time of atmospheric carbon dioxide doubling, in a climate model simulation in which $CO_2$ increases at 1% per year. It is a measure of the strength and rapidity of the surface temperature response to greenhouse gas forcing. (IPCC AR5 WGI Annex III: Glossary)

### Cloud radiative effect

The radiative effect of clouds relative to the identical situation without clouds (previously called cloud radiative forcing). (drawn from IPCC AR5 WGI Annex III: Glossary)

Clouds can act as a greenhouse ingredient to warm the Earth by trapping outgoing longwave infrared radiative flux at the top of the atmosphere (the **longwave cloud radiative effect [LWCRE]**). Clouds can also enhance the planetary albedo by reflecting shortwave solar radiative flux back to space to cool the Earth (the **shortwave cloud radiative effect [SWCRE]**). The net effect of the two competing processes depends on the height, type, and the optical properties of the clouds. (edited from *NOAA, Geophysical Fluid Dynamics Laboratory*)

### CMIP

The Coupled Model Intercomparison Project is a standard experimental protocol for studying the output of coupled atmosphere–ocean general circulation models (AOGCMs). Phases three and five (CMIP3 and CMIP5, respectively) coordinated and archived climate model simulations based on shared model inputs by modeling groups from around the world. The CMIP3 multi-model data set includes projections using the SRES scenarios drawn from the Intergovernmental Panel on Climate Change's Special Report on Emissions Scenarios. The CMIP5 dataset includes projections using the **Represen-**

tative Concentration Pathways. (edited from IPCC AR5 WGII Annex II: Glossary).

### Compound event

An event that consists of 1) two or more extreme events occurring simultaneously or successively, 2) combinations of extreme events with underlying conditions that amplify the impact of the events, or 3) combinations of events that are not themselves extremes but lead to an extreme event or impact when combined. The contributing events can be of similar or different types. (*CSSR, Ch. 15*, drawing upon *SREX 3.1.3*)

### Critical threshold

A threshold that arises within a system as a result of the amplifying effects of positive **feedbacks**. The crossing of a critical threshold commits the system to a change in state. (*CSSR, Ch. 15*)

### Cryosphere

All regions on and beneath the surface of the Earth and ocean where water is in solid form, including sea ice, lake ice, river ice, snow cover, glaciers and ice sheets, and frozen ground (which includes permafrost). (IPCC AR5 WGI Annex III: Glossary)

### Cyclonic circulation

Fluid motion in the same sense as that of the earth, that is, counterclockwise in the Northern Hemisphere, clockwise in the Southern Hemisphere, undefined at the equator. (AMS glossary).

### Denitrification

As used in this report, refers to the loss of fixed nitrogen in the ocean through biogeochemical processes. (*CSSR, Ch. 13*).

### Deoxygenation

See **hypoxia.**

### Downscaling

A method that derives local- to regional-scale (10–100 km) information from larger-scale models or data analyses. Two main methods exist. **Dynamical downscaling** uses the output of regional climate models, global models with variable spatial resolution, or high-resolution global models. **Empirical/statistical downscal-**

ing methods develop statistical relationships that link the large-scale atmospheric variables with local/regional climate variables. In all cases, the quality of the driving model remains an important limitation on the quality of the downscaled information. (IPCC AR5 WGI Annex III: Glossary)

**Drought**

A period of abnormally dry weather long enough to cause a serious hydrological imbalance. Drought is a relative term; therefore, any discussion in terms of precipitation deficit must refer to the particular precipitation-related activity that is under discussion. For example, shortage of precipitation during the growing season impinges on crop production or ecosystem function in general (due to soil moisture drought, also termed **agricultural drought**), and during the runoff and percolation season primarily affects water supplies (**hydrological drought**). Storage changes in soil moisture and groundwater are also affected by increases in actual evapotranspiration in addition to reductions in precipitation. A period with an abnormal precipitation deficit is defined as a **meteorological drought**. (IPCC AR5 WGI Annex III: Glossary)

**Dynamical downscaling**

See **downscaling**.

**Earth System Model**

A coupled atmosphere–ocean general circulation model in which a representation of the carbon cycle is included, allowing for interactive calculation of atmospheric $CO_2$ or compatible emissions. Additional components (for example, atmospheric chemistry, ice sheets, dynamic vegetation, nitrogen cycle, but also urban or crop models) may be included. (IPCC AR5 WGI Annex III: Glossary)

**Effective radiative forcing**

See **radiative forcing**.

**El Niño–Southern Oscillation**

A natural variability in ocean water surface pressure that causes periodic changes in ocean surface temperatures in the tropical Pacific Ocean. El Niño–Southern Oscillation (ENSO)

has two phases: the warm oceanic phase, El Niño, accompanies high air surface pressure in the western Pacific, while the cold phase, La Niña, accompanies low air surface pressure in the western Pacific. Each phase generally lasts for 6 to 18 months. ENSO events occur irregularly, roughly every 3 to 7 years. The extremes of this climate pattern's oscillations cause extreme weather (such as floods and droughts) in many regions of the world. (*USGCRP*)

**Empirical/statistical downscaling**

See **downscaling**.

**Equivalent carbon dioxide concentration**

The concentration of carbon dioxide that would cause the same radiative forcing as a given mixture of carbon dioxide and other forcing components. Those values may consider only greenhouse gases, or a combination of greenhouse gases and aerosols. Equivalent carbon dioxide concentration is a metric for comparing radiative forcing of a mix of different greenhouse gases at a particular time but does not imply equivalence of the corresponding climate change responses nor future forcing. There is generally no connection between equivalent carbon dioxide emissions and resulting equivalent carbon dioxide concentrations. (IPCC AR5 WGI Annex III: Glossary)

**Eutrophication**

Over-enrichment of water by nutrients such as nitrogen and phosphorus. It is one of the leading causes of water quality impairment. The two most acute symptoms of eutrophication are **hypoxia** (a state of oxygen depletion) and harmful algal blooms. (IPCC AR5 WGII Annex II: Glossary).

**Extratropical cyclone**

A large-scale (of order 1,000 km) storm in the middle or high latitudes having low central pressure and fronts with strong horizontal gradients in temperature and humidity. A major cause of extreme wind speeds and heavy precipitation especially in wintertime. (IPCC AR5 WGI Annex III: Glossary)

## Feedbacks

An interaction between processes in the climate system, in which the result of an initial process triggers changes in a second process that in turn influences the initial one. A **positive feedback** magnifies the original process, while a **negative feedback** attenuates or diminishes it. Positive feedbacks are sometimes referred to as "vicious" or "virtuous" cycles, depending on whether their effects are viewed as harmful or beneficial. (*CSSR, Ch. 15*)

## Geoengineering

A broad set of methods and technologies that aim to deliberately alter the climate system in order to alleviate the impacts of climate change (also known as **climate intervention** (National Academy of Sciences) or climate engineering). Most, but not all, methods seek to either 1) reduce the amount of absorbed solar energy in the climate system (**Solar Radiation Management**) or 2) increase net carbon sinks from the atmosphere at a scale sufficiently large to alter climate (**Carbon Dioxide Removal**). Scale and intent are of central importance. Two key characteristics of geoengineering methods of particular concern are that they use or affect the climate system (e.g., atmosphere, land, or ocean) globally or regionally and/or could have substantive unintended effects that cross national boundaries. (adapted from IPCC AR5 WGI Annex III: Glossary)

## Glacial isostatic adjustment (GIA)

The deformation of the Earth and its gravity field due to the response of the earth–ocean system to changes in ice and associated water loads. It includes vertical and horizontal deformations of the Earth's surface and changes in geoid due to the redistribution of mass during the ice–ocean mass exchange. GIA is currently contributing to relative sea level rise in much of the continental United States. (IPCC AR5 WGI Annex III: Glossary)

## Glacier

A perennial mass of land ice that originates from compressed snow, shows evidence of past or present flow (through internal deformation and/or sliding at the base), and is constrained by internal stress and friction at the base and sides. A glacier is maintained by accumulation of snow at high altitudes, balanced by melting at low altitudes and/or discharge into the sea. An ice mass of the same origin as glaciers, but of continental size, is an **ice sheet**, defined further below. (IPCC AR5 WGI Annex III: Glossary)

## Global mean sea level

The average of relative sea level or of sea surface height across the ocean.

## Global warming potential (GWP)

An index, based on radiative properties of greenhouse gases, measuring the radiative forcing following a pulse emission of a unit mass of a given greenhouse gas in the present-day atmosphere integrated over a chosen time horizon, relative to that of carbon dioxide. The GWP represents the combined effect of the differing times these gases remain in the atmosphere and their relative effectiveness in causing radiative forcing. (truncated from IPCC AR5 WGI Annex III: Glossary)

## Gravimetry

Measurement of the Earth's gravitational field. Using satellite data from the Gravity Recovery and Climate Experiment (GRACE), measurements of the mean gravity field help scientists better understand the structure of the solid Earth and learn about ocean circulation. Monthly measurements of time-variable gravity can be used to study ground water fluctuations, sea ice, sea level rise, deep ocean currents, ocean bottom pressure, and ocean heat flux. (modified from *NASA Earth Observatory on the GRACE project*)

## Greenhouse gas (GHG)

Greenhouse gases are those gaseous constituents of the atmosphere, both natural and anthropogenic, that absorb and emit radiation at specific wavelengths within the spectrum of terrestrial radiation emitted by the Earth's surface, the atmosphere itself, and by clouds. This property causes the greenhouse effect. Water vapor ($H_2O$), carbon dioxide ($CO_2$), nitrous oxide ($N_2O$), methane ($CH_4$), and ozone ($O_3$) are the primary greenhouse gases in the Earth's atmosphere. Moreover, there are a number of entirely human-made greenhouse gases in the atmosphere, such as the halocarbons and other chlorine- and bromine-containing substances, dealt with under the Montreal Protocol. Beside $CO_2$, $N_2O$, and $CH_4$, the Kyoto Protocol dealt with the greenhouse gases sulfur hexafluoride ($SF_6$), hydrofluorocarbons (HFCs), and perfluorocarbons (PFCs). (adapted from IPCC AR5 WGI Annex III: Glossary)

## Hydrological drought
See **drought**.

## Hypoxia
Deficiency of oxygen in water bodies, which can be a symptom of **eutrophication** (nutrient overloading). **Deoxygenation** (the process of removing oxygen) leads to hypoxia, and the expansion of **oxygen minimum zones** (IPCC AR5 WGII Annex II: Glossary supplemented with other sources).

## Ice sheet
A mass of land ice of continental size that is sufficiently thick to cover most of the underlying bed, so that its shape is mainly determined by its dynamics (the flow of the ice as it deforms internally and/or slides at its base). An ice sheet flows outward from a high central ice plateau with a small average surface slope. The margins usually slope more steeply, and most ice is discharged through fast flowing ice streams or outlet glaciers, in some cases into the sea or into ice shelves floating on the sea. There are only two ice sheets in the modern world, one on Greenland and one on Antarctica. During glacial periods there were others, including the Laurentide Ice Sheet in North America, whose loss is the primary driver of **glacial isostatic adjustment** in the United States today. (adapted from IPCC AR5 WGI Annex III: Glossary)

## Ice wedge
Common features of the subsurface in permafrost regions, ice wedges develop by repeated frost cracking and ice vein growth over hundreds to thousands of years. Ice wedge formation causes the archetypal polygonal patterns seen in tundra across the Arctic landscape. (adapted from Liljedal et al., 2016)

## Instantaneous radiative forcing
See **radiative forcing**.

## Irreversible
Changes in components of the climate system that either cannot be reversed, or can only be reversed on timescales much longer than the timescale over which the original forcing occurred. (*CSSR, Ch. 15*)

## Longwave cloud radiative effect (LWCRE)
See **cloud radiative effect**.

## Meridional overturning circulation (MOC)
Meridional (north–south) overturning circulation in the ocean quantified by zonal (east–west) sums of mass transports in depth or density layers. In the North Atlantic, away from the subpolar regions, the Atlantic MOC (AMOC, which is in principle an observable quantity) is often identified with the thermohaline circulation (THC), which is a conceptual and incomplete interpretation. It must be borne in mind that the AMOC is also driven by wind, and can also include shallower overturning cells such as occur in the upper ocean in the tropics and subtropics, in which warm (light) waters moving poleward are transformed to slightly denser waters and subducted equatorward at deeper levels. (adapted from IPCC AR5 WGI Annex III: Glossary)

## Meridional temperature gradient
North–South temperature variation

## Meteorological drought
See **drought**.

## Mode water
Water of exceptionally uniform properties over an extensive depth range, caused in most instances by convection. Mode waters represent regions of water mass formation; they are not necessarily water masses in their own right but contribute significant volumes of water to other water masses. Because they represent regions of deep sinking of surface water, mode water formation regions are atmospheric heat sources. Subantarctic Mode Water is formed during winter in the subantarctic zone just north of the subantarctic front and contributes to the lower temperature range of central water; only in the extreme eastern Pacific Ocean does it obtain a temperature low enough to contribute to Antarctic Intermediate Water. Subtropical

465

Mode Water is mostly formed through enhanced subduction at selected locations of the subtropics and contributes to the upper temperature range of central water. Examples of Subtropical Mode Water are the 18°C water formed in the Sargasso Sea, Madeira Mode Water formed at the same temperature but in the vicinity of Madeira, and 13°C water formed not by surface processes but through mixing in Agulhas Current eddies as they enter the Benguela Current. (AMS glossary).

## Model ability/model skill

Representativeness of the ability of a climate model to reproduce historical climate observational data.

## Model bias

Systematic error in model output that over- or under-emphasizes particular model mechanism or results.

## Model ensemble

Also known as a multimodel ensemble (MME), a group of several different global climate models (GCMs) used to create a large number of climate simulations. An MME is designed to address **structural model uncertainty** between different climate models, rather than **parametric uncertainty** within any one particular model. (*UK Met Office, Climate Projections*, Glossary)

## Model independence

An analysis of the degree to which models are different from one another. Also is used as an interpretation of an ensemble as constituting independent samples of a distribution which represents our collective understanding of the climate system. (summarized based on Annan and Hargreaves, 2017)

## Nationally determined contributions (NDCs)

See **Paris Agreement**.

## Negative feedbacks

See **feedbacks**.

## Nitrogen mineralization

Mineralization/remineralization is the conversion of an element from its organic form to an inorganic form as a result of microbial decomposition. In nitrogen mineralization, organic nitrogen from decaying plant and animal residues (proteins, nucleic acids, amino sugars and urea) is converted to ammonia ($NH_3$) and ammonium ($NH4+$) by biological activity. (IPCC AR5 WGI Annex III: Glossary)

## Ocean acidification

The process by which ocean waters have become more acidic due to the absorption of human-produced carbon dioxide, which interacts with ocean water to form carbonic acid and lower the ocean's pH. Acidity reduces the capacity of key plankton species and shelled animals to form and maintain shells. (*USGCRP*)

## Ocean stratification

The existence or formation of distinct layers or laminae in the ocean identified by differences in thermal or salinity characteristics (e.g., densities) or by oxygen or nutrient content. (adapted from AMS glossary).

## Oxygen minimum zones (OMZs)

The midwater layer (200–1,000 m) in the open ocean in which oxygen saturation is the lowest in the ocean. The degree of oxygen depletion depends on the largely bacterial consumption of organic matter, and the distribution of the OMZs is influenced by large-scale ocean circulation. In coastal oceans, OMZs extend to the shelves and may also affect benthic ecosystems. OMZs can expand through a process of **deoxygenation**. (supplemented version of IPCC AR5 WGII Annex II: Glossary).

## Pacific Decadal Oscillation

The pattern and time series of the first empirical orthogonal function of sea surface temperature over the North Pacific north of 20°N. The PDO broadened to cover the whole Pacific Basin is known as the Interdecadal Pacific Oscillation. The PDO and IPO exhibit similar temporal evolution. (IPCC AR5 WGI Annex III: Glossary)

## Parameterization

In climate models, this term refers to the technique of representing processes that cannot be explicitly resolved at the spatial or temporal resolution of the model (sub-grid scale processes) by relationships between model-resolved larger-scale variables and the area- or time-averaged effect of such subgrid scale processes. (IPCC AR5 WGI Annex III: Glossary)

**Parametric uncertainty**

See **uncertainty**.

**Paris Agreement**

An international climate agreement with the central aim to hold global temperature rise this century well below 2°C above preindustrial levels and to pursue efforts to limit the temperature increase even further to 1.5°C. For the first time, all parties are required to put forward emissions reductions targets, and to strengthen those efforts in the years ahead as the Agreement is assessed every five years. Each country's proposed mitigation target (the intended nationally determined contribution [INDC]) becomes an official **nationally determined contribution (NDC)** when the country ratifies the agreement. The Paris Agreement was finalized on December 12, 2015, at the 21st Conference of Parties (COP 21) of the United National Framework Convention on Climate Change (UNFCCC). "Paris" entered into force on November 4, 2016, after ratification by 55 countries that account for at least 55% of global emissions). The agreement had a total of 125 national parties by early 2017. (summarized/edited from *UNFCCC*)

**Pattern scaling**

A simple and computationally cheap method to produce climate projections beyond the scenarios run with expensive global climate models (GCMs). The simplest technique has known limitations and assumes that a spatial climate anomaly pattern obtained from a GCM can be scaled by the global mean temperature anomaly. (Herger et al., 2015)

**Permafrost**

Ground that remains at or below freezing for at least two consecutive years. (*USGCRP*)

**Permafrost active layer**

The layer of ground that is subject to annual thawing and freezing in areas underlain by permafrost. (IPCC AR5 WGI Annex III: Glossary)

**Petagram**

One petagram (Pg) = $10^{15}$ grams or $10^{12}$ kilograms. A petagram is the same as a gigaton, which is a billion metric tons, where 1 metric ton is 1,000 kg. Estimated 2014 global fossil fuel emissions were 9.855 Pg = 9.855 Gt = 9,855 million metric tons of carbon. (CDIAC – Carbon Dioxide Information Center: Boden et al., 2017)

**Positive feedbacks**

See **feedbacks**.

**Proxy**

A way to indirectly measure aspects of climate. Biological or physical records from ice cores, tree rings, and soil boreholes are good examples of proxy data. (*USGCRP*)

**Radiative forcing**

The change in the net (downward minus upward) radiative flux (expressed in $W/m^{2)}$ at the tropopause or top of atmosphere due to a change in an external driver of climate change, such as a change in the concentration of carbon dioxide or in the output of the Sun. Sometimes internal drivers are still treated as forcings even though they result from the alteration in climate, for example aerosol or greenhouse gas changes in paleoclimates. The traditional radiative forcing is computed with all tropospheric properties held fixed at their unperturbed values, and after allowing for stratospheric temperatures, if perturbed, to readjust to radiative–dynamical equilibrium. Radiative forcing is **instantaneous** if no change in stratospheric temperature is accounted for. The radiative forcing once rapid adjustments are accounted for is the **effective radiative forcing**. Radiative forcing is not to be confused with cloud radiative forcing, which describes an unrelated measure of the impact of clouds on the radiative flux at the top of the atmosphere. (truncated from IPCC AR5 WGI Annex III: Glossary)

**Relative sea level**

The height of the sea surface, measured with respect to the height of the underlying land. Relative sea level changes in response to both changes in the height of the sea surface and changes in the height of the underlying land.

## Representative Concentration Pathways

Scenarios that include time series of emissions and concentrations of the full suite of greenhouse gases and aerosols and chemically active gases, as well as land use/land cover. The word "representative" signifies that each RCP provides only one of many possible scenarios that would lead to the specific radiative forcing characteristics. The term "pathway" emphasizes that not only the long-term concentration levels are of interest, but also the trajectory taken over time to reach that outcome. RCPs usually refer to the portion of the concentration pathway extending up to 2100. Four RCPs produced from Integrated Assessment Models were selected from the published literature for use in the Intergovernmental Panel on Climate Change's Fifth Assessment Report: **RCP2.6**, a pathway where radiative forcing peaks at approximately 3 W/m² before 2100 and then declines; **RCP4.5** and **RCP6.0**, two intermediate stabilization pathways in which radiative forcing is stabilized at approximately 4.5 W/m² and 6.0 W/m², respectively, after 2100; and **RCP8.5**, a high pathway for which radiative forcing reaches greater than 8.5 W/m² by 2100 and continues to rise for some amount of time (truncated and adapted from IPCC AR5 WGI Annex III: Glossary, excluding discussion of extended concentration pathways)

## Rossby waves

Rossby waves, also known as planetary waves, naturally occur in rotating fluids. Within the Earth's ocean and atmosphere, these waves form as a result of the rotation of the planet. These waves affect the planet's weather and climate. Oceanic Rossby waves are huge, undulating movements of the ocean that stretch horizontally across the planet for hundreds of kilometers in a westward direction. Atmospheric Rossby waves form primarily as a result of the Earth's geography. Rossby waves help transfer heat from the tropics toward the poles and cold air toward the tropics in an attempt to return the atmosphere to balance. They also help locate the jet stream and mark out the track of surface low pressure systems. The slow motion of these waves often results in fairly long, persistent weather patterns. (adapted from NOAA National Ocean Service)

## Saffir-Simpson hurricane scale

A classification scheme for hurricane intensity based on the maximum surface wind speed and the type and extent of damage done by the storm. The wind speed categories are as follows: 1) 33–42 m/s (65–82 knots or 74–95 mph); 2) 43–49 m/s (83–95 knots or 96–110 mph); 3) 50–58 m/s (96–113 knots or 111–129 mph); 4) 59–69 m/s (114–134 knots or 130–156 mph); and 5) 70 m/s (135 knots or 156 mph) and higher. These categories are used routinely by weather forecasters in North America to characterize the intensity of hurricanes for the public. (adapted from AMS glossary).

## Saturation

The condition in which vapor pressure is equal to the equilibrium vapor pressure over a plane surface of pure liquid water, or sometimes ice. (AMS glossary).

## Scenarios

Plausible descriptions of how the future may develop based on a coherent and internally consistent set of assumptions about key driving forces (e.g., rate of technological change, prices) and relationships. Note that scenarios are neither predictions nor forecasts, but are useful to provide a view of the implications of developments and actions. (IPCC AR5 WGI Annex III: Glossary)

## Sea level pressure

The atmospheric pressure at mean sea level, either directly measured or, most commonly, empirically determined from the observed station pressure. In regions where the Earth's surface is above sea level, it is standard observational practice to reduce the observed surface pressure to the value that would exist at a point at sea level directly below if air of a temperature corresponding to that actually present at the surface were present all the way down to sea level. In actual practice, the mean temperature for the preceding 12 hours is employed, rather than the current temperature. This "reduction of pressure to sea level" is responsible for many anomalies in the pressure field in mountainous areas on the surface synoptic chart. (AMS glossary).

**Shared Socioeconomic Pathways**

A basis for emissions and socioeconomic scenarios, an SSP is one of a collection of pathways that describe alternative futures of socioeconomic development in the absence of climate policy intervention. The combination of SSP-based socioeconomic scenarios and **Representative Concentration Pathway (RCP)**-based climate projections can provide a useful integrative frame for climate impact and policy analysis. (updated from IPCC AR5 WGIII Annex I: Glossary).

**Shortwave cloud radiative effect (SWCRE)**

See **cloud radiative effect**.

**Snow water equivalent**

The depth of liquid water that would result if a mass of snow melted completely. (IPCC AR5 WGI Annex III: Glossary)

**Solar radiation management (SRM)**

The intentional modification of the Earth's shortwave radiative budget with the aim to reduce climate change according to a given metric (for example, surface temperature, precipitation, regional impacts, etc). Artificial injection of stratospheric aerosols and cloud brightening are two examples of SRM techniques. Methods to modify some fast-responding elements of the longwave radiative budget (such as cirrus clouds), although not strictly speaking SRM, can be related to SRM. See also **geoengineering**. (edited from IPCC AR5 WGI Annex III: Glossary)

**Static-equilibrium (sea level change) fingerprint**

The near-instantaneous pattern of **relative sea level** change associated with changes in the distribution of mass at the surface of the Earth, for example due to the melting of ice on land. Near a shrinking ice sheet (within ~2,000 km of the margin), sea level will fall due to both crustal uplift and the reduction of the gravitational pull on the ocean from the ice sheet. Close to the ice sheet, this fall can be an order of magnitude greater than the equivalent rise in global mean sea level associated with the meltwater addition to the ocean. Far from the ice sheet, sea level will generally rise with greater amplitude as the distance from the ice sheet increases, and this rise

can exceed the global mean value by up to about 30%. (draws on Hay et al., 2012)

**Structural model uncertainty**

See **uncertainty**.

**Teleconnection**

A statistical association between climate variables at widely separated, geographically fixed spatial locations. Teleconnections are caused by large spatial structures such as basin-wide coupled modes of ocean–atmosphere variability, Rossby wave-trains, midlatitude jets and storm tracks, etc. (IPCC AR5 WGI Annex III: Glossary)

**Thermohaline circulation (THC)**

Large-scale circulation in the ocean that transforms low-density upper ocean waters to higher-density intermediate and deep waters and returns those waters back to the upper ocean. The circulation is asymmetric, with conversion to dense waters in restricted regions at high latitudes and the return to the surface involving slow upwelling and diffusive processes over much larger geographic regions. The THC is driven by high densities at or near the surface, caused by cold temperatures and/or high salinities, but despite its suggestive though common name, is also driven by mechanical forces such as wind and tides. Frequently, the name THC has been used synonymously with the **Meridional Overturning Circulation**. (IPCC AR5 WGI Annex III: Glossary)

**Thermokarst**

The process by which characteristic landforms result from the thawing of ice-rich permafrost or the melting of massive ground ice. (IPCC AR5 WGI Annex III: Glossary)

**Threshold**

The value of a parameter summarizing a system, or a process affecting a system, at which qualitatively different system behavior emerges. Beyond this value, the system may not conform to statistical relationships that described it previously. For example, beyond a threshold level of ocean acidification, wide-scale collapse of coral ecosystems may occur. (*CSSR, Ch. 15*)

**Tipping elements**

Systems with critical thresholds, beyond which small perturbations in forcing can—as a result of positive feedbacks—lead to large, nonlinear, and irreversible shifts in state. In the climate system, a tipping element is a subcomponent of the climate system (typically at a spatial scale of approximately 1,000 km or larger). (*CSSR, Ch. 15*)

**Tipping point**

The critical **threshold** of a tipping element. Some limit its use to critical thresholds in which both the commitment to change and the change itself occur without a significant lag, while others also apply it to situations where a commitment occurs rapidly, but the committed change may play out over centuries and even millennia. (*CSSR, Ch. 15*)

**Transient climate response**

See **climate sensitivity**.

**Tropopause**

The boundary between the troposphere and the stratosphere. (IPCC AR5 WGI Annex III: Glossary)

**Uncertainty**

A state of incomplete knowledge that can result from a lack of information or from disagreement about what is known or even knowable. It may have many types of sources, from imprecision in the data to ambiguously defined concepts or terminology, or uncertain projections of human behavior. Uncertainty can therefore be represented by quantitative measures (for example, a probability density function) or by qualitative statements (for example, reflecting the judgment of a team of experts) (cut from IPCC AR5 WGII Annex II: Glossary).

Given that no model can represent the world with complete accuracy, **structural model uncertainty** refers to how well the physical processes of the real world are represented in the structure of a model. Different modeling research groups will represent the climate system in different ways, and to some extent this decision is a subjective judgement. The use of climate **model ensembles** can address the uncertainty of differently structured models. (adapted from *UK Met Office, Climate Projections*, Glossary)

In contrast, **parametric uncertainty** refers to incomplete knowledge about real world processes in a climate model. A parameter is well-specified in that it has a true value, even if this value is unknown. Such empirical quantities can be measured, and the level of uncertainty about them can be represented in probabilistic terms. (adapted from *Morgan and Henrion, 1990, pp 50-52*)

**Urban heat island effect**

The relative warmth of a city compared with surrounding rural areas, associated with changes in runoff, effects on heat retention, and changes in surface albedo. (IPCC AR5 WGI Annex III: Glossary)

**Zonal mean**

Data average along a latitudinal circle on the globe.

71888466R00263

Made in the USA
Lexington, KY
25 November 2017